QUANTUM DISSIPATIVE SYSTEMS

SYSTEMS

Third Edition

SERIES IN MODERN CONDENSED MATTER PHYSICS

Editors-in-charge: I. Dzyaloshinski and Yu Lu

Published

Series in
Modern
Condensed
Matter
Physics

Vol. 13

QUANTUM DISSIPATIVE SYSTEMS

Third Edition

Ulrich Weiss

Institute for Theoretical Physics
University of Stuttgart

World Scientific
Singapore • New Jersey • London • Hong Kong

Published by

World Scientific Publishing Co. Pte. Ltd.

5 Toh Tuck Link, Singapore 596224

USA office: 27 Warren Street, Suite 401-402, Hackensack, NJ 07601

UK office: 57 Shelton Street, Covent Garden, London WC2H 9HE

British Library Cataloguing-in-Publication Data
A catalogue record for this book is available from the British Library.

QUANTUM DISSIPATIVE SYSTEMS (3rd Edition)
Series in Modern Condensed Matter Physics — Vol. 13

ISBN-13 978-981-279-162-7 (pbk)
ISBN-10 981-279-162-0 (pbk)

Printed in Singapore by B & JO Enterprise

Preface

In the fourteen years since the appearance of the first edition, the subject kept freshness. There have been interesting theoretical progress, important new applications and lots of stunning new experiments in the field.

The present edition of *Quantum Dissipative Systems* reflects two endeavours on my part: the improvement and refinement of material contained already in the second edition; the addition of new topics (and the omission of few).

The emphasis and major intensions are still the same, but there are changes, augmentations and additions. The major extensions, altogether about 60 pages, are:

- Chapter 3 contains a more detailed discussion of the quasiclassical Langevin equation and a subsection on Josephson flux and charge qubits.

- Chapter 4 gives wider space to the basics of path integration and to the treatment of an electromagnetic environment.

- Chapter 5 discusses the stochastic unraveling of path integrals for the reduced density matrix.

- Chaper 6 gives an extended discussion of the damped quantum harmonic oscillator. It includes discussion of internal energy, purity, and uncertainty.

- Chapter 15 presents a generalization of the Smoluchowski diffusion equation which includes quantum effects.

- Chapter 20 offers a broader discussion of single-charge tunneling in the weak-tunneling or Coulomb blockade regime.

- Chapter 21 discusses relaxation and decoherence in the spin-boson model at zero temperature. It presents analytical results for the relaxation and decoherence rate at general damping strength which cover the entire regime extending from weak to strong tunneling.

- Chapter 24 includes a discussion of the full counting statistics for Poissonian quantum transport and presents many analytical results available in special cases.

- Chapter 25 presents the scaling-invariant solution of the full counting statistics in diverse limits and discusses application to charge transport in Josephson junctions.

- Chapter 26 gives an extended discussion of charge transport in quantum impurity systems, including full counting statistics. It points out an intimate connection of these systems with models for coherent conductors and with others discussed in the preceding Chapters 23 – 25.

- The bibliography is updated.

This new edition has benefited from comments, suggestions and criticisms from many students and colleagues. Among those to whom I owe specific debt of gratitude are Holger Baur, Pino Falci, Hermann Grabert, Milena Grifoni, Yuli Nazarov, Elisabetta Paladino, Jürgen Stockburger, and Ruggero Vaia.

It is a pleasure to thank the students P. Diemand and A. Herzog for proofreading and tracking down misprints. I also wish to thank the responsible editor Kim Tan for advice and patience until completion of the third edition.

Finally, I am grateful to my wife Christel for her sympathy and constant encouragement.

Stuttgart
October 2007   *Ulrich Weiss*

Preface to the Second Edition

Since the first publication of this book in 1993, there have been enormous research activities in quantum dissipative mechanics both experimentally and theoretically. For this reason, it has been highly desirable after the book has been sold out almost three years ago to undergo a number of extensions and improvements. I have been encouraged by the positive reception of this book by a large community and by many colleagues to write not simply an updated second edition. What came out now after all is almost a new book of roughly double content.

In an extensive rewriting, the 19 Chapters of the First Edition have been expanded by about one third to better meet the desires of both the newcomers to the field and the advanced readership, and I have added 7 new chapters. The most relevant extensions are as follows. In the first part, I have added a section on stochastic dynamics in Hilbert space and I have extended the discussion of relevant microscopic global models considerably. Now, there are also treated acoustic phonons with two-phonon coupling, a microscopic model for tunneling between surfaces, charging and environmental effects in normally conducting and superconducting tunnel junctions, and nonlinear quantum environments. Part II now contains an extended discussion of the damped harmonic oscillator (e.g., a study of the density of states is added), and new chapters on the thermodynamic variational approach and variational perturbation expansion method, and on the quantum decoherence problem. Part III, which deals with quantum-statistical decay, is extended by two chapters. In the new edition, the turnover theory to the energy-diffusion limited regime is discussed, and the treatment of dissipative quantum tunneling has been extended and improved. Ample space is now provided in Part IV to a thorough discussion of the dissipative two-state system. A number of new results on the thermodynamics and dynamics of this archetypal system are presented. An extensive discussion of electron transfer in a solvent, incoherent tunneling in the nonadiabatic regime, and single-charge tunneling is provided in a unified framework. Regarding dynamics, new sections on exact master equations, improved approximation schemes, and recent results on correlation functions have been written, and a new chapter on the driven dissipative two-state is included. Part V, which deals with the dissipative multi-state system, is completely rewritten. It now contains four chapters on quantum Brownian motion in a cosine potential, multi-state dynamics, duality symmetry, and tunneling of charge through an impurity in a quantum wire. Many new results available only very recently are presented. The about 460 references are suggestions for additional reading on particular subjects and are not intended as a comprehensive bibliography.

Stuttgart
December 1998   *Ulrich Weiss*

Acknowledgements

A number of colleagues and friends have made valuable remarks on the first edition. In this regard, I am especially grateful to Theo Costi, Thorsten Dröse, Reinhold Egger, Hermann Grabert, Peter Hänggi, Gert-Ludwig Ingold, Chi Mak, Maura Sassetti, Rolf Schilling, Herbert Spohn, and Wilhelm Zwerger. In writing the new edition, discussions with Pino Falci, Igor Goychuk, Milena Grifoni, Gunther Lang, Elisabetta Paladino, and Manfred Winterstetter have been extremely profitable, and I wish to thank them for their suggestions and for tracking down misprints. I also would like to acknowledge the preparation of a number of figures by Jochen Bauer, Gunther Lang, Jörg Rollbühler, and Manfred Winterstetter. Finally, I would like to thank Mrs. Karen Yeo as the responsible Editor of World Scientific Publishing Co. for her useful advice.

Preface to the First Edition

This book is an outgrowth of a series of lectures which I taught at the ICTP at Trieste and at the University of Stuttgart during the spring and summer of 1991. The purpose of my lectures was to present the approaches and techniques that accurately treat quantum processes in the presence of frictional influences.

The problem of *open* quantum systems has been around since the beginnings of quantum mechanics. Important contributions to this general area have been made by researchers working in fields as diverse as solid-state physics, chemical physics, biophysics, quantum measurement theory, quantum optics, nuclear and particle physics. Often, there has been used, and still is used, a language well known in one context or one field, yet sufficiently different from others that it is not altogether easy to make out the connection. Here, I offer a collection of ideas and examples rather than a comprehensive review of the topic and the history.

The central theme is the space-time functional integral or path integral formulation of quantum theory. This approach is particularly well suited for treating the quantum generalization of friction. Here we are faced to understand the behavior of a system with few quantal degrees of freedom coupled to a thermal reservoir. After integrating out the bath while keeping the system's coordinates fixed we get the influence functional describing the influence of the many bath degrees of freedom on the few relevant ones. This leads to an effective action weighting the paths of the open system in the functional integral description. Indeed, if one wishes to perform numerical computations on a rigorous level, there are no alternatives to this approach at present.

Path integration in condensed matter and chemical physics has become a growth industry in the last one or two decades. A newcomer to this thriving field may not yet be very familiar with the path integral method. Here, I do assume a knowledge of standard text book quantum mechanics and statistical mechanics augmented by a knowledge of Feynman's approach on a first introductory level. The books by Baym [1], Chandler [2], Feynman and Hibbs [3], and by Feynman [4] provide the elementary material in this regard. Further background and supplementary material on the path integral method are contained in the books by Schulman [5] and by Kleinert [6]. However, advanced mastery of these subjects is not necessary.

Some of the more sophisticated concepts, such as preparation functions, propagating functions, and correlation functions, are basic to the development as it is presented here. To cover this material at an introductory level, I make frequent use of simplified models. In this way, I can keep the mathematics relatively simple.

Many of the problems, methods, and ideas which are discussed here have become essential to the current understanding of quantum statistical mechanics. I have made a considerable effort to make the material largely self-contained. Thus, although the theoretical tools are not developed systematically and in its full beauty, the material may be useful to many graduate students to become familiar with the field and learn the methods. For the most part, I refrain from just quoting results without explaining where they come from. With regard to citations, I have tried to give references to the historical development and also to provide a selection of the very recent important ones. But the list is surely not a comprehensive bibliography.

This book exists because of the physics I learned and enjoyed from the fertile collaboration with Hermann Grabert, Peter Hänggi, Gert-Ludwig Ingold, Peter Riseborough, and Maura Sassetti. I am particularly indebted to Maura who took time off her research to weed out points of confusion and who persistently encouraged me to finish this venture. I am also grateful to my students Reinhold Egger, Manfried Milch, Jürgen Stockburger, and Dietmar Weinmann for helpful comments concerning the presentation of many subjects discussed here and for preparing the figures.

In writing this book, I have benefited from the discussion with many companions; in particular Uli Eckern, Enrico Galleani d'Agliano, Anthony J. Leggett, Hajo Leschke, Franco Napoli, Albert Schmid, Gerd Schön, Larry Schulman, Peter Talkner, Valerio Tognetti, Andrei Zaikin, and Wilhelm Zwerger, who helped me in increasing my understanding of many of the subjects which are discussed here.

It is a pleasure to thank my teacher, colleague, and friend Wolfgang Weidlich for many fruitful discussions over the years.

Finally, and most importantly, I am deeply grateful to my wife Christel and my children Ulrike, Jan, and Meike for their infinite patience and omnipresent sympathy.

Stuttgart
October 1992   *Ulrich Weiss*

Contents

1. Introduction

Quantum-statistical mechanics is a very rich and checkered field. It is the theory dealing with the dynamical behavior of spontaneous quantal fluctuations.

When probing dynamical processes in complex many-body systems, one usually employs an external force which drives the system slightly or far away from equilibrium, and then measures the time-dependent response to this force. The standard experimental methods are quasielastic and inelastic scattering of light, electrons, or neutrons off a sample, and the system's dynamics is analyzed from the line shapes of the corresponding spectra. Other experimental tools are, e. g., spin relaxation experiments, studies of the absorptive and dispersive acoustic behavior, and investigations of transport properties. In such experiments, the system's response gives information about the dynamical behavior of the spontaneous fluctuations. Theoretically, the response is rigorously described in terms of time correlation functions. Therefore, time correlation functions are the center of interest in theoretical studies of the relaxation of non-equilibrium systems.

This book deals with the theories of open quantum systems with emphasis on phenomena in condensed matter. An essential ingredient of any particular real dissipative quantum system is the separation of a global quantum system into a subsystem, usually called the relevant part, and the environment, called the irrelevant part. In most cases of practical interest, the environment is thought to be in thermal equilibrium. The coupling to a quantum statistical environment results in a fluctuating force acting on the relevant system and reflecting the characteristics of the heat reservoir. It is the very nature of the fluctuating forces to cause decoherence and damping, and to drive everything to disorder.

While quantum mechanics was conceived as a theory for the microcosm, there is apparently no contradiction with this theory in the mesoscopic and macroscopic world. The understanding of the appearance of classical behavior within quantum mechanics is of fundamental importance. This issue is intimately connected with the understanding of decoherence. Despite the stunning success of quantum theory, there is still no general agreement on the interpretation. The main disputes circle around "measurement" and "observation".

Decoherence is the phenomenon that the superposition of macroscopically distinct states decays on a short time scale. It is omnipresent because information about quantum interference is carried away in some physical form into the surroundings. In a sense, the environmental coupling acts as a continuous measuring apparatus, leading to an incessant destruction of phase correlations. The relevance of this coupling for macroscopic systems is nowadays generally accepted by the respectable community.

This is a book of methods, techniques, and applications. The level is such that anyone with a first course in quantum mechanics and rudimentary knowledge of path integration should not find difficulties. An attempt is made to present the problem of dissipation in quantum mechanics in a unified form. A general framework is developed which can deal with weak and strong dissipation, and with all kinds of memory effects. The reader will find a presentation of the relevant ideas and theoretical concepts, and a discussion of a wide collection of microscopic models. In the models and applications, emphasis is put on condensed matter physics. We have tried to use vocabulary and notation which should be fairly familiar to scientists working in chemical and condensed matter physics.

The book is divided into five parts. The following sequence of topics is adopted. The first part of the book is devoted to the general theory of open quantum systems. In Chapter 2, we review traditional approaches, such as formulations by master equations for weak coupling, operator-valued and quasiclassical Langevin equations. We also discuss attempts to interpret the dynamics of an open quantum system in terms of a stochastic process in the Hilbert space of state vectors pertaining to the reduced system. In Chapter 3, various global models are introduced. They are partly connected with microscopic models which are of relevance in condensed matter physics. Chapter 4 is devoted to the equilibrium statistical mechanics for the relevant subsystems of these models using the imaginary-time path integral approach. Chapter 5 concerns dynamics — quantum-mechanical motion, decoherence and relaxation of macroscopic systems that are far from or close to equilibrium. In particular, we discuss the concepts of preparation functions, propagating functions, and correlation functions and derive the corresponding exact formal solutions using the path integration method.

Part II with Chapters 6 – 9 covers a discussion of exactly solvable damped linear quantum systems (damped harmonic oscillator and free Brownian particle), the useful thermodynamic variational approach with extension to open nonlinear quantum systems, and a semiclassical treatment of the quantum decoherence problem for a particle traveling in a medium.

Part III deals with quantum-statistical metastability: a problem of fundamental importance in chemical physics and reaction theory. After an introduction into the problem in Chapter 10, the relevant theoretical concepts and the characteristic features of the decay are discussed in Chapters 11 to 17 . The treatment mainly relies on a thermodynamic method in which the decay rate is related to the imaginary part of the analytically continued free energy of the damped system. This allows for a uniform theoretical description in the entire temperature range. The discussion extends from high temperatures where thermal activation prevails down to zero temperature where the system can only decay by quantum-mechanical tunneling out of the ground state in the metastable well. Results in analytic form are presented where available.

In Part IV, we consider the thermodynamics and dynamics of the dissipative two-state or spin-boson system. This is the simplest nonlinear system for the study of the interplay between quantum coherence, quantal and thermal fluctuations, and friction. After an introduction into the model in Chapter 18, the discussion in Chapter 19 is focused on equilibrium properties for a general form of the system-reservoir coupling. In particular, the partition function is discussed and the specific heat and static susceptibility are studied. The relationship with Kondo and Ising models is explained.

Chapter 20 is devoted to the electron transfer problem in a solvent, nonadiabatic tunneling under exchange of energy, and single charge tunneling in the presence of an electromagnetic environment. Chapter 21 deals with the dissipative two-state dynamics. Different kinds of initial preparations of the system-plus-reservoir complex are treated and exact formal expressions for the system's dynamics in the form of series expressions and generalized master equations are derived. Ample space is devoted to the discussion of non-equilibrium and equilibrium correlation functions, and to adequate approximate treatments in the various regions of the parameter space. Part IV closes with a chapter on the dynamics of the dissipative two-state system under exposure to time-dependent external fields.

The last part reviews dissipative quantum transport of a quantum Brownian particle in a tilted cosine potential. In Chapters 23 and 24, we introduce the respective global models in the weak- and strong-tunneling representations. We present the appropriate nonequilibrium quantum transport formalism and derive exact formal expressions describing the system's dynamics for factorizing and thermal initial states. Furthermore, explicit solutions in analytic form in various limits are given. Chapter 25 provides a discussion of a duality symmetry between the weak- and strong-tunneling representations which becomes an exact self-duality in the so-called Ohmic scaling limit. We show that self-duality offers the possibility to construct the exact scaling function for the nonlinear mobility at zero temperature. In addition, the full counting statistics of these generic quantum transport models is discussed. The book closes with a chapter on quantum transport of charge in quantum impurity models. Both the weak- and strong-tunneling representations are discussed. In addition, the close relationship of the quantum impurity model with the Brownian particle model, and with charge transport in a coherent conductor and in Josephson systems is pointed out. This allows us to to translate results obtained for one of these system in corresponding results for related other systems.

We have tried to concentrate on models which are simple enough to be largely tractable by means of analytical methods. There are, however, important examples where numerical computations have given clues to the analytical solution of a problem. If one wishes to calculate the full dynamics of the global system, one is faced with the problem that the number of basis states is growing exponentially. Therefore, even on supercomputers, the number of reservoir modes which can be treated numerically exactly, is rather limited. When the number of bath modes is above ten or even tends to infinity, an inclusive description of the environmental effects, e.g., in terms of the influence functional method (cf. Chapters 4 and 5) is indispensable. Various numerical schemes developed within the framework of the influence functional approach are available. The most valuable numerical tool in many-body quantum theory is probably the path integral Monte Carlo simulation method. Unfortunately, in simulations of the real-time quantum dynamics, the numerical stability of long-time propagation is spoilt by the destructive interference of different paths contributing to the path sum. This so-called *dynamical sign problem* is intrinsic in real-time quantum

mechanics, and is characterized by an exponential drop of the signal-to-noise ratio
with increasing propagation time.

In recent years, considerable progress in reducing the sign problem has been
achieved by implementation of blocking algorithms in quantum Monte Carlo simulations based on a Trotter split-up of the elementary propagator. A possible strategy
consists in sampling "blocks" of which the corresponding average sign is nonzero,
instead of single states. This method always reduces the sign problem [C. H. Mak
and R. Egger, Adv. Chem. Phys. **93**, 39 (1996)]. Alternatively, one may use iterative procedures which are based on systematic approximations. In the so-called
tensor-propagator approach, a maximal correlation time of the influence functional
interactions is introduced [N. Makri, J. Math. Phys. **36**, 2430 (1995)]. In the so-
termed path class approach, the exact summation of a class of paths is approximated
by a low-order cumulant expansion of averages of the path class history [M. Winterstetter and W. Domcke, Chem. Phys. Letters **236**, 455 (1995)]. As an alternative
method, Stockburger and Grabert proposed to unravel the Feynman-Vernon influence functional into a stochastic one, and to solve numerically the ensuing stochastic
Liouville-von Neumann equations, or the related stochastic Schrödinger equations,
which are free of quantum memory effects [J.T. Stockburger and H. Grabert, Phys.
Ref. Lett. **88**, 170407 (2002)]. We have refrained from adding sections which deal
with numerical methods in detail. Where appropriate, we give relevant information
and literature.

After all, the reader may not find a comprehensive account of what interests him
most. Since the number of articles in this general field has become enormous in recent
years, a somewhat arbitrary choice among the various efforts is inevitable. My choice
of topics is just one possibility. It reflects, to some extent, the author's personal
valuation of an active and rapidly developing area in science.

PART I

GENERAL THEORY OF OPEN QUANTUM SYSTEMS

2. Diverse limited approaches: a brief survey

Often in condensed phases, a rather complex physical situation can adequately be described by a global model system consisting of only one or few relevant dynamical variables in contact with a huge environment, of which the number of degrees of freedom is very large or even infinity. If we are interested in the physical properties of the *small* relevant system alone, we have to handle this system as an *open system* which exchanges energy with its surroundings in a random manner. In recent years, a variety of theoretical methods for open quantum systems has been developed and employed. The emphasis in this book is on the functional integral approach to open quantum systems. This method has turned out to be very powerful and has found broad application. Nevertheless, I find it appropriate to begin with a brief survey of various other formalisms. Clearly, the short discussion given subsequently can not do justice to all of them. However, I hope that the interested reader will be able to get a line along the given references for deeper studies. I find it appropriate to begin with a brief discussion of the classical regime.

2.1 Langevin equation for a damped classical system

It is our everyday experience that the motion of any macroscopic physical system comes to a stop when supply of energy is cut off. The reason for this is energy dissipation: in fact, no system can be completely isolated from the surroundings. Therefore, energy accumulated in the system is inevitably given away to the environment, and the system is trapped in a local minimum of the potential energy. The loss of energy is phenomenologically described in terms of a frictional force. In this place we anticipate that the open system also is subject to a fluctuating or noise force and that dissipation and fluctuations are related.

Consider for simplicity an open system with a single degree of freedom, which we associate with the coordinate $q(t)$ of a particle with mass M. The simplest assumptions about the dissipative process one can make is that dissipation is state-independent. Then the frictional force is a *linear* functional of the history of the velocity $\dot{q}(t)$, and the stochastic force $\xi(t)$ obeys stationary Gaussian statistics. It is fully characterized by the ensemble averages

$$\langle \xi(t) \rangle = 0 , \qquad \langle \xi(t)\xi(0) \rangle_{\mathrm{cl}} \equiv \mathcal{X}_{\mathrm{cl}}(t) . \qquad (2.1)$$

A classical heat reservoir at temperature T with zero memory time constitutes a white noise source. Then the frictional force is local in time, $F_{\text{fric}}(t) = -M\gamma\dot{q}(t)$, where γ is the damping rate, and the stochastic force is δ-correlated according to

$$\mathcal{X}_{\text{cl}}(t) = 2M\gamma k_{\text{B}}T\,\delta(t)\,. \tag{2.2}$$

The dynamics of the damped particle is described by the Langevin equation

$$M\ddot{q}(t) + M\gamma\,\dot{q}(t) + V'(q) = \xi(t)\,. \tag{2.3}$$

Time-local friction proportional to the velocity is usually called Ohmic because of the correspondence with a series resistor in an electrical circuit. The dynamical equation (2.3) describes, for example, a heavy Brownian particle with mass M immersed in a fluid of light particles and driven by a systematic force $-V'(q)$, where $V(q)$ is an externally applied potential. Equation (2.3) together with the relations (2.1) and (2.2) forms the basis for the theory of Brownian motion since the seminal studies by Einstein, Langevin and Smoluchowski. The early work on Brownian motion theory was reviewed in an excellent article by Chandrasekhar [7].

In many cases of practical interest, the heat reservoir exhibits retardation. Then it is a source of coloured noise, which has finite memory time, and friction depends on the velocity in the past. The corresponding dynamical equation is the generalized Langevin equation

$$M\ddot{q}(t) + M\int_{-\infty}^{t} dt'\,\gamma(t-t')\,\dot{q}(t') + V'(q) = \xi(t)\,. \tag{2.4}$$

Since the random force $\xi(t)$ has zero mean, the effect of the reservoir on average is in the memory-friction kernel $\gamma(t)$, which obeys causality, $\gamma(t) = 0$ for $t < 0$. Hence the frequency-dependent damping function[1]

$$\tilde{\gamma}(\omega) \equiv \tilde{\gamma}'(\omega) + i\,\tilde{\gamma}''(\omega) = \int_{-\infty}^{\infty} dt\,\gamma(t)\,e^{i\omega t} \tag{2.5}$$

satisfies Kramers-Kronig relations. It is connected with the force autocorrelation function $\langle \xi(t)\xi(t')\rangle_{\text{cl}} = \langle \xi(t-t')\xi(0)\rangle_{\text{cl}}$ by the Green-Kubo formula [8]

$$\tilde{\gamma}(\omega) = \frac{1}{Mk_{\text{B}}T}\int_{0}^{\infty} dt\,\langle \xi(t)\xi(0)\rangle_{\text{cl}}\,e^{i\omega t}\,. \tag{2.6}$$

From this we infer that the power spectrum of the classical stochastic force

$$\tilde{\mathcal{X}}_{\text{cl}}(\omega) = \int_{-\infty}^{\infty} dt\,\mathcal{X}_{\text{cl}}(t)\cos(\omega t) \tag{2.7}$$

is related to the real part $\tilde{\gamma}'(\omega)$ of the damping function $\tilde{\gamma}(\omega)$ by the relation

---

[1]Throughout this book, we use for the Fourier transform the normalization and sign convention in the exponent as in Eq. (2.5).

$$\widetilde{\mathcal{X}}_{\mathrm{cl}}(\omega) \; = \; 2Mk_{\mathrm{B}}T\,\widetilde{\gamma}'(\omega)\,. \tag{2.8}$$

This is a version of the classical fluctuation-dissipation or Nyquist theorem.

The Langevin equation (2.4) will be derived from a global system-plus-reservoir Hamiltonian in Subsection 3.1.2.

Evidently, the equation (2.4) with (2.8) is limited to the classical domain. We expect that at sufficiently low temperatures all types of quantum effects should occur. Since the standard procedure of quantization relies upon the existence of a Lagrangian or a Hamiltonian function for the system, the question arises: how can one reconcile dissipation with the canonical scheme of quantization?

2.2 New schemes of quantization

The equation of damped motion (2.3) can not be obtained from the application of Hamilton's principle unless the Lagrangian has an explicit time dependence. The use of time-dependent Lagrangians or Hamiltonians would permit us to use the standard schemes of quantization directly. Historically, the first researchers taking this path were Caldirola [9] and Kanai [10] who employed a time-dependent mass chosen in such a way that a friction term appears in the corresponding classical equation of motion. However, this procedure does not properly handle the uncertainty principle [11]. As the resumé of diverse studies, it is generally accepted meanwhile that dissipation cannot be described adequately by simply introducing a time-dependent mass.

Many approaches to open quantum systems were introduced over the last thirty years. The variety of attempts falls into three main categories. One either modifies the procedure of quantization, or one uses a so-called stochastic Schrödinger equation for state vectors, or the system-plus-reservoir approach is employed.

Among the first group, Dekker [12] proposed a theory with a canonical quantization procedure for complex variables, thereby reproducing the Fokker-Planck equation for the Wigner distribution function. However, some *ad hoc* assumptions in the theory seem questionable, such as the introduction of noise sources in the canonical equations for position and momentum. Kostin [13] introduced a theory with a nonlinear Schrödinger equation. The same equation was found later by Yasue [14] using Nelson's stochastic quantization procedure [15]. However, this theory violates the superposition principle, and also yields some dubious results such as stationary damped states. Apart from the fact that the theoretical foundations are completely unclear, these approaches can reproduce, at best, known results only for very limited cases, such as weakly damped linear systems. Therefore, all attempts of the first group can be assessed to have failed. We shall not consider them further here. The second category is discussed in some detail in Section 2.4.

The more natural, and also more successful approach has been to view the dissipative system as a relevant system with a single or few significant degrees of freedom, which is in contact with (infinitely) many degrees of freedom. These additional degrees of freedom are commonly referred to as bath, or reservoir or environment. Both the relevant system and the reservoir are the constituents of an energy-conserving global system which obeys the standard rules of quantization. In this picture, friction comes about by the transfer of energy from the "small" system to the "huge" environment. The energy, once transferred, dissipates into the environment and is not given back within any physically relevant period of time.

2.3   Traditional system–plus–reservoir methods

Common approaches to open quantum systems based on system-plus-reservoir models are generally divided into two classes. Working in the Schrödinger picture, the dynamics is conventionally described in terms of generalized quantum master equations for the reduced density matrix or density operator [16] – [18]. Working in the Heisenberg picture, the description is given in terms of generalized Langevin equations for the relevant set of operators of the reduced system [19] – [21].

2.3.1   Quantum-mechanical master equations for weak coupling

The starting point of this method is the familiar Liouville equation of motion for the density operator $W(t)$ of the global system,

$$\dot{W} = -\frac{i}{\hbar}[H, W(t)] \equiv \mathcal{L}W(t), \qquad (2.9)$$

where H is the Hamiltonian of the total system, and where the second equality defines the Liouville operator $\mathcal{L}$. Next, assume that the Hamiltonian H and the Liouvillian $\mathcal{L}$ of the total system are decomposed as

$$H = H_{\mathrm{S}} + H_{\mathrm{R}} + H_{\mathrm{I}} ; \qquad \mathcal{L} = \mathcal{L}_{\mathrm{S}} + \mathcal{L}_{\mathrm{R}} + \mathcal{L}_{\mathrm{I}} . \qquad (2.10)$$

The individual parts refer to the free motion of the relevant system and of the reservoir, and to the interaction term, respectively. Employing a certain projection operator P, chosen as to project on the relevant part of the density matrix, the full density operator is reduced to an operator acting only in the space of the relevant variables,

$$\rho(t) = PW(t) . \qquad (2.11)$$

The operator $\rho(t)$ is usually called *reduced density* operator. For systems with the Hamiltonian form (2.10), the projection operator contains a trace operation over the reservoir coordinates. By means of the projection operator P, the density operator can be decomposed into the relevant part $\rho(t)$ and the irrelevant part $(1 - P)W(t)$,

$$W(t) = \rho(t) + (1 - P)W(t) ; \qquad P^2 = P . \qquad (2.12)$$

Upon substituting the decomposition (2.12) into Eq. (2.9), and acting on the resulting equation from the left with the operator P and with the operator $1 - P$, respectively, we obtain two coupled equations for the relevant part $\rho(t)$ and the irrelevant part $(1 - P)W(t)$. A closed equation for $\rho(t)$ is obtained by inserting the formal integral for $(1 - P)W(t)$ into the first equation. We then finally arrive at the formally exact generalized quantum master equation, the *Nakajima-Zwanzig equation* [16, 17]

$$\dot{\rho}(t) = P\mathcal{L}\rho(t) + \int_0^t dt' \, P\mathcal{L} \exp[(1-P)\mathcal{L}t'] \, (1-P)\mathcal{L}\rho(t-t')$$
$$+ P\mathcal{L} \exp[(1-P)\mathcal{L}t] \, (1-P)W(0) \, . \qquad (2.13)$$

The generalized master equation is an inhomogeneous integro-differential equation in time. It describes the dynamics of the open (damped) system in contact with the reservoir $\mathcal{R}$. Observe that the inhomogeneity in Eq. (2.13) still depends on the initial value of the irrelevant part $(1 - P)W(0)$. In applications, it is attempted to choose the projection operator in such a way that the irrelevant part of the initial state $(1 - P)W(0)$ can be disregarded. Assuming further that P commutes with $\mathcal{L}_S$ one then finds the homogeneous time-retarded quantum master equation

$$\dot{\rho}(t) = P(\mathcal{L}_S + \mathcal{L}_I)\rho(t) + \int_0^t dt' \, P\mathcal{L}_I e^{(1-P)\mathcal{L}t'} \, (1-P)\mathcal{L}_I\rho(t-t') \, . \qquad (2.14)$$

The first (instantaneous) term describes the reversible motion of the relevant system while the second (time-retarded) term brings on irreversibility. It describes all possible effects the reservoir may exert on the system, such as relaxation, decoherence and energy shifts. Equation (2.14) is still too complicated for explicit evaluation. First, the kernel of (2.14) contains any power of $\mathcal{L}_I$. Secondly, the dynamics of ρ at time t depends on the whole history of the density matrix. In order to surmount these difficulties, one usually considers the kernel of Eq. (2.14) only to second order in $\mathcal{L}_I$. Disregarding also retardation effects, one finally arrives at the quantum master equation in Born-Markov approximation

$$\dot{\rho}(t) = P(\mathcal{L}_S + \mathcal{L}_I)\rho(t) + \int_0^t dt' \, P\mathcal{L}_I e^{(1-P)(\mathcal{L}_S+\mathcal{L}_R)t'} \, (1-P)\mathcal{L}_I\rho(t) \, . \qquad (2.15)$$

Master equations of this form were successfully used to describe weak-damping phenomena, for instance in quantum optics or spin dynamics. Various excellent reviews of this sort of approach including many applications are available in Refs. [22] – [30]. While the Markov assumption can easily be dropped, the more severe limitation of this method is the Born approximation for the kernel. The truncation of the Born series at second order in the interaction $\mathcal{L}_I$ effectively restricts the application of these generalized master equations to weakly damped systems with relaxation times that are large compared to the relevant time scales of the reversible dynamics.

When the Born-Markov quantum master equation (2.15) is given in the energy eigenstate basis of H_S, it is usually referred to as *Redfield* equation [31, 26, 32]

$$\dot{\rho}_{nm}(t) = -i\,\omega_{nm}\,\rho_{nm}(t) - \sum_{k,l} R_{nmkl}\,\rho_{kl}(t) \, . \qquad (2.16)$$

The first term represents the reversible motion in terms of the transition frequencies ω_{nm}, and the second term describes relaxation. The Redfield relaxation tensor reads

$$R_{nmkl} = \delta_{lm}\sum_r \Gamma^{(+)}_{nrrk} + \delta_{nk}\sum_r \Gamma^{(-)}_{lrrm} - \Gamma^{(+)}_{lmnk} - \Gamma^{(-)}_{lmnk} \, . \tag{2.17}$$

The rates are given by the Golden Rule expressions

$$\Gamma^{(+)}_{lmnk} = \hbar^{-2}\int_0^\infty dt\, e^{-i\omega_{nk}t}\langle \widetilde{H}_{I,\,lm}(t)\widetilde{H}_{I,\,nk}(0)\rangle_R \, ,$$

$$\Gamma^{(-)}_{lmnk} = \hbar^{-2}\int_0^\infty dt\, e^{-i\omega_{lm}t}\langle \widetilde{H}_{I,\,lm}(0)\widetilde{H}_{I,\,nk}(t)\rangle_R \, . \tag{2.18}$$

Here, $\widetilde{H}_I(t) = \exp(iH_R t/\hbar)H_I\exp(-iH_R t/\hbar)$ is the interaction in the interaction picture, and the bracket denotes thermal average of the bath degrees of freedom.

The Redfield equations (2.16) are well-established in wide areas of physics and chemistry, e.g., in nuclear magnetic resonance (NMR), in optical spectroscopy, and in laser physics. In NMR, one deals with the externally driven dynamics of the density matrix for the nuclear spin [33, 34, 35]. In optical spectroscopy, a variant of the Redfield equations are the optical Bloch equations [36, 37].

Multilevel Redfield theory has been applied to the electron transfer dynamics in condensed phase reactions by several authors [38].

Markovian reduced density matrix (RDM) theory has been also utilized in the diabatic state representation [39]. Denoting electronic-vibrational direct-product states in the diabatic representation[2] by α, β, $\cdots$, the RDM equations of motion read

$$\dot{\rho}_{\alpha\beta}(t) = -i\,\omega_{\alpha\beta}\,\rho_{\alpha\beta}(t) - \frac{i}{\hbar}\sum_\nu [\,V_{\alpha\nu}\rho_{\nu\beta}(t) - V_{\nu\beta}\rho_{\alpha\nu}(t)\,] - \sum_{\nu,\sigma} R_{\alpha\beta\nu\sigma}\,\rho_{\nu\sigma}(t) \, . \tag{2.19}$$

The $\omega_{\alpha\beta}$ are the transition frequencies between unperturbed diabatic surfaces, the $V_{\alpha\beta}$ are the matrix elements of the diabatic interstate coupling, and the relaxation tensor $R_{\alpha\beta\nu\sigma}$ describes relaxation of the diabatic electronic-vibrational states. For electron transfer processes, the diabatic surfaces can often be taken as harmonic, which simplifies the calculation of the relaxation tensor drastically. The master equation (2.19) in the diabatic basis is especially useful when the interstate coupling is weak or moderate. In contrast to Redfield theory in the system's eigenstate basis, the computation of the Redfield tensor in the diabatic basis, Eq. (2.19), is without difficulties even for multimode vibronic-coupling systems [42]. When the RDM approach is applied to complex systems, the dimension N of the Hilbert space of the relevant system H_S is possibly very large. Since the density matrix scales with N^2 and the relaxation tensor with N^4, the computational problem may become easily nontrivial. The so-called Monte-Carlo wave function propagation or quantum jump method discussed below in Section 2.4, which is equivalent to the Born-Markov RDM approach, provides a considerably more favorable scaling of the computational costs with the number of states than the direct integration of the Redfield equations.

Within the conception of quantum mechanics, the time evolution of the density operator of a closed system is a unitary map. If the system is open, the possible

---

[2]The concepts of diabatic states are discussed in recent reviews [40, 41, 42].

transformations are thought to be "completely positive" [43, 29], $\rho \to \sum_n O_n \rho O_n^\dagger$. Here $\{O_n\}$ is a set of linear operators on the reduced state space, restricted only by $\sum_n O_n^\dagger O_n = 1$, which guarantees that $\mathrm{tr}\,\rho$ does not change. The most general form of generators $\mathcal{L}$, $\dot\rho(t) = \mathcal{L}\rho(t)$, preserving complete positivity of density operators and conveying time-directed irreversibility is established by the Lindblad theory [43]. The Lindblad form of the quantum master equation for an open system reads

$$\frac{d\rho(t)}{dt} = -\frac{i}{\hbar}[\,H_{\mathrm{S}}, \rho(t)\,] + \frac{1}{2}\sum_j \left\{ [\,L_j\rho(t), L_j^\dagger\,] + [\,L_j, \rho(t)L_j^\dagger\,] \right\} . \qquad (2.20)$$

The first term represents the reversible dynamics of the relevant system. The Lindblad operators L_j describe the effect of the environment on the system in Born-Markov approximation. In concrete applications of Eq. (2.20), the L_j transmit emission and absorption processes. For linear dissipation, the simplest form of a Lindblad operator is a linear combination of a raising and lowering operator, or equivalently of a coordinate and a momentum operator [44],

$$L = \mu q + i\nu p , \qquad L^\dagger = \mu q - i\nu p , \qquad (2.21)$$

where μ and ν are c-numbers. Explicit temperature-dependent expressions for μ and ν are obtained for the damped harmonic quantum oscillator by matching the dissipation terms of the Lindblad master equation with corresponding expressions of this exactly solvable model (cf. Chapter 6) in the Born-Markov limit [45].

Complete positivity of reduced density operators is considered as a strict guideline by researchers working on stochastic Schrödinger equations (see Section 2.4). However, as argued by Pechukas [46, a], complete positivity of the RDM is an artefact of product initial conditions $W^{(i)} = \rho^{(i)} \otimes \rho_{\mathrm{R}}^{(i)}$ for the global density matrix. Product initial conditions are only appropriate in the weak-coupling limit. In general, reduced dynamics need not be completely positive. It is known that the Redfield-Bloch master equation may break the positivity of the density matrix. Violation of positivity may occur if memory effects are not adequately taken into account at the initial stage of the time evolution. Then some "slippage" of the initial conditions must occur before the reduced dynamics looks Markovian [46, b]. The slippage captures the effects of the actual non-Markovian evolution in a short transient regime of the order of the reservoir's memory time. (see also [46, c]). The Markovian master equation holds only in a subspace in which rapidly decaying components of the density matrix are disregarded, whereas Lindblad theory requires validity for *any* reduced density matrix. These findings have been confirmed in a study of the exactly solvable damped harmonic oscillator [47]. Using the exact path integral solution, it has been shown that in general there is no exact dissipative Liouville operator describing the dynamics of the oscillator in terms of an exact master equation that is independent of the initial preparation. Exact non-stationary Liouville operators can be found only for particular preparations. Time-independent Liouville operators which are valid for arbitrary preparation of the initial state can be extracted only in a sub-space where the fast transient components have already decayed. However, the Liouville operators are still not of the Lindblad form. The Lindblad master equation is obtained only when the weak-coupling limit is performed and a coarse graining in time is carried out. It is perfectly obvious from the study of this exactly solvable model that one should not attribute fundamental significance to the Lindblad master equation.

Besides the Markov approximation and the weak-coupling limit, there is another severe limitation of the usefulness of the above quantum master equations. Namely, the relaxation dynamics towards the equilibrium state $\rho_\beta = P W_\beta$, where W_β is the thermal equilibrium density matrix of the global system, is correctly described only when the system's relaxation times are large compared to the thermal time $\hbar\beta$ (see, e.g., the discussion in Chapters 6 and 17).

In conclusion, the Born-Markov quantum master equation approach provides a reasonable description in many cases, e. g. in NMR, in quantum optics, and in a variety of chemical reactions. However, this method is not applicable in most problems of solid state physics at low temperatures for which neither the Born approximation is valid nor the Markov assumption holds.

2.3.2 *Operator Langevin equations for weak coupling*

Just as we projected the density matrix $W(t)$ of the global system onto the relevant part $\rho(t)$, we may proceed in the Heisenberg picture by projecting the operators of the global system on the set of macroscopically relevant operators. The various efforts in studying the dynamics of these operators have been described by Gardiner [49].

Let us denote the set of operators of the global system by $\{X\}$ and the set of macroscopically relevant operators governing the open system by $\{Y\}$. Now consider the operators X_i and Y_μ as elements $|X_i)$ and $|Y_\mu)$ in the Liouville space Λ.[3] At time $t = 0$, the operators $\{Y\}$ span a subspace Λ_Y of the Liouville space Λ. We use the convention that an operator in the Heisenberg representation without time argument denotes the operator at time zero. Next, it is convenient to define a time-independent projection operator which projects onto the subspace Λ_Y,

$$\mathcal{P} = \sum_{\mu,\nu} |Y_\mu)\, g_{\mu\nu}\, (Y_\nu| \; ; \qquad \mathcal{P}^2 = \mathcal{P} \,. \tag{2.22}$$

The metric $g_{\mu\nu}$ is the inverse of the scalar product $(Y_\mu|Y_\nu)$ which has to be chosen appropriately in practical calculations. For quantum statistical linear response and relaxation problems, a suitable form is the Mori scalar product [21, 48]

$$(Y_\mu|Y_\nu) \equiv \frac{1}{\beta} \int_0^\beta d\lambda \, \langle \mathrm{e}^{-\lambda H}\, Y_\mu^\dagger\, \mathrm{e}^{\lambda H}\, Y_\nu \rangle \,. \tag{2.23}$$

The angular brackets denote average with respect to the canonical ensemble of the global system $W_\beta = Z_\beta^{-1}\, \mathrm{e}^{-\beta H}$, and $\beta = 1/k_\mathrm{B}T$. The superoperator $\mathcal{P}$ projects onto the subspace Λ_Y according to

---

[3]See Ref. [48] for a review of the formulation of quantum mechanics in Liouville space.

$$\mathcal{P}\,|X_i) \;=\; \sum_{\mu,\nu} |Y_\mu)\, g_{\mu\nu}\,(Y_\nu|X_i)\,. \tag{2.24}$$

Acting now from the left with $\mathcal{P}$ and with $1-\mathcal{P}$ on the Heisenberg equation of motion

$$|\dot{X}_i) \;=\; \mathcal{L}\,|X_i)\,, \tag{2.25}$$

where $\mathcal{L}$ is the Liouville superoperator, and eliminating $(1-\mathcal{P})|X_i)$ with the aid of the exact formal solution, it is straightforward to derive for $Y_\mu(t)$ a system of coupled integro-differential equations, which have been popularized as the Mori equations.

$$\dot{Y}_\mu(t) \;=\; i\sum_\nu Y_\nu(t)\,\Omega_{\nu\mu}(t) - \sum_\nu \int_0^t ds\, Y_\nu(t-s)\,\gamma_{\nu\mu}(s) + \xi_\mu(t)\,. \tag{2.26}$$

The generally temperature-dependent drift matrix $\Omega_{\nu\mu}(t)$ is given by

$$i\,\Omega_{\nu\mu}(t) \;=\; \sum_\rho g_{\nu\rho}\,(Y_\rho|\dot{Y}_\mu(t))\,. \tag{2.27}$$

The stochastic force $\xi_\nu(t)$ is a functional of the operators of the *irrelevant* part,

$$\xi_\nu(t) \;=\; \exp[(1-\mathcal{P})\mathcal{L}\,t]\,(1-\mathcal{P})\dot{X}_\nu(t)\,. \tag{2.28}$$

Finally, the memory matrix is expressed in terms of the correlation function of the stochastic force as

$$\gamma_{\nu\mu}(t) \;=\; \sum_\rho g_{\nu\rho}(\xi_\rho|\xi_\mu(t))\,. \tag{2.29}$$

The actual computation of the fluctuating force and of the memory matrix is again restricted to weak coupling. Altogether, this approach is subject to exactly the same limitations we encountered above in the master equation method.

In conclusion, it is important in the Mori formalism that the complete set of macrovariables spans the subspace. Otherwise, the fluctuating force contains slowly varying components, and the separation of time scales is incomplete.

2.3.3 Quantum and quasiclassical Langevin equation

One further approach consists in attempting to generalize the classical Langevin equation for a Brownian particle to the quantum case [20, 50, 51]. The quantum mechanical version of the Langevin equation for the coordinate operator reads[4]

$$M\frac{d^2\hat{q}(t)}{dt^2} + M\int_{t_0}^t dt'\,\gamma(t-t')\frac{d\hat{q}(t')}{dt'} + V'(\hat{q}) \;=\; \hat{\xi}(t)\,, \tag{2.30}$$

where $\hat{\xi}(t)$ is the Gaussian random force operator with autocorrelation

$$\langle\hat{\xi}(t)\hat{\xi}(0)\rangle \;=\; \frac{\hbar M}{\pi}\int_0^\infty d\omega\,\omega\tilde{\gamma}(\omega)\Big(\coth(\beta\hbar\omega/2)\cos(\omega t) - i\sin(\omega t)\Big)\,. \tag{2.31}$$

---

[4]Here we dropped a term which depends on the initial condition and which decays on a time scale given by the memory time of the reservoir (See the discussion in Subsection 3.1.3).

The equation (2.30) can be derived, e. g., for the linear response oscillator model discussed below in Section 3.1 [50]. One can even show that the form (2.31) of the correlation is a general result of the fluctuation-dissipation theorem (cf. Section 6.2) and is therefore independent of the model.[5] Benguria and Kac [52], and Ford and Kac [50] argued that the system approaches the correct equilibrium state for the form (2.30) with (2.31) and Gaussian noise. Recently, the quantum Langevin equation (2.30) has been derived for the white-noise case from the Feynman-Vernon forward-backward path integral discussed below in Section 5.1 [53].

In the *quasiclassical* Langevin equation (QLE) [19, 54], the operator-valued quantities in Eq. (2.30) are replaced by c-numbers, but the Gaussian property of the stochastic force and the correlation (2.31) are retained. We then end up at the generalized classical Langevin equation (2.4) in which the power spectrum of the force autocorrelation function has the quantum mechanical form

$$\widetilde{\mathcal{X}}(\omega) \equiv \int_{-\infty}^{\infty} dt \, \langle \xi(t)\xi(0) \rangle \cos(\omega t) = M\hbar\omega \coth\left(\frac{\hbar\omega}{2k_{\mathrm{B}}T}\right) \widetilde{\gamma}'(\omega) . \qquad (2.32)$$

The derivation of this expression from a quantum statistical ensemble average is given in Subsection 3.1.4. In addition, we briefly sketch in Section 5.5 the derivation of the QLE within the path integral method following the approach by Schmid [54]. There, it will turn out that the QLE is exact (apart from disregarding a term describing the initial transient behaviour) if the external force is harmonic (see also Ref. [19]). The QLE gives a reasonable description for systems which are nearly harmonic [55, 56]. However, the predictions of the QLE are unreliable when the anharmonicity of the potential is of crucial importance like, for instance, in quantum tunneling. The mere insertion of the quantum noise (2.32) into a classical equation is insufficient to render a proper description of the quantum statistical decay of a metastable state [55].

When the unharmonicity of the potential is relevant, the most successful approach is the functional integral description. Like in the classical regime, the dissipative system is considered to interact with a complex environment, and the "complete universe" formed by the system plus environment is assumed to be energy-conserving so that it can be quantized in the standard way. For equations of motion which are linear in the bath coordinates, the environment can easily be eliminated. Thus one obtains closed equations for the damped system alone. In the path integral description, the environment reveals itself through an influence functional depending on the spectral properties of the environmental coupling and on temperature. A general discussion of the influence functional method is presented in Chapters 4 and 5.

2.3.4 Phenomenological methods

Often, a physical or chemical system cannot be characterized by a simple model Hamiltonian of the form (2.10), or the Hamiltonian is simply unknown. In such sit-

---

[5]The Gaussian property of the force operator does not seem to follow from such general considerations, but is implied by a harmonic oscillator reservoir.

uations, it is sometimes useful to describe the dissipative quantum dynamics on a phenomenological level. For instance, one may introduce a dynamical description of the system in terms of occupation probabilities $p_n(t)$ of energy levels or of spatially localized states, rather than in terms of complex probability amplitudes or wave functions, or the full reduced density matrix $\rho_{nn'}$. Then, the relaxation dynamics of a macroscopic system is described by a *Pauli master equation* for $p_n(t) \equiv \rho_{nn}(t)$,

$$\dot{p}_n(t) = \sum_m [A_{nm} p_m(t) - A_{mn} p_n(t)] . \qquad (2.33)$$

The first term on the r.h.s. describes the gain and the second term the loss of probability to occupy the state n. In this formulation, knowledge of the full set of transition rates $\{A_{nm}\}$ is requisite in order to have a complete description of the relaxation process. The transition rates may be inferred, e. g., from standard quantum mechanical perturbation theory, or from experimental data, or they may be chosen by a phenomenological ansatz. The Pauli master equation (2.33) has found widespread application to the study of rate dynamics in physics, chemical kinetics, and biology.

In the quantum coherence regime, off-diagonal matrix elements of the density matrix become relevant. Nevertheless, the full coherent dynamics of the populations can still be formulated in terms of dynamical equations for diagonal matrix elements. However, the corresponding master equation is time-nonlocal (see Section 21.2.6).

2.4 Stochastic dynamics in Hilbert space

In recent years, there have been made numerous attempts to postulate non-Hamiltonian dynamics as a fundamental modification of the Schrödinger equation in order to explain the spontaneous stochastic collapse of the wave function and the appearance of a "classical world" (cf. for a survey the article by I.-O. Stamatescu in Ref. [57]). In these approaches, the non-unitary dynamics of open quantum systems is interpreted in terms of a fundamental stochastic process in the Hilbert space of state vectors pertaining to the system. To retain the standard probability rules, the respective dynamical equations become inevitably nonlinear. The evolution of state vectors is considered as a stochastic Markov process, and the covariance matrix of the state vector is taken as the density operator. The stochastic process is usually constructed in such a way that the equation of motion of the density operator is the familiar Markovian quantum master equation in Lindblad form (2.20), e.g., the optical Bloch equations [36, 37]. First of all, the approaches of this type were introduced on phenomenological grounds. Later on, a fundamental significance has been allocated to the "stochastic Schrödinger equations" by several authors, in particular to propose explanation of the omnipresent decoherence phenomena observed in real quantum systems. The above scheme does not lead to a definite stochastic representation of the dynamics of the reduced system in Hilbert space, even though the Markov approximation is made, since the stochastic process is not unambiguously determined by merely fixing the covariance. An infinity of different realizations is possible. Basically, one may distinguish two classes of stochastic models for the evolution of state vectors in the Schrödinger picture. In the first class of stochastic Schrödinger equations, the stochastic increment is a diffusion process, the so-called Wiener process. In the second class, the evolution of the state vector is represented as a stochastic process of which the realizations are piecewise deterministic paths, and the smooth segments are interruped by stochastic sudden jump processes [58].

The *quantum-state diffusion* (QSD) method proposed by Gisin [59] and developed further by Gisin, Percival and coworkers [60, 61] belongs to the first class. The QSD method is based upon a correspondence between the solutions of the master equation for the ensemble density operator ρ and the solutions of a Langevin-Itô diffusion equation for the normalized pure state vector $|\psi>$ of an individual system of the ensemble. If the master equation has the Lindblad form (2.20), then the corresponding QSD equation is the nonlinear stochastic differential equation

$$|d\psi> = -\frac{i}{\hbar}H|\psi> dt + \sum_j (<L_j^\dagger> L_j - \tfrac{1}{2}L_j^\dagger L_j - \tfrac{1}{2}<L_j^\dagger><L_j>)|\psi> dt$$
$$+ \sum_j (L_j - <L_j>)|\psi> d\xi_j \,, \tag{2.34}$$

where $<L_j> \equiv <\psi|L_j|\psi>$ is the quantum expectation.[6] The first sum describes the nonlinear drift of the state vector in the state space and the second sum the random fluctuations. The $d\xi_j$ are complex differential variables of a Wiener process satisfying

$$\langle d\xi_j \rangle = 0 \,, \qquad \langle d\xi_j \, d\xi_k \rangle = 0 \,, \qquad \langle d\xi_j^* \, d\xi_k \rangle = \delta_{j,k} \, dt \,, \tag{2.35}$$

where $\langle \cdots \rangle$ represents a mean over the ensemble. The density operator is given by the mean over the projectors onto the quantum states of the ensemble

$$\rho = \langle |\psi><\psi| \rangle \,. \tag{2.36}$$

A relativistic quantum state diffusion model has been proposed in Ref. [62].

Attempts have also been made to describe the stochastic evolution of the state vector in terms of a stochastic differential equation with a linear drift [63].

The Monte-Carlo wave function simulation or *quantum jump methods* proposed by Diósi [64], by Dalibard, Castin, and Mølmer [65], by Zoller and coworkers [66], and by Carmichael *et al.* [67] belong to the second class. In these related methods, the Schrödinger equation is supplemented by a non-Hermitean term and by a stochastic term undergoing a Poisson jump process. Because of the non-unitary time evolution of the state vector under a non-Hermitean Hamiltonian, the trace of the density operator is no more conserved. Conservation of probability is restored again and again by imposing stochastically chosen quantum jumps (see Refs. [65, 66, 68]). In the Monte Carlo algorithm by Mølmer *et al.* [69], the deviation of the norm δp of the wave function from unity after a certain time step is compared with a number ϵ,

---

[6]We constantly use the symbol $|\cdots>$ for a pure state, and $\langle \cdots \rangle$ for an ensemble average.

which is randomly chosen from the interval $[0, 1]$. If $\delta p > \epsilon$, a quantum jump occurs by which the wave function is renormalized to unity. A comparison of some of the quantum jump and state diffusion models was given in Ref. [70].

Recently, Breuer and Petruccione showed that a unique stochastic process in Hilbert space for the dynamics of the open system may be derived directly from the underlying microscopic system-plus-reservoir model [71, 72]. They employed a description of quantum mechanical ensembles in terms of probability distributions on projective Hilbert space. In order to eliminate the reservoir, they made the Markovian approximation, and they employed second-order perturbation theory in the system-reservoir coupling. They then obtained a Liouville-master functional equation for the reduced probability distribution. The Liouville part of this equation corresponds to a deterministic Schrödinger-type equation with a non-Hermitean Hamiltonian which is intrinsically nonlinear in order to preserve the norm. The master part of this equation describes gain and loss of the probabilities for individual states due to discontinuous quantum jumps. In this description, the realization of the stochastic process is very similar to those generated by the piecewise deterministic quantum jump method [65] – [67]. Therefore, the stochastic simulation algorithms of all these approaches are very similar likewise. The equation of motion for the reduced density matrix derived from the Liouville-master equation is in the Lindblad form (2.20).

In the first place, the stochastic wave function methods are computational tools with which the solution of the Born-Markov master equation is simulated by using Monte Carlo importance-sampling techniques [68]. The stochastic methods are numerically superior to the conventional integration of the master equation when the rank of the reduced density matrix is large. Over and above the computational advantage of stochastic wave function methods for Born-Markov processes, some groups are presuming to claim that the instantaneous discontinuous processes are *real* and provide a natural description of individual quantum jump events (and not only of their statistics) as observed, e.g., in experiments with single ions in radio-frequency traps [73, 74]. Against that, we wish to point out that the assignment of definite states to a subsystem is *incompatible* with standard quantum theory, and has been proven wrong, e.g., in Einstein-Rosen-Podolsky experiments. Moreover, there is no experimental indication for non-standard phenomena (e.g., spontaneous collapse) in connection with the explanation of classical properties. Hence there is no phenomenological necessity for the introduction of a stochastic equation for state vectors. Besides computational advantages in the simulation of Born-Markov processes, the quantum-state diffusion method provides an alternative approach to measurement theory. In this method, a continuous measurement process, by which the system is steadily reduced within a certain time period to an eigenstate, is an integral part of the dynamical description. In conclusion, the stochastic wave function approaches provide efficient numerical simulations of quantum master equations which are of the Lindblad form. However, since the Markov and the Born approximation are made, the application of these methods to solid state physics problems is as limited as the Born-Markov quantum master equation approach.

In the sequel, we move on firm ground taking the conservative view that the Hamiltonian dynamics of a global system induces a non-unitary dynamics for a subsystem. Upon performing a reduction of the global system to the relevant subsystem, all effects of the environmental coupling are put into in an influence functional. Non-Markovian generalizations of quantum state diffusion and related stochastic Schrödinger equations can be found by a stochastic unraveling of the influence functional. The related discussion is given in Section 5.6.

3. System–plus–reservoir models

For many complex quantum systems we do not have a clear understanding of the microscopic origin of damping. In some systems, however, it is possible to track down the power spectrum of the stochastic force in the classical regime, and hence the spectral damping function $\tilde{\gamma}(\omega)$. Therefore, it is very important to have phenomenological system-plus-reservoir models which on the one hand open up the full quantum mechanical treatment, and on the other hand reduce in the classical limit to a description of the stochastic process in terms of a Langevin equation of the form (2.4).

The simplest model of a dissipative quantum mechanical system that one can envisage is a damped quantum mechanical linear oscillator: a central harmonic oscillator is coupled linearly via its displacement coordinate q to a fluctuating dynamical reservoir or bath. If the equilibrium state of the reservoir is only weakly perturbed by the central oscillator, its classical dynamics can be represented by linear equations. Therefore, it can be described in terms of a bosonic field or a (infinite) set of harmonic oscillators. Then the noise statistics of the stochastic process induced by the reservoir is strictly Gaussian. This simple system–plus–reservoir model has been introduced and discussed in a series of four papers by Ullersma [75]. Zwanzig generalized the model to the case in which the central particle moves in an anharmonic potential and studied the classical regime [76]. Caldeira and Leggett [77] were among the first who applied this model to a study of quantum mechanical tunneling of a macroscopic variable. The relevant model is considered in this chapter.

In the first section of this chapter we introduce the model and we track down the relations between the parameters of the model and the quantities appearing on the classical phenomenological level. Subsequent to this, we discuss a number of physically important particle–plus–reservoir systems for which we can base the description to some extent on a microscopic footing. In this chapter, we cannot deal with these systems in any detail. We shall outline only the underlying Hamiltonians and postpone the path integral formulation of the quantum statistical mechanics until the next chapter. In the last section, we touch on the discussion of nonlinear quantum environments.

3.1 Harmonic oscillator bath with linear coupling

In this section, we first introduce the most general Hamiltonian underlying a dissipative system obeying Eq. (2.4) in the classical limit, and we explain the important generalization to the case where the viscosity is state-dependent. After that, we determine the relation between the parameters of the global model and the phenomenological frequency-dependent friction coefficient $\tilde{\gamma}(\omega)$.

3.1.1 The Hamiltonian of the global system

Consider a system with one or few degrees of freedom which is coupled to a huge environment and imagine that the environment is represented by a bath of harmonic excitations above a stable ground state. The interaction of the system with each individual degree of freedom of the reservoir is proportional to the inverse of the volume of the reservoir. Hence, *the coupling to an individual bath mode is weak for a geometrically macroscopic environment*. Therefore, it is physically very reasonable for macroscopic global systems to assume that the system-reservoir coupling is a *linear* function of the bath coordinates. This property is favourable since it allows to eliminate the environment exactly. Importantly, the weak perturbation of any individual bath mode does not necessarily mean that the dissipative influence of the reservoir on the system is weak as well since the couplings of the individual bath modes add up and the number of modes can be very large.

The general form of the Hamiltonian for the global system complying with these properties (barring pathological cases) is (see Ref. [77], Appendix C)

$$H = H_S + H_R + H_I . \tag{3.1}$$

Here, H_S is the Hamiltonian of the relevant system. For simplicity, we imagine a particle of mass M moving in a potential $V(q)$,

$$H_S = p^2/2M + V(q) . \tag{3.2}$$

The reservoir consists of a set of harmonic oscillators,

$$H_R = \sum_{\alpha=1}^{N} \left(\frac{p_\alpha^2}{2m_\alpha} + \frac{1}{2} m_\alpha \omega_\alpha^2 x_\alpha^2 \right) , \tag{3.3}$$

and the system-bath interaction H_I is assumed to be linear in the bath coordinates,

$$H_I = -\sum_{\alpha=1}^{N} F_\alpha(q) x_\alpha + \Delta V(q) . \tag{3.4}$$

For specific purpose we have added a counter-term $\Delta V(q)$ which depends on $F_\alpha(q)$, and on the parameters m_α, ω_α of the reservoir, but not on its dynamical variables $x_\alpha(t)$. The additional potential term $\Delta V(q)$ is introduced in order to compensate a renormalization of the potential $V(q)$ which is caused by the coupling term linear in x_α in the interaction H_I. In the absence of $\Delta V(q)$, the minimum of the potential surface of the global system for fixed q in x_α-direction is at $x_\alpha = F_\alpha(q)/m_\alpha\omega_\alpha^2$. Thus, the "effective" potential renormalized by the coupling is given by

$$V_{\text{eff}}(q) = V(q) - \sum_{\alpha=1}^{N} \frac{F_\alpha^2(q)}{2m_\alpha\omega_\alpha^2} . \tag{3.5}$$

In the special case $F_\alpha(q) = c_\alpha q$, the second term in Eq. (3.5) causes a negative shift $(\Delta\omega)^2 = -\sum_\alpha c_\alpha^2/Mm_\alpha\omega_\alpha^2$ in the squared circular frequency ω_0^2 of small oscillations about the minimum. This coupling-induced renormalization of the potential can be very large, and, if $\omega_{\text{eff}}^2 = \omega_0^2 + (\Delta\omega)^2 < 0$, it changes the potential even qualitatively.

If we wish that the coupling of the relevant system to the reservoir solely introduces dissipation — and not in addition a renormalization of the potential $V(q)$ — we must compensate the second term in Eq. (3.5) by a suitable choice of $\Delta V(q)$. Full compensation of the coupling-induced potential distorsion is achieved if we put

$$\Delta V(q) = \sum_{\alpha=1}^{N} \frac{F_\alpha^2(q)}{2m_\alpha\omega_\alpha^2} . \tag{3.6}$$

The specific choice of *separable* interaction

$$F_\alpha(q) = c_\alpha F(q) , \tag{3.7}$$

where $F(q)$ is independent of α, is of particular interest. Thus, under the assumptions specified above, the most general translational invariant Hamiltonian with a separable interaction is

$$H = \frac{p^2}{2M} + V(q) + \frac{1}{2}\sum_{\alpha=1}^{N}\left[\frac{p_\alpha^2}{m_\alpha} + m_\alpha\omega_\alpha^2\left(x_\alpha - \frac{c_\alpha}{m_\alpha\omega_\alpha^2}F(q)\right)^2\right] . \tag{3.8}$$

The case of a nonlinear function $F(q)$ occurs e.g. in rotational tunneling systems, in polaron systems and in Josephson systems.

It has been argued [78] that the periodicity of a hindering potential in a *rotational tunneling system* is an exact symmetry which cannot be destroyed whatever the external influences are. The argument applies when several identical particles are tunneling at the same time, as e.g., in a rotating molecule complex. If there are N identical particles coherently tunneling [e.g., $N = 3$ for a methyl-(CH_3-)group], both the potential $V(\varphi)$ and the coupling function $F(\varphi)$, where φ is the dynamical angular variable, belong to the same symmetry group C_N, i.e., $V(\varphi) = V(\varphi + 2\pi/N)$ and $F(\varphi) = F(\varphi + 2\pi/N)$.

In a polaron system, the particle's interaction energy due to linear lattice distorsions is nonlinear in the particle's coordinate according to $F_{\mathbf{k}}(\mathbf{q}) = e^{i\mathbf{k}\cdot\mathbf{q}}$ [cf. Sec. 3.3].

Quasiparticle tunneling in Josephson systems is another important case. Then the coordinate q is again identified with a phase variable φ. In a phenomenological modeling of charge tunneling between superconductors the interaction term is [79]

$$H_I = \sin(\varphi/2) \sum_\alpha c_\alpha^{(1)} x_\alpha^{(1)} + \cos(\varphi/2) \sum_\alpha c_\alpha^{(2)} x_\alpha^{(2)} , \tag{3.9}$$

where $\{x_\alpha^{(1)}\}$ and $\{x_\alpha^{(2)}\}$ represent two independent sets of oscillators. For a discussion of the microscopic theory, see Subsection 4.2.10.

If we require that the dissipation be *strictly* linear, we must constrain $F_\alpha(q)$ as

$$F_\alpha(q) = c_\alpha q . \tag{3.10}$$

This is the case of state-independent dissipation. The form (3.10) describes in von Neumann's sense an ideal measurement of the particle's position by the reservoir.

Substituting Eq. (3.10) into Eq. (3.1), the Hamiltonian takes the form

$$H = \frac{p^2}{2M} + V(q) + \frac{1}{2} \sum_{\alpha=1}^{N} \left[\frac{p_\alpha^2}{m_\alpha} + m_\alpha \omega_\alpha^2 \left(x_\alpha - \frac{c_\alpha}{m_\alpha \omega_\alpha^2} q \right)^2 \right] . \tag{3.11}$$

For later convenience, we rewrite the Hamiltonian (3.11) as

$$H = \frac{p^2}{2M} + \sum_{\alpha=1}^{N} \frac{p_\alpha^2}{2m_\alpha} + V(q, \boldsymbol{x}) , \tag{3.12}$$

$$V(q, \boldsymbol{x}) = V(q) + \frac{1}{2} \sum_{\alpha=1}^{N} m_\alpha \omega_\alpha^2 \left(x_\alpha - \frac{c_\alpha}{m_\alpha \omega_\alpha^2} q \right)^2 . \tag{3.13}$$

Here, $V(q, \boldsymbol{x})$ is the potential of the global system, and $\boldsymbol{x}$ represents the set of bath coordinates $\{x_\alpha\}$. The Hamiltonian (3.11) has been used to model dissipation for about thirty years. Early studies were limited to a harmonic potential $V(q)$. Probably the first who showed that Eq. (3.11) leads to dissipation was Magalinskiĭ [80]. Shortly later studies include the work by Rubin [81] for classical systems, and Senitzky [19], Ford *et al.* [20], and Ullersma [75] for quantum systems. In the more recent literature, the model described by Eq. (3.11) is usually referred to as the Caldeira-Leggett model.

3.1.2 The road to the classical generalized Langevin equation

The equations of motion for a global system described by the Hamiltonian (3.11) read

$$M\ddot{q} + V'(q) + \sum_\alpha (c_\alpha^2 / m_\alpha \omega_\alpha^2) q = \sum_\alpha c_\alpha x_\alpha , \tag{3.14}$$

$$m_\alpha \ddot{x}_\alpha + m_\alpha \omega_\alpha^2 x_\alpha = c_\alpha q , \tag{3.15}$$

where $V'(q) = \partial V / \partial q$. The dynamical equation for the oscillator position $x_\alpha(t)$ is an ordinary second order linear differential equation with inhomogeneity $c_\alpha q(t)$. This equation is solved by standard Green's function techniques. In Fourier space, a particular solution of the inhomogeneous equation reads

$$\tilde{x}_\alpha(\omega) = \tilde{\chi}_\alpha(\omega) c_\alpha \tilde{q}(\omega) , \quad \text{where} \quad \tilde{\chi}_\alpha(\omega) = \lim_{\epsilon \to 0^+} \frac{1}{m_\alpha} \frac{1}{\omega_\alpha^2 - \omega^2 - i\omega\epsilon} \tag{3.16}$$

is the dynamical susceptibility of an individual bath oscillator. Of particular later interest is the absorptive part $\operatorname{Im}\widetilde{\chi}_\alpha(\omega) \equiv \widetilde{\chi}''_\alpha(\omega) = (\pi/m_\alpha)\operatorname{sgn}(\omega)\,\delta(\omega_\alpha^2 - \omega^2)$. For simplicity we fix the initial time to $t_0 = 0$. The solution evolving from the initial values $x_\alpha^{(0)}$ and $p_\alpha^{(0)}$ reads

$$x_\alpha(t) = x_\alpha^{(0)}\cos(\omega_\alpha t) + \frac{p_\alpha^{(0)}}{m_\alpha\omega_\alpha}\sin(\omega_\alpha t) + \frac{c_\alpha}{m_\alpha\omega_\alpha}\int_0^t dt'\sin[\,\omega_\alpha(t - t')\,]\,q(t') . \quad (3.17)$$

It is convenient to rewrite Eq. (3.17) as a functional of the particle's velocity. Integrating the last term by parts, we get

$$\begin{aligned}
x_\alpha(t) = {} & x_\alpha^{(0)}\cos(\omega_\alpha t) + \frac{p_\alpha^{(0)}}{m_\alpha\omega_\alpha}\sin(\omega_\alpha t) + \frac{c_\alpha}{m_\alpha\omega_\alpha^2}\Big(q(t) - \cos(\omega_\alpha t)q(0)\Big) \\
& - \frac{c_\alpha}{m_\alpha\omega_\alpha^2}\int_0^t dt'\cos[\,\omega_\alpha(t - t')\,]\,\dot{q}(t') .
\end{aligned} \quad (3.18)$$

Next, we eliminate the bath degrees of freedom by substituting Eq. (3.18) into Eq. (3.14). Then, the last term of the l.h.s. in Eq. (3.14) originating from the counter term in Eq. (3.4) cancels out. The dynamical equation for $q(t)$ alone is found to read

$$M\ddot{q}(t) + M\int_0^t dt'\,\gamma(t - t')\dot{q}(t') + V'(q) = \zeta(t) - M\gamma(t - t_0)q(0) . \quad (3.19)$$

Here we have introduced the memory-friction kernel

$$\gamma(t) = \Theta(t)\,k(t) , \qquad \text{where} \qquad k(t) = \frac{1}{M}\sum_\alpha \frac{c_\alpha^2}{m_\alpha\omega_\alpha^2}\cos(\omega_\alpha t) , \quad (3.20)$$

and the force

$$\zeta(t) = \sum_\alpha c_\alpha\left(x_\alpha^{(0)}\cos(\omega_\alpha t) + \frac{p_\alpha^{(0)}}{m_\alpha\omega_\alpha}\sin(\omega_\alpha t)\right) . \quad (3.21)$$

Taking the average of the initial values $x_\alpha^{(0)}$, $p_\alpha^{(0)}$ with respect to the canonical classical equilibrium density of the unperturbed reservoir

$$\rho_R^{(0)} = Z^{-1}\exp\left[-\beta\sum_\alpha\left(\frac{p_\alpha^{(0)\,2}}{2m_\alpha} + \frac{m_\alpha\omega_\alpha^2}{2}x_\alpha^{(0)\,2}\right)\right] , \quad (3.22)$$

the force $\zeta(t)$ becomes a stochastic force with stationary Gaussian statistics,

$$\langle\zeta(t)\rangle_{\rho_R^{(0)}} = 0 , \qquad \langle\zeta(t)\zeta(0)\rangle_{\rho_R^{(0)}} \equiv \mathcal{X}_{cl}(t) = Mk_BT\,k(t) . \quad (3.23)$$

The second relation is the classical fluctuation-dissipation theorem Upon Fourier-transformation of the second relation we obtain the power spectrum of the random force as $\widetilde{\mathcal{X}}(\omega) = 2Mk_BT\widetilde{\gamma}'(\omega)$, given already earlier in Eq. (2.8). From this we see that in the case of memory-friction the stochastic force $\xi(t)$ is not anymore delta-correlated. Since $\widetilde{\mathcal{X}}(\omega)$ is frequency-dependent, this case is usually referred to as coloured noise.

The dynamical equation (3.19) is a Langevin equation with a linear memory-friction force and a random force representing additive noise. The generalization to quantum-thermal noise is given in Subsection 3.1.4.

The stochastic equation of motion (3.19) still contrasts with the usual form of the Langevin equation by the spurious term $-M\gamma(t)q(0)$. This term is a transient depending on the initial position $q(0)$.

Upon strengthening an observation by Bez [82], it has been argued [83] that the initial transient slippage is an artefact of the decoupled thermal initial state (3.22).

The annoying term in Eq. (3.19) can be eliminated by a suitable definition of the thermal average as follows. In the first step of the elimination, the term $-M\gamma(t)q(0)$ is absorbed by a corresponding shift of the random force $\zeta(t)$,

$$\xi(t) = \zeta(t) - M\gamma(t)q(0) . \tag{3.24}$$

The modified random force $\xi(t)$ does not vanish on average when the mean value is taken with respect to $\rho_R^{(0)}$. In the second step, $\xi(t)$ is reconciled with the usual properties of a Gaussian random force by performing the thermal average in the initial state of the reservoir with the shifted canonical equilibrium distribution

$$\rho_R = Z^{-1} \exp\left\{ -\beta \sum_\alpha \left[\frac{p_\alpha^{(0)\,2}}{2m_\alpha} + \frac{m_\alpha \omega_\alpha^2}{2} \left(x_\alpha^{(0)} - \frac{c_\alpha}{m_\alpha \omega_\alpha^2} q(t_0) \right)^2 \right] \right\} . \tag{3.25}$$

Upon averaging $x_\alpha^{(0)}$ and $p_\alpha^{(0)}$ with the weight function (3.25) $\xi(t)$ becomes a random force with the same statistical properties as the random force $\zeta(t)$,

$$\langle \xi(t) \rangle_{\rho_R} = 0 , \qquad \langle \xi(t)\xi(0) \rangle_{\rho_R} = Mk_B T\, k(t) . \tag{3.26}$$

Thus we arrive at the standard form of the generalized Langevin equation,

$$M\ddot{q}(t) + V'(q) + M \int_0^t dt'\, \gamma(t - t')\dot{q}(t') = \xi(t) , \tag{3.27}$$

anticipated in Section 2.1.

From the above we see that care has to be taken if one refers to statistical properties of a random force [83]. We have been able to remove the initial transient slippage by taking the thermal average in the initial state with respect to bath modes which are shifted by the coupling to the particle.

If we had started at the Hamiltonian (3.8), we would have reached the form

$$M\ddot{q}(t) + MF'[q(t)] \int_0^t dt'\, \gamma(t - t')F'[q(t')]\,\dot{q}(t') + V'(q) = F'[q(t)]\,\xi(t) . \tag{3.28}$$

In this modified Langevin equation the damping force describes nonlinear (state-dependent) memory friction and the random force represents multiplicative noise.

As important examples for state-dependent dissipation, we mention quasiparticle tunneling through a barrier between superconductors (Josephson contact) and strong electron tunneling through a mesoscopic junction [cf. Subsections 4.2.10 and 5.5.3].

3.1.3 Phenomenological modeling

If the number N of the bath oscillators is small, the period of time, in which the energy transferred to the bath is fed back to the central system, is of the order of other relevant time scales. However, when N is about 20 or larger, the Poincaré recurrence time is found to be practically infinity. In such cases it is appropriate to replace the sum over the discrete bath modes by a frequency integral with a continuous spectral density of the reservoir coupling.

With these preliminary remarks it is now straightforward to establish the connection of the frequency-dependent damping function $\widetilde{\gamma}(\omega)$ defined in Eq. (2.5) with the parameters of the Hamiltonian (3.8) or (3.11). The Fourier transform of the retarded memory-friction kernel (3.20) is

$$\widetilde{\gamma}(\omega) \;=\; \lim_{\epsilon \to 0^+} -i \, \frac{\omega}{M} \sum_{\alpha=1}^{N} \frac{c_\alpha^2}{m_\alpha \omega_\alpha^2} \, \frac{1}{\omega_\alpha^2 - \omega^2 - i \, \omega \epsilon} \,. \tag{3.29}$$

Next we introduce the spectral density of the environmental coupling

$$J(\omega) \;=\; \frac{\pi}{2} \sum_\alpha \frac{c_\alpha^2}{m_\alpha \omega_\alpha} \delta(\omega - \omega_\alpha) \,. \tag{3.30}$$

For a set of discrete modes, the spectral density consists of a sequence of δ-peaks. In order to work on a genuine heat bath, we assume that the eigenfrequencies ω_α are so dense as to form a continuous spectrum. In the continuum limit, $J(\omega)$ becomes a smooth function of ω, and the sum in Eq. (3.29) is replaced by the integral

$$\widetilde{\gamma}(\omega) \;=\; \lim_{\epsilon \to 0^+} \frac{-i\omega}{M} \frac{2}{\pi} \int_0^\infty d\omega' \, \frac{J(\omega')}{\omega'} \, \frac{1}{\omega'^{\,2} - \omega^2 - i \, \omega \epsilon} \,. \tag{3.31}$$

By this subtle distinction, the function $\widetilde{\gamma}(\omega)$ acquires a smooth real part,

$$\widetilde{\gamma}'(\omega) \;=\; J(\omega)/M\omega \,. \tag{3.32}$$

The real and imaginary parts of $\widetilde{\gamma}(\omega)$ are related by Kramers-Kronig relations, as required for a response function.

Sometimes it is convenient to consider the Laplace transform $\hat{\gamma}(z)$ of the damping kernel $\gamma(t)$. The damping function $\hat{\gamma}(z)$ is related to the Fourier transform $\widetilde{\gamma}(\omega)$ by analytic continuation,

$$\hat{\gamma}(z) \;=\; \widetilde{\gamma}(\omega = iz) \,; \qquad \widetilde{\gamma}(\omega) \;=\; \lim_{\epsilon \to 0^+} \hat{\gamma}(z = -i\omega + \epsilon) \,. \tag{3.33}$$

Upon using Eqs. (3.29) and (3.31) we then get

$$\hat{\gamma}(z) \;=\; \frac{z}{M} \sum_{\alpha=1}^{N} \frac{c_\alpha^2}{m_\alpha \omega_\alpha^2} \, \frac{1}{(\omega_\alpha^2 + z^2)} \;=\; \frac{z}{M} \frac{2}{\pi} \int_0^\infty d\omega' \, \frac{J(\omega')}{\omega'} \, \frac{1}{\omega'^{\,2} + z^2} \,. \tag{3.34}$$

We may express $\widetilde{\gamma}(\omega)$ and $J(\omega)$ in terms of the dynamical susceptibilities of the individual reservoir modes given in Eq. (3.16). The resulting expressions are

$$\tilde{\gamma}(\omega) \;=\; \frac{1}{-i\omega M} \sum_{\alpha=1}^{N} c_\alpha^2 \Big(\tilde{\chi}_\alpha(0) - \tilde{\chi}_\alpha(\omega) \Big) , \qquad (3.35)$$

and

$$J(\omega) \;=\; \Theta(\omega) \sum_{\alpha=1}^{N} c_\alpha^2\, \tilde{\chi}_\alpha''(\omega) . \qquad (3.36)$$

We see from Eq. (3.35) that the aforementioned counter term serves to eliminate a contribution from the static susceptibility. These forms pave the way for situations in which the bath modes are effectively nonlinear, as discussed below in Section 3.5.

As far as we are interested in properties of the particle alone, the dynamics is fully determined by the mass M, the potential $V(q)$, and the spectral density $J(\omega)$. We see from Eq. (3.32) that $J(\omega)$ is uniquely determined by the classical frequency-dependent damping coefficient $\tilde{\gamma}(\omega)$. With the roles reversed, we can also express the microscopic characteristics in terms of the phenomenological real-time damping kernel $\gamma(t)$. Using Eq. (3.31), we can give the damping kernel $\gamma(t)$ in terms of the spectral density $J(\omega)$ of the environmental coupling,

$$\gamma(t) \;=\; \Theta(t)\frac{1}{M}\frac{2}{\pi}\int_0^\infty d\omega\, \frac{J(\omega)}{\omega}\, \cos(\omega t) . \qquad (3.37)$$

The inversion of the Fourier integral (3.37) gives

$$J(\omega) \;=\; M\omega \int_0^\infty dt\, \gamma(t)\cos(\omega t) . \qquad (3.38)$$

The spectral density $J(\omega)$ may also be expressed in terms of the Laplace transform of the damping kernel. From Eq. (3.38) we find

$$J(\omega) \;=\; \lim_{\epsilon \to 0^+} M\omega[\,\hat{\gamma}(\epsilon + i\omega) + \hat{\gamma}(\epsilon - i\omega)\,]/2 . \qquad (3.39)$$

We conclude this subsection with an important remark: The spectral density $J(\omega)$, and later on quantum mechanics, is determined by quantities that appear already in the classical phenomenological equation of motion [84, 85]. This property holds exactly in the case of strict linear (i.e. state-independent) dissipation. For instance, a conventional molecular dynamics simulation may be used to compute $\tilde{\gamma}'(\omega)$, which then determines $J(\omega)$ by Eq. (3.32). Therefore, this relation plays a fundamental role in the phenomenological modeling of dissipative quantum systems.

3.1.4 Quasiclassical Langevin equation

The quasiclassical Langevin equation looks formally the same as the classical Langevin equation (3.27). The generalization is that the random force $\xi(t)$ bears the quantum statistical fluctuations of the reservoir instead of the pure classical ones. To account for quantum statistical behaviour in a simple way, we assign to the random force $\xi(t)$ operator character via $x_\alpha^{(0)}$ and $p_\alpha^{(0)}$. Next, we write these operators in terms of creation and annihilation operators,

$$x_\alpha^{(0)} = \left(\frac{\hbar}{2m_\alpha\omega_\alpha}\right)^{1/2}\left(b_\alpha + b_\alpha^\dagger\right), \qquad p_\alpha^{(0)} = i\left(\frac{m_\alpha\omega_\alpha\hbar}{2}\right)^{1/2}\left(b_\alpha^\dagger - b_\alpha\right), \qquad (3.40)$$

which obey the usual commutator relations

$$\left(b_\alpha, b_\gamma\right) = \left(b_\alpha^\dagger, b_\gamma^\dagger\right) = 0, \qquad \text{and} \qquad \left(b_\alpha, b_\gamma^\dagger\right) = \delta_{\alpha\gamma}. \qquad (3.41)$$

In the language of these operators, the random force (3.21) reads[1]

$$\xi(t) = \sum_\alpha c_\alpha\left(\frac{\hbar}{2m_\alpha\omega_\alpha}\right)^{1/2}\left(e^{i\omega_\alpha t}b_\alpha^\dagger + e^{-i\omega_\alpha t}b_\alpha\right). \qquad (3.42)$$

The relevant quantum statistical equilibrium averages of these operators are

$$\langle b_\alpha\rangle_\beta = \langle b_\alpha^\dagger\rangle_\beta = \langle b_\alpha b_\gamma\rangle_\beta = \langle b_\alpha^\dagger b_\gamma^\dagger\rangle_\beta = 0,$$

$$\langle b_\alpha^\dagger b_\gamma\rangle_\beta = \delta_{\alpha\gamma}\, n(\omega_\alpha), \qquad \langle b_\alpha b_\gamma^\dagger\rangle_\beta = \delta_{\alpha\gamma}\left[1 + n(\omega_\alpha)\right], \qquad (3.43)$$

where $n(\omega)$ is the single-particle Bose distribution

$$n(\omega) = \frac{1}{e^{\beta\hbar\omega} - 1}. \qquad (3.44)$$

With these ensemble averages the random force obeys Gaussian statistics

$$\langle\xi(t)\rangle_\beta = 0, \qquad \langle\xi(t)\xi(0)\rangle_\beta = \mathcal{X}(t), \qquad (3.45)$$

and the force correlator takes the form

$$\mathcal{X}(t) = \hbar\sum_\alpha \frac{c_\alpha^2}{2m_\alpha\omega_\alpha}\left\{e^{-i\omega_\alpha t}\left[1 + n(\omega_\alpha)\right] + e^{i\omega_\alpha t}n(\omega_\alpha)\right\}. \qquad (3.46)$$

From this expression we see that the force correlator conveys absorption to and emission from the reservoir of a single quantum of energy $\hbar\omega_\alpha$. Since we have

$$\frac{n(\omega)}{1 + n(\omega)} = e^{-\beta\hbar\omega}, \qquad (3.47)$$

emission and absorption are related by detailed balance. Upon introducing the continuous spectral density of the coupling (3.30), the force correlator is found to read

$$\mathcal{X}(t) = \frac{\hbar}{\pi}\int_0^\infty d\omega\, J(\omega)\left[\coth(\beta\hbar\omega/2)\cos(\omega t) - i\sin(\omega t)\right]. \qquad (3.48)$$

From this we find with Eq. (3.32) that the power spectrum of the random force is

$$\widetilde{\mathcal{X}}(\omega) \equiv \int_{-\infty}^\infty dt\,\mathcal{X}(t)\cos(\omega t) = M\hbar\omega\coth(\beta\hbar\omega/2)\widetilde{\gamma}'(\omega), \qquad (3.49)$$

which is a version of the quantum-mechanical fluctuation-dissipation theorem. This form has been anticipated above in Eq. (2.32).

One remark is appropriate. We see from the classical fluctuation-dissipation theorem (3.23), or from the classical limit of Eq. (3.49), that classical noise is δ-correlated for strict Ohmic friction. On the other hand, Eq. (3.49) tells us that quantum statistical noise stays coloured even when friction becomes Ohmic.

---

[1] For simplicity we disregard here the initial transient slippage $-M\gamma(t)q(0)$.

3.1.5 Ohmic and frequency–dependent damping

In the strict *Ohmic* (instant response) limit, damping is frequency-independent,

$$\widetilde{\gamma}(\omega) = \gamma \,. \tag{3.50}$$

We see from Eq. (3.38) or from Eq. (3.39) that strict Ohmic damping is described by the model (3.11) in which the spectral density $J(\omega)$ is chosen as [77]

$$J(\omega) = \eta\omega = M\gamma\omega \tag{3.51}$$

for all frequencies ω. In the first equality, we have introduced the familiar viscosity coefficient η. The relation (3.51) implies memoryless friction $\gamma(t) = 2\gamma\Theta(t)\delta(t)$.[2] This, of course, is an idealized situation. In reality, any particular spectral density $J(\omega)$ of physical origin falls off in the limit $\omega \to \infty$, because there is always a microscopic memory time setting the time scale for inertia effects in the reservoir. If $J(\omega)$ would grow steadily with ω, certain physical quantities, e.g. the momentum dispersion, would be divergent (cf. the discussion in Section 7.3).

In the simplest form, the damping kernel $\gamma(t)$ in the classical equation of motion (2.4) is regularized with a Drude memory time $\tau_D = 1/\omega_D$,

$$\gamma(t) = \gamma\omega_D\Theta(t)\exp(-\omega_D t) \,, \tag{3.52}$$

which is known as *Drude regularization*. We then have in the frequency domain

$$\widetilde{\gamma}(\omega) = \gamma/[1 - i\omega/\omega_D] \,, \quad \text{and} \quad \widehat{\gamma}(z) = \gamma/[1 + z/\omega_D] \,. \tag{3.53}$$

In the Drude case, the imaginary part of $\widetilde{\gamma}(\omega)$ differs from the real part by a factor of ω/ω_D. Hence it is small well below the Drude frequency.

The Drude form (3.53) originates from a spectral density with algebraic cutoff,

$$J(\omega) = M\gamma\omega/[1 + \omega^2/\omega_D^2] \,. \tag{3.54}$$

The damping kernel brings in memory-friction on the time scale $\tau_D = \omega_D^{-1}$. When the relevant frequencies of the system are much lower than the Drude cutoff frequency ω_D, the reservoir described by (3.54) behaves like an Ohmic heat bath with effective damping strength $\gamma_{\text{eff}} = \int_0^\infty dt\,\gamma(t) = \gamma$.

Next, consider the extension to general frequency-dependent damping. It is convenient to assume (though it is not strictly necessary) that the function $J(\omega)$ has a power law form at low frequencies, $J(\omega) \propto \omega^s$. It will become clear later on that the dissipative influences can be classified by the power s. Negative values of s are excluded since otherwise the counter term (3.6) or the quantity $\gamma(0)$ would diverge which is pathological. The power-law form is assumed to hold in the frequency range $0 \leq \omega \lesssim \omega_c$, where ω_c is much less than a characteristic cutoff frequency ω_{ch}, which is of the order of the Drude, Debye, or Fermi frequency etc., depending on the model.

---

[2]In the integral in Eq. (2.4) the δ-function counts only half, and thus the damping term reduces to the Ohmic form $M\gamma\dot{q}(t)$ in Eq. (2.3).

For frequencies of the order of or greater than ω_{ch}, the behaviour of $J(\omega)$ may be complicated and may not easily be inferable from the classical motion or from microscopic considerations. The important point, however, is that, as long as we are interested in times $t \gg \omega_c^{-1}$, the effect of the environmental modes with $\omega \gtrsim \omega_c$ can be absorbed into a renormalization of parameters appearing in the Hamiltonian H_S of the system. The only additional property we need to postulate is that $J(\omega)$ falls off at least with some negative power of ω in the limit $\omega \to \infty$. The minimal decrease of $J(\omega)$ required depends on the physical quantity under consideration.

To make the discussion quantitative, we decompose $J(\omega)$ into a low-frequency and a high-frequency contribution,

$$J(\omega) = J_{\mathrm{lf}}(\omega) + J_{\mathrm{hf}}(\omega) , \tag{3.55}$$

$$J_{\mathrm{lf}}(\omega) = J(\omega) f(\omega/\omega_c) , \tag{3.56}$$

$$J_{\mathrm{hf}}(\omega) = J(\omega) [1 - f(\omega/\omega_c)] . \tag{3.57}$$

Here, $f(\omega/\omega_c)$ is a cutoff function defined in such a way that $J_{\mathrm{hf}}(\omega)$ is negligibly small for $\omega \ll \omega_c$ and $J_{\mathrm{lf}}(\omega)$ is negligibly small for $\omega \gg \omega_c$. According to convenience, we may choose either a sharp cutoff $f(\omega/\omega_c) = \Theta(1 - \omega/\omega_c)$, or a smooth cutoff. Expedient forms are a Gaussian, an exponential, or the rational Drude form $f(\omega/\omega_c) = 1/[1 + (\omega/\omega_c)^2]$.

We shall consider in most applications the case in which $J_{\mathrm{lf}}(\omega)$ has a power-law form with an exponential cutoff,

$$J_{\mathrm{lf}}(\omega) = \eta_s \omega_{\mathrm{ph}}^{1-s} \omega^s \, \mathrm{e}^{-\omega/\omega_c} , \qquad \eta_s = M\gamma_s . \tag{3.58}$$

Here we have introduced for $s \neq 1$ a "phononic" reference frequency ω_{ph}, so that the coupling constant η_s has dimension of viscosity for all s. We shall distinguish the frequency ω_{ph} from the cutoff frequency ω_c. Where convenient, we shall use the respective temperature scales

$$T_{\mathrm{ph}} \equiv \hbar\omega_{\mathrm{ph}}/k_{\mathrm{B}} , \qquad T_c \equiv \hbar\omega_c/k_{\mathrm{B}} . \tag{3.59}$$

In many physical situations, the frequency ω_c at which the power law $J_{\mathrm{lf}}(\omega) \propto \omega^s$ is cut off is very large compared with all other relevant frequencies of the system. A number of analytic results will be available in this limit.

The partial spectral density $J_{\mathrm{hf}}(\omega)$ adds to the damping function $\widetilde{\gamma}(\omega)$ in the regime $\omega \ll \omega_c$ the contribution

$$\widetilde{\gamma}_{\mathrm{hf}}(\omega) = -i\omega \, \Delta M_{\mathrm{hf}}/M , \tag{3.60}$$

$$\Delta M_{\mathrm{hf}} = \frac{2}{\pi} \int_0^\infty d\omega' \, \frac{J_{\mathrm{hf}}(\omega')}{\omega'^3} , \tag{3.61}$$

as follows from Eq. (3.31). With (3.60), the Fourier transform of the memory-friction force $-i\omega\widetilde{\gamma}_{\mathrm{hf}}(\omega)\widetilde{q}(\omega)$ reduces at low frequencies to the form $-\Delta M_{\mathrm{hf}} \, \omega^2 \widetilde{q}(\omega)$. This term, however, simply adds to the kinetic term $-M\omega^2\widetilde{q}(\omega)$. Thus for times $t \gg \omega_c^{-1}$, the partial spectral density $J_{\mathrm{hf}}(\omega)$ manifests itself just as mass renormalization,

$$M_r = M + \Delta M_{\mathrm{hf}} . \tag{3.62}$$

From this we conclude that if we are not interested in the time regime $t < 1/\omega_c$, we may treat $J_{\mathrm{hf}}(\omega)$ in the adiabatic approximation. If not stated differently, we will eliminate $J_{\mathrm{hf}}(\omega)$ from the explicit description, and we will regard the mass M as a renormalized mass which is already dressed by the reservoir's high-frequency modes.

The integral (3.61) is convergent also for the spectral density $J_{\mathrm{lf}}(\omega)$ when the power s in Eq. (3.56) is larger than 2. For this case, the full spectral density results in no more than mass renormalization for asymptotic times. Further discussion of this point is given in Section 7.4.

Upon inserting the form (3.56) of the spectral density with a sharp cutoff into the representation (3.34), the spectral damping function takes the exact form [86]

$$\widehat{\gamma}(z) = \frac{2\gamma_s}{\pi s}\left(\frac{\omega_c}{\omega_{\mathrm{ph}}}\right)^{s-1}\frac{\omega_c}{z}\,{}_2\mathrm{F}_1(1, \tfrac{1}{2}s; 1 + \tfrac{1}{2}s; -\omega_c^2/z^2), \tag{3.63}$$

where ${}_2\mathrm{F}_1(a, b; c; x)$ is a hypergeometric function [87, 88]. The explicit form of the spectral damping function $\widehat{\gamma}(z)$ for $z \ll \omega_c$ is found from the asymptotic expansion of the hypergeometric function. It may be calculated upon using standard transformation formulas of the hypergeometric function. Putting $\lambda_s \equiv \gamma_s/\sin(\pi s/2)$, we find

$$\widehat{\gamma}(z) = \begin{cases} \lambda_s\left(z/\omega_{\mathrm{ph}}\right)^{s-1}, & 0 < s < 2, \\[1.5ex] (\gamma_2/\pi)(z/\omega_{\mathrm{ph}})\ln\left(1 + \omega_c^2/z^2\right), & s = 2, \\[1.5ex] z\left[\Delta M/M + (\lambda_s/\omega_{\mathrm{ph}})(z/\omega_{\mathrm{ph}})^{s-2} + \mathcal{O}\left(\frac{z^2}{\omega_c^2}\right)\right], & 2 < s < 4, \\[1.5ex] z\left[\Delta M/M + \mathcal{O}\left(\frac{z^2}{\omega_c^2}\right)\right], & s \geq 4. \end{cases} \tag{3.64}$$

In the regime $0 < s < 2$, we have disregarded terms depending on ω_c. From these results we see that in the most interesting regime $0 < s < 2$ the damping function $\widehat{\gamma}(z)$ varies as z^{s-1}. In the case $2 < s < 4$, we have also given the next to leading order term which determines the leading behaviour beyond trivial mass renormalization. This term is relevant in the transient time regime.

As we have discussed already, the case $s = 1$ in Eq. (3.56) describes Ohmic friction. The cases $0 < s < 1$ and $s > 1$ have been dubbed *sub-Ohmic* and *super-Ohmic*, respectively [89]. We shall present in Sections 3.3 and 4.2 various microscopic models and consider the respective spectral densities of the coupling. We anticipate that an Ohmic friction contribution is almost ubiquitously met in real physical systems at low temperatures. The Ohmic case is important, e.g., for tunneling systems in a metallic environment. A phonon bath in d spatial dimensions corresponds to the case $s = d$ or $s = d + 2$, depending on the underlying symmetry of the strain field. Irrational values of s may occur for environments with fractal dimension.

$$M \quad f \quad m$$

Figure 3.1: Pictorial sketch of the mechanical analogue of the Rubin model.

3.1.6 Rubin model

In Rubin's model, a heavy particle of mass M and coordinate q is bilinearly coupled to a half-infinite chain of harmonic oscillators with masses m and spring constants $f = m\omega_R^2/4$ [90], as sketched in Fig. 3.1. The Hamiltonian of the global model is

$$H = \frac{p^2}{2M} + V(q) + \sum_{n=1}^{\infty} \left(\frac{p_n^2}{2m} + \frac{f}{2}(x_{n+1} - x_n)^2 \right) + \frac{f}{2}(q - x_1)^2 . \qquad (3.65)$$

The model is not yet of the standard form (3.11) since the harmonic bath modes are coupled with each other. The bath is diagonalized with the transformation

$$x_n = \sqrt{2/\pi} \int_0^\pi dk \, \sin(kn) \, X(k) . \qquad (3.66)$$

In the normal mode representation of the reservoir, the Hamiltonian reads

$$H = \frac{p^2}{2M} + V(q) + \frac{f}{2}q^2 + \int_0^\pi dk \left(\frac{P^2(k)}{2m} + \frac{m}{2}\omega^2(k)X^2(k) - c(k)X(k)\,q \right) , \qquad (3.67)$$

where the eigenfrequencies $\omega(k)$ and the coupling function $c(k)$ are given by

$$\omega(k) = \omega_R \sin(k/2) , \qquad c(k) = \sqrt{2/\pi}\,[\,m\omega_R^2/4\,]\sin(k) . \qquad (3.68)$$

The frequency ω_R is the highest frequency of the reservoir modes. Substituting Eq. (3.68) into Eq.(3.30), the spectral density of the coupling is found as

$$J(\omega) = \frac{m\omega_R^3}{16} \int_0^\pi dk \, \frac{\sin^2(k)}{\sin(\tfrac{1}{2}k)} \, \delta\,[\,\omega - \omega(k)\,] = \frac{m\omega_R}{2}\,\omega\left(1 - \frac{\omega^2}{\omega_R^2}\right)^{1/2} \Theta(\omega_R - \omega) ,$$

which is Ohmic for $\omega \ll \omega_R$. With this form, the damping kernel (3.37) emerges as

$$\gamma(t) = \frac{m}{2M}\,\omega_R \frac{J_1(\omega_R t)}{t} , \qquad (3.69)$$

where $J_1(z)$ is a Bessel function of the first kind. Thus, the damping kernel of the Rubin model oscillates with time, and for asymptotic times $t \gg 1/\omega_R$, it decays algebraically,

$$\gamma(t) = \frac{m}{2M}\,\sqrt{2\omega_R/\pi}\,\frac{\sin(\omega_R t - \pi/4)}{t^{3/2}} . \qquad (3.70)$$

This behaviour drastically differs from the exponential decay of the Drude kernel (3.52). Finally, observe that the memory time of the Rubin kernel is of the order of $1/\omega_R$.

3.2 The Spin–Boson model

Many physical and chemical systems can be described by a generalized coordinate with which is associated an effective potential energy function with two separate minima at roughly the same energy. At thermal energies which are very small compared with the level spacing of the low-lying states, only the ground states of the two wells are involved. Then the essential dynamics goes off in a two-dimensional Hilbert space. For a high barrier, the two states are localized in the left and right well, respectively, and they are spatially well-separated. The states are weakly coupled through a transfer or tunneling matrix element. The two-state system (TSS) is the simplest system showing constructive and destructive quantum interference effects, e.g., clockwise oscillations of the occupation of the left and right well. There is a great amount of interest in the thermodynamical and dynamical behaviour of this model due to the diversity of physical realizations, as well as to theoretical advances in recent years.

The basic element in a quantum computer is a *qubit*. Principally, it can be formed by any physical system, whose motion is effectively restricted to a two-dimensional Hilbert space. It is often convenient to represent such a system as a "particle" of spin $\frac{1}{2}$. In any real physical situation, the spin is in contact with the environment. Since the surroundings can often be imagined as a bath of bosons, the global model system has been dubbed *spin-boson model*. Most interesting in this system, in particular in the context of quantum computing, is the extent to which the phase relation between the "spin-up" and "spin-down" components of the wave function is preserved.

Other examples of the situation just described include the motion of defects in crystalline solids (e.g., impurity ions in alkali halides), tunneling of light particles (e.g., hydrogen, muon, proton) in metals [91, 92, 93], the tunneling entities believed to be responsible for the anomalous low-temperature thermal and acoustic properties of oxide glasses and amorphous metals [94] and for the anomalous conductance in mesoscopic wires [95, 96], and also some types of chemical reactions involving electron transfer processes [97, 98].

3.2.1   The model Hamiltonian

We begin with considering the reduction of a spatially expanded double-well system to a two-state system. We characterize the asymmetric double well potential by a "detuning" energy $\hbar\epsilon$ between the two potential minima and by a coupling energy $\hbar\Delta_0$ representing the tunnel splitting of the symmetric system, as sketched in Fig. 3.2. Throughout the book, we shall use the convention that the right well is the lower one for positive bias, $\epsilon > 0$. The tunneling matrix element Δ_0 may be calculated using standard WKB or instanton methods (see Refs. [99, 100]).

Consider a symmetric double well potential $V(q)$ with minima at $q = \pm\frac{1}{2}q_0$ and potential energy $V(\pm\frac{1}{2}q_0) = 0$. The frequency of small oscillations in the wells is $\omega_0 = \sqrt{V''(\pm q_0/2)/M}$. In the semiclassical limit, the tunnel splitting for a particle of mass M is determined by properties of the so-called instanton path $q_{\text{inst}}(\tau)$. This

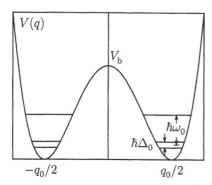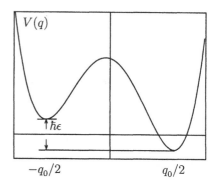

Figure 3.2: Symmetric double well potential (left) with barrier height V_b. The spacing between the first excited state and the ground state in each well is $\hbar\omega_0$, and the tunnel splitting is $\hbar\Delta_0$. The biased double well with detuning energy $\hbar\epsilon$ is sketched on the right.

path describes motion in the *upside-down* potential $-V(q)$ with zero total energy,

$$\tfrac{1}{2} M \dot{q}_{\text{inst}}^2(\tau) = V[q_{\text{inst}}(\tau)] \,, \tag{3.71}$$

and satisfies the boundary conditions $q_{\text{inst}}(\tau \to \pm\infty) = \pm\tfrac{1}{2}q_0$. The instanton is a kink-like path centered at the time where the path rushes with maximal velocity through the minimum of the upside-down potential $-V(q)$. The action of this path is

$$S_{\text{inst}} = M \int_{-\infty}^{\infty} d\tau \, \dot{q}_{\text{inst}}^2(\tau) \,. \tag{3.72}$$

The instanton with center at $\tau = 0$ approaches the boundary values $\pm q_0/2$ as

$$q_{\text{inst}}(\tau \to \pm\infty) = \pm \frac{q_0}{2} \mp C_0 \left(\frac{S_{\text{inst}}}{2M\omega_0} \right)^{1/2} e^{-\omega_0|\tau|} \,, \tag{3.73}$$

where we have chosen the preexponential factor conveniently. The factor C_0 is a numerical constant of order unity which depends on the shape of the potential barrier. In the semiclassical limit, the tunneling matrix element Δ_0 for any particular symmetric double well is determined by the instanton action S_{inst}, the frequency ω_0, and by the numerical constant C_0 occuring in the asymptotic behaviour (3.73). The semiclassical expression for the tunneling matrix element expressed in terms of instanton parameters reads [99, 100]

$$\Delta_0 = 2\omega_0 C_0 (S_{\text{inst}}/2\pi\hbar)^{1/2} e^{-S_{\text{inst}}/\hbar} \,, \tag{3.74}$$

For the archetypal double-well potential of quartic form

$$V(q) = M\omega_0^2 \left(q^2 - q_0^2/4 \right)^2 \Big/ 2q_0^2 \,, \tag{3.75}$$

which has barrier height $V(q = 0) \equiv V_{\mathrm{b}} = M\omega_0^2 q_0^2/32$, the instanton trajectory reads

$$q_{\mathrm{inst}}(\tau) = \tfrac{1}{2}q_0 \tanh\left(\tfrac{1}{2}\omega_0\tau\right) . \tag{3.76}$$

For this particular path, the action S_{inst} and the numerical factor C_0 are given by

$$S_{\mathrm{inst}} = M\omega_0 q_0^2/6 = 16V_{\mathrm{b}}/3\omega_0 , \quad \text{and} \quad C_0 = 2\sqrt{3} . \tag{3.77}$$

Thus we find from Eq. (3.74) for the tunneling matrix element the expression

$$\Delta_0 = 8(2V_{\mathrm{b}}/\pi\hbar\omega_0)^{1/2}\omega_0 \exp\left(-16V_{\mathrm{b}}/3\hbar\omega_0\right) . \tag{3.78}$$

For later convenience, we employ in the sequel instead of the bare Δ_0 the renormalized tunnel matrix element Δ, which is dressed by a Franck-Condon factor representing the polarization cloud of the high-frequency environmental modes in adiabatic approximation (see the discussion in Subsection 18.1.2). In the parameter regime

$$V_{\mathrm{b}} \gg \hbar\omega_0 \gg \hbar\Delta, \ \hbar|\epsilon|, \ k_{\mathrm{B}}T , \tag{3.79}$$

where V_{b} is the barrier height and $\hbar\omega_0$ is the separation of the first excited state from the ground state in either well, the system will be effectively restricted to a two-dimensional Hilbert space spanned, e.g., by the states $|R> \,\hat{=}\, |\uparrow>$ (right) and $|L> \,\hat{=}\, |\downarrow>$ (left) localized in the right and left well, respectively. The two-state Hamiltonian may be expressed in terms of the Pauli matrices in the pseudospin form[3]

$$H_{\mathrm{TSS}} = -\tfrac{1}{2}\hbar\Delta\,\sigma_x - \tfrac{1}{2}\hbar\epsilon\,\sigma_z = \frac{\hbar}{2}\begin{pmatrix} -\epsilon & -\Delta \\ -\Delta & +\epsilon \end{pmatrix}, \tag{3.80}$$

where the basis is formed by the localized states $|R>$ and $|L>$ which are eigenstates of σ_z with eigenvalues $+1$ and -1, respectively. The position operator is $\hat{q} = \tfrac{1}{2}q_0\sigma_z$ and the eigenvalues $\pm\tfrac{1}{2}q_0$ of $\hat{q}$ are the positions of the localized states. In this representation, the Pauli operators may equivalently be written in the form

$$\sigma_z = |R><R| - |L><L| ,$$
$$\sigma_x = |R><L| + |L><R| , \tag{3.81}$$
$$\sigma_y = i\,|L><R| - i\,|R><L| .$$

The localized states $|R>$ and $|L>$ are related to the eigenstates $|g>$ (ground) and $|e>$ (excited) of the Hamiltonian (3.80) by the orthogonal transformation

$$|R> = \cos\frac{\varphi}{2}\,|g> + \sin\frac{\varphi}{2}\,|e> ,$$
$$|L> = \sin\frac{\varphi}{2}\,|g> - \cos\frac{\varphi}{2}\,|e> , \tag{3.82}$$

---

[3]We use the spin representation

$$\sigma_x = \begin{pmatrix} 0 & 1 \\ 1 & 0 \end{pmatrix}, \quad \sigma_y = \begin{pmatrix} 0 & -i \\ i & 0 \end{pmatrix}, \quad \sigma_z = \begin{pmatrix} 1 & 0 \\ 0 & -1 \end{pmatrix} .$$

where

$$\sin\varphi = \frac{\Delta}{\Delta_{\mathrm{b}}}, \qquad \cos\varphi = \frac{\epsilon}{\Delta_{\mathrm{b}}}, \qquad \tan\varphi = \frac{\Delta}{\epsilon}. \tag{3.83}$$

The tunnel splitting energy of the biased TSS is

$$E_{\mathrm{e}} - E_{\mathrm{g}} = \hbar\Delta_{\mathrm{b}} = \hbar\sqrt{\Delta^2 + \epsilon^2}. \tag{3.84}$$

Consider the coupling to a heat bath which is sensitive to the position of the TSS and which is represented by a collective bath mode, the polarization energy $\mathfrak{E}(t)$,

$$H_{\mathrm{I}} = -\sigma_z \mathfrak{E}(t), \qquad \mathfrak{E}(t) \equiv \frac{q_0}{2} \sum_{\alpha=1}^{N} c_\alpha x_\alpha(t). \tag{3.85}$$

The $x_\alpha(t)$ are the individual modes of the reservoir. A dipole–local-field interaction provides a simple physical model for this kind of coupling. As already discussed in some detail in Subsection 3.1.2, the coupling to the heat bath introduces a fluctuating force $\xi(t) = \sum_\alpha c_\alpha x_\alpha(t)$ which causes a fluctuating bias or polarization energy $\mathfrak{E}(t)$. Assuming strict Gaussian statistics, the heat bath can again be modelled exactly by a harmonic bath of bosons. Thus the essential physics is captured by a model Hamiltonian which has become known in the literature as the "spin-boson" Hamiltonian,

$$\begin{aligned} H_{\mathrm{SB}} &= -\tfrac{1}{2}\hbar\Delta\,\sigma_x - \tfrac{1}{2}\hbar\epsilon\,\sigma_z - \sigma_z \mathfrak{E}(t) + H_{\mathrm{R}} \\ &= -\tfrac{1}{2}\hbar\Delta\,\sigma_x - \tfrac{1}{2}\hbar\epsilon\,\sigma_z + \frac{1}{2}\sum_\alpha \left(\frac{p_\alpha^2}{m_\alpha} + m_\alpha\omega_\alpha^2 x_\alpha^2 - q_0\sigma_z c_\alpha x_\alpha \right). \end{aligned} \tag{3.86}$$

In the eigenbasis of the spin part, the Hamiltonian takes the form

$$H_{\mathrm{SB}} = -\tfrac{1}{2}\hbar\Delta_{\mathrm{b}}\,\sigma_z + (\cos\varphi\,\sigma_z - \sin\varphi\,\sigma_x)\,\mathfrak{E}(t) + H_{\mathrm{R}}. \tag{3.87}$$

In this basis, the coupling to the reservoir comprises a transverse ($\propto \sin\varphi$) and a longitudinal ($\propto \cos\varphi$) coupling. Only the transverse part can induce spin flips.

The Hamiltonian (3.86) can equivalently be expressed in terms of the annihilation and creation operators b_α and $b_\alpha^\dagger$ introduced in Eq. (3.40). We then have

$$H_{\mathrm{SB}} = H_{\mathrm{TSS}} - \frac{1}{2}\sigma_z \sum_{\alpha=1}^{N} \hbar\lambda_\alpha \left(b_\alpha + b_\alpha^\dagger \right) + \sum_{\alpha=1}^{N} \hbar\omega_\alpha b_\alpha^\dagger b_\alpha. \tag{3.88}$$

Here we have dropped the irrelevant zero-point energy. The environmental effects are again administered by a spectral density of the coupling,

$$G(\omega) = \sum_{\alpha=1}^{N} \lambda_\alpha^2 \delta(\omega - \omega_\alpha) = \frac{q_0^2}{2\hbar} \sum_{\alpha=1}^{N} \frac{c_\alpha^2}{m_\alpha\omega_\alpha} \delta(\omega - \omega_\alpha). \tag{3.89}$$

The second equality relates λ_α to c_α. The spin-boson spectral density (3.89) is related to the spectral density of the continuous model $J(\omega)$ defined in Eq. (3.30) by

$$G(\omega) = (q_0^2/\pi\hbar)\,J(\omega). \tag{3.90}$$

The spectral density $J(\omega)$ of the continuous model (3.8) or (3.11) has dimension mass times frequency squared whereas $G(\omega)$ has dimension frequency.

3.2.2   Josephson two-state systems: flux and charge qubit

Superconducting circuits based on Josephson tunnel junctions are considered as candidates for quantum computing devices. Making qubits from such electrical elements might be advantageous because coupling of qubits, electrical control, and scaling to large numbers should be feasible using integrated-circuit fabrication technology.

The simplest design is an rf SQUID which is formed by a superconducting ring interrupted by a Josephson junction. The relevant macroscopic degree of freedom is the magnetic flux ϕ threading the ring. It is related to the phase difference ψ of the Cooper pair wave function[4] across the junction (cf. Subsection 4.2.10) by the relation

$$\psi = 2\pi(n - \phi/\phi_0)\,, \tag{3.91}$$

where $2\pi n$ is the change of the phase of the pair wave function per cycle around the ring, and where $\phi_0 = h/2e$ is the flux quantum. The total flux ϕ through the ring consists of the externally applied flux ϕ_{ext} and the flux induced by the Josephson supercurrent I resulting from Cooper pair tunneling, $\phi = \phi_{\text{ext}} + LI$. Here L is the self-inductance of the ring and the supercurrent is $I = I_c \sin \psi$. The critical supercurrent I_c depends on the junction properties (see Subsection 4.2.10). The relation between the Josephson coupling energy E_J and the critical current I_c is $E_J = I_c \phi_0/2\pi = I_c \hbar/2e$. In the regime $E_J \gg E_c$, the flux ϕ is the appropriate quantum degree of freedom.

According to the phenomenological resistively shunted junction (RSJ) model [101] (see Subsection 3.3.4), the Josephson element is shunted by an Ohmic resistance R and by a capacitance C, which bears charging energy $Q^2/2C$. Correspondingly, a current through the loop splits into three pieces,

$$I = I_c \sin \psi + U/R + C\dot{U}\,. \tag{3.92}$$

Josephson's second relation
$$\dot{\psi} = 2eU/\hbar \tag{3.93}$$

relates the change of ψ per unit time with the voltage drop U across the junction. Substituting Eqs. (3.91) and (3.93) into Eq. (3.92), one obtains the deterministic equation of motion for the total flux ϕ threading the ring,

$$C\ddot{\phi} + \dot{\phi}/R + \partial V(\phi)/\partial\phi = 0\,, \tag{3.94}$$

in which the potential $V(\phi)$ is a sinusoidally modulated parabola,

$$V(\phi) = \frac{1}{2L}\left(\phi - \phi_{\text{ext}}\right)^2 - E_J \cos\left(\frac{2\pi\phi}{\phi_0}\right)\,. \tag{3.95}$$

Apart from the missing stochastic force, the equation of motion (3.94) is homologous with the Langevin equation of a damped particle, Eq. (2.3).

If the self-inductance L is so large that $\beta_L \equiv 2\pi L I_c/\phi_0 = 4\pi^2 L E_J/\phi_0^2 > 1$, then the SQUID is hysteretic, i.e., the potential $V(\phi)$ has one or several relative

---

[4]We repeatedly use for the phase difference across a Josephson junction the variable ψ and across a normal junction the variable φ.

minima in which the flux may be trapped. When the flux is trapped in a well, the flux state is metastable with respect to a neighbouring flux state of lower energy and fluxoid macroscopic transitions to the lower state may occur. At high temperature, the transition comes about by thermal activation. At sufficiently low temperature, the transition predominantly occurs via quantum tunneling.

Josephson systems have been employed since the mid 1980s in order to study new specific quantum effects, such as macroscopic quantum tunneling (MQT) of the phase (or flux) [see Part III], as well as indirect (spectroscopic) evidence for quantum superpositions of macroscopic states [102, 103].

Flux qubit

Upon tuning the external flux to $\phi_{ext} = \phi_0/2$, the two wells lowest in energy become degenerate, so that at sufficiently low T, the system is effectively reduced to a two-state system. The two minima correspond to the two different senses of rotation of the supercurrent in the ring. The symmetric double well potential reads as function of $\varphi = 2\pi\phi/\phi_0$ in quartic approximation

$$V_{quart}(\varphi) = \frac{\phi_0^2}{4\pi^2 L} \left[\beta_L - \tfrac{1}{2}(\beta_L - 1)(\varphi - \pi)^2 + \tfrac{1}{24}\beta_L(\varphi - \pi)^4 \right] , \qquad (3.96)$$

The height of the barrier between the two wells of this potential is

$$V_b = \frac{\phi_0^2}{4\pi^2 L} \frac{3}{2} \frac{(\beta_L - 1)^2}{\beta_L} . \qquad (3.97)$$

The bias energy of the TSS can be adjusted by an external magnetic field. When the external flux is put out of tune, $\phi_{ext} \neq \tfrac{1}{2}\phi_0$, the double well becomes asymmetric. For $\beta_L - 1 \ll 1$, the bias energy is $\hbar\epsilon = 4\pi\sqrt{6(\beta_L - 1)}(\phi_{ext}/\phi_0 - \tfrac{1}{2})E_J$. When, in addition, the single junction is replaced by a dc SQUID ring, then the trapped flux $\tilde{\phi}_{ext}$ in this ring serves as an additional independent control parameter by which also the coupling energy between the two localized states can be controlled.

The condition $\beta_L > 1$ requires a relatively large loop, which makes the system very vulnerable to external noise. To overcome this difficulty, one may use smaller superconducting loops with three or four junctions [104, 105].

In these systems, the environment usually consists of resistive elements in the circuits needed for manipulations and measurements. They produce voltage and current noise. In the usual case, the noise is Gaussian. Friction and noise can be introduced in the model by coupling the flux ϕ dynamically to a harmonic heat bath, as we have discussed already in Section 3.1. Under these conditions, the flux qubit is well described by the spin boson model with tunable bias and coupling energy.

Charge qubit

The simplest form of a Josephson charge qubit is a superconducting charge box populated by excess Cooper pair charges. It consists of a small superconducting island ("box") connected to a superconducting electrode by a tunnel junction with capacitance C_J and Josephson coupling energy E_J [106]. A control gate voltage V_g is coupled to the system via a gate capacitor C_g. When the gap energy $\hbar \Delta_g$ is the largest relevant energy of the system, quasiparticle tunneling is suppressed so that quasiparticle excitations on the island are negligible. Then the system is described by the Hamiltonian

$$H = 4E_C (N - N_g)^2 - E_J \cos \psi \,, \tag{3.98}$$

where $E_C = e^2/2(C_g + C_J)$ is the single-electron charging energy and N is the number operator of excess Cooper pair charges on the island, which is conjugate to the phase of the order parameter, $N = -i\partial/\partial\psi$. The gate charge $N_g = C_g V_g/2e$ is controlled by the gate voltage. In the regime $E_C \gg E_J$, the charge on the island is almost sharp. It is then convenient to introduce a basis of charge states, parametrized by the number N of Cooper pairs on the island,

$$Q = 2e \sum_N N |N\rangle\langle N| \,, \qquad e^{i\psi} = \sum_N |N+1\rangle\langle N| \,. \tag{3.99}$$

In this basis, the Hamiltonian (3.98) reads

$$H = \sum_N \left[4E_C(N - N_g)^2 |N\rangle\langle N| - \tfrac{1}{2} E_J \left(|N+1\rangle\langle N| + |N\rangle\langle N+1| \right) \right] \,. \tag{3.100}$$

When the gate voltage is tuned to the symmetry point, $V_g = e/C_g$, the gate charge N_g is $\tfrac{1}{2}$, so that the states with $N = 0$ and $N = 1$ are degenerate. When the charging energy E_C is large enough, all other charge states are elevated to much higher energy, so that they can be ignored. The charge box then reduces to a two-state or spin-$\tfrac{1}{2}$ quantum system (qubit) of the form (3.80). The charge states with $N = 0$ and $N = 1$ correspond to the localized states $|R\rangle$ and $|L\rangle$ of the TSS. The coupling of the states is described by E_J, and the bias energy is

$$\delta E_{\rm ch}(V_g) = 4E_C[1 - 2N_g(V_g)] \,. \tag{3.101}$$

When the single Josephson junction is replaced by two identical junctions (each with coupling energy E_J^0) in a loop, and when the loop is penetrated by an external flux $\phi_{\rm ext}$, the effective Josephson coupling energy of the modified device is

$$E_J(\phi_{\rm ext}) = E_J^0 \cos(\pi \phi_{\rm ext}/\phi_0) \,. \tag{3.102}$$

The SQUID-controlled charge qubit is thus a TSS with tunable coupling energy and tunable bias energy,

$$H = -\frac{1}{2} E_J(\phi_{\rm ext})\, \sigma_x - \frac{1}{2} \delta E_{\rm ch}(V_g)\, \sigma_z \,. \tag{3.103}$$

In this system, the most serious source of decoherence is noise of the biasing voltage ("charge noise"). Fluctuations of the voltage entail fluctuations of the phase ψ, which are described by the phase autocorrelation function given below in Eq. (3.179) with (3.180). In charge-qubit experiments, the major source of decoherence was found to be low frequency $1/f$ noise which originates from background charge fluctuators [107, 108]. When the system is affected by $1/f$ noise in the σ_z or longitudinal ($\propto \cos\varphi$) coupling, this source of decoherence outstrips Ohmic noise at very low temperature. There are experimental indications for a connection between high-frequency Ohmic noise, responsible for relaxation, and low frequency $1/f$ noise accounting for dephasing. This may be explained with a distribution of coherent two-level background charges which is log-uniform in the tunnel splitting and linear in the bias, like the distribution of two-level tunneling systems in amorphous materials, which has been introduced in order to understand anomalous low-temperature properties [110, 111, 112].

With use of the control parameters V_g and ϕ_{ext} one can tune the longitudinal linear couplings to the charge and flux noise to zero: tuning the gate voltage to the degeneracy point $E_{ch}(V_g) = 0$ yields $\cos\varphi = 0$ in the Hamiltonian (3.87) [109]. Further, by tuning ϕ_{ext} to the point of maximal $E_J(\phi_{ext})$, also linear couplings of the qubit to flux fluctuations can be suppressed.

Nakamura $et~al.$ were the first who achieved control of coherent evolution of a macroscopic quantum state in a charge qubit [113].

Operation of the charge qubit at this optimal point may increase the coherence time by 2-3 orders of magnitude as shown for a charge-phase qubit ($E_c \approx E_J$) [114]. Meanwhile several types of superconducting circuits based on Josephson junctions have given evidence that control of coherent quantum state evolution is feasible. Overall, they have proven to be promising candidates for qubit implementation [113] - [117]. There has been made also substantial progress in the control of the interactions between individual flux qubits operating at the optimal point while retaining quantum coherence [118].

3.3   Microscopic models

We now describe physical systems for which the Hamiltonian of the global system can be determined from microscopic considerations.

In unpolar crystals like semiconductors and metals, a charged particle (electron, heavy interstitial particle, tunneling defect particle) distorts the lattice in its neighbourhood, and when the particle moves, the lattice distortion moves with it. In the underlying microscopic description, the particle interacts with acoustic phonons. In the conceptually simplest case, the interaction is described by a scalar deformation potential. The particle together with its attached vibrating environment is called *acoustic polaron.*

An electron in an *ionic* crystal interacts with surrounding ions and thus polarizes the lattice. The interaction lowers the energy of the electron, and when the electron moves, the polarization cloud moves with it. The accompanying cloud of vibrating displacements of the ionic sublattices originates from the electron's coupling to longitudinal optical phonons. The electron together with its polarized environment is called *optical polaron.*

Even without lattice vibrations, there is a rigid periodic potential $H_{\text{rig}}(\boldsymbol{q})$ acting on the particle at position $\boldsymbol{q}$. The Hamiltonian in the absence of lattice vibrations is

$$H_{\text{S}} = p^2/2M + H_{\text{rig}}(\boldsymbol{q}) \,. \tag{3.104}$$

In the literature, one distinguishes between *small* and *large* polaron, depending on whether it is described by localized states in a discrete lattice (small polaron) or by extended states in a continuous medium (large polaron).

In the *large-polaron* system, the rigid potential $H_{\text{rig}}(\boldsymbol{q})$ provides the electron with an effective mass M_{eff}. This is the usual case for ionic crystals, semiconductors and metals. For a moving electron, the lattice must adjust the local distortions. Therefore, the actual mass of the polaron is even higher than the effective mass M_{eff} for a rigid lattice. This dynamical effect is the subject of Feynman's polaron problem [4].

In the opposite *small-polaron* limit, the particle moves in a discrete tight-binding lattice. The rigid lattice potential leads to spatially localized states. The overlap of these states is represented by a transfer matrix describing transitions of the system between the localized states. The dynamics of the particle in the absence of lattice vibrations is then described by the tight-binding Hamiltonian

$$H_{\text{S}} = -\frac{1}{2}\hbar\Delta\sum_n [a_{n+1}^\dagger a_n + a_n^\dagger a_{n+1}] \,, \tag{3.105}$$

where we have used the language of creation and annihilation operators. To this class belong motion of a single excess electron (or hole) in a molecular crystal [119, 120], reorientation processes of dipolar defects [121], and phonon assisted transport [122] and quantum diffusion of light interstitials and defects in crystals [123] – [125].

The particle's interaction energy due to linear lattice distortions is described by the Fröhlich interaction [126]. The Hamiltonian of the particle-phonon system reads

$$H = H_{\text{S}} + \sum_{\boldsymbol{k},\lambda} \hbar\omega_{\boldsymbol{k},\lambda} b_{\boldsymbol{k},\lambda}^\dagger b_{\boldsymbol{k},\lambda} + \sum_{\boldsymbol{k},\lambda} W_{\boldsymbol{k},\lambda}\, e^{i\boldsymbol{k}\cdot\boldsymbol{q}} \left(b_{\boldsymbol{k},\lambda} + b_{-\boldsymbol{k},\lambda}^\dagger \right) \,, \tag{3.106}$$

where the $b_{\boldsymbol{k},\lambda}$ and $b_{\boldsymbol{k},\lambda}^\dagger$ are the annihilation and creation operators for the normal mode with wave vector $\boldsymbol{k}$ and branch index λ. In order that the interaction is Hermitean, we need to have $W_{\boldsymbol{k},\lambda}^* = W_{-\boldsymbol{k},\lambda}$. The linear coupling to the phonon modes leads to a shift of the equilibrium position and to a shift of the potential energy as we have discussed already in Subsection 3.1.1 (cf. Eq. 3.5). We have

$$\Delta V_{\text{relax}} = -\sum_{\boldsymbol{k},\lambda} |W_{\boldsymbol{k},\lambda}|^2/\hbar\omega_{\boldsymbol{k},\lambda} \,. \tag{3.107}$$

In connection with optical polarons, ΔV_{relax} is called the *relaxation energy*. The relaxation energy has no effect on the dynamics of the polaron. It is a matter of convenience whether ΔV_{relax} is subtracted from the Hamiltonian (3.106).

In the following, the form factor $W_{\boldsymbol{k},\lambda}$ is calculated from a microscopic consideration for acoustic and optical polarons. In many particular cases, the form factors are practically independent of the lattice coordinates. In this case, the Hamiltonian is translational invariant. The Fröhlich Hamiltonian has been of basic importance for numerous branches of solid state physics over the last four decades.

In metals, the polaron is also interacting with the conduction electrons. The electromagnetic interaction of the polaron distorts the electron gas or Fermi liquid. This results in a screening cloud of virtual electron-hole excitations around the particle which is dragged along as the particle moves. As we shall see, this leads to various singular behaviours which are known as Fermi surface effects [127].

3.3.1 Acoustic polaron: one-phonon and two-phonon coupling

The interaction of the charged particle with the lattice shifts the position of the atom n from the equilibrium value $R^{(n)}$ to the actual position $X^{(n)}$ by the displacement vector $u^{(n)}$. We have $X^{(n)} = R^{(n)} + u^{(n)}$. Since the displacement $u^{(n)}$ is usually small, it is convenient to expand the particle-lattice potential in a power series in the displacement. We assume that the interaction potential depends only on the distance of the particle from the host atoms, and we express this property by the short-hand notation $H_{\text{latt}}(q - \{X\})$. We then have [128]

$$H_{\text{latt}}(q - \{X\}) = H_{\text{rig}}(q) + H_{\text{lin}}(q) + H_{\text{quadr}}(q) + \cdots , \qquad (3.108)$$

where

$$H_{\text{rig}}(q) = H_{\text{latt}}(q - \{R\}) ,$$

$$H_{\text{lin}}(q) = \sum_n \left(\nabla_{X^{(n)}} H_{\text{latt}}(q - \{X\}) \Big|_{u^{(n)}=0} \right) \cdot u^{(n)}(q) , \qquad (3.109)$$

$$H_{\text{quadr}}(q) = \frac{1}{2} \sum_{n,m} u^{(n)}(q) \cdot \left(\nabla_{X^{(n)}} \nabla_{X^{(m)}} H_{\text{latt}}(q - \{X\}) \Big|_{u^{(n/m)}=0} \right) \cdot u^{(m)}(q) .$$

The first term is the rigid lattice potential introduced already in Eq. (3.104). The displacement is expressed in terms of the usual bosonic annihilation and creation operators. We find it convenient to choose the form [129]

$$u^{(n)}(q) = i \sum_{k,\lambda} \left(\frac{\hbar}{2V \varrho \omega_{k,\lambda}} \right)^{1/2} e(k, \lambda) \exp\left(i k \cdot (q - R^{(n)}) \right) \left(b_{k,\lambda} + b^\dagger_{-k,\lambda} \right) , \qquad (3.110)$$

where V is the volume of the lattice, ϱ is the mass density, and $e(k, \lambda)$ is the polarization vector. Hermitecity of the Hamiltonian requires that $-i\, e^*(k, \lambda) = i\, e(-k, \lambda)$. We choose the polarization vectors $e(k, \lambda)$ real but change sign with k direction,

$$e(-k, \lambda) = -e(k, \lambda) . \qquad (3.111)$$

For the sake of clarity, we use from now on the abridged notation

$$k := (k, \lambda) \qquad \text{and} \qquad -k := (-k, \lambda) . \qquad (3.112)$$

It is convenient to introduce the scalar coupling functions

$$\kappa_k^{(n)}(\boldsymbol{q}) = -i\nabla_{\boldsymbol{X}^{(n)}} H_{\text{latt}}\left(\boldsymbol{q} - \{\boldsymbol{X}\}\right)\Big|_{\boldsymbol{u}^{(n)}=0} \cdot \boldsymbol{e}(k) \equiv \boldsymbol{g}^{(n)}(\boldsymbol{q}) \cdot \boldsymbol{e}(k) , \quad (3.113)$$

$$\gamma_{k,k'}^{(n,m)}(\boldsymbol{q}) = -\boldsymbol{e}(k) \cdot \left(\nabla_{\boldsymbol{X}^{(n)}}\nabla_{\boldsymbol{X}^{(m)}} H_{\text{latt}}\left(\boldsymbol{q} - \{\boldsymbol{X}\}\right)\Big|_{\boldsymbol{u}^{(n/m)}=0}\right) \cdot \boldsymbol{e}(k') . \quad (3.114)$$

The interaction which is linear in the lattice diplacement reads

$$H_{\text{lin}}(\boldsymbol{q}) = -\left(\frac{\hbar}{2V\varrho}\right)^{1/2} \sum_n \sum_k \frac{\kappa_k^{(n)}(\boldsymbol{q})}{\sqrt{\omega_k}} \left(b_k + b_{-k}^\dagger\right) \exp\left(i\boldsymbol{k}\cdot(\boldsymbol{q} - \boldsymbol{R}^{(n)})\right) , \quad (3.115)$$

and the interaction which is quadratic in the displacement is given by

$$\begin{aligned} H_{\text{quadr}}(\boldsymbol{q}) = {} & \frac{\hbar}{4V\varrho} \sum_{n,m} \sum_{k,k'} \frac{\gamma_{k,k'}^{(n,m)}(\boldsymbol{q})}{\sqrt{\omega_k\omega_{k'}}} \left(b_k + b_{-k}^\dagger\right)\left(b_{k'} + b_{-k'}^\dagger\right) \\ & \times \exp\left(i\boldsymbol{k}\cdot(\boldsymbol{q} - \boldsymbol{R}^{(n)})\right)\exp\left(i\boldsymbol{k}'\cdot(\boldsymbol{q} - \boldsymbol{R}^{(m)})\right) . \end{aligned} \quad (3.116)$$

The elementary process of the interaction $H_{\text{lin}}(\boldsymbol{q})$ is the absorption or emission of a single acoustic phonon. In the nth order of the perturbation series in $H_{\text{lin}}(\boldsymbol{q})$, the polaron absorbs and emits altogether n *uncorrelated* phonons. In contrast, the elementary process of the interaction $H_{\text{quadr}}(\boldsymbol{q})$, which is of second order in the lattice displacement, describes simultaneous absorption and emission of two phonons. In different terms, the contribution which is quadratic in $H_{\text{lin}}(\boldsymbol{q})$ is a single-phonon process of second order while the first order in $H_{\text{quadr}}(\boldsymbol{q})$ describes a genuine two-phonon process. We shall see in Section 4.2 that the effects of acoustic phonons on the tunneling of atoms between surfaces, as in a scanning-tunneling microscope, are qualitatively different from atom tunneling in the bulk. The first case corresponds to Ohmic dissipation while dissipation in the second case belongs to the super-Ohmic variety. The correlated two-phonon contribution (3.116) is disregarded in most studies because it is usually small compared to the contributions of $H_{\text{lin}}(\boldsymbol{q})$. However, as we shall see in Subsection 4.2.5, the two-phonon coupling $H_{\text{quadr}}(\boldsymbol{q})$ corresponds to Ohmic dissipation for tunneling in the bulk, but the viscosity coefficient is strongly temperature dependent.

3.3.2   Optical polaron

In *polar* crystals, such as sodium chloride, the ions of positive charge and the ions of negative charge oscillate with opposite phase. By the relative displacement of the ionic sublattices, polarization charges are set up which act as a source of a polarization field. The polarization field scatters the electrons. The one displacement mode causing strong polarization is the *longitudinal optical* (LO) mode [$\lambda = 1$ in Eq. (3.106)] for which the polarization vector is a unit vector in $\boldsymbol{k}$ direction,

$$e(\boldsymbol{k}, \mathrm{l}) = \hat{\boldsymbol{k}} . \tag{3.117}$$

We assume that the LO mode is independent of the positions of the individual ions and that the frequency of the optical phonon is constant. We put $\omega_{\boldsymbol{k},\mathrm{l}} = \omega_{\mathrm{LO}}$. The relative displacement of the ionic sublattices by the longitudinal optical mode is then obtained from Eq. (3.110) as

$$\boldsymbol{u}(\boldsymbol{q}) = i \left(\frac{\hbar}{2V \varrho \, \omega_{\mathrm{LO}}} \right)^{1/2} \sum_{k} \hat{\boldsymbol{k}} \, \exp \left(i \boldsymbol{k} \cdot \boldsymbol{q} \right) \left(b_{\boldsymbol{k}} + b^{\dagger}_{-\boldsymbol{k}} \right) . \tag{3.118}$$

The optical mode induces a polarization field $\boldsymbol{P}(\boldsymbol{q})$ which is proportional to the displacement in Eq. (3.118). We write

$$\boldsymbol{P}(\boldsymbol{q}) = U \boldsymbol{u}(\boldsymbol{q}) , \tag{3.119}$$

where the constant U has dimension charge density and is to be determined yet. The polarization charge density

$$\rho_{\mathrm{pol}}(\boldsymbol{q}) = -\nabla \cdot \boldsymbol{P}(\boldsymbol{q}) \tag{3.120}$$

is a source for an electrical field which itself is a gradient field of a scattering potential for the electron. Thus, the scattering potential $V_{\mathrm{scatt}}(\boldsymbol{q})$ satisfies the Poisson equation

$$\nabla^2 V_{\mathrm{scatt}}(\boldsymbol{q}) = 4\pi e \rho_{\mathrm{pol}}(\boldsymbol{q}) = -4\pi e U \nabla \cdot \boldsymbol{u}(\boldsymbol{q}) , \tag{3.121}$$

where the electron has charge $-e$. The solution of Eq. (3.121) reads

$$V_{\mathrm{scatt}}(\boldsymbol{q}) = -4\pi e U \left(\frac{\hbar}{2V \varrho \, \omega_{\mathrm{LO}}} \right)^{1/2} \sum_{k} \frac{1}{|\boldsymbol{k}|} \exp \left(i \boldsymbol{k} \cdot \boldsymbol{q} \right) \left(b_{\boldsymbol{k}} + b^{\dagger}_{-\boldsymbol{k}} \right) . \tag{3.122}$$

The interaction of the electron with the modes of the polar crystal, Eq. (3.122), is exactly of the Fröhlich form (3.106) [4, 126, 129]. Upon comparing Eq. (3.122) with the coupling term in Eq. (3.106), the form factor of the longitudinal optical polaron $W_{\boldsymbol{k},\mathrm{l}}$ reads

$$W_{\boldsymbol{k},\mathrm{l}} = -4\pi e U \left(\frac{\hbar}{2V \varrho \, \omega_{\mathrm{LO}}} \right)^{1/2} \frac{1}{|\boldsymbol{k}|} . \tag{3.123}$$

Finally, the charge density parameter U is found from the consideration of the interaction potential between two electrons at fixed distance $|\boldsymbol{q}|$. The interaction energy originates from the relaxation of the ionic environment in which the charges are embedded. For a finite distance $\boldsymbol{q}$ of the two charges, we find from Eq. (3.107) the relaxation contribution

$$V_{\mathrm{relax}}(\boldsymbol{q}) = -2 \sum_{k} \frac{|W_{\boldsymbol{k},\mathrm{l}}|^2}{\hbar \omega_{\mathrm{LO}}} \exp \left(i \boldsymbol{k} \cdot \boldsymbol{q} \right) . \tag{3.124}$$

Inserting Eq. (3.123), we obtain

$$V_{\mathrm{relax}}(\boldsymbol{q}) = -\frac{(4\pi e U)^2}{\varrho \, \omega_{\mathrm{LO}}^2} \int \frac{d^3 k}{(2\pi)^3} \frac{1}{|\boldsymbol{k}|^2} \exp \left(i \boldsymbol{k} \cdot \boldsymbol{q} \right) , \tag{3.125}$$

which is the Coulomb potential

$$V_{\text{relax}}(\boldsymbol{q}) = -\gamma \frac{e^2}{|\boldsymbol{q}|} \quad \text{with} \quad \gamma = \frac{4\pi U^2}{\varrho \omega_{\text{LO}}^2}. \tag{3.126}$$

The relaxation potential $V_{\text{relax}}(\boldsymbol{q})$ represents the contribution from the optical phonons to the dielectric screening of the Coulomb potential. This is the difference in screening between the cases of low and high frequencies. Thus we have

$$\frac{e^2}{\epsilon_0 |\boldsymbol{q}|} = \frac{e^2}{|\boldsymbol{q}|} \left(\frac{1}{\epsilon_\infty} - \gamma \right), \quad \text{yielding} \quad \gamma = \frac{1}{\epsilon_\infty} - \frac{1}{\epsilon_0}, \tag{3.127}$$

where ϵ_∞ and ϵ_0 are the high-frequency (optical) and static dielectric constants, respectively. The charge density U is gathered from Eqs. (3.126) and (3.127). We find

$$U^2 = \frac{\varrho \omega_{\text{LO}}^2}{4\pi} \left(\frac{1}{\epsilon_\infty} - \frac{1}{\epsilon_0} \right). \tag{3.128}$$

Thus, U is entirely given in terms of measurable quantities. In the literature, it is customary to use the dimensionless polaron coupling constant

$$\alpha \equiv \left(\frac{1}{\epsilon_\infty} - \frac{1}{\epsilon_0} \right) \frac{e^2}{\hbar \omega_{\text{LO}}} \left(\frac{M \omega_{\text{LO}}}{2\hbar} \right)^{1/2}. \tag{3.129}$$

Expressed in terms of the parameter α, the form factor (3.123) reads for an optical polaron

$$W_{\boldsymbol{k},\text{l}} = - \left(\frac{4\pi \alpha}{V} \right)^{1/2} \left(\frac{\hbar}{2M \omega_{\text{LO}}} \right)^{1/4} \frac{\hbar \omega_{\text{LO}}}{|\boldsymbol{k}|}. \tag{3.130}$$

With the coupling parameter α given by Eq. (3.129), the self-energy of the polaron takes a particular simple form [4]. Typical values of α run from about 1 to 20.

3.3.3 Interaction with fermions (normal and superconducting)

As a result of the Pauli principle, the response of the noninteracting electron gas to a time-dependent local potential shows interesting characteristic features. The transient perturbations of the Fermi gas or Fermi liquid lead to a screening cloud of virtual electron-hole excitations which turns out to be very singular in character. The singular behaviour can be seen, e. g., in the energy dependence of soft X-ray absorption or emission of metals (see, e.g., the review by Mahan [130], and by Ohtaka and Tanabe [131]). The singular response of the electron gas is also reflected in the electrical resistivity when dilute magnetic impurities are dissolved in a non-magnetic metallic host crystal. The understanding of these peculiar properties constitutes the crux of the Kondo problem (cf. the review by Tsvelik and Wiegmann [132]).

It has been shown by Kondo that quantum diffusion of charged interstitials in normally conducting metals exhibits also singular behaviour which is closely related to the above phenomema. The singular transient response of the conduction electrons results from the high density of electron-hole excitations around the Fermi surface (the Fermi surface effects are reviewed by Kondo in Ref. [127]). When the particle moves, it drags behind it a screening cloud of electron-hole pairs which have bosonic character. These excitations shall be found to be adequately represented by a spectral density of a virtual coupling to a bosonic reservoir which is of Ohmic form. This particular coupling leads, e.g., to self-trapping at zero temperature above a critical value of the coupling strength [133] – [138], and to anomalous temperature dependence such as the increase of the diffusion coefficient with decreasing temperature for weak damping (see below and Ref. [127]). Ohmic friction fades away when the fermionic system changes from the normally conducting to the superconducting state. We shall discuss the respective properties of the coupling in some detail in Subsection 4.2.8.

In this subsection we introduce the underlying Hamiltonian for the normal and superconducting state of the fermions and postpone the derivation of the effective action of the particle until Subsection 4.2.8. The description of the fermionic effects on the charged interstitial is based on the Hamiltonian

$$H = H_S(\boldsymbol{q}) + H_R + H_I(\boldsymbol{q}) , \tag{3.131}$$

where H_S is the Hamiltonian of the system in the absence of the environment, and H_R and H_I are the terms due to the reservoir and the interaction, respectively. For our purposes it is convenient to formulate H_R and H_I in the scheme of second quantization. We shall assume that the particle couples to local fluctuations of the electronic density via a potential $U(\boldsymbol{r})$.

For fermions in the *normally conducting* state, the reservoir Hamiltonian has the standard harmonic form

$$H_R^{(nc)} = \sum_{\boldsymbol{k},\sigma} \hbar\omega_{\boldsymbol{k}} c_{\boldsymbol{k}\sigma}^\dagger c_{\boldsymbol{k}\sigma} , \tag{3.132}$$

where $c_{\boldsymbol{k}\sigma}^\dagger$ is the creation operator for a conduction electron with wave vector $\boldsymbol{k}$ and spin polarization σ. The energy $\hbar\omega_{\boldsymbol{k}}$ is measured relative to the Fermi energy E_F.

The interaction of the conduction electrons with the defect is a charge density-density interaction. For a point particle at position $\boldsymbol{q}$, we have

$$H_I^{(nc)}(\boldsymbol{q}) = \int d^3r\, U(\boldsymbol{r} - \boldsymbol{q})\Psi^\dagger(\boldsymbol{r})\Psi(\boldsymbol{r}) . \tag{3.133}$$

Upon using the normal mode expansion $\Psi^\dagger(\boldsymbol{r}) = \sum_{\boldsymbol{k},\sigma} \exp(-i\boldsymbol{k}\cdot\boldsymbol{r}) c_{\boldsymbol{k},\sigma}^\dagger$, we obtain

$$H_I^{(nc)}(\boldsymbol{q}) = \sum_{\boldsymbol{k},\boldsymbol{k}',\sigma,\sigma'} <\boldsymbol{k},\sigma|U|\boldsymbol{k}',\sigma'> \exp\left(i(\boldsymbol{k}'-\boldsymbol{k})\cdot\boldsymbol{q}\right) c_{\boldsymbol{k}\sigma}^\dagger c_{\boldsymbol{k}'\sigma'} , \tag{3.134}$$

where $<\boldsymbol{k},\sigma|U|\boldsymbol{k}',\sigma'>$ is the matrix element of the interaction. The potential $U(\boldsymbol{q})$ depends on the particular case. For charged particles in metals, it is the Coulomb potential, while in superconductors, it is a screened Coulomb potential. Other choices such as dipolar, or multipolar couplings for neutral particles are also possible.

When the fermions are in the *superconducting* state, the reservoir Hamiltonian describes BCS quasi-particles,

$$H_R^{(\text{sc})} = \sum_{k,\alpha} \hbar\Omega_k \, \gamma_{k\alpha}^\dagger \gamma_{k\alpha} \, . \tag{3.135}$$

Here, $\gamma_{k\alpha}^\dagger$ is the creation operator for a quasi-particle with wave vector k, spin polarization α, and excitation energy $\hbar\Omega_k$, where

$$\hbar\Omega_k = \hbar(\omega_k^2 + \Delta_g^2)^{1/2} \, , \tag{3.136}$$

and $\hbar\Delta_g$ is the temperature dependent gap energy of the superconducting state. The operators $\gamma_{k\alpha}^\dagger$ are connected with $c_{k\sigma}^\dagger$ by a Bogoliubov transformation [139]

$$\begin{aligned}
\gamma_{k\uparrow}^\dagger &= u_k c_{k\uparrow}^\dagger - v_k c_{-k\downarrow} \, , \\
\gamma_{k\downarrow}^\dagger &= u_k c_{k\downarrow}^\dagger + v_k c_{-k\uparrow} \, .
\end{aligned} \tag{3.137}$$

Since the screening of the electrons in the superconducting state is the same as in the normal state [140], the interaction is the same as in Eq. (3.134). We now wish to express the interaction as a function of the quasi-particle operators. The inversion of Eq. (3.137) gives

$$c_{k\alpha}^\dagger = u_k \gamma_{k\alpha}^\dagger + \sum_\beta \rho_{\alpha\beta} v_k \gamma_{-k\beta} \, , \tag{3.138}$$

where

$$\rho = \begin{pmatrix} 0 & -1 \\ 1 & 0 \end{pmatrix} \, . \tag{3.139}$$

Inserting Eq. (3.138) into Eq. (3.134) we obtain

$$\begin{aligned}
H_I(q) = & \sum_{k,k',\alpha,\alpha'} < k,\alpha|U|k',\alpha' > \exp\left(i(k'-k)\cdot q\right) \\
& \times \left\{ u_k u_{k'} \gamma_{k\alpha}^\dagger \gamma_{k'\alpha'} + v_k v_{k'} \sum_{\sigma,\sigma'} \rho_{\alpha\sigma}\rho_{\alpha'\sigma'} \gamma_{-k\sigma} \gamma_{-k'\sigma'}^\dagger \right. \\
& \left. + u_k v_{k'} \sum_{\sigma'} \rho_{\alpha'\sigma'} \gamma_{k\alpha}^\dagger \gamma_{-k'\sigma'}^\dagger + u_{k'} v_k \sum_\sigma \rho_{\alpha\sigma} \gamma_{-k\sigma} \gamma_{k'\alpha'} \right\} \, .
\end{aligned} \tag{3.140}$$

We are interested in the first two terms of the interaction since they describe scattering of quasi-particles. The last two terms are irrelevant at low temperatures as they create and destroy two quasi-particles. Thus we have

$$H_I^{(\text{sc})}(q) = \sum_{k,k',\alpha,\alpha'} M(k,\alpha|k',\alpha') \exp\left(i(k'-k)\cdot q\right) \gamma_{k\alpha}^\dagger \gamma_{k'\alpha'} \, , \tag{3.141}$$

where $M(k,\alpha|k',\alpha')$ is determined by the first two terms of Eq. (3.140).

Upon using $\gamma_i \gamma_j^\dagger = -\gamma_j^\dagger \gamma_i$ for $i \neq j$, we find [139]

$$M(k,\alpha|k',\alpha') = u_k u_{k'} < k,\alpha|U|k',\alpha' > -v_k v_{k'} \sum_{\sigma,\sigma'} \rho_{\sigma'\alpha'}\rho_{\sigma\alpha} < -k',\sigma'|U|-k,\sigma > \, .$$

The term $\sum_{\sigma\sigma'} \rho_{\sigma'\alpha'}\rho_{\sigma\alpha} < -\boldsymbol{k}', \sigma'|U| - \boldsymbol{k}, \sigma >$ is essentially the matrix element of U in which the polarizations and momenta of the electrons have the opposite sign. Accordingly, the sign of this term depends on the behaviour of the interaction under time-reversal. Thus we may write

$$M(\boldsymbol{k}, \alpha|\boldsymbol{k}', \alpha') = \left\{ u_{\boldsymbol{k}} u_{\boldsymbol{k}'} - \zeta\, v_{\boldsymbol{k}} v_{\boldsymbol{k}'} \right\} < \boldsymbol{k}, \alpha|\, U\,|\boldsymbol{k}', \alpha' >, \qquad (3.142)$$

where $\zeta = \pm 1$. The curly bracket in Eq. (3.142) is called the coherence factor of the transition. If the external perturbation does not break the time-reversal symmetry, which is the case of our interest, we have $\zeta = +1$. By using the definitions

$$u_{\boldsymbol{k}} v_{\boldsymbol{k}} = \frac{\Delta_{\mathrm{g}}}{2\Omega_{\boldsymbol{k}}}\,; \quad u_{\boldsymbol{k}}^2 = \frac{1}{2}\left(1 + \frac{\omega_{\boldsymbol{k}}}{\Omega_{\boldsymbol{k}}}\right)\,; \quad v_{\boldsymbol{k}}^2 = \frac{1}{2}\left(1 - \frac{\omega_{\boldsymbol{k}}}{\Omega_{\boldsymbol{k}}}\right), \qquad (3.143)$$

we obtain

$$|\,M(\boldsymbol{k}, \sigma|\boldsymbol{k}', \sigma')\,|^2 = \frac{1}{2}\left\{ 1 + \frac{\omega_{\boldsymbol{k}}\omega_{\boldsymbol{k}'}}{\Omega_{\boldsymbol{k}}\Omega_{\boldsymbol{k}'}} - \zeta\frac{\Delta_{\mathrm{g}}^2}{\Omega_{\boldsymbol{k}}\Omega_{\boldsymbol{k}'}} \right\} |< \boldsymbol{k}, \sigma|\, U|\boldsymbol{k}', \sigma' >|^2. \qquad (3.144)$$

Since to each value of $\Omega_{\boldsymbol{k}}$ there belong the values $\pm\omega_{\boldsymbol{k}}$, the term $\omega_{\boldsymbol{k}}\omega_{\boldsymbol{k}'}/\Omega_{\boldsymbol{k}}\Omega_{\boldsymbol{k}'}$ cancels out when Eq. (3.144) is summed over these two values.

Our subsequent study of the dissipative influences of the fermionic environment in Subsection 4.2.8 will be based on the above microscopic Hamiltonians. For normally conducting electrons, this is Eq. (3.131) with Eq. (3.132) and Eq. (3.134), whereas in the superconducting case the underlying model is defined by Eq. (3.131) with Eq. (3.135) and Eq. (3.141).

3.3.4 Superconducting tunnel junction

The collective variable which determines the dynamics of a Josephson junction is the phase difference of the order parameter of the superconductor on the two sides of the barrier. The phase difference couples to the electronic degrees of freedom, which act again as a heat bath and source of phase fluctuations. A well-known phenomenological description of this coupling is given by the *resistively shunted junction* (RSJ) model in which the barrier acts as an Ohmic resistor shunted in parallel to the displacement and to the super-current [101].

For Josephson junctions or weak links, we need not resort to a phenomenological modeling. We rather may start from the microscopic theory, as shown by Ambegaokar *et al.* [141], and by Larkin and Ovchinnikov [142]. As we shall see, the microscopic theory may lead to a model Hamiltonian of the form (3.8) with linear or nonlinear Ohmic friction, depending on the magnitude of the phase difference across the junction. The description of the bulk superconductor is based on the grand canonical Hamiltonian [143]

$$H_{\mathrm{bulk}} = \int d^3r\, \Psi_\sigma^\dagger(\boldsymbol{r}) \left[-\frac{\hbar^2}{2m}\left(\boldsymbol{\nabla} - \frac{ie}{\hbar}\boldsymbol{A}\right)^2 - \mu + U(\boldsymbol{r}) \right] \Psi_\sigma(\boldsymbol{r}) \qquad (3.145)$$

$$-\frac{1}{2}\int d^3r\, \Psi_\sigma^\dagger(\boldsymbol{r})\,\Psi_{-\sigma}^\dagger(\boldsymbol{r})\, g(\boldsymbol{r})\, \Psi_{-\sigma}(\boldsymbol{r})\, \Psi_\sigma(\boldsymbol{r}) + \frac{1}{8\pi}\int d^3r\, \left(\boldsymbol{h} - \boldsymbol{h}_{\mathrm{ext}}\right)^2.$$

Here a summation over spin polarizations is implied. The vector field $\boldsymbol{A}$ represents the vector potential, and μ is the chemical potential. The Hamiltonian describes conduction electrons in a potential $U(\boldsymbol{r})$ which may also account for impurities and boundaries. The second line describes an effective attractive BCS interaction of strength $g(\boldsymbol{r})$, and in the last line we have added the magnetic field contribution, where $\boldsymbol{h}$ and $\boldsymbol{h}_{\text{ext}}$ are the magnetic field and the externally applied magnetic field, respectively.

A superconducting tunnel junction consists of two superconductors to the left (L) and right (R) of an oxide barrier. It is conveniently described by the tunneling Hamiltonian

$$H = H_{\text{bulk,L}} + H_{\text{bulk,R}} + H_{\text{T}} + H_{\text{Q}} . \qquad (3.146)$$

The coupling is due to the transfer of electrons through the barrier and due to the Coulomb interaction term H_{Q}. The former is described by the tunneling term

$$H_{\text{T}} = \int_{\boldsymbol{r} \in \text{R}} d^3 r \int_{\boldsymbol{r}' \in \text{L}} d^3 r' \left[T_{\boldsymbol{r} \boldsymbol{r}'} \Psi_{\text{R},\sigma}^{\dagger}(\boldsymbol{r}) \, \Psi_{\text{L},\sigma}(\boldsymbol{r}') + \text{h.c.} \right] . \qquad (3.147)$$

The range of the tunneling matrix element $T_{\boldsymbol{r} \boldsymbol{r}'}$ is limited to the vicinity of the barrier. As long as we are interested in frequencies small compared with the plasma frequency, the Coulomb interaction across the barrier effectively behaves as a capacitive interaction depending on the Cooper pair charges Q_{L} and Q_{R} stored on the electrodes and on the capacitor with capacitance C. The latter is determined by the geometry and by properties of the insulating barrier. We then have

$$H_{\text{Q}} = (Q_{\text{L}} - Q_{\text{R}})^2 / 8C , \qquad (3.148)$$

$$Q_{\text{L/R}} = e \int_{\boldsymbol{r} \in \text{ins}} d^3 r \, \Psi_{\text{L/R},\sigma}^{\dagger}(\boldsymbol{r}) \, \Psi_{\text{L/R},\sigma}(\boldsymbol{r}) . \qquad (3.149)$$

The model also describes inhomogeneous systems including normally conducting domains when $g(\boldsymbol{r}) = 0$, constrictions and tunnel junctions, as well as superconducting rings with a weak link in which the electrodes L and R are joined up in a loop.

The relevant quantum statistical properties of the system can be extracted from the generating functional

$$Z(\xi) = \text{tr}_{A,\Psi} \left\{ T_\tau \exp\left[-\frac{1}{\hbar} \int_0^{\hbar\beta} d\tau \left(H(\tau) - \hbar\xi(\tau)B(\tau) \right) \right] \right\} , \qquad (3.150)$$

where $B(\tau)$ is the quantity of interest and $\xi(\tau)$ is the usual source term. Here, T_τ is the time ordering operator for the imaginary-time variable τ (cf. Chapter 4). Further, the trace is taken over the fermion fields Ψ and the vector potential $\boldsymbol{A}$. The effective action of this model is discussed in Subsection 4.2.10.

3.4   Charging and environmental effects in tunnel junctions

With the enormous progress in microfabrication techniques, it has become possible to fabricate metallic tunnel junctions with capacitances C of 10^{-15} F or less. The corresponding charging energy of an electron, $E_c = e^2/2C$, is larger than the thermal energy for temperatures of 1 K or smaller. Therefore, it is nowadays possible to study charge transport through barriers in a regime where charging effects play an important role. Here we are interested in a quantum mechanical treatment of these effects.

Charging effects are also relevant in semiconductor nanostructures, e.g. quantum dots in a 2-D electron gas with typical capacitances of 10^{-15} F. In the still smaller molecular electron transfer systems, the charging energy can be so large that single-electron tunneling becomes observable even at room temperatures. On the one hand, the systems are small, on the other hand, they are still large enough that they can be connected to macroscopic current and voltage sources or probes. Just for this reason, the systems are sensitive to the electromagnetic environment.

The concepts introduced here and in Chapter 20.3 are generally important for metallic, semiconductor, and molecular systems. Here we concentrate the attention on metallic systems with a high electron density of states.

A collection of research papers in the field of single-charge tunneling is published as a special issue of Zeitschrift für Physik B [144]. A series of nine tutorial articles by renowned experts which deal with particular aspects of the field are compiled in the book *Single Charge Tunneling* [145]. We now introduce the global system for single-electron tunneling and present various examples of electromagnetic environments. For a further study, we refer the reader to the review by Ingold and Nazarov [146].

For an ideal current-biased junction at zero temperature, a charge can only tunnel if the balance of the electrostatic charging energy before and after the charge transfer is positive,

$$\Delta E_c = \frac{Q^2}{2C} - \frac{(Q-e)^2}{2C} = e[V_a - e/2C] > 0 . \tag{3.151}$$

This condition is satisfied if $Q > e/2$, which in turn implies that the voltage V_a across the junction must be larger than $V_c = e/2C$. For a nonzero current through the junction, the junction capacitor charge Q will increase until the threshold charge $e/2$ is reached. Then, a tunneling event may occur leaving a charge $Q = -e/2$ on the capacitor. Afterwards, a new charging cycle starts again. Altogether, the junction voltage performs sawtooth single electron tunneling (SET) oscillations [147, 143] with a frequency $\nu_{\mathrm{SET}} = I/e$. [5]

Alternatively, we may employ an ideal voltage source for driving the current through the junction. According to the above argument, there is no current in the regime $-e^*/2C < V_a < e^*/2C$, where e^* is the charge e in the normal state and is $2e$ in the superconducting state. This is the *Coulomb blockade* phenomenon for a single junction. For a voltage larger than $e/2C$ one finds for a normal state junction at $T = 0$ an Ohmic (linear) current-voltage characteristics which is shifted by the Coulomb gap $e/2C$, $I(V_a) = (V_a - e/2C)/R_T$ [cf. Subsection 20.3.2].

---

[5]By a similar argument, analogous so-called Bloch oscillations occur for a superconducting Josephson junction with a fundamental frequency $\nu_{\mathrm{Bloch}} = I/2e$.

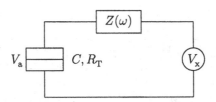

Figure 3.3: A voltage-biased tunnel junction with tunnel resistance R_T and capacitance C. The impedance $Z(\omega)$ models the frequency response of the electromagnetic environment which is dominated by the effects of the leads attached to the junction.

3.4.1 The global system for single electron tunneling

In reality, the single-electron device is influenced by stray capacitances, inductances, and resistances, as well as by quantum and thermal fluctuations. The simplest model in which we can study these influences on charge tunneling is a single junction in series with a general impedance $Z(\omega)$ and connected to an ideal voltage source V_x as sketched in Fig. 3.3. The impedance $Z(\omega)$ represents the frequency response of the electromagnetic environment. Generally, the dc voltage drop V_a across the junction differs from the applied voltage V_x by the dc voltage drop at the impedance. Assuming that we have calculated the $I(V_a)$ characteristics of the junction, the applied voltage V_x can be reconstructed from the relation $V_x = V_a + Z(0)\, I(V_a)$.

To derive a phenomenological "electromagnetic" Hamiltonian for the environment, we recall the well-known correspondence between electrical and mechanical quantities assembled in Table I [146]. We introduce the phase difference at the junction

$$\varphi_V(t) = \frac{e}{\hbar} \int_{-\infty}^{t} dt'\, V(t') , \qquad (3.152)$$

where $V(t) = V_a + \delta V(t)$ is the voltage drop at the junction. The impedance $Z(\omega)$ introduces voltage fluctuations $\delta V(t)$ and charge fluctuations $\mathcal{Q}(t)$ which vanish on average. The charge $Q(t)$ on the junction is the canonical conjugate variable to $(\hbar/e)\varphi_V(t)$. In equilibrium, the average charge on the capacitor is CV_a. Next, we introduce the fluctuations of the charge $\mathcal{Q}(t)$ and of the phase $\varphi(t)$ around the mean values determined by the voltage V_a,

$$\mathcal{Q} = Q - CV_a , \qquad \varphi(t) = \varphi_V(t) - eV_a t/\hbar , \qquad \text{with} \qquad [\varphi, \mathcal{Q}] = ie . \quad (3.153)$$

In a phenomenological modeling analogous to a harmonic oscillator bath, the electromagnetic reservoir is formed by LC-circuits with eigenfrequencies $\omega_\alpha = 1/\sqrt{L_\alpha C_\alpha}$, and the coupling of the junction to the circuits is bilinear in the phase variables. Thus, the electromagnetic environment consists of the capacitance of the junction, the LC-circuits, and the coupling between the junction and the circuits. It is described by the Hamiltonian [the charge $\mathcal{Q}$ is conjugate to $(\hbar/e)\,\varphi$]

Table I. Correspondence between mechanical and electrical quantities

Mechanical quantity	Electrical quantity
mass M	capacitance C
momentum p	charge Q
velocity $v = p/M$	voltage $V = Q/C$
coordinate x	phase φ
$[x, p] = i\hbar$	$[\varphi, Q] = ie$
spring constant f	inverse inductance $1/L$
harmonic oscillator	LC-circuit

$$H_{\text{env}} = \frac{Q^2}{2C} + \sum_{\alpha=1}^{N}\left[\frac{q_\alpha^2}{2C_\alpha} + \left(\frac{\hbar}{e}\right)^2\frac{1}{2L_\alpha}(\varphi - \varphi_\alpha)^2\right]. \tag{3.154}$$

It is straightforward to write down the Heisenberg equations of motion and to eliminate the variables of the LC-circuits and the charge variable $\mathcal{Q}(t)$. We then obtain the Langevin equation for $\varphi(t)$ in the form of the equation of motion for a Brownian particle with frequency-dependent damping, Eq. (3.27),

$$C\ddot{\varphi}(t) + \int_0^t dt'\, Y(t-t')\dot{\varphi}(t') = \frac{e}{\hbar} I_{\text{noise}}(t). \tag{3.155}$$

The damping kernel describing relaxation of the charge into equilibrium is given by

$$Y(t) = \sum_\alpha \frac{1}{L_\alpha}\cos(\omega_\alpha t), \qquad Y^*(\omega) = \lim_{\epsilon\to 0^+}\sum_\alpha \frac{1}{L_\alpha}\frac{-i\omega}{\omega_\alpha^2 - \omega^2 - i\epsilon\,\text{sgn}(\omega)}. \tag{3.156}$$

The Fourier transform of $Y(t)$ is the admittance $Y^*(\omega)$, and $Z^*(\omega) = 1/Y^*(\omega)$ is the impedance function.[6] The quantum mechanical noise current $I_{\text{noise}}(t)$ obeys stationary Gaussian statistics with correlations analogous to Eq. (2.31) and correspondence $M\tilde\gamma'(\omega) \leftrightarrow Y'(\omega)$. Switching by integration from the current-current to the charge-charge correlation function $Q_{\mathcal{Q}}(t) = \langle[\mathcal{Q}(0) - \mathcal{Q}(t]\mathcal{Q}(0)\rangle_\beta$, we find

$$Q_{\mathcal{Q}}(t) = \int_0^\infty d\omega\, \frac{G_{\mathcal{Q}}(\omega)}{\omega^2}\left(\coth(\tfrac{1}{2}\beta\hbar\omega)[1 - \cos(\omega t)] + i\sin(\omega t)\right), \tag{3.157}$$

where

$$G_{\mathcal{Q}}(\omega) = 2\left(\frac{e}{2\pi}\right)^2 R_{\text{K}} Y'(\omega)\,\omega, \tag{3.158}$$

---

[6]We use the quantum mechanical convention $e^{-i\omega t}$ for the kernels, whereas impedances are expressed in terms of the engineering convention $e^{+i\omega t}$. This is the reason for complex conjugation.

and where $Y'(\omega) = \mathrm{Re}\, Y(\omega)$. We can easily determine the dynamical susceptibility $\widetilde{\chi}(\omega)$ describing the response of the phase to the current $I_{\mathrm{noise}}(t)$. Upon Fourier-transforming Eq. (3.155) we get $\widetilde{\varphi}(\omega) = \widetilde{\chi}(\omega)\, e\widetilde{I}_{\mathrm{noise}}(\omega)/\hbar$, where

$$\widetilde{\chi}(\omega) = \frac{1}{-\omega^2 C - i\omega Y^*(\omega)} = \frac{Z_t^*(\omega)}{-i\omega}, \qquad Z_t(\omega) = \frac{1}{i\omega C + Z^{-1}(\omega)}. \qquad (3.159)$$

Thus, the total impedance of the circuit seen by the junction consists of the external impedance $Z(\omega)$ in parallel with the capacitance C of the junction. Anticipating the fluctuation-dissipation theorem discussed in Section 6.1, in particular Eq. (6.24), we can express the phase autocorrelation function $Q_\varphi(t) = \langle [\,\varphi(0) - \varphi(t)\,]\varphi(0)\rangle_\beta$ in terms of the absorptive part of $\widetilde{\chi}(\omega)$. The resulting expression is

$$Q_\varphi(t) = \int_0^\infty d\omega\, \frac{G_\varphi(\omega)}{\omega^2}\Big(\coth(\beta\hbar\omega/2)[\,1 - \cos(\omega t)\,] + i\sin(\omega t)\Big). \qquad (3.160)$$

Here we have introduced the spectral density $G_\varphi(\omega)$ of the electromagnetic coupling, which for the present circuit model is

$$G_\varphi(\omega) = (e^2/\pi\hbar)\,\omega^2\widetilde{\chi}''(\omega) = 2\omega\, Z_t'(\omega)/R_K = 2\omega R_K \bar{Y}'(\omega), \qquad (3.161)$$

where $R_K = 2\pi\hbar/e^2$ is the resistance quantum. In the third form, we have introduced for later convenience the admittance $\bar{Y}(\omega) = Z_t(\omega)/R_K^2$ [cf. Subsection 25.3]. We have normalized $G_\varphi(\omega)$ in such a way that it directly corresponds to the spectral density $G(\omega)$ introduced in Eq. (3.89), as will become obvious in Section 20.3. An alternative derivation of the integral expression (3.160) is given in Subsection 4.2.11. There we calculate the imaginary-time correlation function, $W_\varphi(\tau) = Q_\varphi(t = i\tau)$ upon performing average with the Euclidean influence action $S_{\mathrm{infl}}^{(\mathrm{E})}[\varphi]$ of the electromagnetic environment.

Apparently, any particular harmonic electromagnetic environment can be modelled by a suitable choice of the parameters L_α and C_α in the expresssion (3.154). Eventually, the sum in Eq. (3.156) turns into an integral over a continuous distribution, as discussed in Subsection 3.1.3.

So far, we have treated the junction merely as a capacitor. Now we admit tunneling of charge through the junction by adding the tunneling Hamiltonian

$$H'_T = \sum_{k,k',\sigma} T_{k,k'} c_{k',\sigma}^\dagger c_{k,\sigma}\, e^{-i\varphi_V} + \text{h.c.}, \qquad (3.162)$$

where the operators $c_{k,\sigma}$ and $c_{k,\sigma}^\dagger$ are annihilation and creation operators for quasiparticles with wave vector k, energy E_k, and spin polarization σ. The wave vectors k and k' correspond to quasiparticles in the left and right electrode of the tunnel barrier, respectively. Finally, the model is completed by a Hamiltonian describing the quasiparticles in the left and right electrode, respectively,

$$H'_{\mathrm{qp}} = \sum_{k,\sigma} E_k c^\dagger_{k,\sigma} c_{k,\sigma} + \sum_{k',\sigma} E_{k'} c^\dagger_{k',\sigma} c_{k',\sigma} \, . \qquad (3.163)$$

In the next step we wish to substitute the phase φ for the phase φ_V in the tunneling term. This can be achieved by performing a time-dependent transformation with the unitary operator acting on quasiparticle operators in the left electrode

$$U = \prod_{k,\sigma} \exp \left(i \frac{e}{\hbar} V_{\mathrm{a}} t \, c^\dagger_{k,\sigma} c_{k,\sigma} \right) \, . \qquad (3.164)$$

We have

$$H_{\mathrm{T}} = U^\dagger H'_{\mathrm{T}} U = \sum_{k,k',\sigma} T_{k,k'} c^\dagger_{k',\sigma} c_{k,\sigma} \, \mathrm{e}^{-i\varphi} + \mathrm{h.c.} \, , \qquad (3.165)$$

where φ is given in Eq. (3.153). While the corresponding transformation has no effect on H_{env}, it shifts the quasiparticle energies in the left electrode by eV_{a},

$$H_{\mathrm{qp}} = U^\dagger H'_{\mathrm{qp}} U - i\hbar U^\dagger \partial U / \partial t = \sum_{k,\sigma} [E_k + eV_{\mathrm{a}}] c^\dagger_{k,\sigma} c_{k,\sigma} + \sum_{k',\sigma} E_{k'} c^\dagger_{k',\sigma} c_{k',\sigma} \, . \quad (3.166)$$

The Hamiltonian of the global system takes the form

$$H = H_{\mathrm{T}} + H_{\mathrm{qp}} + H_{\mathrm{env}} \, , \qquad (3.167)$$

in which the tunneling Hamiltonian (3.165) couples the quasiparticle Hamiltonian (3.166) to the reservoir part (3.154).

For weak tunneling, the charge on the junction is nearly sharp. It is then natural to choose as basis the discrete charge representation

$$Q = e \sum_N N |N\rangle\langle N| \, , \qquad \mathrm{e}^{i\varphi} = \sum_N |N+1\rangle\langle N| \, . \qquad (3.168)$$

In this limit, the phase fluctuations given by the correlation function (3.160) are large. In Section 20.3, single-electron tunneling is considered to lowest order in the tunneling coupling. As we shall, the energy-dependence of the Golden Rule tunneling rate is governed by the probability density $P(E)$ for exchange of energy E with the environment. The function $P(E)$ will turn out as the Fourier transform of $\mathrm{e}^{-Q_\varphi(t)}$, where $Q_\varphi(t)$ is the phase correlation function given in Eq. (3.160).

On the other hand, for strong tunneling, the jump of the phase at the junction is nearly sharp. Then, it is favorable to switch to the discrete phase representation

$$\varphi = 2\pi \sum_n n |n\rangle\langle n| \, , \qquad \mathrm{e}^{i2\pi Q/e} = \sum_n |n\rangle\langle n+1| \, . \qquad (3.169)$$

In this representation, the probability for exchange of energy with the environment will turn out as the Fourier transform of the function $\mathrm{e}^{-Q_Q(t)}$.

These two different representation are dual to each other. We shall discuss the charge-phase duality in the context of Cooper pair tunneling in Subsection 25.3.

The full weak- and strong-tunneling series expansions of charge transport across a weak link and a weak constriction, respectively, will be discussed in Section 26.1.

3.4.2 Resistor, inductor and transmission lines

An important example for an electromagnetic environment is the case where the impedance $Z(\omega)$ is represented by an ideal Ohmic resistor, $Z(\omega) = R$. Then we obtain from Eqs. (3.159) and (3.161)

$$G_\varphi(\omega) = \frac{2\alpha\,\omega}{1 + (\omega/\omega_R)^2}\,, \tag{3.170}$$

where we have introduced a dimensionless resistance α and a cutoff frequency ω_R,

$$\alpha = R/R_K = R e^2/2\pi\hbar\,, \qquad \omega_R = 1/RC = E_c/\pi\alpha\hbar\,. \tag{3.171}$$

Here, $E_c = e^2/2C$ is the charging energy. The spectral density is Ohmic, and the junction capacitance provides a Drude form with cutoff frequency ω_R.

Consider next the case where the impedance $Z(\omega)$ is inductive, $Z(\omega) = i\omega L$. Now, the total impedance is given by $Z_t(\omega) = (i\omega/C)/[\omega_L^2 - (\omega - i0^+)^2]$, where $\omega_L = 1/\sqrt{LC}$ is the resonance frequency of the LC circuit. Thus, the junction is coupled to a single environmental mode,

$$G_\varphi(\omega) = (E_c/\hbar)\,\omega_L\,\delta(\omega - \omega_L)\,. \tag{3.172}$$

Putting a resistor in series with the inductor, the resonance at the frequency ω_L is broadened because of the finite quality factor $Q_{\text{qual}} \equiv \omega_R/\omega_L$. With $Z(\omega) = R + i\omega L$, and Eqs. (3.159) and (3.161), we obtain the spectral density

$$G_\varphi(\omega) = \frac{2\alpha\,\omega}{[1 - (\omega/\omega_R)^2 Q_{\text{qual}}^2]^2 + (\omega/\omega_R)^2}\,. \tag{3.173}$$

This form captures the Ohmic case (3.170) in the limit $Q_{\text{qual}} \to 0$, and the single mode case (3.172) in the limit $Q_{\text{qual}} \to \infty$. Thus, by variation of the quality factor, we can tune the properties of the environment between resonant and resistive.

Until now, we considered impedances which are built by one or two lumped circuit elements. A more realistic model describing the coaxial leads attached to a junction in a real experiment is a resistive transmission line model with distributed resistors, inductors, and capacitors. The resistive transmission line model for $Z(\omega)$ can be thought of as a ladder of discrete building blocks, in which the inductor and the resistor are in series along one stringboard of the ladder, and the capacitor is on the rung. Putting the resistance, inductance, and capacitance per building block as R_0, L_0, and C_0, one finds for an infinitely long transmission line the impedance [146]

$$Z(\omega) = \left(\frac{R_0 + i\omega L_0}{i\omega C_0}\right)^{1/2}\,. \tag{3.174}$$

A LC transmission line is characterized by a resistive impedance $Z_{\text{LC}}(0) \approx \sqrt{L_0/C_0}$. This form results from Eq. (3.174) by disregarding R_0. The corresponding spectral density $G_\varphi(\omega)$ is of the Ohmic form (3.170) in which $\sqrt{L_0/C_0}$ is substituted for R.

Qualitatively different effects of the environment occur when the impedance is not approximately constant in the relevant frequency range. This is the case for a RC

transmission line, which is the limit $L_0 \to 0$ in Eq. (3.174). Substituting the form $Z_{RC}(\omega) = \sqrt{R_0/(i\omega C_0)}$ into Eq. (3.159), the spectral density (3.161) takes the form

$$G_\varphi(\omega) \;=\; 2\delta\,\omega_{\mathrm{ph}}^{1/2}\,\omega^{1/2} \qquad \text{for} \qquad \omega \ll \omega_c\,, \tag{3.175}$$

where

$$\delta \;=\; \frac{R_0}{R_K}\,, \qquad \omega_{\mathrm{ph}} \;=\; \frac{1}{2R_0 C_0}\,, \qquad \omega_c \;=\; \frac{C_0^2}{C^2}\,\omega_{\mathrm{ph}}\,. \tag{3.176}$$

Thus, the spectral density modeling the coupling to a RC transmission line is sub-Ohmic with the power $s = \frac{1}{2}$. Observe that the spectral density does not depend on the junction properties in the frequency regime $\omega \ll \omega_c$.

We shall study the effect of these electromagnetic environments on single-electron tunneling in Section 20.3.

3.4.3 Charging effects in Josephson junctions

In Josephson tunnel junctions there are two kinds of charge carriers traversing the barrier, namely Cooper pairs and fermionic quasiparticles. The effective action governing the equilibrium properties of the junction including quasiparticle and Cooper pair tunneling is considered in Subsection 4.2.10. At temperatures well below the critical temperature of the superconductor and applied voltage far below the gap voltage, the tunneling of quasiparticles may be neglected. Then the only ingredient we have to anticipate is the Josephson potential [cf. Eq. (3.95)]

$$H_J \;=\; -E_J \cos\psi\,. \tag{3.177}$$

Here, ψ is the phase difference of the Cooper pair wave function across the junction, and $E_J = (\hbar/2e)I_c$ is the Josephson energy, where I_c is the critical current. In view of the operator relation $e^{i\psi}Q\,e^{-i\psi} = Q - 2e$, the Josephson coupling term (3.177) describes Cooper pair tunneling changing the charge on the junction by $2e$. We assume that the junction has capacitance C and is coupled to an ideal voltage source V_x through a general impedance $Z(\omega)$, as sketched earlier in Fig. 3.3.

Since the voltage drop V_R across the resistor is $V_R = V_x - V_a = V_x - \hbar\dot\psi/2e$ [cf. Eq. (3.93)], we may associate with V_R the phase $\psi_R = 2eV_xt/\hbar - \psi$. Thus the coupling of the junction to the electromagnetic environment is described by a Hamiltonian of the form (3.153) with (3.154) with the substitutions $e \to 2e$ and $V_a \to V_x$. The Hamiltonian of the junction-plus-environment system may be written in the form

$$H \;=\; \frac{Q^2}{2C} + \sum_{\alpha=1}^{N}\left[\frac{q_\alpha^2}{2C_\alpha} + \left(\frac{\hbar}{2e}\right)^2\frac{1}{2L_\alpha}\Big(\psi - 2eV_xt/\hbar - \psi_\alpha\Big)^2\right] - E_J\cos\psi\,. \tag{3.178}$$

The charge q_α is conjugate to $(\hbar/2e)\psi_\alpha$. For $E_J \ll E_c$, where $E_c = 2e^2/C$ is the charging energy for Cooper pairs, the Josephson coupling term can be treated as a perturbation. We then may use for convenience the charge representation (3.99), which is analogous to Eq. (3.168) with $2e$ substituted for e. The effects of the environment are again described in terms of the capacitance C and the admittance function $Y(\omega)$ given in Eq. (3.156). The charge autocorrelation function $Q_Q(t) = \langle [\, Q(0) - Q(t)\,]Q(0)\rangle_\beta$ is again given by Eq. (3.157) with (3.158).

The phase autocorrelation function $Q_\psi(t) = \langle [\, \psi(0) - \psi(t)\,]\psi(0)\rangle_\beta$ takes the form

$$Q_\psi(t) \;=\; \int_0^\infty d\omega\, \frac{G_\psi(\omega)}{\omega^2}\Big(\coth(\beta\hbar\omega/2)[\,1 - \cos(\omega t)\,] + i\sin(\omega t)\Big) \qquad (3.179)$$

with the spectral density

$$G_\psi(\omega) \;=\; 2\omega\,\mathrm{Re}\,Z_t^*(\omega)/R_Q\,, \qquad (3.180)$$

where $R_Q = R_K/4 = 2\pi\hbar/4e^2$ is the resistance quantum for Cooper pairs. The phase autocorrelation function for a Josephson junction differs from that for a normal junction, Eq.(3.160) with (3.161), by a factor four. This is because the charge of the Cooper pair is twice the electron charge. We shall study in Subsection 20.3.4 the case of weak Josephson coupling in which the charge on the junction is sharp.

In the opposite limit of large Josephson coupling, $E_J \gg E_c$, the Josephson junction is in a localized phase state and the environmental influences are described in terms of charge fluctuations. We shall discuss in Subsection 25.3 an interesting duality symmetry between the charge and phase representation of the Josephson junction.

3.5 Nonlinear quantum environments

Generally, the low-energy physics of complex systems coupled to their also complex surroundings can be studied using effective Hamiltonians. The reduction usually leads to one of only few canonical forms. For instance, a double well reduces to a two-state system, and an environmental mode with a whole ladder of roughly equidistant excited states is well described by a harmonic oscillator. The opposite extreme case is a nonlinear bath mode with only one or few accessible excited states. Such a mode behaves like a spin. If there is a whole distribution of such systems, the environment effectively acts like a bath of spins. As opposed to the harmonic bath, the populations of the spin bath are saturated at temperatures above the available excitation energies. Every nonlinear bath is situated between these two idealized cases.

Consider a global Hamiltonian analogous to Eq. (3.1) consisting of a system part, a nonlinear reservoir part, and an interaction term with a counter term,

$$H \;=\; \frac{p^2}{2M} + V(q) + H_R(\{x\}) - F(q)\sum_\alpha c_\alpha x_\alpha + \frac{1}{2}F^2(q)\sum_\alpha c_\alpha^2 \tilde\chi_\alpha(0)\,. \qquad (3.181)$$

The counter-term is defined in terms of the static susceptibility $\tilde\chi_\alpha(\omega = 0)$ for the individual modes x_α. It is again easy to write down the equations of motion and to eliminate formally the reservoir coordinates. This is most conveniently performed in frequency space upon introducing the dynamical susceptibility for the reservoir mode x_α. We then find an equation of motion of the form (3.19) where the Fourier transform of the damping kernel has the form (3.35). The corresponding expression for the spectral density of the coupling is as in Eq. (3.36),

$$J(\omega) = \sum_\alpha c_\alpha^2 \, \widetilde{\chi}_\alpha''(\omega) \,. \tag{3.182}$$

It is obvious from the derivation that this form is generally valid for any global system in which the coupling is *linear* in the bath coordinate. For a nonlinear bath, the dynamical susceptibility may drastically differ from the form (3.16) and generally depends on $\hbar$ and on temperature. Hence the spectral density of the coupling for a nonlinear environment is also temperature-dependent.

A "nanomagnet", i.e. a monodomain particle, may contain as many as 10^8 magnetically ordered spins. Such a giant spin, denoted by $\boldsymbol{S}$, is usually coupled to surrounding nuclear spins and paramagnetic impurity spins. The global Hamiltonian for a central spin bilinearly coupled to a spin bath in a magnetic field and with internal interactions has the canonical form

$$H(\boldsymbol{S}, \{\boldsymbol{\sigma}^\alpha\}) = H_S + \frac{1}{S} \sum_\alpha E_\alpha \boldsymbol{S} \cdot \boldsymbol{\sigma}^\alpha + \sum_\alpha \boldsymbol{h}^\alpha \cdot \boldsymbol{\sigma}^\alpha + \tfrac{1}{2} \sum_{\alpha, \alpha'; i, j} V_{ij}^{\alpha, \alpha'} \sigma_i^\alpha \sigma_j^{\alpha'} \,.$$

Many forms for the giant spin Hamiltonian H_S have been discussed in the literature (cf. the book [148] and the review [149], and references therein). A simple example of the giant spin Hamiltonian H_S is the biaxial (easy axis/easy plane) form

$$H_S = [\, - K_\parallel S_z^2 + K_\perp S_x^2 \,]/S + g\mu_B H_z S_z \,, \tag{3.183}$$

in which tunneling between the classical minima at $S_z = \pm S$ is accomplished by the symmetry-breaking transverse term $K_\perp$ and we have added a magnetic field term. The low-energy physics of the Hamiltonian (3.183) can be understood in terms of the spin-$\frac{1}{2}$ Hamiltonian (3.80). The passage from the form (3.183) to the form (3.80) for $S \gg 1$ has been studied using WKB methods [151] and instanton methods [152].

Single molecule magnets like Mn$_{12}$ac are conveniently described by similar spin Hamiltonians [153].

For giant spin systems, there are two scenarios. In the usual case, the coupling between the giant spin and the nuclear spins is very large. Then the nuclear spins are slaved to the giant spin, and the spectrum of the spin environment is drastically altered. The coupling to the nuclear spins may introduce three different effects [149]. Transitions of the giant spin may cause a change of the phase in the nuclear bath state, which then reacts back upon the giant spin and induces phase randomization ("topological decoherence"). Secondly, transitions of the giant spin may be hindered by a mismatch between the initial and final bath state ("orthogonality blocking"). This leads to a Franck-Condon type dressing factor (cf. Subsection 18.1.3). Thirdly, the spread of coupling energies of the nuclear spins leads to a spread of bias energies which may bring the initial state of the tunneling giant spin out of resonance with the final state. This results in a Landau-Zener type suppression of tunneling transitions ("degeneracy blocking").

The strong-coupling case contrasts with the usual weakly-coupled harmonic oscillator environment. It is impossible to integrate out the spin bath using the functional averaging method described in the next chapter, and one has to study the global system directly, e.g., the instanton trajectory in the full system-bath space [154].

In the other scenario, the coupling to an individual bath spin is weak. Then, the treatment is similar to a boson bath. The resulting spectral density $J(\omega)$ for the spin bath is of the form (3.182), where $\tilde{\chi}_\alpha(\omega)$ is the dynamical susceptibility for a spin degree. Consider as example a global system with a system part $H_S(p, q)$, a spin-$\frac{1}{2}$ bath with excitation frequencies $\{\omega_\alpha\}$, a bilinear coupling term and a counter term,

$$H = H_S(p, q) - \frac{1}{2}\sum_\alpha \hbar\omega_\alpha \sigma_x^\alpha - \frac{1}{2}\sum_\alpha c_\alpha q_0^{(\alpha)} \sigma_z^\alpha q + \frac{1}{2}\sum_\alpha c_\alpha^2 \tilde{\chi}_\alpha(0) q^2 . \qquad (3.184)$$

The coupling terms in Eq. (3.11) and (3.184) correspond to each other at $T = 0$ if we identify the length $\frac{1}{2}q_0^{(\alpha)}$ with the position spread of the equivalent bath oscillator with eigenfrequency ω_α in the ground state, $\frac{1}{2}q_0^{(\alpha)} = \sqrt{\hbar/2m_\alpha\omega_\alpha}$.

The spin correlation function is $\mathrm{Re}\,\langle \sigma_z^\alpha(t)\sigma_z^\alpha(0)\rangle = \cos(\omega_\alpha t)$. Taking the Fourier transform and anticipating the fluctuation dissipation theorem (6.24), we obtain

$$\tilde{\chi}_{\alpha,\,\mathrm{spin}}''(\omega) = (\pi/\hbar)\,\tanh(\beta\hbar\omega_\alpha/2)\delta(\omega - \omega_\alpha) . \qquad (3.185)$$

With the equivalent oscillator susceptibility, $\chi_\alpha(\omega) = (\frac{1}{2}q_0^{(\alpha)})^2\chi_{\alpha,\,\mathrm{spin}}(\omega)$, we find

$$J_{\mathrm{spin}}(\omega) = \sum_\alpha c_\alpha^2\,\tilde{\chi}_\alpha''(\omega) = \frac{\pi}{2}\sum_\alpha \frac{c_\alpha^2}{m_\alpha\omega_\alpha}\,\tanh(\beta\hbar\omega_\alpha/2)\delta(\omega - \omega_\alpha) . \qquad (3.186)$$

The same form is found by calculating the influence function (see Section 4.2) for a spin bath up to second order in c_α [155]. Because of the non-Gaussian nature of the spin-bath, the cumulant expansion does not break off at this order. Cumulants of higher order in c_α are not considered here. We see from Eq. (3.186) that, to order c_α^2, the spectral densities of a spin-$\frac{1}{2}$ bath and a boson bath are related by

$$J_{\mathrm{spin}}(\omega) = J_{\mathrm{boson}}(\omega)\tanh(\beta\hbar\omega/2) . \qquad (3.187)$$

The two reservoirs have the same dissipative influences at $T = 0$. At finite temperatures, the spin bath has a smaller effect on the system because of the possibility for saturation of the populations in the bath.

The striking counter-example to a spin bath is when the ladder of excited states of a nonlinear bath mode is narrowing with increasing energy. In this case, we find enhancement of $J(\omega)$ compared with the spectral density of a harmonic bath.

The findings of this section can be summarized as follows. The effects of a nonlinear bath are captured, to second order in the coupling, by a spectral density of the coupling. For every nonlinear bath, this function is temperature dependent. In conclusion, whenever the concept of a spectral density for the environmental coupling is applicable, we can use the methods expounded here for a linear environment.

4. Imaginary–time path integrals and statistical mechanics

4.1 The density matrix: general concepts

In quantum-statistical mechanics, the state of a system at a particular time is not perfectly known. When one has incomplete information about a system, one usually appeals to the concept of probability. Typically, the incomplete information about a system presents itself, in quantum mechanics, as a *statistical mixture*: the state of the system may be either the state $|\psi_1>$ with a probability p_1, or the state $|\psi_2>$ with a probability p_2, etc. We have $0 \leq p_1, p_2, \cdots, p_k, \cdots \leq 1$, and $\sum_k p_k = 1$. Under these conditions, the system is fully characterized by the density operator

$$W(t) \;=\; \sum_k p_k \,|\psi_k(t)><\psi_k(t)| \;. \tag{4.1}$$

We have $\operatorname{tr} W = \sum_k p_k = 1$, whereas $\operatorname{tr} W^2 = \sum_k p_k^2 \leq 1$. The equality holds for a pure state. An important example is the equilibrium density operator in which the p_k are canonically distributed, and the $|\psi_k><\psi_k|$ are projectors on energy eigenstates,

$$W^{(\mathrm{eq})} \;=\; W_\beta \;=\; \frac{1}{Z}\,\mathrm{e}^{-\beta H} \;=\; \frac{1}{Z}\sum_k \mathrm{e}^{-\beta E_k}\,|\psi_k><\psi_k| \;, \tag{4.2}$$

where $Z = \operatorname{tr} \mathrm{e}^{-\beta H} = \sum_k \mathrm{e}^{-\beta E_k}$. In the orthonormal basis $\{|\varphi_n>\}$, the density matrix reads

$$W_{n,m} \;=\; \sum_k p_k <\varphi_n|\psi_k><\psi_k|\varphi_m> \;. \tag{4.3}$$

The diagonal matrix elements $W_{n,n}$ of W in the $\{|\varphi_n>\}$ basis are

$$W_{n,n} \;=\; \sum_k p_k\,|<\varphi_n|\psi_k>|^2 \;, \tag{4.4}$$

where $|<\varphi_n|\psi_k>|^2$ is a real positive number with the following physical interpretation: if the state of the system is $|\psi_k>$, it is the probability of finding, in a measurement, this system in the state $|\varphi_n>$. Because of the uncertainty of the state $|\psi_k>$ before the measurement, $W_{n,n}$ represents the average probability of finding the system in the state $|\varphi_n>$. For this reason, the diagonal matrix element $W_{n,n}$ is called the *population* of the state $|\varphi_n>$. Evidently, $W_{n,n}$ is a real positive number. The off-diagonal element

$$W_{n,m} \;=\; \sum_k p_k <\varphi_n|\psi_k><\psi_k|\varphi_m> \tag{4.5}$$

is the average of the cross terms which express interference effects between the states $|\varphi_n>$ and $|\varphi_m>$ appearing when the state $|\psi_k>$ is a superposition of these states. While $W_{n,n}$ is a sum of real positive numbers, $W_{n,m}$ is a sum of complex numbers. If $W_{n,m}$ is zero, the average in Eq. (4.5) has abolished interference effects between the states $|\varphi_n>$ and $|\varphi_m>$. If $W_{n,m}$ is nonzero, coherence effects between the states $|\varphi_n>$ and $|\varphi_m>$ are present. For this reason, the off-diagonal elements of the density matrix are called *coherences*. It is easy to proof that

$$W_{n,n}W_{m,m} \geq |W_{n,m}|^2 . \qquad (4.6)$$

We remark that the distinction between populations and coherences depends on the basis $\{|\varphi_n>\}$ chosen in the state space. The density matrix concept for coupled few-state systems in $SU(n)$ representation is neatly presented in Ref. [156].

Within the density matrix approach, expectation values are expressed as

$$\langle A \rangle = \text{tr}\{WA\}, \qquad \langle A \rangle^{(\text{eq})} = \text{tr}\{W^{(\text{eq})}A\} . \qquad (4.7)$$

The density operator of the global system acts in the full state space $\mathcal{H} = \mathcal{H}_S \otimes \mathcal{H}_R$. For observables of which the operators act only in the system's space $\mathcal{H}_S$, we can perform in Eq. (4.7) the trace with respect to the reservoir. It is then natural to construct a so-called *reduced* density operator ρ acting only in the system's space $\mathcal{H}_S$,

$$\rho = \text{tr}_R W . \qquad (4.8)$$

The operation (4.8) is called *partial trace*. All physical predictions about measurements bearing only on the system S are captured by the reduced density ρ,

$$\langle A^{(\text{S})} \rangle = \text{tr}_S\{\rho A^{(\text{S})}\} . \qquad (4.9)$$

For a normalized reduced density matrix, $\text{tr}\,\rho = 1$, we have

$$0 < \text{tr}\,\rho^2 \leq 1 . \qquad (4.10)$$

The maximum value $\text{tr}\,\rho^2 = 1$ corresponds to a pure quantum state, whereas small values of $\text{tr}\,\rho^2$ describe almost classical situations.

Our main goal in this chapter is to derive for the global systems discussed in the preceding chapter the path integral for the reduced density matrix of the open system in thermal equilibrium. The reduced density operator of the open system in thermal equilibrium is defined as the partial trace of the canonical density operator W_β of the total system with respect to the reservoir coordinates, $\rho_\beta \equiv \text{tr}_R W_\beta$. We shall concentrate the discussion on the imaginary-time path integral representation. The essential element of the path integral is the effective action which is found by tracing out the reservoir coordinates in the path integral for the equilibrium density matrix W_β of the global system.

In the next section of this chapter, we shall calculate the effective actions for the various system-plus-reservoir models introduced previously. We then study the partition function of these systems and discuss several methods and concepts for the approximate evaluation of imaginary-time path integrals.

We are interested in studying thermodynamic properties both at zero and at finite temperature. The equilibrium properties of the full system are determined by the canonical density operator $W_\beta = Z^{-1}\,e^{-\beta H}$, where H is the underlying Hamiltonian.

The canonical operator $e^{-\beta H}$ is related to the time-evolution operator $e^{-iHt/\hbar}$ by analytic continuation, known in field theory as *Wick rotation*, $t = -i\hbar\beta$. Feynman taught us that the coordinate matrix elements of $e^{-iHt/\hbar}$ and $e^{-\beta H}$ can be written as a sum over histories of real- and imaginary-time paths in configuration space, respectively. From a contempory point of view, path integrals represent not only an approach alternative to canonical quantization of classical mechanics, but are basic to the foundation and interpretation of quantum mechanics. Besides that, they form a perfect basis for numerical simulations of quantum thermodynamics and dynamics employing Monte-Carlo methods.

Consider briefly the derivation of the path integral for the imaginary time propagator $\langle q''| e^{-\tau H/\hbar}|q'\rangle$ for the simple Hamiltonian $H = p^2/2M + V(q)$. The starting point of the derivation is a group property of the imaginary-time evolution,

$$\langle q''| e^{-\tau H/\hbar}|q'\rangle = \langle q''| e^{-\tau H/N\hbar} \, \mathbf{1} \, e^{-\tau H/N\hbar} \, \mathbf{1} \, e^{-\tau H/N\hbar} \, \cdots \, e^{-\tau H/N\hbar}|q'\rangle . \qquad (4.11)$$

Upon insertion of the completeness property of the orthonormal position eigenstates

$$\mathbf{1} = \int_{-\infty}^{\infty} dq \, |q\rangle\langle q| , \qquad (4.12)$$

the thermal propagator $\langle q''| e^{-\tau H/\hbar}|q'\rangle$ becomes a multiple integral representing succession of N short-time propagators over imaginary time intervals of length $\epsilon = \tau/N$,

$$\langle q''| e^{-\tau H/\hbar}|q'\rangle = \lim_{N\to\infty} \int_{-\infty}^{\infty} dq_{N-1} \cdots dq_1 \prod_{n=1}^{N} \langle q_n| e^{-\epsilon H/\hbar}|q_{n-1}\rangle , \qquad (4.13)$$

where $q'' = q_N$ and $q' = q_0$, and where we eventually take the limit $N \to \infty$. The short-time propagator with initial condition $\langle q_n|q_{n-1}\rangle = \delta(q_n - q_{n-1})$ reads

$$\langle q_n| e^{-\epsilon H/\hbar}|q_{n-1}\rangle = \sqrt{\frac{M}{2\pi\hbar\epsilon}} \, e^{-S_n/\hbar} \qquad (4.14)$$

where S_n is the the Euclidean short-time action,

$$S_n = \epsilon \left[\frac{M}{2}\left(\frac{q_n - q_{n-1}}{\epsilon}\right)^2 + V(q_n)\right] . \qquad (4.15)$$

With the normalization in Eq. (4.14) the limit $N \to \infty$ exists and we may write

$$\langle q''| e^{-\tau H/\hbar}|q'\rangle = \lim_{N\to\infty} \int_{-\infty}^{\infty} \frac{d^{N-1}q}{(2\pi\hbar\epsilon/M)^{N/2}} e^{-S_N} , \qquad (4.16)$$

where $S_N = \sum_{n=1}^{N} S_n$. As $N \to \infty$, S_N turns into the Euclidean action integral[1]

---

[1]Throughout the book, the center dot in "$q(\cdot)$" is used to express that the path q is a functional of time (here imaginary-time), while $q(\tau)$ means the value the path takes at a particular time τ. The notation $q(\tau)$ for the path – though standard in the literature – is not used here since the path $q(\cdot)$ and the value $q(\tau)$ it takes at a particular time will often appear in the same formula.

$$\lim_{N\to\infty} S_N \;\to\; S^{(\mathrm{E})}[q(\cdot)] \;=\; \int_0^\tau d\tau' \Big(\frac{M}{2}\dot{q}^2(\tau') + V[q(\tau')] \Big) \,. \qquad (4.17)$$

With the short-hand notation for the well-behaved multiple-integral measure

$$\int_{q'}^{q''} \mathcal{D}[q(\cdot)] \cdots \;=\; \lim_{N\to\infty} \int_{-\infty}^{\infty} \frac{d^{N-1}q}{(2\pi\hbar\epsilon/M)^{N/2}} \cdots \,, \qquad (4.18)$$

the propagator takes the suggestive form of a path sum or path integral

$$\langle q'' | e^{-\tau H/\hbar} | q' \rangle \;=\; \int_{q'}^{q''} \mathcal{D}[q(\cdot)]\, e^{-S^{(\mathrm{E})}[q(\cdot)]/\hbar} \,. \qquad (4.19)$$

As already emphasized by Feynman [3], the concept of the sum over all paths, like the concept of an ordinary integral, is independent of a special definition of the path sum. It may be chosen according to convenience. In the representation (4.16), the path is a zigzag of straight segments. Evaluation of path integrals by Fourier series is discussed in Subsection 4.3.4. The notion of a path integral can be extended to discrete systems, for which no classical analogue exists. Since the position operator has a discrete spectrum for a tight-binding system, paths are now piecewise constant functions of time with a countable number of sudden jumps between the localized TB states. The respective path sum is discussed in Chapters 19, 21 and 24.

This concludes the brief introduction into path integrals. The reader interested in getting familiarized with this field is referred to Refs. [3] – [6].

Let us first consider a one-dimensional system described by a coordinate q and a Lagrangian (in view of correlation functions, we add a source term)

$$\mathcal{L}_{\mathrm{S}}[q(t),\dot{q}(t)] \;=\; \tfrac{1}{2}M\dot{q}^2(t) - V[q(t)] - \mathcal{J}(t)q(t) \,. \qquad (4.20)$$

For simplicity, we assume that there is no velocity-dependent force. The canonical density matrix $W_\beta(q'',q')$ of the system may be expressed in terms of the eigenstates $\psi_n(q)$ and eigenvalues E_n of the Hamiltonian H_{S}. We then have

$$< q'' | W_\beta | q' > \equiv W_\beta(q'',q') \;=\; Z_{\mathrm{S}}^{-1} \sum_n \psi_n(q'')\,\psi_n^*(q')\, \exp(-\beta E_n) \,, \qquad (4.21)$$

where Z_{S} is the partition function which normalizes W_β as $\mathrm{tr}\,W_\beta = 1$. The canonical density matrix can be represented as an imaginary-time path integral (see e.g. [3, 4])

$$W_\beta(q'',q') \;=\; Z_{\mathrm{S}}^{-1} \int_{q(0)=q'}^{q(\hbar\beta)=q''} \mathcal{D}q(\cdot) \exp\Big(-\int_0^{\hbar\beta} d\tau\, \mathcal{L}_{\mathrm{S}}^{(\mathrm{E})}[q(\tau),\dot{q}(\tau)]/\hbar \Big) \,. \qquad (4.22)$$

Here, $\hbar\beta$ is the *imaginary* or *thermal* time. The path integral in Eq. (4.22) runs over all paths $q(\tau)$ in imaginary time τ which leave position q' at time zero and arrive at position q'' at imaginary time $\hbar\beta$. In the absence of velocity-dependent forces, the "classical dynamics" in imaginary time is described by the *Euclidean* Lagrangian $\mathcal{L}_{\mathrm{S}}^{(\mathrm{E})}(q,\dot{q})$. As a result of the analytical continuation, the Euclidean Lagrangian differs from the real-time Lagrangian (4.20) by the reverse sign in the potential term,

$$\mathcal{L}_{\mathrm{S}}^{(\mathrm{E})}[q(\tau), \dot{q}(\tau)] \;=\; \tfrac{1}{2} M \dot{q}^2(\tau) + V[q(\tau)] + \mathcal{J}(\tau) q(\tau) \,. \qquad (4.23)$$

Here, we use the convention that in real-time quantities, as e. g. in Eq. (4.20), the overdot denotes differentiation with respect to t while in *Euclidean* quantities the overdot denotes differentiation with respect to the imaginary time.

4.2 Effective action and equilibrium density matrix

We now generalize the discussion to the case in which the system is coupled to a thermal reservoir. The equilibrium density matrix of the "universe" (system plus environment) is then represented by

$$W_\beta(q'', \boldsymbol{x}''; q', \boldsymbol{x}') \;=\; Z_{\mathrm{tot}}^{-1} \sum_{\{n\}} \Psi_{\{n\}}(q'', \boldsymbol{x}'') \, \Psi_{\{n\}}^*(q', \boldsymbol{x}') \, \exp(-\beta E_{\{n\}}) \,. \qquad (4.24)$$

The N-component vector $\boldsymbol{x}$ stands for $x_1, x_2, \ldots, x_N$, $\Psi_{\{n\}}$ denotes the eigenstates of the Hamiltonian of the global system, and $\{n\}$ is the full set of quantum numbers. Again, Z_{tot} is a normalization factor so that $\mathrm{tr}\, W_\beta = 1$. We shall assume that the forces in the Hamiltonian of the global system are velocity-independent. Then, the canonical density matrix (4.24) can be written as a path integral in the usual way,

$$W_\beta(q'', \boldsymbol{x}''; q', \boldsymbol{x}') \;=\; Z_{\mathrm{tot}}^{-1} \int_{q(0)=q'}^{q(\hbar\beta)=q''} \mathcal{D}q(\cdot) \int_{\boldsymbol{x}(0)=\boldsymbol{x}'}^{\boldsymbol{x}(\hbar\beta)=\boldsymbol{x}''} \mathcal{D}\boldsymbol{x}(\cdot) \, \exp\left(-S^{(\mathrm{E})}[q(\cdot), \boldsymbol{x}(\cdot)]/\hbar\right) \,. \qquad (4.25)$$

As is explicitly indicated, the functional integrations run over all paths taken by the coordinates $q(\tau)$ and $\boldsymbol{x}(\tau)$ with endpoints $q(0) = q'$, $\boldsymbol{x}(0) = \boldsymbol{x}'$, $q(\hbar\beta) = q''$ and $\boldsymbol{x}(\hbar\beta) = \boldsymbol{x}''$. The Euclidean action contains contributions from the system (S), the reservoir (R), and the interaction (I),

$$S^{(\mathrm{E})} \;=\; S_{\mathrm{S}}^{(\mathrm{E})} + S_{\mathrm{R}}^{(\mathrm{E})} + S_{\mathrm{I}}^{(\mathrm{E})} \;=\; \int_0^{\hbar\beta} d\tau \left(\mathcal{L}_{\mathrm{S}}^{(\mathrm{E})} + \mathcal{L}_{\mathrm{R}}^{(\mathrm{E})} + \mathcal{L}_{\mathrm{I}}^{(\mathrm{E})}\right) \,. \qquad (4.26)$$

From here on, we focus our interest on a reduced description in which the reservoir coordinates are eliminated. To this end, we introduce the *reduced* equilibrium density matrix describing the open (damped) system

$$\rho_\beta(q'', q') \;=\; \mathrm{tr}\, W_\beta \;=\; \int_{-\infty}^{+\infty} d\boldsymbol{x}' \, W_\beta(q'', \boldsymbol{x}'; q', \boldsymbol{x}') \,, \qquad (4.27)$$

and the *reduced* partition function of the damped system alone,

$$Z(\hbar\beta) \;=\; \frac{Z_{\mathrm{tot}}(\hbar\beta)}{Z_{\mathrm{R}}(\hbar\beta)} \,, \qquad (4.28)$$

where Z_{tot} is the partition function of the global system, and Z_{R} is the partition function of the unperturbed reservoir consisting of N harmonic oscillators,

$$Z_{\text{R}} \equiv \prod_{\alpha=1}^{N} Z_{\text{R}}^{(\alpha)} = \prod_{\alpha=1}^{N} \frac{1}{2\sinh(\beta\hbar\omega_{\alpha}/2)} \ . \qquad (4.29)$$

Upon employing Eqs. (4.25) – (4.28), the reduced equilibrium density matrix can be written in the form

$$\rho_{\beta}(q'', q') = \frac{1}{Z} \int_{q(0)=q'}^{q(\hbar\beta)=q''} \mathcal{D}q(\cdot) \exp\left\{-S_{\text{S}}^{(\text{E})}[q(\cdot)]/\hbar\right\} \mathcal{F}^{(\text{E})}[q(\cdot)] \ , \qquad (4.30)$$

where

$$\mathcal{F}^{(\text{E})}[q(\cdot)] \equiv \exp\left\{-S_{\text{infl}}^{(\text{E})}[q(\cdot)]/\hbar\right\} = \frac{1}{Z_{\text{R}}} \oint \mathcal{D}x(\cdot) \exp\left\{-S_{\text{R,I}}^{(\text{E})}[q(\cdot), x(\cdot)]/\hbar\right\} \quad (4.31)$$

with

$$S_{\text{R,I}}^{(\text{E})}[q, x] = S_{\text{R}}^{(\text{E})}[x] + S_{\text{I}}^{(\text{E})}[q, x] \ . \qquad (4.32)$$

Because of the trace operation in Eq. (4.27), the path integral in Eq. (4.31) runs over all *periodic* paths with period $\hbar\beta$ taken by the coordinate x. The *influence functional* $\mathcal{F}^{(\text{E})}[q]$ captures the influences of the environment on the equilibrium properties of the open system. The factor Z_{R}^{-1} normalizes $\mathcal{F}^{(\text{E})}[q]$, so that $\mathcal{F}^{(\text{E})} = 1$ when the coupling $S_{\text{I}}^{(\text{E})}[q, x]$ is switched off, and the factor Z^{-1} normalizes ρ_{β} as $\text{tr}\,\rho_{\beta} = 1$.

In the following subsections, we calculate the Euclidean influence action $S_{\text{infl}}^{(\text{E})}[q(\cdot)]$ for particular systems.

4.2.1 Open system with bilinear coupling to a harmonic reservoir

For the model described by the Hamiltonian (3.11), the Euclidean Lagrangian $\mathcal{L}_{\text{S}}^{(\text{E})}$ is given in Eq. (4.23), and

$$
\begin{aligned}
\mathcal{L}_{\text{R}}^{(\text{E})} &= \frac{1}{2} \sum_{\alpha=1}^{N} m_{\alpha}\left(\dot{x}_{\alpha}^2 + \omega_{\alpha}^2 x_{\alpha}^2\right) \ , \\
\mathcal{L}_{\text{I}}^{(\text{E})} &= \sum_{\alpha=1}^{N} \left(-c_{\alpha} x_{\alpha} q + \frac{1}{2} \frac{c_{\alpha}^2 q^2}{m_{\alpha}\omega_{\alpha}^2}\right) \ .
\end{aligned}
\qquad (4.33)
$$

The stationary paths of the action (4.26) with (4.23) and (4.33), which we denote by $\bar{q}$ and by $\bar{x}_{\alpha}$, obey the Euclidean classical equations of motion

$$
\begin{aligned}
M\ddot{\bar{q}} - \frac{\partial V(\bar{q})}{\partial \bar{q}} + \sum_{\alpha=1}^{N} c_{\alpha}\left(\bar{x}_{\alpha} - \frac{c_{\alpha}}{m_{\alpha}\omega_{\alpha}^2}\bar{q}\right) &= 0 \ , \\
m_{\alpha}\ddot{\bar{x}}_{\alpha} - m_{\alpha}\omega_{\alpha}^2\bar{x}_{\alpha} + c_{\alpha}\bar{q} &= 0 \ .
\end{aligned}
\qquad (4.34)
$$

Since the Lagrangian is a quadratic form in $\boldsymbol{x}$ and $\dot{\boldsymbol{x}}$, the functional integrations over the periodic paths $x_\alpha(\tau)$ can be performed in closed form. The procedure is entirely straightforward. We choose for convenience to periodically continue the paths $q(\tau)$ and $x_\alpha(\tau)$ outside the range $0 \leq \tau < \hbar\beta$ by writing them as Fourier series

$$x_\alpha(\tau) = \frac{1}{\hbar\beta} \sum_{n=-\infty}^{+\infty} x_{\alpha,n}\, e^{i\nu_n \tau} ,$$

$$q(\tau) = \frac{1}{\hbar\beta} \sum_{n=-\infty}^{+\infty} q_n\, e^{i\nu_n \tau} ,$$

(4.35)

where $x_{\alpha,n} = x^*_{\alpha,-n}$, $q_n = q^*_{-n}$, and $\nu_n = 2\pi n/\hbar\beta$ is a bosonic Matsubara frequency. Substituting Eq. (4.35) into Eq. (4.32) with Eq. (4.33), we obtain

$$S^{(\mathrm{E})}_{\mathrm{R,I}}[q,\boldsymbol{x}] = \sum_{\alpha=1}^{N} \frac{1}{\hbar\beta} \sum_{n=-\infty}^{+\infty} \frac{m_\alpha}{2} \left(\nu_n^2 |x_{\alpha,n}|^2 + \omega_\alpha^2 \left| x_{\alpha,n} - \frac{c_\alpha}{m_\alpha \omega_\alpha^2} q_n \right|^2 \right) .$$

(4.36)

Next, we decompose $x_{\alpha,n}$ into the classical term $\bar{x}_{\alpha,n}$ and a deviation $y_{\alpha,n}$ describing quantum fluctuations,

$$x_{\alpha,n} = \bar{x}_{\alpha,n} + y_{\alpha,n} = \frac{c_\alpha}{m_\alpha(\nu_n^2 + \omega_\alpha^2)} q_n + y_{\alpha,n} .$$

(4.37)

In the second form, we have used the solution of the oscillator mode $\bar{x}_{\alpha,n}$ following from Eq. (4.34). Since $\bar{x}_\alpha(\tau)$ is a stationary point of the action, the term linear in the deviation $y_{\alpha,n}$ is absent, and we get an expression in which the quadratic forms in $\boldsymbol{y}$ and q are decoupled,

$$S^{(\mathrm{E})}_{\mathrm{R,I}}[q, \bar{\boldsymbol{x}} + \boldsymbol{y}] = S^{(\mathrm{E})}_{\mathrm{R}}[\boldsymbol{y}] + S^{(\mathrm{E})}_{\mathrm{infl}}[q] ,$$

(4.38)

$$S^{(\mathrm{E})}_{\mathrm{R}}[\boldsymbol{y}] = \sum_{\alpha=1}^{N} \frac{1}{\hbar\beta} \sum_{n=-\infty}^{+\infty} \frac{m_\alpha}{2}(\nu_n^2 + \omega_\alpha^2)|y_{\alpha,n}|^2 = \sum_{\alpha=1}^{N} \int_0^{\hbar\beta} d\tau \, \frac{m_\alpha}{2}(\dot{y}_\alpha^2 + \omega_\alpha^2 y_\alpha^2) ,$$

$$S^{(\mathrm{E})}_{\mathrm{infl}}[q] = \sum_{\alpha=1}^{N} \frac{c_\alpha^2}{2m_\alpha} \frac{1}{\hbar\beta} \sum_{n=-\infty}^{+\infty} \left(\frac{|q_n|^2}{\omega_\alpha^2} - \frac{|q_n|^2}{\nu_n^2 + \omega_\alpha^2} \right) .$$

(4.39)

The first term in $S^{(\mathrm{E})}_{\mathrm{infl}}[q]$ originates from the potential counter term in $\mathcal{L}^{(\mathrm{E})}_{\mathrm{I}}$ given in Eq. (4.33). The path integral over the $\hbar\beta$-periodic quantum fluctuations $\boldsymbol{y}(\tau)$ gives just the partition function of the reservoir,

$$Z_{\mathrm{R}} = \oint \mathcal{D}\boldsymbol{y}(\cdot) \exp\left\{ -S^{(\mathrm{E})}_{\mathrm{R}}[\boldsymbol{y}(\cdot)]/\hbar \right\} ,$$

(4.40)

thus cancelling the factor Z_R^{-1} in Eq. (4.31). In Fourier series representation, the influence action (4.39) takes the compact form

$$S_{\text{infl}}^{(E)}[q] = \frac{M}{2} \frac{1}{\hbar\beta} \sum_{n=-\infty}^{+\infty} \xi_n |q_n|^2 , \tag{4.41}$$

$$\xi_n = \frac{1}{M} \sum_{\alpha=1}^{N} \frac{c_\alpha^2}{m_\alpha \omega_\alpha^2} \frac{\nu_n^2}{(\nu_n^2 + \omega_\alpha^2)} = \frac{2}{M\pi} \int_0^\infty d\omega \, \frac{J(\omega)}{\omega} \frac{\nu_n^2}{\nu_n^2 + \omega^2} . \tag{4.42}$$

In the second form, we have introduced the spectral density (3.30). The Fourier coefficient ξ_n of the influence action is expressed in terms of the spectral damping functions $\tilde{\gamma}(\omega)$ and $\hat{\gamma}(z)$ defined in Eq. (3.31) and (3.33), respectively, by the relation

$$\xi_n = |\nu_n| \tilde{\gamma}(i|\nu_n|) = |\nu_n| \hat{\gamma}(|\nu_n|) . \tag{4.43}$$

Returning to the time representation, the influence action (4.41) reads

$$S_{\text{infl}}^{(E)}[q(\cdot)] = \int_0^{\hbar\beta} d\tau \int_0^\tau d\tau' \, k(\tau - \tau') \, q(\tau) \, q(\tau') , \tag{4.44}$$

$$k(\tau) = \frac{M}{\hbar\beta} \sum_{n=-\infty}^{+\infty} \xi_n \, e^{i\nu_n \tau} . \tag{4.45}$$

The kernel $k(\tau)$ satisfies the symmetry relations

$$k(\tau) = k(\hbar\beta - \tau) = k(\tau + \hbar\beta) . \tag{4.46}$$

Because of the property

$$\int_0^{\hbar\beta} d\tau \, k(\tau) = M\xi_0 = 0 , \tag{4.47}$$

the influence action (4.44) can be rewritten as

$$S_{\text{infl}}^{(E)}[q(\cdot)] = -\frac{1}{2} \int_0^{\hbar\beta} d\tau \int_0^\tau d\tau' \, k(\tau - \tau') \left(q(\tau) - q(\tau') \right)^2 . \tag{4.48}$$

This form clearly shows that the influence action $S_{\text{infl}}^{(E)}[q]$ is fully *nonlocal* and therefore does not cause potential renormalization.

Alternatively, we may express the influence action (4.48) in terms of the kernel

$$K(\tau) = \mu : \delta(\tau): - \, k(\tau) . \tag{4.49}$$

Various forms of μ are

$$\mu = \sum_{\alpha=1}^{N} \frac{c_\alpha^2}{m_\alpha \omega_\alpha^2} = \frac{2}{\pi} \int_0^\infty d\omega \, \frac{J(\omega)}{\omega} = \lim_{t \to 0^+} M\gamma(t) = M\gamma(0^+) , \tag{4.50}$$

and $:\delta(\tau):$ is the periodically continued δ-function

$$:\delta(\tau): \equiv \frac{1}{\hbar\beta} \sum_{n=-\infty}^{+\infty} e^{i\nu_n \tau} = \sum_{n=-\infty}^{+\infty} \delta(\tau - n\hbar\beta) . \tag{4.51}$$

We then have

$$S_{\text{infl}}^{(E)}[q(\cdot)] = \frac{1}{2} \int_0^{\hbar\beta} d\tau \int_0^{\tau} d\tau' \, K(\tau - \tau') \left(q(\tau) - q(\tau')\right)^2 . \tag{4.52}$$

The Matsubara sum or Fourier series representation of the kernel $K(\tau)$ reads

$$K(\tau) = \frac{M}{\hbar\beta} \sum_{n=-\infty}^{+\infty} \zeta_n \, e^{i\nu_n \tau} , \tag{4.53}$$

$$\zeta_n = \frac{\mu}{M} - \xi_n = \frac{1}{M} \sum_{\alpha=1}^{N} \frac{c_\alpha^2}{m_\alpha} \frac{1}{\nu_n^2 + \omega_\alpha^2} = \frac{2}{\pi M} \int_0^{\infty} d\omega \, J(\omega) \frac{\omega}{\nu_n^2 + \omega^2} . \tag{4.54}$$

Interchanging the frequency integral with the Matsubara sum, we find the form

$$K(\tau) = \frac{1}{\pi} \int_0^{\infty} d\omega \, J(\omega) \, D_\omega(\tau) , \tag{4.55}$$

where $D_\omega(\tau)$ is the temperature Green function of a free boson

$$D_\omega(\tau) = \frac{1}{\hbar\beta} \sum_{n=-\infty}^{+\infty} \frac{2\omega}{\nu_n^2 + \omega^2} e^{i\nu_n \tau} . \tag{4.56}$$

The series in Eq. (4.56) is easily summed, yielding in the principal interval $0 \le \tau < \hbar\beta$

$$D_\omega(\tau) = \frac{\cosh[\omega(\hbar\beta/2 - \tau)]}{\sinh(\omega\hbar\beta/2)} = [1 + n(\omega)] \, e^{-\omega\tau} + n(\omega) \, e^{\omega\tau} . \tag{4.57}$$

The second form is written in terms of the single-particle Bose distribution (3.44), showing that $D_\omega(\tau)$ is the imaginary-time boson propagator in thermal equilibrium. Outside the principal interval $0 \le \tau < \hbar\beta$, the thermal Green function $D_\omega(\tau)$ is periodically continued with cusps at the points $\tau = n\hbar\beta$ ($n = 0, \pm1, \pm2, \cdots$). We shall refer to $D_\omega(\tau)$ as the imaginary-time boson propagator. The kernel $K(\tau)$ can also be directly related to the real-time memory-friction kernel $\gamma(t)$. Substituting the expression (3.38) for $J(\omega)$ into Eq. (4.55), reversing the order of integrations, and performing the ω integral with the form (4.57) for $D_\omega(\tau)$, we get

$$K(\tau) = \frac{M}{\hbar\beta} \int_0^{\infty} dt \, \gamma(t) \frac{\partial}{\partial t} \left(\frac{\sinh(\nu t)}{\cosh(\nu t) - \cos(\nu\tau)} \right) , \tag{4.58}$$

where $\nu \equiv \nu_1 = 2\pi/\hbar\beta$ is the lowest Matsubara frequency.

Another form of the influence action $S_{\text{infl}}^{(E)}[q]$ is found if we agree to periodically continue the path $q(\tau)$ outside the range $0 \leq \tau < \hbar\beta$ according to $q(\tau + n\hbar\beta) = q(\tau)$. By extension of the integration regime, the expression (4.52) is transformed into [77]

$$S_{\text{infl}}^{(E)}[q(\cdot)] = \frac{1}{4} \int_0^{\hbar\beta} d\tau \int_{-\infty}^{\infty} d\tau' \, K_0(\tau - \tau') \left(q(\tau) - q(\tau') \right)^2, \qquad (4.59)$$

where the kernel $K_0(\tau - \tau')$ is the the kernel $K(\tau)$ for zero temperature,

$$K_0(\tau - \tau') = \sum_{\alpha=1}^{N} \frac{c_\alpha^2}{2m_\alpha\omega_\alpha} e^{-\omega_\alpha|\tau-\tau'|} = \frac{1}{\pi} \int_0^{\infty} d\omega \, J(\omega) \, e^{-\omega|\tau-\tau'|}. \qquad (4.60)$$

While $K(\tau - \tau')$ in Eq. (4.52) has effect only within the principal interval of length $\hbar\beta$, the function $K_0(\tau - \tau')$ in Eq. (4.59) introduces also correlations between the path in the principal interval and the periodically continued path.

Finally, we substitute Eq. (4.31) with Eq. (4.52) into Eq. (4.30). Then the reduced equilibrium density matrix of the open system can be written in the compact form

$$< q'' |\rho_\beta| q' > \equiv \rho_\beta(q'', q') = Z^{-1} \int_{q(0)=q'}^{q(\hbar\beta)=q''} \mathcal{D}q(\cdot) \, \exp\left\{ -S_{\text{eff}}^{(E)}[q(\cdot)]/\hbar \right\}, \qquad (4.61)$$

where the effective action $S_{\text{eff}}^{(E)}[q(\cdot)] = S_S^{(E)}[q(\cdot)] + S_{\text{infl}}^{(E)}[q(\cdot)]$ is given by the expression

$$S_{\text{eff}}^{(E)}[q(\cdot)] = \int_0^{\hbar\beta} d\tau \left(\tfrac{1}{2}M\dot{q}^2 + V(q) \right) + \frac{1}{2} \int_0^{\hbar\beta} d\tau \int_0^{\tau} d\tau' \, K(\tau - \tau') \left(q(\tau) - q(\tau') \right)^2$$

$$= \int_0^{\hbar\beta} d\tau \left(\tfrac{1}{2}M\dot{q}^2 + V(q) \right) + \int_0^{\hbar\beta} d\tau \int_0^{\tau} d\tau' \, k(\tau - \tau')q(\tau)q(\tau'). \qquad (4.62)$$

The action (4.62), as well as the trivial generalization to many dimensions, describes any system exposed to a linear dissipative process. It is the essential component in the path integral approach to the thermodynamics of open quantum systems.

4.2.2 State-dependent memory-friction

The treatment of the previous subsection is easily generalized to nonlinear state-dependent friction provided that the coupling is linear in the bath degrees of freedom. For the global system described by the Hamiltonian (3.8), we finally obtain the path integral expression (4.61) for the reduced equilibrium density matrix, but with a modified influence action in $S_{\text{eff}}^{(E)}[q] = S_S^{(E)}[q] + S_{\text{infl}}^{(E)}[q]$,

$$S_{\text{infl}}^{(E)}[q(\cdot)] = \frac{1}{2} \int_0^{\hbar\beta} d\tau \int_0^{\tau} d\tau' \, K(\tau - \tau') \left(F[q(\tau)] - F[q(\tau')] \right)^2$$

$$= \int_0^{\hbar\beta} d\tau \int_0^{\tau} d\tau' \, k(\tau - \tau')F[q(\tau)]F[q(\tau')]. \qquad (4.63)$$

With a nonlinear function for $F(q)$ in this action, the dissipative process is intrinsically nonlinear. This form is relevant, e.g., in rotational tunneling systems, in which the coupling function $F(q)$ must obey the same periodicity symmetry as the hindering potential [78] [cf. the discussion following Eq. (3.8)] and for quasiparticle tunneling between superconductors.

For the phenomenological coupling (3.9), the influence action takes the form

$$S_{\text{infl}}^{(\text{E})}[q(\cdot)] = \sum_{m=1}^{2} \int_{0}^{\hbar\beta} d\tau \int_{0}^{\tau} d\tau' \, k^{(m)}(\tau - \tau') F^{(m)}[q(\tau)] \, F^{(m)}[q(\tau')] , \qquad (4.64)$$

where $F^{(1)}[q(\tau)] = \sin[q(\tau)/2]$ and $F^{(2)}[q(\tau)] = \cos[q(\tau)/2]$. By straightforward manipulation, the action (4.64) can be brought into the form (4.186) of the microscopic model for quasiparticle tunneling in a Josephson junction [79] (cf. Subsection 4.2.10).

Another important example of state-dependent friction is when the relevant system interacts with individual bath modes within a spatially restricted region [157, 158]. The case of interaction with localized modes is discussed in the context of the decoherence problem in Section 9.3. With the modifications discussed in Sec. 3.5, the results of this subsection are also valid for a weakly coupled nonlinear bath.

4.2.3 Spin-boson model

In the dissipative two-state or spin-boson model, Eq. (3.86) or (3.88), the transitions between the two states are sudden. Correspondingly, the path $q(\tau)$ for the TSS is piecewise constant and it occasionally jumps between the positions $+\frac{1}{2}q_0$ and $-\frac{1}{2}q_0$. It is now convenient to put

$$q(\tau) = \tfrac{1}{2} q_0 \, \sigma(\tau) , \qquad (4.65)$$

where $\sigma(\tau)$ is a spin path in the model (3.88). The path $\sigma(\tau)$ jumps forth and back between the values $+1$ and -1, as sketched for 8 moves (spin flips, or kinks or instantons with zero width) in Fig. 4.1. A path with $2m$ alternating spin orientations (m kink–anti-kink pairs) is given by (the centers are in chronological order, $s_{j+1} > s_j$)

$$\sigma^{(m)}(\tau) = 1 + 2 \sum_{j=1}^{2m} (-1)^j \Theta(\tau - s_j) . \qquad (4.66)$$

For the shifted canonical distribution (3.25) with initial state $q(0) = \frac{1}{2}q_0\sigma_i$, the mean value of the polarization energy is $\mathfrak{E}_0 = \frac{1}{2}M\gamma(0)q(0)q_0 \equiv \frac{1}{2}\Lambda_{\text{cl}}\sigma_i$, as follows with Eqs. (3.85) and (3.24). The quantity $\Lambda_{\text{cl}} = \frac{1}{2}q_0^2 M\gamma(0)$ is called the classical *reorganization energy* of the bath [cf. Subsection 20.1.1],

$$\Lambda_{\text{cl}} = \frac{q_0^2}{2} \sum_{\alpha} \frac{c_\alpha^2}{m_\alpha \omega_\alpha^2} = \hbar \int_{0}^{\infty} d\omega \, \frac{G(\omega)}{\omega} . \qquad (4.67)$$

The correlation function for the polarization fluctuations $\delta\mathfrak{E}(\tau) \equiv \mathfrak{E}(\tau) - \mathfrak{E}_0$ reads

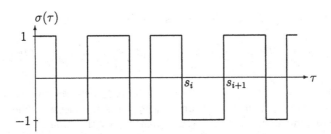

Figure 4.1: A typical multi-kink path with kink centers at s_j.

$$\frac{4}{\hbar^2}\Big\langle \delta\mathfrak{E}(\tau)\delta\mathfrak{E}(0)\Big\rangle_\beta \equiv \mathcal{K}(\tau) = \frac{q_0^2}{\hbar}K(\tau) = \int_0^\infty d\omega\, G(\omega)D_\omega(\tau), \qquad (4.68)$$

where $K(\tau)$ is defined in Eq. (4.55). In the last form, we have introduced the spectral density $G(\omega)$ of the TSS, Eq. (3.89) or (3.90). Substituting Eq. (4.65) into Eq. (4.52), the Euclidean influence functional (4.31) takes the form

$$\mathcal{F}^{(E)}[\sigma(\cdot)] = \exp\left\{-\frac{1}{2}\int_0^{\hbar\beta} d\tau \int_0^\tau d\tau'\, \mathcal{K}(\tau-\tau')\Big(\sigma(\tau)-\sigma(\tau')\Big)^2\Big/4\right\}. \qquad (4.69)$$

With the form (4.66) for $\sigma(\tau)$, the time integrals can be carried out. We then obtain a representation in terms of the kink interaction $W(\tau)$ which is a second integral of $\mathcal{K}(\tau)$, $\ddot{W}(\tau) = -\mathcal{K}(\tau)$. For a path with m kink–anti-kink pairs, we find

$$\mathcal{F}_m^{(E)}[\{s_j\}] = \exp\left\{\sum_{j=2}^{2m}\sum_{i=1}^{j-1}(-1)^{i+j}\,W(s_j-s_i)\right\},$$
$$W(\tau) = \int_0^\infty d\omega\, \frac{G(\omega)}{\omega^2}\Big(D_\omega(0)-D_\omega(\tau)\Big). \qquad (4.70)$$

The form (4.69) is the state representation of the influence function, whereas the form (4.70) is usually referred to as the charge representation. The charges are stringed at the kink positions with alternating sign. The function $W(\tau)$ describes the interaction between two charges, and the exponent of the influence function (4.70) represents the sum of all pair interactions of the $2m$ alternating charges.

4.2.4 Acoustic polaron and defect tunneling: one-phonon coupling

Consider a particle coupled to acoustic phonons by the interaction term $H_{\text{lin}}(q)$ given in Eq. (3.115). Since the global Hamiltonian is harmonic in the phonon variables, the thermal average over the phonon degrees of freedom can be performed exactly. Rather than performing the average in the path integral representation, here we choose, for a change, to proceed directly in the operator formulation. The influence functional describing the effects of the coupling H_{lin} is formally defined by

$$F_{\mathrm{lin}}^{(\mathrm{E})}[\boldsymbol{q}(\cdot)] \equiv \exp\left(-S_{\mathrm{lin}}^{(\mathrm{E})}[\boldsymbol{q}(\cdot)]/\hbar\right) = \left\langle T_\tau \exp\left(-\int_0^{\hbar\beta} d\tau\, \tilde{H}_{\mathrm{lin}}[\boldsymbol{q}(\tau)]/\hbar\right)\right\rangle_\beta . \quad (4.71)$$

Here, T_τ is the imaginary-time ordering operator, $\langle\cdots\rangle_\beta$ denotes the thermal average with respect to $\exp(-\sum_k \beta\hbar\omega_k b_k^\dagger b_k)$, and $\tilde{H}_{\mathrm{lin}}[\boldsymbol{q}(\tau)]$ is the electron-phonon interaction, Eq. (3.115), in the interaction picture (denoted by the tilde),

$$\tilde{H}_{\mathrm{lin}}[\boldsymbol{q}(\tau)] = \exp(H_{\mathrm{R}}\tau/\hbar)H_{\mathrm{lin}}[\boldsymbol{q}(\tau)]\exp(-H_{\mathrm{R}}\tau/\hbar) . \quad (4.72)$$

Since we are in imaginary time, we have

$$\tilde{b}_k(\tau) = b_k\, e^{-\omega_k\tau}, \qquad \tilde{b}_k^\dagger(\tau) = b_k^\dagger\, e^{\omega_k\tau} . \quad (4.73)$$

Here we use again the notation introduced in Eq. (3.112).

The influence functional $F_{\mathrm{lin}}^{(\mathrm{E})}[\boldsymbol{q}(\cdot)]$ is calculated by expanding the formal expression (4.71) into a power series in $\tilde{H}_{\mathrm{lin}}[\boldsymbol{q}]$,

$$\mathcal{F}_{\mathrm{lin}}^{(\mathrm{E})}[\boldsymbol{q}(\cdot)] = 1 + \sum_{n=1}^{\infty}\left(\frac{-1}{\hbar}\right)^n \int_0^{\hbar\beta} d\tau_n \cdots \int_0^{\tau_3} d\tau_2 \int_0^{\tau_2} d\tau_1$$
$$\times \left\langle \tilde{H}_{\mathrm{lin}}[\boldsymbol{q}(\tau_n)]\cdots\tilde{H}_{\mathrm{lin}}[\boldsymbol{q}(\tau_1)]\right\rangle_\beta , \quad (4.74)$$

and by employing *Wick's theorem* for thermodynamic averages (see, e.g., Ref. [159]). The influence action is then given by the respective cumulant expansion. Observing that the interaction (3.115) is linear in the phonon field, we have $\langle\tilde{H}_{\mathrm{lin}}[\boldsymbol{q}(\tau)]\rangle = 0$. Because of the Gaussian statistics, the cumulants of third and higher order in $\tilde{H}_{\mathrm{lin}}[\boldsymbol{q}]$ are exactly zero. The only non-vanishing cumulant contraction is $\langle\tilde{H}_{\mathrm{lin}}[\boldsymbol{q}(\tau)]\tilde{H}_{\mathrm{lin}}[\boldsymbol{q}(\tau')]\rangle_\beta$. Thus we find for the influence action

$$S_{\mathrm{lin}}^{(\mathrm{E})}[\boldsymbol{q}(\cdot)] = -\frac{1}{\hbar}\int_0^{\hbar\beta} d\tau \int_0^{\tau} d\tau' \left\langle \tilde{H}_{\mathrm{lin}}[\boldsymbol{q}(\tau)]\tilde{H}_{\mathrm{lin}}[\boldsymbol{q}(\tau')]\right\rangle_\beta . \quad (4.75)$$

Upon inserting Eq. (3.115) with Eq. (4.73), the contraction takes the form

$$\left\langle \tilde{H}_{\mathrm{lin}}[\boldsymbol{q}(\tau)]\tilde{H}_{\mathrm{lin}}[\boldsymbol{q}(\tau')]\right\rangle_\beta = \sum_{n,m}\sum_{k,\lambda} \frac{\hbar\kappa_k^{(n)}[\boldsymbol{q}(\tau)]\kappa_k^{(m)}[\boldsymbol{q}(\tau')]}{2V\varrho\omega_k}\, e^{-i\boldsymbol{k}\cdot[\boldsymbol{R}^{(n)}-\boldsymbol{R}^{(m)}]}$$
$$\times \exp\left(i\boldsymbol{k}\cdot[\boldsymbol{q}(\tau)-\boldsymbol{q}(\tau')]\right)D_{\omega_k}(\tau-\tau') . \quad (4.76)$$

In the derivation of the expression (4.76) we have used the relations (3.43), and we have identified the resulting linear combination of $\langle b_k^\dagger b_k\rangle_\beta$ and $\langle b_k b_k^\dagger\rangle_\beta$ with the boson propagator $D_{\omega_k}(\tau-\tau')$ defined in Eq. (4.57).

Quantum diffusion of light interstitials and defect tunneling in solids usually takes place by incoherent tunneling transitions between localized states in the rigid multi-well potential of the host crystal. At thermal energy small compared with the level spacing of the low-lying states in the wells, only the lowest state in each local minimum remains relevant [160].[2]

Consider the case of a defect particle which tunnels between two spatially separated interstitial positions located at $q_1 = \frac{1}{2}q_0$ and $q_2 = -\frac{1}{2}q_0$. We then have

$$q(\tau) = \tfrac{1}{2}q_0\,\sigma(\tau)\,, \qquad (4.77)$$

where $\sigma(\tau) = \pm 1$, as sketched in Fig. 4.1. The eigenvalues $\pm\frac{1}{2}q_0$ are the two energetically accessible interstitial positions of the defect. It is convenient to introduce

$$\kappa_k^{(n)}(\sigma) \equiv \kappa_k^{(n)}[q = \sigma\tfrac{1}{2}q_0]\,. \qquad (4.78)$$

Then the thermal average (4.76) takes the form

$$\left\langle \tilde{H}_{\mathrm{lin}}[\sigma(\tau)]\tilde{H}_{\mathrm{lin}}[\sigma(\tau')]\right\rangle_\beta = \sum_{n,m}\sum_{k,\lambda} \frac{\hbar\kappa_k^{(n)}(\sigma)\kappa_k^{(m)}(\sigma')}{2V\varrho\omega_k}\,e^{-i\boldsymbol{k}\cdot[\boldsymbol{R}^{(n)}-\boldsymbol{R}^{(m)}]}D_{\omega_k}(\tau - \tau')$$

$$\times \left[\cos^2\left(\frac{\boldsymbol{k}\cdot\boldsymbol{q_0}}{2}\right) + \sin^2\left(\frac{\boldsymbol{k}\cdot\boldsymbol{q_0}}{2}\right)\sigma(\tau)\sigma(\tau')\right]\,. \qquad (4.79)$$

For defect tunneling, we are not interested in the energy shift which is caused by the linear coupling $H_{\mathrm{lin}}(\boldsymbol{q})$ [cf. the discussion after Eq. (3.4)]. Restricting the attention to the change of the action when the particle hops, we define the influence action with a suitable subtraction for a constant path σ. We write

$$S_{\mathrm{lin}}^{(\mathrm{E})}[\sigma(\cdot)] = \frac{1}{\hbar}\int_0^{\hbar\beta}d\tau\int_0^\tau d\tau'\left\langle \tilde{H}_{\mathrm{lin}}[\sigma]\tilde{H}_{\mathrm{lin}}[\sigma] - \tilde{H}_{\mathrm{lin}}[\sigma(\tau)]\tilde{H}_{\mathrm{lin}}[\sigma(\tau')]\right\rangle_\beta\,. \qquad (4.80)$$

Putting $[\kappa_k^{(n)}(+)]^2 = [\kappa_k^{(n)}(-)]^2$, the influence action may be written in the form (4.69),

$$S_{\mathrm{lin}}^{(\mathrm{E})}[\sigma(\cdot)]/\hbar = \frac{1}{2}\int_0^{\hbar\beta}d\tau\int_0^\tau d\tau'\,\mathcal{K}_{\mathrm{lin}}(\tau - \tau')\Big(\sigma(\tau) - \sigma(\tau')\Big)^2\Big/4\,, \qquad (4.81)$$

where the kernel $\mathcal{K}_{\mathrm{lin}}(\tau)$ is defined in terms of the integral representation

$$\mathcal{K}_{\mathrm{lin}}(\tau) = \int_0^\infty d\omega\, G_{\mathrm{lin}}(\omega)D_\omega(\tau)\,, \qquad (4.82)$$

and where $D_\omega(\tau)$ is the imaginary-time phonon propagator (4.56) or (4.57). The spectral density is given by

---

[2]The dissipative two- and multi-state system are discussed in Part IV and Part V, respectively.

$$G_{\text{lin}}(\omega) = \sum_{\boldsymbol{k},\lambda} \frac{\delta(\omega - \omega_{\boldsymbol{k},\lambda})}{2V\varrho\,\hbar\omega_{\boldsymbol{k},\lambda}} \left(\cos^2\left(\frac{\boldsymbol{k}\cdot\boldsymbol{q}_0}{2}\right) \left| \sum_n e^{i\boldsymbol{k}\cdot\boldsymbol{R}^{(n)}} \left(\kappa^{(n)}_{\boldsymbol{k},\lambda}(+) - \kappa^{(n)}_{\boldsymbol{k},\lambda}(-) \right) \right|^2 \right.$$

$$\left. + \sin^2\left(\frac{\boldsymbol{k}\cdot\boldsymbol{q}_0}{2}\right) \left| \sum_n e^{i\boldsymbol{k}\cdot\boldsymbol{R}^{(n)}} \left(\kappa^{(n)}_{\boldsymbol{k},\lambda}(+) + \kappa^{(n)}_{\boldsymbol{k},\lambda}(-) \right) \right|^2 \right) . \qquad (4.83)$$

Consider now the Debye model with linear dispersion relation $\omega_{\boldsymbol{k},\lambda} = v_\lambda|\boldsymbol{k}|$. We then have for space dimension $d = 3$

$$\frac{1}{V}\sum_{\boldsymbol{k},\lambda} \cdots \; \to \; \frac{1}{2\pi^2}\sum_\lambda \frac{1}{v_\lambda^3} \int \frac{d\Omega}{4\pi} \int_0^\infty d\omega_{\boldsymbol{k},\lambda}\, \omega_{\boldsymbol{k},\lambda}^2 \Theta(\omega_{\text{D}} - \omega_{\boldsymbol{k},\lambda}) \cdots . \qquad (4.84)$$

We have assumed for simplicity that the Debye frequency ω_{D} is the same for the longitudinal and transversal phonon branches. Denoting the number of atoms per volume V by N, we find from Eq. (4.84)

$$\omega_{\text{D}} = \left(\frac{2\pi^2 N}{V}\right)^{1/3} \bar{v} \qquad \text{with} \qquad \frac{1}{\bar{v}^3} = \frac{1}{3}\left(\frac{1}{v_l^3} + \frac{2}{v_t^3}\right) . \qquad (4.85)$$

Here we are interested in the long wavelength limit. Therefore, we expand the exponential and trigonometric functions in Eq. (4.83) in $\boldsymbol{k}$. Then we obtain for the spectral density of the coupling the low frequency expansion

$$G_{\text{lin}}(\omega) = \left(2\alpha_1\omega + 2\alpha_3\frac{\omega^3}{\omega_{\text{D}}^2} + 2\alpha_5\frac{\omega^5}{\omega_{\text{D}}^4} + \mathcal{O}\!\left(\omega^7\right) \right)\Theta(\omega_{\text{D}} - \omega) . \qquad (4.86)$$

The coupling parameter of the Ohmic contribution $G_{\text{lin}}(\omega) \propto \omega$ is given by

$$\alpha_1 = \frac{1}{8\pi^2\varrho\hbar}\sum_\lambda \frac{1}{v_\lambda^3}\int \frac{d\Omega}{4\pi} \left| \sum_n \left(\kappa^{(n)}_{\boldsymbol{k},\lambda}(+) - \kappa^{(n)}_{\boldsymbol{k},\lambda}(-) \right) \right|^2 . \qquad (4.87)$$

The dimensionless coupling coefficients α_3 and α_5 are conveniently expressed in terms of the symmetrized and antisymmetrized components

$$\mathbb{P}_{\text{s}} = \frac{1}{2}\Big(\mathbb{P}(+) + \mathbb{P}(-)\Big) \qquad \text{and} \qquad \mathbb{P}_{\text{a}} = \frac{1}{2}\Big(\mathbb{P}(+) - \mathbb{P}(-)\Big) \qquad (4.88)$$

of the deformation potential tensor ($\mathbb{P}$ has dimension energy, κ has dimension force)

$$\mathbb{P}(\pm) = \sum_n \boldsymbol{R}^{(n)} \otimes \boldsymbol{g}^{(n)}(\pm) , \qquad (4.89)$$

where $\boldsymbol{g}^{(n)}(\sigma)$ is the Kanzaki force defined in Eq. (3.113). Equation (4.83) yields [128]

$$\alpha_3 = \frac{\omega_{\text{D}}^2}{2\pi^2\varrho\hbar}\sum_\lambda \frac{1}{v_\lambda^5}\int \frac{d\Omega}{4\pi} \left| \hat{\boldsymbol{k}}\cdot\mathbb{P}_{\text{a}}\cdot\boldsymbol{e}(\hat{\boldsymbol{k}},\lambda) \right|^2 , \qquad (4.90)$$

and

$$\alpha_5 = \frac{\omega_D^4}{32\pi^2\varrho\hbar}\sum_\lambda \frac{1}{v_\lambda^7}\int \frac{d\Omega}{4\pi}\left\{\left|\sum_n\left(\hat{\boldsymbol{k}}\cdot\boldsymbol{R}^{(n)}\right)^2\left(\kappa_{\hat{k},\lambda}^{(n)}(+) - \kappa_{\hat{k},\lambda}^{(n)}(-)\right)\right|^2\right.$$

$$\left. + 4\left(\hat{\boldsymbol{k}}\cdot\boldsymbol{q}_0\right)^2\left(\left|\hat{\boldsymbol{k}}\cdot\mathbb{P}_s\cdot\boldsymbol{e}(\hat{\boldsymbol{k}},\lambda)\right|^2 - \left|\hat{\boldsymbol{k}}\cdot\mathbb{P}_a\cdot\boldsymbol{e}(\hat{\boldsymbol{k}},\lambda)\right|^2\right)\right\} , \tag{4.91}$$

where $\hat{\boldsymbol{k}}$ is a unit vector. The amplitude factors $|\sum_n(\boldsymbol{R}^{(n)})^{(\nu-1)/2}\cdots|$ in the coefficients α_ν are moments of the positions of the atoms. Therefore, we may regard α_1 as the coefficient of a monopole force, α_3 as the coefficient of a dipole force, etc. [161].

For a crystalline environment, the defect does not cause a *monopole* force because of the symmetry of the interstitial positions. Therefore, the coupling coefficient α_1 is zero. Hence dissipation from acoustic phonons is super-Ohmic rather than Ohmic.

The tensor character of $\mathbb{P}$ in Eqs. (4.90) and (4.91) may lead to strong selection rules for the coefficients α_3 and α_5. The particular case depends on the symmetry of the interstitial sites [123, 125, 128].

In bcc lattices, such as Nb and Fe, the octahedral and tetrahedral interstitial sites have tetragonal symmetry so that a selection rule is absent. Denoting the eigenvalues of the deformation potential tensor (4.89) by p_1 and p_3, where p_1 is double and p_3 belongs to the tetragonal axis, we obtain for an isotropic elastic medium

$$\alpha_3 = \frac{\omega_D^2(p_1 - p_3)^2}{2\pi^2\varrho\hbar}\left(\frac{1}{10}\frac{1}{v_t^5} + \frac{1}{15}\frac{1}{v_l^5}\right) . \tag{4.92}$$

For niobium, we have $\alpha_3 \approx 1.3\,(p_1 - p_3)^2/[eV]^2$ (Ref. [128]), and tabulated values of p_1 and p_3 can be found in Ref. [162]. In bcc metals, the coefficient α_5 is estimated to be smaller than or at most of the same order of magnitude as the coefficient α_3.

In fcc lattices, such as Cu and Al, the octahedral and tetrahedral sites have cubic symmetry. Therefore, the deformation potential tensor is degenerate, $p_1 = p_2 = p_3 \equiv p$. As a result of this symmetry, the dipole force is absent. Hence the coefficient α_3 is zero. Furthermore, there follows from Eq. (4.91) that only the longitudinal phonon branch contributes to the coefficient α_5. One finds [128]

$$\alpha_5 = \frac{\omega_D^4 q_0^2}{24\pi^2\varrho\hbar v_l^7}p^2 . \tag{4.93}$$

For tetrahedral sites in Al, Eq. (4.93) gives the numerical value $\alpha_5 = 0.1\,p^2/[eV]^2$, and for octahedral sites the value $\alpha_5 = 0.2\,p^2/[eV]^2$.

For tunneling centers in amorphous solids, the monopole force term is again missing for symmetry reasons, i.e., the coupling parameter α_1 of the Ohmic spectral density vanishes again. However, there are generally no selection rules for the dipole force. It is convenient to introduce the coupling strength as in Eq. (4.86),

$$G_{\text{lin}}(\omega) = 2\alpha_3\frac{\omega^3}{\omega_D^2} , \qquad \text{where} \qquad \alpha_3 = \frac{\omega_D^2}{2\pi^2\varrho\hbar}\left(\frac{p_l^2}{v_l^5} + \frac{2p_t^2}{v_t^5}\right) . \tag{4.94}$$

The elastic energies or deformation potentials p_l and p_t describe again the coupling to longitudinal and transversal acoustic phonons, respectively.

For acoustic phonons in d dimensions, the spectral density of states per unit volume takes the form

$$\sum_{k,\lambda} \delta(\omega - \omega_{k,\lambda}) \equiv D(\omega) = D_0 \left(\frac{\omega}{\omega_D}\right)^{d-1}, \qquad (4.95)$$

where D_0 is a normalization constant. Assuming a frequency-dependent coupling of the form $[\lambda_\alpha \to \lambda(\omega)$ in Eq. (3.89)]

$$\lambda(\omega) = \lambda_0 (\omega/\omega_D)^\nu \qquad \text{for} \qquad \omega \lesssim \omega_D, \qquad (4.96)$$

where λ_0 is proportional to the deformation potential, we obtain for the spectral density $G_{\text{lin}}(\omega)$ in Eq. (3.89) the power-law form [cf. Eq. (3.58)]

$$G_{\text{lin}}(\omega) \equiv D(\omega)\lambda^2(\omega) = 2\alpha_s \omega_D^{1-s} \omega^s, \qquad (4.97)$$

where $2\alpha_s = D_0\lambda_0^2/\omega_D$. The power s of the spectral algebraic law is

$$s = d - 1 + 2\nu. \qquad (4.98)$$

The relation (4.98) connects the power s of the power-law form (4.97) for $G_{\text{lin}}(\omega)$ with the dimension d of the lattice and with the power ν of the frequency-dependent coupling function $\lambda(\omega)$ in Eq. (4.96). In the absence or presence of a cubic symmetry, we have $\nu = 1/2$ and $\nu = 3/2$, respectively.

A concluding thought is appropriate. When the polaron moves coherently in the crystal, the two-state assumption breaks down. In this case, it is reasonable to eliminate the dependence of the expression (4.76) on the individual positions $R^{(n)}$ of the atoms by averaging over many atoms. Then the angular integrals in Eq. (4.76) are easily carried out. We then find

$$S_{\text{lin}}^{(E)}[\boldsymbol{q}(\cdot)] = \frac{3}{\pi} \int\limits_0^\infty d\omega \, \frac{J(\omega)}{k^2(\omega)} \int\limits_0^{\hbar\beta} d\tau \int\limits_0^\tau d\tau' \, D_\omega(\tau - \tau') \left\{ 1 - \frac{\sin[k(\omega)|\boldsymbol{q}(\tau) - \boldsymbol{q}(\tau')|]}{k(\omega)|\boldsymbol{q}(\tau) - \boldsymbol{q}(\tau')|} \right\},$$
$$(4.99)$$

where the function $k(\omega)$ is the dispersion relation solved for $k = |\boldsymbol{k}|$, and the spectral density $J(\omega)$ is related to the function $G(\omega)$ given in Eq. (4.86) by Eq. (3.90), where q_0 is a characteristic length. For $k|\boldsymbol{q}(\tau) - \boldsymbol{q}(\tau')| \ll 1$ in the relevant integration regime, we may expand the sine in Eq. (4.99). Then, the expression (4.99) reduces to the standard form given in Eq. (4.52).

Generally, the influence action (4.52) overestimates the dissipative influences of the environment as compared with the more accurate expression (4.99). The influence action (4.99) has been used repeatedly in studies of the ground-state energy and the effective mass of the acoustic polaron.[3]

Particular interest has been devoted over about forty years to the investigation of the possibility of a phonon-induced self-trapping phase transition. In recent years, this question has been answered in the negative. For a clarification of this issue we refer to a review by Gerlach and Löwen [165].

---

[3]Cf., e.g., Refs. [163, 164]. See also the review in Ref. [165].

4.2.5 Acoustic polaron: two-phonon coupling

We now turn to the discussion of the two-phonon coupling (3.116). Again, the influence action $\mathcal{F}^{(E)}_{\text{quadr}}[q]$ for the two-phonon process is formally given by the series expansion (4.74) in which $\tilde{H}_{\text{lin}}[q(\tau)]$ is replaced by $\tilde{H}_{\text{quadr}}[q(\tau)]$. Using Eqs. (4.73), (3.43), and (4.57), the thermal expectation value of the lowest order term takes the form

$$\left\langle \tilde{H}_{\text{quadr}}[q(\tau)] \right\rangle_\beta = \frac{\hbar}{4V\varrho} \sum_{n,m} \sum_{k,\lambda} \frac{\gamma^{(n,m)}_{k,-k}[q(\tau)]}{\omega_k} e^{i k \cdot [R^{(m)} - R^{(n)}]} \coth(\hbar\beta\omega_k/2) \,. \quad (4.100)$$

Now assume that the particle-lattice potential in Eq. (3.108) or Eq. (3.114) is a sum of pair potentials. If we disregard umklapp processes, which is a reasonable assumption for low temperatures, we then find [128]

$$\sum_{n,m} \gamma^{(n,m)}_{k,-k}[q(\tau)] e^{i k \cdot [R^{(m)} - R^{(n)}]} = 0 \,. \quad (4.101)$$

and thus

$$\langle \tilde{H}_{\text{quadr}}[q(\tau)] \rangle_\beta = 0 \,. \quad (4.102)$$

The computation of the second cumulant is straightforward. Using relations for expectation values with four phonon operators holding for Gaussian statistics, e.g.,

$$\langle b_k b^\dagger_{-k'} b^\dagger_{-k''} b_{k'''} \rangle_\beta = \langle b_k b^\dagger_{-k'} \rangle_\beta \langle b^\dagger_{-k''} b_{k'''} \rangle_\beta + \langle b_k b^\dagger_{-k''} \rangle_\beta \langle b^\dagger_{-k'} b_{k'''} \rangle_\beta \,, \quad (4.103)$$

and Eqs. (4.102), (3.43), (4.57), and the relation $\gamma^{(n,m)}_{k,k'} = \gamma^{(n,m)}_{-k,-k'}$, we find

$$\left\langle \tilde{H}_{\text{quadr}}[q(\tau)] \tilde{H}_{\text{quadr}}[q(\tau')] \right\rangle_\beta = \sum_{n,m} \sum_{n',m'} \sum_{k,k'} \frac{\hbar^2 \gamma^{(n,m)}_{k,k'}[q(\tau)] \gamma^{(n',m')}_{k,k'}[q(\tau')]}{2(2V\varrho)^2 \omega_k \omega_{k'}} \quad (4.104)$$

$$\times\ e^{i k \cdot [R^{(n')} - R^{(n)}]} e^{i k' \cdot [R^{(m')} - R^{(m)}]} e^{i(k+k') \cdot [q(\tau) - q(\tau')]}$$

$$\times\ \left\{ 2\Theta(\omega_k - \omega_{k'}) \Big(n(\omega_{k'}) - n(\omega_k) \Big) D_{\omega_k - \omega_{k'}}(\tau - \tau') \right.$$

$$\left. + \Big(1 + n(\omega_k) + n(\omega_{k'}) \Big) D_{\omega_k + \omega_{k'}}(\tau - \tau') \right\} \,.$$

We now restrict the attention to defect tunneling between two interstitial sites. We use again the parametrization (4.77) for the path and introduce the notation

$$\gamma^{(n,m)}_{k,k'}(\sigma) \equiv \gamma^{(n,m)}_{k,k'}[q = \sigma q_0/2] \,. \quad (4.105)$$

We then have

$$\text{Re}\, e^{i(k+k') \cdot [q(\tau) - q(\tau')]} = \cos^2\left(\frac{(k+k') \cdot q_0}{2}\right) + \sin^2\left(\frac{(k+k') \cdot q_0}{2}\right) \sigma(\tau)\sigma'(\tau) \,. \quad (4.106)$$

As in the previous analysis of linear phonon coupling, we are not interested in the constant energy shift induced by the coupling for $\sigma = \sigma'$. Therefore, as before in Eq.(4.80), we subtract this term from the expression given in Eq. (4.104). Upon using Eq. (4.106) and performing the subtraction, we obtain the influence action in the form

$$
S_{\text{quadr}}^{(E)}[\sigma(\cdot)]/\hbar \;=\; \frac{1}{2}\int_0^{\hbar\beta} d\tau \int_0^\tau d\tau' \, \mathcal{K}_{\text{quadr}}(\tau - \tau')\Big(\sigma(\tau) - \sigma(\tau')\Big)^2 \Big/ 4 \tag{4.107}
$$

with the kernel

$$
\mathcal{K}_{\text{quadr}}(\tau) \;=\; \int_0^\infty d\omega \, G_{\text{quadr}}(\omega) D_\omega(\tau) . \tag{4.108}
$$

The spectral density for two-phonon coupling is given by

$$
\begin{aligned}
G_{\text{quadr}}(\omega) \;=\; \frac{1}{2(2V\varrho)^2} \sum_{k,k'} \Bigg\{ & 2\delta(\omega - \omega_k + \omega_{k'})\frac{n(\omega_{k'}) - n(\omega_k)}{\omega_k \omega_{k'}} \\
& + \delta(\omega - \omega_k - \omega_{k'})\frac{1 + n(\omega_k) + n(\omega_{k'})}{\omega_k \omega_{k'}} \Bigg\} f(k,k') ,
\end{aligned} \tag{4.109}
$$

where

$$
\begin{aligned}
f(k,k') \;=\; & \cos^2\left(\frac{(k + k')\cdot q_0}{2}\right) \left| \sum_{n,m} e^{ik\cdot R^{(n)}} e^{ik'\cdot R^{(m)}} \left(\gamma_{k,k'}^{(n,m)}(+) - \gamma_{k,k'}^{(n,m)}(-)\right) \right|^2 \\
& + \sin^2\left(\frac{(k + k')\cdot q_0}{2}\right) \left| \sum_{n,m} e^{ik\cdot R^{(n)}} e^{ik'\cdot R^{(m)}} \left(\gamma_{k,k'}^{(n,m)}(+) + \gamma_{k,k'}^{(n,m)}(-)\right) \right|^2 .
\end{aligned}
$$

Next, we expand Eq. (4.109) in the frequency ω. To linear order in ω, we find

$$
G_{\text{quadr}}(\omega) \;=\; 2\alpha_{\text{quadr}}\, \omega + \mathcal{O}\left((\omega/\omega_D)^3\right) , \tag{4.110}
$$

where the prefactor α_{quadr} is the dimensionless function

$$
\alpha_{\text{quadr}} \;=\; \frac{\hbar\beta}{2(2V\varrho)^2} \sum_{k,k'} \delta\Big(\omega_k - \omega_{k'}\Big)\frac{n(\omega_k)[1 + n(\omega_{k'})]}{\omega_k \omega_{k'}} f(k,k') . \tag{4.111}
$$

We are not interested in the term $\propto \omega^3$, since this contribution can be absorbed into the spectral density $G_{\text{lin}}(\omega)$ in Eq. (4.86), thereby giving a small temperature dependent contribution to the coefficient α_3.

Turning to the Debye model (4.84) and expanding the trigonometric functions in k and k', we obtain the low temperature expansion for α_{quadr}. We find

$$
\alpha_{\text{quadr}} \;=\; \kappa_1\left(\frac{k_B T}{\hbar\omega_D}\right)^6 + \kappa_2\left(\frac{k_B T}{\hbar\omega_D}\right)^8 + \mathcal{O}\left(\left(\frac{k_B T}{\hbar\omega_D}\right)^{10}\right) . \tag{4.112}
$$

The coupling parameter κ_1 is given by

$$
\kappa_1 = \frac{9\pi^2 \omega_D^6}{370\varrho^2} \sum_{\lambda,\lambda'} \frac{1}{v_\lambda^5 v_{\lambda'}^5} \int \frac{d\Omega}{4\pi} \int \frac{d\Omega'}{4\pi}
$$
$$
\times \left| \sum_{n,m} \left(\gamma_{\mathbf{k},\lambda}^{(n,m)}(+) - \gamma_{\mathbf{k},\lambda}^{(n,m)}(-) \right) \left(\hat{\mathbf{k}} \cdot \mathbf{R}^{(n)} \right) \left(\hat{\mathbf{k}}' \cdot \mathbf{R}^{(m)} \right) \right|^2 .
\tag{4.113}
$$

It is also straightforward to calculate the coefficient κ_2. Since the resulting expression is quite lengthy, it is not given here, and we refer to Ref. [128].

In the absence of a cubic symmetry for the interstitial sites, e.g., for tetrahedral sites in bcc crystals, the coefficient κ_1 is estimated to be about 10^4. For cubic symmetry of the interstitial sites, there holds $\gamma^{(n)}(+) = \gamma^{(n)}(-)$, and therefore the coefficient κ_1 vanishes. The coefficient κ_2 is estimated to be about 10^6.

The spectral density (4.110) resulting from the two-phonon process is Ohmic. While the Ohmic viscosity due to electron-hole excitations in metals is temperature-independent (see the discussion in Subsection 4.2.8), the present viscosity coefficient varies strongly with T as T^6 or T^8, depending on the particular symmetry of the interstitial sites.

Higher-order correlated n-phonon processes with $n \geq 3$, which arise from the terms of order u^n in the expansion (3.108), give again an Ohmic spectral density. The respective viscosity coefficient behaves as T^{2n+2} in the absence of cubic symmetry, and as T^{2n+4} for cubic symmetry. At low temperatures therefore, the multi-phonon processes with $n > 2$ merely gives a weak renormalization of the parameters κ_1 and κ_2. The two-phonon process has been studied in Refs. [124, 166, 168, 169, 128].

In conclusion, even if one finds, for some model, a non-Ohmic form for $G(\omega)$, in most cases an Ohmic spectral density contribution is found if one pursues the calculation to higher orders. However, the respective viscosity will usually be strongly *temperature-dependent*.

4.2.6 Tunneling between surfaces: one-phonon coupling

One of the most interesting developments in surface science in the last fifteen years has been the possibility to manipulate atoms and molecules at a surface on an atomic scale by a scanning tunneling microscope (STM) [170]. In the atomic switch realized by Eigler *et al.* [171], a Xe atom has been reversibly transferred between a Ni surface and a tungsten tip. Thus, scanning-tunneling microscopy can be used not only for imaging, but also to test fundamental aspects of quantum mechanics [172]. The crystalline environment can have strong influence on the transfer of the atom. Louis and Sethna have emphasized that linear coupling of acoustic phonons to atoms tunneling between surfaces corresponds to Ohmic dissipation, as opposed to the bulk case where dissipation is super-Ohmic [161]. With the groundwork already done in Subsection 4.2.4, the case of surface tunneling is easily understood.

The potential for the surface-tip system of a STM or atomic-force microscope (AFM) has a double well shape. Consider the parameter regime (3.79), allowing us to truncate the Hilbert space to two states. The tunneling atom exerts a force on the surface which changes by an amount $\Delta \boldsymbol{F}$ when the atom hops from the surface to the tip. In response to the atom being on the surface or on the tip, the surface atoms will switch their equilibrium positions. Since the displacement is small, we may consider the interaction of the tunneling atom with the surface in linear order in the displacement (one-phonon coupling). The atom on the surface or tip together with the relaxation of the atoms on the surface can thus be modelled by the spin-boson Hamiltonian (3.88). The possibility of varying many parameters is very appealing. By applying an electrical field and by varying the tip position, the bias energy and the tunnel matrix element can be varied in a wide range.

All we need to connect tunneling between surfaces with defect tunneling in the bulk is to identify the force $\Delta \boldsymbol{F}$ with the change of the sum of the Kanzaki forces $\boldsymbol{g}^{(n)}$ when the atom moves from the surface $(+)$ to the tip $(-)$. We have

$$\Delta \boldsymbol{F} = \sum_n \left(\boldsymbol{g}^{(n)}(+) - \boldsymbol{g}^{(n)}(-) \right) , \qquad (4.114)$$

where the Kanzaki forces are defined in Eq. (3.113). The link missing yet is easily gathered from the findings in Subsection 4.2.4, which for the present purpose are Eq. (4.86) with Eq. (4.87). Thus, the tip-plus-atom system coupled to acoustic phonons is represented by an Ohmic spectral density[4]

$$G(\omega) = \left(2K\omega + \mathcal{O}(\omega^3) \right) \Theta(\omega_D - \omega) \qquad (4.115)$$

with the dimensionless coupling parameter

$$K = \frac{1}{8\pi^2 \varrho \hbar} \sum_\lambda \frac{1}{v_\lambda^3} \int \frac{d\Omega}{4\pi} \left| \Delta \boldsymbol{F} \cdot \boldsymbol{e}(\hat{\boldsymbol{k}}, \lambda) \right|^2 . \qquad (4.116)$$

While for defect tunneling in the bulk of a solid the monopole force is absent and hence super-Ohmic dissipation prevails, atom tunneling between a tip and a surface has lower symmetry, which results in a nonzero monopole force $\Delta \boldsymbol{F}$ on the surface and hence in Ohmic dissipation. Assuming a point force $\Delta \boldsymbol{F}/2$ on a semi-infinite isotropic medium, the Ohmic coupling parameter K is estimated as [161]

$$K = \frac{\delta}{8\pi^2 \varrho \hbar v_t^3} (\Delta \boldsymbol{F})^2 , \qquad (4.117)$$

where v_t is the tranverse sound velocity and the numerical constant δ varies between 0.3 and 0.8, depending on the Lamé constants of the medium. By varying the distance between the tip and the surface, the force $\Delta \boldsymbol{F}$ (and thus the coupling parameter K)

---

[4]We deliberately denote the coupling parameter by K in order to emphasize the similarity with the Kondo parameter K occuring for a fermionic environment [see Eqs. (4.153) and (18.18)].

could be tuned besides the tunneling coupling Δ and the bias ϵ. Thus, atom tunneling in a STM or AFM is obviously a prominent test ground for studying *macroscopic quantum tunneling* (MQT) and *macroscopic quantum coherence* (MQC) phenomena in different parameter regimes. In conclusion, we have studied the effects of phonons on the tunneling of an atom between two surfaces, and we have found, because of the lower symmetry compared with tunneling in the bulk, Ohmic dissipation.

4.2.7 Optical polaron

Consider an electron interacting with longitudinal optical phonons in a polar crystal as described by the Hamiltonian (3.106) with Eq. (3.130). In the interaction picture, the interaction term reads

$$\tilde{H}_{\rm I}[\boldsymbol{q}(\tau)] = \sum_{k} W_{k,l}\, e^{i\boldsymbol{k}\cdot\boldsymbol{q}(\tau)} \left(e^{-\omega_{\rm LO}\tau}\, b_{k} + e^{\omega_{\rm LO}\tau}\, b_{k}^{\dagger} \right). \tag{4.118}$$

The influence functional $F_{\rm LO}^{(\rm E)}[\boldsymbol{q}(\cdot)]$ is calculated as in the two preceding subsections. We expand the formal expression for $\mathcal{F}_{\rm LO}^{(\rm E)}[\boldsymbol{q}(\cdot)]$ into a power series in $\tilde{H}_{\rm I}$, as in Eq. (4.74), and employ *Wick's theorem* for thermodynamic Gaussian averages. For the simple form (4.118) of $\tilde{H}_{\rm I}[\boldsymbol{q}(\tau)]$, the only nonvanishing cumulant contraction is again the two-point function $\langle \tilde{H}_{\rm I}[\boldsymbol{q}(\tau)]\tilde{H}_{\rm I}[\boldsymbol{q}(\tau')]\rangle_{\beta}$. The influence action is exactly given by [cf. Eq. (4.75)]

$$S_{\rm LO}^{(\rm E)}[\boldsymbol{q}(\cdot)] = -\frac{1}{\hbar}\int_{0}^{\hbar\beta} d\tau \int_{0}^{\tau} d\tau'\, \langle \tilde{H}_{\rm I}[\boldsymbol{q}(\tau)]\tilde{H}_{\rm I}[\boldsymbol{q}(\tau')]\rangle_{\beta}. \tag{4.119}$$

The contraction takes the form

$$\langle \tilde{H}_{\rm I}[\boldsymbol{q}(\tau)]\tilde{H}_{\rm I}[\boldsymbol{q}(\tau')]\rangle_{\beta} = \sum_{k} |W_{k,l}|^2\, e^{i\boldsymbol{k}\cdot[\boldsymbol{q}(\tau)-\boldsymbol{q}(\tau')]} \tag{4.120}$$

$$\times \left[e^{-\omega_{\rm LO}(\tau-\tau')}\langle b_{k}b_{k}^{\dagger}\rangle_{\beta} + e^{\omega_{\rm LO}(\tau-\tau')}\langle b_{k}^{\dagger}b_{k}\rangle_{\beta} \right],$$

in which
$$\langle b_{k}^{\dagger}b_{k}\rangle_{\beta} = n(\omega_{\rm LO})\,, \qquad \langle b_{k}b_{k}^{\dagger}\rangle_{\beta} = 1 + n(\omega_{\rm LO})\,, \tag{4.121}$$

where $n(\omega)$ is the single-particle Bose distribution (3.44). Next, we observe that the square bracket in Eq. (4.120) is just the boson propagator $D_{\omega_{\rm LO}}(\tau - \tau')$ defined in Eq. (4.57). Thus we have

$$\langle \tilde{H}_{\rm I}[\boldsymbol{q}(\tau)]\tilde{H}_{\rm I}[\boldsymbol{q}(\tau')]\rangle_{\beta} = \sum_{k} |W_{k,l}|^2\, e^{i\boldsymbol{k}\cdot[\boldsymbol{q}(\tau)-\boldsymbol{q}(\tau')]}\, D_{\omega_{\rm LO}}(\tau - \tau')\,. \tag{4.122}$$

The effective action of a longitudinal optical polaron has two contributions,

$$S_{\rm eff}^{(\rm E)}[\boldsymbol{q}] = \frac{M}{2}\int_{0}^{\hbar\beta} d\tau\, \dot{\boldsymbol{q}}^2 + S_{\rm LO}^{(\rm E)}[\boldsymbol{q}]\,, \tag{4.123}$$

where the influence action is found upon using the form (3.130) for $W_{k,l}$ to read

$$S_{\text{LO}}^{(\text{E})}[\boldsymbol{q}(\cdot)] = -\frac{\hbar}{V}\frac{4\pi\alpha\omega_{\text{LO}}^2\sqrt{\hbar}}{(2M\omega_{\text{LO}})^{1/2}}\int_0^{\hbar\beta}d\tau\int_0^{\tau}d\tau'\,D_{\omega_{\text{LO}}}(\tau-\tau')\sum_{k}\frac{e^{i\boldsymbol{k}\cdot[\boldsymbol{q}(\tau)-\boldsymbol{q}(\tau')]}}{|\boldsymbol{k}|^2}\,. \quad (4.124)$$

Next, we write the sum over $\boldsymbol{k}$ as an integral, $\sum_{\boldsymbol{k}} \to \int d^d\boldsymbol{k}/(2\pi)^d$, and calculate the integral for an isotropic medium with dimension $d=3$. We then obtain the expression

$$S_{\text{LO}}^{(\text{E})}[\boldsymbol{q}(\cdot)] = -\alpha\hbar\left(\frac{\hbar}{2M\omega_{\text{LO}}}\right)^{1/2}\omega_{\text{LO}}^2\int_0^{\hbar\beta}d\tau\int_0^{\tau}d\tau'\,\frac{D_{\omega_{\text{LO}}}(\tau-\tau')}{|\boldsymbol{q}(\tau)-\boldsymbol{q}(\tau')|}\,. \quad (4.125)$$

The polaron problem represents one of the simplest examples of the interaction of a particle and a field. With the influence action (4.125) however, we are left with a considerably complicated path integral for the optical polaron. Fortunately, one can tackle the polaron path integral with a variational approach (cf. Subsection 8.2). The variational method has been successfully utilized in diverse applications and gives reliable results for arbitrary strength of the coupling.

4.2.8   Heavy particle in a metal

We now turn our attention to the evaluation of the response of a noninteracting Fermi liquid to a local time-dependent perturbation. The elimination procedure adequate for bosons is generally not suitable for fermions. The anticommutation relations for fermions create difficulties which are absent in the case of bosons [173]. Fortunately, the low-energy excitations of the Fermi liquid have bosonic character. Therefore, it is possible to circumvent these difficulties and give a consistent formulation following previous lines. To determine the effective action for a heavy particle interacting with conduction electrons, we have to evaluate the influence functional (4.31) for the Hamiltonian (3.131). It is convenient to work out $\mathcal{F}^{(\text{E})}[\boldsymbol{q}(\cdot)]$ using the operator formulation for the fermionic degrees of freedom. We have

$$\begin{aligned}\mathcal{F}^{(\text{E})}[\boldsymbol{q}(\cdot)] &= Z_{\text{R}}^{-1}\text{tr}_{\text{R}}\left\{e^{-\beta H_{\text{R}}}\,T_\tau\exp\left(-\int_0^{\hbar\beta}d\tau\,\tilde{H}_{\text{I}}[\boldsymbol{q}(\tau)]/\hbar\right)\right\}\\ &\equiv \left\langle T_\tau\exp\left(-\int_0^{\hbar\beta}d\tau\,\tilde{H}_{\text{I}}[\boldsymbol{q}(\tau)]/\hbar\right)\right\rangle_\beta,\end{aligned} \quad (4.126)$$

where T_τ is the imaginary-time ordering operator for fermion fields. In the second form, $\langle\cdots\rangle_\beta$ denotes thermal average with respect to $\exp(-\beta H_{\text{R}})$. The tilde indicates again that we are in the interaction representation [cf. Eq. (4.72)]. We proceed by expanding $\mathcal{F}^{(\text{E})}[\boldsymbol{q}]$ into a power series in $\tilde{H}_{\text{I}}$ as given in Eq. (4.74).

Consider first *normally conducting electrons*. We find from Eq. (3.134)

$$\tilde{H}_{\text{I}} = \sum_{\boldsymbol{k},\boldsymbol{k}',\sigma,\sigma'}<\boldsymbol{k},\sigma|\,U\,|\boldsymbol{k}',\sigma'>\exp[i(\boldsymbol{k}'-\boldsymbol{k})\cdot\boldsymbol{q}(\tau)]\exp[-(\omega_{\boldsymbol{k}'}-\omega_{\boldsymbol{k}})\tau]\,c_{\boldsymbol{k}\sigma}^\dagger\,c_{\boldsymbol{k}'\sigma'}\,. \quad (4.127)$$

Working in perturbation theory, all terms with odd powers of $\tilde{H}_I$ vanish because $\langle \tilde{H}_I[q(\tau)] \rangle_\beta = 0$. The lowest-order term $(n = 2)$ in the series expression analogous to (4.74) is written in terms of the two-time electron-hole contraction

$$\mathcal{H}_2[q(\tau), q(\tau')] \equiv \langle \tilde{H}_I[q(\tau)] \, \tilde{H}_I[q(\tau')] \rangle_\beta \qquad (4.128)$$

as

$$\mathcal{F}_2^{(E)}[q(\cdot)] = \int_0^{\hbar\beta} d\tau \int_0^\tau d\tau' \, \mathcal{H}_2[q(\tau), q(\tau')]/\hbar^2 . \qquad (4.129)$$

Upon substituting Eq. (4.127) we find

$$\begin{aligned}
\mathcal{H}_2[q(\tau), q(\tau')] &= \sum_{k,k',\sigma,\sigma'} |<k,\sigma|U|k',\sigma'>|^2 \exp\Big(i(k'-k)\cdot[q(\tau) - q(\tau')]\Big) \\
&\times \exp\Big(-(\omega_{k'} - \omega_k)(\tau - \tau')\Big) f_k(1 - f_{k'}) .
\end{aligned} \qquad (4.130)$$

Here we have used the relations

$$\langle c_{k\sigma}^\dagger c_{k'\sigma'} \rangle_\beta = \delta_{\sigma\sigma'}\delta_{kk'} f_k , \qquad \langle c_{k\sigma} c_{k'\sigma'}^\dagger \rangle_\beta = \delta_{\sigma\sigma'}\delta_{kk'}(1 - f_k) , \qquad (4.131)$$

where f_k is the Fermi distribution function,

$$f_k \equiv f(\omega_k) = \frac{1}{\exp(\beta\hbar\omega_k) + 1} , \qquad (4.132)$$

and where the energy $\hbar\omega_k$ is measured relatively to the Fermi energy.

The expression (4.129) with Eq. (4.130) is a simple density response function of the noninteracting Fermi gas. The function $\mathcal{H}_2[q(\tau), q(\tau')]$ may be interpreted in terms of an electron-hole pair injected at time τ' and removed at a later imaginary time τ. This is indicated by diagram (a) in Fig. 4.2. All those diagrams representing multiple self-energy insertions arising from uncorrelated electron-hole loops, e. g., the two shown in (b) of Fig. 4.2, can be summed up and expressed as an exponential of the simple diagram (a). As we shall see immediately, the spectral density of the particle's coupling to the boson-like electron-hole excitations implicit in Eq. (4.129) is of the Ohmic form $J(\omega) \propto \omega$, this being due to the constant density of states around the Fermi surface. Contractions of higher order in the cumulant expansion for the action resulting from Eq. (4.126) [cf. Eq. (4.74)], e.g., diagram (c) in Fig. 4.2, correspond to coherent excitation of two or more electron-hole pairs. Since they depend on a convolution of the densities of pairs, they are represented by powers of ω larger than one in the spectral density $J(\omega)$ at low frequencies. Therefore, they contribute negligibly small corrections to the response of the electron gas at low temperatures.

Within the approximation of uncorrelated electron-hole pairs, we thus obtain the influence functional in the form

$$\mathcal{F}^{(E)}[q(\cdot)] = \exp\left(\int_0^{\hbar\beta} d\tau \int_0^\tau d\tau' \, \mathcal{H}_2[q(\tau), q(\tau')]\Big/\hbar^2\right) . \qquad (4.133)$$

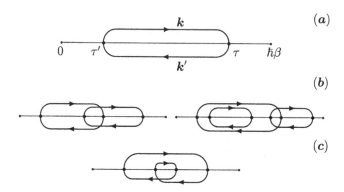

Figure 4.2: Some virtual electron-hole diagrams appearing in the expansion in $\tilde{H}_{\mathrm{I}}[\boldsymbol{q}(\tau)]$ are shown. The oval (thick) lines represent electron and hole propagators, respectively. Diagrams (a) and (b) are included in the expression (4.133), while diagram (c) is neglected.

The evaluation of Eq. (4.130) is simplified when the impurity potential is a contact potential. Then the potential matrix is constant, $< \boldsymbol{k}, \sigma|U|\boldsymbol{k}', \sigma' > = \delta_{\sigma\sigma'}U_0$. The coupling strength U_0 is called *deformation potential*. This form is a reasonable approximation also for extended potentials, since the Fermi functions effectively restrict $\hbar|\omega_{\boldsymbol{k}}|$ and $\hbar|\omega_{\boldsymbol{k}'}|$ in Eq. (4.130) to small values compared to the Fermi energy E_{F}.
Computation of the angular integrals for an isotropic medium ($d = 3$) gives

$$\mathcal{H}_2[\Delta q(\tau,\tau')] = 2U_0^2 \sum_{\boldsymbol{k},\boldsymbol{k}'} \frac{\sin[k\Delta q(\tau,\tau')]}{k\Delta q(\tau,\tau')} \frac{\sin[k'\Delta q(\tau,\tau')]}{k'\Delta q(\tau,\tau')} \, \mathrm{e}^{[\omega_{\boldsymbol{k}}-\omega_{\boldsymbol{k}'}](\tau-\tau')} f_k(1-f_{k'})\,,$$

(4.134)

with

$$\Delta q(\tau,\tau') \equiv |\boldsymbol{q}(\tau) - \boldsymbol{q}(\tau')|\,. \tag{4.135}$$

Next, we take the continuum limit

$$\sum_{\boldsymbol{k}} \cdots \quad \rightarrow \quad \hbar\varrho \int d\omega\, \mathcal{N}(\omega) \cdots\,, \tag{4.136}$$

and we count the energy relative to the Fermi energy E_{F}. The quantity $\varrho\mathcal{N}(\omega)$ is the number of states per energy interval $\hbar\,d\omega$, where ϱ is the density of states at the Fermi energy and has dimension inverse energy. Thus by definition

$$\mathcal{N}(\omega = 0) = 1\,. \tag{4.137}$$

As we are interested in the low energy excitations around the Fermi surface, we may replace k and k' in the fractions of Eq. (4.134) by the Fermi wave number k_{F}. At this point it is convenient to absorb various factors into the "viscosity" coefficient

$$\eta = 4\pi\, U_0^2 \varrho^2 \hbar k_{\mathrm{F}}^2/3\,, \tag{4.138}$$

which has the usual dimension mass times frequency. Then we may write Eq. (4.134) as

$$\mathcal{H}_2[\Delta q(\tau, \tau')] = \frac{3\hbar}{2k_F^2} \frac{\sin^2[k_F \Delta q(\tau, \tau')]}{[k_F \Delta q(\tau, \tau')]^2} K(\tau - \tau'), \quad (4.139)$$

$$K(\tau) = \frac{\eta}{\pi} \int_{-\infty}^{\infty} d\omega' \int_{-\infty}^{\infty} d\omega'' \, \mathcal{N}(\omega') \mathcal{N}(\omega'') \exp[(\omega'' - \omega')\tau] f(\omega'') f(-\omega'). \quad (4.140)$$

With the substitution $\omega'' = \omega' - \omega$, we may write $K(\tau)$ as

$$K(\tau) = \frac{1}{\pi} \int_{-\infty}^{\infty} d\omega \, \frac{\exp(-\omega\tau)}{1 - \exp(-\omega\hbar\beta)} J(\omega) \quad (4.141)$$

with

$$J(\omega) = \eta \left[1 - e^{-\omega\hbar\beta} \right] \int_{-\infty}^{\infty} d\omega' \, \mathcal{N}(\omega') \mathcal{N}(\omega' - \omega) f(\omega' - \omega) f(-\omega'). \quad (4.142)$$

Upon substituting the explicit form of the Fermi function, it is straightforward to rewrite the expression (4.142) as

$$J(\omega) = \frac{\eta}{2} \int_{-\infty}^{\infty} d\omega' \, \mathcal{N}(\omega') \mathcal{N}(\omega' - \omega) \left\{ \tanh(\tfrac{1}{2}\beta\hbar\omega') - \tanh[\tfrac{1}{2}\beta\hbar(\omega' - \omega)] \right\}, \quad (4.143)$$

from where we see that $J(\omega)$ is antisymmetric in ω. The antisymmetric part of the fraction in Eq. (4.141) is just the phonon propagator (4.57). Thus we may write Eq. (4.141) in the form

$$K(\tau) = \frac{1}{\pi} \int_0^{\infty} d\omega \, J(\omega) D_\omega(\tau). \quad (4.144)$$

Hence the function $K(\tau)$ is found as the familiar bosonic heat bath kernel (4.55). We now wish again that the influence functional $\mathcal{F}^{(E)}[q(\cdot)]$ is unity for a constant path. Therefore, we subtract from $\mathcal{H}_2[\Delta q(\tau, \tau')]$ in Eq. (4.139) the expression $\mathcal{H}_2[\Delta q = 0]$. We then find $\mathcal{F}^{(E)}[q(\cdot)]$ in the form (4.31) with the action

$$S_{\text{infl}}^{(E)}[q(\cdot)] = \frac{3}{2 k_F^2} \int_0^{\hbar\beta} d\tau \int_0^{\tau} d\tau' \, K(\tau - \tau') \left(1 - \frac{\sin^2[k_F|q(\tau) - q(\tau')|]}{[k_F|q(\tau) - q(\tau')|]^2} \right). \quad (4.145)$$

The influence action (4.145) arising from the coupling of the heavy particle to the conduction electrons holds without restriction on the numerical value of $k_F \Delta q$.

For defect tunneling in metals between two interstitial sites of distance q_0, we may use again the spin path (4.65). We then obtain from Eq. (4.145) the influence action exactly in the spin-boson form [cf. Eq. (4.69) with Eq. (4.68)]

$$S_{\text{infl}}^{(E)}[\sigma(\cdot)]/\hbar = \frac{1}{2} \int_0^{\hbar\beta} d\tau \int_0^{\tau} d\tau' \, \mathcal{K}(\tau - \tau') [\sigma(\tau) - \sigma(\tau')]^2 / 4 \quad (4.146)$$

with

$$\mathcal{K}(\tau) = \frac{q_0^2}{\hbar} K(\tau) = \int_0^{\infty} d\omega \, G(\omega) D_\omega(\tau). \quad (4.147)$$

The spin-boson spectral density $G(\omega)$ is expressed in terms of the density $J(\omega)$ by

$$G(\omega) \;=\; \frac{3}{(k_F q_0)^2} \left(1 - \frac{\sin^2(k_F q_0)}{(k_F q_0)^2} \right) \frac{q_0^2}{\pi\hbar} J(\omega) \,. \tag{4.148}$$

For conduction electrons, the density of states around the Fermi level is constant,

$$\mathcal{N}(\omega) \;=\; 1 \,. \tag{4.149}$$

For simplicity, we now disregard that the width of the conduction band is actually finite. Substituting the constant electron density into Eq. (4.143), the ω'-integral is easily performed. We then find

$$J(\omega) \;=\; \eta\omega \,, \tag{4.150}$$

which is just in the Ohmic form, Eq. (3.51). Thus, the electron-hole excitations behave as a bosonic heat reservoir with Ohmic spectral density.

The expression (4.138) gives the viscosity in the Born approximation, $\eta \propto U_0^2$. In a nonperturbative treatment, Sassetti et $al.$ [174] showed by calculating the single-particle propagator for a contact potential that the friction coefficient is expressed in terms of the s-wave scattering phase shift δ_0 at the Fermi energy by the relation

$$\eta \;=\; 4\hbar\, k_F^2 \sin^2\delta_0 \,/3\pi \,, \qquad \text{where} \qquad \tan\delta_0 \;=\; -\pi U_0 \varrho \,. \tag{4.151}$$

The same form for the viscosity is found by calculating the force auto-correlation function for a heavy particle in a free electron gas. Generalizing to a spin-independent spherically symmetric finite-range potential, one finds [175]

$$\eta \;=\; \frac{4\hbar}{3\pi} k_F^2 \sum_{l=0}^{\infty} (l+1) \sin^2(\delta_{l+1} - \delta_l) \,, \tag{4.152}$$

where δ_l represents the scattering phase shift of the lth partial wave. Thus also for a finite-range potential, the electronic excitations behave collectively as if they were bosons with an Ohmic spectral density of the coupling. Similar conclusions were drawn by Chang and Chakravarty [176] who studied Dyson's equation for the fermion propagator in presence of the impurity within a real-time formulation. They solved the problem by employing the long-time approximation of the free fermion propagator, a method originally proposed by Nozières and De Dominicis [177]. It has been shown by a direct bosonization of the fermion operators [178] that the low-energy excitations of a fermionic bath can be mapped onto a bosonic bath. We remark that the scattering phase of a contact potential depends on the regularization scheme employed. Only the leading Born term is universal, i.e., does not depend on the particular regularization prescription. For a discussion of this point we refer to Ref. [178].

The spin-boson spectral density $G(\omega)$ takes the form

$$G(\omega) \;=\; 2K\omega \,, \tag{4.153}$$

where the dimensionless Ohmic damping strength K is usually called the Kondo parameter. For interstitials in metals with a tunneling distance q_0 of the order of the lattice constant, we have $k_F q_0 \ll 1$. In this important limit, we have

$$K = \frac{\eta q_0^2}{2\pi\hbar} + \mathcal{O}\left((k_F q_0)^4\right) . \tag{4.154}$$

The relation (4.154) is exactly in the form as conjectured by Sols and Guinea [179] from a linear response calculation. Recently, Schönhammer [180] has shown by taking all the scattering phase shifts into account that the relation (4.154) is generally valid in the limit $k_F q_0 \ll 1$ for any spherically symmetric potential independent of the strength of the coupling constant.

Substituting Eq. (4.150) with (4.151) into Eq. (4.148), the Kondo parameter reads

$$K = (2/\pi^2) \sin^2 \delta_0 \left[1 - \sin^2(k_F q_0)/(k_F q_0)^2 \right] , \tag{4.155}$$

which has the limiting cases

$$K = 2(U_0 \rho)^2 [1 - \sin^2(k_F q_0)/(k_F q_0)^2] , \qquad \delta_0 \ll \pi/2 , \tag{4.156}$$

$$K = \frac{2}{3\pi^2}(k_F q_0)^2 \sin^2 \delta_0 + \mathcal{O}[(k_F q_0)^4] , \qquad k_F q_0 \ll 1 . \tag{4.157}$$

The parameter K is equivalent to a particular overlap parameter introduced by Kondo (cf. Ref. [127] and references therein). The nature of the electronic screening cloud around an impurity in a metal has been the subject of intense studies. Anderson [181] considered the overlap of the ground state $|\phi>$ of the fermions in the absence of the impurity with the corresponding state $|\psi>$ in the presence of the impurity and found that the overlap integral for a state with N conduction electrons behaves as

$$<\psi|\phi> \propto N^{-K_+} , \tag{4.158}$$

where K_+ is a positive number. This relation is known as Anderson's *orthogonality theorem*. For a contact potential, one finds

$$K_+ = (\delta_0/\pi)^2 , \tag{4.159}$$

where δ_0 is the s-wave phase for scattering of the electrons off the impurity.

Next, consider the overlap of the ground state $|\psi_1>$ of the fermions for the impurity at position q_1 with the ground state $|\psi_2>$ for the impurity at position q_2. A special version of the orthogonality theorem now reads

$$<\psi_1|\psi_2> \propto N^{-K} . \tag{4.160}$$

The Kondo parameter is a function of the phase shifts and the distance $q_0 = |q_1 - q_2|$. A general calculation of K as a function of the distance is rather difficult [182, 183, 180], since each partial wave centered at q_1 mixes with *all* partial waves centered at q_2. Even when the impurity potential is spherically symmetric, the problem lacks spherical symmetry.

For s-wave scattering and spin degeneracy, one finds [182]

$$K = \frac{2}{\pi^2} \left\{ \arctan\left(\frac{\sqrt{1 - x(q_0)} \tan \delta_0}{\sqrt{1 + x(q_0) \tan^2 \delta_0}} \right) \right\}^2 . \tag{4.161}$$

The phase shift δ_0 is given in Eq. (4.151). Equation (4.161) gives the bound $K \leq 1/2$. The function $x(q_0)$ depends on the space dimension d,

$$x(q_0) = \sin^2(k_F q_0)/(k_F q_0)^2 \qquad \text{for} \qquad d = 3 , \qquad (4.162)$$
$$x(q_0) = \sin^2(k_F q_0) \qquad\qquad \text{for} \qquad d = 1 . \qquad (4.163)$$

For $d = 3$, the expression (4.161) reduces in the limit $k_F q_0 \ll 1$ to the form (4.157). For weak coupling, $\delta_0 \ll \pi/2$, Eq. (4.161) corresponds to the expression (4.156).

The *short-distance* form of K, Eq. (4.154), does not depend on the spatial dimension d [180], as opposed to the large distance behaviour which depends on d. For $d = 1$, where Eq. (4.163) applies, the function $x(q_0)$ is oscillating around the mean value $\frac{1}{2}$. For $d \geq 2$, the function $x(q_0)$ goes to zero as $k_F q_0 \to \infty$. Putting $x = 0$ in Eq. (4.161) and using Eq. (4.159), we obtain $K = 2K^+$.

For s-wave scattering, the overlap parameter K has the upper bound $1/2$ for arbitrary distances as follows from (4.161). In the general case of an extended potential, K is *not* bounded in principle by this value in two or three dimensions, since the exponent K_+ is not bounded [180]. Experimentally, K turns out to be a small parameter, typically about 0.1 for charged interstitials in metals. The smallness of K is due to a screening of the impurity potential by surrounding charges.

For the Ohmic spectral density (4.153), the kernel $\mathcal{K}(\tau)$ in the retarded action (4.146) can be calculated from (4.147) in analytic form. We obtain

$$\mathcal{K}(\tau) = 2K \frac{(\pi/\hbar\beta)^2}{\sin^2(\pi\tau/\hbar\beta)} . \qquad (4.164)$$

The influence action (4.145) with the kernel (4.164) conveys the effects of a normal-state metallic environment at thermal energies well below the Fermi energy.

4.2.9   Heavy particle in a superconductor

It is straightforward to generalize the discussion of the preceding subsection to the *superconducting* state. The underlying Hamiltonian is given by Eqs. (3.135) and (3.141) with (3.144). According to the BCS theory, the electronic spectrum is modified and the coherence factor is as specified in Eq. (3.144). The "dissipative" part of the action is found again in the form (4.145) or in the form (4.146). Also the kernel $\mathcal{K}(\tau)$ takes again the earlier form (4.147), and the spectral density of the coupling (4.143) is adapted to BCS-quasiparticle excitations,

$$G_{qp}(\omega) = K \int_{-\infty}^{\infty} d\omega' \, \mathcal{N}_{qp}(\omega') \, \mathcal{N}_{qp}(\omega' - \omega) \left\{ 1 - \frac{\Delta_g^2}{\omega'(\omega' - \omega)} \right\}$$
$$\times \left\{ \tanh(\tfrac{1}{2}\hbar\beta\omega') - \tanh[\tfrac{1}{2}\hbar\beta(\omega' - \omega)] \right\} . \qquad (4.165)$$

Here $\mathcal{N}_{qp}(\omega)$ is the density of states for the quasi-particle excitations, and the coherence factor $1 - \Delta_g^2/[\omega'(\omega' - \omega)]$ corresponds to a time-reversal-invariant interaction.

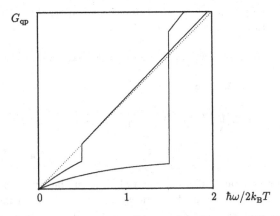

Figure 4.3: Quasiparticle spectral density G_{qp} as a function of $\hbar\omega/2k_B T$. The solid curves correspond to $\hbar\Delta_g/k_B T = 0.5$ and $\hbar\Delta_g/k_B T = 1.5$, respectively. The dotted straight line represents the spectral density in the normally conducting state.

Within the BCS theory, the density of states is given by

$$\mathcal{N}_{qp}(\omega) = \Theta(|\omega| - \Delta_g) \frac{|\omega|}{\sqrt{\omega^2 - \Delta_g^2}} . \tag{4.166}$$

At low frequencies $\omega \ll \Delta_g$, the function $G_{qp}(\omega)$ takes the Ohmic form

$$G_{qp}(\omega) = 2K_{qp}\,\omega \tag{4.167}$$

with the temperature-dependent dimensionless Ohmic coupling constant

$$K_{qp} = 2K /[1 + \exp(\beta\hbar\Delta_g)] . \tag{4.168}$$

Thus, K_{qp} is suppressed by the gap and vanishes at $T = 0$. The function $G_{qp}(\omega)$ is discontinuous at the excitation threshold $\omega = 2\Delta_g$, with a jump of height

$$\Delta G_{qp} = 2K\pi\Delta_g \tanh\left(\frac{\hbar\Delta_g}{2k_B T}\right) . \tag{4.169}$$

Above the threshold, $G_{qp}(\omega)$ increases with increasing frequency and approaches in the regime $\omega \gg \Delta_g$ the Ohmic form $G_{qp}(\omega) = 2K\omega$. Fig. 4.3 shows plots of $G_{qp}(\omega)$ for different temperatures.

At zero temperature, the integral in Eq. (4.165) can be expressed in terms of the complete elliptic integral of the second kind $E(\pi/2, k)$ [87]. We obtain

$$G_{qp}(\omega) = \Theta\left(\omega - 2\Delta_g\right) 2K\omega\, E\left(\pi/2, \sqrt{1 - 4\Delta_g^2/\omega^2}\right) , \tag{4.170}$$

where $\Delta_g = \Delta_g(T = 0)$. The kernel $\mathcal{K}(\tau)$ may be given for $T = 0$ in terms of the modified Bessel functions $K_0(z)$ and $K_1(z)$. We find

$$\mathcal{K}(\tau) \;=\; 2K\Delta_g^2 \left\{ K_1^2(\Delta_g|\tau|) + K_0^2(\Delta_g|\tau|) \right\} . \qquad (4.171)$$

The limiting behaviour of $\mathcal{K}(\tau)$ is

$$\mathcal{K}(\tau) \;=\; 2K \begin{cases} \dfrac{1}{\tau^2} + \Delta_g^2 \Big(\ln(\Delta_g|\tau|/2) - C_{\mathrm{E}} \Big)^2 , & \text{for} \quad \Delta_g|\tau| \ll 1 , \\[3mm] \dfrac{\pi}{|\tau|}\Delta_g\, \mathrm{e}^{-2\Delta_g|\tau|} , & \text{for} \quad \Delta_g|\tau| \gg 1 , \end{cases} \qquad (4.172)$$

where C_{E} is Euler's constant. The exponential drop of the kernel at long times is qualitatively different from the algebraic decay $\propto 1/\tau^2$ in the normally conducting state at zero temperature [cf. Eq. (4.164)]. The qualitative difference originates from the gap in the low frequency spectrum of the superconducting environment.

4.2.10   Effective action for a Josephson junction

To find the effective action for a Josephson junction, we start out from the microscopic Hamiltonian (3.146) put down in Subsection 3.3.4 and follow the procedure outlined by Ambegaokar *et al.* [141]. In the first step, the quartic interactions in Eqs. (3.145) and (3.148) are eliminated in the functional integral representation of the expression (3.150) by means of a Gaussian identity which has become known as Hubbard-Stratonovich transformation. In this method, new fields are introduced in exchange for the quartic pair interaction and for the quartic Coulomb interaction, respectively. The former can be identified with the complex order parameter fields $\Delta_{\mathrm{L}}(\boldsymbol{r}, \tau)$ and $\Delta_{\mathrm{R}}(\boldsymbol{r}, \tau)$, the latter with the (real) voltage field $U(\tau)$. The resulting action is only quadratic in the fermion field, so that the trace over this field can readily be performed explicitly. The partition function takes the form

$$Z \;=\; \int \mathcal{D}^2\Delta_{\mathrm{L}}(\boldsymbol{r}, \tau)\, \mathcal{D}^2\Delta_{\mathrm{R}}(\boldsymbol{r}, \tau)\, \mathcal{D}U(\tau)\, \mathcal{D}\boldsymbol{A}(\boldsymbol{r}, \tau) \exp(-\mathcal{A}[\Delta_{\mathrm{L}}, \Delta_{\mathrm{R}}, U, \boldsymbol{A}]/\hbar) , \quad (4.173)$$

with the effective action

$$\begin{aligned} \mathcal{A} \;=\;\; & \hbar \operatorname{tr}\ln \widehat{\underline{G}}^{-1} + \frac{1}{2}\int_0^{\hbar\beta} d\tau\, CU^2(\tau) + \frac{1}{8\pi}\int_0^{\hbar\beta} d\tau \int d^3r \left(\boldsymbol{h}(\boldsymbol{r}, \tau) - \boldsymbol{h}_{\mathrm{ext}} \right)^2 \\ & + \int_0^{\hbar\beta} d\tau \left\{ \int_{r\in\mathrm{L}} d^3r\, g_{\mathrm{L}}^{-1}(\boldsymbol{r})\, |\Delta_{\mathrm{L}}(\boldsymbol{r}, \tau)|^2 + \int_{r\in\mathrm{R}} d^3r\, g_{\mathrm{R}}^{-1}(\boldsymbol{r})\, |\Delta_{\mathrm{R}}(\boldsymbol{r}, \tau)|^2 \right\} . \quad (4.174) \end{aligned}$$

Here, $\widehat{\underline{G}}$ is a 4×4 matrix Green function in the space spanned by (L) and (R) and in the Nambu pseudo-spin space.[5] We indicate matrices in Nambu space by carets,

---

[5]The two-dimensional Nambu space facilitates a compact formulation of BCS superconductivity.

and matrices describing the (L) and (R) superconductors by underlines. The inverse of the Green function is given by

$$\underline{\widehat{G}}^{-1}(r,\tau;r',\tau') = \begin{pmatrix} \widehat{G}_{\mathrm{L}}^{-1}(r,\tau;r',\tau') & \widehat{T}_{rr'}\,\delta(\tau-\tau') \\ \widehat{T}_{rr'}^{\dagger}\,\delta(\tau-\tau') & \widehat{G}_{\mathrm{R}}^{-1}(r,\tau;r',\tau') \end{pmatrix}. \tag{4.175}$$

The diagonal elements in the L-R space are

$$\widehat{G}_{\mathrm{L/R}}^{-1}(r,\tau;r',\tau') = \left\{ \hbar\frac{\partial}{\partial\tau}\widehat{1} + \left[-\frac{\hbar^2}{2m}\left(\boldsymbol{\nabla} - \frac{ie}{\hbar c}A\widehat{\tau}_3\right)^2 - \mu + ie\,U_{\mathrm{L/R}}(\tau)\right]\widehat{\tau}_3 \right.$$

$$\left. + \widehat{\Delta}_{\mathrm{L/R}}(r,\tau)\right\}\delta(r-r')\,\delta(\tau-\tau'), \tag{4.176}$$

where $\widehat{\tau}_3$ is the Pauli matrix in the Nambu space, and where $U(\tau) = U_{\mathrm{L}}(\tau) - U_{\mathrm{R}}(\tau)$. The order parameter field in Nambu space is

$$\widehat{\Delta}_{\mathrm{L/R}}(r,\tau) = \begin{pmatrix} 0 & \Delta_{\mathrm{L/R}}(r,\tau) \\ \Delta_{\mathrm{L/R}}^{*}(r,\tau) & 0 \end{pmatrix} = |\Delta_{\mathrm{L/R}}|\exp(-i\,\varphi_{\mathrm{L/R}}\widehat{\tau}_3)\,\widehat{\tau}_1. \tag{4.177}$$

The two superconducters (L) and (R) are coupled by the transfer matrix

$$\widehat{T}_{rr'} = \begin{pmatrix} T_{rr'} & 0 \\ 0 & -T_{rr'}^{*} \end{pmatrix}. \tag{4.178}$$

It is useful to perform a gauge transformation with the aim of making the off-diagonal elements in Eq. (4.177) real. This is achieved with a transformation induced by the unitary matrix

$$\underline{\widehat{U}} = \begin{pmatrix} e^{i\varphi_{\mathrm{L}}(r,\tau)\widehat{\tau}_3/2} & 0 \\ 0 & e^{i\varphi_{\mathrm{R}}(r,\tau)\widehat{\tau}_3/2} \end{pmatrix}. \tag{4.179}$$

Then the diagonal element of $\underline{\widehat{G}}'^{-1} = \underline{\widehat{U}}\,\underline{\widehat{G}}^{-1}\underline{\widehat{U}}^{-1}$ in the L-space becomes

$$\widehat{G}_{\mathrm{L}}'^{-1}(r,\tau;r',\tau') = \left\{ \hbar\frac{\partial}{\partial\tau}\widehat{1} - i\hbar\left(v_{\mathrm{S,L}}\cdot\boldsymbol{\nabla}\right)\widehat{1} + \widehat{\Delta}_{\mathrm{L}}(r,\tau) \right. \tag{4.180}$$

$$\left. + \left[-\frac{\hbar^2\boldsymbol{\nabla}^2}{2m} - \mu + \frac{m}{2}v_{\mathrm{SL}}^2 - i\left(\frac{\hbar}{2}\frac{\partial\varphi_{\mathrm{L}}}{\partial\tau} - eU_{\mathrm{L}}(\tau)\right)\right]\widehat{\tau}_3 \right\}\delta(r-r')\,\delta(\tau-\tau'),$$

while the left-right off-diagonal elements pick up a phase,

$$\widehat{T}_{rr'}' = \begin{pmatrix} T_{rr'}\,e^{i\psi(\tau)/2} & 0 \\ 0 & -T_{rr'}^{*}\,e^{-i\psi(\tau)/2} \end{pmatrix}, \tag{4.181}$$

where $\psi(\tau) = \varphi_{\rm L}(\boldsymbol{r},\tau) - \varphi_{\rm R}(\boldsymbol{r}',\tau)$. Since $T_{\boldsymbol{r}\,\boldsymbol{r}'}$ differs essentially from zero only for $\boldsymbol{r}$, $\boldsymbol{r}'$ near to the right and left of the barrier, the phase difference across the junction $\psi(\tau) = \varphi_{\rm L}(\boldsymbol{r} \to 0_-,\tau) - \varphi_{\rm R}(\boldsymbol{r}' \to 0_+,\tau)$ enters in Eq. (4.181). Above we have introduced the gauge-invariant superfluid velocity

$$\boldsymbol{v}_{\rm S,L/R} = -\frac{\hbar}{2m}\left(\boldsymbol{\nabla}\varphi_{\rm L/R} + \frac{2e}{\hbar c}\boldsymbol{A}\right). \qquad (4.182)$$

In Eq. (4.180), the order parameter matrix in Nambu space $\widehat{\Delta}_{\rm L}$ has *real* matrix elements instead of the complex ones in Eq. (4.177).

The action $\mathcal{A}$ depends on the independent collective variables $|\Delta_{\rm L/R}|$, $\boldsymbol{A}$, U and ψ. The partition function is a functional integral over all these variables. For bulk superconductors, the quantities $|\Delta_{\rm L/R}|$, $\boldsymbol{A}$, and U have insignificant uncertainties, so that we can disregard fluctuations of these variables. This simplifies the further discussion of Z considerably. We proceed by observing the following three points.

(i) For uncoupled homogeneous bulk superconductors, the extremal condition $\delta\mathcal{A}/\delta|\Delta_{\rm L/R}| = 0$ reproduces the gap equation of the BCS theory, which yields the mean field value $|\Delta_{\rm L/R}^{\rm BCS}|$. Since the second variation $\delta^2\mathcal{A}/(\delta|\Delta_{\rm L/R}|)^2$ is proportional to the density N_0 of free electron states at $E_{\rm F}$ and proportional to the third power of the BCS coherence length, fluctuations of $|\Delta_{\rm L/R}|$ about $|\Delta_{\rm L/R}^{\rm BCS}|$ are suppressed except in a narrow temperature range just below the transition temperature $T_{\rm c}$.

(ii) Expanding the action to second order in the superfluid velocity, we obtain

$$\mathcal{A} = \mathcal{A}_0 + \int_0^{\hbar\beta} d\tau \int d^3\boldsymbol{r} \left(\frac{m}{2}\rho_{\rm S}\boldsymbol{v}_{\rm S}^2 + \frac{1}{8\pi}\left(\boldsymbol{h}(\boldsymbol{r},\tau) - \boldsymbol{h}_{\rm ext}\right)^2\right), \qquad (4.183)$$

where $\rho_{\rm S}$ is the superfluid density [141]. From this one finds that the least action condition $\delta\mathcal{A}/\delta\boldsymbol{A}(\boldsymbol{r},\tau) = 0$ reproduces the familiar law $\boldsymbol{j} = \boldsymbol{j}_{\rm ext} + \boldsymbol{j}_{\rm S}$, where $\boldsymbol{j}_{\rm ext}$ is the current density related to the external field, and where $\boldsymbol{j}_{\rm S}$ is the supercurrent

$$\boldsymbol{j}_{\rm S} = \frac{e\hbar}{2m}\rho_{\rm S}\left(\boldsymbol{\nabla}\psi + \frac{2e}{\hbar c}\boldsymbol{A}\right). \qquad (4.184)$$

In a bulk superconductor, fluctuations of $\boldsymbol{j}$ are suppressed, as follows from the study of the second variation of $\mathcal{A}$ with respect to the vector potential $\boldsymbol{A}$.

(iii) The action is extremal if the phase difference across the junction satisfies the Josephson relation (3.93). We may disregard again deviations from this mean field relation since fluctuations are suppressed by bulk energies.

Evidently, the only remaining fluctuating variable is the phase difference ψ. In a SQUID geometry where the two superconductors are joined in a loop, the corresponding fluctuating variable is the total flux ϕ through the loop. The flux ϕ is related to the phase variable ψ as given in Eq. (3.91). When the quantities $|\Delta_{\rm L/R}|$, U and $\boldsymbol{j}$ are fixed to their mean values, the multiple functional integral (4.173) is reduced to a single one over the phase variable ψ. The partition function of a junction then reads

$$Z = \oint \mathcal{D}\psi(\cdot) \exp\left\{-\frac{C}{2\hbar}\int_0^{\hbar\beta} d\tau \left(\frac{\hbar}{2e}\frac{\partial\psi}{\partial\tau}\right)^2 - \frac{1}{\hbar}\mathcal{A}_{\mathrm{T}}[\psi(\cdot)]\right\}, \qquad (4.185)$$

where $\mathcal{A}_{\mathrm{T}}[\psi(\cdot)]$ represents the tunneling contribution to the action. Working out $\mathcal{A}_{\mathrm{T}}[\psi(\cdot)]$ to second order in the (averaged) tunnel matrix element from the first term on the right hand side of (4.174), one finds[6] [141] – [143]

$$\mathcal{A}_{\mathrm{T}} = 2\int_0^{\hbar\beta} d\tau \int_0^{\hbar\beta} d\tau' \left\{\alpha(\tau-\tau')\left(1 - \cos\frac{\psi(\tau)-\psi(\tau')}{2}\right) + \beta(\tau-\tau')\cos\frac{\psi(\tau)+\psi(\tau')}{2}\right\}.$$
$$(4.186)$$

For later convenience, we extracted a factor 2 in front of the integrals, and we added a ψ-independent term in such a way that the first contribution to $\mathcal{A}_{\mathrm{T}}$ vanishes for a constant path $[\psi(\tau) = \psi(\tau')]$. The kernels $\alpha(\tau)$ and $\beta(\tau)$ are given in terms of the diagonal $(1,1)$ and off-diagonal $(1,2)$ components of the Green function $\widehat{G}$ in Nambu space, usually denoted by $G(\tau,\boldsymbol{p})$ and $F(\tau,\boldsymbol{p})$,

$$\alpha(\tau) = -\frac{|T_{\mathrm{t}}|^2}{\hbar}\int\frac{d^3\boldsymbol{p}_{\mathrm{L}}}{(2\pi\hbar)^3}\int\frac{d^3\boldsymbol{p}_{\mathrm{R}}}{(2\pi\hbar)^3} G_{\mathrm{L}}(\tau,\boldsymbol{p}_{\mathrm{L}}) G_{\mathrm{R}}(-\tau,\boldsymbol{p}_{\mathrm{R}}),$$
$$(4.187)$$
$$\beta(\tau) = -\frac{|T_{\mathrm{t}}|^2}{\hbar}\int\frac{d^3\boldsymbol{p}_{\mathrm{L}}}{(2\pi\hbar)^3}\int\frac{d^3\boldsymbol{p}_{\mathrm{R}}}{(2\pi\hbar)^3} F_{\mathrm{L}}(\tau,\boldsymbol{p}_{\mathrm{L}}) F_{\mathrm{R}}(-\tau,\boldsymbol{p}_{\mathrm{R}}).$$

Here we have assumed that the tunnel matrix element T_{t} is independent of the momentum near to the Fermi momentum p_{F}. The Green functions are given by

$$\int\frac{d^3\boldsymbol{p}}{(2\pi\hbar)^3}\binom{G}{F}(\tau,\boldsymbol{p}) = -N_0\int_{-\infty}^{\infty} d\omega\, \mathcal{N}_{\mathrm{qp}}(\omega)\binom{1}{\Delta_{\mathrm{g}}/\omega}$$
$$\times e^{-\omega\tau}\left(f(-\omega)\Theta(\tau) - f(\omega)\Theta(-\tau)\right). \qquad (4.188)$$

Here $\mathcal{N}_{\mathrm{qp}}(\omega)$ is the density of states of BCS quasi-particles given in Eq. (4.166), $f(\omega)$ denotes the Fermi distribution function and Δ_{g} is the mean field gap frequency $\Delta_{\mathrm{g}} = \Delta^{\mathrm{BCS}}/\hbar$. For notational simplicity, we now assume that $\mathcal{N}_{\mathrm{qp}}^{(\mathrm{L})}(\omega) = \mathcal{N}_{\mathrm{qp}}^{(\mathrm{R})}(\omega) = \mathcal{N}_{\mathrm{qp}}$ and $\Delta_{\mathrm{g,L}} = \Delta_{\mathrm{g,R}} = \Delta_{\mathrm{g}}$. Substituting Eq. (4.188) into Eq. (4.187) and assuming that both superconductors are described by the same mean field parameters, we obtain

$$\binom{\alpha(\tau)}{\beta(\tau)} = \frac{|T_{\mathrm{t}}|^2}{\hbar}N_0^2\int_{-\infty}^{\infty} d\omega'\int_{-\infty}^{\infty} d\omega''\, \mathcal{N}_{\mathrm{qp}}(\omega')\mathcal{N}_{\mathrm{qp}}(\omega'')\binom{1}{\Delta_{\mathrm{g}}^2/\omega'\omega''}$$
$$\times e^{(\omega''-\omega')|\tau|} f(\omega'')f(-\omega'). \qquad (4.189)$$

---

[6]This form corresponds to the leading order term of a cumulant expansion.

These expressions are similar to the previous form (4.140). Following the steps described below Eq. (4.140), we find in analogy to the expression (4.55)

$$\begin{pmatrix} \alpha(\tau) \\ \beta(\tau) \end{pmatrix} = \frac{1}{\pi} \int_0^\infty d\omega \begin{pmatrix} J_\alpha(\omega) \\ J_\beta(\omega) \end{pmatrix} D_\omega(\tau) . \qquad (4.190)$$

Here, $D_\omega(\tau)$ is the boson propagator (4.57), and $J_{\alpha(\beta)}(\omega)$ is the spectral density

$$\begin{pmatrix} J_\alpha(\omega) \\ J_\beta(\omega) \end{pmatrix} = \frac{\pi}{2} \frac{|T_t|^2}{\hbar} N_0^2 \int_{-\infty}^\infty d\omega' \, \mathcal{N}_{qp}(\omega') \mathcal{N}_{qp}(\omega' - \omega) \begin{pmatrix} 1 \\ \frac{\Delta_g^2}{\omega'(\omega'-\omega)} \end{pmatrix}$$
$$\times \left\{ \tanh(\tfrac{1}{2}\beta\hbar\omega') - \tanh[\tfrac{1}{2}\beta\hbar(\omega' - \omega)] \right\} . \qquad (4.191)$$

The physics of the terms involving $\alpha(\tau)$ and $\beta(\tau)$ is quite different. The first term in Eq. (4.186) describes dissipation because of quasiparticle tunneling. The second term describes tunneling of Cooper pairs (Josephson tunneling). In this term, the kernel $\beta(\tau - \tau')$ appears in combination with $\cos[(\psi(\tau) + \psi(\tau'))/2]$. Assuming that $\psi(\tau)$ varies slowly on the time scale on which the kernel $\beta(\tau)$ decays, the β-term reduces to an ordinary potential term, the Josephson "washboard potential" action

$$\mathcal{A}_T^{(\beta)}[\psi(\cdot)] = -I_c \frac{\hbar}{2e} \int_0^{\hbar\beta} d\tau \, \cos\psi(\tau) , \qquad (4.192)$$

where

$$I_c = -\frac{4e}{\hbar} \int_{-\hbar\beta/2}^{\hbar\beta/2} d\tau \, \beta(\tau) = -\frac{8e}{\pi\hbar} \int_0^\infty d\omega \, \frac{J_\beta(\omega)}{\omega} \qquad (4.193)$$

is the critical current of the junction. The nonlocal correction in the β-term represents the nonlocal supercurrent found by Wertheimer [184]. This contribution is usually referred to as the "cos ψ" term or quasi-particle pair interference current [141, 142]. In order to keep the discussion simple, we shall ignore this time-nonlocal current apart from a contribution to the effective capacitance (see below).

In cases where the superconducting tunnel junction is biased by an externally applied current I_{ext}, we have to add an additional potential contribution linear in ψ which tilts the Josephson washboard. At this level of approximation, the effective action of a current biased tunnel junction with capacitance C is

$$\mathcal{A}_{eff}[\psi(\cdot)] = \int_0^{\hbar\beta} d\tau \left[\frac{C}{2} \left(\frac{\hbar}{2e} \frac{\partial\psi}{\partial\tau} \right)^2 + V(\psi) \right] + \mathcal{A}_T^{(\alpha)}[\psi] ,$$
$$\mathcal{A}_T^{(\alpha)}[\psi(\cdot)] = 2 \int_0^{\hbar\beta} d\tau \int_0^{\hbar\beta} d\tau' \, \alpha(\tau - \tau') \left(1 - \cos\frac{\psi(\tau) - \psi(\tau')}{2} \right) , \qquad (4.194)$$

where $V(\psi)$ is the tilted washboard potential

$$V(\psi) = -\frac{\hbar}{2e} I_c \cos\psi - \frac{\hbar}{2e} I_{ext}\psi . \qquad (4.195)$$

The potential energy of the total flux ϕ in a SQUID geometry is given in Eq. (3.95).

The first term in $\mathcal{A}_{\text{eff}}$ accounts for the charging energy, and the potential $V(\psi)$ for the Josephson coupling and a bias term. The term $\mathcal{A}_T^{(\alpha)}$ describes state-dependent damping through quasiparticle tunneling in superconducting junctions across the barrier, or through single electrons in normal junctions. The trigonometric dependence on the phase difference describes discrete quasiparticle or single electron tunneling rather than continuous flow of charge. Only when the phase deviates weakly from an equilibrium value, we may linearize the function $\sin[(\psi(\tau) - \psi(\tau'))/2]$. Then we recover the α term exactly in the standard quadratic form (4.52), namely

$$\mathcal{A}_T^{(\alpha)}[\psi(\cdot)] = \frac{1}{2} \int_0^{\hbar\beta} d\tau \int_0^\tau d\tau' \, \alpha(\tau - \tau') \, [\psi(\tau) - \psi(\tau')]^2 \, . \tag{4.196}$$

In consideration of the analogy $\psi(\tau)/2\pi \,\hat{=}\, q(\tau)/q_0$ we see that the correspondence of the dissipation kernel for quasiparticle tunneling with that of the phenomenological model discussed in Subsection 4.2.1 is $4\pi^2\alpha(\tau) \,\hat{=}\, q_0^2 K(\tau)$. Accordingly, the correspondence of the spectral densities is $4\pi^2 J_\alpha(\omega) \,\hat{=}\, q_0^2 J(\omega)$.

Important limiting cases are:
(i) If the junction is formed by ideal BCS superconductors, as assumed in Eq. (4.191), the excitation spectrum has a gap, and $J_\alpha(\omega)$ has a behaviour similar to the one of $G_{\text{qp}}(\omega)$ sketched in Fig. 4.3. At zero temperature, we find [cf. Eq. (4.171)]

$$\alpha(\tau) = \frac{\hbar^2 \Delta_{\text{g,L}} \Delta_{\text{g,R}}}{4\pi e^2 R_N} K_1(\Delta_{\text{g,L}}|\tau|) \, K_1(\Delta_{\text{g,R}}|\tau|) \, , \tag{4.197}$$

where

$$R_N^{-1} = 4\pi N_{0,L} N_{0,R} \frac{|T|^2}{\hbar^2} \frac{e^2}{\hbar} \tag{4.198}$$

denotes the normal state conductance of the tunnel junction. For equal gaps, $\Delta_{\text{g,L}} = \Delta_{\text{g,R}} = \Delta_{\text{g}}$, the limiting cases are

$$\alpha(\tau) = \frac{\hbar^2}{4\pi e^2 R_N} \begin{cases} \dfrac{1}{\tau^2} & \text{for} \quad \Delta_{\text{g}}|\tau| \ll 1 \, , \\[2mm] \dfrac{\pi\Delta_{\text{g}}}{2|\tau|} e^{-2\Delta_{\text{g}}|\tau|} & \text{for} \quad \Delta_{\text{g}}|\tau| \gg 1 \, . \end{cases} \tag{4.199}$$

If the phase varies only slowly with time on the time scale given by $1/\Delta_{\text{g}}$, we may expand $\psi(\tau) - \psi(\tau')$ about $\tau = \tau'$. This gives

$$\mathcal{A}_T^{(\alpha)}[\psi(\cdot)] = \frac{1}{2} \int_0^{\hbar\beta} d\tau \left(\frac{\partial\psi}{\partial\tau}\right)^2 \int_0^{\hbar\beta/2} d\tau' \, \alpha(\tau')\tau'^2 \, . \tag{4.200}$$

Hence under these conditions, the α term acts as a kinetic term so that in the end the effects of the quasiparticles result in an increase of the effective capacitance. At $T = 0$ this change can easily be calculated using Eq. (4.197) and is

$$\delta C^{(\alpha)} = \frac{3\pi}{32} \frac{1}{\Delta_{\text{g}} R_N} \, . \tag{4.201}$$

The increase of the capacitance is similar to the mass renormalization in a mechanical analogue for super-Ohmic dissipation ($s > 2$) discussed in Sections 3.1.5 and 7.4. Another correction results from the β-term in Eq. (4.186), if we include the quadratic term of the expansion in $\tau - \tau'$. This gives the phase-dependent contribution

$$\delta C^{(\beta)} = -\tfrac{1}{3}\delta C^{(\alpha)} \cos \psi . \tag{4.202}$$

(ii) In a normal junction ($\Delta_{\mathrm{g}} = 0$), the kernel $\alpha(\tau)$ is Ohmic [see Eq. (4.164)]

$$\alpha(\tau) = \frac{\hbar^2}{4\pi e^2 R_{\mathrm{N}}} \frac{(\pi/\hbar\beta)^2}{\sin^2(\pi\tau/\hbar\beta)} . \tag{4.203}$$

However even in this case, there exists an important difference between dissipation by quasiparticle tunneling and by an Ohmic resistor, namely the trigonometric dependence on the phase in the former case [Eq. (4.194)] versus the quadratic dependence in the latter case [Eq. (4.196)]. Indeed one finds from microscopic transport theory for electrons in the presence of impurities that an Ohmic resistor is well described by the Caldeira-Leggett type action term (4.196) [185]. The difference reflects different physics. In the former case we have a discrete transfer of charge in units of e, while in the latter case charge is flowing continuously.

(iii) In a non-ideal junction we may have pair-breaking effects, spatial variation of the order parameter, or even locally non-perfect energy gaps leading to a finite sub-gap conductance down to $T = 0$ [143]. These effects may be accounted for either by smearing the density of states $\mathcal{N}_{\mathrm{qp}}(\omega)$ in Eq. (4.191), or by writing the kernel $\alpha(\tau)$ as a linear combination of Eq. (4.199) and Eq. (4.203).

Given the action $\mathcal{A}_{\mathrm{eff}}[\psi(\cdot)]$, one may ask whether ψ is an extended variable defined in the range $-\infty \leq \psi \leq \infty$, or a compact variable restricted either to the interval $0 < \psi \leq 2\pi$ or to $0 < \psi \leq 4\pi$. In the first case, the minima of the potential $U(\psi)$ are distinguishable, and $Q = (2e/i)\partial/\partial\psi$ is the usual charge operator with continuous eigenvalues, while in the other two cases Q is a quasicharge operator with discrete eigenvalues similar to quasimomentum of a particle in a periodic potential [143]. All these interpretations are possible and which case applies depends on the experimental conditions. If charge transport is dominated by Cooper pair tunneling, the 2π-periodic potential $U(\psi)$ determines the symmetry of the problem and it is convenient to choose a basis of corresponding Bloch states. If quasiparticle tunneling is relevant, the symmetry is reduced to 4π periodicity, which is the symmetry of $\mathcal{A}_{\mathrm{T}}^{(\alpha)}[\psi]$. Finally, if there is a continuous flow of charge, as in the Ohmic case, or in a SQUID, or when the junction is coupled to an external circuit, Bloch states are inappropriate and we have to choose a basis of continuous charge states. When ψ is defined on a ring, the path integral for the partition function generally includes a summation over the winding numbers. In many cases it is convenient to work with the restricted space of discrete charges and allow for a continuous change of charges perturbatively [143].

When quasiparticle tunneling is suppressed by the superconducting gap, higher-order processes may become relevant, e.g., correlated tunneling of two electrons across a junction with a normal and a superconducting electrode. A diagram taking into account two diagonal (G) propagators on the normal metal side and two off-diagonal (F) propagators on the superconductor side describes Andreev scattering across the interface. The resulting action is similar in form to the quasiparticle contribution $A_{\mathrm{T}}^{(\alpha)}$ in Eq. (4.194) except that it is of order $|T|^4$ and that the factor $\frac{1}{2}$ in the argument of the cosine function is missing because, instead of the charge e, the charge $2e$ is transferred in the Andreev scattering process. This implies that dissipation by Andreev reflection is very similar to dissipation by quasiparticle or single-electron tunneling [186, 187]. If the normal electrode is made of a dirty metal, an impurity averaging must be performed leading to a Cooperon propagator. The respective analysis for different junction geometries is given in Ref. [188].

4.2.11 Electromagnetic environment

It is straightforward to calculate the influence action due to an electromagnetic environment formed by LC-circuits. Upon comparing the Hamiltonian (3.154) with the oscillator model Hamiltonian (3.11) and the respective influence action (4.41) with (4.43), we find the Euclidean influence action in the form

$$S_{\mathrm{infl}}^{(\mathrm{E})}[\varphi]/\hbar = \frac{R_{\mathrm{K}}}{4\pi}\frac{1}{\hbar\beta}\sum_n\left[|\nu_n|^2 C + |\nu_n|\hat{Y}(\nu_n)\right]|\varphi_n|^2, \qquad (4.204)$$

where $\hat{Y}(z)$ is the Laplace transform of the admittance function $Y(\tau)$ and φ_n is the Fourier coefficient of the phase function in imaginary time, $\varphi(\tau) = \sum_n e^{i\nu_n\tau}\varphi_n/\hbar\beta$. For a resistive environment $\hat{Y}(z) = 1/R$ this reduces in the Ohmic scaling limit $|\nu_n| \ll 1/RC$ to the form

$$S_{\mathrm{infl}}^{(\mathrm{E})}[\varphi]/\hbar = \frac{1}{4\pi}\frac{R_{\mathrm{K}}}{R}\frac{1}{\hbar\beta}\sum_n|\nu_n|\,|\varphi_n|^2. \qquad (4.205)$$

Consider next the phase correlator $\langle e^{i[\varphi(0)-\varphi(\tau)]}\rangle_\beta$ in the imaginary-time regime $0 \le \tau < \hbar\beta$. Here, $\langle\cdots\rangle_\beta$ denotes average with the weight function $\exp[-S_{\mathrm{infl}}^{(\mathrm{E})}[\varphi]/\hbar]$. With the form (4.204), we get

$$\left\langle e^{i[\varphi(0)-\varphi(\tau)]}\right\rangle_\beta = e^{-\langle[\varphi(0)-\varphi(\tau)]\varphi(0)\rangle_\beta} = e^{-W_\varphi(\tau)} \qquad (4.206)$$

Upon completing the square in Eq. (4.206) the phase-autocorrelation function is found to be given by

$$W_\varphi(\tau) = \frac{1}{R_{\mathrm{K}}}\frac{2\pi}{\hbar\beta}\sum_n\frac{1}{\nu_n^2 C + |\nu_n|\hat{Y}(\nu_n)}\left[1 - e^{i\nu_n\tau}\right]. \qquad (4.207)$$

The Matsubara sum may be transformed by standard action into the frequency integral

$$W_\varphi(\tau) \;=\; \int_0^\infty d\omega \, \frac{G_\varphi(\omega)}{\omega^2} \, \frac{\cosh(\omega\hbar\beta/2) - \cosh[\omega(\hbar\beta/2 - \tau)]}{\sinh(\omega\hbar\beta/2)} \;, \qquad (4.208)$$

where $G_\varphi(\omega)$ is given in Eq. (3.161). The equivalence of the epression (4.208) with (4.207) follows upon writing the fraction of the hyperbolic functions as the Matsubara sum given in Eq. (5.68). For a resistive environment described by an Ohmic impedance R, this expression simplifies for $\tau \gg RC$ to the form

$$W_\varphi(\tau) \;=\; \frac{R}{R_{\rm K}} \frac{2\pi}{\hbar\beta} \sum_n \frac{1}{|\nu_n|} \left[1 - e^{i\nu_n\tau} \right] \;. \qquad (4.209)$$

We shall employ the expressions derive here in Section 20.3 and in Subsection 26.1.4, where we discuss quantum transport of charge through a weak link and a weak constriction under influence of an electromagnetic environment.

4.3   Partition function of the open system

4.3.1   General path integral expression

Quantum statistical equilibrium properties can be computed directly if the partition function Z is known as a function of temperature, volume, external field, etc. The standard thermodynamic quantities are internal energy, entropy, pressure, specific heat, and susceptibility. They are obtained by differentiation of Z with respect to the control parameters. Since the partition function is the Laplace transform of the density of states, it also contains relevant information about the spectrum of the Hamiltonian (cf. the discussion of the density of states of the damped harmonic oscillator in Sect. 6.5).

The thermodynamical key quantity of open quantum systems is the reduced partition function. The subsequent discussion is based on the Euclidean path integral

$$Z(\hbar\beta) \;=\; \oint \mathcal{D}q(\cdot) \, \exp\left(- S_{\rm eff}^{\rm (E)}[q(\cdot)]/\hbar \right) , \qquad (4.210)$$

where the symbol $\oint$ indicates that the path sum is over all periodic paths $q(\tau) = q(\tau + \hbar\beta)$. Here, $S_{\rm eff}^{\rm (E)}[q]$ is the effective action of the open system under consideration.

If not stated otherwise, we shall restrict our attention in the remainder of this chapter to the simple phenomenological model discussed previously in Subsection 4.2.1. The exact formal expression (4.210) with the effective action (4.62) finds application in diverse open quantum systems. We may use it as a starting point, e. g., for the calculation of thermodynamic properties, or, in problems connected with tunneling, for the calculation of quantum statistical decay rates of metastable states. The discussion of quantum statistical decay is postponed until Part III.

The "time-retarded" Euclidean action $S_{\text{eff}}^{(E)}[q(\cdot)]$ determines the path probability in the functional integral expression for the partition function of the open quantum system. Unfortunately, the path sum can be performed exactly only when the exponent $S_{\text{eff}}^{(E)}[q]$ is a quadratic form in the dynamical variable q. Therefore, I find it useful to expound a variety of different approaches which yield reasonable approximations over the whole temperature range. Among them are the semiclassical approximation and the variational approach to quantum-statistical mechanics. I find it convenient to defer the discussion of the latter approach until Chapter 8, which is after the discussion of the damped harmonic oscillator.

4.3.2 Semiclassical approximation

In the semiclassical limit, $S/\hbar \to \infty$, the path sum (4.61) or (4.210) is dominated by the regions about the stationary points of the action. The first variation of the action vanishes for the *extremal* path $\bar{q}(\tau)$ which obeys the equation of motion

$$-M\ddot{\bar{q}}(\tau) + V'(\bar{q}) + \int_0^{\hbar\beta} d\tau'\, k(\tau - \tau')\,\bar{q}(\tau') = 0\,. \qquad (4.211)$$

The sign of the inertia term is negative because the motion is in imaginary time. The time-nonlocal third term originates from the projection of the original energy-conserving path $\{\bar{q}(\tau),\, \bar{x}(\tau)\}$ in the $(N+1)$-dimensional $\{q, x\}$ coordinate space [see Eqs. (4.34)] onto the q-axis. Since here we are interested in the partition function, the solution of Eq. (4.211) must be periodic $\bar{q}(\tau + \hbar\beta) = \bar{q}(\tau)$. To evaluate the contributions of paths in the vicinity of $\bar{q}(\tau)$, we expand the action (4.62) about $\bar{q}(\tau)$ up to terms of second order in the fluctuations about the classical path. Putting

$$q(\tau) = \bar{q}(\tau) + y(\tau)\,, \qquad (4.212)$$

we find

$$S_{\text{eff}}^{(E)}[q(\cdot)] = S_{\text{eff}}^{(E)}[\bar{q}(\cdot)] + \frac{1}{2}M\int_0^{\hbar\beta} d\tau\, y(\tau)\Lambda[\bar{q}(\tau)]\,y(\tau) + \mathcal{O}(y^3)\,, \qquad (4.213)$$

where $\Lambda[\bar{q}(\tau)]$ is a linear operator acting in the space of $\hbar\beta$-periodic functions,

$$\Lambda[\bar{q}(\tau)]y(\tau) \equiv \left(-\frac{\partial^2}{\partial\tau^2} + \frac{1}{M}V''[\bar{q}(\tau)]\right)y(\tau) + \frac{1}{M}\int_0^{\hbar\beta} d\tau'\, k(\tau - \tau')\,y(\tau')\,. \qquad (4.214)$$

The path integral for Z is now rewritten as a functional integral over the deviations $y(\tau)$ with Gaussian weight,

$$Z = \exp\left(-S_{\text{eff}}^{(E)}[\bar{q}]/\hbar\right) \oint \mathcal{D}y(\cdot)\, \exp\left(-\frac{M}{2\hbar}\int_0^{\hbar\beta} d\tau\, y(\tau)\Lambda[\bar{q}(\tau)]y(\tau)\right)\,. \qquad (4.215)$$

Performing all the Gaussian integrals in a suitable representation, we obtain

$$Z = \frac{N}{\sqrt{D[\bar{q}]}} \exp\left(-S_{\text{eff}}^{(\text{E})}[\bar{q}]/\hbar\right) , \qquad (4.216)$$

where N is a normalization factor, and where $D[\bar{q}]$ is the determinant of the fluctuation operator $\Lambda[\bar{q}]$ on the space of $\hbar\beta$-periodic functions,

$$D[\bar{q}] \equiv \det(\Lambda[\bar{q}]) . \qquad (4.217)$$

The yet undetermined normalization factor N is universal in the sense that it neither depends on the specific form of the potential nor on the specific form of the dissipative mechanism. This is because the normalization factor N is entirely controlled by the kinetic part $-\partial^2/\partial\tau^2$ of $\Lambda[\bar{q}]$ (see Subsection 4.3.4). Therefore, in ratios of partition functions the normalization factor N drops out. Thus we have

$$Z = Z_0 \left(\frac{D_0}{D[\bar{q}]}\right)^{1/2} \exp\left(-S_{\text{eff}}^{(\text{E})}[\bar{q}]/\hbar\right) . \qquad (4.218)$$

Here for convenience, Z_0 is chosen as the partition function of a harmonic system which is free of zero modes, e.g., the undamped or damped harmonic oscillator, and D_0 is the determinant of the associated Gaussian fluctuation operator Λ_0. With the form (4.218), the remaining computational problem is reduced to the calculation of ratios of determinants. In practical computations of the ratio $D_0/D[\bar{q}]$ it is favorable to choose for D_0 the same dissipative mechanism which underlies $D[\bar{q}]$.

The semiclassical form (4.218) is a valid expression when all of the fluctuation modes can be treated in Gaussian approximation and when all eigenvalues of $\Lambda[\bar{q}]$ are positive. The appropriate treatment of modes with negative eigenvalues and with zero or quasi-zero eigenvalues is given in Part III.

4.3.3 Partition function of the damped harmonic oscillator

Consider a single-well potential with the minimum located at $q = 0$, and with the energy counted from the bottom of the well, i.e., $V(0) = 0$. We denote the frequency of small oscillations about the minimum of the well by ω_0. For this potential, $V(q) = \frac{1}{2}M\omega_0^2 q^2$, Eq. (4.211) admits only the trivial periodic solution where the particle sits for ever at the barrier top of the upside-down potential $-V(q)$. Hence we have $\bar{q}(\tau) = 0$ and $S_{\text{eff}}^{(\text{E})}[\bar{q} = 0] = 0$. The eigenvalues of the fluctuation operator

$$\Lambda_0 y(\tau) \equiv \left(-\frac{\partial^2}{\partial\tau^2} + \omega_0^2\right) y(\tau) + \frac{1}{M}\int_0^{\hbar\beta} d\tau'\, k(\tau - \tau') y(\tau') \qquad (4.219)$$

on the space of $\hbar\beta$-periodic functions are found with Eqs. (4.45) and (4.43) as

$$\Lambda_n^{(0)} = \nu_n^2 + \omega_0^2 + |\nu_n|\widehat{\gamma}(|\nu_n|) , \qquad (4.220)$$

where $\nu_n = 2\pi n/\hbar\beta$ $(n = 0, \pm 1, \pm 2, \cdots)$ is a bosonic Matsubara frequency, and where $\widehat{\gamma}(z)$ is the Laplace transform of the friction kernel $\gamma(t)$. The determinant of the fluctuation operator Λ_0 now reads

$$D_0 = \prod_{n=-\infty}^{\infty} \Lambda_n^{(0)} = \omega_0^2 \prod_{n=1}^{\infty} \left(\Lambda_n^{(0)} \right)^2 , \tag{4.221}$$

by which the partition function gets the semiclassical form

$$Z = \frac{N}{\sqrt{D_0}} = \frac{1}{\hbar\beta\omega_0} \prod_{n=1}^{\infty} \frac{\nu_n^2}{\Lambda_n^{(0)}} . \tag{4.222}$$

In the second expression, we have fixed the normalization factor N by matching (4.222) for $\widehat{\gamma} = 0$ with the partition function of the undamped oscillator

$$Z_0 = \frac{1}{2\sinh(\beta\hbar\omega_0/2)} = \frac{1}{\beta\hbar\omega_0} \prod_{n=1}^{\infty} \frac{\nu_n^2}{\omega_0^2 + \nu_n^2} . \tag{4.223}$$

For a harmonic system, all the quantum fluctuations about the classical path are exactly of Gaussian form. Therefore, we end up at the exact expression of the partition function of the damped harmonic oscillator

$$Z = \frac{1}{\beta\hbar\omega_0} \prod_{n=1}^{\infty} \frac{\nu_n^2}{\omega_0^2 + \nu_n^2 + \nu_n\widehat{\gamma}(\nu_n)} . \tag{4.224}$$

4.3.4 Functional measure in Fourier space

Lagrangians which consist of a kinetic term $\frac{1}{2}M\dot{q}^2$ and a general potential $V(q)$ are usually referred to as standard Lagrangians. Open systems described by Eqs. (4.26), (4.23) and (4.33) also belong to this class, whereas magnetic systems are non-standard. Consider now the normalized functional measure for standard Lagrangians in Fourier space. Upon expanding the $\hbar\beta$-periodic deviation $y(\tau)$ into a Fourier series

$$y(\tau) = \sum_{n=-\infty}^{\infty} y_n e^{i\nu_n\tau} , \tag{4.225}$$

the exponent of the Gaussian weight function becomes

$$\frac{M}{2\hbar} \int_0^{\hbar\beta} d\tau\, y(\tau)\Lambda_0\, y(\tau) = \frac{M\beta}{2} \sum_{n=-\infty}^{\infty} \Lambda_n^{(0)} y_n y_{-n} , \tag{4.226}$$

where Λ_0 is defined in (4.219). We now can see that the result (4.224) can be found directly from (4.215) if we define the functional measure as

$$\oint \mathcal{D}y(\cdot) \cdots \equiv \int_{-\infty}^{\infty} \frac{dy_0}{\sqrt{2\pi\hbar^2\beta/M}} \prod_{n=1}^{\infty} \left[\int_{-\infty}^{\infty} \int_{-\infty}^{\infty} \frac{dy_n\, dy_{-n}}{2\pi/(iM\beta\nu_n^2)} \right] \cdots$$

$$= \int_{-\infty}^{\infty} \frac{dy_0}{\sqrt{2\pi\hbar^2\beta/M}} \prod_{n=1}^{\infty} \left[\int_{-\infty}^{\infty} \int_{-\infty}^{\infty} \frac{d\mathrm{Re}\, y_n\, d\mathrm{Im}\, y_n}{\pi/(M\beta\nu_n^2)} \right] \cdots . \tag{4.227}$$

The measure (4.227) in quantum statistical path integrals is generally valid for standard Lagrangians. It is appropriate for Gaussian fluctuations, while modifications are necessary for modes with zero or quasi-zero eigenvalues (cf. Chapters 16 and 17).

4.3.5 Partition function of the damped harmonic oscillator revisited

The expression (4.224) for the partition function of the damped harmonic oscillator is exact, as we have already emphasized. Additional insight into the structure and specific form of the partition function of the open harmonic system is achieved by showing that the concise expression (4.224) is rediscovered via a reduction of the partition function $Z^{(\text{tot})}$ of the full system-plus-reservoir complex [189]. We have

$$ Z \;=\; Z^{(\text{tot})}/Z_{\text{R}} \,, \tag{4.228}$$

where Z_{R} is the partition function of the unperturbed reservoir given in Eq. (4.29).

To this purpose we put $V(q) = \frac{1}{2} m \omega_0^2 q^2$ in the Hamiltonian (3.11), and we introduce mass weighted coordinates,

$$ u_0 \;=\; M^{1/2} q \,, \qquad u_\alpha \;=\; m_\alpha^{1/2} x_\alpha \qquad (\alpha = 1, 2, \cdots, N) \,. \tag{4.229}$$

The resulting form of the Hamiltonian of the global system describing $N+1$ coupled stable oscillators reads

$$ H \;=\; \frac{1}{2} \sum_{\alpha=0}^{N} \sum_{\alpha'=0}^{N} [\, \delta_{\alpha\alpha'} \, \dot{u}_\alpha \dot{u}_{\alpha'} + U_{\alpha\alpha'} \, u_\alpha u_{\alpha'} \,] \,. \tag{4.230}$$

The force constant matrix U with spring coefficients $U_{\alpha\alpha'}$ is given by

$$ U \;=\; \begin{pmatrix} \widetilde{C}_0 & \widetilde{C}_1 & \widetilde{C}_2 & \cdots & \widetilde{C}_N \\ \widetilde{C}_1 & \omega_1^2 & 0 & \cdots & 0 \\ \widetilde{C}_2 & 0 & \omega_2^2 & \cdots & 0 \\ \vdots & \vdots & \vdots & \ddots & \vdots \\ \widetilde{C}_N & 0 & 0 & \cdots & \omega_N^2 \end{pmatrix} , \tag{4.231}$$

where we have put

$$ \widetilde{C}_0 \;\equiv\; \omega_0^2 + \sum_{\alpha'=1}^{N} \frac{\widetilde{C}_{\alpha'}^2}{\omega_{\alpha'}^2} \,, \qquad \widetilde{C}_\alpha \;\equiv\; -\frac{c_\alpha}{(m_\alpha M)^{1/2}} \,, \qquad (\alpha = 1, 2, \cdots, N) \,. \tag{4.232}$$

The eigenvalues of the matrix U, denoted by μ_α^2 ($\alpha = 0, 1, \cdots, N$), represent the squared eigenfrequencies of the harmonic system, and they are positive real. In the normal mode representation, the Hamiltonian (4.230) takes the form

$$ H \;=\; \frac{1}{2} \sum_{\alpha=0}^{N} \left(\dot{\rho}_\alpha^2 + \mu_\alpha^2 \rho_\alpha^2 \right) \,. \tag{4.233}$$

Expressed in terms of the set of frequencies of free vibration $\{\mu_\alpha\}$, the partition function of the global system reads

$$Z^{(\text{tot})} = \prod_{\alpha=0}^{N} \frac{1}{2 \sinh(\beta \hbar \mu_\alpha / 2)} = \prod_{\alpha=0}^{N} \left(\frac{1}{\beta \hbar \mu_\alpha} \prod_{n=1}^{\infty} \frac{\nu_n^2}{\mu_\alpha^2 + \nu_n^2} \right) , \qquad (4.234)$$

where in the second equality we have used again the infinite product representation of the hyperbolic sine function.

Consider next the dynamical matrix $\boldsymbol{A}$ for $\hbar\beta$-periodic functions in imaginary time associated with the Hamiltonian (4.230),

$$\boldsymbol{A}(\nu_n) \equiv \boldsymbol{U} + \nu_n^2 \boldsymbol{1} = \begin{pmatrix} \overline{C}_0 & \tilde{C}_1 & \tilde{C}_2 & \cdots & \tilde{C}_N \\ \tilde{C}_1 & \Omega_1^2 & 0 & \cdots & 0 \\ \tilde{C}_2 & 0 & \Omega_2^2 & \cdots & 0 \\ \vdots & \vdots & \vdots & \ddots & \vdots \\ \tilde{C}_N & 0 & 0 & \cdots & \Omega_N^2 \end{pmatrix} , \qquad (4.235)$$

where

$$\overline{C}_0 = \tilde{C}_0 + \nu_n^2 , \qquad \text{and} \qquad \Omega_\alpha^2 = \omega_\alpha^2 + \nu_n^2 . \qquad (4.236)$$

In diagonal basis, the determinant of $\boldsymbol{A}(\nu_n)$ has the form

$$\det \boldsymbol{A}(\nu_n) = \prod_{\alpha=0}^{N} (\mu_\alpha^2 + \nu_n^2) . \qquad (4.237)$$

On the other hand, we may calculate $\det \boldsymbol{A}(\nu_n)$ by employing the matrix representation (4.235). We then find

$$\det \boldsymbol{A}(\nu_n) = \left(\overline{C}_0 - \sum_{\alpha'=1}^{N} \frac{\tilde{C}_{\alpha'}^2}{\Omega_{\alpha'}^2} \right) \prod_{\alpha=1}^{N} \Omega_\alpha^2 . \qquad (4.238)$$

By means of the definitions (4.232) and (4.236), this expression is transformed into

$$\det \boldsymbol{A}(\nu_n) = \left(\omega_0^2 + \nu_n^2 + \frac{\nu_n^2}{M} \sum_{\alpha'=1}^{N} \frac{c_{\alpha'}^2}{m_{\alpha'} \omega_{\alpha'}^2} \frac{1}{(\omega_{\alpha'}^2 + \nu_n^2)} \right) \prod_{\alpha=1}^{N} \left(\omega_\alpha^2 + \nu_n^2 \right)$$

$$= \left(\omega_0^2 + \nu_n^2 + |\nu_n| \hat{\gamma}(|\nu_n|) \right) \prod_{\alpha=1}^{N} \left(\omega_\alpha^2 + \nu_n^2 \right) . \qquad (4.239)$$

In the second form, we have used the representation (3.34) for the Laplace transform $\hat{\gamma}(\nu_n)$ of the damping kernel $\gamma(t)$. Upon equating (4.237) with (4.239), we get

$$\prod_{\alpha=0}^{N} (\mu_\alpha^2 + \nu_n^2) = \left(\omega_0^2 + \nu_n^2 + |\nu_n| \hat{\gamma}(|\nu_n|) \right) \prod_{\alpha=1}^{N} \left(\omega_\alpha^2 + \nu_n^2 \right) . \qquad (4.240)$$

Putting $n = 0$ in Eq. (4.240) and observing that $\nu_0 = 0$, we see that the product of the eigenfrequencies is equal to the product of the original system and bath frequencies,

$$\prod_{\alpha=0}^{N} \mu_\alpha^2 = \omega_0^2 \prod_{\alpha=1}^{N} \omega_\alpha^2 . \tag{4.241}$$

Upon using the relations (4.240) and (4.241), we find from Eq. (4.234)

$$Z^{(\text{tot})} = \prod_{\alpha=1}^{N} \left\{ \frac{1}{\beta\hbar\omega_\alpha} \prod_{n=1}^{\infty} \frac{\nu_n^2}{\omega_\alpha^2 + \nu_n^2} \right\} \frac{1}{\beta\hbar\omega_0} \prod_{n=1}^{\infty} \frac{\nu_n^2}{\omega_0^2 + \nu_n^2 + \nu_n\widehat{\gamma}(\nu_n)} . \tag{4.242}$$

Now, the N-fold product of the curly bracket term is the partition function Z_R of the reservoir, Eq. (4.29), and the residual set of factors is just the product representation of the reduced partition function Z of the damped quantum oscillator. Thus indeed, we find the partition function of the global system in the form

$$Z^{(\text{tot})} = Z_\text{R} Z . \tag{4.243}$$

In conclusion, we have shown by two different lines of reasoning that the partition function of the damped system is given by the expression (4.224).

4.4   Quantum statistical expectation values in phase space

We may ask ourselves if there is a phase space representation of quantum statistical expectation values[7] for general operator functions $\hat{A}(\hat{p}, \hat{q})$. A classical particle is described by a probability density function in phase space $f^{(\text{cl})}(p, q)$. The average of a function of momentum and position can then be expressed as

$$\langle A \rangle_\text{cl} = \iint \frac{dp\, dq}{2\pi\hbar} A(p, q) f^{(\text{cl})}(p, q) . \tag{4.244}$$

All integrations in this section will be from $-\infty$ to $+\infty$. Because of the uncertainty principle, one cannot define a true phase space probability distribution for a quantum mechanical particle. Nevertheless, "quasiprobability distribution functions" which bear resemblance to classical phase space distribution functions are useful not only as a calculational tool but can also elucidate the nature of quantum mechanics. With the use of a quasiprobability distribution $f^{(\text{qm})}(p, q)$, the quantum statistical average takes the form

$$\langle A \rangle_\text{qm} \equiv \operatorname{tr}(\hat{A}\hat{\rho}) = \iint \frac{dp\, dq}{2\pi\hbar} A(p, q) f^{(\text{qm})}(p, q) . \tag{4.245}$$

First of all, the question arises whether an operator function $\hat{A}(\hat{p}, \hat{q})$ which is defined in terms of a certain ordering prescription can unambiguously be represented as

---

[7]There is a neat monography by W. P. Schleich on quantum physics in phase space [190].

an ordinary function in phase space. One positive answer is the generalized Weyl correspondence. This is discussed in the following subsection.

Since the pioneering work by Weyl [191], the association of distribution functions in phase space with operator ordering rules for operators $\hat{p}$ and $\hat{q}$ has been the subject of many studies (cf. the review Ref. [192]). The quasiprobability distribution which is appropriate to the Weyl correspondence between operator-valued and ordinary functions is the Wigner distribution $f^{(W)}(p, q)$ [193].

4.4.1 Generalized Weyl correspondence

We want to establish the correspondence between quantum mechanical operators $\hat{A}(\hat{p}, \hat{q})$ and ordinary functions $A(p, q)$ in phase space. As a specific example, let us consider a particle in one space dimension. The generalization to a coordinate space with d dimensions and curvature is discussed in Ref. [194]. We start with expanding the operator $\hat{A}(\hat{p}, \hat{q})$ in position and momentum eigenstates,

$$\hat{A}(\hat{p}, \hat{q}) = \int dp' \, dp'' \, dq' \, dq'' \, |p''><p''|q''><q''|\hat{A}(\hat{p}, \hat{q})|q'><q'|p'><p'| \, , \quad (4.246)$$

and consider a one-parameter family of transformations of the phase space variables with Jacobian unity,

$$\begin{aligned} p'' &= p + \tfrac{1}{2}(1 - \gamma)\eta \,; & p' &= p - \tfrac{1}{2}(1 + \gamma)\eta \, , \\ q'' &= q + \tfrac{1}{2}(1 + \gamma)y \,; & q' &= q - \tfrac{1}{2}(1 - \gamma)y \, . \end{aligned} \quad (4.247)$$

The range of the parameter γ is $-1 \leq \gamma \leq 1$. Upon using

$$< q|q' > = \delta(q - q') \,; \quad < p|p' > = \delta(p - p') \,; \quad < q|p > = (2\pi\hbar)^{-1/2} e^{ipq/\hbar} \, , \quad (4.248)$$

we then get

$$\hat{A}(\hat{p}, \hat{q}) = \int\int \frac{dp \, dq}{2\pi\hbar} \, A_\gamma(p, q) \hat{\Delta}(p, q) \, , \quad (4.249)$$

where

$$A_\gamma(p, q) = \int dy \, e^{-ipy/\hbar} < q + \tfrac{1}{2}(1 + \gamma)y \, | \, \hat{A}(\hat{p}, \hat{q}) \, | \, q - \tfrac{1}{2}(1 - \gamma)y > \, , \quad (4.250)$$

$$\hat{\Delta}(p, q) = \int d\eta \, e^{-iq\eta/\hbar} |p + \tfrac{1}{2}(1 - \gamma)\eta > < p - \tfrac{1}{2}(1 + \gamma)\eta| \, . \quad (4.251)$$

The function $A_\gamma(p, q)$ is the generalized Weyl transform of the operator-valued function $\hat{A}(\hat{p}, \hat{q})$. The correspondence is one-to-one and is denoted by the symbol $\longleftrightarrow$. The functional form of $A_\gamma(p, q)$ depends on the ordering of the operators $\hat{p}$ and $\hat{q}$ in $\hat{A}(\hat{p}, \hat{q})$, and on the parameter γ. By calculation one finds, e.g., the correspondence

$$\hat{p}^m \hat{F}(\hat{q}) \hat{p}^n \quad \longleftrightarrow \quad \left(p - i\hbar \frac{1 - \gamma}{2} \frac{\partial}{\partial q}\right)^m \left(p + i\hbar \frac{1 + \gamma}{2} \frac{\partial}{\partial q}\right)^n F(q) \, . \quad (4.252)$$

Thus, the ordinary function in phase space usually differs from the corresponding operator function. It is useful to define a one-parameter family of operator ordering schemes in such a way that the functional form of the operator function coincides exactly with the ordinary function in phase space. Denoting this family symbolically by $\mathcal{Z}_\gamma\{\hat{A}(\hat{p},\hat{q})\}$, we thus have the direct correspondence

$$\mathcal{Z}_\gamma\{\hat{A}(\hat{p},\hat{q})\} \longleftrightarrow A(p,q) . \tag{4.253}$$

By definition, $\mathcal{Z}_\gamma$ is an operator which orders the operators $\hat{p}$ and $\hat{q}$ in an operator function $\hat{A}(\hat{p},\hat{q})$ regardless of the commutation relations such that the correspondence (4.253) holds. There follows with Eq. (4.252) that the operator corresponding to the transform $p^m F(q)$ is given by

$$\mathcal{Z}_\gamma\{\hat{p}^m \hat{F}(\hat{q})\} =: \sum_{l=0}^{m} \binom{m}{l} \left(\frac{1+\gamma}{2}\hat{p}\right)^{m-l} \hat{F}(\hat{q}) \left(\frac{1-\gamma}{2}\hat{p}\right)^{l} . \tag{4.254}$$

With this form, we can find $\mathcal{Z}_\gamma\{\hat{A}(\hat{p},\hat{q})\}$ for any operator function $\hat{A}(\hat{p},\hat{q})$ of which the Weyl transform can be represented in the form of a power series in the momentum, $A(p,q) = \sum_j F_j(q)\,p^j$,

$$\mathcal{Z}_\gamma\{\hat{A}(\hat{p},\hat{q})\} =: \sum_{j}\sum_{l=0}^{j} \binom{j}{l} \left(\frac{1+\gamma}{2}\hat{p}\right)^{j-l} \hat{F}_j(\hat{q}) \left(\frac{1-\gamma}{2}\hat{p}\right)^{l} . \tag{4.255}$$

Equation (4.254) or (4.255) defines the ordering prescription for a continuous parameter γ in the range $-1 \leq \gamma \leq 1$. The special case $\gamma = 0$ is the case of Weyl ordering. We have, e.g., for $m = 2$ in Eq. (4.254)

$$\mathcal{Z}_0\{\hat{p}^2 \hat{F}(\hat{q})\} =: \frac{1}{4}\left(\hat{p}^2 \hat{F}(\hat{q}) + 2\hat{p}\hat{F}(\hat{q})\hat{p} + \hat{F}(\hat{q})\hat{p}^2\right) . \tag{4.256}$$

The cases $\gamma = 1$ and $\gamma = -1$ represent the antistandard ($\hat{p}\hat{q}$) and standard ($\hat{q}\hat{p}$) ordering prescription, respectively. We have from Eqs. (4.251) and (4.248) in the coordinate representation

$$< q''|\hat{\Delta}_\gamma(p,x)|q' > = \delta(x-q)\,e^{ip(q''-q')/\hbar} , \tag{4.257}$$

where
$$q = \tfrac{1}{2}(q''+q') - \tfrac{1}{2}\gamma(q''-q') . \tag{4.258}$$

Notice that q depends on the endpoints q'' and q', and on the continuous parameter γ. We find from Eq. (4.249) with (4.257) and (4.253) the coordinate representation

$$< q''|\mathcal{Z}_\gamma\{\hat{A}(\hat{p},\hat{q})\}|q' > = (2\pi\hbar)^{-1}\int dp\, A(p,q)\,e^{ip(q''-q')/\hbar} . \tag{4.259}$$

This may be inverted to yield

$$A(p,q) = \int dy\, e^{-ipy/\hbar} < q + \tfrac{1}{2}(1+\gamma)y\,|\mathcal{Z}_\gamma\{\hat{A}(\hat{p},\hat{q})\}|\,q - \tfrac{1}{2}(1-\gamma)y > . \qquad (4.260)$$

Thus, the generalized Weyl transform $A(p,q)$ is calculated by regarding the matrix element $< q''|\mathcal{Z}_\gamma\{\hat{A}(\hat{p},\hat{q})\}|q' >$ as a function of q and y, and then taking the Fourier transform with respect to y. For symmetric (Weyl) ordering, $\gamma = 0$, we have

$$A(p,q) = \int dy\, e^{-ipy/\hbar} < q + \tfrac{1}{2}y|\mathcal{Z}_0\{\hat{A}(\hat{p},\hat{q})\}|q - \tfrac{1}{2}y > . \qquad (4.261)$$

In summary of the results obtained so far, we have found for the operator ordering prescription (4.254) a unique representation in phase space.

4.4.2 Generalized Wigner function and expectation values

Wigner's function $f^{(W)}(p,q)$ is a phase space representation of the reduced density matrix (4.27) in thermal equilibrium. Wigner's function is derived by regarding $\rho_\beta(q'',q')$ as a function of $y = q'' - q'$ and $q = \tfrac{1}{2}(q'' + q')$, and then taking the Fourier transform with respect to y

$$f^{(W)}(p,q) = \int dy\, e^{-ipy/\hbar} \rho_\beta(q + \tfrac{1}{2}y,\, q - \tfrac{1}{2}y) . \qquad (4.262)$$

The quasiprobability distribution $f^{(W)}(p,q)$ is the appropriate density function in phase space for operator functions with Weyl ordering.

Since we aim at the phase space representation of quantum statistical expectation values for general operator functions $\mathcal{Z}_\gamma\{\hat{A}(\hat{p},\hat{q})\}$, we have to introduce a slight generalization of Wigner's function (4.262). It is convenient to define

$$f^{(\gamma)}(p,\bar{q}) \equiv \int dy\, e^{-ipy/\hbar}\, \rho_\beta\left(\bar{q} + \tfrac{1}{2}(1-\gamma)y,\, \bar{q} - \tfrac{1}{2}(1+\gamma)y\right) . \qquad (4.263)$$

Again $y = q'' - q'$, but instead of q given in (4.258) we use as second variable

$$\bar{q} = \tfrac{1}{2}(q'' + q') + \tfrac{1}{2}\gamma(q'' - q') , \qquad (4.264)$$

which differs from the expression for q by the replacement $\gamma \rightarrow -\gamma$. The reason for this subtle change will become clear shortly. For $\gamma = 0$, the generalized Wigner function $f^{(\gamma)}(p,\bar{q})$ coincides with $f^{(W)}(p,q)$. The expression (4.263) is inverted to yield

$$\rho_\beta(q'',q') = (2\pi\hbar)^{-1} \int dp\, f^{(\gamma)}(p,\bar{q})\, e^{ip(q''-q')/\hbar} . \qquad (4.265)$$

The quasiprobability distributions $f^{(W)}(p,q)$ and $f^{(\gamma)}(p,\bar{q})$ have properties similar to the classical density function $f(p,q)$ in phase space,

$$(2\pi\hbar)^{-1}\int dp\, f^{(W)}(p,q) \;=\; (2\pi\hbar)^{-1}\int dp\, f^{(\gamma)}(p,q) \;=\; <q|\hat{\rho}_{\beta}|q> \;\equiv\; P(q)\,,$$

$$\int dq\, f^{(W)}(p,q) \;=\; \int d\bar{q}\, f^{(\gamma)}(p,\bar{q}) \;=\; <p|\hat{\rho}_{\beta}|p> \;\equiv\; \tilde{P}(p)\,. \qquad (4.266)$$

The diagonal elements $P(q')$ and $\tilde{P}(p)$ represent the probability for finding the damped particle in thermal equilibrium at the point q' and with momentum p, respectively. Although the quasiprobability distribution $f^{(W)}(p,q)$ $[f^{(\gamma)}(p,\bar{q})]$ satisfies (4.266), it does not describe the probability for finding the particle at the position q $[\bar{q}]$ and with the momentum p, because the function can become negative for some values of p and q $[\bar{q}]$.

We are now ready to consider the phase space representation of the quantum statistical average of the operator $\mathcal{Z}_{\gamma}\{\hat{A}(\hat{p},\hat{q})\}$,

$$\langle A_{\gamma}\rangle =: \left\langle \mathcal{Z}_{\gamma}\{\hat{A}(\hat{p},\hat{q})\}\right\rangle \;=\; \int dq'' \int dq' <q''|\mathcal{Z}_{\gamma}\{\hat{A}(\hat{p},\hat{q})\}|q'> \rho_{\beta}(q',q'')\,. \qquad (4.267)$$

Because $\rho_{\beta}(q',q'')$, instead of $\rho_{\beta}(q'',q')$, appears in Eq. (4.267), we used Eq. (4.264) in Eq. (4.265), instead of Eq. (4.258). As a result, the Wigner function related to $\rho_{\beta}(q',q'')$ is a function of q instead of $\bar{q}$. Substituting (4.259) and (4.265) we find

$$\langle A_{\gamma}\rangle \;=\; \int\int \frac{dp'\,dq}{2\pi\hbar}\int\int \frac{dp\,dy}{2\pi\hbar}\, A(p,q)f^{(\gamma)}(p',q)\,\mathrm{e}^{iy(p-p')/\hbar}\,, \qquad (4.268)$$

where we have interchanged q'' with q' in Eq. (4.265) and observed Eqs. (4.264) and (4.258). Since the y integral gives $\delta(p-p')$, we find in the end

$$\langle A_{\gamma}\rangle \;=\; \int\int \frac{dp\,dq}{2\pi\hbar}\, A(p,q)f^{(\gamma)}(p,q)\,. \qquad (4.269)$$

Thus, in the case where the quasiprobability distribution in Eq. (4.245) is chosen to be $f^{(\gamma)}(p,q)$, the correspondence between $A(p,q)$ and $\hat{A}(\hat{p},\hat{q})$ is that given by Eq. (4.253). Notice that Eq. (4.269) is true for the family of operator ordering prescriptions $\mathcal{Z}_{\gamma}\{\hat{A}(\hat{p},\hat{q})\}$ among which Weyl ordering, standard ordering, and antistandard ordering are special cases. The main criterion for the choice of a quasiprobability distribution function for a particular problem is convenience.

The phase space representation of quantum statistical expectation values is especially convenient if the density matrix in coordinate representation is already known. Then the remaining problem is limited to integrations of ordinary functions. We will make use of the result (4.269) in Section 8.2.

5. Real–time path integrals and dynamics

In the preceding chapter, I have formulated quantum statistical mechanics for a number of systems in contact with a thermal reservoir using the imaginary-time or Euclidean path integral representation. I have also shown some techniques useful in the approximate evaluation of path integrals. In this chapter we are concerned with the description of non-equilibrium time-dependent phenomena. I shall outline the general formalism with the simple phenomenological model introduced in Section 3.1. The proceeding will also illuminate the appropriate modifications for the various microscopic models introduced previously. Thus, I take them up only briefly in the summary at the end of the chapter. Since we are interested in the evolution of mixed states, the proper vehicle is the density matrix representation.

In the first section, I study the evolution of the damped system under the assumption that it was prepared initially in a product state of the system-plus-reservoir complex. In the second section, I squeeze in a brief discussion on the physical significance of the influence functional obtained in the first section. The third section deals with several other initial preparations which are appropriate in specific experiments. The general concepts will be applied in the fourth section to describe the evolution of a damped system for any form of initial preparation. This leads, in general, to a complex-time triple path integral. In the last section, I show that for ergodic systems we can regain a real time path integral, but for paths evolving from the infinite past. I also discuss the semiclassical limit of the path integral. This will end up in a Langevin-type description of the stochastic process with a quantum mechanical power spectrum of the random force.

Let us examine the possibility of obtaining closed expressions for the dynamics of a dissipative quantum mechanical system. We assume that the underlying global system is governed by the Hamiltonian (3.11). We shall see that the Hamiltonian dynamics of the global system induces a non-unitary dynamics for the reduced density matrix. Like in the static problem discussed in the preceding section, it is convenient to carry out the reduction within the functional integral description, a technique introduced by Feynman and Vernon [195] already in 1963. Our starting point in performing the reduction is the operator relation for the density operator of the global system

$$\hat{W}(t) = e^{-i\hat{H}t/\hbar} \, \hat{W}(0) \, e^{i\hat{H}t/\hbar} \, . \tag{5.1}$$

In coordinate representation we may write

$$< q_f, \boldsymbol{x}_f | \, \hat{W}(t) \, | q'_f, \boldsymbol{x}'_f > \; = \; \int dq_i \, dq'_i \, d\boldsymbol{x}_i \, d\boldsymbol{x}'_i \, K(q_f, \boldsymbol{x}_f, t; q_i, \boldsymbol{x}_i, 0) \tag{5.2}$$

$$\times \; < q_i, \boldsymbol{x}_i | \, \hat{W}(0) \, | q'_i, \boldsymbol{x}'_i > K^*(q'_f, \boldsymbol{x}'_f, t; q'_i, \boldsymbol{x}'_i, 0) \, .$$

Here, the N-component vector $\boldsymbol{x}_{i/f}$ stands again for $(x_{i/f,1}, \ldots, x_{i/f,N})$, and K is the coordinate representation of the time evolution operator

$$K(q_f, \boldsymbol{x}_f, t; q_i, \boldsymbol{x}_i, 0) \; = \; < q_f, \boldsymbol{x}_f | \, e^{-i\hat{H}t/\hbar} | q_i, \boldsymbol{x}_i > \, , \tag{5.3}$$

which may be represented as a path integral,

$$
\begin{aligned}
K(q_f, \boldsymbol{x}_f, t; q_i, \boldsymbol{x}_i, 0) &= \int \mathcal{D}q\, \mathcal{D}\boldsymbol{x}\ e^{iS[q, \boldsymbol{x}]/\hbar}\ , \\
K^*(q'_f, \boldsymbol{x}'_f, t; q'_i, \boldsymbol{x}'_i, 0) &= \int \mathcal{D}q'\, \mathcal{D}\boldsymbol{x}'\ e^{-iS[q', \boldsymbol{x}']/\hbar}\ .
\end{aligned}
\tag{5.4}
$$

The functional integrations in Eq. (5.4) extend over all paths with endpoints

$$
q(0) = q_i\ , \quad q(t) = q_f\ ; \quad q'(0) = q'_i\ , \quad q'(t) = q'_f\ , \tag{5.5}
$$

and

$$
\boldsymbol{x}(0) = \boldsymbol{x}_i\ , \quad \boldsymbol{x}(t) = \boldsymbol{x}_f\ ; \quad \boldsymbol{x}'(0) = \boldsymbol{x}'_i\ , \quad \boldsymbol{x}'(t) = \boldsymbol{x}'_f\ . \tag{5.6}
$$

The action in Eq. (5.4) is given by

$$
S = S_{\mathrm{S}} + S_{\mathrm{R}} + S_{\mathrm{I}} = \int_0^t dt'\left(\mathcal{L}_{\mathrm{S}}(t') + \mathcal{L}_{\mathrm{R}}(t') + \mathcal{L}_{\mathrm{I}}(t')\right)\ , \tag{5.7}
$$

where [we add a source term in $\mathcal{L}_{\mathrm{S}}(t)$, as in Eq. (4.20)]

$$
\begin{aligned}
\mathcal{L}_{\mathrm{S}}(t) &= \frac{1}{2}M\dot{q}^2(t) - V[q(t)] - \mathcal{J}(t)q(t)\ , \\
\mathcal{L}_{\mathrm{R}}(t) &= \frac{1}{2}\sum_{\alpha=1}^N m_\alpha\left[\dot{x}_\alpha^2(t) - \omega_\alpha^2 x_\alpha^2(t)\right]\ , \\
\mathcal{L}_{\mathrm{I}}(t) &= \sum_{\alpha=1}^N \left(c_\alpha x_\alpha(t)q(t) - \frac{1}{2}\frac{c_\alpha^2}{m_\alpha \omega_\alpha^2}q^2(t)\right)\ .
\end{aligned}
\tag{5.8}
$$

The expression (5.2) for the density matrix describes the dynamics of the system-plus-environment complex as a whole. However, in most cases of interest the only information we wish to have is the system's dynamics under the reservoir's influence. Then the quantity we are really interested in is the reduced density matrix [11, 23, 22]

$$
\begin{aligned}
\rho(q_f, q'_f; t) &\equiv \int d\boldsymbol{x}_f < q_f, \boldsymbol{x}_f|\,\hat{W}(t)\,|q'_f, \boldsymbol{x}_f > \\
&= \int dq_i\, dq'_i\, d\boldsymbol{x}_i\, d\boldsymbol{x}'_i\, d\boldsymbol{x}_f\, K(q_f, \boldsymbol{x}_f, t; q_i, \boldsymbol{x}_i, 0)
\end{aligned}
\tag{5.9}
$$

$$
\times\ < q_i, \boldsymbol{x}_i|\,\hat{W}(0)\,|q'_i, \boldsymbol{x}'_i > K^*(q'_f, \boldsymbol{x}_f, t; q'_i, \boldsymbol{x}'_i, 0)\ .
$$

We are now ready to work out the path integral for the reduced density matrix.

5.1   Feynman–Vernon method for a product initial state

Suppose that the density operator of the global system at the initial time $t = 0$ is in product form, and assume that the system and the bath are decoupled, and the unperturbed bath is in thermal equilibrium. We then have

$$
\hat{W}_{\mathrm{fc}}(0) = \hat{\rho}(0) \otimes \hat{W}_{\mathrm{R}}(0) = \hat{\rho}(0) \otimes Z_{\mathrm{R}}^{-1}\, e^{-\beta \hat{H}_{\mathrm{R}}}\ . \tag{5.10}
$$

where $\hat{\rho}$ is the reduced density operator. The initial state (5.10) is free of correlations between the system and the bath. The product form is very convenient for computations. However, since the system-bath coupling is not at the disposal of the experimentalist, the product initial state is somewhat artificial in many applications.

Assume that the system-bath coupling is suddenly switched on at time $t = 0^+$ and consider the dynamics of $\rho(t)$ for $t \geq 0$. Substituting Eq. (5.10) in Eq. (5.9) and using Eqs. (5.4) – (5.8), we obtain for the reduced density matrix the expression

$$\rho(q_f, q_f'; t) = \int dq_i \, dq_i' \, J_{FV}(q_f, q_f', t; q_i, q_i', 0) \, \rho(q_i, q_i'; 0) , \tag{5.11}$$

where J_{FV} is a *propagating function* describing the time evolution of the reduced density matrix. The path integral for the propagating function is

$$J_{FV}(q_f, q_f', t; q_i, q_i', 0) = \int \mathcal{D}q \, \mathcal{D}q' \, \exp\left\{\frac{i}{\hbar}\left(S_S[q] - S_S[q']\right)\right\} \mathcal{F}_{FV}[q, q'] . \tag{5.12}$$

The still formal expression

$$\begin{aligned}
\mathcal{F}_{FV}[q, q'] &\equiv \int d\boldsymbol{x}_f \, d\boldsymbol{x}_i \, d\boldsymbol{x}_i' \, W_R(\boldsymbol{x}_i, \boldsymbol{x}_i') \int_{\boldsymbol{x}(0)=\boldsymbol{x}_i}^{\boldsymbol{x}(t)=\boldsymbol{x}_f} \mathcal{D}\boldsymbol{x} \int_{\boldsymbol{x}'(0)=\boldsymbol{x}_i'}^{\boldsymbol{x}'(t)=\boldsymbol{x}_f} \mathcal{D}\boldsymbol{x}' \\
&\times \exp\left\{\frac{i}{\hbar}\left(S_R[\boldsymbol{x}] + S_I[\boldsymbol{x}, q] - S_R[\boldsymbol{x}'] - S_I[\boldsymbol{x}', q']\right)\right\}
\end{aligned} \tag{5.13}$$

is the so-called Feynman-Vernon influence functional [195]. This can be rewritten as

$$\mathcal{F}_{FV}[q, q'] = \int d\boldsymbol{x}_f \, d\boldsymbol{x}_i \, d\boldsymbol{x}_i' \, W_R(\boldsymbol{x}_i, \boldsymbol{x}_i') \, F(q; \boldsymbol{x}_f, \boldsymbol{x}_i) \, F^*(q'; \boldsymbol{x}_f, \boldsymbol{x}_i') , \tag{5.14}$$

where

$$F[q(\cdot); \boldsymbol{x}_f, \boldsymbol{x}_i] = \int_{\boldsymbol{x}(0)=\boldsymbol{x}_i}^{\boldsymbol{x}(t)=\boldsymbol{x}_f} \mathcal{D}\boldsymbol{x}(\cdot) \, \exp\left\{\frac{i}{\hbar}\left(S_R[\boldsymbol{x}(\cdot)] + S_I[\boldsymbol{x}(\cdot), q(\cdot)]\right)\right\} \tag{5.15}$$

is a real-time functional integral over the path $\boldsymbol{x}(t')$ with endpoints $\boldsymbol{x}(0) = \boldsymbol{x}_i$ and $\boldsymbol{x}(t) = \boldsymbol{x}_f$. For the reservoir model described by $\mathcal{L}_R$ in Eq. (5.8), both the canonical density matrix W_R and the functional F are in product form,

$$\begin{aligned}
W_R(\boldsymbol{x}_i, \boldsymbol{x}_i') &= \prod_{\alpha=1}^{N} W_R^{(\alpha)}(x_{i,\alpha}, x_{i,\alpha}') , \\
F[q(\cdot); \boldsymbol{x}_f, \boldsymbol{x}_i] &= \prod_{\alpha=1}^{N} F_\alpha[q(\cdot); x_{f,\alpha}, x_{i,\alpha}] ,
\end{aligned} \tag{5.16}$$

where $W_{\rm R}^{(\alpha)}$ is the canonical density matrix for the αth oscillator, and F_α is the propagator of this oscillator under the influence of an external force $c_\alpha q(t')$. A standard calculation, which is an exercise in doing Gaussian integrals, gives [195, 86]

$$W_{\rm R}^{(\alpha)}(x_{i,\alpha}, x_{i,\alpha}') = \left(\frac{m_\alpha \omega_\alpha}{2\pi\hbar \sinh(\beta\hbar\omega_\alpha)}\right)^{1/2} \frac{1}{Z_{\rm R}^{(\alpha)}} \tag{5.17}$$

$$\times \ \exp\left\{-\frac{m_\alpha \omega_\alpha}{2\hbar \sinh(\beta\hbar\omega_\alpha)}\left[(x_{i,\alpha}^2 + x_{i,\alpha}'^2)\cosh(\beta\hbar\omega_\alpha) - 2x_{i,\alpha}x_{i,\alpha}'\right]\right\},$$

$$F_\alpha[q(\cdot); x_{f,\alpha}, x_{i,\alpha}] = \left(\frac{m_\alpha \omega_\alpha}{2\pi i\hbar \sin(\omega_\alpha t)}\right)^{1/2} \exp\left(\frac{i}{\hbar}\phi_\alpha[q(\cdot); x_{f,\alpha}, x_{i,\alpha}]\right), \tag{5.18}$$

where

$$\phi_\alpha[q(\cdot); x_{f,\alpha}, x_{i,\alpha}] = \frac{m_\alpha \omega_\alpha}{2\sin(\omega_\alpha t)}\left[(x_{i,\alpha}^2 + x_{f,\alpha}^2)\cos(\omega_\alpha t) - 2x_{i,\alpha}x_{f,\alpha}\right] \tag{5.19}$$

$$+ \ \frac{x_{i,\alpha}c_\alpha}{\sin(\omega_\alpha t)}\int_0^t dt' \ \sin\left(\omega_\alpha(t - t')\right)q(t')$$

$$+ \ \frac{x_{f,\alpha}c_\alpha}{\sin(\omega_\alpha t)}\int_0^t dt' \ \sin(\omega_\alpha t')q(t') \ - \ \frac{c_\alpha^2}{2m_\alpha\omega_\alpha^2}\int_0^t dt' \ q^2(t')$$

$$- \ \frac{c_\alpha^2}{m_\alpha\omega_\alpha \sin(\omega_\alpha t)}\int_0^t dt'\int_0^{t'} dt'' \sin[\omega_\alpha(t - t')]\sin(\omega_\alpha t'')\,q(t')\,q(t'')$$

is the action in the presence of the external force. Further, the partition function of the unperturbed reservoir is given by

$$Z_{\rm R} = \prod_{\alpha=1}^N Z_{\rm R}^{(\alpha)} = \prod_{\alpha=1}^N \left\{2\sinh(\beta\hbar\omega_\alpha/2)\right\}^{-1}. \tag{5.20}$$

Substituting Eqs. (5.16) – (5.19) into Eq. (5.14), we obtain Gaussian integrals over the endpoints that are easily evaluated. In the end, we find that the pre-exponential factors complement each other to a factor unity. The resulting expression for the Feynman-Vernon influence functional is conveniently written as

$$\mathcal{F}_{\rm FV}[q(\cdot), q'(\cdot)] = \exp\left\{-\mathcal{S}_{\rm FV}[q(\cdot), q'(\cdot)]/\hbar\right\}, \tag{5.21}$$

where the influence action is given by

$$\mathcal{S}_{\rm FV}[q, q'] = \int_0^t dt'\int_0^{t'} dt'' \left\{q(t') - q'(t')\right\}\left\{L(t' - t'')\,q(t'') - L^*(t' - t'')\,q'(t'')\right\}$$

$$+ \ i\frac{\mu}{2}\int_0^t dt' \left\{q^2(t') - q'^2(t')\right\}. \tag{5.22}$$

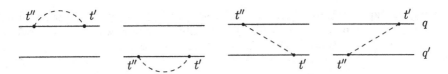

Figure 5.1: Graphical representation of the four contributions to the influence function $S_{\mathrm{FV}}[q, q']$. The dashed line is the propagator $L(t' - t'')$ in the first and third graph, and $L^*(t' - t'')$ in the second and fourth graph.

Here, the last contribution is the potential renormalization term. The quantity $\mu = M\gamma(0^+)$ has been encountered already in Eq. (4.50). The kernel L has for complex time $z = t - i\tau$ and real time t, respectively, the spectral representations

$$L(z) = \sum_{\alpha=1}^{N} \frac{c_\alpha^2}{2m_\alpha\omega_\alpha} \frac{\cosh[\omega_\alpha(\hbar\beta/2 - iz)]}{\sinh(\omega_\alpha\hbar\beta/2)} = \frac{1}{\pi} \int_0^\infty d\omega\, J(\omega) \frac{\cosh[\omega(\hbar\beta/2 - iz)]}{\sinh(\omega\hbar\beta/2)},$$

$$L(t) \equiv L'(t) + iL''(t) = \frac{1}{\pi} \int_0^\infty d\omega\, J(\omega) \left(\coth(\beta\hbar\omega/2) \cos(\omega t) - i\sin(\omega t) \right). \quad (5.23)$$

We see from Eq. (4.55) with Eq. (4.57) that the kernel $L(z)$ is the analytic continuation of the imaginary-time kernel $K(\tau)$ in the influence action (4.52),

$$L(z) = K(\tau = iz). \quad (5.24)$$

The Feynman-Vernon influence action (5.22) can be split up into three contibutions

$$S_{\mathrm{FV}}[q, q'] = S_{\mathrm{FV, self}}[q] - S_{\mathrm{FV, self}}^*[q'] + S_{\mathrm{FV, cross}}[q, q'], \quad (5.25)$$

where

$$S_{\mathrm{FV, self}}[q] = \int_0^t dt' \int_0^{t'} dt''\, q(t')\, L(t' - t'')\, q(t'') + i\frac{\mu}{2} \int_0^t dt'\, q^2(t'),$$

$$S_{\mathrm{FV, cross}}[q, q'] = \int_0^t dt' \int_0^{t'} dt'' \left\{ q'(t')L(t' - t'') q(t'') + q(t')\, L^*(t' - t'')\, q'(t'') \right\}.$$

The four diagrams contributing to $S_{\mathrm{FV}}[q, q']$ are shown in Fig. 5.1. The first two graphs are the selfinteractions of the paths q and q', respectively, and the other two describe cross interactions between the paths. The exponentiation of $S_{\mathrm{FV}}[q, q']$ in $\mathcal{F}_{\mathrm{FV}}[q, q']$ gives all kinds of irreducible and reducible diagrams with any number of exchanged lines made up of the four fundamental contributions.

The imaginary part $L''(t)$ of the Feynman-Vernon kernel is related to the damping kernel $\gamma(t)$ in the classical equation of motion [cf. Eq. (3.37)] by

$$L''(t) = \frac{M}{2} \frac{d\gamma(t)}{dt}. \quad (5.26)$$

Substituting Eq. (5.26) into Eq. (5.22), we can simplify the imaginary part of $\mathcal{S}_{\mathrm{FV}}[q, q']$ with integration by parts. Then a boundary contribution occurs which exactly cancels the potential renormalization term. The resulting expression is

$$\mathrm{Im}\, \mathcal{S}_{\mathrm{FV}}[q, q'] = \frac{M}{2} \int_0^t dt' \int_0^{t'} dt'' \left(q(t') - q'(t') \right) \gamma(t' - t'') \left(\dot{q}(t'') + \dot{q}'(t'') \right)$$

$$+ \frac{M}{2} \left(q(0) + q'(0) \right) \int_0^t dt'\, \gamma(t') \left(q(t') - q'(t') \right) . \tag{5.27}$$

At this point, it is useful to introduce symmetric and antisymmetric coordinates,

$$r(t) = \tfrac{1}{2} \left[q(t) + q'(t) \right], \qquad \text{and} \qquad y(t) = q(t) - q'(t) . \tag{5.28}$$

The path $r(t)$ measures propagation along the diagonal of the density matrix and is therefore termed *quasiclassical* path. The path $y(t)$ is book-keeping the system's off-diagonal excursions while propagating. This path describes quantum fluctuations. Using Eqs. (5.22) and (5.27), we arrive at the convenient form

$$\mathcal{S}_{\mathrm{FV}}[q, q'] = \mathcal{S}^{(\mathrm{N})}[y] + i\mathcal{S}^{(\mathrm{F})}[r, y] ,$$

$$\mathcal{S}^{(\mathrm{N})}[y] = \int_0^t dt' \int_0^{t'} dt''\, y(t')\, L'(t' - t'')\, y(t'') , \tag{5.29}$$

$$\mathcal{S}^{(\mathrm{F})}[r, y] = M \int_0^t dt' \int_0^{t'} dt''\, y(t') \gamma(t' - t'') \dot{r}(t'') + Mr(0) \int_0^t dt'\, \gamma(t') y(t') ,$$

where $\mathcal{S}^{(\mathrm{N})}[y]$ is the noise action, and $\mathcal{S}^{(\mathrm{F})}[r, y]$ is the friction action. The expression (5.29) is the influence action for a product initial state. This form has been given first in Ref. [86].[1] Using the influence functional formalism, the explicit solution for the time evolution of the density matrix for a product initial state can be obtained in the case of a damped linear oscillator[196, 86].

The two diagrams representing $\mathcal{S}^{(\mathrm{N})}[y]$ and $\mathcal{S}^{(\mathrm{F})}[r, y]$ are sketched in Fig. 5.2. Exponentiation of the action leads to graphs which contain any number of the two fundamental diagrams. Before turning to more general initial conditions, let us briefly reflect the significance of the influence functional and the effects caused by it.

5.2   Decoherence and friction

The coupling of the system to the reservoir manifests itself in two different ways. On the one hand, the influence action $\mathcal{S}^{(\mathrm{F})}[r, y]$ brings friction into the system as it provides the damping force $\int_0^t dt'\, \gamma(t - t') \dot{r}(t')$ in the equation of motion for the quasiclassical path $r(t)$. This we shall see below in Subsection 5.5.1 from a stationary

---

[1]The last term in Eq. (5.29) is missing in the work by Caldeira and Leggett [196].

Figure 5.2: Graphical representation of the two contributions to $S_{\mathrm{FV}}[r, y]$. The left diagram is the selfinteraction of the off-diagonal path $y(t)$ and represents $S^{(\mathrm{N})}[y]$. The exchanged line is the propagator $L'(t'-t'')$. The right diagram illustrates the correlations between $y(t)$ and $\dot{r}(t)$ in $S^{(\mathrm{F})}[r, y]$. The exchanged line represents the friction kernel $\gamma(t'-t'')$.

phase analysis. Pictorially, friction is conveyed by the right diagram in Fig. 5.2. On the other hand, the noise action $S^{(\mathrm{N})}[y]$ is responsible for loss of coherence since it provides random pumping of energy back and forth between system and reservoir.

Quantum interference is a phenomenon directly related to the fact that transition amplitudes representing two different paths contributing to the propagation of a system may accumulate a phase difference φ. The interference terms of the squared amplitude come with phase factors $e^{\pm i\varphi}$. In the presence of the bath coupling, the phase φ is fluctuating. Then the relevant object is the statistical average of it, the "decoherence" factor $\langle e^{\pm i\varphi}\rangle$ [197, 198]. The extinction of quantum coherence for the pair of paths $q(t)$ and $q'(t)$ is determined by the noise action $S^{(\mathrm{N})}[y]$,

$$\langle e^{\pm i\varphi}\rangle = e^{-S^{(\mathrm{N})}[y]/\hbar} , \tag{5.30}$$

which is independent of the quasiclassical path $r(t)$. The noise functional acts as a Gaussian filter which quenches off-diagonal quantum fluctuations. Physically, the environment is continuously measuring the position of the system. It therefore suppresses quantum interference between different position eigenstates and makes the system behave more classical. Diagrammatically, $S^{(\mathrm{N})}[y]$ is represented by the left diagram in Fig. 5.2. To elucidate in some more detail the way $S^{(\mathrm{N})}[y]$ works, consider the Ohmic case, $J(\omega) = \eta\omega$, at high temperatures. In this white noise limit, we have $L'(t) = (2\eta/\hbar\beta)\,\delta(t)$, as follows from Eq. (5.23). Thus we find for two different localized states with spatial separation q_0 the decoherence factor

$$\langle e^{-i\varphi}\rangle = \exp(-\gamma_{\mathrm{decoh}}t) , \qquad \text{where} \qquad \gamma_{\mathrm{decoh}} = \eta q_0^2 k_{\mathrm{B}}T/\hbar^2 . \tag{5.31}$$

Thus, the rate γ_{decoh} is the inverse time scale for decoherence of these two states.

On the other hand, the effects of friction are characterized by the damping rate

$$\gamma_{\mathrm{damping}} = \eta/M . \tag{5.32}$$

The ratio of the two rates is

$$\gamma_{\mathrm{decoh}}/\gamma_{\mathrm{damping}} = q_0^2 M k_{\mathrm{B}}T/\hbar^2 = q_0^2/\lambda_{\mathrm{th}}^2 , \tag{5.33}$$

where $\lambda_{\mathrm{th}} = \hbar/p_{\mathrm{th}} = \hbar/\sqrt{M k_{\mathrm{B}}T}$ is the thermal wave length. The ratio (5.33) relates the range of spatial coherence with the thermal wave length of the particle. For a typical macroscopic situation, $M = 1\,\mathrm{g}$, $T = 300\,\mathrm{K}$, $q_0 = 1\,\mathrm{mm}$, a value of 10^{38} for the ratio $\gamma_{\mathrm{decoh}}/\gamma_{\mathrm{damping}}$ results. Thus for practically all macroscopic systems, friction is fully negligible on time scales where quantum coherence occurs. The ratio between the inverse time scales of decoherence and friction, Eq. (5.33), is generally valid for all systems which are in contact with an Ohmic heat bath with thermal energy large compared to the relevant internal energy scales of the system.[2] We shall discuss decoherence in some more detail in Chapter 9.

Upon comparing the expression (3.48) with (5.23) we see that the kernel $L(t)$ in the influence action (5.22) is related to the correlator of the random force $\xi(t)$ and of the collective bath mode $\mathfrak{E}(t)$, defined in Eq. (3.85), by

$$L(z) = \frac{1}{\hbar} \langle \xi(z)\xi(0)\rangle_\beta = \frac{4}{q_0^2\,\hbar} \langle \mathfrak{E}(z)\mathfrak{E}(0)\rangle_\beta \, . \tag{5.34}$$

We shall refer to $L(z)$ as the *bath correlation function*. Here we have tacitly assumed that the fluctuations $\xi(z)$ are the deviations from the equilibrium value for the shifted canonical distribution (3.25).

For a linearly responding bath, the random force obeys Gaussian statistics. Therefore, all influences of the reservoir exerted on the relevant system are fully captured by the autocorrelation function $\langle \xi(t)\xi(0)\rangle_\beta$. Hence a conventional molecular dynamics simulation may be used to determine this correlation function in the classical limit, which then defines $J(\omega)$ via Eq. (2.6) and Eq. (3.32). The quantum mechanical response of the bath is then obtained from Eq. (3.48) as a function of $J(\omega)$.

After having calculated the Feynman-Vernon influence functional (5.29) within the model (5.8), we have all pieces together. We call to the reader's mind that the evolution of the reduced density matrix is given by the relation (5.11) with Eq. (5.12). The double functional integral in Eq. (5.12) is over paths $q(t')$ and $q'(t')$ with the endpoint constraints given by Eq. (5.5).

Finally, let us consider the case when the behaviour of the coupled system-reservoir complex is ergodic. For such systems, effects of the initial preparation die completely out in the course of time, and the system relaxes to the thermal equilibrium state $\rho_\beta(q_f, q_f')$ independently of the particular initial state $\rho(q_i, q_i'; 0)$ chosen,

$$
\begin{aligned}
\rho_\beta(q_f, q_f') &= \lim_{t\to\infty} \rho(q_f, q_f'; t) \, , \\
&= \lim_{t\to\infty} \int dq_i\, dq_i'\, J_{\mathrm{FV}}(q_f, q_f', t; q_i, q_i', 0)\, \rho(q_i, q_i'; 0) \, ,
\end{aligned}
\tag{5.35}
$$

where we have employed the relation (5.11). This is a remarkably useful property. Namely, instead of calculating the thermal equilibrium state of the damped system from the imaginary-time path integral (4.61), we may get the equilibrium density matrix $\rho_\beta(q_f, q_f')$ from the long-time limit in the dynamical approach.

The dynamical approach to the equilibrium density matrix has considerable advantage over the imaginary-time method in applications to specific systems. The detailed discussion is deferred until Section 21.9.

---

[2] See Sections 21.3 and 21.4 for the discussion of decoherence in the dissipative two-state system.

5.3 General initial states and preparation function

Often in experiments, the "small" system is prepared, say at $t = 0$, by first letting it equilibrate with the bath and then measuring certain dynamical variables of it. The measurement leads to a reduction of the equilibrium density matrix of the system-plus-reservoir complex. Initial states of this type are generated by a linear transformation of the equilibrium density matrix of the global system. We write

$$< q_i, x_i| \hat{W}(0) |q_i', x_i' > \; = \; \int d\bar{q}\, d\bar{q}'\, \lambda(q_i, q_i'; \bar{q}, \bar{q}') < \bar{q}, x_i| \hat{W}_\beta |\bar{q}', x_i' > , \qquad (5.36)$$

where $\hat{W}_\beta = e^{-\beta\hat{H}}/Z$. The linking function $\lambda(q, q'; \bar{q}, \bar{q}')$ is a *preparation function* depending on the specific measurement apparatus at $t = 0$ [86].

Initial states of the form (5.36) are not in the product form (5.10). Thus we cannot apply the procedure described in Section 5.1. Before explaining the corresponding generalization for correlated initial states, we briefly discuss specific initial preparations which have certain experimental relevance.

Let the effect of the measuring device be described by operators $\hat{O}_j, \hat{O}_j'$ which act on the system only and leave the reservoir unaffected. Then the preparation function is in the form

$$\lambda(q_i, q_i'; \bar{q}, \bar{q}') \; = \; \sum_j < q_i|\hat{O}_j|\bar{q} >< \bar{q}'|\hat{O}_j'|q_i' > . \qquad (5.37)$$

Consider briefly three important cases. First, if we are interested in the response of a system in thermal equilibrium to an external force switched on at time $t = 0$, we have $\hat{W}(0) = \hat{W}_\beta$. Then the preparation function $\lambda(q_i, q_i'; \bar{q}, \bar{q}')$ takes the simple form

$$\lambda_\beta(q_i, q_i'; \bar{q}, \bar{q}') \; = \; \delta(q_i - \bar{q})\, \delta(q_i' - \bar{q}') . \qquad (5.38)$$

The second important example is that we perform initially a measurement of a dynamical variable of the particle. In accordance with the principles of quantum measurement theory (see, for instance, Refs. [199, 57, 200]), the effect of an "observation" is to project the density matrix that described the system just prior to observation onto the manifold corresponding to the measured interval of eigenvalues. For an ideal measurement at time $t = 0^-$, the initial density matrix is then given by

$$\hat{W}(0) \; = \; \hat{P}\hat{W}_\beta\hat{P} , \qquad (5.39)$$

where $\hat{P}$ is the projection operator corresponding to the measurement. For example, a measurement of the position of the particle with the outcome q_0 with uncertainty Δq is described by the projection operator

$$\hat{P}_q \; = \; \int_{q_0-\Delta q/2}^{q_0+\Delta q/2} dq\, |q >< q| . \qquad (5.40)$$

Thirdly, the structure factor investigated in scattering experiments is related to the Fourier transform of the equilibrium coordinate autocorrelation function of the scatterer [201]. Now, an equilibrium correlation function, say $\langle A(t)\,B(0)\rangle$, may be formally looked upon as the expectation value of $\hat{A}$ at time t for a system starting out from the initial state $\hat{W}(0) = \hat{B}W_\beta$. In this case, the specific form of the preparation function follows directly from Eq. (5.36). For the coordinate autocorrelation function in thermal equilibrium $C_{qq}(t) = \operatorname{tr}\{q(t)q(0)W_\beta\}$, the preparation function takes the form

$$\lambda_q(q_i, q_i'; \bar{q}, \bar{q}') \;=\; q_i\,\lambda_\beta(q_i, q_i'; \bar{q}, \bar{q}')\,, \tag{5.41}$$

while one finds for the equilibrium coordinate-momentum correlation function $C_{qp}(t) = \operatorname{tr}\{q(t)\,p(0)W_\beta\}$

$$\lambda_p(q_i, q_i'; \bar{q}, \bar{q}') \;=\; \frac{\hbar}{i}\frac{\partial}{\partial q_i}\lambda_\beta(q_i, q_i'; \bar{q}, \bar{q}')\,. \tag{5.42}$$

The two cases correspond to $\hat{W}(0) = \hat{q}\hat{W}_\beta$ and $\hat{W}(0) = \hat{p}\hat{W}_\beta$, respectively. These expressions are not proper density matrices since they are not positive definite. However, they belong to the class of initial conditions specified by Eq. (5.37).

5.4 Complex–time path integral for the propagating function

For the general class of initial states discussed in the preceding section, the quantum mechanical stochastic process is characterized by a *preparation function* $\lambda(q_i, q_i'; \bar{q}, \bar{q}')$ and by a *propagating function* $J(q_f, q_f', t; q_i, q_i'; \bar{q}, \bar{q}')$. The propagation function describes the time evolution of the reduced system starting out from an initial state which is introduced by the preparation function. The time evolution out of the initially prepared state is described by the relation

$$\rho(q_f, q_f'; t) \;=\; \int dq_i\, dq_i'\, d\bar{q}\, d\bar{q}'\; J(q_f, q_f', t; q_i, q_i'; \bar{q}, \bar{q}')\,\lambda(q_i, q_i'; \bar{q}, \bar{q}')\,. \tag{5.43}$$

An explicit expression for J is easily found when we insert Eq. (5.36) into Eq. (5.9) and compare the result with Eq. (5.43). We then get

$$
\begin{aligned}
J(q_f, q_f'; t; q_i, q_i'; \bar{q}, \bar{q}') \;=\;& \int d\boldsymbol{x}_i\, d\boldsymbol{x}_i'\, d\boldsymbol{x}_f\; K(q_f, \boldsymbol{x}_f, t; q_i, \boldsymbol{x}_i, 0) \\
&\times <\bar{q}, \boldsymbol{x}_i|W_\beta|\bar{q}', \boldsymbol{x}_i'> K^*(q_f', \boldsymbol{x}_f, t; q_i', \boldsymbol{x}_i', 0)\,.
\end{aligned}
\tag{5.44}
$$

Using the functional integral expressions (4.25) and (5.4) for the density matrix and for the propagators, we can write the propagating function (5.44) in the form

$$J(q_f, q_f'; t; q_i, q_i'; \bar{q}, \bar{q}') \;=\; Z^{-1} \int \mathcal{D}q\, \mathcal{D}q'\, \mathcal{D}\bar{q} \tag{5.45}$$

$$\times\, \exp\left\{\frac{i}{\hbar}\Big(S_{\mathrm{S}}[q] - S_{\mathrm{S}}[q']\Big) - \frac{1}{\hbar}S_{\mathrm{S}}^{(\mathrm{E})}[\bar{q}]\right\}\, \mathcal{F}_{\mathrm{I}}[q, q', \bar{q}]\,,$$

where Z is the reduced partition function of the damped system, and where

$$
\mathcal{F}_I[q, q', \bar{q}] = \int d\boldsymbol{x}_f \, d\boldsymbol{x}_i \, d\boldsymbol{x}_i' \frac{1}{Z_R} \int \mathcal{D}\boldsymbol{x} \, \mathcal{D}\boldsymbol{x}' \, \mathcal{D}\bar{\boldsymbol{x}}
$$

$$
\times \exp\left\{ i\left(S_R[\boldsymbol{x}] + S_I[\boldsymbol{x}, q] - S_R[\boldsymbol{x}'] - S_I[\boldsymbol{x}', q'] \right)/\hbar \right. \tag{5.46}
$$

$$
\left. - \left(S_R^{(E)}[\bar{\boldsymbol{x}}] + S_I^{(E)}[\bar{\boldsymbol{x}}, \bar{q}] \right)/\hbar \right\}
$$

represents the influence functional. The functional integrations in Eq. (5.46) are evaluated for paths $\boldsymbol{x}(t')$, $\boldsymbol{x}'(t')$, $\bar{\boldsymbol{x}}(\tau)$ which form a closed path on a contour in the complex-time plane according to the endpoint conditions

$$
\boldsymbol{x}(t) = \boldsymbol{x}'(t) = \boldsymbol{x}_f \;; \quad \boldsymbol{x}(0) = \bar{\boldsymbol{x}}(\hbar\beta) = \boldsymbol{x}_i \;; \quad \bar{\boldsymbol{x}}(0) = \boldsymbol{x}'(0) = \boldsymbol{x}_i' \,. \tag{5.47}
$$

We can now proceed as in Section 5.1 by writing Eq. (5.46) in the form

$$
\mathcal{F}_I[q(\cdot), q'(\cdot), \bar{q}(\cdot)] = \prod_{\alpha=1}^{N} \frac{1}{Z_R^{(\alpha)}} \int dx_{f,\alpha} \, dx_{i,\alpha} \, dx_{i,\alpha}' \tag{5.48}
$$

$$
\times F_\alpha[q(\cdot), x_{f,\alpha}, x_{i,\alpha}] \, F_\alpha^*[q'(\cdot), x_{f,\alpha}, x_{i,\alpha}'] \, F_\alpha^{(E)}[\bar{q}(\cdot), x_{i,\alpha}, x_{i,\alpha}'] \,,
$$

in which F_α, F_α^* and $F_\alpha^{(E)}$ are represented by sums over the real-time paths $x_\alpha(t')$, $x_\alpha'(t')$, and over the imaginary-time path $\bar{x}_\alpha(\tau)$, respectively. The functional dependence on $q(t')$, $q'(t')$ and $\bar{q}(\tau)$ is in the external forces. The resulting expression for $F_\alpha(q, x_{f,\alpha}, x_{i,\alpha})$ has been given in Eq. (5.18) with Eq. (5.19), and the result for $F_\alpha^{(E)}(q, x_{f,\alpha}, x_{i,\alpha})$ follows from this with the substitution $t = -i\hbar\beta$. Hence the integrand of Eq. (5.48) is a Gaussian in $x_{f,\alpha}, x_{i,\alpha}$ and $x_{i,\alpha}'$, so that the evaluation of Eq. (5.48) is straightforward. In the end one finds [86]

$$
\mathcal{F}_I[q, q', \bar{q}] = \exp\left\{ -S_I[q, q', \bar{q}]/\hbar \right\}, \tag{5.49}
$$

where

$$
S_I[q, q', \bar{q}] = -\int_0^{\hbar\beta} d\tau \int_0^{\tau} d\tau' \, K(\tau - \tau') \, \bar{q}(\tau) \, \bar{q}(\tau') + \frac{M\gamma(0^+)}{2} \int_0^{\hbar\beta} d\tau \, \bar{q}^2(\tau)
$$

$$
- i \int_0^{\hbar\beta} d\tau \int_0^{t} dt' \, L^*(t' - i\tau) \, \bar{q}(\tau) \left\{ q(t') - q'(t') \right\}
$$

$$
+ \int_0^{t} dt' \int_0^{t'} dt'' \left\{ q(t') - q'(t') \right\} \left\{ L(t' - t'')q(t'') - L^*(t' - t'') \, q'(t'') \right\}
$$

$$
+ i \frac{M\gamma(0^+)}{2} \int_0^{t} dt' \left\{ q^2(t') - q'^2(t') \right\}, \tag{5.50}
$$

and where $K(\tau)$ and $L(t)$ are defined in Eqs. (4.55) and (5.23), respectively. The triple path integral (5.45) with Eqs. (5.49) and (5.50) is the exact formal solution for the propagating function $J(q_f, q'_f; t; q_i, q'_i; \bar{q}, \bar{q}')$. The paths $q(t')$, $q'(t')$ and $\bar{q}(\tau)$ are subject to the endpoint conditions

$$
\begin{aligned}
q(0) &= q_i, & q'(0) &= q'_i, & \bar{q}(0) &= \bar{q}', \\
q(t) &= q_f, & q'(t) &= q'_f, & \bar{q}(\hbar\beta) &= \bar{q}.
\end{aligned}
\tag{5.51}
$$

The influence functional (5.49) with Eq. (5.50) can be written compactly in terms of a path $\tilde{q}(z)$ for which the time z goes by along a contour $\mathcal{C}_{\mathrm{I}}$ or a contour $\mathcal{C}_{\mathrm{II}}$ in the complex-time plane (see Fig. 5.3). The path $\tilde{q}(z)$ on the contour $\mathcal{C}_{\mathrm{I}}$ is defined by

$$
\tilde{q}(z) = \begin{cases}
q'(s) & \text{for} \quad z = s - i\,0^+, & 0 \le s \le t, \\
\bar{q}(\tau) & \text{for} \quad z = -i\tau, & 0 \le \tau \le \hbar\beta, \\
q(s) & \text{for} \quad z = -i\,(\hbar\beta - 0^+) + s, & 0 \le s \le t,
\end{cases}
\tag{5.52}
$$

while along the contour $\mathcal{C}_{\mathrm{II}}$

$$
\tilde{q}(z) = \begin{cases}
q(s) & \text{for} \quad z = s + i\,0^+, & 0 \le s \le t, \\
q'(s) & \text{for} \quad z = s - i\,0^+, & 0 \le s \le t, \\
\bar{q}(\tau) & \text{for} \quad z = -i\tau, & 0 \le \tau \le \hbar\beta.
\end{cases}
\tag{5.53}
$$

With these definitions the expression (5.50) can be rewritten as

$$
\mathcal{S}[\tilde{q}(\cdot)] = \int dz \int_{z>z'} dz'\, L(z - z')\, \tilde{q}(z)\, \tilde{q}(z') + i\, \frac{M\gamma(0^+)}{2} \int dz\, \tilde{q}^2(z),
\tag{5.54}
$$

where the integrals are *time-ordered* along the contours $\mathcal{C}_{\mathrm{I}}$ and $\mathcal{C}_{\mathrm{II}}$ in the direction indicated by the arrows in Fig. 5.3. The equivalence of the expression (5.54) with Eq. (5.50) (both for the contour $\mathcal{C}_{\mathrm{I}}$ and $\mathcal{C}_{\mathrm{II}}$) follows upon using symmetry relations resulting from the definitions (5.23) and (5.24),

$$
\begin{aligned}
L\left(t + i\tau - i\hbar\beta\right) &= L^*(t - i\tau) = L(-t - i\tau), \\
L\left(-i\tau\right) &= K(\tau).
\end{aligned}
\tag{5.55}
$$

Defining the particle's (Minkowskian) action along the path $\tilde{q}(z)$,

$$
S_{\mathrm{S}}[\tilde{q}(\cdot)] = \int dz \left[\frac{M}{2} \left(\frac{\partial \tilde{q}}{\partial z} \right)^2 - V(\tilde{q}) \right],
\tag{5.56}
$$

we can write the propagating function in the compact form

$$
J(q_f, q'_f; t; q_i, q'_i; \bar{q}, \bar{q}') = Z^{-1} \int \mathcal{D}\tilde{q}(\cdot)\, e^{iS_{\mathrm{S}}[\tilde{q}(\cdot)]/\hbar - \mathcal{S}[\tilde{q}(\cdot)]/\hbar},
\tag{5.57}
$$

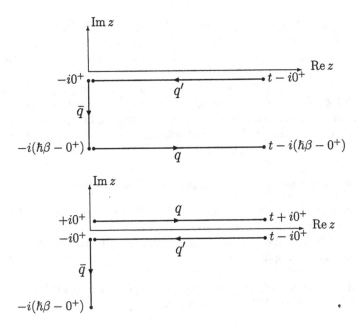

Figure 5.3: Schematic depiction of the contours $\mathcal{C}_{\mathrm{I}}$ (top) and $\mathcal{C}_{\mathrm{II}}$ (bottom) for propagators in thermal correlations functions. The dots represent the endpoint conditions (5.51). The Feynman-Vernon influence functional for product initial states is defined on a contour $\mathcal{C}_{\mathrm{FV}}$, which differs from the contours $\mathcal{C}_{\mathrm{I}}$ or $\mathcal{C}_{\mathrm{II}}$ by omission of the path segment $\bar{q}(z)$.

where the history of the path $\widetilde{q}(z)$ goes by either along the contour $\mathcal{C}_{\mathrm{I}}$ or along $\mathcal{C}_{\mathrm{II}}$.

The contour $\mathcal{C}_{\mathrm{I}}$ has been discussed and applied to quantum Brownian motion in Ref. [86], while $\mathcal{C}_{\mathrm{II}}$ is the standard Kadanoff-Baym contour [202]. The latter formulation is related to the Green function technique introduced by Keldysh [203]. The closed-time-path Green function approach has been developed further and has been applied to various equilibrium and nonequilibrium problems. For recent reviews we refer to the work by Chou et al. [204], and by Rammer and Smith [205].

The results obtained in this section shed light on the evolution of a general initial state. The dynamics is specified by a preparation function $\lambda(q_i, q_i'; \bar{q}, \bar{q}')$ and by a propagating function $J(q_f, q_f', t; q_i, q_i'; \bar{q}, \bar{q}')$. The preparation function fixes the particular initial state $\rho_0 = \rho(0)$ within the preparation class through the relation

$$\rho_0(q_i, q_i') \;=\; \int d\bar{q}\, d\bar{q}'\, \lambda(q_i, q_i'; \bar{q}, \bar{q}')\, \rho_\beta(\bar{q}, \bar{q}') \,, \tag{5.58}$$

where $\rho_\beta(\bar{q}, \bar{q}')$ is the equilibrium reduced density matrix given in Eq. (4.61). The relation (5.58) is obtained by tracing out the reservoir coordinates in Eq. (5.36). It also follows from (5.43) if we put $t = 0$ and use the initial condition [cf. Eq. (5.44)]

$$J(q_f, q_f', 0; q_i, q_i'; \bar{q}, \bar{q}') \;=\; \delta(q_f - q_i)\,\delta(q_f' - q_i')\rho_\beta(\bar{q}, \bar{q}') \,. \tag{5.59}$$

The preparation function $\lambda(q_i, q_i'; \bar{q}, \bar{q}')$ represents the quantum analogue of the initial classical phase space distribution. Note that it is not uniquely specified through the relation (5.58). It is rather determined by the initial density matrix of the total system through relation (5.36). In quantum mechanics, the propagating function $J(q_f, q_f', t; q_i, q_i'; \bar{q}, \bar{q}')$ has taken over the role of the classical conditional probability. From the considerations above it is clear that the propagating function is determined by the equilibrium properties of the reduced system. This is in correspondence with the classical stationary conditional probability.

5.5   Real–time path integral for the propagating function

In the formal solution (5.45) with Eqs. (5.49) and (5.50) for the propagating function, one faces difficulties which arise from the correlations of the forward and backward paths $q(t')$ and $q'(t')$ with the imaginary-time path $\bar{q}(\tau)$. The complications caused by this coupling can be simplified to some extent in actual calculations for systems with ergodic environments, as we explain in the sequel.

We start with the observation that the propagating function (5.44) has the initial condition (5.59). Using the relation (5.43), we then find

$$\rho(q_f, q_f'; 0) \;=\; \int d\bar{q}\,d\bar{q}'\,\lambda(q_f, q_f'; \bar{q}, \bar{q}')\,\rho_\beta(\bar{q}, \bar{q}') \,. \tag{5.60}$$

Suppose that the system has been prepared at some large negative time $-|t_p|$ in a particular initial state and recall that the system relaxes for an ergodic system-plus-environment complex to thermal equilibrium independently of the chosen initial preparation (cf. the discussion given at the end of Section 5.1). In this case, the system will have equilibrated with the reservoir at time $t = 0$, if it started out in the infinite past from some nonequilibrium initial state. Therefore, we may choose in the infinite past without loss of generality a product initial state of the form

$$\lim_{t_p \to -\infty} \hat{W}(t_p) \;=\; \hat{\rho}_p \otimes \mathrm{e}^{-\beta \hat{H}_\mathrm{R}}/Z_\mathrm{R} \,. \tag{5.61}$$

Now, we may follow closely the lines explained in Section 5.1, with the result [cf. Eq. (5.35)] that at time $t = 0^-$

$$\rho(\bar{q}, \bar{q}'; 0^-) \;=\; \rho_\beta(\bar{q}, \bar{q}') \;=\; \lim_{t_p \to -\infty} \int dq_p\,dq_p'\,J_\mathrm{FV}(\bar{q}, \bar{q}', 0^-; q_p, q_p'; t_p)\,\rho_p(q_p, q_p') \,, \tag{5.62}$$

where the propagating function J_FV is given by Eq. (5.12) with (5.22) and (5.5).

After having determined the correlated equilibrium state of the damped particle at time $t = 0^-$, it is straightforward to describe the evolution of the reduced density matrix. We have

$$\rho(q_f, q_f'; t) \;=\; \int dq_i\,dq_i'\,d\bar{q}\,d\bar{q}'\;J(q_f, q_f', t; q_i, q_i', 0^+; \bar{q}, \bar{q}', 0^-)\,\lambda(q_i, q_i'; \bar{q}, \bar{q}') \,. \tag{5.63}$$

The propagating function J is given by the real-time double path integral

$$J(q_f, q'_f, t; q_i, q'_i, 0^+; \bar{q}, \bar{q}', 0^-) \;=\; \int dq_p \, dq'_p \, \rho_p(q_p, q'_p) \lim_{t_p \to -\infty} \int \mathcal{D}q \, \mathcal{D}q'$$
$$\times \exp \left\{ i \left(S_S[q] - S_S[q'] \right)/\hbar \right\} \mathcal{F}_{\mathrm{II}}[q, q']$$

(5.64)

with the influence functional

$$\mathcal{F}_{\mathrm{II}}[q, q'] \;=\; \exp \left\{ -\, \mathcal{S}_{\mathrm{II}}[q, q']/\hbar \right\} ,$$

(5.65)

where

$$\mathcal{S}_{\mathrm{II}}[q, q'] \;=\; \int_{t_p}^{t} dt' \int_{t_p}^{t'} dt'' \left\{ q(t') - q'(t') \right\} \left\{ L(t' - t'')\, q(t'') - L^*(t' - t'')\, q'(t'') \right\}$$
$$+\, i\, \frac{M\gamma(0^+)}{2} \int_{t_p}^{t} dt' \left\{ q^2(t') - q'^2(t') \right\} .$$

(5.66)

The functional integrations in (5.64) are over paths $q(t')$ and $q'(t')$ which are subject to the constraints

$$q(t_p) = q_p, \qquad q(0^-) = \bar{q}, \qquad q(0^+) = q_i, \qquad q(t) = q_f,$$
$$q'(t_p) = q'_p, \qquad q'(0^-) = \bar{q}', \qquad q'(0^+) = q'_i, \qquad q(t) = q'_f.$$

(5.67)

Observe that the paths $q(t')$ and $q'(t')$ are discontinuous at time $t = 0$. In order to write the exponent of the influence functional in compact form, we introduce a path $\tilde{q}(z)$ for which the progression of z is schematically depicted by the contour $\mathcal{C}_{\mathrm{III}}$ or by $\mathcal{C}_{\mathrm{IV}}$ in Fig. 5.4. The path $\tilde{q}(z)$ with z along the contour $\mathcal{C}_{\mathrm{III}}$ is given by

$$\tilde{q}(z) = \begin{cases} q'(s) & \text{for} \quad z = s - i\,0^+, & t_p \leq s \leq t, \\ q(s) & \text{for} \quad z = -i\,(\hbar\beta - 0^+) + s, & t_p \leq s \leq t, \end{cases}$$

(5.68)

and with z along the contour $\mathcal{C}_{\mathrm{IV}}$ by

$$\tilde{q}(z) = \begin{cases} q(s) & \text{for} \quad z = s + i\,0^+, & t_p \leq s \leq t, \\ q'(s) & \text{for} \quad z = s - i\,0^+, & t_p \leq s \leq t. \end{cases}$$

(5.69)

Then the influence action may be written as

$$S[\tilde{q}(\cdot)] = \int dz \int_{z>z'} dz'\, L(z - z')\, \tilde{q}(z)\, \tilde{q}(z') + i\, \frac{M\gamma(0^+)}{2} \int dz\, \tilde{q}^2(z) ,$$

(5.70)

and the propagating function J takes the compact form

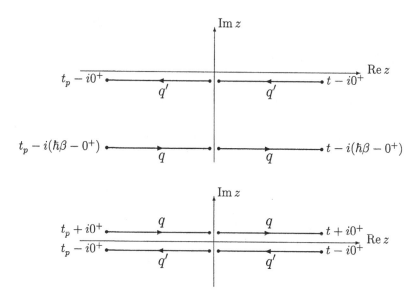

Figure 5.4: Schematic representation of the contours $\mathcal{C}_{\mathrm{III}}$ (top) and $\mathcal{C}_{\mathrm{IV}}$ (bottom). The dots symbolize the endpoint conditions (5.67).

$$J(q_f, q_f'; t; q_i, q_i', 0^+; \bar{q}, \bar{q}', 0^-) = \int dq_p \, dq_p' \, \rho_p(q_p, q_p') \tag{5.71}$$

$$\times \lim_{t_p \to -\infty} \int \mathcal{D}\widetilde{q}[z] \, \exp\left(i \, S_{\mathrm{S}}[\widetilde{q}]/\hbar - \mathcal{S}[\widetilde{q}]/\hbar \right),$$

where $S_{\mathrm{S}}[\widetilde{q}]$ is defined in Eq. (5.56). The z-integrals are time-ordered along the contours $\mathcal{C}_{\mathrm{III}}$ and $\mathcal{C}_{\mathrm{IV}}$ in the direction indicated by the arrows (see Fig. 5.4). When the system is ergodic, the expression (5.71) for the propagating function is exactly equivalent to the former expression (5.57). An explicit demonstration of the equivalence has been given for specific models such as the damped harmonic oscillator [206] and the spin-boson system [207].

At this stage of the discussion, it is convenient to introduce again the symmetric and antisymmetric variables $r(t)$ and $y(t)$ defined in Eq. (5.28). The expression (5.43) for the reduced density matrix is rewritten as

$$\rho(r_f, y_f; t) = \int dr_i \, dy_i \, d\bar{r} \, d\bar{y} \, J(r_f, y_f, t; r_i, y_i, 0^+; \bar{r}, \bar{y}, 0^-) \, \lambda(r_i, y_i; \bar{r}, \bar{y}), \tag{5.72}$$

and the propagating function (5.64) is expressed by the path integral

$$J(r_f, y_f, t; r_i, y_i, 0^+; \bar{r}, \bar{y}, 0^-) = \int dr_p \, dy_p \, \rho_p(r_p, y_p) \tag{5.73}$$

$$\times \lim_{t_p \to -\infty} \int \mathcal{D}r \, \mathcal{D}y \, e^{i\Sigma[r,y]/\hbar - S_{\mathrm{II}}[r,y]/\hbar},$$

where $\Sigma[r, y] = S_S[\tilde{q}]$ is the reversible action and $S_{II}[r, y]$ is the influence action. Using Eqs. (5.7), (5.8) and (5.66) we find

$$\Sigma[r, y] = \int_{t_p}^{t} dt' \left(M\dot{r}\dot{y} - V(r + y/2) + V(r - y/2) - M\gamma(0^+)ry \right), \quad (5.74)$$

$$S_{II}[r, y] = \int_{t_p}^{t} dt' \int_{t_p}^{t'} dt'' \left(y(t')L'(t' - t'')y(t'') + i2y(t')L''(t' - t'')r(t'') \right),$$

where $L'(t)$ and $L''(t)$ are the real and imaginary part of the function $L(t)$ defined in Eq. (5.23). The endpoints of the paths $y(t')$ and $r(t')$ are

$$
\begin{aligned}
r(t_p) &= r_p, & r(0^-) &= \bar{r}, & r(0^+) &= r_i, & r(t) &= r_f, \\
y(t_p) &= y_p, & y(0^-) &= \bar{y}, & y(0^+) &= y_i, & y(t) &= y_f.
\end{aligned}
\quad (5.75)
$$

The limit $t_p \to -\infty$ in Eq. (5.73) is quite subtle. The reason that effects of the arbitrary initial distribution $\rho_p(r_p, y_p)$ at $t = t_p$ die out completely is the following. As $t_p \to -\infty$, the double path integral expression in Eq. (5.73) becomes a δ–distribution in the variable y_p which is centered at $y_p = 0$, and at the same time it becomes independent of the variable r_p. On the other hand, it remains as a smooth function of all the other endpoint variables given in Eq. (5.75). Thus we may write

$$J(r_f, y_f, t; r_i, y_i, 0^+; \bar{r}, \bar{y}, 0^-) = \int dr_p \, \rho_p(r_p, y_p = 0) \quad (5.76)$$

$$\times \lim_{t_p \to -\infty} \int dy_p \int \mathcal{D}r \, \mathcal{D}y \, e^{i\Sigma[r,y]/\hbar - S_{II}[r,y]/\hbar}.$$

Now observe that the integral over r_p is just tr $\rho_p = 1$. Hence the propagating function J is indeed independent of the particular initial state chosen in the infinite past.

It is worth closing this section with the remark that we ended up with a pure real-time path integral expression for the propagating function for the class of preparations discussed in Section 5.2. In this formulation, the effects of the system-reservoir correlations of the equilibrium state at time $t = 0^+$ are described by interactions between the positive and negative time branches of the functional integral expression (5.64). The interactions are contained in the influence functional (5.65) with (5.66). The real-time approach has been applied successfully, e. g., to the calculation of equilibrium correlation functions for dissipative two-state systems [207] and to the investigation of universal low temperature properties [208] (see Chapter 21).

5.5.1 Extremal paths

In the semiclassical limit, the functional integral is dominated by paths for which the action is stationary. Stationarity of the action (5.74) under variation of the paths $y(s)$

and $r(s)$ with fixed endpoints gives for $s > 0$ the equations of motion

$$M\ddot{r}(s) + 2 \int_{t_p}^{s} dt'\, L''(s - t')r(t') + M\gamma(0^+)\, r(s)$$

$$+ \frac{d}{dy}\Big(V(r + y/2) - V(r - y/2)\Big) \; - \; i \int_{t_p}^{t} dt'\, L'(s - t')y(t') \; = \; 0 \,,$$

$$\tag{5.77}$$

$$M\ddot{y}(s) + 2 \int_{s}^{t} dt'\, y(t')L''(t' - s) + M\gamma(0^+)\, y(s)$$

$$\tag{5.78}$$

$$+ \frac{d}{dr}\Big(V(r + y/2) - V(r - y/2)\Big) \; = \; 0 \,.$$

We see that the positive and negative time branches of the paths $r(t')$ and $y(t')$ are coupled with each other via the bath correlation function $L(t)$. Next, it is convenient to introduce the damping kernel $\gamma(s)$ via the relation (5.26). Integrating by parts, one finds that the terms with $M\gamma(0^+)$ in Eqs. (5.77) and (5.78) cancel out. The final form of the equations of motion is

$$\ddot{r}(s) \; + \; \int_{t_p}^{s} dt'\, \gamma(s - t')\dot{r}(t') + \frac{1}{M}\frac{d}{dy}\Big(V(r + y/2) - V(r - y/2)\Big)$$

$$= \; i \int_{t_p}^{t} dt'\, L'(s - t')y(t') - r_p\gamma(s - t_p) - \Theta(s)\gamma(s)\Big(r_i - \bar{r}\Big) \,,$$

$$\tag{5.79}$$

$$\ddot{y}(s) \; - \; \int_{s}^{t} dt'\, \dot{y}(t')\gamma(t' - s) + \frac{1}{M}\frac{\partial}{\partial r}\Big(V(r + y/2) - V(r - y/2)\Big)$$

$$= \; -y_f\gamma(t - s) + \Theta(-s)\gamma(-s)\Big(y(0_+) - y(0_-)\Big) \,,$$

$$\tag{5.80}$$

where $\Theta(s)$ is the unit step function. The last terms in Eqs. (5.79) and (5.80) are due to the discontinuity of the paths at $s = 0$.

5.5.2 Classical limit

In the classical limit, the off-diagonal elements of the density matrix vanish, and the diagonal elements merge into the probability distribution $w(r; t) = \rho(r, y = 0; t)$. Then we may restrict our attention to initial states of the form

$$\rho(r_i, y = 0; t = 0) \; = \; w(r_i; t = 0) \,. \tag{5.81}$$

At time t we are again only interested in a probability distribution. Thus we may choose for the path $y(s)$ the boundary conditions $y(t_0) = y(0^-) = y(0^+) = y(t) = 0$ and for the path $r(s)$ we have $r(0^-) = r(0^+)$, i.e., $\bar{r} = r_i$. For these conditions, Eq. (5.80) has the trivial constant solution $y(s) = 0$. With this and the fact that $\gamma(s - t_0) \to 0$ as $t_0 \to -\infty$, the equation for $r(s)$ becomes homogeneous,

$$\ddot{r}(s) + \int_{-\infty}^{s} dt'\, \gamma(s - t')\dot{r}(t') + \frac{1}{M}\frac{dV(r)}{dr} = 0 \,. \tag{5.82}$$

Thus we recover the standard form of the deterministic equation of motion for a damped particle. In the next step, we may go beyond the deterministic description by taking into account the Gaussian fluctuations about the stationary path. This is discussed next.

5.5.3 Semiclassical limit: quasiclassical Langevin equation

We now consider Gaussian fluctuations about the minimal action path described by $y(s)$. Again, we are interested only in the probability distribution $w(r_f; t)$ evolving from the initial distribution $w(r_i; 0)$. We wish that the evolution of $w(r_f; t)$ is represented by a double path integral of the form

$$w(r_f; t) = \int dr_i\, w(r_i; t) \int \mathcal{D}r\, \mathcal{D}y\, \exp\left(\frac{i}{\hbar}\Sigma^{(\mathrm{sc})}[r, y]\right) , \tag{5.83}$$

in which the yet to be determined action $\Sigma^{(\mathrm{sc})}$ is quadratic in y. The functional integral extends over all paths $y(t')$ and $r(t')$ that satisfy $y(0) = y(t) = 0$, $r(0) = r_i$ and $r(t) = r_f$. We proceed with the observation that in the functional integral (5.73) the last term in the expression (5.74) for the action $\Sigma[r, y]$ acts as the weight function

$$p[y(\cdot)] = \exp\left\{ -\frac{1}{2\hbar} \int_{t_p}^{t} dt' \int_{t_p}^{t} dt''\, y(t')L'(t' - t'')y(t'') \right\} . \tag{5.84}$$

In the semiclassical limit, $p[y(\cdot)]$ represents a Gaussian filter function which suppresses large deviations from the classical path. Since the fluctuations $y(t')$ are confined to small values, it is appropriate to expand the potential term in the action into powers in y and to keep the lowest nontrivial order only,

$$-V(r + y/2) + V(r - y/2) = -V'(r)\, y + \mathcal{O}(y^3) \,. \tag{5.85}$$

Next, we substitute Eq. (5.85) into Eq. (5.74) and insert the expression (5.26) for $L''(t' - t'')$. Integrating the latter term by parts we find that the term with the factor $M\gamma(0^+)$ in the action cancels out. It is also convenient to integrate the kinetic term in the action by parts. We then obtain the effective action in semiclassical approximation

$$\begin{aligned}
\Sigma^{(\mathrm{sc})}[r, y] = &- \int_{0}^{t} ds\, M\, y(s) \left[\ddot{r}(s) + \int_{0}^{s} du\, \gamma(s - u)\,\dot{r}(u) + \frac{1}{M}V'(r) \right] \\
&+ \frac{i}{2} \int_{0}^{t} ds \int_{0}^{t} du\, y(s)L'(s - u)y(u) ,
\end{aligned} \tag{5.86}$$

where we have dropped an irrelevant constant stemming from the contribution of the negative time branch. Now, the expression for the path probability in the region from $r(t)$ to $r(t) + \mathcal{D}r(t)$,

$$W[r(t)]\,\mathcal{D}r(t) \;\equiv\; \left\{ \int \mathcal{D}y\,\exp\left(\frac{i}{\hbar}\Sigma^{(sc)}[r,y]\right)\right\}\mathcal{D}r(t) \tag{5.87}$$

is a functional integral of Gaussian type which can be evaluated by "completing the square". Omitting a normalization constant, we find

$$W[r(t)]\,\mathcal{D}r(t) \;=\; \exp\left(-\frac{1}{2\hbar}\int_0^t ds \int_0^t du\,\xi[r(s)]\,L'^{-1}(s-u)\,\xi[r(u)]\right)\mathcal{D}r(t)\,. \tag{5.88}$$

Here, $L'^{-1}(s-u)$ is the inverse of $L'(s-u)$ and the functional $\xi[r(s)]$ is defined by

$$\xi[r(s)] \;\equiv\; M\ddot{r}(s) + M\int_0^s du\,\gamma(s-u)\,\dot{r}(u) + V'[r(s)]\,. \tag{5.89}$$

The form of the weight function (5.88) suggests to introduce a measure of path integration $\mathcal{D}\xi(t)$ associated with the $\xi(t)$ fluctuations. The functional Jacobian of the transformation

$$\mathcal{D}r(t) \;=\; \mathcal{J}^{-1}\mathcal{D}\xi(t) \tag{5.90}$$

is given by

$$\mathcal{J} \;=\; \det\left[\left(M\frac{\partial^2}{\partial t^2} + V''(r(t))\right)\delta(t-t') + M\int_0^t dt'\,\gamma(t-t')\frac{\partial}{\partial t'}\right]\,. \tag{5.91}$$

Schmid [54] has argued by discretization of the differential operator on the sliced time axis that the Jacobian is a constant independent of the choice of the potential $V(r)$.[3] Thus, we may consider $\xi(t)$ as an independent stochastic variable. As $\xi(t)$ in fact does not depend on the path $r(t)$, we may regard Eq. (5.89) as a standard Langevin equation with memory-friction and coloured-noise force,

$$M\ddot{r}(t) + M\int_0^t dt'\,\gamma(t-t')\dot{r}(t') + V'(r(t)) \;=\; \xi(t)\,. \tag{5.92}$$

The fluctuating force $\xi(t)$ undergoes the Gaussian stochastic process

$$W[\xi(t)]\,\mathcal{D}\xi(t) \;=\; \exp\left(-\frac{1}{2\hbar}\int_0^t ds \int_0^t du\,\xi(s)L'^{-1}(s-u)\xi(u)\right)\mathcal{D}\xi(t)\,, \tag{5.93}$$

from where we infer the properties

$$\langle\xi(t)\rangle \;=\; 0\,, \tag{5.94}$$

$$\mathcal{X}(t) \;\equiv\; \operatorname{Re}\langle\xi(t)\xi(0)\rangle \;=\; \hbar L'(t) \;=\; \frac{\hbar}{\pi}\int_0^\infty d\omega\,J(\omega)\coth(\beta\hbar\omega/2)\cos(\omega t)\,.$$

The power spectrum of the correlation function $\mathcal{X}(t)$ is found to read

$$\widetilde{\mathcal{X}}(\omega) \;\equiv\; \int_{-\infty}^\infty dt\,\mathcal{X}(t)\cos(\omega t) \;=\; M\hbar\omega\coth(\beta\hbar\omega/2)\,\gamma'(\omega)\,, \tag{5.95}$$

---

[3]The criticism stated in Ref. [6], p.1300, on the slicing in Ref. [54] does not spoil our reasoning.

where we have employed the relation (3.32). The noise characteristics of the Langevin force derived here agrees with that derived above in Subsection 3.1.4.

The relation (5.95) is the quantum mechanical version of Kubo's "Second Fluctuation-Dissipation Theorem". As $\hbar \to 0$, it becomes equivalent to the classical FDT of the second kind, Eq. (2.8). Further, since the quadratic form in y of the action $\Sigma^{(sc)}[r, y]$ becomes exact in the classical limit, the Langevin equation (5.89) becomes exact as well.

In conclusion, in the semiclassical limit, the Brownian particle can be described by a Langevin equation in which the quantum mechanical stochastic nature of the Langevin force is given by Eq. (5.94). From the above derivation it is clear that the QLE is exact for a Brownian particle with Gaussian statistics when the external force is linear in the coordinate. The results of the QLE are also reliable when the anharmonic part of the potential can be treated as a perturbation. On this level of approximation, squeezing of thermal and quantum fluctuations by time-dependent external forces has been studied in Ref. [209]. For a discussion of the validity of the QLE for anharmonic potentials we refer to Refs. [54] and [55], and to the remarks given in Subsection 2.3.3.

General quantum kinetic equations were derived in Ref. [210]. For weak dissipation, they can be reduced, upon employing a generalized Born approximation, to relaxation equations for the diagonal and off-diagonal components of the reduced density matrix, as familiar from nuclear magnetic relaxation theory [31]. In the classical limit, one is led to the energy-diffusion version of the Fokker-Planck equation studied below in Section 11.4. On the other hand, when coarse grained, the kinetic equation gives the full Fokker-Planck equation capturing the relevant physics in the classical limit.

The quasiclassical Langevin equation is easily generalized to state-dependent damping [141, b]. Particular cases are described by the action given in Eq. (4.64) or Eq. (4.186), and the relevant quasiclassical Langevin equation is of the form (3.19). Important examples in mesoscopic physics are quasiparticle tunneling between superconductors and strong electron tunneling through small junctions or metallic grains. When the fluctuations of the electronic tunneling current are strong, fluctuations of the conjugate phase variable φ are weak, and therefore the phase dynamics can well be described in terms of a Langevin equation for the phase variable φ with a state-dependent stochastic force $\xi_1(t) \cos \varphi + \xi_2(t) \sin \varphi$, where $\xi_1(t)$ and $\xi_2(t)$ are two independent stochastic variables obeying Gaussian statistics with a correlator of the form (5.94) with Eq. (5.95). For a recent application of a quasiclassical Langevin equation with state-dependent noise to "strong electron tunneling", see Ref. [211].

5.6 Stochastic unraveling of influence functionals

Present exact numerical techniques such as quantum Monte Carlo (QMC) methods keep struggling with the so-called dynamical sign problem, which is due to the fact that complex-valued probability amplitudes with varying phases rather than probabilities have to be summed up in a quantum mechanical computation.

Path integral representations of the reduced density matrix are intimately connected to stochastic processes for pure quantum states. It has been shown in the recent literature that there is a close connection between influence functionals and non-Markovian generalizations of quantum state diffusion [212] as well as Langevin-like stochastic dynamics of the reduced density matrix [213].

Quantum state diffusion (QSD) has been established as an alternative approach to quantum dissipation in the perturbative weak-coupling regime [65, 66, 60, 61]. In this method, *pure* quantum states are propagated stochastically. This yields an appealing picture of individual quantum trajectories for open systems and facilitates effective numerical calculations. Extension of this method to the case of nonperturbative, non-Markovian dynamics paves the way to overcome many of the above-mentioned numerical limitations. A corresponding generalization of quantum state diffusion has been proposed by Strunz, Diósi, and Gisin [212]. In their approach, the cross term $\exp\{-\mathcal{S}_{\mathrm{FV,cross}}[q, q']/\hbar\}$ of the Feynman-Vernon influence functional (5.21) with (5.25) is unraveled into a stochastic influence functional. However, the resulting propagation amplitude $G[q]$ still experiences quantum memory effects arising from the selfinteraction term $\exp\{-\mathcal{S}_{\mathrm{FV,self}}[q]/\hbar\}$. As a result of that, the respective stochastic Schrödinger equation is equipped with a time-nonlocal functional derivative, for which a general solution strategy, even numerically, is not known, except in restrictive limits. Stockburger and Grabert overcame these difficulties by unraveling the full influence functional into a stochastic one [214]. Following these lines, they derived exact c-number representations of non-Markovian quantum dissipation in terms of stochastic Schrödinger equations or equivalent stochastic Liouville–von Neumann equations, which are free of quantum memory effects.

To study the principle of stochastic unraveling of the influence functional we take the contour path integral (5.57) for the propagating function J as starting point, where the contour is chosen to be the contour $\mathcal{C}_{\mathrm{I}}$ specified in Eq. (5.52) and in Fig. 5.3. To proceed, we define a complex-valued Gaussian noise force $\xi(z)$, where $z \in \mathcal{C}_{\mathrm{I}}$, and a stochastic noise action (here we disregard the potential renormalization term)

$$S_\xi[\widetilde{q}] = S_{\mathrm{S}}[\widetilde{q}] + \int_{\mathcal{C}_{\mathrm{I}}} dz\, \xi(z)\widetilde{q}(z) , \qquad (5.96)$$

where the path $\widetilde{q}(z)$ is specified in Eq. (5.52). For a given realization $\xi(z)$, the time evolution of the propagating function J_ξ and of the density matrix ρ_ξ is unitary,

$$J_\xi(q_f, q_f'; t; q_i, q_i'; \bar{q}, \bar{q}') = \frac{1}{Z} \int_{\mathcal{C}_{\mathrm{I}}} \mathcal{D}\widetilde{q}(\cdot)\, e^{iS_{\mathrm{S}}[\widetilde{q}(\cdot)]/\hbar} \exp\left(\frac{i}{\hbar} \int_{\mathcal{C}_{\mathrm{I}}} dz\, \xi(z)\widetilde{q}(z)\right) . \qquad (5.97)$$

The propagating function $J(q_f, q_f'; t; q_i, q_i'; \bar{q}, \bar{q}')$, as well as the density matrix $\rho(q_f, q_i; t)$, is the average over all the noise realizations,

$$J(q_f, q_f'; t; q_i, q_i'; \bar{q}, \bar{q}') = \big\langle J_\xi(q_f, q_f'; t; q_i, q_i'; \bar{q}, \bar{q}') \big\rangle_\beta . \qquad (5.98)$$

For Gaussian noise, the average over the noise action factor is an exponential of the second cumulant of $\xi(t)$,

$$\left\langle \exp\left(\frac{i}{\hbar}\int_{C_{\mathrm{I}}} dz\, \xi(z)\tilde{q}(z)\right)\right\rangle_\beta = \exp\left(-\frac{1}{\hbar^2}\int_{C_{\mathrm{I}}} dz \int_{z>z'} dz'\, \tilde{q}(z)\,\langle\xi(z)\xi(z')\rangle_\beta\, \tilde{q}(z')\right).$$

$$(5.99)$$

From this we see that the noise-averaged propagating function (5.98) coincides with the former expression (5.57), if we identify [cf. Eq. (5.34)]

$$\langle\xi(z)\xi(z')\rangle_\beta = \hbar L(z - z'), \qquad \text{for} \qquad z \geq z'. \tag{5.100}$$

For the case of a factorizing initial condition described by the Feynman-Vernon contour C_{FV} in Fig. 5.3, we can replace the noise force $\xi(z)$ with complex-time argument z by two noise forces with real-time argument, one along the forward path, $\xi_1(\tau) = \xi(\tau - i\hbar\beta)$, and the other along the backward path, $\xi_2(\tau) = \xi^*(z = \tau)$. Assuming that the random forces have correlations

$$\langle\xi_{1/2}(\tau)\xi_{1/2}(0)\rangle_\beta = \hbar L(\tau), \qquad \text{and} \qquad \langle\xi_1(\tau)\xi_2(0)\rangle_\beta = \hbar L^*(\tau), \tag{5.101}$$

the stochastic influence functional (5.97) for the contour C_{FV} leads to an unraveling of the reduced dynamics

$$\rho(q_f, q_f'; t) = \int dq_i \int dq_i' \left\langle G_{\xi_1}(q_f, q_i; t)\, G_{\xi_2}^*(q_f', q_i'; t)\right\rangle_\beta \rho(q_i, q_i'; 0), \tag{5.102}$$

where $G_{\xi_1}(q_f, q_i; t)$ is the stochastic propagator

$$G_{\xi_j}(q_f, q_i; t) = \int_{q_i}^{q_f} \mathcal{D}[q(\cdot)]\, e^{iS_{\mathrm{S}}[q(\cdot)]/\hbar} \exp\left(\frac{i}{\hbar}\int_0^t d\tau\, \xi_j(\tau)\, q(\tau)\right), \tag{5.103}$$

which is free of quantum mechanical memory. Since the propagator is different for the forward and backward paths, the propagation of a stochastic sample of the density matrix with initial state

$$\rho_{\mathrm{ini}} = |\psi_1> <\psi_2| \tag{5.104}$$

is determined by two different stochastic Schrödinger equations,

$$i\hbar\frac{d}{dt}|\psi_{1/2}> = H_{\mathrm{S}}|\psi_{1/2}> - \xi_{1/2}(t)\, q\, |\psi_{1/2}> . \tag{5.105}$$

These equations are local in time and their solution renders the propagation amplitudes (5.103). Hence all effects of the quantum mechanical memory are contained in the averaging procedure in Eq. (5.102).

The dynamics governed by the stochastic Schrödinger equations (5.105) can be rewritten in the form of a stochastic Liouville–von Neumann (SLN) equation. Upon introducing the linear combinations $\zeta(t) = \frac{1}{2}[\xi_1(t) + \xi_2^*(t)]$ and $\nu(t) = \xi_1(t) - \xi_2^*(t)$

we obtain the SLN equation with the noise term in the form of a commutator and an anticommutator,

$$\frac{d}{dt}\rho_\xi(t) = -\frac{i}{\hbar}[H_{\rm S}, \rho_\xi(t)] + \frac{i}{\hbar}\zeta(t)[q, \rho_\xi(t)] + \frac{i}{2\hbar}\nu(t)\{q, \rho_\xi(t)\}. \qquad (5.106)$$

The correlations of the new random forces are

$$\langle\zeta(t)\zeta(0)\rangle_\beta = \hbar L'(t), \quad \langle\zeta(t)\nu(0)\rangle_\beta = 2i\hbar L''(t), \quad \langle\nu(t)\nu(0)\rangle_\beta = 0, \quad (5.107)$$

where $L(t) = L'(t) + i\,L''(t)$ is given in Eq. (5.23), whereas the correlations $\langle\zeta(t)\zeta^*(0)\rangle_\beta$, $\langle\zeta(t)\nu^*(0)\rangle_\beta$ and $\langle\nu(t)\nu^*(0)\rangle_\beta$ are left undetermined by the physical model, so that they can be chosen according to computational convenience.

The stochastic linear Schrödinger equations (5.105), as other stochastic linear Schrödinger equations, do not preserve the norm of the quantum states. The transition from linear QSD approaches to stochastic processes with normalized quantum states proved to be a decisive step towards robust numerical simulations. Likewise, the SLN approach can be transformed into a stochastic approach for a normalized density matrix [214].

5.7   Brief summary and outlook

So far we have presented the general real-time formalism for the density matrix with the phenomenological model described by the Hamiltonian (3.11). In the first section we have familiarized ourselves with the general formalism by studying the evolution of a product initial state. Then we have addressed the important question of the preparation of initial states. We have introduced a preparation function which describes the reduction of the equilibrium density matrix of the global system due to a measurement of certain variables of the system carried out at time $t = 0$. The subsequent evolution is determined by a propagating function.

In the third section of this chapter we have studied the propagating function and found the exact formal solution in the compact form of a path integral where the progression in time is along a certain contour in the complex-time plane. We then have discussed a real-time formulation of the propagation function in which the system is evolving from the infinite past and is subject to a measurement of its coordinates at time zero, and again at time t.

The appropriate real-time modifications required for the microscopic models introduced in Section 3.3 are easy to understand, and the implementation is left to the reader. The crucial point is the observation that the action $-iS_{\rm S}[\tilde{q}] + \mathcal{S}[\tilde{q}]$ in Eq. (5.57) or (5.71) for the various microscopic models results from the corresponding imaginary-time action $S_{\rm eff}^{\rm (E)}[q]$ by the substitution $\tau = iz$ and $q(\tau) = \tilde{q}(z)$ and subsequent deformation of the contour in the complex-z plane as indicated in Figs. 5.3 and 5.4 [cf. also the analytic continuation in Eq. (5.55)].

The methods of this chapter will be applied below *inter alia* to the dissipative two-state and multi-state system.

PART II
FEW SIMPLE APPLICATIONS

Up to this point, we have employed functional integral methods in order to set out exact formal expressions which describe the thermodynamics and dynamics of damped quantum systems. Now we move to new grounds and consider the physical properties of exactly solvable damped linear systems, i.e. the damped quantum oscillator and quantum Brownian particle. Subsequently, we discuss a variational method which makes it possible for us to treat nonlinear damped quantum systems in much the same way as linear systems. We conclude this part with a brief general discussion of quantum decoherence.

6. Damped harmonic oscillator

In this chapter, I study static and dynamical properties of a quantum mechanical linear oscillator subject to linear dissipation of arbitrary frequency dependence.

The understanding of the quantum-mechanical damped oscillator is very important because of its universal relevance. Whenever a macroscopic quantum system is (slightly) displaced from a stable local potential minimum, this model applies. Applications are, e.g., current oscillations in circuits with impedances (cf. Sect. 3.4).

The problem of a damped harmonic quantum mechanical oscillator has been studied intensively since the early sixties [19, 20, 25, 27, 75]. In subsequent works [54, 56, 196], [215] – [221], attention was put on the asymptotic low temperature characteristics which show anomalous behaviour.

In the presence of a time dependent external force $f_{\text{ext}}(t)$ coupled linearly to the coordinate of the particle, the Hamiltonian of the global system reads

$$H(t) = \frac{p^2}{2M} + \frac{M\omega_0^2 q^2}{2} - q f_{\text{ext}}(t) + \frac{1}{2} \sum_{\alpha=1}^{N} \left[\frac{p_\alpha^2}{m_\alpha} + m_\alpha \omega_\alpha^2 \left(x_\alpha - \frac{c_\alpha}{m_\alpha \omega_\alpha^2} q \right)^2 \right] . \quad (6.1)$$

It describes a central harmonic oscillator with eigenfrequency ω_0 in a spatially constant force field and bilinearly coupled to a bath of harmonic oscillators. The Hamiltonian (6.1) gives linear equations of motion, which can be solved in analytic form. Also, the Hamiltonian of the global system can be diagonalized directly, as I have already discussed briefly in Section 4.3.5. Path integrals for (thermo)-dynamical quantities, e.g. the partition function or the propagating function, can also be calculated exactly since they are in Gaussian form. The respective treatment is similar to the ones given in Sections 4.2 and 5.1 and is given in Ref. [86] (see also the more recent work in Ref. [47]).

Fortunately, a damped linear oscillator is a system simple enough to enable a study of the quantum mechanical stochastic process in the entire range of parameters by using only phenomenological considerations. Here, I neither pursue direct diagonalization nor use the path integral approach. I rather consider a simple stochastic modeling of the quantum particle affected by thermal and quantum stochastic fluctuations. This is possible since the quantum mechanical equation for $q(t)$ is in the form of a classical Langevin equation with quantum mechanical coloured noise. Before embarking on the specific problem, I consider an important theorem which is generally valid for any particular linear and nonlinear quantum system.

6.1   Fluctuation–dissipation theorem

The fluctuation-dissipation theorem (FDT) is the cornerstone of linear response theory. The theorem relates relaxation of a weakly perturbed system to the spontaneous fluctuations in thermal equilibrium. Well-known special cases of the FDT in the classical regime are the Einstein relation [222], which relates the diffusion constant to the viscosity of a Brownian particle, and the Johnson-Nyquist formula [223, 224], which relates thermal current fluctuations in an electrical circuit to the impedance.

Let us now see how to derive the FDT from the principles of quantum statistical mechanics. Consider a system described by a Hamiltonian H which has reached the canonical equilibrium state. Imagine that at a particular time a weak time-dependent external force $\delta f_{\text{ext}}(t)$ is switched on. By the perturbation the system is thrown off equilibrium.

With the additional force, the Hamiltonian is time-dependent,

$$\mathcal{H}(t) = H + \delta H(t) = H - q\,\delta f_{\text{ext}}(t)\,. \tag{6.2}$$

The time evolution of $\langle q(t)\rangle$ for a mixed initial state is carried by the density matrix $\rho(t)$ according to

$$\langle q(t)\rangle = \operatorname{tr}\{\rho(t)\,q\}\,, \tag{6.3}$$

where $\operatorname{tr}\{\rho(t)\} = 1$. Here, $\operatorname{tr}\{\cdots\}$ denotes the usual quantum mechanical trace operation, and $\rho(t)$ obeys the equation of motion

$$i\hbar\,\dot{\rho}(t) = [\mathcal{H}(t),\,\rho(t)] = [H,\,\rho(t)] + [\delta H(t),\,\rho(t)]\,, \tag{6.4}$$

where $[A, B] = AB - BA$. We assume that the system is in thermal equilibrium at the time where the perturbation $\delta H(t)$ is turned on,

$$\lim_{t\to-\infty} \rho(t) = \rho_{\text{eq}} = e^{-\beta H}/Z\,,$$

$$\lim_{t\to-\infty} \langle q(t)\rangle = \langle q\rangle_\beta = \operatorname{tr}\{e^{-\beta H}q\}/Z\,, \tag{6.5}$$

$$Z = \operatorname{tr}\{e^{-\beta H}\}\,.$$

For weak external perturbation $\delta H(t)$, the average displacement of the particle from the equilibrium position is a linear functional of the history of the time-dependent external force $\delta f_{\text{ext}}(t)$,

$$\delta \langle q(t) \rangle \equiv \langle q(t) \rangle - \langle q \rangle_\beta = \int_{-\infty}^{+\infty} dt' \, \chi(t - t') \, \delta f_{\text{ext}}(t') \, . \tag{6.6}$$

The causal response function $\chi(t)$ is a retarded temperature Green function called generalized susceptibility. To determine $\chi(t)$, we put

$$\rho(t) = \rho_{\text{eq}} + \delta \rho(t) \, . \tag{6.7}$$

The change of the density matrix $\delta \rho(t)$ caused by the perturbation obeys the equation

$$i\hbar \, \delta \dot{\rho}(t) = [\, \delta H(t), \rho_{\text{eq}}\,] + [\, H, \delta \rho(t)\,] \, . \tag{6.8}$$

This equation is easily solved and gives

$$\delta \rho(t) = \frac{1}{i\hbar} \int_{-\infty}^{t} dt' \, e^{-iH(t-t')/\hbar} [\, \delta H(t'), \, \rho_{\text{eq}}\,] \, e^{iH(t-t')/\hbar} \, . \tag{6.9}$$

Upon inserting the expression (6.9) into $\delta \langle q(t) \rangle = \text{tr} \{ \delta \rho(t) \, q \}$, we find that $\chi(t)$ is connected with the antisymmetric part of the position auto-correlation function in thermal equilibrium by the relation

$$\chi(t) = \frac{i}{\hbar} \Theta(t) \, \langle \, [\, q(t)q(0) - q(0)q(t)\,] \, \rangle_\beta \, . \tag{6.10}$$

The function $\chi(t)$ describes the linear response of the equilibrated system to an external force. The step function implicit in Eq. (6.10) saves *causality*. Since we did not use specific properties of the system, the expression (6.10) is a general result for any quantum system within linear response.

It is convenient to define the position autocorrelation function $C^{\pm}(t)$ with a suitable subtraction in order that the Fourier transform is well-defined,

$$\begin{aligned}
C^{+}(t) &\equiv C_{qq}(t) &&\equiv \langle q(t)q(0) \rangle_\beta - \langle q(t) \rangle_\beta \langle q(0) \rangle_\beta \, , \\
C^{-}(t) &\equiv C_{qq}(-t) &&\equiv \langle q(0)q(t) \rangle_\beta - \langle q(0) \rangle_\beta \langle q(t) \rangle_\beta \, .
\end{aligned} \tag{6.11}$$

The thermal expectation value $\langle q(t) \rangle_\beta$ is independent of time, $\langle q(t) \rangle_\beta = \langle q(0) \rangle_\beta$. With the subtraction in Eq. (6.11), we ensure that $C^{\pm}(t) \to 0$ as $t \to \infty$. Since $C^{+*}(t) = C^{-}(t)$, we may write

$$C^{\pm}(t) = S(t) \pm i \, A(t) \, . \tag{6.12}$$

The real part $S(t)$ is the symmetrized correlation function

$$S(t) = S(-t) = \frac{1}{2} \langle \, [\, q(t)q(0) + q(0)q(t)\,] \, \rangle_\beta - \langle q(0) \rangle_\beta^2 \, , \tag{6.13}$$

and the imaginary part is expressed in terms of the expectation value of the anti-commutator as

$$A(t) = -A(-t) = \frac{1}{2i} \langle [q(t)q(0) - q(0)q(t)] \rangle_\beta . \tag{6.14}$$

We see from Eqs. (6.10) and (6.14) that the response function $\chi(t)$ and the function $A(t)$ are related by

$$\chi(t) = -\frac{2}{\hbar}\Theta(t)A(t) . \tag{6.15}$$

Switching to the Fourier transforms

$$\begin{aligned}\widetilde{C}^\pm(\omega) &\equiv \int_{-\infty}^{\infty} dt\, C^\pm(t)\, e^{i\omega t} = \widetilde{S}(\omega) \pm i\,\widetilde{A}(\omega) , \\ \widetilde{\chi}(\omega) &\equiv \int_0^\infty dt\, \chi(t)\, e^{i\omega t} ,\end{aligned} \tag{6.16}$$

the relation (6.15) is transformed into

$$\widetilde{\chi}''(\omega) = \frac{i}{\hbar}\widetilde{A}(\omega) = \frac{1}{2\hbar}\left(\widetilde{C}^+(\omega) - \widetilde{C}^-(\omega)\right) . \tag{6.17}$$

We now prepare the ground to proof the fluctuation-dissipation theorem. This can be done in two different ways. In the first, we utilize that the canonical operator $e^{-\beta H}$ is an imaginary-time translational operator, $e^{-\beta H}q(0)\,e^{\beta H} = q(i\hbar\beta)$, and employ invariance of the trace under cyclic permutation. This yields

$$\text{tr}\{e^{-\beta H}q(0)\,q(t)\} = \text{tr}\{q(i\hbar\beta)\,e^{-\beta H}q(t)\} = \text{tr}\{e^{-\beta H}q(t)\,q(i\hbar\beta)\} . \tag{6.18}$$

Since the correlations in thermal equilibrium are time translation invariant,

$$\langle q(t)q(t') \rangle_\beta = \langle q(t-t')q(0) \rangle_\beta , \tag{6.19}$$

we immediately find from Eq. (6.18) the relation

$$C^-(t) = C^+(t - i\hbar\beta) . \tag{6.20}$$

Upon using the relation (6.20), we then obtain in Fourier space as a central result

$$\widetilde{C}^-(\omega) = e^{-\omega\hbar\beta}\,\widetilde{C}^+(\omega) . \tag{6.21}$$

In the alternative second way, we write the correlation functions (6.11) in energy representation and then switch to the respective Fourier transforms. This yields

$$\begin{aligned}\widetilde{C}^+(\omega) &= \frac{2\pi\hbar}{Z}\sum_{n,m} e^{-\beta E_m} <n|q|m><m|q|n> \delta(E_m - E_n + \hbar\omega) , \\ \widetilde{C}^-(\omega) &= \frac{2\pi\hbar}{Z}\sum_{n,m} e^{-\beta E_n} <n|q|m><m|q|n> \delta(E_m - E_n + \hbar\omega) .\end{aligned} \tag{6.22}$$

Since we have the constraint $E_n = E_m + \hbar\omega$, the expressions (6.22) are indeed in agreement with relation (6.21).

Insertion of Eq. (6.21) into Eq. (6.17) completes the derivation and yields the fluctuation-dissipation theorem discovered first by Callen and Welton in 1951 [225].

$$\widetilde{\chi}''(\omega) \;=\; \frac{1}{2\hbar}\left(1 - e^{-\omega\hbar\beta}\right)\widetilde{C}^+(\omega)\,. \qquad (6.23)$$

We can also relate the Fourier transform of the symmetrized autocorrelation function to $\widetilde{\chi}''(\omega)$. Upon using the relation $\widetilde{S}(\omega) = \frac{1}{2}[\widetilde{C}^+(\omega) + \widetilde{C}^-(\omega)]$ and Eq. (6.21), we find

$$\widetilde{S}(\omega) \;=\; \hbar\coth(\omega\hbar\beta/2)\,\widetilde{\chi}''(\omega)\,. \qquad (6.24)$$

There are two important limits of the quantum-mechanical fluctuation-dissipation theorem. In the extreme quantum limit $k_BT \ll \hbar\omega$, we have pure quantum fluctuations,

$$\widetilde{S}(\omega) \;=\; \hbar\,|\widetilde{\chi}''(\omega)|\,. \qquad (6.25)$$

Since the function $\widetilde{S}(\omega)$ in (6.25) is nonanalytic at the origin, the position autocorrelation function decays algebraically with time at zero temperature.

For low frequencies or high temperatures, $\hbar\omega << k_BT$, the spectral function $\widetilde{S}(\omega)$ becomes classical,

$$\widetilde{S}(\omega) \;=\; 2k_BT\,\frac{\widetilde{\chi}''(\omega)}{\omega}\,. \qquad (6.26)$$

Special cases of this formula are the Einstein relation, which relates the diffusion coefficient D to the mobility μ, $D = k_BT\,\mu$ (see Section 7.3 for linear systems and Subsection 24.4.2 for a nonlinear system), and the Johnson-Nyquist formula, which relates the power spectrum of the current fluctuations $S_{\mathrm{II}}(\omega)$ in an electrical circuit to the admittance $Y(\omega)$, $S_{\mathrm{II}}(\omega) = 2k_BT\,\mathrm{Re}\,Y(\omega)$.

It is appropriate to emphasize that at no step of the derivation we used particular properties of a linear system. Thus, the FDT relations (6.23) and (6.24) are generally valid for any open quantum system in thermal equilibrium. Therefore, they have broad implications far beyond the particular linear model systems discussed here.

6.2 Stochastic modeling

Direct knowledge about many important properties of linear dissipative quantum systems is easily gathered from a simple stochastic modeling [215] which is based upon the following three principles:

i. The mean values obey the classical equations of motion (*Ehrenfest's theorem*).

ii. The response function and the equilibrium autocorrelation function are related by the *fluctuation-dissipation theorem*.

iii. The stochastic process is a *stationary Gaussian process*.

While the first two principles are generally valid, the assumption of a strict Gaussian process is limited to linear systems. For *linear* systems, the response to an external perturbation is *linear* for arbitrary strength of the perturbation. Thus, all we said in the previous subsection for an infinitesimal perturbation $\delta f_{\text{ext}}(t)$ is generally valid for an external force $f_{\text{ext}}(t)$ of any strength in Eq. (6.1).

By virtue of Ehrenfest's theorem, the average position $\langle q(t) \rangle$ of a damped quantum oscillator obeys the equation of motion

$$\langle \ddot{q}(t) \rangle + \int_{-\infty}^{t} dt' \, \gamma(t - t') \langle \dot{q}(t') \rangle + \omega_0^2 \langle q(t) \rangle = \frac{1}{M} f_{\text{ext}}(t) \,. \tag{6.27}$$

To be general, we have permitted memory-friction. The response of $\langle q(t) \rangle$ to a general force $f_{\text{ext}}(t)$ is given by [cf. Eq. (6.6)]

$$\langle q(t) \rangle = \int_{-\infty}^{+\infty} dt' \, \chi(t - t') f_{\text{ext}}(t') \,, \tag{6.28}$$

where $\chi(t)$ is the causal response function, $\chi(t) = 0$ for $t \leq 0$. Causality requires that the dynamical susceptibility

$$\tilde{\chi}(\omega) = \int_{-\infty}^{\infty} dt \, \chi(t) \, e^{i\omega t} \tag{6.29}$$

is an analytic function of ω in the upper complex ω–half-plane ($\text{Im}\,\omega > 0$), which is the origin of the significant Kramers–Kronig relations

$$\tilde{\chi}'(\omega) = \mathcal{P} \int_{-\infty}^{\infty} \frac{d\nu}{\pi} \frac{\tilde{\chi}''(\nu)}{\nu - \omega} \,, \qquad \tilde{\chi}''(\omega) = -\mathcal{P} \int_{-\infty}^{\infty} \frac{d\nu}{\pi} \frac{\tilde{\chi}'(\nu)}{\nu - \omega} \,. \tag{6.30}$$

These relations are widely employed, e.g. in optics, since they express dispersion in terms of absorption and vice versa.

It is a general feature for any linear quantum system that the Ehrenfest equation (6.27) coincides exactly with the classical equation of motion for the position $q(t)$. Therefore, the quantum mechanical response function is identical with the classical response function, $\chi(t) = \chi_{\text{cl}}(t)$.

For the linear equation of motion (6.27), the dynamical susceptibility reads

$$\tilde{\chi}(\omega) \equiv \tilde{\chi}'(\omega) + i\,\tilde{\chi}''(\omega) = \frac{1}{M} \frac{1}{\omega_0^2 - \omega^2 - i\omega\tilde{\gamma}(\omega)} \,, \tag{6.31}$$

where $\tilde{\gamma}(\omega) = \int_0^{\infty} dt\, \gamma(t)\, e^{i\omega t}$ is the frequency-dependent damping coefficient. To complete the picture, we also give the response function in Laplace representation,

$$\chi(t) = \frac{1}{2\pi i} \int_{-i\infty + c}^{i\infty + c} dz\, \hat{\chi}(z)\, e^{zt} \,, \tag{6.32}$$

$$\hat{\chi}(z) = \frac{1}{M[\omega_0^2 + z^2 + z\hat{\gamma}(z)]} \,. \tag{6.33}$$

The constant c in the definition of the integration contour in Eq. (6.32) is chosen such that all singularities of $\hat{\chi}(z)$ are lying to the left of the integration path.

Since we wish to calculate the thermal position autocorrelation function[1]

$$C_{qq}(t) \equiv \langle q(t)q(0)\rangle_\beta \tag{6.34}$$

with the help of the fluctuation-dissipation theorem, we require the absorptive part of the dynamical susceptibility. From (6.31) we get

$$\tilde{\chi}''(\omega) = \frac{1}{M} \frac{\omega\tilde{\gamma}'(\omega)}{[\omega^2 - \omega_0^2 - \omega\tilde{\gamma}''(\omega)]^2 + \omega^2\tilde{\gamma}'^2(\omega)} . \tag{6.35}$$

For frequency-independent damping, $\tilde{\gamma}(\omega) = \gamma$, the expression (6.35) can be decomposed into two Lorentzians centered about $\pm\,\omega_0\sqrt{1 - (\gamma/2\omega_0)^2}$. In the limit $\gamma \to 0$, the Lorentzians reduce to δ-functions located at $+\omega_0$ and $-\omega_0$. This is the correct result for the bare oscillator.

Looking back we come to an important conclusion. Since for linear quantum systems the dynamical susceptibility $\tilde{\chi}(\omega)$ is purely classical, quantum mechanics enters into the correlation functions solely by the fluctuation dissipation theorem, Eq. (6.24).

In a stationary stochastic process, the time correlations for dynamical variables do *not* depend on absolute times. Time translation invariance of correlations is reflected by the property (6.19). Thus, once we have gained knowledge about the function $C_{qq}(t)$, the other relevant pair correlation functions can be found by differentiation. Upon relating the momentum to the time derivative of the position and using Eq. (6.19), we obtain the simple relations

$$
\begin{aligned}
C_{pq}(t) &\equiv \langle p(t)q(0)\rangle_\beta &=& \quad M\frac{d}{dt}\langle q(t)q(0)\rangle_\beta , \\[6pt]
C_{qp}(t) &\equiv \langle q(t)p(0)\rangle_\beta &=& \quad -M\frac{d}{dt}\langle q(t)q(0)\rangle_\beta , \\[6pt]
C_{pp}(t) &\equiv \langle p(t)p(0)\rangle_\beta &=& \quad -M^2\frac{d^2}{dt^2}\langle q(t)q(0)\rangle_\beta .
\end{aligned}
\tag{6.36}
$$

It follows from the Gaussian property (iii) that correlation functions with an odd number of variables (position or momentum) vanish whereas correlation functions with an even number of variables can be written as a sum of combinations of factorized pair correlation functions. Within each pair, the original order of the variables is preserved [226]. For illustration, we give an example,

$$\langle q(t)p(t')p(t'')\,q(0)\rangle_\beta = \langle q(t)p(t')\rangle_\beta\langle p(t'')q(0)\rangle_\beta \tag{6.37}$$

$$+ \langle q(t)p(t'')\rangle_\beta\langle p(t')q(0)\rangle_\beta + \langle q(t)q(0)\rangle_\beta\langle p(t')p(t'')\rangle_\beta .$$

Thus, all correlation functions with an even number of dynamical variables can be expressed in terms of the position autocorrelation function and of its time derivatives. In this way, the complete quantum mechanical stochastic process of the damped harmonic system can be determined.

---

[1] Note that there is no subtraction in $C_{qq}(t)$ since the equilibrium average $\langle q\rangle_\beta$ is zero.

6.3   Susceptibility for Ohmic friction and Drude damping

6.3.1   Strict Ohmic friction

For *strict Ohmic* or frequency-independent damping, the dynamical susceptibility
(6.31) has two poles in the lower ω half-plane at $\omega = -i\,\lambda_{1/2}$. In Laplace space, we
have

$$\widehat{\chi}(z) \; = \; \frac{1}{M(\omega_0^2 + z^2 + \gamma z)} \; = \; \frac{1}{M(z + \lambda_1)(z + \lambda_2)}\,, \qquad (6.38)$$

where $\lambda_1 + \lambda_2 = \gamma$, and $\lambda_1 \lambda_2 = \omega_0^2$. The poles are located at $z = -\lambda_{1/2}$, where

$$\lambda_{1/2} \; = \; \frac{\gamma}{2} \pm i\,\Omega\,, \qquad \Omega \; = \; \sqrt{\omega_0^2 - \tfrac{1}{4}\gamma^2}\,. \qquad (6.39)$$

Upon closing the contour in Eq. (6.32) to pick up the two poles at $z = -\lambda_{1/2}$, we find

$$\begin{aligned} \chi(t) &= -(2/\hbar)\,\Theta(t)\,A(t)\,, \\ A(t) &= -(\hbar/2M\Omega)\sin(\Omega t)\,\mathrm{e}^{-\gamma|t|/2}\,. \end{aligned} \qquad (6.40)$$

The function $A(t)$ and the response function $\chi(t)$ describe damped oscillations when
$0 < \gamma < 2\omega_0$. In the overdamped regime $\gamma > 2\omega_0$, the frequency Ω is imaginary, and
both $A(t)$ and $\chi(t)$ decay exponentially with relaxation rates $\frac{\gamma}{2} \pm \sqrt{\frac{\gamma^2}{4} - \omega_0^2}$.

6.3.2   Drude damping

Explicit expressions are also found for other forms of the spectral density. When $\widehat{\gamma}(z)$
is rational in z, the characteristic equation determining the poles of $\widehat{\chi}(z)$ is algebraic.
For the familiar Drude damping (3.53), the susceptibility (6.33) takes the form

$$\widehat{\chi}(z) \; = \; \frac{z + \omega_{\mathrm{D}}}{M(z + \lambda_1)(z + \lambda_2)(z + \lambda_3)}\,. \qquad (6.41)$$

The λ_k are the roots of the cubic equation

$$\lambda^3 - \omega_{\mathrm{D}}\lambda^2 + (\omega_0^2 + \gamma\omega_{\mathrm{D}})\lambda - \omega_{\mathrm{D}}\omega_0^2 \; = \; 0\,, \qquad (6.42)$$

and they satisfy the Vieta relations

$$\begin{aligned} \lambda_1 + \lambda_2 + \lambda_3 &= \omega_{\mathrm{D}}\,, \\ \lambda_1\lambda_2 + \lambda_2\lambda_3 + \lambda_3\lambda_1 &= \omega_0^2 + \gamma\omega_{\mathrm{D}}\,, \\ \lambda_1\lambda_2\lambda_3 &= \omega_{\mathrm{D}}\omega_0^2\,. \end{aligned} \qquad (6.43)$$

The λ_k can be given in explicit form by employing Cardano's formula. For nonzero
damping, the roots have positive real parts. The discriminant D vanishes for the
critical damping $\gamma = \gamma_{\mathrm{c}}$. For $\gamma < \gamma_{\mathrm{c}}$, one of the roots is real while the other two
are complex conjugate. For $\gamma > \gamma_{\mathrm{c}}$, they are all real. In the first case, the oscillator

is *underdamped*, otherwise *overdamped*. For $\omega_{\rm D}/\omega_0 \gg 1$ and $\gamma < 2\omega_0$, the λ_k are approximately given by (we use the parameter $\alpha = \gamma/2\omega_0$, where appropriate)

$$\lambda_{1/2} = \frac{\gamma}{2}\left(1 + \frac{\gamma}{\omega_{\rm D}}\right) \pm i\Omega\left(1 + \frac{1 - 2\,\alpha^2}{2(1 - \alpha^2)}\frac{\gamma}{\omega_{\rm D}}\right),$$

$$\lambda_3 = \omega_{\rm D}\left(1 - \frac{\gamma}{\omega_{\rm D}} - \frac{\gamma^2}{\omega_{\rm D}^2}\right),$$

(6.44)

where Ω is defined in Eq. (6.39), and where terms of higher order in $\gamma/\omega_{\rm D}$ have been disregarded. The generalization of the expressions (6.40) in the Drude case is straight and left to the attentive reader.

6.4 The position autocorrelation function

With the aid of the FDT (6.23), we obtain for the position autocorrelation function $C^+(t)$ defined in Eq. (6.11) the Fourier integral representation

$$C^+(t) = \frac{\hbar}{\pi}\int_{-\infty}^{\infty} d\omega\,\widetilde{\chi}''(\omega)\frac{e^{-i\omega t}}{1 - e^{-\omega\hbar\beta}}.$$

(6.45)

For the damped harmonic oscillator, the absorptive part of the dynamical susceptibility is given in Eq. (6.35). With this we have

$$C^+(t) = \frac{\hbar}{\pi M}\int_{-\infty}^{\infty} d\omega\,\frac{\omega\widetilde{\gamma}'(\omega)}{[\omega^2 - \omega_0^2 - \omega\widetilde{\gamma}''(\omega)]^2 + \omega^2\widetilde{\gamma}'^2(\omega)}\frac{e^{-i\omega t}}{1 - e^{-\omega\hbar\beta}}.$$

(6.46)

From this expression we see that $C^+(t)$ can be analytically continued to complex times $z = t - i\tau$ in the strip $0 < \tau < \hbar\beta$. Since $\widetilde{\chi}''(\omega)$ is an odd function of ω, the Euclidean or imaginary-time correlation function

$$C^{({\rm E})}(\tau) \equiv C^+(z = -i\tau)$$

(6.47)

is real and periodic, $C^{({\rm E})}(0) = C^{({\rm E})}(\hbar\beta)$, so that it can be Fourier expanded,

$$C^{({\rm E})}(\tau) = \frac{1}{\hbar\beta}\sum_{n=-\infty}^{+\infty} c_n\,e^{-i\nu_n\tau},$$

(6.48)

where the frequencies ν_n are the bosonic Matsubara frequencies

$$\nu_n = \frac{2\pi}{\hbar\beta}n, \qquad n = \pm 1, \pm 2, \cdots.$$

(6.49)

The Fourier coefficients c_n of the imaginary-time correlation function can be related to the dynamical susceptibility $\widetilde{\chi}(\omega)$. Upon using Eqs. (6.47) and (6.45) we get

$$
\begin{aligned}
c_n &= \int_0^{\hbar\beta} d\tau\,C^{({\rm E})}(\tau)\,e^{i\nu_n\tau} \\
&= \frac{\hbar}{\pi}\int_{-\infty}^{\infty} d\omega\,\frac{\widetilde{\chi}''(\omega)}{\omega - i\nu_n} = \frac{\hbar}{\pi}{\rm Im}\int_{-\infty}^{\infty} d\omega\,\frac{\omega\,\widetilde{\chi}(\omega)}{\omega^2 + \nu_n^2}.
\end{aligned}
$$

(6.50)

The frequency integral is conveniently evaluated by contour integration. Observing that $\widetilde{\chi}(\omega)$ is analytic in the upper half-plane, we then obtain[2]

$$c_n = c_{-n} = \hbar\widetilde{\chi}(\omega = i\,|\nu_n|) = \hbar\widehat{\chi}(z = |\nu_n|)\,, \qquad (6.51)$$

and finally with the explicit form (6.31),

$$c_n = \frac{\hbar}{M}\frac{1}{\omega_0^2 + \nu_n^2 + |\nu_n|\,\widehat{\gamma}(|\nu_n|)}\,. \qquad (6.52)$$

Upon dividing up the expression (6.46) into real and imaginary part, we find for the symmetrized and antisymmetrized correlation functions (6.13) and (6.14) the forms

$$S(t) = \frac{\hbar}{2\pi}\int_{-\infty}^{\infty}d\omega\,\widetilde{\chi}''(\omega)\coth(\hbar\beta\omega/2)\,e^{-i\omega t}\,, \qquad (6.53)$$

$$A(t) = -\frac{\hbar}{2\pi}\int_{-\infty}^{\infty}d\omega\,\widetilde{\chi}''(\omega)\sin(\omega t)\,. \qquad (6.54)$$

Using the relation (6.15), we then obtain for the response function the useful representation

$$\chi(t) = \Theta(t)\frac{1}{\pi}\int_{-\infty}^{\infty}d\omega\,\widetilde{\chi}''(\omega)\sin(\omega t)\,. \qquad (6.55)$$

Explicit expressions for the functions $S(t)$, $A(t)$ and $\chi(t)$ can be given for special forms of the damping function $\widetilde{\gamma}(\omega)$.

6.4.1 Ohmic damping

For strict Ohmic damping, $\widehat{\gamma}(|\nu_n|) = \gamma$, explicit expressions for $\chi(t)$ and $A(t)$ have been given already above in Eq. (6.40). Consider next the symmetrized position autocorrelation function $S(t)$. Evaluation of the representation (6.53) by closed-contour integration gives two different contributions,

$$S(t) = S_1(t) + S_2(t)\,. \qquad (6.56)$$

The one, say $S_1(t)$, originates from the poles of $\widetilde{\chi}''(\omega)$ at $\omega = -i\lambda_1$ and $\omega = -i\lambda_2$, where the frequencies $\lambda_{1/2}$ are given in Eq. (6.39). This contribution is

$$S_1(t) = \frac{\hbar}{2Mi(\lambda_2 - \lambda_1)}\left(\coth(i\lambda_2\hbar\beta/2)\,e^{-\lambda_2|t|} - \coth(i\lambda_1\hbar\beta/2)\,e^{-\lambda_1|t|}\right)\,. \qquad (6.57)$$

Since λ_1 and λ_2 are either real or complex conjugate, the expression (6.57) is real. In the underdamped limit, $\gamma < 2\omega_0$, it can be rewritten as

$$S_1(t) = \frac{\hbar}{2M\Omega}\frac{[\sinh(\hbar\beta\Omega)\cos(\Omega t) + \sin(\hbar\beta\gamma/2)\sin(\Omega|t|)]}{\cosh(\hbar\beta\Omega) - \cos(\hbar\beta\gamma/2)}\,e^{-\gamma|t|/2}\,. \qquad (6.58)$$

---

[2]We recall that the Fourier transform is denoted by a tilde and the Laplace transform by a hat.

The other contribution, $S_2(t)$, stems from the infinite sequence of simple poles of the function $\coth(\hbar\beta\omega/2)$ located at $\omega = -i\,\nu_n$ $(n = 1, 2, \ldots)$, where the ν_n are the positive bosonic Matsubara frequencies (6.49). We find

$$S_2(t) = -\frac{2\gamma}{M\beta} \sum_{n=1}^{\infty} \frac{\nu_n\, e^{-\nu_n|t|}}{(\omega_0^2 + \nu_n^2)^2 - \gamma^2\nu_n^2}. \tag{6.59}$$

Decomposing the rational fraction in Eq. (6.59) into partial fractions, we find that $S_2(t)$ can be expressed as a linear combination of four hypergeometric functions [86].

Expanding the coth terms in Eq. (6.57) in a series of simple fractions, we find that the initial value of $S(t)$ is

$$S(0) = \frac{1}{M\beta} \sum_{n=-\infty}^{\infty} \frac{1}{\omega_0^2 + \nu_n^2 + \gamma|\nu_n|}, \tag{6.60}$$

which agrees with $C^{(E)}(\tau = 0)$, Eq. (6.48) with (6.52). For $k_B T >> \hbar\omega_0$, we have

$$|S_2(0)|/S_1(0) = \zeta(3)\,(2\gamma/\omega_0)(\hbar\omega_0/2\pi k_B T)^3,$$

where $\zeta(3)$ is a Riemann number. Hence at high temperature, $S_2(t = 0)$ is very small compared to $S_1(t = 0)$. In addition, when $k_B T \gg \hbar\gamma/4\pi$, the function $S_2(t)$ drops to zero much faster than $S_1(t)$. For these two reasons, the contribution $S_2(t)$ is negligibly small at high temperatures for all t. In the high temperature limit, Planck's constant drops out in Eq. (6.58). Hence the correlations are classical in this limit,

$$S(t) = \frac{1}{M\beta\omega_0^2} \left[\cos(\Omega t) + \left(\frac{\gamma}{2\Omega}\right) \sin(\Omega t)\right] e^{-\gamma|t|/2}. \tag{6.61}$$

As the temperature is lowered, quantum effects appear, and also the term $S_2(t)$ becomes more important. In the temperature regime $0 < k_B T < \hbar\gamma/4\pi$, the $n = 1$ term in Eq. (6.59) governs $S(t)$ at long times, $\lim_{t\to\infty} S(t) \propto e^{-\nu_1 t}$. In this regime, the decay of the correlations is determined by the lowest Matsubara frequency, and not any more by the damping rate γ. In the limit $T \to 0$, the series of exponentials in Eq. (6.59) merges into the integral expression

$$S_2(t) = -\frac{\hbar\gamma}{M\pi} \int_0^{\infty} d\nu \frac{\nu\, e^{-\nu|t|}}{(\omega_0^2 + \nu^2)^2 - \gamma^2\nu^2}, \tag{6.62}$$

which may be expressed as a linear combination of exponential-integral functions. At long times, $t \gg 1/\omega_0$, γ/ω_0^2, the integral (6.62) may be evaluated asymptotically. From this we find algebraic decay with leading order

$$\lim_{t\to\infty} S(t) = -\frac{\hbar\gamma}{\pi M\omega_0^4} \frac{1}{t^2} = -\frac{\hbar M\gamma}{\pi} \frac{\chi_0^2}{t^2}. \tag{6.63}$$

In the second form, we have introduced the static susceptibility $\chi_0 = 1/(M\omega_0^2)$.

The t^{-2} power law in the algebraic long-time tail of the correlation function is characteristic for Ohmic damping. The second form in Eq. (6.63) is universally valid for all linear or nonlinear systems with Ohmic damping which have a finite static susceptibility at zero temperature. The parameters of the particular system under consideration are in the static susceptibility χ_0. This form may be compared with the corresponding expression for the dissipative two-state system given below in Eq. (21.251). With the proper normalization of the respective susceptibilities, they are identical in form. We anticipate that, in frequency space, we have a Shiba relation analogous to Eq. (21.253), as follows with Eq. (6.35).

We remark that the weak-coupling assumption is not valid when $k_{\mathrm{B}}T \lesssim \hbar\gamma/4$. Therefore, the standard weak-coupling theories fail to describe this regime.

We refrain from writing down the corresponding expressions for the Drude model (3.53) since they can easily be found upon using Eq. (6.41). We conclude this subsection with the remark that a Drude cutoff ω_{D} does not change the low-frequency behaviour of the spectral density. Hence the $1/t^2$ tail in Eq. (6.63) remains unaffected.

6.4.2 Algebraic spectral density

Consider the case in which the power s of the spectral density $J(\omega)$ in Eq. (3.58) is in the range $0 < s < 2$. We then have $\widehat{\gamma}(z) \propto z^{s-1}$, as given in Eq. (3.64). The Laplace transform $\widehat{\chi}(z)$ in Eq. (6.33) has a branch point at $z = 0$ for $s \neq 1$. The complex z-plane is cut along the negative real axis, and in the cut plane $\widehat{\chi}(z)$ is single-valued. The principal branch of z^s is taken in the cut plane. The two complex conjugate poles of $\widehat{\chi}(z)$ move away from the points $\pm i\omega_0$ when friction is switched on, and, depending on the particular values of the parameters chosen, they may or may not lie in the cut plane, depending on the parameters. The location of the poles can not be given in analytic form for general s. If they lie in the cut plane, they lead to a damped oscillatory contribution. In addition, there is an incoherent contribution from the branch cut,

$$\chi_{\mathrm{cut}}(t) = \frac{1}{M\pi}\,\mathrm{Im}\int_0^\infty d\nu\, \mathrm{e}^{-\nu t}\frac{1}{\omega_0^2 + \nu^2 + \lambda_s\omega_{\mathrm{ph}}^{1-s}\nu^s\mathrm{e}^{-i\pi s}}\,, \qquad (6.64)$$

where $\lambda_s = \gamma_s/\sin(\pi s/2)$. From Eq. (6.64) we find that the response function decays algebraically at long times with power t^{-1-s},

$$\lim_{t\to\infty}\chi(t) \approx \Gamma(1+s)\cos\left(\frac{\pi s}{2}\right)\frac{2\gamma_s}{\pi M\omega_0^4\omega_{\mathrm{ph}}^{s-1}}\frac{1}{t^{1+s}}\,. \qquad (6.65)$$

The algebraic decay of the response function of the damped harmonic oscillator in Eq. (6.65) does not depend on temperature and is therefore a general feature for $s \neq 1$. For the Ohmic case $s = 1$, the prefactor vanishes and thus the algebraic tail is absent. Using the relation (6.15), we find that the antisymmetrized position autocorrelation function at long times takes the form

$$\lim_{t\to\infty} A(t) = -\,\Gamma(1+s)\cos\left(\frac{\pi s}{2}\right)\frac{\hbar\gamma_s}{\pi M\omega_0^4\omega_{\text{ph}}^{s-1}}\frac{1}{t^{1+s}}\ . \tag{6.66}$$

Next, consider the symmetrized position autocorrelation function at zero temperature. We obtain from Eq. (6.53)

$$S(t) = \frac{\hbar}{2\pi}\int_{-\infty}^{\infty} d\omega\,\chi''(\omega)\,\text{sgn}(\omega)\,e^{-i\omega t}\ , \tag{6.67}$$

from where we find with Eq. (6.35) at long times

$$\lim_{t\to\infty} S(t) = \frac{\hbar}{\pi M\omega_0^4}\,\text{Re}\int_0^{\infty} d\omega\,\omega\tilde{\gamma}'(\omega)\,e^{-i\omega t}\ . \tag{6.68}$$

Substituting $\tilde{\gamma}'(\omega) = \gamma_s(\omega/\omega_{\text{ph}})^{s-1}$, we obtain the asymptotic decay law

$$\lim_{t\to\infty} S(t) = -\,\Gamma(1+s)\sin\left(\frac{\pi s}{2}\right)\frac{\hbar M\gamma_s}{\pi\omega_{\text{ph}}^{s-1}}\frac{\chi_0^2}{t^{1+s}}\ . \tag{6.69}$$

The expressions (6.69) and (6.66) may be combined to yield at zero temperature and asymptotic times in the regime $s < 2$

$$\lim_{t\to\infty} C^{+}(t) = \Gamma(1+s)\frac{\hbar M\,\gamma_s}{\pi\omega_{\text{ph}}^{s-1}}\frac{\chi_0^2}{(it)^{1+s}}\ . \tag{6.70}$$

Finally, we remark that we would have obtained the same expression if we had computed the imaginary-time correlation function $C^{(\text{E})}(\tau)$, Eq. (6.48) with (6.52) and (3.64) for $T = 0$ and $t \to \infty$, and had performed the analytical continuation $\tau \to it$.

6.5 Partition function, internal energy and density of states

6.5.1 Partition function and internal energy

The partition function $Z(\beta)$ of the damped linear quantum oscillator has been calculated with the imaginary-time path integral method in Section 4.3. There we obtained the partition function of the composite system in the factorized form $Z^{(\text{tot})} = Z_{\text{R}}Z$, where Z_{R} and Z are the partition functions of the reservoir and of the damped oscillator, respectively. The evaluation of the fluctuation modes resulted in the Matsubara infinite-product representation [cf. Eq.(4.224)]

$$Z(\beta) = \frac{1}{\hbar\beta\omega_0}\prod_{n=1}^{\infty}\frac{\nu_n^2}{\omega_0^2 + \nu_n^2 + \nu_n\widehat{\gamma}(\nu_n)}\ . \tag{6.71}$$

This expression is generally valid for arbitrary frequency-dependent damping. In order that the infinite product is convergent, we must have $\lim_{z\to\infty}\widehat{\gamma}(z) \to 0$. Hence the strict Ohmic case is excluded. For the Drude-regularized damping function (3.53), the partition function of the damped oscillator takes the form

$$Z(\beta) = \frac{1}{\hbar\beta\omega_0} \prod_{n=1}^{\infty} \frac{\nu_n^2 (\nu_n + \omega_D)}{(\nu_n + \lambda_1)(\nu_n + \lambda_2)(\nu_n + \lambda_3)} , \tag{6.72}$$

where the λ_j $(j = 1, 2, 3)$ are the roots of Eq. (6.42) satisfying the Vieta relations (6.43). By means of an infinite-product representation of the gamma function [87, 88], the expression (6.72) can be compactly written as

$$Z(\beta) = \frac{\hbar\beta\omega_0}{4\pi^2} \frac{\Gamma(\hbar\beta\lambda_1/2\pi)\Gamma(\hbar\beta\lambda_2/2\pi)\Gamma(\hbar\beta\lambda_3/2\pi)}{\Gamma(\hbar\beta\omega_D/2\pi)} . \tag{6.73}$$

This form can now be used to calculate various thermodynamic quantities.

Consider first the internal energy $U = \langle E \rangle_\beta$. Using the eigenfrequency representation (4.234) of the composite system, the mean total energy is found in the form

$$\langle E_{\text{tot}} \rangle_\beta \equiv -\frac{\partial}{\partial\beta} \ln Z^{(\text{tot})} = \sum_{\alpha=0}^{N} \frac{\hbar\mu_\alpha}{2} \coth\left(\frac{\hbar\beta\mu_\alpha}{2}\right) . \tag{6.74}$$

To divide the internal energy into the reservoir's and the relevant system's part, we start out from the representation (4.242) of the product form $Z^{(\text{tot})} = Z_R Z$. We then get $\langle E_{\text{tot}} \rangle_\beta = \langle E_R \rangle_\beta + \langle E \rangle_\beta$, where $\langle E_R \rangle_\beta$ is the internal energy of the reservoir

$$\langle E_R \rangle_\beta = \sum_{\alpha=1}^{N} \frac{\hbar\omega_\alpha}{2} \coth\left(\frac{\hbar\beta\omega_\alpha}{2}\right) , \tag{6.75}$$

and where $\langle E \rangle_\beta = -\frac{\partial}{\partial\beta} \ln Z$ is the internal energy of the open system. In the Drude case, we find with use of expression (6.73) that the internal energy of the damped oscillator can be analytically expressed in terms of the digamma function $\psi(z)$,

$$\langle E \rangle_\beta = \frac{1}{\beta} + \frac{\hbar}{2\pi}\left\{\omega_D\,\psi(1 + \hbar\beta\omega_D/2\pi) - \sum_{i=1}^{3} \lambda_i\,\psi(1 + \hbar\beta\lambda_i/2\pi)\right\}. \tag{6.76}$$

When the thermal energy $k_B T$ is large compared to the other energy scales of the damped oscillator, we may expand the digamma functions $\psi(z)$ about $z = 1$ in the expression (6.76). Then we get the high temperature expansion as

$$\langle E \rangle_\beta = k_B T \left\{1 + \frac{1 + 2\alpha\omega_D/\omega_0}{24} \frac{1}{\vartheta^2} + \mathcal{O}(1/\vartheta^4)\right\} . \tag{6.77}$$

where we have introduced a dimensionless temperature and damping strength

$$\vartheta = \frac{k_B T}{\hbar\omega_0} , \qquad \alpha = \frac{\gamma}{2\omega_0} . \tag{6.78}$$

Thus, the internal energy of the damped oscillator approaches in the classical limit the equipartition value $k_B T$ as expected.

At zero temperature, the Matsubara sum resulting from the logarithm of the infinite product representation (6.71) turns into an integral. The resulting expression for the ground-state energy $E_0 = \lim_{\beta\to\infty}\langle E \rangle_\beta$ is

$$E_0 \equiv \hbar\varepsilon_0 = \frac{\hbar}{2\pi} \int_0^\infty d\nu \, \ln\left(\frac{\omega_0^2 + \nu^2 + \nu\hat{\gamma}(\nu))}{\nu^2}\right). \tag{6.79}$$

This form holds for arbitrary frequency-dependent damping. In the Drude case, we find either from Eq. (6.79) or from the asymptotic limit of the expression (6.76)

$$
\begin{aligned}
\hbar\varepsilon_0 &= \frac{\hbar}{2\pi} \left[\lambda_1 \ln(\omega_D/\lambda_1) + \lambda_2 \ln(\omega_D/\lambda_2) + \lambda_3 \ln(\omega_D/\lambda_3)\right], \\
&= \frac{\hbar\omega_0}{2} \left\{ (1-\alpha^2) f(\alpha) + \frac{2\alpha}{\pi} \left[1 + \ln(\omega_D/\omega_0)\right] + \mathcal{O}\left(\frac{\ln\omega_D}{\omega_D}\right) \right\},
\end{aligned}
\tag{6.80}
$$

where

$$
f(\alpha) =
\begin{cases}
\dfrac{1}{\sqrt{1-\alpha^2}} \left(1 - \dfrac{2}{\pi} \arcsin\alpha\right), & \text{for} \quad \alpha < 1, \\[2ex]
\dfrac{2}{\pi} \dfrac{\ln\left(\alpha + \sqrt{\alpha^2-1}\right)}{\sqrt{\alpha^2-1}}, & \text{for} \quad \alpha > 1.
\end{cases}
\tag{6.81}
$$

The energy of the ground state increases monotonously with α and with ω_D. For strong damping $\alpha \gg 1$, we have $\varepsilon_0 = (\gamma/2\pi)\ln(\omega_D/\gamma)$.

With use of the asymptotic expansion of the digamma function [88], the low temperature expansion of the internal energy is found to read

$$\langle E \rangle_\beta = \hbar\varepsilon_0 + a_1\hbar\omega_0\vartheta^2 + a_2\hbar\omega_0\vartheta^4 + \mathcal{O}(\vartheta^6), \tag{6.82}$$

where $a_1 = \pi\alpha/3$ and $a_2 = 2\pi^3(3-4\alpha^2)/15 + \mathcal{O}(\omega_0/\omega_D)$. The corresponding expansion for the free energy $F = -k_B T \ln Z$ reads

$$F = \hbar\varepsilon_0 - a_1\hbar\omega_0\vartheta^2 - \tfrac{1}{3}a_2\hbar\omega_0\vartheta^4 + \mathcal{O}(\vartheta^6). \tag{6.83}$$

The leading thermal variation $\propto T^2$ of $\langle E \rangle_\beta$ or of F at low T is a characteristic feature of Ohmic dissipation. This contribution can be rewritten in terms of the static susceptibility $\chi_0 = 1/M\omega_0^2$ as $\delta\langle E \rangle_\beta = (\pi/6\hbar)\gamma M\chi_0(k_B T)^2$. It turns out that this form of the T^2-term universally holds for any nonlinear Ohmic system which has a nonvanishing static susceptibility at zero temperature.

The corresponding analysis for a spectral density $J(\omega) \propto \omega^s$ gives the leading temperature dependence both of the internal energy and the free energy at low T as T^{1+s}. From this we find that the specific heat $c = \partial U/\partial T = -T\partial^2 F/\partial T^2$ varies as T^s. We remark that the discussion about the Wilson ratio given below in Subsection 19.2.1 and in Section 21.9 also applies to the linear model discussed here.

The power law in T of the internal and free energy indicates that the energy gap above the ground state is dissolved. This conjecture is confirmed by the analysis of the energy spectrum given in the following subsection for the Ohmic case.

6.5.2 Spectral density of states

Knowledge about the spectrum of the damped linear quantum mechanical oscillator is gained from the spectral density of states $\rho(\varepsilon)$. Observing that the partition function is the Laplace transform of the density of states, the latter is given by the standard inversion contour integral

$$\rho(\varepsilon) = \frac{\hbar}{2\pi i} \int_{c-i\infty}^{c+i\infty} d\beta \, Z(\beta) \, e^{\beta\hbar\varepsilon} , \qquad (6.84)$$

where c is a real constant that exceeds the real part of all singularities of $Z(\beta)$. Equipped with the integral expression (6.84) with (6.73), we can now study how the sharp lines of the discrete spectrum of the undamped oscillator,

$$\rho_0(\varepsilon) = \sum_{n=0}^{\infty} \delta\left(\varepsilon - \left(n + \tfrac{1}{2}\right)\omega_0\right) , \qquad (6.85)$$

spread out and melt into another when damping is turned on and increased.

Direct numerical computation of the integral in Eq. (6.84) is difficult to perform since the integrand is rapidly oscillating along the integration contour. Fortunately, some valuable insights can be obtained by analytic methods.

First, it is immediately obvious from the low temperature expansion of the free energy, Eq. (6.83) that the ground state remains as a separate δ-function at the frequency ε_0,

$$\rho(\varepsilon) = \delta(\varepsilon - \varepsilon_0) + \Phi(\varepsilon) . \qquad (6.86)$$

The δ-peak is shifted with increasing damping strength towards higher frequency as specified by the expression (6.80). The function $\Phi(\varepsilon)$ is zero for $\varepsilon < \varepsilon_0$. It represents the density of states above the ground state. The series expansion of $\Phi(\varepsilon)$ about ε_0 for $\varepsilon > \varepsilon_0$ is found from the asymptotic expansion of F, Eq. (6.83), as

$$\Phi(\varepsilon_0 + \sigma) = \frac{1}{\omega_0}\left\{ a_1 + \frac{a_1^2 \, \sigma}{2 \, \omega_0} + \left(\frac{a_1^3}{12} + \frac{a_2}{6}\right)\left(\frac{\sigma}{\omega_0}\right)^2 + \mathcal{O}\left[(\sigma/\omega_0)^3\right] \right\} . \qquad (6.87)$$

From this we draw the following conclusions. First, we see that the gap above the ground state in the energy spectrum is dissolved by the dissipative coupling. The function $\Phi(\varepsilon)$ makes a jump at $\varepsilon = \varepsilon_0$ of height $\pi\alpha/3\omega_0$. The leading thermal enhancement $\propto T^2$ of the internal energy is a direct effect of this jump. Secondly, the derivative of $\Phi(\varepsilon)$ at $\varepsilon = \varepsilon_0^+$ is positive for all α, whereas the curvature of $\Phi(\varepsilon)$ at $\varepsilon = \varepsilon_0^+$ changes from positive to negative at $\alpha = 0.88$.

Well above ε_0 the density $\Phi(\varepsilon)$ is given by the sum of contributions collected from the infinite series of simple poles in the expression (6.73) [227],

$$\Phi(\varepsilon) = \frac{1}{\omega_0} + \sum_{k=1}^{3}\sum_{n=1}^{\infty} R_{n,k} \, e^{-2\pi n\varepsilon/\lambda_k} , \qquad (6.88)$$

where the residues are given by

$$R_{n,1} = \frac{(-1)^{n-1}}{(n-1)!} \frac{\Gamma(-n\lambda_2/\lambda_1)\Gamma(-n\lambda_3/\lambda_1)}{\Gamma(-n\omega_D/\lambda_1)} \frac{\omega_0}{\lambda_1^2}, \tag{6.89}$$

and analogous expressions for $R_{n,2}$ and $R_{n,3}$ by cyclic permutation of the indices.

The expression (6.88) approaches at high frequency the proper classical limit $1/\omega_0$. The analysis gives that the two series (6.87) and (6.88) have an overlapping region of convergence and that $\Phi(\varepsilon)$ for $\alpha > 0$ is an absolutely continuous function of ε in the interval $\varepsilon_0 < \varepsilon < \infty$. Therefore, apart from the ground state, there are no other discrete states embedded in the continuum. In the underdamped regime $\gamma < \gamma_c$, where two of the λ_k are complex conjugate to each other, the density of states shows roughly equidistant resonances with exponentially reduced amplitudes towards higher frequencies. The transition from the bumpy to the smooth behaviour occurs roughly at the aperiodic limit of the classical oscillator. For $\alpha > 1$, the density increases shortly from the threshold value $\pi\alpha/3\omega_0$ to a maximum and then decreases monotonously to the classical value $1/\omega_0$ reached at high T. The characteristic features of the density of states in the underdamped and overdamped regime are shown in Fig. 6.1.

Experimentally, the density of states can be found by measuring the absorbed microwave power of the externally driven oscillator. The linear absorption of an external field with frequency ω is proportional to the density of states at frequency $\varepsilon_0 + \omega$.

Finally, we remark that for a spectral density of the coupling $J(\omega) \propto \omega^s$ or damping function $\widetilde{\gamma}(\omega) \propto \omega^{s-1}$ there is again a δ-peak related to the ground state, but now we have $\Phi(\varepsilon) \propto (\varepsilon - \varepsilon_0)^{s-1}$ for ε slightly above ε_0. Hence the density of states above the ground state grows smoothly with increasing frequency from zero in the super-Ohmic case, while it becomes singular at the threshold in the sub-Ohmic case.

6.6 Mean square of position and momentum

6.6.1 General expressions for coloured noise

The dispersion of the position in the equilibrium state is given by

$$\langle q^2 \rangle = S(0) = \frac{\hbar}{\pi} \int_0^\infty d\omega \, \widetilde{\chi}''(\omega) \coth(\hbar\beta\omega/2), \tag{6.90}$$

as follows from Eq. (6.53). A different form is found from the imaginary-time correlation function (6.48) with Eq. (6.52),

$$\langle q^2 \rangle = C^E(0) = \frac{1}{M\beta} \sum_{n=-\infty}^{\infty} \frac{1}{\omega_0^2 + \nu_n^2 + |\nu_n|\widehat{\gamma}(|\nu_n|)}. \tag{6.91}$$

Upon extending the integral in Eq. (6.90) to a contour integral encircling the upper ω half-plane and observing that $\chi(\omega)$ is analytic for $\mathrm{Re}\,\omega > 0$, one finds that Eq.

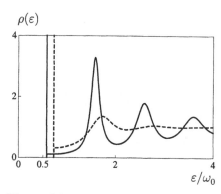
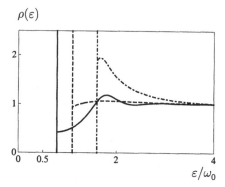

Figure 6.1: The density of states $\rho(\varepsilon)$ is plotted versus ε/ω_0 for $\omega_D = 10\omega_0$ for weak and strong damping. The left plot shows the cases $\alpha = 0.1$ (full curve) and $\alpha = 0.3$ (dashed curve), and the right plot the cases $\alpha = 0.4$ (full curve), $\alpha = 0.9$ (dashed curve) and $\alpha = 1.8$ (dashed-dotted curve). Both the ground-state energy and the height of the jump increase monotonously with α. The respective ground-state values are $\varepsilon/\omega_0 = 0.576,\ 0.721,\ 0.790,\ 1.108,$ and 1.607.

(6.90) coincides with Eq. (6.91). The expression (6.91) is conveniently split into the classical dispersion $\langle q^2 \rangle_{\mathrm{cl}}$ and the quantum-mechanical dispersion $\langle q^2 \rangle_{\mathrm{qm}}$. We have

$$\langle q^2 \rangle \;=\; \langle q^2 \rangle_{\mathrm{cl}} + \langle q^2 \rangle_{\mathrm{qm}} \;=\; \frac{1}{M\beta\omega_0^2} + \frac{2}{M\beta} \sum_{n=1}^{\infty} \frac{1}{\omega_0^2 + \nu_n^2 + |\nu_n|\widehat{\gamma}(|\nu_n|)} \;. \tag{6.92}$$

With use of the expressions (6.36) and (6.45) the mean value of p^2 is found as

$$\langle p^2 \rangle \;=\; \frac{\hbar M^2}{\pi} \int_0^{\infty} d\omega\, \omega^2 \widetilde{\chi}''(\omega) \coth(\omega\hbar\beta/2) \;. \tag{6.93}$$

A Matsubara representation of $\langle p^2 \rangle$ is easily obtained from the imaginary-time correlation function $\langle p(\tau)p(0) \rangle$. Using Eqs. (6.36), (6.47), (6.48) and (6.52), we find

$$\langle p(\tau)p(0) \rangle \;=\; -\hbar M :\!\delta(\tau)\!: + \frac{M}{\beta} \sum_{n=-\infty}^{\infty} \frac{\omega_0^2 + |\nu_n|\widehat{\gamma}(|\nu_n|)}{\omega_0^2 + \nu_n^2 + |\nu_n|\widehat{\gamma}(|\nu_n|)} e^{i\nu_n\tau} \;, \tag{6.94}$$

where $:\!\delta(\tau)\!:$ is the periodically continued δ-function (4.51). The Fourier series (6.94) continues $\langle p(\tau)p(0) \rangle$ periodically beyond the fundamental strip $0 < \mathrm{Re}\,\tau < \hbar\beta$. As a consequence of the continuation, the expression (6.94) has δ-function singularities at $\tau = \hbar\beta m$ $(m = 0, \pm1, \pm2, \cdots)$. Since $\langle p(\tau)p(0) \rangle$ is an analytic continuation of the real-time momentum autocorrelation function only when τ is in the strip $0 < \mathrm{Re}\,\tau < \hbar\beta$, we perform the limit $\tau \to 0^+$, thereby excluding the δ-peak at $\tau = 0$. Thus we find for the momentum dispersion in thermal equilibrium the expression

$$\langle p^2 \rangle = \lim_{\tau \to 0^+} \langle p(\tau)p(0) \rangle_{\mathrm{reg}} \;=\; \frac{M}{\beta} \sum_{n=-\infty}^{\infty} \frac{\omega_0^2 + |\nu_n|\widehat{\gamma}(|\nu_n|)}{\omega_0^2 + \nu_n^2 + |\nu_n|\widehat{\gamma}(|\nu_n|)} \;. \tag{6.95}$$

Since $\widetilde{\chi}(\omega)$ is analytic in the upper half-plane, the expression (6.95) can also directly be found from Eq. (6.93) by contour integration.

The dispersions of position and momentum can also be related to the partition function $Z(\beta)$ of the damped linear oscillator. It is obvious from the product representation (6.71) that the dispersions (6.91) and (6.95) may be expressed in the form

$$\langle q^2 \rangle = -\frac{1}{M\beta\omega_0}\frac{d}{d\omega_0}\ln Z(\beta), \tag{6.96}$$

$$\langle p^2 \rangle = -\frac{M}{\beta}\left(\omega_0\frac{d}{d\omega_0} + 2\gamma\frac{d}{d\gamma}\right)\ln Z(\beta), \tag{6.97}$$

where in the second expression we have employed the representation $\widehat{\gamma}(z) = \gamma g(z)$. The above expressions are generally valid for any form of linear memory-friction.

6.6.2 Strict Ohmic case

For strict Ohmic friction $\widehat{\gamma}(|\nu_n|) = \gamma$, the expression (6.91) may be rewritten as

$$\langle q^2 \rangle = \frac{1}{M\beta\omega_0^2} + \frac{2}{M\beta(\lambda_2 - \lambda_1)}\sum_{n=1}^{\infty}\left(\frac{1}{\nu_n + \lambda_1} - \frac{1}{\nu_n + \lambda_2}\right), \tag{6.98}$$

where λ_1 and λ_2 are the characteristic frequencies defined in Eq. (6.39). The sum in Eq. (6.98) can be rewritten as a linear combination of digamma functions,

$$\langle q^2 \rangle = \frac{1}{M\beta\omega_0^2} + \frac{\hbar}{M\pi(\lambda_2 - \lambda_1)}\Big(\psi(1 + \lambda_2/\nu) - \psi(1 + \lambda_1/\nu)\Big), \tag{6.99}$$

where again $\nu \equiv \nu_1 = 2\pi/\hbar\beta$. In the absence of damping, this reduces to the familiar expression $\langle q^2 \rangle = (\hbar/2m\omega_0)\coth(\hbar\beta\omega_0/2)$.

In the high-temperature regime $k_B T \gg \hbar|\lambda_{1,2}|/2\pi$, the quantum mechanical contribution $\langle q^2 \rangle_{\text{qm}}$ is found from the expression (6.99) to read

$$\langle q^2 \rangle_{\text{qm}} = \frac{1}{12}\frac{\hbar^2}{Mk_B T}\left[1 + \mathcal{O}\Big(\frac{\hbar\gamma}{k_B T}\Big)\right]. \tag{6.100}$$

This shows again that damping becomes irrelevant at high temperature.

On the other hand, in the low temperature limit the asymptotic expansion of the digamma functions applies. Then the term $\langle q^2 \rangle_{\text{cl}}$ is cancelled by a corresponding counter-term. The resulting series is

$$\langle q^2 \rangle = \frac{\hbar}{2M\omega_0}\left\{f(\alpha) + \frac{2\pi}{3}\frac{\gamma}{\omega_0}\Big(\frac{k_B T}{\hbar\omega_0}\Big)^2 + \mathcal{O}\left[\Big(\frac{k_B T}{\hbar\omega_0}\Big)^4\right]\right\}. \tag{6.101}$$

The function $f(\alpha)$ is defined in Eq. (6.81).

The expression (6.101) reproduces the correct form $\langle q^2 \rangle_{T=0} = \hbar/2M\omega_0$ for the dispersion of an undamped oscillator in the ground state.

At finite T, the dispersion of the position is enhanced compared with the zero temperature case. The corrections grow algebraically with T for a damped system. The leading T^2 power law at low T is again a signature of Ohmic friction. For a spectral density $J(\omega \to 0) \propto \omega^s$, the leading thermal enhancement grows with T^{1+s}. For large friction $\gamma \gg \omega_0$ and temperature in the range

$$k_{\rm B}T \gg \hbar\omega_0^2/2\pi\gamma \,, \tag{6.102}$$

the leading quantum mechanical contribution to the coordinate dispersion is

$$\langle q^2 \rangle_{\rm qm} \equiv \langle q^2 \rangle - \langle q^2 \rangle_{\rm cl} = \frac{\hbar}{\pi M\gamma} \left[\psi(1 + \hbar\beta\gamma/2\pi) - \psi(1) \right] \equiv \lambda \,. \tag{6.103}$$

Interestingly enough, the quantum mechanical part of the coordinate dispersion does not depend in leading order on properties of the potential for strong friction and temperature in the regime (6.102). At temperature $T \gg \hbar\gamma/2\pi k_{\rm B}$, the expression (6.103) reduces to the form (6.100).

In the temperature regime $\hbar\omega_0^2/2\pi\gamma \ll k_{\rm B}T \ll \hbar\gamma/2\pi$, the leading quantum contribution to the coordinate dispersion is found from the expression (6.103) as

$$\lambda = \frac{\hbar}{\pi M\gamma} \ln\left(\frac{\hbar\beta\gamma}{2\pi} \right) \,. \tag{6.104}$$

In classical physics, the evolution of a Brownian particle in phase space effectively reduces for very large friction to a time evolution in position space. The corresponding dynamics is well described by the Smoluchowski diffusion equation, which is discussed below in Section 11.3. In the temperature range (6.102), quantum effects are important. However it is still possible to describe the time-evolution of the position distribution in terms of a diffusion equation. The corresponding quantum Smoluchowski equation (QSE) is discussed in Subsection 15.3. We shall see that the strength of the quantum effects in the QSE is controlled by the quantity λ.

6.6.3 Ohmic friction with Drude regularization

For strict Ohmic damping, the integral in Eq. (6.93) as well as the sum in Eq. (6.95) has a logarithmic ultra-violet divergence. The divergence of the momentum dispersion shows that the often made Markov assumption, i.e. the assumption of a memoryless reservoir is unphysical. In any physical system there is always a "microscopic" time scale below which the inertia of the environment becomes relevant. In its simplest form, this is represented by a high-frequency cutoff in $\tilde{\gamma}(\omega)$. By this, the thermal average of $\langle p^2 \rangle$ is regularized. For the Drude model (3.53), we find

$$
\begin{aligned}
\langle q^2 \rangle &= \frac{1}{M\beta} \sum_{n=-\infty}^{+\infty} \frac{1}{\omega_0^2 + \nu_n^2 + \gamma\omega_{\rm D}|\nu_n|/(\omega_{\rm D} + |\nu_n|)} \,, \\
\langle p^2 \rangle &= \frac{M}{\beta} \sum_{n=-\infty}^{+\infty} \frac{\omega_0^2 + \gamma\omega_{\rm D}|\nu_n|/(\omega_{\rm D} + |\nu_n|)}{\omega_0^2 + \nu_n^2 + \gamma\omega_{\rm D}|\nu_n|/(\omega_{\rm D} + |\nu_n|)} \,.
\end{aligned}
\tag{6.105}
$$

Employing a partial fraction decomposition, we obtain

$$\langle q^2 \rangle = \frac{1}{\beta M \omega_0^2} + \frac{\hbar}{M\pi} \sum_{j=1}^{3} q_j \psi(1 + \lambda_j/\nu) , \qquad (6.106)$$

$$\langle p^2 \rangle = M^2 \omega_0^2 \langle q^2 \rangle + \Pi^2 , \qquad (6.107)$$

$$\Pi^2 = \frac{M\gamma\hbar\omega_D}{\pi} \sum_{j=1}^{3} p_j \psi(1 + \lambda_j/\nu) , \qquad (6.108)$$

where

$$q_1 = \frac{\lambda_1 - \omega_D}{(\lambda_1 - \lambda_2)(\lambda_1 - \lambda_3)} , \qquad p_1 = \frac{\lambda_1}{(\lambda_1 - \lambda_2)(\lambda_1 - \lambda_3)} , \qquad (6.109)$$

and where q_2, q_3 and p_2, p_3 are defined by cyclic permutation of the indices.

Use of the approximate roots (6.44) shows that the expression for $\langle q^2 \rangle$, Eq. (6.106), differs from the previous result (6.99) by terms of order ω_0/ω_D and γ/ω_D.

For extremely high temperatures, $k_B T \gg \hbar\omega_D$, we find for the contribution Π^2 to $\langle p^2 \rangle$ upon expanding the digamma functions in Eq. (6.108) about 1

$$\Pi^2 = \frac{M\gamma\hbar\omega_D}{12} \frac{\hbar}{k_B T} . \qquad (6.110)$$

From this we see that Π^2 is negligibly small at high temperatures, and the proper classical limit $\langle p^2 \rangle \to M^2 \omega_0^2 \langle q^2 \rangle \to M k_B T$ is found from the expression (6.107).

When $\hbar\omega_D \gg k_B T$, the leading contribution to Π^2 comes from the term with $j = 3$ in Eq. (6.108), yielding

$$\Pi^2 = (\hbar\gamma M/\pi) \ln(\omega_D \hbar\beta) . \qquad (6.111)$$

At very low temperatures where $k_B T \ll \hbar\omega_0$ and $k_B T \ll \hbar\gamma$, we may use the asymptotic expansion of the digamma function for all three arguments in Eq. (6.108). We then find

$$\Pi^2 = \frac{\hbar\gamma M}{\pi} \ln\left(\frac{\omega_D}{\omega_0}\right) + i \frac{\hbar\gamma^2 M}{4\pi\zeta} \ln\left(\frac{\lambda_1}{\lambda_2}\right) - \frac{\pi\hbar\gamma M}{3} \left(\frac{k_B T}{\hbar\omega_0}\right)^2 , \qquad (6.112)$$

where terms of order ω_0/ω_D, γ/ω_D, and $(k_B T/\hbar\omega_0)^4$ have been disregarded. Now, Eq. (6.112) combines with Eq. (6.101) to give for $\langle p^2 \rangle$ at zero temperature

$$\langle p^2 \rangle_{T=0} = M^2 \omega_0^2 (1 - 2\alpha^2) \langle q^2 \rangle_{T=0} + (2\alpha\hbar\omega_0 M/\pi) \ln(\omega_D/\omega_0) . \qquad (6.113)$$

Since, in a sense, the reservoir is continuously measuring the position of the particle, the oscillator gets more localized, whereas the momentum spread and the mean kinetic energy becomes larger as damping is increased.[3] For large damping $\gamma \gg \omega_0$ we have

---

[3] If the reservoir would couple to the momentum of the oscillator, the features of coordinate and momentum dispersion would perform role reversal.

$$\langle q^2 \rangle_{T=0} = \frac{2\hbar}{M\pi\gamma} \ln(\gamma/\omega_0) \,, \qquad \langle p^2 \rangle_{T=0} = \frac{M\hbar\gamma}{\pi} \ln(\omega_D/\gamma) \,. \tag{6.114}$$

For $\alpha = 0$, the expression (6.113) reduces to the momentum dispersion in the ground state of the undamped system, $\langle p^2 \rangle_{T=0} = M\hbar\omega_0/2$. In Fig. 6.3 (left diagram) the normalized dispersions $2M\omega_0\langle q^2 \rangle/\hbar$ (dashed-dotted line) and $2\langle p^2 \rangle/M\hbar\omega_0$ (dashed line), and the normalized uncertainty $2\sqrt{\langle q^2 \rangle \langle p^2 \rangle}/\hbar$ (full line) in the ground state are plotted as functions of the damping parameter α . The uncertainty reaches a plateau fairly above $\alpha = 2$. There, it has only weak logarithmic dependence on α.

Since $\tilde{\chi}''(\omega)$ is odd in ω, the asymptotic low temperature expansions of both the position and momentum dispersion are power series expansions in T^2, as follows from Eq. (6.90) and Eq. (6.93). For the Drude model, the leading term in the position dispersion is T^2. In the momentum dispersion, the T^2 correction in Eq. (6.112) cancels a corresponding contribution from the term $M^2\omega_0^2\langle q^2 \rangle$ in Eq. (6.107). Thus the leading enhancement in $\langle p^2 \rangle$ varies as T^4.

The temperature dependence of the dispersions $\langle q^2 \rangle$ and $\langle p^2 \rangle$ for the Drude model is depicted in Fig. 6.2. The findings are as expected by intuition. For fixed temperature, the dispersion of the position decreases as the damping is increased while the dispersion of the momentum gets larger. At high temperature, the variances are independent of the damping strength, and they vary linearly with T. This is in agreement with the equipartition law.

One might have guessed that the energy of the oscillator in thermal equilibrium

$$\bar{E} = \frac{\langle p^2 \rangle}{2M} + \frac{1}{2}M\omega_0^2\langle q^2 \rangle \tag{6.115}$$

coincides with the expression (6.76) for $\langle E \rangle_\beta$. One finds that they actually differ to leading order in γ/ω_D by a temperature-independent shift, $\langle E \rangle_\beta - \bar{E} = \hbar\omega_0\alpha/\pi$.

6.7 Equilibrium density matrix

For a Gaussian process, the normalized equilibrium density matrix is completely specified by the position dispersion $\langle q^2 \rangle$ and the momentum dispersion $\langle p^2 \rangle$. Therefore, the correct form for any linear dissipative mechanism can be inferred from the density matrix of the undamped oscillator. Thus we have

$$< q|\hat{\rho}_\beta|q' > = \frac{1}{\sqrt{2\pi\langle q^2 \rangle}} \exp\left(-\frac{(q+q')^2}{8\langle q^2 \rangle} - \frac{\langle p^2 \rangle}{2\hbar^2}(q-q')^2 \right) \,, \tag{6.116}$$

where we have used the normalization tr $\hat{\rho}_\beta = 1$. Let us now see how the exponent in Eq. (6.116) emerges from the action (4.62), which in the present case is

$$S[q] = \int_0^{\hbar\beta} d\tau \left(\frac{M}{2}\dot{q}^2 + \frac{1}{2}M\omega_0^2 q^2 \right) + \int_0^{\hbar\beta} d\tau \int_0^\tau d\tau' \, k(\tau - \tau')q(\tau)q(\tau') \,. \tag{6.117}$$

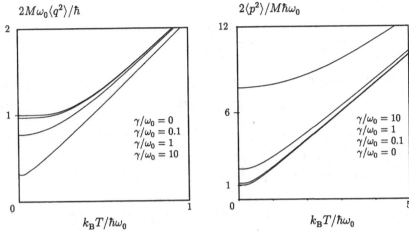

Figure 6.2: The normalized equilibrium variances of position and momentum are shown for a Drude model as a function of temperature. The cutoff frequency ω_D is chosen as $\omega_D = 10\,\omega_0$ and γ/ω_0 varies between 0 and 10.

The Fourier expansion method is somewhat complicated by the fact that the end-points of the path $q(\tau)$ are different and therefore the periodically continued path

$$q(\tau) = \frac{1}{\hbar\beta} \sum_{n=-\infty}^{\infty} q_n e^{i\nu_n \tau} \qquad (6.118)$$

involves periodically repeated jumps from q back to q' and periodically repeated cusps, both of them at times $\tau = m\hbar\beta$ ($m = 0, \pm1, \cdots$). It is convenient to write

$$q(\tau) = q^{(1)}(\tau) + q^{(2)}(\tau) , \qquad (6.119)$$

where $q^{(1)}(\tau)$ makes jumps, and $q^{(2)}(\tau)$ has cusps with jumps in the velocity,

$$q^{(1)}(0^- + m\hbar\beta) - q^{(1)}(0^+ + m\hbar\beta) = q - q' , \qquad (6.120)$$

$$\dot{q}^{(2)}(0^- + m\hbar\beta) = -\dot{q}^{(2)}(0^+ + m\hbar\beta) = v . \qquad (6.121)$$

Away from the discontinuities, the equation of motion for the extremal path reads

$$-M\ddot{q}(\tau) + M\omega_0^2 q(\tau) + \int_0^\tau d\tau'\, k(\tau - \tau')q(\tau') = 0 . \qquad (6.122)$$

Taking into account the jump conditions (6.120) and (6.121), we have in Fourier space

$$\left(\nu_n^2 + \omega_0^2 + |\nu_n|\hat{\gamma}(|\nu_n|)\right)q_n^{(1)} = i\nu_n(q - q') ,$$

$$\left(\nu_n^2 + \omega_0^2 + |\nu_n|\hat{\gamma}(|\nu_n|)\right)q_n^{(2)} = 2v ,$$
$$\qquad (6.123)$$

where v is determined by the requirement

$$q^{(2)}(\tau = m\hbar\beta) = \frac{1}{\hbar\beta} \sum_{n=-\infty}^{\infty} q_n^{(2)} = (q+q')/2 \, . \tag{6.124}$$

This gives upon identifying the resulting Matsubara sum with the expression (6.91)

$$v = \hbar(q+q')/4M\langle q^2 \rangle \, . \tag{6.125}$$

Thus we get

$$q_n = \frac{1}{\hbar\beta} \sum_{n=-\infty}^{\infty} \left\{ \left(-\frac{1}{i\nu_n} + \frac{1}{i\nu_n} \frac{\omega_0^2 + |\nu_n||\widehat{\gamma}(|\nu_n|)}{\omega_0^2 + \nu_n^2 + |\nu_n||\widehat{\gamma}(|\nu_n|)} \right) (q-q') \right.$$

$$\left. + \frac{1}{\nu_n^2 + \omega_0^2 + |\nu_n||\widehat{\gamma}(|\nu_n|)} \frac{\hbar}{2M\langle q^2 \rangle}(q+q') \right\} \, ,$$

yielding

$$q(0^+) = q' \, ; \qquad q(0^-) = q \, ,$$

$$\dot{q}(0^\pm) = \frac{\langle p^2 \rangle}{M\hbar}(q-q') \pm \frac{\hbar}{4M\langle q^2 \rangle}(q+q') \, . \tag{6.126}$$

Since the extremal path satisfies the equation of motion (6.122), the harmonic action (6.117) along this path may be transformed into

$$S[q] = \int_0^{\hbar\beta} d\tau \, \frac{M}{2} \frac{d}{d\tau}\Big(q(\tau)\dot{q}(\tau) \Big) = \frac{M}{2}\Big(q(0^-)\dot{q}(0^-) - q(0^+)\dot{q}(0^+) \Big) \, , \tag{6.127}$$

where we have used $q(\hbar\beta - 0^+) = q(0^-)$. This combines with Eq. (6.126) to yield

$$S[q]/\hbar = \frac{1}{8\langle q^2 \rangle}(q+q')^2 + \frac{\langle p^2 \rangle}{2\hbar^2}(q-q')^2 \, , \tag{6.128}$$

which is just (modulo a minus sign) the exponent in Eq. (6.116).

6.7.1 Purity

After having found the explicit form of the equilibrium density matrix of the damped oscillator, we can now check the purity, Eq. (4.10), which is

$$\text{pur}\,(\alpha) = \text{tr}\,\hat{\rho}^2 = \int dx\, dx'\, \rho(x,x')\,\rho(x',x) = \frac{\hbar}{2\sqrt{\langle p^2 \rangle \langle q^2 \rangle}} \, . \tag{6.129}$$

Using the expressions for $\langle q^2 \rangle_{T=0}$ and $\langle p^2 \rangle_{T=0}$ derived in previous subsections, we see that already weak damping makes the ground state rather impure. Hence, pure quantum states never exist in real life. On the other hand, for γ somewhat above ω_0 the purity reaches a plateau which is still of the order of one even for large α. There,

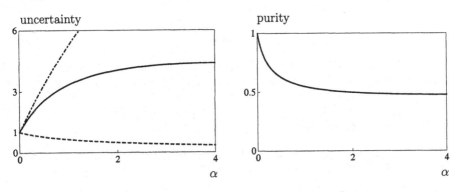

Figure 6.3: The left figure shows the normalized variances $2M\omega_0\langle q^2\rangle/\hbar$ (dashed curve) and $2\langle p^2\rangle/M\hbar\omega_0$ (dashed-dotted curve), and the normalized uncertainty $2\sqrt{\langle q^2\rangle\,\langle p^2\rangle}/\hbar$ (full curve) of the ground state as functions of the damping parameter α. In the right figure, the purity $\mathrm{pur}(\alpha)$ at zero temperature is plotted as a function of α. The Drude cutoff is $\omega_D = 100\,\omega_0$.

the dependence of the purity on α is only logarithmic [cf. Fig 6.3]. This indicates that the damped oscillator stays quantum mechanical even for very large damping.

Let us now study what the effective mass and effective frequency of a fictitious linear quantum oscillator with a discrete equidistant spectrum would be, if it had the same coordinate and momentum dispersions as specified in the preceding chapter. The respective statistical operator with normalization $\mathrm{tr}\,\hat{\rho}_\beta = 1$ is

$$\hat{\rho}_\beta = \widetilde{Z}^{-1}(\beta)\,e^{-\beta\hat{H}_{\text{eff}}} , \tag{6.130}$$

$$\hat{H}_{\text{eff}} = \frac{1}{2M_{\text{eff}}}\hat{p}^2 + \frac{1}{2}M_{\text{eff}}\omega_{\text{eff}}^2\hat{q}^2 . \tag{6.131}$$

Here the parameters ω_{eff} and M_{eff} are fixed by the requirement that Eq. (6.130) has the coordinate representation (6.116). The density matrix can be represented as

$$\hat{\rho}_\beta = \frac{1}{\widetilde{Z}(\beta)}\sum_{n=0}^{\infty} e^{-\beta E_n} |n><n| , \tag{6.132}$$

where $|n>$ is an eigenstate of the effective Hamiltonian

$$\hat{H}_{\text{eff}} |n> = \hbar\omega_{\text{eff}}(n + \tfrac{1}{2})|n> . \tag{6.133}$$

We then have

$$\widetilde{Z}(\beta) = [\,2\sinh(\omega_{\text{eff}}\hbar\beta/2)\,]^{-1} . \tag{6.134}$$

In coordinate representation, we have

$$<q|n> = \left(\frac{1}{\pi}\right)^{1/4}\left(\frac{c}{2^n n!}\right)^{1/2}\exp\left(-\frac{c^2 q^2}{2}\right)H_n(cq) , \tag{6.135}$$

where the $H_n(x)$ are Hermite polynomials of degree n, and where

$$c = \sqrt{M_{\text{eff}}\omega_{\text{eff}}/\hbar} \,. \tag{6.136}$$

Using Eqs. (6.132), (6.133), (6.135) and the generating function of the Hermite polynomials, we find the coordinate representation of the density matrix in the form [4]

$$< q|\hat{\rho}_\beta|q' > = \frac{c}{\widetilde{Z}\sqrt{2\pi \sinh\Omega}} \exp\left\{ -\frac{c^2\,[(q^2 + q'^2)\cosh\Omega - 2qq']}{2\sinh\Omega} \right\} \,, \tag{6.137}$$

where $\Omega = \omega_{\text{eff}}\hbar\beta$. The quantities ω_{eff} and M_{eff} can be determined upon comparing Eq. (6.116) with Eq. (6.137). This gives for the effective parameters the expression

$$\omega_{\text{eff}} = \frac{1}{\hbar\beta} \ln \frac{\sqrt{\langle p^2\rangle\langle q^2\rangle} + \hbar/2}{\sqrt{\langle p^2\rangle\langle q^2\rangle} - \hbar/2} \,, \tag{6.138}$$

$$M_{\text{eff}} = \sqrt{\langle p^2\rangle/\langle q^2\rangle}\Big/\omega_{\text{eff}} \,. \tag{6.139}$$

At high temperatures, the effective quantities ω_{eff} and M_{eff} approach their bare values ω_0 and M. At low temperatures, ω_{eff} and M_{eff} strongly differ from their bare values. Substituting the zero temperature values for $\langle q^2\rangle$ and $\langle p^2\rangle$, we find

$$\hbar\omega_{\text{eff}} = 2 k_{\text{B}}T \operatorname{arcoth} \sqrt{f(\alpha)\,[\,(1 - 2\alpha^2)\,f(\alpha) + (4\alpha/\pi)\,\ln(\omega_{\text{D}}/\omega_0)\,]} \,, \tag{6.140}$$

where the function $f(\alpha)$ is given in Eq. (6.81). Observe that ω_{eff} vanishes linearly with T as $T \to 0$. Hence the discrete energy levels in Eq. (6.133) become very narrowly spaced near zero temperature. In fact, the occupation probabilities $p_n = \widetilde{Z}^{-1} \exp(-\beta E_n)$ remain finite in the limit $T \to 0$. From this we infer that the ground state of the fictitious oscillator is not a pure state but a mixture.

We conclude with the remark that the partition function of the damped oscillator Z and the partition function of the fictitious oscillator $\widetilde{Z}$ drastically differ at low temperatures for $\alpha > 0$. They approach each other as temperature is increased and become equal in the classical limit.

7. Quantum Brownian free motion

Historically, the Langevin equation (2.3) was used to analyze the irregular motion (Zitterbewegung) of a heavy particle moving in a thermally equilibrated molecular medium. This phenomenon is known as Brownian motion, named after Robert Brown, who observed the random motion of pollen grains immersed in a fluid. In many cases, the systematic external force can be neglected, and we are left with the problem of Brownian free motion. Often, the memory times of the fluctuating forces are negligibly short compared to the time scale over which one observes the Brownian particle. The assumption of memoryless friction is usually called Markovian approximation.

Here we discuss quantum Brownian free motion for general frequency-dependent damping $\tilde{\gamma}(\omega)$. Since the equation of motion is linear, the particle is classical, and there are no intrinsic quantum effects. Planck's constant enters merely via the fluctuation-dissipation theorem, as already emphasized after Eq. (6.35). A brief discussion of genuine quantum effects resulting from state-dependent dissipation is postponed to Subsection 9.3.1. Intrinsic quantum effects are pertinent to nonlinear systems, and they are decisive for the transport properties, as we shall see in Part V.

The simplest microscopic model for a Brownian particle is described by the Hamiltonian (3.11) with $V(q) = 0$. The classical equation of motion is given by

$$M\ddot{q}(t) + M \int_{-\infty}^{t} dt'\, \gamma(t - t')\, \dot{q}(t') \;=\; \xi(t)\,, \tag{7.1}$$

where $\xi(t)$ is the fluctuating force with properties as discussed in Section 3.1.2. As emphasized previously, the dynamical equation (7.1) holds also exactly in the quantum regime, but the fluctuating force is characterized by the quantum mechanical force auto-correlation function (5.34). Additional light is shed on the underlying physics by introducing new coordinates y_α and masses μ_α for the bath oscillators [228],

$$y_\alpha \;=\; \frac{m_\alpha \omega_\alpha^2}{c_\alpha}\, x_\alpha\,; \qquad \mu_\alpha \;=\; \frac{c_\alpha^2}{m_\alpha \omega_\alpha^4}\,. \tag{7.2}$$

With these definitions the Hamiltonian (3.11) with $V(q) = 0$ becomes

$$H \;=\; \frac{p^2}{2\,M_0} + \frac{1}{2}\sum_{\alpha=1}^{N} \mu_\alpha \left(\dot{y}_\alpha^2 + \omega_\alpha^2\,(y_\alpha - q)^2 \right)\,, \tag{7.3}$$

where we have distinguished M_0 from M for reasons that will shortly become apparent. We see that the model describes a particle of mass M_0 with all of the reservoir's effective masses μ_α attached with springs to its position coordinate q. This form elucidates the translational invariance of the model.

7.1 Spectral density, damping function and mass renormalization

In this section, we briefly discuss renormalization of the mass of the Brownian particle induced by the environmental coupling.

With the substitution (7.2) the spectral density of the coupling takes the form

$$J(\omega) \;=\; \frac{\pi}{2}\sum_{\alpha=1}^{N} \mu_\alpha \omega_\alpha^3\, \delta(\omega - \omega_\alpha)\,. \tag{7.4}$$

We can still model any desired frequency dependence of $J(\omega)$ by an appropriate choice of the spectral distribution of the oscillators.

If $\sum_\alpha \mu_\alpha$ is finite, the total energy associated with the Hamiltonian (7.3) is the sum of an internal energy, which is independent of the total momentum P, and the kinetic energy of the center of mass. The latter is the square of the total momentum divided by twice the mass of the total system. The total or renormalized mass is given by

$$M_{\rm r} \;=\; M_0 + \sum_{\alpha=1}^{N} \mu_\alpha \;=\; M_0 + \frac{2}{\pi} \int_0^\infty d\omega \, \frac{J(\omega)}{\omega^3} \;. \tag{7.5}$$

In the second form, we have taken the continuum limit. Clearly, the total mass $M_{\rm r}$ is only defined when the integral in Eq. (7.5) is finite.

Consider now the case when the spectral density $J(\omega)$ has the form (3.55) with (3.56). For the purpose of this section, details of the high-frequency behaviour of $J(\omega)$ are irrelevant. It is then convenient to choose a sharp cutoff $f(\omega/\omega_{\rm c}) = \Theta(\omega_{\rm c}-\omega)$. Thus for times $t \gg \omega_{\rm c}^{-1}$, the only effect of the high frequency modes is mass renormalization,

$$M \;=\; M_0 + \frac{2}{\pi} \int_{\omega_{\rm c}}^\infty d\omega \, \frac{J_{\rm hf}(\omega)}{\omega^3} \;. \tag{7.6}$$

Now, we are left with the problem of a Brownian particle of mass M coupled to a heat bath environment described by the spectral density $J_{\rm lf}(\omega)$ given in Eq. (3.56).

If the exponent s in Eq. (3.56) exceeds 2, the integral in Eq. (7.5) is infrared-convergent. Then all environmental modes down to zero frequency simply contribute to a polaronic cloud which increases inertia of the particle. The mass of inertia due to the low-frequency part $J_{\rm lf}(\omega)$ is

$$\Delta M \;\equiv\; \frac{2}{\pi} \int_0^{\omega_{\rm c}} d\omega \, \frac{J(\omega)}{\omega^3} \;=\; \frac{2}{\pi(s-2)} \, \frac{\gamma_s}{\omega_{\rm ph}} \left(\frac{\omega_{\rm c}}{\omega_{\rm ph}} \right)^{s-2} M \;. \tag{7.7}$$

Here we have substituted the form (3.58) for $J(\omega)$ with a sharp cutoff at $\omega = \omega_{\rm c}$.

With the form (7.7), the damping coefficient vanishes linearly with ω as $\omega \to 0$,

$$\lim_{\omega \to 0} \widetilde{\gamma}(\omega)/\omega \;=\; -i\,\Delta M/M \;, \tag{7.8}$$

as follows from Eq. (3.31). The dynamical susceptibility of the Brownian particle

$$\widetilde{\chi}(\omega) \;=\; -\frac{1}{M[\omega^2 + i\,\omega\widetilde{\gamma}(\omega)]} \;, \tag{7.9}$$

following from Eq. (6.31) by setting $\omega_0 = 0$, has then the low-frequency behaviour

$$\widetilde{\chi}(\omega \to 0) \;=\; -\frac{1}{(M + \Delta M)\omega^2} \;. \tag{7.10}$$

From this we conclude that, for $s > 2$, the damping effectively vanishes for long times, and the Brownian particle behaves as a free particle with the renormalized mass

$$M_{\rm r} \;=\; M + \Delta M \;. \tag{7.11}$$

In contrast, when $s \leq 2$, the integral in Eq. (7.5) or Eq. (7.7) is infrared-divergent. This indicates that the low-energy excitations of the reservoir have deeper effects than simple mass renormalization. In this parameter regime, we have [cf. Eq. (3.64)]

$$\widetilde{\gamma}(\omega) \;=\; [\gamma_s/\sin(\pi s/2)]\,(-i\omega/\omega_{\rm ph})^{s-1} \;.$$

For $s < 2$, $\widetilde{\gamma}(\omega) \propto \omega^{s-1}$ is qualitatively stronger than linear behaviour in ω as $\omega \to 0$. Therefore, in this regime damping is effective also at long times.

7.2 Displacement correlation and response function

If we take the limit $\omega_0 \to 0$ in the expression (6.91) or Eq. (6.98), the mean square of the position of the damped oscillator diverges. This singularity is natural since the position of a free particle is not bounded. Subtracting from the position autocorrelation function $C^+(t)$ the divergent initial value, we get the displacement correlation function

$$D^+(t) \equiv C^+(t) - C^+(0) = \langle [q(t) - q(0)] \, q(0) \rangle_\beta \, . \tag{7.12}$$

The function $D^+(t)$ describes correlations between the displacement $q(t) - q(0)$ of the particle at time t with the initial position $q(0)$. Next, we write

$$D^+(t) = \overline{S}(t) + i \, A(t) \, , \tag{7.13}$$

and introduce the mean square displacement

$$\sigma^2(t) \equiv \langle [q(t) - q(0)]^2 \rangle_\beta \, . \tag{7.14}$$

We then have

$$\sigma^2(t) = -2\overline{S}(t) = 2[S(0) - S(t)] \, , \tag{7.15}$$

where $S(t)$ is the symmetrized correlation function given in Eq. (6.13).

The imaginary part of $D^+(t)$ is connected with the thermal expectation value of the commutator in the usual way,

$$A(t) = \frac{1}{2i} \langle [\, q(t), q(0) \,] \rangle_\beta \, , \tag{7.16}$$

so that the dynamical susceptibility is given by

$$\chi(t) = -\frac{2}{\hbar} \Theta(t) \, A(t) \, . \tag{7.17}$$

To obtain an explicit expression for the mean square displacement $\sigma^2(t)$, we use Eq. (7.15) and Eq. (6.53) and insert the imaginary part of $\widetilde{\chi}(\omega)$ from Eq. (7.9). We then find the expressions

$$
\begin{aligned}
\sigma^2(t) &= \frac{\hbar}{\pi} \int_{-\infty}^{+\infty} d\omega \, \widetilde{\chi}''(\omega) \coth(\omega \hbar \beta / 2) \, [1 - \cos(\omega t)] \\
&= \frac{\hbar}{\pi M} \int_{-\infty}^{+\infty} \frac{d\omega}{\omega} \, \frac{\widetilde{\gamma}'(\omega) \coth(\omega \hbar \beta / 2) \, [1 - \cos(\omega t)]}{[\omega + \widetilde{\gamma}''(\omega)]^2 + \widetilde{\gamma}'^2(\omega)} \, .
\end{aligned}
\tag{7.18}
$$

Similarly, the representation (6.55) of the response function takes the form

$$\chi(t) = \Theta(t) \frac{1}{\pi M} \int_{-\infty}^{+\infty} \frac{d\omega}{\omega} \, \frac{\widetilde{\gamma}'(\omega)}{[\omega + \widetilde{\gamma}''(\omega)]^2 + \widetilde{\gamma}'^2(\omega)} \sin(\omega t) \, . \tag{7.19}$$

From Eqs. (7.18) and (7.19) we see that, in the long-time limit, the response function $\chi(t)$ is connected with the change of $\sigma^2(t)$ per unit time by the simple relation

$$\lim_{t\to\infty} \frac{1}{\chi(t)} \frac{d}{dt} \sigma^2(t) = 2k_{\mathrm{B}}T . \tag{7.20}$$

It will soon become clear that the relation (7.20) is a version of the *Einstein relation*. It is interesting to note that in the derivation of Eq. (7.20) we made no specific assumptions on $\widetilde{\gamma}(\omega)$. The significance of the response function may be illustrated by the effect of a constant force F applied to the Brownian particle for $t \geq 0$. There follows from Eq. (6.6) with the substitution $\delta f_{\mathrm{ext}}(t) \to \Theta(t)\,F$, where F need not be infinitesimal, that the drift velocity of the particle is related to $\chi(t)$ according to

$$\langle \dot{q}(t) \rangle = \chi(t)\,F . \tag{7.21}$$

We shall use this relation when we discuss the Brownian dynamics at long times.

7.3   Ohmic damping

Consider now the case of Ohmic damping in some detail. It is practical knowledge from daily experience that for a constant force F and Ohmic friction, the drift velocity becomes asymptotically constant,

$$\lim_{t\to\infty} \langle \dot{q}(t) \rangle = \mu\,F , \tag{7.22}$$

where μ is the dc mobility of the Brownian particle. The mobility μ is related to the response function $\chi(t)$ by

$$\mu = \lim_{t\to\infty} \chi(t) = \lim_{\omega\to 0} \omega\,\widetilde{\chi}''(\omega = 0) , \tag{7.23}$$

as follows from Eqs. (7.21) and (6.55). Putting $\widetilde{\gamma}(\omega) = \gamma$ in Eq. (7.19), we obtain for the response function of a free Brownian particle the familiar expression

$$\chi(t) = \frac{1}{M\gamma}\Big(1 - \mathrm{e}^{-\gamma t}\Big) . \tag{7.24}$$

Thus we have

$$\mu = \frac{1}{M\gamma} = \frac{1}{\eta} . \tag{7.25}$$

The time-dependence of the mean square displacement can be found either by direct integration of Eq. (7.18), or by use of the relation (7.15) with Eqs. (6.57) and (6.59) in the limit $\omega_0 \to 0$. In the end we find

$$\sigma^2(t) = \frac{2}{M\beta\gamma}t - \frac{\hbar}{M\gamma}\cot(\hbar\beta\gamma/2)\,[1 - \mathrm{e}^{-\gamma t}] + \frac{4\gamma}{M\beta}\sum_{n=1}^{\infty}\frac{1 - \mathrm{e}^{-\nu_n t}}{\nu_n(\gamma^2 - \nu_n^2)} . \tag{7.26}$$

In the high-temperature limit, the expression (7.26) simplifies to the classical mean square displacement

$$\sigma_{\text{cl}}^2(t) = \frac{2}{M\beta\gamma^2}\left(\gamma t - 1 + e^{-\gamma t}\right). \tag{7.27}$$

At short times, $\sigma_{\text{cl}}^2(t)$ grows $\propto t^2$. We see from Eq. (7.26) that the quantum mechanical contributions to $\sigma^2(t)$ are relevant only at low temperatures where $\hbar\beta\gamma/2\pi > 1$, and for short to intermediate times. With decreasing temperature, the regime where quantum fluctuations are relevant extends to longer times. However, at all finite temperatures, the dynamics at long times is dominated by the first term in Eq. (7.26), which is purely classical. We find for the diffusion coefficient

$$D \equiv \frac{1}{2}\lim_{t\to\infty}\frac{d}{dt}\sigma^2(t) = \frac{1}{M\beta\gamma} = \frac{k_{\text{B}}T}{M\gamma}. \tag{7.28}$$

Thus, the diffusion coefficient vanishes at zero temperature. This indicates that the growth of the mean square displacement is slowing down when T approaches zero.

The remaining task is to link the (linear) mobility μ to the diffusion coefficient. Upon inserting Eqs. (7.23) with (7.25) and (7.28) into (7.20), or by comparison of Eq. (7.25) with Eq. (7.28), we obtain the Einstein relation in the familiar form

$$D = k_{\text{B}}T\mu. \tag{7.29}$$

At $T = 0$, the Matsubara sum in Eq. (7.26) turns into an integral,

$$\lim_{T\to 0}\frac{1}{\beta}\sum_{n=1}^{\infty}f(\nu_n) = \frac{\hbar}{2\pi}\int_0^\infty d\nu\, f(\nu). \tag{7.30}$$

With this, the time derivative of $\sigma_{T=0}^2(t)$ takes the form

$$\frac{d}{dt}\sigma_{T=0}^2(t) = \frac{2\hbar\gamma}{\pi M}\int_0^\infty d\nu\,\frac{e^{-\nu t}}{\gamma^2 - \nu^2}. \tag{7.31}$$

The integral in Eq. (7.31) can be expressed in terms of the exponential integral function Ei(x). For x not on the positive real axis, Ei(x) is defined by the integral representation

$$\text{Ei}(x) = \int_{-\infty}^{x} dx'\,\frac{e^{x'}}{x'}. \tag{7.32}$$

The function is analytically continued to the positive real axis according to

$$\overline{\text{Ei}}(x) = \lim_{\epsilon\to 0}\frac{1}{2}\Big[\text{Ei}(x + i\epsilon) + \text{Ei}(x - i\epsilon)\Big]. \tag{7.33}$$

Upon fixing the integration constant by the initial condition $\sigma_{T=0}^2(0) = 0$, one then finds the integral representation[86]

$$\sigma_{T=0}^2(t) = \frac{\hbar}{\pi M}\int_0^t ds\left(\overline{\text{Ei}}(\gamma s)\,e^{-\gamma s} - \text{Ei}(-\gamma s)\,e^{\gamma s}\right). \tag{7.34}$$

For long times $t \gg \gamma^{-1}$, the integral may be evaluated asymptotically. We then find that in the absence of thermal fluctuations the mean square displacement grows only logarithmically with time,

$$\sigma_{T=0}^2(t) = [2\hbar/\pi M\gamma] \ln(\gamma t) \qquad \text{for} \qquad t \to \infty . \tag{7.35}$$

The sluggish logarithmic growth of $\sigma_{T=0}^2(t)$ at long times was obtained first by Hakim and Ambegaokar [228]. We note in passing that the logarithmic law (7.35) is not confined to the displacement correlation function of a free Brownian particle. It rather holds for all systems for which the linear mobility at zero temperature is finite,

$$\mu_0 = \mu(T=0) = \lim_{\omega \to 0} \omega \, \tilde{\chi}''(\omega, T=0) . \tag{7.36}$$

For all systems with finite linear mobility at $T = 0$, we get from Eq. (7.18) the long-time behaviour (see also Subsection 24.7.2)

$$\sigma_{T=0}^2(t \to \infty) = (2\hbar\mu_0/\pi) \ln(t/t_0) , \tag{7.37}$$

where t_0 is a reference time. It is obvious from the derivation that the logarithmic growth of the mean square displacement at zero temperature has its origin in the *fluctuation-dissipation theorem*.

We wish to emphasize that the $\hbar$-dependence in Eq. (7.37) comes from the noise and is not a quantum mechanical property of the particle.

With the absorptive part of the dynamical susceptibility taken from Eq. (7.9), we get from Eq. (6.93) for the mean value of p^2 the expression

$$\langle p^2 \rangle = \frac{\hbar M}{\pi} \int_0^\infty d\omega \, \frac{\omega \, \tilde{\gamma}'(\omega)}{[\omega + \tilde{\gamma}''(\omega)]^2 + \tilde{\gamma}'^2(\omega)} \coth(\omega\hbar\beta/2) . \tag{7.38}$$

Alternatively, putting $\omega_0 = 0$ in Eq. (6.95), we may express $\langle p^2 \rangle$ in terms of the series

$$\langle p^2 \rangle = \frac{M}{\beta} \left(1 + 2 \sum_{n=1}^\infty \frac{\widehat{\gamma}(\nu_n)}{\nu_n + \widehat{\gamma}(\nu_n)} \right) . \tag{7.39}$$

In the Ohmic case, $\langle p^2 \rangle$ is logarithmically ultra-violet divergent, as before for the harmonic oscillator in Eq. (6.95), and must be regularized by a cutoff in the spectral density. The cutoff frequency of the bath is actually a theoretically measurable quantity. To be definite, we obtain for the Drude model (3.53) the exact expression

$$\langle p^2 \rangle = \frac{M}{\beta} + \frac{M\hbar\gamma}{\pi} \frac{\omega_D}{\sqrt{\omega_D^2 - 4\gamma\omega_D}} \left[\psi(1 + \lambda_1/\nu) - \psi(1 + \lambda_2/\nu) \right] , \tag{7.40}$$

where

$$\lambda_{1/2} = \omega_D/2 \pm \sqrt{\omega_D^2/4 - \gamma\omega_D} . \tag{7.41}$$

The first term in Eq. (7.40) dominates in the high-temperature limit. The residual terms become increasingly important as temperature is decreased. At zero temperature, we have in the limit $\omega_D \gg 4\gamma$

$$\langle p^2 \rangle_{T=0} = (M\hbar\gamma/\pi) \ln(\omega_D/\gamma) , \tag{7.42}$$

which depends logarithmically on the cutoff ω_D.

7.4 Frequency–dependent damping

For general frequency-dependent damping, the response function can still be expressed in terms of known functions. To my knowledge, this is not possible for the mean square displacement, except for long times where the dominant contributions to $\sigma^2(t)$ come from frequencies near $\omega = 0$.

7.4.1 Response function and mobility

We begin with the discussion of the response function, which for linear systems is temperature-independent. Instead of using the integral expression (7.19), we shall base the discussion on the study of the Laplace integral representation of the response function. Putting $\omega_0 = 0$ in Eq. (6.32), we obtain for the time derivative of $\chi(t)$

$$\dot{\chi}(t) \;=\; \frac{1}{2\pi i} \int_{-i\infty+c}^{i\infty+c} dz \, \frac{e^{zt}}{M[z + \widehat{\gamma}(z)]} \, . \tag{7.43}$$

Now consider the behaviour of $\chi(t)$ for the various cases of $\widehat{\gamma}(z)$ in Eq. (3.64).

In the regime $s < 2$, the integral in Eq. (7.43) is a contour integral representation of a known special function, namely the Mittag-Leffler function $E_\alpha(x)$ [229],

$$E_\alpha(x) \;=\; \frac{1}{2\pi i} \int_{-i\infty+c}^{i\infty+c} dy \, \frac{y^{\alpha-1} \, e^y}{y^\alpha - x} \;=\; \sum_{k=0}^{\infty} \frac{x^k}{\Gamma(\alpha k + 1)} \, , \tag{7.44}$$

which is a particular case of the generalized Mittag-Leffler function, $E_\alpha(x) = E_{\alpha,1}(x)$,

$$E_{\alpha,\beta}(x) \;=\; \frac{1}{2\pi i} \int_{-i\infty+c}^{i\infty+c} dy \, \frac{y^{\alpha-\beta} \, e^y}{y^\alpha - x} \;=\; \sum_{k=0}^{\infty} \frac{x^k}{\Gamma(\alpha k + \beta)} \, . \tag{7.45}$$

In the present case we have $\alpha = 2 - s$ and $x = -(\omega_s t)^{2-s}$,

$$\dot{\chi}(t) \;=\; E_{2-s}[- (\omega_s t)^{2-s}]/M \tag{7.46}$$

with

$$\omega_s \;=\; \Big(\frac{\gamma_s}{\sin(\pi s/2)\omega_{\mathrm{ph}}} \Big)^{1/(2-s)} \omega_{\mathrm{ph}} \, . \tag{7.47}$$

Similarly, we find that the response function is a generalized Mittag-Leffler function $E_{\alpha,\beta}(x)$ with $\alpha = 2 - s$ and $\beta = 2$,

$$\chi(t) \;=\; t \, E_{2-s,2}[- (\omega_s t)^{2-s}]/M \, . \tag{7.48}$$

For $s < 1$, the integrand $1/[z + \widehat{\gamma}(z)]$ in Eq. (7.43) has three singularities:

(1) A complex conjugate pair of simple poles on the principal sheet at $z = -\Gamma_s \pm i\Omega_s$,

$$\Omega_s \;=\; \omega_s \cos\Big(\frac{\pi}{2} \frac{s}{(2-s)} \Big) \; ; \qquad \Gamma_s \;=\; \omega_s \sin\Big(\frac{\pi}{2} \frac{s}{(2-s)} \Big) \, . \tag{7.49}$$

The residues of these poles give a "coherent" contribution to $\dot\chi(t)$ describing damped oscillations,

$$\dot\chi_{\text{coh}}(t) \approx \frac{\cos(\Omega_s t - \varphi_s)\, e^{-\Gamma_s t}}{\cos\varphi_s}, \qquad \varphi_s = \arctan\left(\frac{\Gamma_s}{\Omega_s}\right). \qquad (7.50)$$

(2) A branch point at $z = 0$ with a cut in the complex plane along the negative real axis. The branch cut gives an "incoherent" contribution to $\dot\chi(t)$, which is in the form of a power series in $(\omega_{\text{ph}}t)^{s-2}$. The contribution of the cut dominates the behaviour of $\dot\chi(t)$ at long times.

For $s = 1$, the branch cut is absent and Eq. (7.48) reduces to the familiar form (7.24). In the parameter regime $1 < s < 2$, the poles of the integrand in Eq. (7.45) are not anymore on the principal sheet, and $\dot\chi(t)$ is entirely given by the cut contribution. In this parameter regime, there follows from the integral representation (7.45) that $\dot\chi(t)$ is a completely monotonous function of t. We find from Eq. (7.48) at long times

$$\chi(t) = \frac{\sin(\pi s/2)}{M\gamma_s\Gamma(s)}(\omega_{\text{ph}}t)^{s-1}\left(1 + \mathcal{O}\left[(\gamma_s/\omega_{\text{ph}})(\omega_{\text{ph}}t)^{s-2}\right]\right), \qquad (7.51)$$

which includes the Ohmic result $\chi(t \to \infty) = 1/M\gamma$ given above as a special case.

The behaviour (7.51) of $\chi(t)$ for long times suggests to define a generalized linear dc mobility by

$$\mu_\ell^{(s)} \equiv \lim_{t\to\infty} \chi(t)/(\omega_{\text{ph}}t)^{s-1}. \qquad (7.52)$$

With the use of (7.51) we then obtain

$$\mu_\ell^{(s)} = \frac{\sin(\pi s/2)}{M\gamma_s\Gamma(s)} \qquad \text{for} \qquad 0 < s < 2. \qquad (7.53)$$

Alternatively and more commonly, we may introduce the linear ac mobility

$$\widetilde{\mu}(\omega) = -i\omega\widetilde{\chi}(\omega). \qquad (7.54)$$

For $s < 2$, we find upon using Eq. (3.64) the low-frequency form

$$\widetilde{\mu}(\omega) = \frac{1}{M\widetilde{\gamma}(\omega)} = \frac{\sin(\pi s/2)}{M\gamma_s}\left(\frac{\omega_{\text{ph}}}{-i\omega}\right)^{s-1}. \qquad (7.55)$$

For $s = 2$, the ac mobility takes the form

$$\widetilde{\mu}(\omega) = \frac{\pi}{2M\gamma_2}\frac{i\omega_{\text{ph}}}{\omega\ln(\omega_{\text{c}}/\omega)} \qquad \text{for} \qquad \omega \ll \omega_{\text{c}}. \qquad (7.56)$$

Accordingly, the response function at times $\omega_c t \gg 1$ reads

$$\chi(t) = \frac{\pi}{2M\gamma_2}\frac{\omega_{\text{ph}}t}{\ln(\omega_c t)}, \qquad (7.57)$$

which shows logarithmic modification of the free particle behaviour.

For $s > 2$, the main effect of the heat bath on the dynamics at long times is to renormalize the particle's mass M into $M_r = M + \Delta M$, where ΔM is defined in Eq. (7.7). In the regime $2 < s < 4$, the transient behaviour of $\dot{\chi}(t)$ and $\chi(t)$ is described again in terms of a Mittag-Leffler and generalized Mittag-Leffler function, respectively. We find

$$\dot{\chi}(t) = \left(1 - E_{s-2}[(M_r/M)(\omega_s t)^{s-2}]\right)\Big/M_r, \tag{7.58}$$

$$\chi(t) = t\left(1 - E_{s-2,2}[(M_r/M)(\omega_s t)^{s-2}]\right)\Big/M_r, \tag{7.59}$$

where M_r is given in Eq. (7.11) and ω_s in Eq. (7.47). At long times, this reduces to

$$\chi(t) = \frac{t}{M_r}\left(1 + \frac{M}{M_r}\frac{\gamma_s \omega_{\text{ph}}^{1-s} t^{2-s}}{\Gamma(4-s)|\sin(\pi s/2)|}\right). \tag{7.60}$$

7.4.2 Mean square displacement

Let us now turn to the discussion of the long-time behaviour of the mean square displacement function $\sigma^2(t)$ introduced in Section 7.2. At finite temperature, we may infer the leading dependence of $\sigma^2(t)$ at long times from the first integral of the relation (7.20), which is

$$\sigma^2(t) = \frac{2}{\beta}\int_0^t dt'\,\chi(t'). \tag{7.61}$$

Using Eqs. (7.51), (7.57) and (7.60), we then find for $T > 0$ and large t

$$\sigma^2(t) = \begin{cases} \dfrac{2\sin(\pi s/2)}{M\beta\gamma_s\omega_{\text{ph}}\Gamma(s+1)}(\omega_{\text{ph}}t)^s, & \text{for} \quad s < 2, \\[3mm] \dfrac{\pi}{2M\beta\gamma_2\omega_{\text{ph}}}\dfrac{(\omega_{\text{ph}}t)^2}{\ln(\omega_c t)}, & \text{for} \quad s = 2, \\[3mm] \dfrac{2}{M_r\beta}t^2, & \text{for} \quad s > 2. \end{cases} \tag{7.62}$$

These results can be summarized as follows. For sub-Ohmic damping ($s < 1$), the ac mobility, the response function, and the mean square displacement in thermal equilibrium show subdiffusive behaviour, i.e., the dynamics is slower than in the diffusive regime. In the super-Ohmic regime $1 < s < 2$, the dynamics is faster than in the diffusive regime. For instance, the ac mobility $\tilde{\mu}(\omega)$ vanishes for $s < 1$ and diverges for $s > 1$ in the zero frequency limit. In the case $s = 2$ we find logarithmic corrections to the free particle behaviour, while for $s > 2$ the dynamics at long times is that of a free particle with a renormalized mass.

At zero temperature, the mean square displacement is given by

$$\sigma^2_{T=0}(t) = \frac{2\hbar}{\pi}\int_0^\infty d\omega\,\tilde{\chi}''(\omega)\,[1 - \cos(\omega t)]. \tag{7.63}$$

We can again relate this function to the response function by use of the identity [87]

$$\frac{2}{\pi} \int_0^\infty dx \, \frac{\sin(x\omega t)}{x\left(1 - x^2\right)} = 1 - \cos(\omega t) \,. \tag{7.64}$$

We then find

$$\sigma_{T=0}^2(t) = \frac{2\hbar}{\pi} \int_0^\infty dx \, \frac{1}{x\left(1 - x^2\right)} \chi(xt) \,, \tag{7.65}$$

where we have used Eq. (7.19). Given the long-time behaviour of $\chi(t)$ by Eqs. (7.51), (7.57) and (7.60), the behaviour of $\sigma_{T=0}^2(t)$ for $t \to \infty$ follows from Eq. (7.65) by integration. The results in the various damping regimes are as follows [86]

$$\sigma_{T=0}^2(t \to \infty) = \begin{cases} \sigma_\infty^2 \,, & \text{for} \quad 0 < s < 1 \,, \\ a_1 \ln(\gamma_1 t) \,, & \text{for} \quad s = 1 \,, \\ a_s \left(\omega_{\text{ph}} t\right)^{s-1} \,, & \text{for} \quad 1 < s < 2 \,, \\ a_2 \, \omega_{\text{ph}} t / \ln^2(\omega_{\text{c}} t) \,, & \text{for} \quad s = 2 \,, \\ a_s \left(\omega_{\text{ph}} t\right)^{3-s} \,, & \text{for} \quad 2 < s < 3 \,, \\ a_s \ln(\omega_{\text{c}} t) \,, & \text{for} \quad s = 3 \,, \\ \overline{\sigma}_\infty^2 \,, & \text{for} \quad s > 3 \,. \end{cases} \tag{7.66}$$

For sub-Ohmic friction, $\chi(t)$ goes to zero as $t \to \infty$. Thus, after substituting $u = xt$ in Eq. (7.65), we can readily carry out the limit $t \to \infty$ in the integrand and find that $\sigma_{T=0}^2(t)$ approaches the constant

$$\sigma_\infty^2 = \frac{2\hbar}{\pi} \int_0^\infty du \, \frac{\chi(u)}{u} = \frac{2\hbar}{\pi} \int_0^\infty dz \, \widehat{\chi}(z) \,. \tag{7.67}$$

Upon inserting $\widehat{\chi}(z) = M^{-1}[z^2 + z\,\hat{\gamma}(z)]^{-1}$ and the expression (3.64) for $\hat{\gamma}(z)$, the integral can be evaluated exactly. We then find

$$\sigma_\infty^2 = \frac{2}{(2-s) \sin[\pi/(2-s)]} \left(\frac{\omega_{\text{ph}} \sin(\pi s/2)}{\gamma_s} \right)^{1/(2-s)} \frac{\hbar}{M\omega_{\text{ph}}} \,. \tag{7.68}$$

For $1 \leq s \leq 3$, we can directly calculate the x-integral in Eq. (7.65) with the asymptotic expression for $\chi(xt)$. From this we obtain the behaviour (7.66), where the coefficients are given by

$$a_s = \begin{cases} \dfrac{2}{\pi} \dfrac{\hbar}{M\gamma_1} \,, & \text{for} \quad s = 1 \,, \\[2ex] \dfrac{\sin^2(\pi s/2)}{|\cos(\pi s/2)|\Gamma(s)} \dfrac{\hbar}{M\gamma_s} \,, & \text{for} \quad 1 < s < 2 \,, \\[2ex] \dfrac{\pi^2}{4} \dfrac{\hbar}{M\gamma_2} \,, & \text{for} \quad s = 2 \,, \\[2ex] \dfrac{1}{|\cos(\pi s/2)|\Gamma(s)} \dfrac{\hbar M\gamma_s}{M_{\text{r}}^2 \omega_{\text{ph}}^2} \,, & \text{for} \quad 2 < s < 3 \,, \\[2ex] \dfrac{2}{\pi} \dfrac{\hbar M\gamma_3}{M_{\text{r}}^2 \omega_{\text{ph}}^2} \,, & \text{for} \quad s = 3 \,. \end{cases} \tag{7.69}$$

For $s = 1$, we recover the previous result (7.35). In the region $1 < s < 2$, the mean square displacement for $T = 0$ grows as t^{s-1}, whereas for $2 < s < 3$ we have the behaviour $\propto t^{3-s}$. In the cases $s = 2$ and $s = 3$, $\sigma^2_{T=0}(t \to \infty)$ is not described by a simple power-law. For $s > 3$, the leading contribution to $\chi(t)$ grows linearly with t. This term does not contribute to $\sigma^2_{T=0}(t \to \infty)$. Observing that the subleading term $\propto t^{3-s}$ in $\chi(t)$ vanishes at long times for $s > 3$, the situation is similar to the sub-Ohmic case. As a result, we find that $\sigma^2_{T=0}(t)$ approaches a constant $\overline{\sigma}^2_\infty$ as $t \to \infty$,

$$\overline{\sigma}^2_\infty = \frac{2\hbar}{\pi} \int_0^\infty dz \left(\widehat{\chi}(z) - \frac{1}{M_{\rm r} z^2} \right). \tag{7.70}$$

In difference to the sub-Ohmic case, the constant $\overline{\sigma}^2_\infty$ is not determined by the low-frequency behaviour of $\widehat{\gamma}(z)$. Rather we find for $s > 3$ that, besides mass renormalization, also the behaviour of the mean square displacement at long times depends on the high-frequency properties of the memory-friction.

8. The thermodynamic variational approach

It is appealing to study quantum statistical expectation values with concepts familiar from classical statistical mechanics. The phase space representation of the reduced density matrix in terms of Wigner's function discussed in Subsection 4.4.2 is just one example. Another important concept which goes back to Feynman [230] is the *effective classical potential*.

In the effective classical potential method, all physical effects related to nonlinearity are exactly taken into account in the classical regime. In the quantum regime, the effective potential is optimized by a variational principle with a trial action which is quadratic in the dynamical variables. Feynman originally used a "free particle" trial action. Considerable improvement has been achieved by Giachetti and Tognetti [231], and independently by Feynman and Kleinert [232], by adding a harmonic potential term in the trial action, thereby introducing a second variational parameter. Further generalization is obvious: since linear dissipation is rendered by a "quadratic" action functional, it can be included in the variational approach with minor ancillary efforts.

In the following we present the main ideas. To be general, the discussion will be given for open systems with arbitrary linear dissipation.

8.1 Centroid and the effective classical potential

8.1.1 Centroid

A useful path decomposing concept has been developed by Feynman [3, 4]. The basic idea is to divide up the paths into equivalence classes and to rewrite the original path sum as a sum over paths belonging to the same equivalence class and an additional sum over the equivalence classes. In connection with the *effective classical potential*, it is convenient to consider as equivalence class all those paths which have the same mean position $q_c[q(\cdot)]$, usually referred to as the *centroid*,[1]

$$q_c[q(\cdot)] \;=\; \frac{1}{\hbar\beta} \int_0^{\hbar\beta} d\tau\, q(\tau) \,. \tag{8.1}$$

Within the centroid method, the path sum (4.61) for the reduced density matrix in thermal equilibrium is rearranged as a sum over all paths which are constrained to the same centroid and an additional ordinary integral over the centroid coordinate. We then have (in this chapter we use the normalization tr $\rho = Z$)

$$\rho(q'', q') \;=\; \int_{-\infty}^{\infty} dq_c\, \rho(q'', q'; q_c) \,,$$
$$\rho(q'', q'; q_c) \;\equiv\; \int_{q(0)=q'}^{q(\hbar\beta)=q''} \mathcal{D}q(\cdot)\, \delta\Big(q_c - q_c[q(\cdot)] \Big) \exp\Big(-S_{\mathrm{eff}}^{\mathrm{E}}[q(\cdot)]/\hbar \Big) \,. \tag{8.2}$$

Correspondingly, we may also write the partition function $Z = \int dq'\, \rho(q', q')$ in the form of an integral over the centroid density,

$$Z \;=\; \int_{-\infty}^{\infty} dq_c\, \rho_c(q_c) \,, \tag{8.3}$$

$$\rho_c(q_c) \;\equiv\; \oint \mathcal{D}q(\cdot)\, \delta\Big(q_c - q_c[q(\cdot)] \Big) \exp\Big(-S_{\mathrm{eff}}^{\mathrm{E}}[q(\cdot)]/\hbar \Big) \,. \tag{8.4}$$

To proceed, we Fourier-expand the periodic path $q(\tau)$ with period $\hbar\beta$,

$$q(\tau) \;=\; q_c + \tilde{q}(\tau) \;=\; Q_0 + \sum_{\substack{n=-\infty \\ n\neq 0}}^{\infty} Q_n\, e^{i\nu_n\tau} \,. \tag{8.5}$$

The Fourier coefficient Q_0 represents the centroid q_c, as follows from definition (8.1). Next, we use the functional integral measure (4.227) and write the path sum in (8.2) as an infinite product of ordinary integrals. We then get the centroid density as

$$\rho_c(q_c) \;=\; \Big(M/2\pi\hbar^2\beta \Big)^{1/2} \exp\Big(-\beta\mathcal{V}(q_c) \Big) \,, \tag{8.6}$$

in which $\mathcal{V}(q_c)$ is the *effective classical potential*,

$$e^{-\beta\mathcal{V}(q_c)} \;=\; \oint \mathcal{D}\tilde{q}\, \exp\Big(-M\beta \sum_{n=1}^{\infty} [\,\nu_n^2 + \nu_n\widehat{\gamma}(\nu_n)\,]\, |Q_n|^2 - \int_0^{\hbar\beta} \frac{d\tau}{\hbar}\, V[\,q_c + \tilde{q}(\tau)\,] \Big) \,. \tag{8.7}$$

The functional integral measure is limited to pure quantum fluctuations. It reads

$$\oint \mathcal{D}\tilde{q} \times \cdots \equiv \prod_{n=1}^{\infty} \int_{-\infty}^{\infty} \int_{-\infty}^{\infty} \frac{d\mathrm{Re}\, Q_n\, d\mathrm{Im}\, Q_n}{\pi/M\beta\nu_n^2} \times \cdots \,. \tag{8.8}$$

The expression (8.6) with (8.7) and (8.8) is still exact. Now, all quantum effects in the partition function are included in the effective classical potential $\mathcal{V}(q_c)$.

---

[1]The meaning of the central dot in $q(\cdot)$ is explained in the footnote to Eq. (4.22).

8.1.2 The effective classical potential

Due to the kinetic action term $M\beta\nu_n^2|Q_n|^2$ in the exponent of (8.7), the Q_n are effectively of order $1/(\nu_n\sqrt{M\beta}) \propto 1/\sqrt{T}$. Therefore, the quantum fluctuations $\tilde{q}(\tau)$ in the potential term in Eq. (8.7) decrease when temperature is increased. In the classical limit, $V[q(\tau)]$ becomes independent of $\tilde{q}(\tau)$, so that the integrals over $\mathrm{Re}\,Q_n$ and $\mathrm{Im}\,Q_n$ in Eq. (8.7) take Gaussian form. Since $\widehat{\gamma}(\nu_n)/\nu_n \to 0$ in this limit, the fluctuation integrals simply give a factor unity. Thus we have in the classical limit

$$\lim_{T\to\infty} \mathcal{V}(q_c) \to V(q_c) \,, \qquad (8.9)$$

and hence

$$\lim_{T\to\infty} Z \to Z^{(\mathrm{cl})} = \left(M/2\pi\hbar^2\beta\right)^{1/2} \int_{-\infty}^{\infty} dq_c \, \exp\left(-\beta V(q_c)\right) \,. \qquad (8.10)$$

This consideration shows that the quantum statistical partition function of any particular smooth potential reduces to the classical partition function in the high temperature limit. Because of the similarity of Eq. (8.6) with (8.10), the function $\mathcal{V}(q_c)$ has been dubbed by Feynman and Kleinert [232] the *effective classical potential*.

To obtain first insights, consider the effective classical potential $\mathcal{V}(q_c)$ for the damped harmonic oscillator. Now, all the integrals contained in Eq. (8.7) with (8.8) are in strict Gaussian form at any temperature. We then get

$$\mathcal{V}_{\mathrm{osc}}(q_c) = V_{\mathrm{osc}}(q_c) - (1/\beta)\ln\mu_{\mathrm{osc}}(\omega_0) \,, \qquad (8.11)$$

where μ_{osc} is the pure quantum part of the partition function (6.71),

$$\mu_{\mathrm{osc}}(\omega_0) = \prod_{n=1}^{\infty} \frac{\nu_n^2}{\nu_n^2 + \nu_n\widehat{\gamma}(\nu_n) + \omega_0^2} = \hbar\beta\omega_0 Z_{\mathrm{osc}}(\beta) \,. \qquad (8.12)$$

For the Drude model (3.53), μ_{osc} can be expressed in terms of gamma functions,

$$\mu_{\mathrm{osc}}(\omega_0) = \left(\frac{\hbar\beta\omega_0}{2\pi}\right)^2 \frac{\Gamma(\lambda_1/\nu_1)\,\Gamma(\lambda_2/\nu_1)\,\Gamma(\lambda_3/\nu_1)}{\Gamma(\omega_D/\nu_1)} \,, \qquad (8.13)$$

where λ_1, λ_2, and λ_3 are the three roots of the cubic equation (6.42). In the absence of damping, we have

$$\mu_{\mathrm{osc}}(\omega_0) = \frac{\hbar\omega_0/2k_B T}{\sinh(\hbar\omega_0/2k_B T)} \,. \qquad (8.14)$$

Alternatively, we may express $\mathcal{V}_{\mathrm{osc}}(q_c)$ in terms of the free energy $F_{\mathrm{osc}} = -\ln Z_{\mathrm{osc}}/\beta$,

$$\mathcal{V}_{\mathrm{osc}}(q_c) = V_{\mathrm{osc}}(q_c) - (1/\beta)\ln(\hbar\beta\omega_0) + F_{\mathrm{osc}} \,. \qquad (8.15)$$

Thus for the damped harmonic oscillator, $\mathcal{V}_{\mathrm{osc}}(q_c)$ differs from $V_{\mathrm{osc}}(q_c)$ only by a temperature-dependent energy shift. At high temperatures, the shift vanishes as $1/T$, $\mathcal{V}_{\mathrm{osc}}(q_c) - V_{\mathrm{osc}}(q_c) = \hbar^2[\omega_0^2 + \gamma\omega_D]/12k_B T + \mathcal{O}(1/T^3)$. In the opposite limit $T \to 0$, we have

$$\mathcal{V}_{\mathrm{osc}}(q_c) = V_{\mathrm{osc}}(q_c) + \hbar\varepsilon_0 \,, \qquad (8.16)$$

where $\hbar\varepsilon_0$ is the ground state energy of the oscillator given in Eq. (6.79) or (6.80).

Utilization of the effective classical potential $\mathcal{V}(q_c)$ is physically very appealing. Use of $\mathcal{V}(q_c)$ for anharmonic potentials may seem of little aid, however, since it is impossible to carry out the integrals over all Q_n in Eq. (8.7) in analytic form. Fortunately, there is a variational method in which the path integral in question is compared with the path integral of a simpler problem solvable in analytic form. There is a minimum principle which leads to an upper bound for the effective classical potential. Upon choosing a suitable trial action, the partition function can be calculated quite accurately in the whole temperature range for many anharmonic systems.

8.2 Variational method

We now discuss a method which combines the path integral approach with a variational principle. The minimum principle is based on *Jensen's inequality* for convex functions $f(x)$, e.g. $f(x) = e^{-x}$, in probability theory.[2] For x being random, the average value of $f(x)$ always exceeds or equals the function of the average value of x, as long as x is real and the distribution of the underlying stochastic process is positive,

$$\langle f(x) \rangle \geq f(\langle x \rangle) . \tag{8.17}$$

This inequality for convex functions can be generalized to an exponential functional. Therefore it is very powerful in quantum statistical mechanics.

8.2.1 Variational method for the free energy

Suppose we have a trial action S_{tr} that is easier to work with. Then we may write

$$Z \equiv e^{-\beta F} = \oint \mathcal{D}q \, e^{-(S-S_{\mathrm{tr}})/\hbar} e^{-S_{\mathrm{tr}}/\hbar} = e^{-\beta F_{\mathrm{tr}}} \left\langle e^{-(S-S_{\mathrm{tr}})/\hbar} \right\rangle_{\mathrm{tr}} , \tag{8.18}$$

where $Z_{\mathrm{tr}} \equiv \exp(-\beta F_{\mathrm{tr}}) = \oint \mathcal{D}q \exp(-S_{\mathrm{tr}}/\hbar)$ is the partition function connected with S_{tr}, and $\langle O \rangle_{\mathrm{tr}}$ is the average of the observable $O[q]$ with weight $\exp(-S_{\mathrm{tr}}[q]/\hbar)$,

$$\langle O \rangle_{\mathrm{tr}} \equiv \frac{\oint \mathcal{D}q(\cdot) \, O[q(\cdot)] \exp\left(-S_{\mathrm{tr}}[q(\cdot)]/\hbar\right)}{\oint \mathcal{D}q(\cdot) \exp\left(-S_{\mathrm{tr}}[q(\cdot)]/\hbar\right)} . \tag{8.19}$$

Using (8.17), we obtain from (8.18) the lower bound for the partition function

$$\exp(-\beta F) \geq \exp(-\beta F_{\mathrm{tr}}) \exp\left(-\langle S - S_{\mathrm{tr}} \rangle_{\mathrm{tr}}/\hbar\right) . \tag{8.20}$$

This implies Feynman's upper bound for the free energy [3, 4],

$$F \leq F_{\mathrm{tr}} + \langle S - S_{\mathrm{tr}} \rangle_{\mathrm{tr}}/\hbar\beta . \tag{8.21}$$

The minimum principle (8.21) has been used to determine the "best" free energy from a trial action $S_{\mathrm{tr}}[q]$ for which the path integral can be worked out.

---

[2]For a discussion of bounds in quantum partition functions, cf. Ref. [233].

8.2.2 Variational method for the effective classical potential

Similar to the derivation of a global bound for the free energy, we can also deduce a local bound for the effective classical potential $\mathcal{V}(q_c)$, which is somewhat simpler to study. Since we wish to perform the variational principle in the quantum fluctuation part, we introduce a trial potential $V_{\text{tr}}(\tilde{q}; q_c)$ depending on $\tilde{q} = q - q_c$, and also the corresponding trial action $S_{\text{tr}}[\tilde{q}; q_c]$. Here the extra variable q_c is to indicate parametric dependence of the trial potential and of the trial action on the centroid q_c. By this ansatz, both the trial potential and the trial action become *nonlocal*.

For notational distinction from the full average $\langle O \rangle$, we introduce the double bracket symbol $\langle\!\langle O[q(\cdot)] \rangle\!\rangle_{q_c}^{(\text{tr})}$ in order to denote the average of the observable $O[q(\cdot)]$ over the pure quantum fluctuations $\tilde{q}(\cdot)$ with weight $\exp\{-S_{\text{tr}}[\tilde{q}(\cdot); q_c]/\hbar\}$,

$$\langle\!\langle O[q(\cdot)] \rangle\!\rangle_{q_c}^{(\text{tr})} \equiv \frac{\oint \mathcal{D}\tilde{q}(\cdot)\, O[q_c + \tilde{q}(\cdot)] \exp\left\{ -S_{\text{tr}}[\tilde{q}(\cdot); q_c]/\hbar \right\}}{\oint \mathcal{D}\tilde{q}(\cdot) \exp\left\{ -S_{\text{tr}}[\tilde{q}(\cdot); q_c]/\hbar \right\}}. \tag{8.22}$$

With the definition (8.22), the effective classical potential may be written as

$$e^{-\beta\mathcal{V}(q_c)} = \mu_{\text{tr}}(q_c) \left\langle\!\!\left\langle \exp\left\{ -\int_0^{\hbar\beta} d\tau \left(V[q(\tau)] - V_{\text{tr}}[\tilde{q}(\tau); q_c] \right)\Big/\hbar \right\} \right\rangle\!\!\right\rangle_{q_c}^{(\text{tr})}, \tag{8.23}$$

where $\mu_{\text{tr}}(q_c)$ is the pure quantum contribution to the trial partition function,

$$\mu_{\text{tr}}(q_c) \equiv e^{-\beta F_{\text{tr}}(q_c)} = \oint \mathcal{D}\tilde{q}(\cdot) \exp\left\{ -S_{\text{tr}}[\tilde{q}(\cdot); q_c]/\hbar \right\}. \tag{8.24}$$

Jensen's inequality (8.17) yields $\exp[-\beta\mathcal{V}(q_c)] \geq \exp[-\beta\mathcal{V}_{\text{tr}}(q_c)]$, which implies for the effective classical potential the upper bound

$$\mathcal{V}(q_c) \leq \mathcal{V}_{\text{tr}}(q_c). \tag{8.25}$$

The trial effective potential is given by

$$\mathcal{V}_{\text{tr}}(q_c) = \left\langle\!\!\left\langle \int_0^{\hbar\beta} d\tau \left(V[q(\tau)] - V_{\text{tr}}[\tilde{q}(\tau); q_c] \right)\Big/\hbar\beta \right\rangle\!\!\right\rangle_{q_c}^{(\text{tr})} - \frac{1}{\beta} \ln \mu_{\text{tr}}(q_c). \tag{8.26}$$

Consider first an undamped system, and use the trial action of a free particle,

$$S_{\text{tr}}[\tilde{q}; q_c] \quad \rightarrow \quad S_0[q] = S_0[\tilde{q}] = M\hbar\beta \sum_{n=1}^{\infty} \nu_n^2 |Q_n|^2. \tag{8.27}$$

The simple trial action (8.27) does not depend on the centroid. Using the functional integration measure (8.8), we find from Eq. (8.24)

$$\mu_0 = 1, \tag{8.28}$$

showing that μ is purely classical for a free particle trial action. The computation of the average in Eq. (8.26) is straightforward by writing $V(q)$ as Fourier integral and by completing the square of the Q_n in the exponent. The resulting upper bound for the effective potential is a fuzzy copy of the original potential $V(q)$ [4],

$$\mathcal{V}_0(q_c) \;=\; \langle\!\langle V[q(\cdot)] \rangle\!\rangle_{q_c}^{(0)} \,. \tag{8.29}$$

Substituting the form (8.27) into Eq. (8.22), we obtain

$$\langle\!\langle V[q(\cdot)] \rangle\!\rangle_{q_c}^{(0)} \;\equiv\; \frac{1}{\sqrt{2\pi\langle\xi^2\rangle_0}} \int_{-\infty}^{\infty} dq\, V(q) \, \exp\left(-\frac{(q-q_c)^2}{2\langle\xi^2\rangle_0} \right) . \tag{8.30}$$

The convolution integral smears the original potential $V(q)$ out over a length scale $\sqrt{\langle\xi^2\rangle_0}$. The mean square spread of the Gaussian weight function in Eq. (8.30) is

$$\langle\xi^2\rangle_0 \;\equiv\; \langle\tilde{q}^2\rangle_0 \;=\; \frac{2}{M\beta} \sum_{n=1}^{\infty} \frac{1}{\nu_n^2} \;=\; \frac{\hbar^2\beta}{12M} \,. \tag{8.31}$$

The Gaussian smearing of the original potential in the trial effective classical potential $\mathcal{V}_0(q_c)$ on the scale (8.31) is determined by the quantum fluctuations of a free particle.

A considerable improvement on Feynman's original variational method has been put forward by Giachetti and Tognetti [231], and independently by Feynman and Kleinert [232]. The essence of the refined variational ansatz is to include a harmonic potential in the trial action,

$$V_{\rm h}[q(\tau); q_c] \;=\; \frac{M}{2}\Omega^2(q_c)\Big(q(\tau) - q_c\Big)^2 \;=\; \frac{M}{2}\Omega^2(q_c)\,\tilde{q}^2(\tau) \,, \tag{8.32}$$

where $\Omega(q_c)$ is a local trial frequency to be optimally chosen. The trial potential $V_{\rm h}[q; q_c]$ is nonlocal because of the parametric dependence on the centroid q_c.

At this stage, I find it appropriate to generalize the discussion to include dissipation. For damped systems, it is natural to incorporate the dissipative action into the trial action. Thus we choose as trial action the general harmonic form

$$S_{\rm tr}[\tilde{q}; q_c]/\hbar \;\;\to\;\; S_{\rm h}[\tilde{q}; q_c]/\hbar \;=\; M\beta \sum_{n=1}^{\infty} \Big(\nu_n^2 + \nu_n\widehat{\gamma}(\nu_n) + \Omega^2(q_c) \Big)|Q_n|^2 \,. \tag{8.33}$$

Substituting this form into Eq. (8.24), we find

$$\mu_{\rm h}[\Omega(q_c)] \;\equiv\; e^{-\beta F_{\rm h}[\Omega(q_c)]} \;=\; \prod_{n=1}^{\infty} \frac{\nu_n^2}{\nu_n^2 + \nu_n\widehat{\gamma}(\nu_n) + \Omega^2(q_c)} \;=\; \hbar\beta\Omega(q_c) Z_{\rm h}(\beta) \,. \tag{8.34}$$

The quantity $F_{\rm h}[\Omega(q_c)]$ is the quantum part of the free energy, and $Z_{\rm h}(\beta)$ is the full partition function of the trial damped harmonic oscillator [cf. Eq.(6.71)]. Performing the average in Eq. (8.26) with the form (8.33) in the exponent of the weight function, we find the variational classical effective potential

$$\mathcal{V}_{\rm h}(q_c) \;=\; \langle\!\langle V[q(\cdot)] \rangle\!\rangle_{q_c}^{(\rm h)} + F_{\rm h}[\Omega(q_c)] - \frac{M}{2}\Omega^2(q_c)\langle\xi^2\rangle_{\rm h} \,. \tag{8.35}$$

Here $\langle\!\langle V[q(\cdot)]\rangle\!\rangle_{q_c}^{(\mathrm{h})}$ is the original potential $V(q)$ smeared with a Gaussian of half-width $\langle\xi^2\rangle_{\mathrm{h}}$ analogous to Eq. (8.30). The mean spread is given by

$$\langle\xi^2\rangle_{\mathrm{h}} \equiv \langle\tilde{q}^2\rangle_{\mathrm{h}} = \frac{2}{M\beta}\sum_{n=1}^{\infty}\frac{1}{\nu_n^2 + \nu_n\hat{\gamma}(\nu_n) + \Omega^2(q_c)}\,. \tag{8.36}$$

Comparing the expression (8.36) with Eq. (6.91) we see that $\langle\xi^2\rangle_{\mathrm{h}}$ differs from the full coordinate dispersion by the missing $n = 0$ term, which is the classical dispersion $\langle q^2\rangle_{\Omega}^{(\mathrm{cl})} = 1/M\beta\Omega^2(q_c)$. While the missing term grows linearly with T, the fluctuation width $\langle\xi^2\rangle_{\mathrm{h}}$ shrinks with $1/T$ at large T, as can be seen from Eq. (6.100). The width $\langle\xi^2\rangle_{\mathrm{h}}$ is the pure quantum coordinate dispersion of the damped harmonic oscillator which is finite at all T,

$$\langle\xi^2\rangle_{\mathrm{h}} = \langle q^2\rangle_{\Omega} - \langle q^2\rangle_{\Omega}^{(\mathrm{cl})}\,. \tag{8.37}$$

The optimal trial function $\Omega^2(q_c)$ is determined by minimization of $\mathcal{V}_{\mathrm{h}}(q_c)$,

$$\frac{d\mathcal{V}_{\mathrm{h}}(q_c)}{d\Omega^2(q_c)} = \frac{\partial\mathcal{V}_{\mathrm{h}}(q_c)}{\partial\Omega^2(q_c)} + \frac{\partial\mathcal{V}_{\mathrm{h}}(q_c)}{\partial\langle\xi^2\rangle_{\mathrm{h}}}\frac{\partial\langle\xi^2\rangle_{\mathrm{h}}}{\partial\Omega^2(q_c)} = 0\,. \tag{8.38}$$

Using the expressions (8.34) and (8.36), we obtain $\partial F_{\mathrm{h}}/\partial\Omega^2 = \frac{1}{2}M\langle\xi^2\rangle_{\mathrm{h}}$. With this we find from the defining expression (8.35) that $\partial\mathcal{V}_{\mathrm{h}}(q_c)/\partial\Omega^2(q_c)$ vanishes. Thus Eq. (8.38) reduces to $\partial\mathcal{V}_{\mathrm{h}}(q_c)/\partial\langle\xi^2\rangle_{\mathrm{h}} = 0$. From this we see that the best choice is

$$\Omega^2(q_c) = \frac{2}{M}\frac{\partial\langle\!\langle V[q(\cdot)]\rangle\!\rangle_{q_c}^{(\mathrm{h})}}{\partial\langle\xi^2\rangle_{\mathrm{h}}}\,. \tag{8.39}$$

Combining Eq. (8.39) with (8.36), we get in general a transcendental equation for the function $\Omega^2(q_c)$, which has to be solved numerically. The expressions (8.36) and (8.39) represent the optimal choice for the functions $\langle\xi^2\rangle_{\mathrm{h}}$ and $\Omega^2(q_c)$ within the trial ansatz (8.33). With the optimal choice for $\langle\xi^2\rangle_{\mathrm{h}}$ and $\Omega(q_c)$, the function $\mathcal{V}_{\mathrm{h}}(q_c)$ gives the best upper local bound for the true effective classical potential $\mathcal{V}(q_c)$, $\mathcal{V}(q_c) \leq \mathcal{V}_{\mathrm{h}}(q_c)$. Accordingly, we find for the partition function the best lower bound

$$Z \geq \left(\frac{M}{2\pi\hbar^2\beta}\right)^{1/2}\int_{-\infty}^{\infty}dq_c\,\exp[-\beta\mathcal{V}_{\mathrm{h}}(q_c)]\,. \tag{8.40}$$

For anharmonic systems at high temperature, the centroid integral has to be treated in most cases numerically. The second derivative of $\mathcal{V}_{\mathrm{h}}(q_c)$ with respect to $\Omega^2(q_c)$ is non-negative, as can be easily demonstrated. This implies that the extremal effective potential $\mathcal{V}_{\mathrm{h}}(q_c)$ is a local minimum. Clearly, for the particular case of the damped harmonic oscillator, $V(q) \propto q^2$, the variational method reproduces the exact partition function. On the other hand, in the limit $\Omega \to 0$, Feynman's original (non-optimal) choice is recovered. By construction, the variational result $\mathcal{V}_{\mathrm{h}}(q_c)$ for $\mathcal{V}(q_c)$ becomes more accurate as the temperature is increased. Janke has shown [234] that the high temperature expansion of $\mathcal{V}_{\mathrm{h}}(q_c)$ agrees with the Wigner expansion of $\mathcal{V}(q_c)$ up to terms of order β^3. In the classical limit, the mean spread $\langle\xi^2\rangle_{\mathrm{h}}$ becomes zero and

$\mu_{\mathrm{h}}(q_c) \to 1$, so that $\langle\!\langle\, V[q(\cdot)]\,\rangle\!\rangle_{q_c}^{(\mathrm{h})}$ is reduced to the original potential $V(q_c)$ and (8.40) maps on the classical partition function (8.10). In the opposite limit $T \to 0$, the variational method corresponds to a Gaussian trial ansatz for the ground state [232]. The Gaussian function is known to give extremely good ground-state energies for smooth single-well potentials. Thus, the effective classical potential turns out to be a good concept both in the high- and in the low-temperature limit, and in many cases at any temperature. Numerically, the approach gives quite reliable results for many smooth potentials in the whole temperature range as emphasized in Refs. [6], [234] – [236]. The variational method works surprisingly well for hard-core potentials as well as for singular potentials except at very low temperatures where the cusp-like shape of the true wave functions becomes more important. The method has also been applied successfully to nonlinear field theories [237]. A friction term has been included in the variational method in Ref. [238].

In the path integral approach, imaginary-time position correlation functions are treated by adding the source term $\int_0^{\hbar\beta} d\tau'\, j(\tau')q(\tau')$ to the Euclidean action and performing functional differentiations with respect to the source $j(\tau)$. Using this method, we can also study position correlation functions within the variational method. The corresponding treatment is straightforward [6]. Analytic continuation to real time may be performed in the end.

A different view on the effective classical potential method has been put forward recently. It has been shown in Ref. [239] that, for standard Hamiltonians, the variational method is equivalent to a scheme in which the pure quantum fluctuations are treated selfconsistently in a harmonic approximation (PQSCHA method). For non-standard Hamiltonians like those describing magnetic systems, the Feynman-Jensen inequality is generally not applicable. The PQSCHA in phase space representation (see Subsection 8.2.4) is still meaningful for nonstandard systems and allows to construct an effective classical potential. For a recent review and survey of applications of the PQSCHA, see Ref. [239, b].

8.2.3 Variational perturbation theory

A systematic improvement of the variational approach consists in performing a variational perturbation expansion [240, 6]. In the first step, the anharmonic action is expanded in powers of the fluctuations about the centroid, $\tilde{q}(\tau) = q(\tau) - q_c$,

$$S[q(\cdot)] \;=\; \hbar\beta V(q_c) + S_{\mathrm{h}}[\tilde{q}(\cdot); q_c] + S_{\mathrm{int}}[\tilde{q}(\cdot); q_c]\,, \qquad (8.41)$$

$$S_{\mathrm{int}}[\tilde{q}; q_c] \;=\; \int_0^{\hbar\beta} d\tau\, V_{\mathrm{int}}[\tilde{q}(\tau); q_c]\,, \qquad (8.42)$$

where $V_{\mathrm{int}}(\tilde{q}; q_c)$ is the deviation of the full potential from the trial harmonic potential,

$$V_{\mathrm{int}}(\tilde{q}; q_c) \;\equiv\; \frac{1}{2!}\Big(V''(q_c) - M\Omega^2(q_c)\Big)\tilde{q}^2(\tau) + \frac{1}{3!}\,V'''(q_c)\,\tilde{q}^3(\tau) + \cdots\,, \qquad (8.43)$$

and where the prime denotes differentiation with respect to the coordinate. Substituting the expansion (8.43) into Eq. (8.23), we obtain the variational perturbation expansion for the effective classical potential in the form of a cumulant expansion,

$$\mathcal{V}(q_c) = \mathcal{V}_h(q_c) - \frac{1}{2!\,\hbar^2\beta} \left[\langle\!\langle S_{\text{int}}^2 \rangle\!\rangle_{q_c}^{(\text{h})} - \left(\langle\!\langle S_{\text{int}} \rangle\!\rangle_{q_c}^{(\text{h})} \right)^2 \right] \tag{8.44}$$

$$+ \frac{1}{3!\,\hbar^3\beta} \left[\langle\!\langle S_{\text{int}}^3 \rangle\!\rangle_{q_c}^{(\text{h})} - 3 \langle\!\langle S_{\text{int}}^2 \rangle\!\rangle_{q_c}^{(\text{h})} \langle\!\langle S_{\text{int}} \rangle\!\rangle_{q_c}^{(\text{h})} + 2 \left(\langle\!\langle S_{\text{int}} \rangle\!\rangle_{q_c}^{(\text{h})} \right)^3 \right] \pm \cdots .$$

The expression $\langle\!\langle S_{\text{int}}^m \rangle\!\rangle_{q_c}^{(\text{h})}$ involves the average

$$\langle\!\langle V_{\text{int}}[\tilde{q}(\tau_1); q_c]\, V_{\text{int}}[\tilde{q}(\tau_2); q_c] \cdots V_{\text{int}}[\tilde{q}(\tau_m); q_c]\, \rangle\!\rangle_{q_c}^{(\text{h})} \tag{8.45}$$

over the quantum fluctuations $\tilde{q}(\cdot)$ with weight function $\exp\{-S_h[\tilde{q}(\cdot); q_c]/\hbar\}$, where $S_h[\tilde{q}(\cdot); q_c]$ is given in Eq. (8.33). The smearing formula can be found upon spatial Fourier expansion of $V_{\text{int}}[\tilde{q}(\tau); q_c]$. The average of the Fourier factors gives

$$\left\langle\!\!\left\langle \prod_{\ell=1}^{m} \exp[ik_\ell \tilde{q}(\tau_\ell)] \right\rangle\!\!\right\rangle_{q_c}^{(\text{h})} = \exp\left[-\frac{1}{2} \sum_{i=1}^{m} \sum_{j=1}^{m} k_i \xi_{\tau_i, \tau_j}^2 k_j \right], \tag{8.46}$$

where we have introduced the pure quantum coordinate autocorrelation function

$$\xi_{\tau, \tau'}^2 \equiv \langle \tilde{q}(\tau)\tilde{q}(\tau') \rangle_h = \frac{2}{M\beta} \sum_{n=1}^{\infty} \frac{\cos[\nu_n(\tau - \tau')]}{\nu_n^2 + \nu_n \widehat{\gamma}(\nu_n) + \Omega^2(q_c)} . \tag{8.47}$$

The Fourier inversion of Eq. (8.46) gives again a Gaussian form, but with the inverse matrix. Thus we find in generalization of Eq. (8.30) the smearing formula [241]

$$\langle\!\langle F_1[q(\tau_1)]\, F_2[q(\tau_2)] \cdots F_m[q(\tau_m)] \rangle\!\rangle_{q_c}^{(\text{h})} = \left(\prod_{k=1}^{m} \int_{-\infty}^{\infty} dq_k\, F_k(q_k) \right) \tag{8.48}$$

$$\times \frac{1}{\sqrt{(2\pi)^m \det[\xi_{\tau, \tau'}^2]}} \exp\left[-\frac{1}{2} \sum_{i=1}^{m} \sum_{j=1}^{m} (q_i - q_c)\xi_{\tau_i, \tau_j}^{-2}(q_j - q_c) \right],$$

where $\xi_{\tau, \tau'}^{-2}$ is the inverse of the $m \times m$-matrix $\xi_{\tau, \tau'}^2$.

Truncation of the series (8.44) after the nth order defines the effective classical potential in order n, $\mathcal{V}^{(n)}(q_c)$. The first order corresponds to the variational effective classical potential, $\mathcal{V}^{(1)}(q_c) = \mathcal{V}_h(q_c)$. The infinite sum (8.44) is independent of the variational frequency $\Omega(q_c)$ by construction. For any finite order n, the optimal choice for $\Omega(q_c)$ is determined by the relation

$$\partial\mathcal{V}^{(n)}(q_c)/\partial\Omega(q_c) = 0 . \tag{8.49}$$

The optimal frequency $\Omega^{(n)}(q_c)$ for a given order n has been dubbed *frequency of least dependence*. The variational perturbation expansion (8.44) is an alternating series, and the last cumulant in each order is positive for odd orders. This property leads to the following behaviour. For odd n, $n = 2m + 1$, $\mathcal{V}_{2m+1}(q_c)$ is a minimum in Ω at the optimal frequency, whereas the second derivative for $n = 2m$ even, $\partial^2 \mathcal{V}^{(2m)}(q_c)/(\partial\Omega)^2$, may change sign at the optimal frequency. With increasing odd or even n, the potential $\mathcal{V}^{(n)}(q_c)$ develops as a function of Ω a flat plateau at the optimal frequency. It has been demonstrated for the anharmonic oscillator and quartic double well that the third order effective classical potential $\mathcal{V}^{(3)}(q_c)$ is by far more accurate than the first order effective classical potential $\mathcal{V}^{(1)}(q_c)$.

A discussion of the smearing formula (8.48) and application of the variational perturbation method to the Coulomb potential for zero friction is given in Ref. [241].

For the anharmonic oscillator $V(q) = M\omega^2 q^2/2 + g\, q^4/4$, the energy eigenvalues $\{E_\nu(g)\}$ possess an essential singularity at $g = 0$. Thus, ordinary perturbation expansion in g has zero radius of convergence. The physical origin of the singular behaviour is the possibility for tunneling when g is negative. Therefore, the form of the singularity may be analyzed using semiclassical methods. The important point now is that the variational perturbation expansion involves convergent infinite-order resummations. This changes the ordinary expansion into a systematic expansion which has uniform convergence for arbitrary coupling strength, including the strong-coupling limit [6].

8.2.4   Expectation values in coordinate and phase space

The results of Subsection 8.2.1 can be used directly to formulate quantum statistical averages of operators being functions of the position operator, $\hat{A}(\hat{q})$, or more generally of the position and momentum operator, $\hat{A}(\hat{p}, \hat{q})$. In the remainder of this subsection, we restrict the attention to Weyl ordering $\mathcal{Z}_0\{\hat{A}(\hat{p}, \hat{q})\}$, which is the case $\gamma = 0$ in Subsection 4.4.1. For the harmonic trial ansatz (8.33) or the PQSCHA, the optimally chosen restricted equilibrium density matrix $\rho(q'', q'; q_c)$ defined in Eq. (8.2) takes the particular form [242]

$$\rho(q'', q'; q_c) = \frac{\sqrt{M}\, e^{-\beta \mathcal{V}_h(q_c)}}{2\pi\hbar\sqrt{\beta\langle\xi^2\rangle_h}} \exp\left(-\frac{(q'' + q' - 2q_c)^2}{8\langle\xi^2\rangle_h} - \frac{\langle p^2\rangle_h}{2\hbar^2}(q'' - q')^2\right). \quad (8.50)$$

The effective classical potential $\mathcal{V}_h(q_c)$ and the mean square variation $\langle\xi^2\rangle_h$ are given in Eqs. (8.35) and (8.36), and the momentum dispersion of the trial damped oscillator is

$$\langle p^2\rangle_h = \frac{M}{\beta} \sum_{n=-\infty}^{\infty} \frac{\Omega^2(q_c) + |\nu_n||\hat{\gamma}(|\nu_n|)}{\Omega^2(q_c) + \nu_n^2 + |\nu_n||\hat{\gamma}(|\nu_n|)}, \quad (8.51)$$

which is discussed in some detail in Section 6.6. The form (8.50) immediately follows from undoing the integration over the centroid in the expression (6.116) [cf. Eq. (8.2)].

The restricted Wigner function is a phase space density distribution which is connected with $\rho(q'', q'; q_c)$ as given by the correspondence (4.262). The restricted Wigner function resulting from Eq. (8.50) reads

$$f^{(W)}(p, q; q_c) = \frac{\exp\left(-\beta \mathcal{V}_h(q_c) - (q - q_c)^2/2\langle\xi^2\rangle_h - p^2/2\langle p^2\rangle_h\right)}{\sqrt{2\pi\hbar^2\beta/M}\sqrt{2\pi\langle\xi^2\rangle_h}\sqrt{2\pi\langle p^2\rangle_h}}. \tag{8.52}$$

In the selfconsistent harmonic approximation (8.50) or (8.52), the quantum statistical average of an observable $A(q)$ takes the form of an integral over the centroid q_c,

$$\langle A \rangle = \frac{(M/2\pi\hbar^2\beta)^{1/2}}{Z_h} \int_{-\infty}^{\infty} dq_c \, \exp\left(-\beta\mathcal{V}_h(q_c)\right) \langle\!\langle A[q(\cdot)]\rangle\!\rangle_{q_c}^{(h)}, \tag{8.53}$$

in which $\langle\!\langle A[q(\cdot)]\rangle\!\rangle_{q_c}^{(h)}$ captures the average over the quantum fluctuations. The function $\langle\!\langle A[q(\cdot)]\rangle\!\rangle_{q_c}^{(h)}$ is a smeared portrayal of the original function $A(q)$ like in Eq. (8.30), but now the smearing width is $\sqrt{\langle\xi^2\rangle_h}$,

$$\langle\!\langle A[q(\cdot)]\rangle\!\rangle_{q_c}^{(h)} = \frac{1}{\sqrt{2\pi\langle\xi^2\rangle_h}} \int_{-\infty}^{\infty} dq \, A(q) \exp\left(-\frac{(q - q_c)^2}{2\langle\xi^2\rangle_h}\right). \tag{8.54}$$

Consider next the quantum statistical average of an observable $A(p, q)$ depending on position and momentum. The average of the operator in Weyl form $\mathcal{Z}_0\{\hat{A}(\hat{p}, \hat{q})\}$ over the pure quantum fluctuations in harmonic approximation turns out with the use of Eq. (8.52) as a double Gaussian average with half-widths $\langle\xi^2\rangle_h$ and $\langle p^2\rangle_h$,

$$\langle\!\langle A[q(\cdot), p(\cdot)]\rangle\!\rangle_{q_c}^{(h)} \equiv \frac{1}{2\pi} \iint_{-\infty}^{\infty} \frac{dq\,dp}{\sqrt{\langle\xi^2\rangle_h\langle p^2\rangle_h}} A(p, q) \exp\left(-\frac{(q - q_c)^2}{2\langle\xi^2\rangle_h} - \frac{p^2}{2\langle p^2\rangle_h}\right).$$

The final integral over the centroid is as given in Eq. (8.53). It should be remarked that in the case of Ohmic dissipation, both the quantum part μ_h of the partition function and the momentum dispersion $\langle p^2\rangle_h$ must be Drude-regularized since otherwise they diverge logarithmically (cf. Section 6.5 and 6.6). A different possibility for regularization of the effective classical potential would be to subtract the contribution of a free Brownian particle $(1/\beta)\ln(1/\mu_{\text{free}})$ with $1/\mu_{\text{free}} = \prod_{n=1}^{\infty}[1 + \hat{\gamma}(\nu_n)/\nu_n]$.

One remark is appropriate. Since the local density matrix $\rho(q'', q'; q_c)$ in Eq. (8.50) has a factor $\exp[-(q'' + q' - 2q_c)^2/8\langle\xi^2\rangle_h]$, the fluctuation width of the centroid is generally small at low and intermediate temperatures, and it becomes zero in the classical limit. Thus the special role of the centroid is less important for the density matrix than for the partition function. In numerical computations of the variational perturbation expansion, it turns out that one may treat the centroid in the same way as the quantum fluctuations. By this, the numerical efforts are reduced enormously.[3]

Historically, the variational method has been introduced by Feynman when he studied the polaron problem. For the action (4.123) with (4.125) there are two major reasons that a potential term in the trial action is not the appropriate representative to master the physical situation. First, the polaron is free to wander around in a

---

[3]I am indebted to A. Pelster for this communication.

crystal without giving preference to any particular place. Second, the polaron at
position $q(\tau)$ interacts with itself at positions taken at any previous imaginary time τ'.

A trial action for the polaron which has these properties (except for a different
spatial dependence) and which is analytically tractable is [3, 4]

$$S_0 = \frac{M}{2} \int_0^{\hbar\beta} d\tau \, \dot{\boldsymbol{q}}^2(\tau) + \frac{C}{2} \int_0^{\hbar\beta} d\tau \int_0^{\tau} d\tau' \, D_\omega(\tau - \tau') |\boldsymbol{q}(\tau) - \boldsymbol{q}(\tau')|^2 \,, \qquad (8.55)$$

where $D_\omega(\tau)$ is given in (4.56), and where C and ω are taken as adjustable parameters.
According to the variational principle (8.21), we have for the ground-state energy

$$E \leq E_0 + \lim_{\beta \to \infty} \frac{1}{\hbar\beta} \langle S - S_0 \rangle_0 \,, \qquad (8.56)$$

where E_0 is the ground-state energy of the model (8.55), and $\langle \cdots \rangle_0$ means average
with weight $\exp(-S_0/\hbar)$. The lowest upper bound found by variation of the parameters ω and C gives for E values that are better than those obtained by of all other
methods, and in fact for all values of the polaron coupling constant α [4]. Unfortunately, Jensen's inequality does *not* hold in general for a polaron in a magnetic field
(magneto-polaron) since the action is complex [243]. A slightly modified variational
method adapted for the magneto-polaron case has been suggested in Ref. [244].

9. Suppression of quantum coherence

There are two main lines of investigations on decoherence in quantum mechanics.
On the one hand, one is interested in fundamental problems in the interpretation
of quantum theory, in particular in questions connected with measurement and the
quantum to classical transition (cf., e.g., Ref. [57]). The importance of decoherence
in the appearance of a classical world in quantum theory is well known [245]. In
cosmology, the decoherence which must have been occurred during the inflationary
era of the Universe has been attributed to quantum fluctuations. On the other hand,
environment-induced decoherence is omnipresent in the microscopic world. It is found,
e.g., for an atom confined in a quantum optical trap, or for electron propagation in a
mesoscopic device. Striking examples for coherence effects are the anomalous magne-
toresistance in disordered systems due to weak localization [246], and Aharonov-Bohm
type oscillations in the resistance of mesoscopic rings [247].

A system prepared in a non-equilibrium state and in contact with the environment
relaxes to equilibrium. The decay of the off-diagonal states of the reduced density matrix (coherences) is denoted as decoherence or loosely speaking as dephasing [cf. the
remark at the end of Section 9.1]. Evidently, the origin of decoherence is entanglement
and interaction with a fluctuating environment. Therefore, decoherence depends on
the spectral density of the environmental coupling, on temperature and on specific
properties of the system. There follows from the formal structure of the influence

functional that response and equilibrium correlation functions decay roughly on the same time scale as the coherences. We shall discuss this in some detail in Chapter 21.

There are experimental indications that dephasing of electrons in mesoscopic conductors persists down to zero temperature [248, 249]. On the theoretical side, Altshuler and coworkers advocated that there is no dephasing at $T = 0$ [250], while Golubev and Zaikin found in their studies on diffusive conductors that dephasing of electrons levels off at low temperatures [251]. There is still an ongoing debate about dephasing of elctrons in diffusive conductors at zero temperature.

In a multitude of model systems, decoherence persists down to zero temperature, as we shall see in Section 9.3 and has been found also by others [158, 252, 253].

To define decoherence quantitatively, it is natural to introduce a time scale τ_φ over which interference effects are suppressed. Basically, phase randomization of the system's wave function occurs through energy exchange processes with environmental modes, e.g., electron-electron or electron-phonon scattering.

Two complementary views about dephasing are possible [197]. Either one may treat the problem by studying the changes induced by the interfering particle in the states of the environment, or one may study the accumulation of the phase uncertainty of the interfering waves due to the quantum fluctuations in the reservoir coupling. In the first section, we discuss the important differences between a nondynamical and a dynamical environment with regard to dephasing. Subsequently, we explain that there is no universal dephasing behaviour in models with delocalized bath modes. In the last section, we introduce a model with spatially localized reservoir modes in which the actual geometry is irrelevant for the decoherence process and we calculate semiclassically the time scale over which interference effects are suppressed.

9.1 Nondynamical versus dynamical environment

An environment with no significant dynamics of its own acts upon the system as if it were a classically fluctuating potential. To study the loss of phase memory, consider a particle whose dynamics is governed by the Hamiltonian $H = H_0 + V_0(q) + \delta V(t)$. We assume that $\delta V(t)$ is a stochastic potential term obeying Gaussian statistics with covariance $\langle \delta V(t) \delta V(0) \rangle_{\mathrm{av}} = \langle \delta V^2 \rangle_{\mathrm{av}} \rho(t)$ where $\rho(0) = 1$. The fluctuating term $\delta V(t)$ gives rise to a phase factor in the propagator,

$$K(q_2, q_1; t) = \mathrm{e}^{-i\varphi(t)} K_0(q_2, q_1; t) \qquad \text{with} \qquad \varphi(t) = \frac{1}{\hbar} \int_0^t d\tau \, \delta V(\tau) \,. \qquad (9.1)$$

Taking the statistical average, we have $\langle K(q_2, q_1; t) \rangle_{\mathrm{av}} = \langle \mathrm{e}^{-i\varphi(t)} \rangle_{\mathrm{av}} K_0(q_2, q_1; t)$ with

$$\langle \mathrm{e}^{-i\varphi(t)} \rangle_{\mathrm{av}} = \exp\left[-\tfrac{1}{2} \left\langle \varphi^2(t) \right\rangle_{\mathrm{av}} \right] \,, \qquad (9.2)$$

$$\left\langle \varphi^2(t) \right\rangle_{\mathrm{av}} = \frac{2}{\hbar^2} \left\langle \delta V^2 \right\rangle_{\mathrm{av}} \int_0^t dt' \int_0^{t'} dt'' \, \rho(t' - t'') \,. \qquad (9.3)$$

We now assume that the observation time t is large compared to the characteristic decay time of $\rho(t)$. Then the phase uncertainty increases linearly with t. We write $\frac{1}{2}\langle\varphi^2(t)\rangle_{\mathrm{av}} = \gamma_{\mathrm{deph}}t$. It is natural to interpret γ_{deph} as the inverse time scale for dephasing. The dephasing rate is given by

$$\gamma_{\mathrm{deph}} = \frac{\langle\delta V^2\rangle_{\mathrm{av}}}{\hbar^2} \int_0^\infty d\tau\, \rho(\tau) . \tag{9.4}$$

In a fully quantized model with *dynamical* fluctuations, the situation is more intricate. Assume that for the global model (3.1) the initial state is factorized into a system state $|\psi_i\rangle$ and a bath state $|\chi\rangle$. The transition probability from state $|\psi_i\rangle$ to final state $|\psi_f\rangle$ is

$$P_{fi} = \sum_{\chi'} |\langle\psi_f|\langle\chi'|\exp[-i\left(H_{\mathrm{R}} + H_{\mathrm{S}} + H_{\mathrm{I}}\right)t/\hbar]|\chi\rangle|\psi_i\rangle|^2 . \tag{9.5}$$

When the coupling is weak enough to leave the bath unchanged, we can substitute the expectation values for the bath operators. Assigning the expectation value $\langle H_{\mathrm{R}}\rangle_\chi$ to the phase of the bath state, we find for the transition amplitude of the system

$$A_{fi} = \langle\psi_f|\exp[-i(H_{\mathrm{S}} + \langle H_{\mathrm{I}}\rangle_\chi)t/\hbar]|\psi_i\rangle . \tag{9.6}$$

The amplitude is conveniently written as modulus times phase factor. Next assume that the initial bath state is given in the form of a mixture with weight c_χ for the state $|\chi\rangle$. Then, in analogy with Eq. (9.2), we take the average of the phase factor as

$$\langle e^{-i\varphi}\rangle = \sum_\chi c_\chi |A_{fi}(\chi)| e^{-i\varphi(\chi)} \Big/ \sum_\chi c_\chi |A_{fi}(\chi)| . \tag{9.7}$$

Evidently, this proceeding breaks down for stronger coupling since the environmental states are not any more dwelling on their initial distribution. For a dynamically fluctuating potential it is therefore not possible to split the total phase into a system and a reservoir part. This demonstrates that in general it is impossible to assign to the subsystem a definite phase. From this we may draw the important conclusion that the whole concept of phase memory generally fails for an open system. Therefore, one should speak of *decoherence* rather than of *dephasing*.

9.2 Suppression of transversal and longitudinal interferences

We recall to the reader's attention the Feynman-Vernon method for the reduced density matrix in which the motion of the particle in contact with the reservoir is described by the propagating function [cf. Eq. (5.12)]

$$J_{\mathrm{FV}}(q_f, q_f', t; q_i, q_i', 0) = \int \mathcal{D}q\,\mathcal{D}q' \exp\left\{\frac{i}{\hbar}\left(S_{\mathrm{S}}[q] - S_{\mathrm{S}}[q']\right)\right\} \mathcal{F}_{\mathrm{FV}}[q, q'] . \tag{9.8}$$

Let us now study whether decoherence depends on the properties of the reservoir coupling alone, or also on the particular geometry of the quantum interference system.

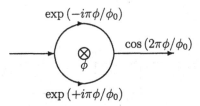

$$\exp\left(-i\pi\phi/\phi_0\right)$$

$$\cos\left(2\pi\phi/\phi_0\right)$$

$$\exp\left(+i\pi\phi/\phi_0\right)$$

Figure 9.1: Schematic sketch of an Aharonov-Bohm device threaded by a magnetic flux ϕ.

Consider first an Aharonov-Bohm device in which a metallic ring is threaded by a magnetic flux ϕ as sketched in Fig. 9.1. The traversing paths are divided into two groups that pass the ring clockwise and anti-clockwise, thereby picking up a phase difference $\varphi = 2\pi\phi/\phi_0$, where ϕ_0 is the flux quantum. Since the paths are spatially separated in transversal direction, we shall refer to this type of interferences as *transversal*. For the global model (3.11), the decoherence factor is given by the noise action, Eq. (5.30) with (5.29). Assuming that the paths $q_1(t)$ and $q_2(t)$ are coupled to statistically independent environments, the cross terms in the noise action $\Phi^{(N)}$ are absent. Then we have

$$\langle e^{-i\varphi(t)} \rangle = \exp\left(-\sum_{j=1}^{2}\frac{1}{\hbar}\int_0^t dt' \int_0^{t'} dt'' \, q_j(t')L'(t'-t'')q_j(t'')\right). \qquad (9.9)$$

The same expression holds for a two-slit device. In the white noise limit, we recover for a slit distance q_0 the previous result (5.31).

The other device is a resonant tunneling structure with reflection and transmission coefficients $\mathcal{R}$ and $\mathcal{T}$, sketched in its simplest form in Fig. 9.2. In this device, the directly through-going path is interfering with a path which has made an additional roundtrip between the barriers. For δ-potentials, the two paths evolve in the scattering region as sketched in Fig. 9.3. Because they travel through the same spatial region and are only separated with respect to time, this interference type has been termed *longitudinal* [157]. Obviously, the interfering paths are influenced by the same environment. Putting $y(t) = q_1(t) - q_2(t)$, the resulting decoherence factor is

$$\langle e^{-i\varphi(t)} \rangle = \exp\left(-\frac{1}{\hbar}\int_0^t dt' \int_0^{t'} dt'' \, y(t')L'(t'-t'')y(t'')\right). \qquad (9.10)$$

The expressions (9.9) and (9.10) have been analyzed in Ref. [157]. One finds that they behave quite differently. For a ballistic path, e.g., the former exponent grows with the third power of t, whereas the latter increases quadratically in t. Observe also that Eq. (9.10) is space translational invariant while Eq. (9.9) is not. In models with delocalized bath modes, the suppression of interferences depends also strongly on spatial dimension, as emphasized in Refs. [197, 254]. In conclusion, in models with delocalized bath modes, the decoherence process depends on the geometry of the interference device. Hence there is *no universal* decoherence behaviour in these models.

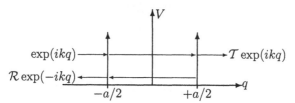

Figure 9.2: A resonant tunneling structure made up of two δ-potentials.

9.3 Localized bath modes and universal decoherence

Physically, a particle moving in a solvent or solid interacts with individual modes of the environment only within a spatially restricted region which we characterize by a length scale λ. When the particle is confined to a region of extension d where $d \ll \lambda$, as it may occur, e.g., in a two- or few-state system, then the actual coupling may be replaced by an effective coupling of the form (3.4) with (3.10). Thus in this case the simplified model with *delocalized* bath modes, Eq. (3.11), is appropriate. However, when the particle travels over distances large compared to the interaction range λ, we have to take into account the *localized* nature of the bath modes.

9.3.1 A model with localized bath modes

In the sequel, we mainly limit ourselves to the case of quasi one-dimensional systems. We assume that the interaction potential is linear in the bath coordinates. In accordance with Eqs. (3.4) and (3.7), we take the interaction in the form

$$H_I = -\left(\sum_\alpha c_\alpha x_\alpha\right)\lambda\sum_{q_0} u\left(\frac{q - q_0}{\lambda}\right) . \qquad (9.11)$$

The form factor $u(x)$ is a dimensionless interaction potential with range of order unity. The form (9.11) describes interaction of the particle with individual bath modes within a range of length λ about the center q_0. To restore translational invariance of the coupling, the scatterers in Eq. (9.11) are assumed to be uniformly distributed in space. Then the interaction is characterized by the spatial autocorrelation function

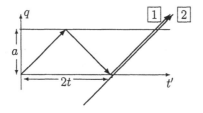

Figure 9.3: Sketch of two interfering path in the resonant tunneling device in Fig. 9.2.

$$\lambda^2 \int_{-\infty}^{\infty} dx'\, u(x')u(q/\lambda - x') = \lambda^2 U(q/\lambda)\,. \tag{9.12}$$

The dimensionless interaction $U(x)$ is an even function of the variable $x = q/\lambda$ and has a range of order one. The elimination of the bath modes may be effected analogously to the case of delocalized modes discussed in Section 5.1. We then get as a generalization of Eq. (5.29) for the present model the influence action

$$\Phi[q_1, q_2] = \frac{\lambda^2}{\hbar} \sum_{j=1}^{2} (-1)^{j-1} \int_0^t dt' \int_0^{t'} dt'' \tag{9.13}$$

$$\times \left\{ L(t' - t'')\, U\!\left(\frac{|q_j(t') - q_1(t'')|}{\lambda}\right) - L^*(t' - t'')\, U\!\left(\frac{|q_j(t') - q_2(t'')|}{\lambda}\right) \right\}.$$

The real and imaginary part are the noise and friction action, respectively.

The model belongs to the general class of models describing state-dependent dissipation discussed in Subsections 3.1 and 4.2.2. It describes, e.g., the interaction with extended field modes. The electron-photon and the electron-phonon interaction, Eq. (3.115), are of the form (9.11). Another important example is the interaction of a charged particle with conduction electrons. The action resulting from the imaginary-time action (4.145) is exactly in the form (9.13), $L(t' - t'')$ is the Ohmic heat bath kernel, and $\lambda^2 U(\Delta q/\lambda) = (3/2k_F^2) \sin^2(k_F \Delta q)/(k_F \Delta q)^2$.

The Zwanzig–Caldeira–Leggett model with delocalized modes, Eq. (3.11), is contained in Eq. (9.13) as the special case $U(|q_j - q_i|) = -\frac{1}{2}(q_j - q_i)^2/\lambda^2$. If we substitute this form, the action (9.13) reduces to the familiar expression (5.22).

The model with localized bath modes has been introduced in Ref. [157] in order to study suppression of interferences without dependence on geometry (see below). The same model has been used in Ref. [158] to study state-dependent quantum effects in Brownian motion. A number of new quantum effects have been found. For instance, the logarithmic law (7.37) for the spatial spreading in the Ohmic case at $T = 0$ holds for this model only on a scale $\sigma^2(t) \ll \lambda^2$. On a larger scale, there is a smooth crossover from a Gaussian to a frozen exponential profile. This genuine quantum effect can be understood from the interplay of the negative time correlations of the quantum noise with the spatial correlations described by the function $U(q/\lambda)$.

In the remainder of this chapter, we base the discussion of the suppression of quantum interferences on the action (9.13). First of all, we observe that the self-terms are added up in the noise action $\Phi^{(N)} = \mathrm{Re}\,\Phi$, whereas they appear with opposite sign in the friction action $\Phi^{(F)} = \mathrm{Im}\,\Phi$. Suppose for simplicity that the self-terms of the two paths give the same contribution. Now assume that the distance between the two interfering paths is large compared to the interaction range λ for most of the time, $\lambda \gg |q_1(t') - q_2(t'')|$. Then the self-terms in Eq. (9.13) are very large compared to the cross terms. Since the self-terms cancel out in the friction action, this action depends only on the cross terms. Therefore we have

$$|\Phi^{(F)}[q_1, q_2]| \ll \Phi^{(N)}[q_1, q_2]\,. \tag{9.14}$$

Thus we have found that friction plays only a secondary role in the decoherence phenomenon. In view of Eq. (9.14), we may disregard the friction action henceforth.[1] The noise action reads

$$\Phi^{(N)}[q] = \frac{2\lambda^2}{\hbar} \int_0^t dt' \int_0^{t'} dt'' \, L'(t' - t'') \, U\left(\frac{|q(t') - q(t'')|}{\lambda}\right) . \qquad (9.15)$$

For this noise action, the actual geometry is irrelevant. Decoherence by scattering events is as effective for small separation as for large separation of the interfering paths. While the action in Eq. (9.9) is proportional to the square of the slit distance q_0, the action (9.15) is proportional to λ^2. Since λ is a microscopic length, we usually have $q_0 \gg \lambda$. In this case, quantum coherence is much better preserved in the model with localized modes than in the model with delocalized modes. Thus we have shown that the suppression of interferences is universal in the model with localized modes.

9.3.2 Statistical average of paths

In a mesoscopic conductor, the electron moves under the concerted influence of static disorder and a dynamical environment. A stochastic static potential for the electron is generated by randomly distributed impurities. Following Chakravarty and Schmid [246], the statistical average of paths in the impurity potential is conveniently performed in a quasiclassical approach. For the pair of trajectories $q(t')$ and $q(t'')$ we define the probability distribution

$$P(y, t' - t'') = \langle \delta\left[y - q(t') + q(t'')\right] \rangle_{\text{imp}} . \qquad (9.16)$$

The dynamical nature of the environment leads to a dephasing factor of the form

$$\langle e^{-i\varphi(t)} \rangle = \exp\left(-\Phi^{(N)}[q]\right) . \qquad (9.17)$$

Taking next the impurity average, we have

$$\langle\langle e^{-i\varphi(t)} \rangle\rangle_{\text{imp}} = \langle \exp\left(-\Phi^{(N)}[q]\right) \rangle_{\text{imp}} \approx \exp\left(-\langle \Phi^{(N)}[q] \rangle_{\text{imp}}\right) . \qquad (9.18)$$

As in the derivation of Eq. (9.4), we now suppose that the observation time t is large compared to the time scale on which the integrand $I(\tau) = I(t' - t'')$ effectively drops to zero. Then $\Phi^{(N)}$ increases linearly with t. Upon averaging the paths in the noise action (9.15) with the distribution (9.16), we obtain the decoherence rate in the form

$$\gamma_\varphi = \frac{2\lambda^2}{\hbar} \int_0^\infty d\tau \, L'(\tau) \int_{-\infty}^\infty dy \, U\left(y/\lambda\right) P(y, \tau) . \qquad (9.19)$$

---

[1]In the model with delocalized bath modes, Eq. (3.11), the situation is different. In this model, also the friction action $\Phi^{(F)}$ may become relevant for dephasing, as noticed in Ref. [198].

It is convenient to write $U(y/\lambda)$ as Fourier integral with transform $\widetilde{U}(\lambda k)$, and $P(y, \tau)$ as double Fourier integral with transform $\widetilde{P}(k, \omega)$. Substituting these forms and the representation (5.23) for $L'(\tau)$ into Eq. (9.19), we obtain

$$\gamma_\varphi = \frac{\lambda^3}{2\pi^2\hbar} \int_0^\infty d\omega \, J(\omega) \coth(\hbar\beta\omega/2) \int_{-\infty}^\infty dk \, \widetilde{U}(\lambda k) \, \widetilde{P}(k, \omega) \,. \qquad (9.20)$$

Since the function $\widetilde{U}(y)$ is substantially different from zero in the range $|y| \lesssim 1$, the k-integration is essentially restricted to the regime

$$|k| \lesssim 1/\lambda \,. \qquad (9.21)$$

For either ballistic or diffusive motion, most of the contribution comes from environmental modes with wave numbers near to $1/\lambda$. We now study this formula in various limits. We suppose that the cutoff ω_c in $J(\omega)$ is the largest frequency of the problem and therefore plays no particular role.

9.3.3 Ballistic motion

The motion of the particle with velocity v is ballistic on the scale of the interaction range when the mean free path ℓ is large compared to λ. With the definition of the diffusion coefficient in 1D, $D = v\ell$, we then have as condition for ballistic motion

$$\kappa \equiv \lambda v/D = \lambda/\ell \ll 1 \,. \qquad (9.22)$$

In the ballistic regime (9.22), the spectral probability function $\widetilde{P}(k, \omega)$ reads

$$\widetilde{P}(k, \omega) = 2\pi\delta(\omega - kv) \,. \qquad (9.23)$$

The characteristic frequency and temperature scale for the ballistic case is given in terms of the inverse time scale for traversing the interaction region,

$$\omega_{\rm b} \equiv v/\lambda \,, \qquad \text{and} \qquad T_{\rm b} \equiv \hbar\omega_{\rm b}/k_{\rm B} \,. \qquad (9.24)$$

With Eq. (9.23), the decoherence rate (9.20) takes the form [157]

$$\gamma_\varphi = \frac{\lambda^2}{\pi\hbar\omega_{\rm b}} \int_0^\infty d\omega \, J(\omega)\widetilde{U}(\omega/\omega_{\rm b}) \coth(\hbar\beta\omega/2) \,. \qquad (9.25)$$

We now suppose that $\omega_{\rm b} \ll \omega_c$. Then only the environmental modes with $\omega \lesssim \omega_{\rm b}$ contribute, and the cutoff in the spectral density $J(\omega)$ is irrelevant. On condition that the environmental spectral density has a low-frequency cutoff $\omega_{\rm min}$, the decoherence rate drops to zero when the particle's velocity v falls below $\lambda\omega_{\rm min}$.

The decoherence rate γ_φ can be calculated in analytic form for an algebraic spectral density and a Lorentzian form of the spatial correlation function (9.12). We put

$$U(x) = 1/[1+x^2] \qquad \text{yielding} \qquad \widetilde{U}(y) = \pi \, e^{-y} \,. \qquad (9.26)$$

With the form (3.58) for $J(\omega)$ the integral can be expressed in terms of Riemann's zeta function $\zeta(q, z)$. We find for the decoherence rate the analytic expression

$$\gamma_\varphi = \left\{ 1 + 2(T/T_{\rm b})^{s+1} \zeta(s+1, 1+T/T_{\rm b}) \right\} \gamma_0 \,, \tag{9.27}$$

where γ_0 is the decoherence rate at zero temperature,

$$\gamma_0 = \Gamma(1+s) \left(\omega_{\rm b}/\omega_{\rm ph}\right)^{s-1} \eta_s \lambda^2 \omega_{\rm b}/\hbar \,. \tag{9.28}$$

Since the spectral density of the environmental coupling is gapless, the environment has a decoherence effect which even persists at zero temperature. The rate saturates at temperatures below $T_{\rm b}$ and varies in this regime with the velocity of the particle as v^s. Well above $T_{\rm b}$, the decoherence rate is determined by classical noise, yielding

$$\gamma_\varphi = \frac{2T}{sT_{\rm b}} \gamma_0 = \Gamma(s) \left(\frac{\omega_{\rm b}}{\omega_{\rm ph}}\right)^{s-1} \frac{2\eta_s \lambda v}{\hbar} \frac{T}{T_{\rm b}} \,. \tag{9.29}$$

The form (9.27) describes the crossover between the limiting cases (9.28) and (9.29).

In the Ohmic case, $s = 1$, the decoherence rate (9.27) takes the form

$$\gamma_\varphi = \left\{ 1 + 2(T/T_{\rm b})^2 \psi'(1 + T/T_{\rm b}) \right\} \eta \lambda^2 \omega_{\rm b}/\hbar \,, \tag{9.30}$$

where $\psi(z)$ is the trigamma function. This reduces in the particular cases $T = 0$ and $T \gg T_{\rm b}$ to the forms

$$\gamma_\varphi = \eta \lambda v/\hbar \,, \qquad\qquad T = 0 \,, \tag{9.31}$$

$$\gamma_\varphi = 2\eta \lambda^2 k_{\rm B} T/\hbar^2 \,, \qquad T \gg T_{\rm b} \,. \tag{9.32}$$

Thus at zero temperature the dephasing rate is proportional to the velocity of the particle. In the white-noise limit, Eq. (9.32), the rate is independent of the velocity and formally agrees with the result for the model with delocalized oscillators, Eq. (5.31). However, the respective length parameters q_0 and λ may differ drastically. Measuring the decoherence rate (9.30) in units of the damping rate $\gamma_{\rm damping} = \eta/M$, we find $\gamma_\varphi/\gamma_{\rm damping} = 2\lambda^2/\lambda_{\rm th}^2$ for $T \gg T_{\rm b}$, where $\lambda_{\rm th} = \hbar/(Mk_{\rm B}T)^{1/2}$ is the thermal wave length. In the opposite limit $T = 0$, we have $\gamma_\varphi/\gamma_{\rm damping} = \lambda/\lambda_{\rm B}$, where $\lambda_{\rm B} = \hbar/Mv$ is the de Broglie wave length of the particle.

9.3.4 Diffusive motion

The motion of the particle is diffusive on the interaction range λ when the condition

$$\kappa \equiv \lambda v/D = \lambda/\ell \gg 1 \tag{9.33}$$

is met. In the diffusive regime, the double transform of $P(y, \tau)$ takes the form

$$\widetilde{P}(k, \omega) = \frac{2Dk^2}{(Dk^2)^2 + \omega^2} \,. \tag{9.34}$$

Substituting the expressions (9.26) and (9.34) into Eq. (9.20), and switching to the dimensionless variable $p = \lambda k$, we obtain the decoherence rate as

$$\gamma_\varphi = \frac{\lambda^2}{\pi \hbar \omega_d} \int_0^{\omega_m} d\omega \, J(\omega) \coth(\hbar \beta \omega / 2) \int_0^\infty dp \, g_1(p) \frac{2p^2}{p^4 + \omega^2/\omega_d^2} \tilde{U}(p) . \qquad (9.35)$$

Here, $g_1(p) = 1/\pi$ is the mode density for one space dimension. The characteristic frequency and temperature scale for diffusive motion is given by

$$\omega_d \equiv D/\lambda^2 = \omega_b/\kappa , \qquad \text{and} \qquad T_d \equiv \hbar \omega_d / k_B = T_b/\kappa . \qquad (9.36)$$

In the diffusive regime, the influence of the environmental modes is effectively limited to the range $\omega < \omega_m$ where $\omega_m \equiv v/\ell = v^2/D$ is the inverse collision time. We have $\omega_m = \kappa^2 \omega_d$ and hence $\omega_m \gg \omega_d$.

For temperatures well above $T > T_d$, we can use the classical noise approximation, $\coth(\hbar \beta \omega / 2) \approx 2/\hbar \beta \omega$. In the regime $s < 2$, we may take the limit $\omega_m \to \infty$, whereas ω_m is a physical high-frequency cutoff when $s \geq 2$. With the form (9.26) we find

$$\gamma_\varphi = \begin{cases} \dfrac{\eta_s \lambda^2}{\pi \hbar} \dfrac{2\pi \Gamma(2s-1)}{\sin(\pi s/2)} \left(\dfrac{\omega_d}{\omega_{ph}} \right)^{s-1} \dfrac{k_B T}{\hbar} , & \text{for} \quad s < 2 , \\[2em] \dfrac{\eta_2 \lambda^2}{\pi \hbar} 16 \ln(\kappa) \left(\dfrac{\omega_d}{\omega_{ph}} \right) \dfrac{k_B T}{\hbar} , & \text{for} \quad s = 2 , \qquad (9.37) \\[2em] \dfrac{\eta_s \lambda^2}{\pi \hbar} \dfrac{8\kappa^{2(s-2)}}{(s-2)} \left(\dfrac{\omega_d}{\omega_{ph}} \right)^{s-1} \dfrac{k_B T}{\hbar} , & \text{for} \quad s > 2 . \end{cases}$$

In the Ohmic case, $s = 1$, we recover the white noise limit, Eq. (9.32).

At temperatures below T_d, the cutoff ω_m becomes relevant for $s \geq 1$. We find from Eq. (9.35) at zero temperature in the sub-Ohmic, Ohmic, and super-Ohmic case the expressions

$$\gamma_\varphi = \begin{cases} \dfrac{\eta_s \lambda^2}{\pi \hbar} \dfrac{\pi \Gamma(1+2s)}{\cos(\pi s/2)} \left(\dfrac{\omega_{ph}}{\omega_d} \right)^{1-s} \omega_d , & \text{for} \quad s < 1 , \\[2em] \dfrac{\eta_1 \lambda^2}{\pi \hbar} 8 \ln(\kappa) \, \omega_d , & \text{for} \quad s = 1 , \qquad (9.38) \\[2em] \dfrac{\eta_s \lambda^2}{\pi \hbar} \dfrac{4}{(s-1)} \left(\dfrac{\omega_m}{\omega_{ph}} \right)^{s-1} \omega_d , & \text{for} \quad s > 1 . \end{cases}$$

Thus, the semiclassical treatment of the above model with localized bath modes yields saturation of the decoherence rate in the diffusive regime at temperature below T_d.

The above semiclassical analysis can easily be generalized to n space dimensions. For an isotropic interaction $U(|\boldsymbol{x}|)$, we find again the form (9.35) with the mode density in n dimensions $g_n(p)$ substituted for $g_1(p)$. We have $g_n(p) = \Omega_n p^{n-1}/(2\pi)^n$, where Ω_n is the total space angle. Now assume that the correlation function $\widetilde{U}(p)$ behaves as $\widetilde{U}(p \to 0) \propto p^{\sigma-n}$. For short-range interaction of Gaussian form, we have $\sigma = n$, whereas $\sigma = n - 2$ in the case of long-range Coulomb interaction. In all cases where $\sigma > 0$, the p-integral in Eq. (9.35) is good-natured so that the decoherence rate takes again the forms (9.37) and (9.38). Only the numerical prefactors depend weakly on the parameters n and σ. The regime $-2 < \sigma \leq 0$ is also well-defined, but requires special treatment since the interaction $U(x)$ diverges at the origin [255].

An expression for γ_φ similar to Eq. (9.35) is discussed in Refs. [158, 252]. Saturation of τ_φ at $T = 0$ is also found in models which are free of form factors in the system-reservoir interaction [256]. Taking into account quantum fluctuations about the classical path leads to additional suppression of quantum interference [198].

In the above semiclassical study of the model with localized bath modes, decoherence persists down to $T = 0$ when $J(\omega)$ is nonzero at frequencies below ω_b for ballistic motion, and below ω_d for diffusive motion. Decoherence endures because the reservoir can randomly absorb arbitrarily small amounts of energy down to $T = 0$ when the spectral density is gapless. Since $\omega_{ph} \gg \omega_m \gg \omega_d$, the dephasing rate at zero temperature is usually very small, and it becomes even smaller as the parameter s is raised from the sub-Ohmic sector $s < 1$ to the super-Ohmic regime $s > 1$. Decoherence decreases at low T with increasing s because of the much weaker super-Ohmic coupling at low frequencies compared to the Ohmic and sub-Ohmic case.

The above treatment relies on a single-particle picture for the electron. In a refined study of dephasing of electrons at low temperature, the multi-particle aspects of the problem and the Pauli principle should be taken into account. This may result in quantitatively if not qualitatively different behaviour of electron dephasing at low T.

There is an ongoing debate about the presence of dephasing at $T = 0$, and about the use of semiclassical methods in strongly disordered systems in which the mean free path is very short. For clarification, accurate measurements of decoherence times at low temperatures for different materials and different degrees of disorder would be extremely helpful.

In Aharonov-Bohm ring devices with a quantum dot (QD) built in one arm of the ring, the phase shift of the electron passing through the quantum dot has been analyzed [257]. The experiments demonstrated that the electron can propagate coherently through a quantum dot. Such device offers the possibility to control dephasing rates by modifying the electromagnetic environment of the QD. An additional wire with a quantum point contact located close to the QD may act as sensitive measurement device to detect when the electron passes through the QD. This opens the possibility to build a "Which Path?" interferometer [258]. The influence of the measurement apparatus on dephasing is twofold. First, creation of real electron-hole pairs in the wire measures which path the electron took around the ring. Secondly, Ohmic damping resulting from the creation of virtual electron-hole pairs (cf. Subsection 4.2.8) leads to power-law suppression of the Aharonov-Bohm oscillations.

PART III
QUANTUM STATISTICAL DECAY

In the second part, I considered the exactly solvable case of linear dissipative quantum systems, the thermodynamic variational approach useful for nonlinear quantum systems, and the quantum decoherence problem. We now turn to a study of the frequent situation in which a metastable state is separated from the outside region by a free energy barrier. The decay of a metastable state plays a central role in many scientific areas including low-temperature physics, nuclear physics, chemical kinetics and transport in biomolecules. At high temperatures, the system escapes from the metastable well predominantly over the barrier by thermal activation. At zero temperature on the other hand, the system is in the localized ground state of the metastable well and can escape only by quantum mechanical tunneling through the barrier. We shall focus the discussion on the influence of friction on the decay process. The treatment will be based on an effective method which allows to study this problem in a unified manner for temperatures ranging from $T = 0$ up to the classical regime.

10. Introduction

There are many processes in physics, chemistry, and biology in which a system makes transitions between different states by traversing a barrier. The theory of rate coefficients for barrier crossing has a long history since the days of Arrhenius [259]. H. A. Kramers' article of 1940 [260] represents a cornerstone in the quantitative analysis of thermally activated rate processes. This work provided a thorough theoretical description in the classical regime both for very *weak* and for *moderate-to-strong damping*. It includes important limiting cases such as the transition state theory or the Smoluchowski model of diffusion controlled processes.

The investigation of quantum mechanics for macroscopic variables has been stimulated considerably by Leggett's discussion concerning the validity of quantum mechanics at the macroscopic level [261, 262]. Strong impulse to the field was given further by the work of Caldeira and Leggett [77], who studied *quantum tunneling in the presence of dissipation* at zero temperature. From both the analytical and numerical point of view, quantum reaction theory has also been one of the most active and challenging areas in chemical physics and theoretical chemistry. In the recent past, the functional integral method has provided a unified description of the thermal quantum decay in the entire temperature range [263]. A comprehensive review of many

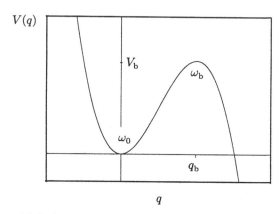

Figure 10.1: A metastable "quadratic-plus-cubic" potential well.

of the theoretical concepts and ideas in reaction rate theory extending from classical rate theory to more recent quantum versions has been given by Hänggi, Talkner and Borkovec [264]. I also refer to a recent account of the semiclassical approach to quantum tunneling in complex systems by Ankerhold [265].

Assume that the metastable system in question can be characterized by a generalized coordinate q. It may be visualized as a particle of mass M moving in an external potential $V(q)$ which has a single metastable minimum at a point which we choose at the origin of q. The bottom of the metastable potential is chosen to lie at zero, i.e., $V(0) = 0$. We assume that the potential $V(q)$ is fairly smooth and has the general form depicted in Fig. 10.1. In particular, $V(q)$ is taken to be negative for all points $q > q_{ex}$, where q_{ex} is the nonzero value of q for which $V(q) = 0$, sometimes called the "exit point" of the barrier. Once the particle has left the metastable well, it will not return in finite time. With this assumption, we can disregard quantum coherence effects. In physical systems, the coordinate q is the tunneling degree of freedom, and the thermal initial state is characterized by the partition function Z_0 of the well region. In chemical reactions, one thinks of q as the reaction coordinate and Z_0 denotes the partition function for reactants. Among the chemical applications are dissociation and recombination reactions and transfer processes of atoms and electrons.

The concept of metastability is useful when the barrier is large enough that the decay time of the metastable state is very long compared with all other characteristic time scales of the system dynamics. There exist many time scales such as the correlation time τ_c of the noise, the time of relaxation in the locally stable well τ_r, the thermal time $\hbar\beta$, and the time scales $\tau_0 = \omega_0^{-1}$ and $\tau_b = \omega_b^{-1}$ which are related to the curvature of the potential at the metastable minimum and at the barrier top. Here, ω_0 is the angular frequency of small oscillations around the metastable minimum

$$\omega_0 = [V''(0)/M]^{1/2} \, , \tag{10.1}$$

$-V(q)$

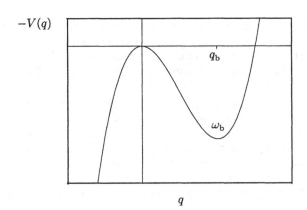

Figure 10.2: The upside-down potential $-V(q)$.

and the unstable barrier frequency

$$\omega_{\text{b}} = [-V''(q_{\text{b}})/M]^{1/2} \tag{10.2}$$

characterizes the width of the parabolic top of the barrier hindering the decay process. It represents the angular frequency of small oscillations around the minimum of the upside-down potential $-V(q)$ (cf. Fig. 10.2). All these various time scales will become important in a precise description of the escape rate. Weak metastability implies that the activation energy V_{b} is by far the largest energy of the problem, in particular

$$V_{\text{b}} \gg k_{\text{B}}T , \qquad \text{and} \qquad V_{\text{b}} \gg \hbar\omega_0 . \tag{10.3}$$

In the sequel, we shall not put restrictions on the numerical value of the ratio $\hbar\omega_0/k_{\text{B}}T$. It is convenient to parameterize the rate of escape from the metastable well in the form

$$k = A\,e^{-B} . \tag{10.4}$$

The quantity B is a dimensionless measure of the massiveness of the barrier the particle has to overcome. The prefactor A is a kind of an attempt frequency of the particle in the well towards the barrier. We shall be concerned with the question on which scale the parameters A and B are influenced by dissipation. The findings are briefly summarized as follows. In the classical regime, the exponent B is independent of damping and only the attempt frequency A is modified. In the quantum regime, both the exponent and the prefactor of the rate expression crucially depend on the strength and on the spectral properties of the dissipation. As the damping is increased, the classical regime extends to lower temperatures.

In the last decade, the problem of quantum tunneling of macroscopic variables has attracted a lot of interest. The research was considerably stimulated by experiments on macroscopic quantum tunneling in Josephson systems (cf., e.g., Ref. [266], and references therein), and by theoretical work put forward by Caldeira and Leggett [77]. As often in science, the problem of tunneling in the presence of coupling to (infinitely) many degrees of freedom, e.g., to phonons, quasiparticles, and electrons, has several precursors. Among them are the work on deep-inelastic collisions of heavy ions by Brink et al. [267], and by Möhring and Smilansky [268]. On the other hand, quantum tunneling in the presence of phonon modes has a long history in solid state physics. The study of multi-phonon effects in the polaron problem was initiated by Holstein [120] and by Emin and Holstein [269]. Further development includes the study of polaron effects on quantum tunneling which received enormous impact through the work by Flynn and Stoneham [123]. A critical review of the early work which relied mainly on the *Condon approximation* has been given by Sethna [160].

Before embarking on a closer look at multidimensional thermal quantum decay, we rove through the classical regime in Chapter 11, and in Chapter 12 we give an overview of the theoretical methods in the various regimes of thermal quantum decay for the one-dimensional case.

11. Classical rate theory: a brief overview

11.1 Classical transition state theory

Let us begin the discussion with the simplified description of thermally activated decay by transition state theory (TST). In its simplest form, which is the case of a single degree of freedom in the absence of a reservoir, one makes two *ad hoc* assumptions. First, thermal equilibrium prevails in the well, e.g., through the action of Maxwell's demon, so that the metastable state is represented by a canonical equilibrium distribution. Secondly, the particle will never return once it has crossed the barrier top even infinitesimally. Upon identifying the rate with the probability current across the barrier, and taking the Boltzmann weight of the normalized current density at the barrier top, we obtain

$$k_{\rm cl}^{\rm (TST)} = \frac{\dfrac{1}{2\pi\hbar}\displaystyle\int dp\,dq\,\exp\left(-\beta\left[p^2/2M + V(q)\right]\right)\delta(q - q_{\rm b})\,\Theta(p)\,p/M}{\dfrac{1}{2\pi\hbar}\displaystyle\int_{q<q_{\rm b}} dp\,dq\,\exp\left(-\beta\left[p^2/2M + V(q)\right]\right)}. \qquad (11.1)$$

Here we have chosen the barrier to the right of the well with the maximum $V_{\rm b} = V(q_{\rm b})$ at $q = q_{\rm b}$. From this we obtain the well-known transition state formula

$$k_{\rm cl}^{\rm (TST)} = \frac{1}{2\pi\hbar}\frac{k_{\rm B}T}{Z_0}\,{\rm e}^{-V_{\rm b}/k_{\rm B}T} = \frac{\omega_0}{2\pi}\,{\rm e}^{-V_{\rm b}/k_{\rm B}T}, \qquad (11.2)$$

where $Z_0 = k_{\rm B}T/\hbar\omega_0$ is the classical partition function of the well region. The Arrhenius law for the escape rate $k \propto {\rm e}^{-V_{\rm b}/k_{\rm B}T}$ reflects the exponentially small tail of the canonical initial state at the threshold energy $E = V_{\rm b}$. This is the reason why the exponential factor in Eq. (11.2) does not depend on the width of the barrier. The attempt frequency $A = \omega_0/2\pi$ is just the frequency of small oscillations in the well. The classical rate vanishes as temperature is lowered to absolute zero.

11.2 Moderate-to-strong-damping regime

In a famous paper published in 1940, Kramers [260] reported his careful study in various limits of the classical escape of a particle from a metastable well. In his treatment of the reaction rate, the starting point is nonlinear Brownian motion in phase space, which is analogous to the Markovian Langevin equation in coordinate space, Eq. (2.3) with (2.1) and (2.2). Following Kramers, the stochastic dynamics for the reaction coordinate q and the velocity $v = \dot{q}$ (or momentum) is conveniently described in terms of a probability density $p(q, v; t)$. The dynamical equation found for $p(q, v; t)$ is of the Fokker-Planck type, nowadays termed Klein-Kramers equation,

$$\frac{\partial p(q, v; t)}{\partial t} = \left(-\frac{\partial}{\partial q} v + \frac{\partial}{\partial v} \left(\frac{1}{M} \frac{\partial V(q)}{\partial q} + \gamma v \right) + \gamma \frac{k_B T}{M} \frac{\partial^2}{\partial v^2} \right) p(q, v; t) . \quad (11.3)$$

A comprehensive review on the Fokker-Planck equation and on the methods of its solution is given in the book by Risken [270]. Equation (11.3) gives a complete description of the stochastic process described by the Langevin equation (2.1) – (2.3).

The Klein-Kramers equation (11.3) has two different stationary solutions satisfying $\partial p(q, v; t)/\partial t = 0$. One solution is the canonical equilibrium state

$$p_{eq}(q, v) = N \exp \left\{ -\left[\tfrac{1}{2} M v^2 + V(q) \right] \big/ k_B T \right\} , \quad (11.4)$$

describing the thermal distribution of q and v in the well region (N is a normalization constant). The population of the well is found in Gaussian approximation as

$$n \equiv \int_{-\infty}^{q_b} dq \int_{-\infty}^{\infty} dv \, p_{eq}(q, v) = \frac{2\pi k_B T}{M \omega_0} N . \quad (11.5)$$

The other time-independent solution of Eq. (11.3) represents a steady state with nonzero probability flux over the barrier. The steady state describes the following situation. Particles are continuously injected at the bottom of the well. Afterwards they stay sufficiently long in the well region so that they thermalize before escaping over the barrier. Outside the barrier region, they are absorbed by a particle sink. To obtain the steady flux state for a potential of the form sketched in Fig. 10.1, we follow Kramers[260] and make the ansatz

$$p_{flux}(q, v) = G[u(q, v)] \, p_{eq}(q, v) , \quad (11.6)$$

where $G[u(q = 0, v)]$ captures the flux property. We impose the boundary conditions

$$G[u(q = 0, v)] = 1 , \quad \text{and} \quad G[u(q = q_+, v)] \approx 0 . \quad (11.7)$$

Here, q_+ is a point far beyond the barrier region ($q_+ \gg q_b$). With the first condition, the steady state (11.6) matches on the equilibrium state (11.4) in the well region. The second condition takes into account that the particles are removed on the other side of the barrier region. The limiting behaviour (11.7) implies that the function $G[u(q,v)]$ depends on q and v only through a linear combination $u(q,v)$ of q and v. If we choose

$$u(q,v) \;=\; q - q_b - \varrho v/\omega_b \,, \tag{11.8}$$

then the Klein-Kramers equation (11.3) with the ansatz (11.6) gives an ordinary differential equation for $G(u)$, i.e. without any additional dependence on v, if we choose the parameter ϱ to be a solution of the quadratic equation

$$\varrho^2 + (\gamma/\omega_b)\varrho - 1 \;=\; 0 \,. \tag{11.9}$$

The resulting differential equation for $G(u)$ reads

$$\kappa u G'(u) + G''(u) \;=\; 0 \tag{11.10}$$

with

$$\kappa \;=\; \frac{M\omega_b^3}{\varrho \gamma k_B T} \,. \tag{11.11}$$

The quantity κ must be positive in order that $G(u)$ vanishes for q near to q_+. This entails that the relevant root of Eq. (11.9) must be positive as well. Thus we have

$$\varrho \;=\; \sqrt{1 + \left(\frac{\gamma}{2\omega_b}\right)^2} - \frac{\gamma}{2\omega_b} \,. \tag{11.12}$$

The solution of Eq. (11.10) with the boundary conditions (11.7) reads

$$G(u) \;=\; \sqrt{\frac{\kappa}{2\pi}} \int_u^\infty du'\, \exp\left(-\frac{\kappa u'^2}{2}\right) \,. \tag{11.13}$$

Consider now the outgoing flux S of the steady state $p_{\text{flux}}(q,v)$ at the barrier top,

$$S(q_b) \;\equiv\; \int_{-\infty}^\infty dv\, v p_{\text{flux}}(q_b, v) \,. \tag{11.14}$$

With the form 11.6) with (11.13), the resulting integral can be transformed upon integration by parts into a standard Gauss integral. We readily get

$$S(q_b) \;=\; \varrho \frac{k_B T}{M} N \exp\left(-V_b/k_B T\right) \,. \tag{11.15}$$

The rate of escape over the barrier equals outgoing flux at the barrier top divided by the population of the well. Upon using the expressions (11.5) and (11.14), we obtain

$$k_{\text{cl}} \;\equiv\; \frac{S(q_b)}{n} \;=\; f_{\text{cl}} \exp\left(-V_b/k_B T\right) \,, \tag{11.16}$$

where the classical attempt frequency f_{cl} is given by

$$f_{cl} = \varrho f_{cl}^{(TST)} = \varrho \frac{\omega_0}{2\pi} . \qquad (11.17)$$

The expression (11.16) with (11.17) is the celebrated form of the classical escape rate found by Kramers for moderate-to-strong damping. Strictly speaking, the form (11.17) holds when damping is strong enough that the particle in the well relaxes to a thermal state before it escapes over the barrier. The attempt frequency (11.17) differs from the transition state value $f_{cl}^{(TST)}$ by the transmission factor ϱ, which captures all effects of damping in the classical regime.

In the light of the derivation given here within the Klein-Kramers equation, or within the equivalent Langevin equation, the transmission factor is reduced, $\rho < 1$, because of multiple re-crossing runs at the barrier which the particle's noisy trajectory makes.

Alternative derivations of the Kramers rate (11.16) with (11.17) from a multi-dimensional point of view are given below in Section 15.1 with the Im F method and in Section 15.4 based on multi-dimensional transition state theory. This will elucidate a different physical interpretation of the transmission factor $\rho < 1$ without recrossing processes of the barrier top (cf. concluding remarks in Section 15.4).

11.3 Strong damping regime

For strong damping or large viscosity, the inertia term in Eq. (2.3) becomes negligibly small. It is then expected that, starting from an arbitrary initial distribution $p(q, v; t)$, a Maxwell velocity distribution will arise within short time for every value of q. After that, a slow diffusion of the density distribution in coordinate space will take place. The time evolution of the reduced probability density

$$p(q, t) \equiv \int_{-\infty}^{\infty} dv \, p(q, v; t) \qquad (11.18)$$

is described in the classical regime by the Smoluchowski diffusion equation [264, 270]

$$\frac{\partial p(q, t)}{\partial t} = \frac{1}{M\gamma} \frac{\partial}{\partial q} \hat{L}_{cl}(q) \, p(q, t) \qquad (11.19)$$

with the Smoluchowski operator

$$\hat{L}_{cl}(q) \equiv \frac{\partial V(q)}{\partial q} + k_B T \frac{\partial}{\partial q} . \qquad (11.20)$$

For state-dependent diffusion, the Smoluchowski operator takes the form

$$\hat{L}_{cl}(q) \equiv \frac{\partial V(q)}{\partial q} + \frac{\partial}{\partial q} D(q) . \qquad (11.21)$$

The particle flux connected with the distribution $p(q, t)$ reads

$$S(q, t) \;=\; -\frac{1}{M\gamma}\left(V'(q) + \frac{\partial}{\partial q}D(q)\right)p(q, t)\,. \tag{11.22}$$

The Langevin equation corresponding to the diffusion equation (11.19) with the Smoluchowski operator (11.21) reads in the Ito representation [270]

$$M\gamma\dot{q}(t) \;=\; -V'(q) + \sqrt{2M\gamma D(q)}\,\xi(t)\,, \tag{11.23}$$

where $\xi(t)$ is Gaussian white noise with zero mean and correlation $\langle \xi(t)\xi(0)\rangle = \delta(t)$.

One stationary solution of the Smoluchowski equation (11.19) with (11.21) is the equilibrium state in the potential well

$$p_{\text{eq}}(q) \;=\; N\,\mathrm{e}^{-\phi(q)}\,, \tag{11.24}$$

where N is a normalization constant, and the effective potential

$$\phi(q) \;=\; \ln D(q) + \psi(q) \qquad \text{with} \qquad \psi(q) \;=\; \int_0^q dq'\,\frac{V'(q')}{D(q')}\,. \tag{11.25}$$

The other stationary solution of the Smoluchowski equation is a steady-flux state with constant current S,

$$p_{\text{flux}}(q) \;=\; M\gamma\, S\,\mathrm{e}^{-\phi(q)}\int_q^{q_+} dq'\,\mathrm{e}^{\psi(q')}\,. \tag{11.26}$$

The state $p_{\text{flux}}(q)$ is exponentially small at the other side of the potential barrier. In the steady-flux state, the population of the well is

$$n \;\equiv\; \int_{-\infty}^{q_{\text{b}}} dq\, p_{\text{flux}}(q)\,. \tag{11.27}$$

Next, we use that the integration over the well and the barrier regime decouple and that they can be done in Gaussian approximation. Observing further that the escape rate k_{cl} is the current divided by the population of the well we obtain the expression

$$k_{\text{cl}} \;=\; \frac{S}{n} \;=\; \frac{D(0)}{2\pi M\gamma}\sqrt{\phi''(0)|\psi''(q_{\text{b}})|}\,\mathrm{e}^{\psi(0)-\psi(q_{\text{b}})}\,, \tag{11.28}$$

where we have assumed $\phi'(0) = \psi'(q_{\text{b}}) = 0$.

In the case of state-independent diffusion $D = k_{\text{B}}T$, this reduces to the expression

$$k_{\text{cl}} \;=\; \frac{\omega_0\omega_{\text{b}}}{2\pi\gamma}\exp\left(-V_{\text{b}}/k_{\text{B}}T\right)\,. \tag{11.29}$$

Thus we find that the escape rate in the Smoluchowski limit is again in the form (11.16) with (11.17), the transmission factor ϱ being inversely proportional to the friction coefficient,

$$\varrho = \omega_{\mathrm{b}}/\gamma \, . \tag{11.30}$$

This form obtained for the transmission factor just represents the strong-friction limit of the Kramers expression (11.12).

Finally, we remark that we take up again the expression (11.29) in Section 15.3, where we discuss quantum corrections of the rate in the Smoluchowski limit.

11.4   Weak-damping regime

For very weak friction, the treatment given in the previous two subsections fails since damping is not strong enough to maintain thermal equilibrium in the well region. In the steady-flux state, the escape over the barrier is limited by the short supply of energy which is required to raise the particle to the barrier top. The escape over the barrier results in a depletion compared with the canonical population in an energy band of width $k_{\mathrm{B}}T$ just below the barrier top.

For very weak friction, the motion of the particle in the well is oscillatory, and the effect of friction is a gradual change of the distribution in energy space. The mean energy loss ΔE during one round trip in the well is found from the deterministic limit of Eq. (2.3) as

$$\Delta E = \gamma I(E) \, , \tag{11.31}$$

where

$$I(E) = M \oint dq \, \dot{q} = 2 \int_{q_1}^{q_2} dq \, \sqrt{2M[E - V(q)]} \tag{11.32}$$

is the abbreviated (Minkowskian) action[1] for one round trip at total energy E in the well with turning points q_1 and q_2.

The energy E is a slowly varying variable undergoing a diffusion process. The diffusion equation for the probability density $p(E, t)$ may be written as [264]

$$\dot{p}(E,t) = \gamma \frac{\partial}{\partial E} I(E) \left(1 + k_{\mathrm{B}}T \frac{\partial}{\partial E} \right) \frac{\omega(E)}{2\pi} p(E,t) \, , \tag{11.33}$$

where $\omega[I(E)] \equiv \omega(E)$ is the angular frequency at abbreviated action $I(E)$.

When the particle has built up energy as large as the threshold energy V_{b}, it escapes from the well with probability 1. The rate is therefore given by the particle flux in energy space through the energy point V_{b}, divided by the population n of the well. The current S associated with the steady-flux state $p_{\mathrm{flux}}(E)$ of Eq. (11.33) is

$$S = -\gamma I(E) \left(1 + k_{\mathrm{B}}T \frac{\partial}{\partial E} \right) \frac{\omega(E)}{2\pi} p_{\mathrm{flux}}(E) \, . \tag{11.34}$$

---

[1]The Minkowskian abbreviated action is denoted by $I(E)$ and the Euclidean abbreviated action by $W(E)$.

Upon using the relation

$$p(E) = (\partial I/\partial E)\,\widetilde{p}(I) = [2\pi/\omega(I)]\,\widetilde{p}(I) ,\qquad (11.35)$$

the current S may be expressed in terms of the steady-flux state $\widetilde{p}_{\text{flux}}(I)$,

$$S = -\gamma I \left(1 + \frac{2\pi k_{\text{B}}T}{\omega(I)}\frac{\partial}{\partial I}\right)\widetilde{p}_{\text{flux}}(I) .\qquad (11.36)$$

Removing the particle at $I = I(E = V_{\text{b}}) \equiv I_{\text{b}}$ from the well region means putting $\widetilde{p}_{\text{flux}}(I = I_{\text{b}}) = 0$. With this boundary condition, the solution of Eq. (11.36) reads

$$\widetilde{p}_{\text{flux}}(I) = \frac{S}{2\pi\gamma k_{\text{B}}T}\,e^{-E(I)/k_{\text{B}}T}\int_{I}^{I_{\text{b}}}dI'\,\frac{\omega(I')}{I'}\,e^{E(I')/k_{\text{B}}T} .\qquad (11.37)$$

The escape rate is again flux over population, this time in abbreviated-action space,

$$k_{\text{cl}} = S\Big/\int_{0}^{I_{\text{b}}}dI\,\widetilde{p}_{\text{flux}}(I) .\qquad (11.38)$$

Upon substituting the expression (11.37) and calculating the integrals in Gaussian approximation, we find the rate k_{cl} again in the form (11.16) with Eq. (11.17), in which the transmission factor ϱ is given by

$$\varrho = \gamma I(V_{\text{b}})/k_{\text{B}}T ,\qquad (11.39)$$

where $I(V_{\text{b}})$ is the abbreviated action for one round trip in the well at energy $E = V_{\text{b}}$.

In summary, the transmission is reduced compared to the transition state value $\varrho = 1$ by two physically different effects. For moderate-to-strong damping, ϱ is diminished because the trajectory of the particle is stochastic and may cross the barrier point q_{b} several times before the particle eventually escapes. The corresponding expression is given in Eq. (11.12). In the opposite weak-damping limit, $\gamma < k_{\text{B}}T/I(V_{\text{b}})$, the stationary distribution near the barrier top, $p(E \approx V_{\text{b}})$, is depleted compared with the canonical distribution. As a result, the transmission factor decreases linearly with γ as in Eq. (11.39).

As already noted by Kramers, the two limiting behaviours (11.12) and (11.39) imply a maximum of the rate at a certain damping value intermediate in strength between the above two limits. The actual transmission factor ϱ undergoes a *turnover* near $\gamma = k_{\text{B}}T/I(V_{\text{b}})$ in the form of a bell-shaped curve as sketched in Fig. 11.1. The turnover theory bridging between the spatial-diffusion-controlled and the energy-diffusion-limited formulas for the activated escape will be discussed in some detail in Section 14.2. There, light is shed also on the multi-dimensional aspects of the classical decay process.

The simplest analytic form of a metastable potential is the "quadratic-plus-cubic" potential sketched in Fig. 10.1,

$$V(q) = \frac{1}{2}M\omega_0^2 q^2 \left(1 - \frac{2q}{3q_{\text{b}}}\right) .\qquad (11.40)$$

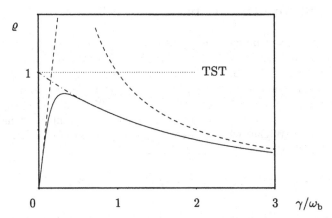

Figure 11.1: Turnover of the transmission factor ϱ from energy-diffusion-limited behaviour [Eq. (11.39), straight dashed line] to spatial-diffusion-controlled behaviour [Eq. (11.12), dashed-dotted curve]. The Smoluchowski limit $\varrho = \omega_b/\gamma$ is approached for large damping.

Here, q_b is the position of the barrier top, and the exit point for $E = 0$ is $q_{ex} = 3q_b/2$. The barrier height and barrier frequency are

$$V_b = M\omega_0^2 q_b^2/6 , \qquad \text{and} \qquad \omega_b = \omega_0 . \tag{11.41}$$

The potential has the symmetry

$$V(q_b - q) = V_b - V(q) . \tag{11.42}$$

For this potential, the round trip path at energy $E = V_b$ leaving the point $q = q_b$ at time $-\infty$ and returning to it at time $+\infty$ reads

$$q_{E=V_b}(t) = q_b - (3q_b/2)\,\text{sech}^2(\omega_0 t/2) , \tag{11.43}$$

and the abbreviated action of this path is

$$I(V_b) = 36V_b/5\omega_0 . \tag{11.44}$$

Using Eq. (11.44), we find that Kramers' formula (11.16) with (11.17) and (11.12) is correct in the classical regime when

$$\gamma/\omega_b \gtrsim 5k_B T/36V_b . \tag{11.45}$$

We see from the conditions (10.3) and (11.45) that the moderate-to-strong damping formula (11.12) is valid down to considerably weak damping.

As the temperature is lowered, the classical escape rate (11.16) decreases exponentially fast so that at very low temperatures the metastable state can only decay via quantum tunneling.

12. Quantum rate theory: basic methods

With the advent of quantum mechanics, the first who introduced quantum tunneling was Friedrich Hund in 1927 when he described the intramolecular rearrangement in ammonia molecules [271]. Shortly later, the tunneling effect was popularized by Oppenheimer [272], who used it to explain the ionization of atoms in strong electric fields, and by Gamow [273], and Gurney and Condon [274] when they explained the radioactive decay of nuclei. Perhaps the oldest guess of a quantum transition state theory was made by Wigner who proposed as a quantum mechanical generalization of the classical TST expression (11.1) the formula [275]

$$k_W = \frac{1}{Z_0} \frac{1}{2\pi\hbar} \int dp\, dq\, \frac{p}{M} \Theta(p)\delta(q - q_b) f^{(W)}(p, q) , \qquad (12.1)$$

where Z_0 is the partition function of the well region, and $f^{(W)}(p, q)$ is the quantum Wigner distribution (4.262). The expression (12.1) gives the correct first order quantum correction to the classical transition state rate expression and has been used by Miller [276] to derive a semiclassical transition state theory that involves a periodic classical trajectory in the upside-down potential surface (see Sections 12.3 and 15.4).

12.1 Formal rate expressions in terms of flux operators

A concise quantum mechanical rate expression in the form of a Boltzmann average of the reactive cross sections can be found from an exact quantum scattering calculation [277, 278]. It reads

$$k = \mathrm{Re}\left\{\mathrm{tr}\left(\mathrm{e}^{-\beta\hat{H}}\hat{F}\hat{\mathcal{P}}\right)\right\}/Z_0 = \mathrm{tr}\left(\mathrm{e}^{-\beta\hat{H}/2}\hat{F}_s\, \mathrm{e}^{-\beta\hat{H}/2}\hat{\mathcal{P}}\right)/Z_0 . \qquad (12.2)$$

Here, $\hat{H}$ is the Hamiltonian, Z_0 is the quantum partition function of the reactants (the partition function of the well region), and tr denotes a quantum mechanical trace. The operator $\hat{F}_s$ is the symmetrized flux through dividing surface operator. For the metastable potential of the form sketched in Fig. 10.1, the flux operator is given by

$$\hat{F}_s = \tfrac{1}{2}(\hat{F} + \hat{F}^\dagger) , \qquad \hat{F} = \delta\left(\hat{q} - q_b\right)\hat{p}/M , \qquad (12.3)$$

where p is the momentum conjugate to q. Finally, $\hat{\mathcal{P}}$ is a projection operator projecting onto states with positive momentum in the infinite future (outgoing states),

$$\hat{\mathcal{P}} = \lim_{t\to\infty} \mathrm{e}^{i\hat{H}t/\hbar}\Theta(\hat{p})\,\mathrm{e}^{-i\hat{H}t/\hbar} = \lim_{t\to\infty} \mathrm{e}^{i\hat{H}t/\hbar}\Theta(\hat{q} - q_b)\,\mathrm{e}^{-i\hat{H}t/\hbar} . \qquad (12.4)$$

The equivalence of both forms was shown in Ref. [277]. A major advantage of the expression (12.2) is that the exact thermal quantum rate expression can be directly given without necessity of first calculating energy-dependent rates and then perform a thermal averaging. The expression (12.2) for the quantum crossing rate involves the Hamiltonian in two ways: first, in the Boltzmann operator $e^{-\beta\hat{H}}$, and secondly through the projection operator $\hat{P}$ in Eq. (12.4). The infinite time limit of this projection can only be handled correctly by a quantum scattering calculation.

The expression for the crossing rate given in Eq. (12.2) is dynamically exact, with the only assumption that the total system is initially in thermal equilibrium. Thus, the formula covers both the spatial-diffusion-limited regime for moderate-to-strong damping and the energy-diffusion-limited regime for weak damping.

Other formally exact expressions for thermal rate constants were derived from Eq. (12.2) in Ref. [277]. An equivalent form of the rate which is similar to a result obtained by Yamamoto using Kubo's linear response formalism is [279]

$$ k = \frac{1}{Z_0} \int_0^\infty dt\, C(t) , \tag{12.5} $$

where $C(t)$ is the flux-flux autocorrelation function

$$ C(t) = \text{tr}\left[e^{i\hat{H}t/\hbar}\, e^{-\beta\hat{H}/2}\hat{F}_s\, e^{-\beta\hat{H}/2}\, e^{-i\hat{H}t/\hbar}\hat{F}_s \right] . \tag{12.6} $$

To determine the rate expression (12.5) with (12.6), it is necessary to deal with the quantum scattering problem. This amounts to the evaluation of the matrix elements of the complex-time evolution operator $< q'|\, e^{-i\hat{H}(t-i\hbar\beta/2)/\hbar}\, |q'' >$. The difficulty in calculating numerically exact rates is a direct consequence of the real-time propagation. Recently, it has been suggested to replace the exact time propagation in Eq. (12.4) by the propagation for a parabolic barrier [280], resulting in

$$ \hat{P} = \Theta\left(\hat{q} - q_b + \hat{p}/M\omega_b\right) . \tag{12.7} $$

With this replacement, the thermal rate is given by the formal approximate expression

$$ k = \text{tr}\left(e^{-\beta\hat{H}/2}\hat{F}_s\, e^{-\beta\hat{H}/2}\Theta(\hat{q} - q_b + \hat{p}/M\omega_b) \right)\Big/ Z_0 , \tag{12.8} $$

which depends only on matrix elements of the thermal density operator $e^{-\beta\hat{H}}$.

The flux operator $\hat{F}_s$ in Eq. (12.6) is of low rank. In one dimension, there are only two nonzero eigenvalues, one positive and one negative, corresponding to flux in forward and backward direction. The eigenvalues of the thermal flux operator $e^{-\beta\hat{H}/2}\, \hat{F}_s\, e^{-\beta\hat{H}/2}$ can be calculated efficiently using the iterative Lanczos scheme [281]. This allows to compute only those eigenvalues which are nonzero and contribute to the rate. The calculated basis functions are then propagated without numerical difficulties. The low rank of $\hat{F}_s$ implies a similar low rank for the full operator in Eq. (12.6). The trace can then be computed in this much smaller basis. Numerical stability of the real-time propagation can be considerably improved by introducing a complex absorbing potential. Following these lines, the numerically exact solution of Eq. (12.5) with (12.6) is feasible for systems with several degrees of freedom [282].

12.2   Quantum transition state theory

The quantum transition state approximation invoking the basic transition state idea in the calculation of the thermal rate corresponds to the replacement [278, 283]

$$\hat{\mathcal{P}} \longrightarrow \Theta(\hat{p}) ,\qquad (12.9)$$

so that in Eq. (12.2)

$$\hat{F}\hat{\mathcal{P}} \overset{\text{TST}}{\longrightarrow} \delta\left(\hat{q}-q_{\mathrm{b}}\right)\frac{\hat{p}}{M}\Theta(\hat{p}) = \frac{1}{2M}\delta\left(\hat{q}-q_{\mathrm{b}}\right)\left(|\hat{p}|+\hat{p}\right) .\qquad (12.10)$$

The projection operator $\Theta(\hat{p})$ projects onto the subspace of states with positive momentum. The expression (12.9) is easier than the form (12.7) since it only involves the momentum. The second term in the bracket of Eq. (12.10) does not contribute in thermal equilibrium because of detailed balance. The physical meaning of the approximation is as in the classical case: any trajectory with positive momentum at the dividing surface (which is the position of the barrier top) does not turn around to make recrossings. The replacement prescription, Eq. (12.9), eliminates the troublesome time-evolved projection operator $\hat{\mathcal{P}}$. Consequently, the semiclassical rate expression involves only matrix elements of the canonical operator $\mathrm{e}^{-\beta\hat{H}}$.

The transition state approximation, Eq. (12.9), circumvents the need to know the full scattering dynamics. The full Hamiltonian is retained however in the canonical operator $\mathrm{e}^{-\beta\hat{H}}$. Upon employing Eq. (12.10) and the identity

$$\mathrm{e}^{-\beta\hat{H}} = \frac{1}{\pi\hbar}\lim_{\epsilon\to 0^{+}}\,\mathrm{Im}\int_{0}^{\infty}dE\,\mathrm{e}^{-\beta E}\int_{0}^{\infty}d\tau\,\mathrm{e}^{(E+i\epsilon-\hat{H})\tau/\hbar} ,$$

we may rewrite the formal expression (12.2) in the form

$$k^{(\text{TST})} = \frac{1}{Z_0}\frac{1}{2\pi\hbar}\int_{0}^{\infty}dE\,p(E)\,\mathrm{e}^{-\beta E} ,\qquad (12.11)$$

where $p(E)$ is the dimensionless distribution function

$$p(E) = \lim_{\epsilon\to 0^{+}}\,\mathrm{Im}\int_{0}^{\infty}d\tau\,\mathrm{e}^{(E+i\epsilon)\tau/\hbar}$$
$$\times \int dq\,\delta(q-q_{\mathrm{b}})\,|\dot{q}\,|_{q=q_{\mathrm{b}}} < q\,|\,\mathrm{e}^{-\hat{H}\tau/\hbar}\,|\,q > .\qquad (12.12)$$

The expression (12.11) with (12.12) represents the quantum mechanical rate expression for a one-dimendional system in the transition state approximation. It is free of subsidiary approximations. The function $p(E)$ describes the transmission probability at energy E. Henceforth, we write k for $k^{(\text{TST})}$.

12.3 Semiclassical limit

In the semiclassical quantum transition state theory [276], the *semiclassical* form for the density matrix $< q \,|\, e^{-\hat{H}\tau/\hbar} \,|\, q' >$ is used in Eq. (12.12). There are different forms for the pre-exponential factor in the semiclassical expression [284]. For the present purpose, it is convenient to write[1]

$$< q'' |\, e^{-\hat{H}\tau/\hbar} \,| q' > \;=\; \frac{e^{-i\phi}}{\sqrt{2\pi\hbar}} \left(\frac{|\partial^2 S_{\mathrm{cl}}/d\tau^2|}{|\dot{q}(0)|\,|\dot{q}(\tau)|} \right)^{1/2} e^{-S_{\mathrm{cl}}(\tau)/\hbar}, \qquad (12.13)$$

where

$$S_{\mathrm{cl}}(\tau) \;=\; \int_0^\tau d\tau' \left(\tfrac{1}{2} M\dot{q}^2(\tau') + V[q(\tau')] \right) \qquad (12.14)$$

is the action of the extremal path in the upside-down potential with boundary conditions $q(0) = q'$ and $q(\tau) = q''$, and obeying $M\ddot{q}(\tau) = \partial V/\partial q$. The phase ϕ is $\pi/2$ times the number of conjugate points along the trajectory from q' to q''.

The second step of the semiclassical approximation consists in calculating the time integral in Eq. (12.12) by the method of steepest descent. The relevant exponential factor $\exp\{-[S_{\mathrm{cl}}(\tau) - E\tau]/\hbar\}$ is stationary for Euclidean time $\tau = \bar{\tau}$,

$$\left. \frac{\partial S_{\mathrm{cl}}}{\partial \tau} \right|_{\tau=\bar{\tau}} \;=\; E. \qquad (12.15)$$

The stationary points are periodic orbits with energy E in the upside-down potential $-V(q)$. Upon expanding the exponent about the stationary points, we obtain

$$S_{\mathrm{cl}}(\tau) - E\tau \;=\; W(E) + \frac{1}{2} \left. \frac{\partial^2 S_{\mathrm{cl}}}{\partial \tau^2} \right|_{\tau=\bar{\tau}} (\tau - \bar{\tau})^2 + \mathcal{O}\!\left((\tau - \bar{\tau})^3 \right), \qquad (12.16)$$

where $W(E)$ is the abbreviated *Euclidean* action[2]

$$W(E) \;=\; 2 \int_{q_1}^{q_2} dq\, \sqrt{2M[V(q) - E]}. \qquad (12.17)$$

The limits q_1 and q_2 are the zeros of the integrand and represent the turning points of the classical motion in the upside-down potential $-V(q)$ with total energy

$$E \;=\; V[q(\tau)] - \tfrac{1}{2} M\dot{q}^2(\tau). \qquad (12.18)$$

Since $\partial^2 S_{\mathrm{cl}}/\partial\tau^2 = 1/(\partial\tau/\partial E) < 0$ at $\tau = \bar{\tau}$, the periodic orbits are unstable against small perturbations. The saddle points are located on the positive-real axis of the

---

[1]For diverse applications of semiclassical propagators and periodic orbit theory to quantum chaos we refer to the monographies by M. Gutzwiller [285] and by F. Haake [286].

[2]The Minkowskian abbreviated action is denoted by $I(E)$ [Eq. (11.32)] and the Euclidean abbreviated action by $W(E)$.

complex-τ plane at $\bar{\tau} \equiv \tau(E) = -\partial W(E)/\partial E$, and the direction of steepest descent is perpendicular to the imaginary-time axis, i.e., along Minkowskian time. The contribution of the periodic orbit with one cycle is

$$p_1(E) = -\mathcal{J}_1\,e^{-W(E)/\hbar}\,. \tag{12.19}$$

The motion of the particle in the well of the upside-down potential is bounded and periodic for $E \geq 0$. For $E \ll V_b$, the path spends most of the time in the vicinity of the inner turning point q_1 and it bounces back from the outer turning point q_2. For this reason, the periodic path satisfying Eq. (12.18) is called *bounce* [99]. We shall denote this path by $q_B(\tau)$. It is convenient to choose the phase of the bounce such that $q_B(\tau) = q_B(-\tau)$, and $q_B(\tau = 0) = q_2$. For a bounce with total energy $E = 0$ we have $q_2 = q_{ex}$ and $q_1 = 0$. The additional factor $\mathcal{J}_1$ originates from the conjugate points. Each turning point along the orbit contributes a phase shift $\pi/2$ to the phase ϕ in Eq. (12.13). For a path with n cycles we thus have $\mathcal{J}_n = e^{-i\pi n} = (-1)^n$.

Next, consider the contribution of a path making n round trips at energy E. Evidently, the abbreviated action is $nW(E)$ and the total time spent is $\tau_n = n\bar{\tau}$. The contributions from the infinite sequence of stationary points can be summed explicitly, yielding

$$p(E) = \sum_{n=1}^{\infty}(-1)^{n-1}\,e^{-nW(E)/\hbar} = \frac{1}{1 + e^{W(E)/\hbar}}\,. \tag{12.20}$$

In the semiclassical limit $\hbar \to 0$ with $T/T_0 < 0$ held fixed, where

$$k_B T_0 \equiv \frac{\hbar\omega_b}{2\pi}\,, \qquad \beta_0 \equiv \frac{2\pi}{\hbar\omega_b}\,, \tag{12.21}$$

the Boltzmann average in Eq. (12.11) is dominated by the periodic orbit with one cycle of period $\tau \equiv -\partial W/\partial E = \hbar\beta$,

$$k = \frac{1}{Z_0}\frac{1}{2\pi\hbar}\int_0^{V_b}dE\,e^{-[\beta E + W(E)/\hbar]}\,. \tag{12.22}$$

Upon calculating the integral by steepest descent, we find

$$k = \frac{1}{Z_0}\frac{1}{\sqrt{2\pi\hbar|\tau_B'|}}\,e^{-S_B(\hbar\beta)/\hbar}\,, \qquad T < T_0\,, \tag{12.23}$$

$$|\tau_B'| = \left|\frac{\partial\tau_B}{\partial E}\right|_{E=E_{\hbar\beta}} = \left|\frac{\partial^2 W(E)}{\partial E^2}\right|_{E=E_{\hbar\beta}}\,, \tag{12.24}$$

and where $S_B(\hbar\beta)$ is the action (12.14) for the bounce trajectory with period $\hbar\beta$ and total conserved energy $E = E_{\hbar\beta}$. Using Eqs. (12.14) and (12.18), we have

$$S_B(\hbar\beta) = W(E_{\hbar\beta}) + E_{\hbar\beta}\hbar\beta\,. \tag{12.25}$$

The expression (12.23) with Eqs. (12.24) and (12.25) represents the semiclassical quantum rate formula in transition state theory [287]. This form is generally valid in the temperature regime $0 \leq k_{\mathrm{B}}T \lesssim \hbar\omega_0$.

For temperatures $T > T_0$, the integral in Eq. (12.11) is dominated by the regime $E \gtrsim V_{\mathrm{b}}$, in which we may use for $p(E)$ the expression (12.20) with $W(E)$ approximated by the abbreviated action of the parabolic barrier,

$$p(E) = \frac{1}{1 + \mathrm{e}^{2\pi(V_{\mathrm{b}}-E)/\hbar\omega_{\mathrm{b}}}} \cdot \qquad (12.26)$$

Upon extending the integration in Eq. (12.11) to $-\infty$, we obtain with Eq. (12.26)

$$k = \frac{\omega_{\mathrm{b}}}{2\pi} \frac{\sinh(\hbar\beta\omega_0/2)}{\sin(\hbar\beta\omega_{\mathrm{b}}/2)} \, \mathrm{e}^{-\beta V_{\mathrm{b}}} , \qquad T > T_0 . \qquad (12.27)$$

As T approaches T_0, the expression (12.27) diverges $\propto 1/(T-T_0)$. The singularity is removed by taking into account that the barrier is actually wider than the parabolic barrier. This leads us for $T \approx T_0$ to the modified abbreviated action

$$W(E) = \frac{2\pi}{\omega_{\mathrm{b}}}\left(V_{\mathrm{b}} - E\right) + \frac{|\tau'(V_{\mathrm{b}})|}{2}\left(V_{\mathrm{b}} - E\right)^2 + \mathcal{O}\!\left((V_{\mathrm{b}} - E)^3\right) . \qquad (12.28)$$

Upon inserting this form into the integral expression

$$k = \frac{1}{Z_0} \frac{1}{2\pi\hbar} \int_{-\infty}^{V_{\mathrm{b}}} dE \; \mathrm{e}^{-[\beta E + W(E)/\hbar]} , \qquad (12.29)$$

we obtain

$$k = \frac{1}{2Z_0} \frac{1}{\sqrt{2\pi\hbar|\tau'|}} \, \mathrm{erfc}\!\left(\left(\frac{\hbar}{2|\tau'|}\right)^{\frac{1}{2}}(\beta_0 - \beta)\right) \exp\!\left(-\beta V_{\mathrm{b}} + \frac{\hbar(\beta - \beta_0)^2}{2|\tau'|}\right) , \qquad (12.30)$$

where $\tau' = \tau'(V_{\mathrm{b}})$, and where $\mathrm{erfc}(z)$ denotes the complementary error function [88]

$$\mathrm{erfc}(z) \equiv \frac{2}{\sqrt{\pi}} \int_z^\infty dt \; \mathrm{e}^{-t^2} . \qquad (12.31)$$

The expression (12.30) smoothly bridges the solution (12.22) for $T < T_0$ to the regime of thermal activation where Eq. (12.27) is valid. The expression (12.30) is only needed in the narrow temperature regime $k_{\mathrm{B}}|T - T_0| \lesssim (k_{\mathrm{B}}T_0)^2\sqrt{2|\tau'|}/\hbar$ about T_0. We shall consider this regime in Section 16.3.

12.4   Quantum tunneling regime

Consider now the decay rate at temperatures $T \ll T_0$ more closely. In the energy regime $E \ll V_{\mathrm{b}}$, Eq. (12.17) may be expanded as

$$W(E) = W_{\mathrm{h}}(E) + W(0) - W_{\mathrm{h}}(0) + E\left[\partial W/\partial E - \partial W_{\mathrm{h}}/\partial E\right]_{E=0} + \cdots , \qquad (12.32)$$

where $W_h(E)$ is the abbreviated action in which the harmonic potential $V_h(q) = \frac{1}{2}M\omega_0^2 q^2$ is substituted for $V(q)$. The action of the bounce with $E = 0$ is

$$S_0 \equiv W(0) = 2 \int_0^{q_0} dq \sqrt{2MV(q)} \,, \qquad (12.33)$$

and the bounce approaches the point $q = 0$ as

$$\lim_{\tau \to \pm\infty} q_B(\tau) = C_0 \sqrt{\frac{S_0}{2M\omega_0}} \, e^{-\omega_0 |\tau|} \,, \qquad (12.34)$$

where we have parametrized the prefactor conveniently. The constant C_0 is a numerical factor which depends on the shape of the barrier. With Eqs. (12.33) and (12.34) we find from Eq. (12.32) the expression

$$W(E) = S_0 - \frac{E}{\omega_0} - \frac{E}{\omega_0} \ln\left(\frac{C_0^2 S_0 \omega_0}{E}\right) + \mathcal{O}\left[(E/V_b)^2\right] \,. \qquad (12.35)$$

We shall use this expression shortly.

If the system is initially localized in the ground state of the metastable potential well, the probability per unit time that it escapes by quantum tunneling from the well is given by the standard WKB formula [99]

$$k = \gamma_0 \,, \qquad \gamma_0 = \omega_0 C_0 \, (S_0/2\pi\hbar)^{1/2} \, e^{-S_0/\hbar} \,, \qquad (12.36)$$

and the rate out of the nth excited state with energy $E_n = \hbar\omega_0(n + 1/2)$ reads [100]

$$\gamma_n = \frac{1}{n!}\left(C_0^2 \frac{S_0}{\hbar}\right)^n \gamma_0 \,. \qquad (12.37)$$

For the cubic metastable potential (11.40), the bounce with zero total energy takes the bell-shaped form

$$q_B(\tau) = \tfrac{3}{2}q_b \, \text{sech}^2(\tfrac{1}{2}\omega_0\tau) \,. \qquad (12.38)$$

Because of the symmetry relation (11.42), the Euclidean action S_0 coincides with the Minkowskian action (11.32) $I(E = V_b)$. The Euclidean action S_0 and the numerical factor C_0 are given by

$$S_0 = I(V_b) = \frac{36}{5}\frac{V_b}{\omega_0} = \frac{6}{5}M\omega_0 q_b^2 \,, \qquad \text{and} \qquad C_0 = \sqrt{60} \,. \qquad (12.39)$$

We now define a thermal rate by averaging the decay rates for the individual energy levels in the well with the canonical distribution,

$$k \equiv \sum_n \gamma_n \, e^{-\beta E_n} \Big/ \sum_m e^{-\beta E_m} \,. \qquad (12.40)$$

Upon inserting the semiclassical expression (12.37), the sum in Eq. (12.40) can be carried out explicitly. We obtain

$$k = \frac{e^{-\hbar\omega_0/2k_{\rm B}T}}{Z_0} \exp\left(C_0^2 \frac{S_0}{\hbar} e^{-\hbar\omega_0/k_{\rm B}T}\right) \gamma_0 \,, \qquad (12.41)$$

where $Z_0^{-1} = 2\sinh(\hbar\beta\omega_0/2)$. The expression (12.41) applies when the thermal energy is very small compared to $\hbar\omega_0$. At higher temperatures, Eq. (12.40) may be replaced by the continuum version, which is

$$k = \frac{1}{Z_0} \int_0^\infty dE\, \rho(E)\gamma(E)\, e^{-\beta E} \,. \qquad (12.42)$$

Here, $\rho(E) = 1/\hbar\omega_0$ is the density of states, and $\gamma(E)$ is the WKB expression of the transmisission probability for well-separated turning points (higher states),

$$\gamma(E) = \frac{\omega_0}{2\pi} e^{-W(E)/\hbar} \,. \qquad (12.43)$$

The rate expression (12.42) with Eq. (12.43) coincides with the expression (12.22). It is somewhat surprising for two reasons that the expression (12.23) is valid down to zero temperature. First, in the initial equation (12.42) the discreteness of the low-lying states is disregarded. Secondly, also the steepest-descent transmission formula (12.43) is not correct for low-lying states [100].[3] Interestingly, the two inaccuracies cancel each other out. Thus, the result Eq. (12.23) matches indeed on the expression (12.41) in the regime $k_{\rm B}T \ll \hbar\omega_0$. Let us briefly show the equivalence. In the first step, we invert the relation $|\partial W/\partial E|_{E=E_{\hbar\beta}} = \hbar\beta$ with $W(E)$ given by Eq. (12.35). We then find for the total energy of the bounce with period $\hbar\beta$

$$E_{\hbar\beta} = C_0^2 S_0\omega_0\, e^{-\beta\hbar\omega_0} \,. \qquad (12.44)$$

With this form and with Eq. (12.35) we obtain

$$S_{\rm B}(\hbar\beta) = \left(1 - C_0^2\, e^{-\beta\hbar\omega_0}\right)S_0 \,, \qquad (12.45)$$

and

$$|\tau_{\rm B}'| = \frac{1}{C_0^2 S_0\omega_0^2} e^{\beta\hbar\omega_0} \,. \qquad (12.46)$$

In the second step, we substitute Eqs. (12.45) and (12.46) into Eq. (12.23). We then recover the previous result (12.41). This verifies that the expression (12.23) is valid down to zero temperature [100, 287].

12.5 Free energy method

A different line of thoughts which has found great popularity in recent years is to calculate the quantum rate expression from a pure thermodynamic equilibrium method. The thermodynamic method was pioneered by Langer [288] in 1967 and reviewed by

---

[3]The formula (12.43) for $\gamma(E_n)$ follows from the expression (12.37) by substituting Stirling's asymptotic formula for the factorial $n!$. This approximation is only justified when n is large.

himself in 1980 in Ref. [289]. In the original treatment, Langer calculated the classical nucleation rate governing the early stage of a first-order phase transition. In this approach the quantity of interest is the free energy of the metastable system. Because of states of lower energy on the other side of the barrier, the partition function can only be defined by means of an analytical continuation from a stable potential to the metastable situation depicted in Fig. 10.1. The procedure of analytic continuation leads to a unique imaginary part of the free energy of the metastable state. This quantity is then related to the decay probability of the system, fully analogous to the interpretation of imaginary energies of resonances in quantum field theory. Interestingly and importantly, this method is not restricted to the classical regime. As we shall see, all the results of Section 12.3 are reproduced by the free energy (Im F) method. In a sense, the Im F method is easier than the method presented in Section 12.3 since the energy and time integrals in Eqs. (12.11) and (12.12) are circumvented.

In this section, we briefly sketch the Im F method for the onedimensional case. The Im F method turns out to be particularly appropriate in the case of a system-plus-environment complex where the number N of bath degrees of freedom is very large or even infinity. The respective treatment is given in Chapters 15 – 17.

The general argument is as follows [100, 287]. For a metastable system, the partition function may be written as

$$Z = \sum_n e^{-\beta z_n} = \sum_n e^{-\beta(E_n - i\hbar\gamma_n/2)} , \qquad (12.47)$$

where the $z_n = E_n - i\hbar\gamma_n/2$ are the complex energies of the individual states, and $\sum_n$ denotes summation over all states. In the limit $V_b/\hbar\omega_0 \gg 1$, we have $\hbar\gamma_n \ll E_n$ for all n. Thus for weak metastability, we have for any temperature

$$Z = Z' + iZ'' \approx \sum_n e^{-\beta E_n} + i(\hbar\beta/2)\sum_n \gamma_n e^{-\beta E_n} . \qquad (12.48)$$

The imaginary part Z'' of the partition function is generally small compared to the real part Z'. The real part Z' is (apart from exponentially small corrections) determined by the properties of the well, and the imaginary part Z'' is determined by the properties of the barrier. We therefore write $Z' = Z_0$, and $Z'' = \mathrm{Im}\, Z_b$. Using the definition of the free energy

$$F = -(1/\beta) \ln Z , \qquad (12.49)$$

we obtain for the imaginary part

$$\mathrm{Im}\, F = -\frac{1}{\beta} \frac{\mathrm{Im}\, Z_b}{Z_0} = -\frac{\hbar}{2} \frac{\sum_n \gamma_n e^{-\beta E_n}}{\sum_m e^{-\beta E_m}} . \qquad (12.50)$$

Thus, the thermal quantum rate (12.40) is related to Im F by the relation

$$k = -(2/\hbar) \,\mathrm{Im}\, F . \qquad (12.51)$$

The formula (12.51) generalizes the decay rate of the ground state, $k(T = 0) = -(2/\hbar)\,\mathrm{Im}E_0$, to finite temperatures. Actually, the relation (12.51) is only valid in the regime $T \leq T_0$, as we shall see shortly.

Let us study $\mathrm{Im}\,F$ in the semiclassical limit. This is done by writing the partition function as a path integral, as in Eq. (4.210). The dominant stationary point for the barrier partition function Z_b in the regime $T < T_0$ is the periodic bounce path $q_B(\tau)$ with period $\hbar\beta$ making one cycle in the upside-down potential $-V(q)$. The second variation operator $-d^2/d\tau^2 + V''[q_B]$ has a periodic zero mode with one node which is $\dot{q}_B(\tau)$. Therefore, there must be a nodeless eigenmode with a negative eigenvalue. This eigenmode appears because the bounce is a saddle point of the action. Here we anticipate the result for the zero mode contribution discussed in Section 17.1. Upon deforming the integration contour for the mode with the negative eigenvalue, we find[4]

$$\mathrm{Im}\,Z_b = \frac{\hbar\beta Z_0}{2}\sqrt{\frac{W_B}{2\pi\hbar}}\left|\frac{\det\{-d^2/d\tau^2 + \omega_0^2\}}{\det'\{-d^2/d\tau^2 + V''[q_B(\tau)]\}}\right|^{1/2} e^{-S_B/\hbar}, \qquad (12.52)$$

where W_B is the abbreviated action of the bounce. The determinants are calculated for eigenfunctions obeying periodic boundary conditions as discussed in Subsection 4.3.2. The prime in $\det'[\cdots]$ indicates that the zero eigenvalue is omitted. Alternatively, we obtain directly from Eq. (12.13) upon substituting the integral $\int_0^{\hbar\beta} d\tau\,\dot{q}(\tau)\cdots$ for the trace integral $\oint dq\cdots$,

$$\mathrm{Im}\,F = -\frac{1}{\beta}\frac{\mathrm{Im}\,Z_b}{Z_0} \quad \text{with} \quad \mathrm{Im}\,Z_b = \frac{1}{2}\frac{\hbar\beta}{\sqrt{2\pi\hbar|\tau_B'|}}\,e^{-S_B/\hbar}. \qquad (12.53)$$

The factor $i/2$ in Z_b results again from the deformation of the contour at the saddle point in the direction of steepest descent. Upon equating Eq. (12.52) with Eq. (12.53), we can express the ratio of the determinants in Eq. (12.52) directly in terms of the classical mechanics of the bounce. We find the relation

$$Z_0^2\left|\frac{\det\{-d^2/d\tau^2 + \omega_0^2\}}{\det'\{-d^2/d\tau^2 + V''[q_B(\tau)]\}}\right| = \frac{1}{W_B|\tau_B'|}, \qquad (12.54)$$

which agrees with the result of a direct computation [291]. Upon substituting Eq. (12.53) into Eq. (12.51), we find that the resulting rate expression is in the form (12.23). This verifies the relation (12.51) in the temperature regime $T \leq T_0$.

For $T > T_0$, the saddle point for the barrier partition function is determined by the constant path $q = q_b$. The zero mode is absent, but there is again a negative eigenvalue which requires analytic continuation of the respective fluctuation mode. The appropriate deformation of the integration contour leads again to an imaginary part of the partition function coming from the barrier region. We obtain

$$\mathrm{Im}\,F = -\frac{1}{\beta}\frac{\mathrm{Im}\,Z_b}{Z_0} \quad \text{with} \quad \mathrm{Im}\,Z_b = \frac{1}{4\sin(\hbar\beta\omega_b/2)}\,e^{-\beta V_b}. \qquad (12.55)$$

---

[4] A detailed study of the bounce method, in particular the analytic continuation scheme related to the deformation of the potential, is reported in Ref. [290].

Comparing Eq. (12.55) with Eq. (12.27), we find consistency provided that

$$k = -(2/\hbar)(\beta/\beta_0)\operatorname{Im} F \qquad \text{for} \qquad T > T_0 \,. \qquad (12.56)$$

These findings which go back to Affleck [287] are summarized as follows. The relation (12.51) is appropriate in the temperature regime $T \leq T_0$, in which $\operatorname{Im} Z_b$ is determined by the periodic bounce trajectory $q_B(\tau)$. Because the phase of the bounce $q_B(\tau)$ is arbitrary, there is a zero mode in the deviations about $q_B(\tau)$. For this reason, $\operatorname{Im} Z_b$ is proportional to β. The factor β cancels out in the expression (12.53) for $\operatorname{Im} F$. Above T_0, the relevant saddle point trajectory is the constant path $q(\tau) = q_b$ which has no fluctuation mode with eigenvalue zero. Because of the absence of the zero mode above T_0, the rate has to be calculated by means of the modified formula (12.56). This is merely a plausible argument and certainly not a proof. Combining the expressions (12.51) and (12.56), the flux rate can be written in the unified form

$$k = -\kappa \,(2/\hbar)\operatorname{Im} F \,, \qquad (12.57)$$

where κ is a piecewise constant numerical factor given by

$$
\begin{aligned}
\kappa &= 1 \,; & \hbar\beta\omega_b &\geq 2\pi \,, \\
\kappa &= \hbar\beta\omega_b/2\pi \,; & \hbar\beta\omega_b &< 2\pi \,.
\end{aligned}
\qquad (12.58)
$$

Later, when we discuss crossing rates in multidimensional systems, the barrier frequency in Eq. (12.58) is actually an effective barrier frequency (cf. Section 14.3).

A general remark is in order. We have seen that quantum tunneling opens new channels for barrier crossing as the temperature is lowered below the classical regime. Therefore, the escape rate is enhanced compared with the classical rate expression. At intermediate temperatures, a rough estimate of the rate is obtained naively by adding to the classical rate a quantum mechanical counterpart, $k = k_{cl} + k_{qm}$ [292]. In a first guess, k_{qm} may be identified with the ground-state tunneling rate (12.36). With similar reasoning, a plausible criterion for the crossover temperature T_0, where roughly the transition from the thermally activated to the quantum mechanical decay occurs, was given by Goldanskii already in 1959 [293]. By equating the Arrhenius exponent with the Gamow factor for the transmission of a parabolic barrier with height V_b and barrier frequency ω_b,

$$\beta_0 V_b = 2\pi V_b/\hbar\omega_b \,, \qquad (12.59)$$

the crossover temperature of an undamped system is found in the form (12.21).

The above considerations shed different light on the significance of the crossover temperature T_0. For $T < T_0$, there is always a classical bounce solution in the upside-down potential $-V(q)$ with a period $\hbar/k_B T$. The shortest possible period is $2\pi/\omega_b$, for which the bounce is a harmonic oscillation about q_b. At the crossover temperature T_0, the periodic bounce coalesces to the harmonic oscillation. At this particular temperature, the respective actions are the same, and they also coincide with the action of the constant path $q = q_b$.

The crossover temperature T_0 can be quite large for very light particles, e.g., above room temperature for electrons. On the other hand, for macroscopic quantum tunneling phenomena in Josephson systems, the effective mass is typically very high and the crossover temperature can well be in the milli-Kelvin region [266].

Since the Im F method is basically a thermodynamic method, it is applicable when there is thermal equilibrium in the well region. This means that the time scale for relaxation in the well should be short compared to the average time for escape from the well. The Im F method may also be applied in certain limits, when the barrier is time-dependent. First, it is applicable in the adiabatic limit of slow-frequency driving [294]. In this case, the escaping particle is subject to the current barrier potential, and the total rate is found from an average over the relevant potential configurations. Secondly, the Im F method is also applicable in the case of fast driving, where the escape occurs in an effective potential resulting from an average over one period of driving. Clearly, the Im F method is not applicable when the driving frequency is of the order of the bare or dressed well frequency.

Let us conclude this section with making a general remark on a recent controversy. Tunneling has been investigated by considering classical real-time trajectories with energies higher than the barrier [295]. In this picture, the smallness of the tunneling rate is due to the rapid oscillations of the amplitudes. Subsequently, it has been shown that the numerical real-time Fourier integral which uses only information from the region above the barrier is inadequate in the deep tunneling regime [296]. In the semiclassical quantum transition state theory discussed above and in subsequent chapters, tunneling is a pure imaginary-time process which can be fully accounted for by considering classical trajectories on the upside-down potential surface.

12.6 Centroid method

Dynamical simulations of barrier crossing processes suffer from the fact that the relevant tunneling paths are rare events. The efficiency of numerically computing reaction rates is enormously enhanced by filtering out irrelevant paths that do not cross the surface dividing the reactant from the reaction product and therefore do not contribute to the crossing rate. An efficient sampling strategy consists in directly sampling the centroid density $\rho_c(q_c)$ discussed in Section 8.1 by using imaginary-time path integral Monte Carlo techniques for the constrained path integral in Eq. (8.4). Upon tuning the centroid variable q_c, the effective potential $\mathcal{V}(q_c)$ can be computed. Generally, the effective potential $\mathcal{V}(q_c)$ may not necessarily have the same shape as the original potential $V(q)$. As noticed by Gillan [297] and subsequently treated more rigorously by Voth $et\ al.$ [298], the reaction rate is directly connected with the change of the effective classical potential $\mathcal{V}(q_c)$ when the centroid is tuned from the reactant state to the barrier top (dividing surface).

Assume that the effective potential has a local minimum at $q_c = q_{min}$ with well frequency ω_{min}, and a maximum at $q_c = q_{max}$ with barrier frequency ω_{max}. Unless

$V(q)$ is symmetric, q_{max} is different from the position q_b of the barrier top of $V(q)$. Following the analysis given in Section 12.5, the point $q_c = q_{min}$ is a stable stationary point, and integration over q_c in the well region [cf. Eq. (8.3) with Eq. (8.6)] gives the partition function of reactants, $Z_0 = (1/\hbar\beta\omega_{min})\exp[-\beta V(q_{min})]$, in the Gaussian approximation. In contrast, the point q_{max} is a saddle point requiring deformation of the integration contour into the complex q_c-plane. Thus we find in steepest descent the barrier contribution

$$\text{Im}\, Z_b = \frac{1}{2\hbar\beta\omega_{max}}\, e^{-\beta V(q_{max})} \,. \qquad (12.60)$$

Using Eq. (12.57), we obtain for the rate the expression

$$k = \frac{\kappa}{\hbar\beta}\frac{\omega_{min}}{\omega_{max}}\, e^{-\beta[V(q_{max})-V(q_{min})]} \,, \qquad (12.61)$$

which for $\kappa = 1$ is Gillan's centroid formula [297]. For temperatures $k_B T \geq \hbar\omega_{max}/2\pi$, Affleck's factor κ in Eq. (12.61) is $\hbar\beta\omega_{max}/2\pi$ [cf. Eq. (12.58)], which is the centroid result obtained in Refs. [298, 299].

Instead of computing the effective classical potential numerically using quantum Monte Carlo techniques, one may estimate the effective potential by employing the variational methods discussed in Subsection 8.2.2 or one may apply semiclassical methods [300]. In the semiclassical limit, we have [cf. Eq. (8.11) with Eq. (8.14), and corresponding expressions for the parabolic barrier]

$$\begin{aligned}
V(q_{min}) &= V(q_{min}) + \ln[\sinh(\hbar\beta\omega_{min}/2)/(\hbar\beta\omega_{min}/2)]/\beta \,, \\
V(q_{max}) &= V(q_{max}) + \ln[\sin(\hbar\beta\omega_{max}/2)/(\hbar\beta\omega_{max}/2)]/\beta \,.
\end{aligned} \qquad (12.62)$$

It is straightforward to see that Eq. (12.61) with Eq. (12.62) agrees for $T > T_0$ with the previous result in Eq. (12.27).

We remark again that the effective potential includes all quantum effects of the partition function. The result (12.61) involves only two Gaussian approximations and therefore seems "less approximate" than the semiclassical expression in which all fluctuation modes are treated in the Gaussian approximation. Generally one finds from numerical studies that the centroid approximation gives reasonable results for a symmetric or almost symmetric barrier and not too low temperatures [300]. It is seemingly superior to semiclassical methods for low barriers. Being nearly exact about T_0, the accuracy of the centroid method deteriorates as temperature is decreased. For very low T, the semiclassical bounce or periodic orbit method appears to be more accurate. With increasing asymmetry, the centroid approximation becomes less reliable. For a metastable potential that is not bounded from below, the centroid density diverges because of the unlimited potential drop beyond the barrier leading to infinite values for the rate. The generalization of the centroid method to the multidimensional case along the lines given in Section 8.1 and Chapters 14 – 16 is straightforward.

13. Multidimensional quantum rate theory

We now generalize the discussion to a system with $N + 1$ degrees of freedom and eventually take the limit $N \to \infty$. The multidimensional system is assumed to be described by the system-plus-reservoir model given in Eq. (3.11). We find it convenient to use the thermodynamic $\mathrm{Im}\, F$ method to study thermal quantum decay. Some of the results, however, shall be confirmed by the multidimensional generalization of the semiclassical method discussed in Section 12.3.

The thermodynamic rate calculation starts out from a functional representation for the partition function of the damped system. The exact formal path integral expression for the reduced partition function Z is written in Eq. (4.210), and the effective action for the motion in the upside-down potential is given in Eq. (4.62). The multi-dimensional upside-down potential landscape $-V(q, \boldsymbol{x})$, where $V(q, \boldsymbol{x})$ is defined in Eq. (3.13) [see also Eq. (13.1)], is sketched in Fig. 13.1.

For $N+1$ degrees of freedom, the potential $-V(q, \boldsymbol{x})$ is concave up in one direction, giving rise to periodic motion, but concave down in N directions and thus unstable with respect to small perturbations in these directions. The total energy

$$
\begin{aligned}
E &= V(q, \boldsymbol{x}) - \frac{M}{2}\dot{q}^2 - \frac{1}{2}\sum_{\alpha=1}^{N} m_\alpha \dot{x}_\alpha^2 \,, \\
V(q, \boldsymbol{x}) &= V(q) + \frac{1}{2}\sum_{\alpha=1}^{N} m_\alpha \omega_\alpha^2 \left(x_\alpha - \frac{c_\alpha}{m_\alpha \omega_\alpha^2}q \right)^2 .
\end{aligned}
\tag{13.1}
$$

is usually the only constant of motion. The periodic orbit is therefore characterized by N stability angular frequencies [301].

The tunneling contribution to the partition function is dominated by the periodic orbit with period $\hbar\beta$. At zero temperature, the total energy is zero and the period is infinity. The condition $V(q, \boldsymbol{x}) = 0$ defines the point P_{m} situated at the maximum of the parabolic hill of the upside-down potential ($q_{\mathrm{m}} = 0$, $\boldsymbol{x}_{\mathrm{m}} = 0$), and also the outer N-dimensional hyperbolic surface $\Sigma_{\mathrm{ex}}(q, \boldsymbol{x})$. The point P_{m} corresponds to the turning point q_1 in Eq. (12.17) for $E = 0$. The surface Σ_{ex} is the generalization of the turning point q_2 for N bath degrees of freedom. The surface separates the well and barrier region from the (classically accessible) outside region. The particle in the periodic orbit stays infinitely long in the (infinite) past at the point P_{m} before it starts moving towards the well of the upside-down potential. Eventually, it rushes through the well (which actually is a saddle point in the presence of the bath coordinates) and then bounces back from the surface Σ_{ex} at time zero. The particle hits the surface Σ_{ex} perpendicularly at a particular point denoted by P_{ex}. On the way back, it rushes again through the well on the same way, and finally comes to rest at time infinity at the starting point P_{m}. The orbit of the particle near the surface Σ_{ex} is parabolic and therefore is very sensitive to a small perturbation. This reflects the fact that the periodic orbit is unstable. If the particle does not approach the surface Σ_{ex} perpendicularly, it would continue near the surface Σ_{ex} on a parabolic path and therefore

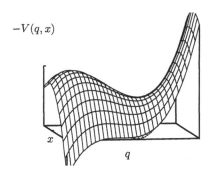
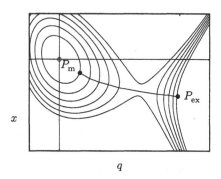

Figure 13.1: Potential landscape of the upside-down potential $-V(q,x)$ for a single bath degree of freedom (left). The contour lines of equal potential height are shown for the potential $-V(q,x)$ (right). The contour lines around the top (symbol $\circ$) are elliptic, while those separating the barrier region from the outside region are hyperbolic. The curved periodic orbit for a particular period with turning points P_{m} on Σ_{m} and P_{ex} on Σ_{ex} is also sketched.

would never return to the starting point. The envelope surface of the parabolic paths is the caustic which touches the surface Σ_{ex} at the point P_{ex}. The particular role of the unstable periodic orbit in multi-dimensional quantum decay has been stressed by Banks et al. already in 1973 [302]. The periodic orbit is often referred to as the most probable escape path (MPEP).

At finite temperatures, the total energy of the periodic orbit is tuned to a positive value for which the period is $\hbar\beta$. Now, the point P_{m} is dissolved into a N-dimensional closed surface Σ_{m} around the point $q = 0$, $\boldsymbol{x} = 0$. The inner surface Σ_{m} as well as the outer surface Σ_{ex} are defined by the condition $E - V(q,\boldsymbol{x}) = 0$. For $E \ll V_{\mathrm{b}}$, the surface Σ_{m} is elliptic. In the periodic orbit or MPEP, the particle leaves the surface Σ_{m} perpendicularly at a particular point (denoted again by P_{m}) at time $-\hbar\beta/2$, rushes through the well of the upside-down potential and hits the outer surface Σ_{ex} perpendicularly at time zero at a particular point P_{ex}. Afterwards, it returns to the starting point on Σ_{m} at time $\hbar\beta/2$. The periodic orbit meets the same points in configuration space on the way forth and back.

In Fig. 13.1, we have sketched the potential landscape (left) and the contour lines of equal potential height (right). We have also drawn the periodic orbit for a particular value of E with turning points on the surfaces Σ_{m} and Σ_{ex}, respectively.

The projection of the periodic orbit or MPEP onto the q-axis is the bounce path $q_{\mathrm{B}}(\tau)$. This path is a stationary point of the action $S_{\mathrm{eff}}^{\mathrm{E}}[q]$ defined in Eq. (4.62) and satisfies the equation of motion

$$-M\,\ddot{q}_{\mathrm{B}}(\tau) + \frac{\partial V[q_{\mathrm{B}}(\tau)]}{\partial q_{\mathrm{B}}(\tau)} + \int_0^{\hbar\beta} d\tau'\, k(\tau - \tau')q_{\mathrm{B}}(\tau') = 0\,. \qquad (13.2)$$

Because of the elimination of the reservoir coordinates $\boldsymbol{x}$, the equation of motion for $q_{\mathrm{B}}(\tau)$ is time-nonlocal. Hence there is *no* constant of motion.

The quantum fluctuations about the bounce $q_{\mathrm{B}}(\tau)$ shall be considered, apart from a zero mode, in Gaussian approximation. Only in a narrow temperature range about the crossover temperature at which the bounce coalesces to the trivial constant path $\bar{q}(\tau) = q_{\mathrm{b}}$ another non-Gaussian fluctuation mode exists. The appropriate treatment shall be discussed in Chapter 16.

For a stable potential, the stationary points are minima of the action. In contrast, the bounce path available in a metastable potential is a saddle point, i.e., there is a direction in function space along which the action gets smaller. This feature reflects the circumstance that there is a fluctuation mode around the bounce with a negative eigenvalue: the nodeless amplitude fluctuation mode. This phenomenon points to the fact that the potential has another local minimum. In order to get a meaningful expression for the bounce contribution to the barrier partition function, the mode with the negative eigenvalue must be treated by analytical continuation. By deforming the respective integration contour from the stable to the unstable situation, the barrier partition function acquires an exponentially small imaginary part which can be assigned to the free energy as in Eq. (12.50).

Since the Im F method relies on equilibrium thermodynamics, it cannot describe effects which are related to nonequilibrium occupation of the states in the well. Such deviations may occur for extremely weak friction where the condition (11.45) is violated and the internal relaxation times for restoring equilibrium are of the order of the escape times or larger. The Im F method gives the proper classical limit of the escape in the damping regime (11.45). In the quantum regime, nonequilibrium effects are weaker. Thus, the Im F method is even valid down to considerably weaker damping. As we shall see, the rate formula for a damped system obtained by the Im F method approaches at high temperatures Kramers' expression (11.16) with Eq. (11.17) and Eq. (11.12). Moreover, the effects of frequency-dependent damping [303] as well as quantum corrections to thermally activated decay calculated by dynamical methods [304] are reproduced *quantitatively* by the thermodynamic method.

Among the first who introduced dissipation in the problem of quantum mechanical decay of a metastable ground state were Caldeira and Leggett in 1983 [77]. They used the "bounce"-technique within the functional integral formulation [99]. Shortly later, the bounce method was found to be also an effective scheme for the calculation of dissipative quantum tunneling at finite temperatures [305] – [308]. A unified description of quantum statistical decay of a metastable state in the entire temperature range was given in Ref. [263].

It is worth noting that some confusion in the literature about whether dissipation increases or decreases the tunneling probability is related to the absence or presence of the potential counter term $\Delta V(q)$, Eq. (3.6) with (3.10), in the system-reservoir coupling term defined in Eq. (3.4). As long as one wishes to study solely the effects of friction and decoherence, and not the addition of potential renormalization effects induced by the coupling, one has to choose $\Delta V(q)$ according to (3.6). Then one finds, as we shall see, that damping *always* decreases the decay probability. This can be simply understood as follows. While the environmental coupling in the Hamiltonian (3.11) does not change the barrier height, the tunneling distance along the curved periodic orbit gets larger with damping (see Fig. 13.1). On the other hand, if we choose $\Delta V(q) = 0$, the barrier is effectively reduced by the coupling. The effect of weakening the barrier overcompensates the opposite curvature effect, so that after all for $\Delta V(q) = 0$ the decay rate is increased by the coupling.

The functional integral method which we shall present in the following four chapters provides a unified approach to weak metastability of a dissipative system in large regions of the parameter space.

14. Crossover from thermal to quantum decay

Under the condition of weak metastability, Eq. (10.3), the functional integral (4.210) for the partition function is dominated by the stationary points of the action. Besides the periodic bounce $q_B(\tau)$, there are two trivial constant solutions of Eq. (4.211). First, we have $\bar{q}(\tau) = q_b$, in which the particle sits in the minimum of the upside-down potential $-V(q)$ [i.e., on the barrier top of the original potential]. Secondly, there is the constant path $\bar{q}(\tau) = 0$, in which the particle sits at the local maximum of the upside-down potential in Fig. 10.2 [i.e., in the minimum of the original potential $V(q)$ in Fig. 10.1]. For temperatures below the crossover temperature T_0, it is always possible to find a periodic bounce solution of Eq. (13.2) with period $\hbar\beta$.

In the absence of damping, the shortest period possible is $\hbar\beta_0 = 2\pi/\omega_b$, for which the bounce is a small harmonic oscillation about q_b. The respective temperature $T_0 = 1/k_B\beta_{min} = \hbar\omega_b/2\pi k_B$ is in agreement with Goldanskii's conjecture, Eq. (12.59). At temperatures below T_0, the action S_B of the bounce $q_B(\tau)$ is generally smaller than the action $\hbar\beta V_b$ of the constant path $\bar{q}(\tau) = q_b$. Hence below T_0, the functional integral for the decay process is dominated by the bounce. From this we may infer that T_0 is the temperature where roughly the transition between thermally activated decay and decay by quantum tunneling occurs.

We now turn to the discussion of a damped particle. Again the crossover temperature is defined as the temperature where the bounce trajectory coalesces with the constant path sitting at the barrier top. To study this regime in some detail, it is useful to consider the normal modes of the global system about the barrier top.

14.1 Normal mode analysis at the barrier top

The analysis of the dynamics near the barrier top is simplified considerably through a normal mode analysis of the system-plus-reservoir Hamiltonian (3.11) [309]. The harmonic approximation of the potential $V(q)$ at the barrier point is

$$V^{(b)}(q) = V_b - \frac{M}{2}\omega_b^2 (q - q_b)^2 . \tag{14.1}$$

The full Hamiltonian may be decomposed into a harmonic part $H^{(b)}$ and a nonlinear potential contribution $V^{(nl)}(q)$,

$$H = H^{(b)} + V^{(nl)}(q) . \tag{14.2}$$

The quadratic Hamiltonian $H^{(b)}$ governing the dynamics near the barrier top may be diagonalized. Introducing mass weighted coordinates as in Eq. (4.229), we find in the diagonal representation

$$H^{(b)} = V_b + \frac{1}{2}\sum_{\alpha=0}^{N}\left(\dot{y}_\alpha^2 + \mu_\alpha^{(b)\,2} y_\alpha^2\right) . \tag{14.3}$$

Here, y_0 describes the unstable normal mode with squared (imaginary) frequency $\mu_0^{(b)\,2} \equiv -\omega_R^2 < 0$, and y_α for $\alpha \neq 0$ is a stable bath mode with real frequency $\mu_\alpha^{(b)}$. The normal mode frequencies are related to the original frequencies by the relations

$$\mu_0^{(b)\,2} = -\omega_b^2 \Big/ \left(1 + \frac{1}{M}\sum_{\alpha'=1}^{N}\frac{c_{\alpha'}^2}{[m_{\alpha'}\omega_{\alpha'}^2(\omega_{\alpha'}^2 - \mu_0^{(b)\,2})]}\right) \equiv -\omega_R^2 , \tag{14.4}$$

$$\mu_\alpha^{(b)\,2} = \omega_\alpha^2 \Big/ \left(1 + \frac{1}{M}\sum_{\alpha'=1}^{N}\frac{c_{\alpha'}^2}{[m_{\alpha'}\omega_{\alpha'}^2(\omega_{\alpha'}^2 - \mu_\alpha^{(b)\,2})]}\right) , \qquad \alpha \neq 0 . \tag{14.5}$$

The force constant matrix at the barrier top $U^{(b)}$ is related to the corresponding matrix in the well regime $U^{(0)}$, which is given in Eq. (4.231), by

$$U_{ij}^{(b)} = U_{ij}^{(0)} - \delta_{i0}\delta_{j0}(\omega_0^2 + \omega_b^2) . \tag{14.6}$$

Following the discussion given in Subsection 4.3.5, we find for the determinant of the force constant matrix $U^{(b)}$

$$\det U^{(b)} = \prod_{\alpha=0}^{N}\mu_\alpha^{(b)\,2} = -\omega_b^2 \prod_{\alpha=1}^{N}\omega_\alpha^2 . \tag{14.7}$$

The original system coordinate q is expressed in terms of the mass weighted normal coordinates $\{y_\alpha\}$ by the orthogonal transformation

$$q = q_b + \frac{1}{\sqrt{M}}u_{00}y_0 + \frac{1}{\sqrt{M}}\sum_{\alpha=1}^{N}u_{\alpha 0}y_\alpha , \tag{14.8}$$

with the matrix elements

$$u_{00} = \left(1 + \frac{1}{M}\sum_{\alpha'=1}^{N}\frac{c_{\alpha'}^2}{m_{\alpha'}(\omega_{\alpha'}^2 + \omega_R^2)^2}\right)^{-1/2}, \tag{14.9}$$

$$u_{\alpha 0} = \left(1 + \frac{1}{M}\sum_{\alpha'=1}^{N}\frac{c_{\alpha'}^2}{m_{\alpha'}(\omega_{\alpha'}^2 - \omega_\alpha^2)^2}\right)^{-1/2}. \tag{14.10}$$

The frequency ω_R of the unstable mode y_0 and the matrix element u_{00} may be equivalently expressed in terms of the spectral density $J(\omega)$ or in terms of the friction kernel $\widehat{\gamma}(z)$. We have from (14.4) upon using the definitions (3.30) and (3.34)

$$\omega_R^2 = \omega_b^2\left(1 + \frac{2}{\pi M}\int_0^\infty d\omega\, \frac{J(\omega)}{\omega}\frac{1}{(\omega^2 + \omega_R^2)}\right)^{-1} = \frac{\omega_b^2}{1 + \widehat{\gamma}(\omega_R)/\omega_R}. \tag{14.11}$$

Since $\widehat{\gamma}(\omega_R) > 0$, the effective barrier frequency ω_R is smaller than the bare barrier frequency ω_b. Similarly, the coefficient u_{00} is given by

$$u_{00}^2 = \left(1 + \frac{2}{\pi}\int_0^\infty d\omega\, \frac{\omega J(\omega)}{(\omega^2 + \omega_R^2)^2}\right)^{-1}. \tag{14.12}$$

The relation (14.11) may be rewritten as

$$\omega_R^2 + \omega_R\widehat{\gamma}(\omega_R) = \omega_b^2. \tag{14.13}$$

Equation (14.13) has a unique real positive solution for ω_R. The renormalized barrier frequency ω_R is the familiar Kramers-Grote-Hynes frequency appearing in the theory of non-Markovian rate processes [303].

14.2 Turnover theory for activated rate processes

The formulation of a unified theory for the classical attempt frequency bridging between the spatial-diffusion limited transmission factor $\varrho = \omega_R/\omega_b$ and the energy-diffusion limited expression (11.39) is known as the *Kramers turnover problem*. Near to the barrier top, the total energy E of the global system becomes the sum of the normal mode energies E_α, $E = \sum_{\alpha=0}^{N} E_\alpha$. Since the normal modes are not coupled, the probability to cross the barrier depends completely on the energy E_0 left in the unstable mode or escape coordinate y_0. When $E_0 < V_b$, the y_0-component of the trajectory goes through a turning point and the particle returns to the well. For $E_0 > V_b$, the particle leaves the well with probability 1. The particular role of the unstable mode has been utilized in Refs. [310, 311]. By calculating the energy left in the unstable mode, rather than in the physical coordinate q, inconsistencies in an earlier theory [312] could be resolved. Following Kramers [260], imagine that particles are constantly injected near to the bottom of the well. Then the system will approach a steady flux state. When the probability is normalized to a single particle in the well, the flux equals the classical escape rate k_{cl}. Let $p(E_0)dE_0$ be the probability per unit time to find the particle near the barrier top with energy E_0 in the unstable mode y_0. Since all particles with $E_0 > V_b$ escape, the rate is given by the expression

$$k_{cl} = \int_{V_b}^{\infty} dE_0 \, p(E_0) \, . \tag{14.14}$$

In the general case, $p(E_0)$ is a nonequilibrium probability.

Now assume as a first crude guess that the Boltzmann distribution remains valid even in the vicinity of the barrier top. It is convenient to write down the canonical distribution in the normal mode representation (14.3) at the barrier top, and the normalization factor in the normal mode representation (4.233) of the well region. Integrating over the coordinates of the stable modes, we find for the equilibrium distribution of the unstable mode coordinates

$$W_{eq}(\dot{y}_0, y_0) = \frac{\prod\limits_{\alpha=1}^{N} \int d\dot{y}_\alpha \, dy_\alpha \exp\left(-\tfrac{1}{2}\beta\left[\dot{y}_\alpha^2 + \mu_\alpha^{(b)\,2} y_\alpha^2\right]\right)}{\prod\limits_{\alpha=0}^{N} \int d\dot{\rho}_\alpha \, d\rho_\alpha \exp\left(-\tfrac{1}{2}\beta\left[\dot{\rho}_\alpha^2 + \mu_\alpha^{(0)\,2} \rho_\alpha^2\right]\right)} \tag{14.15}$$
$$\times \, \exp\left(-\beta\left[\tfrac{1}{2}\dot{y}_0^2 + V_b - \tfrac{1}{2}\omega_R^2 y_0^2\right]\right) \, .$$

Using the identities (4.241) and (14.7), we find with $E_0 = V_b + \tfrac{1}{2}\dot{y}_0^2 - \tfrac{1}{2}\omega_R^2 y_0^2$

$$p_{eq}(E_0) \equiv W_{eq}(\dot{y}_0, y_0) = \frac{\beta\omega_0}{2\pi}\frac{\omega_R}{\omega_b} e^{-\beta E_0} \, . \tag{14.16}$$

Upon replacing $p(E_0)$ in Eq. (14.14) by the equilibrium probability $p_{eq}(E_0)$, we find for the escape rate again the form (11.16)

$$k_{cl} = f_{cl} \, e^{-\beta V_b} \, , \tag{14.17}$$

where the classical attempt frequency is given by

$$f_{cl} = \varrho \frac{\omega_0}{2\pi} \, , \qquad \text{with} \qquad \varrho = \frac{\omega_R}{\omega_b} \, . \tag{14.18}$$

The expression (14.18) is the generalization of the Ohmic result (11.17) with (11.12) to general memory-friction described in terms of the Grote-Hynes frequency ω_R [303].

Consider now the probability distribution $p(E_0)$ in Eq. (14.14) more closely. Because of the steady flux out of the well, the distribution $p(E_0)$ may deviate from the equilibrium form $p_{eq}(E_0)$, Eq. (14.16). In general, $p(E_0)$ obeys the steady-state condition

$$p(E_0) = \int_0^{V_b} dE_0' \, P(E_0|E_0') \, p(E_0') \, , \tag{14.19}$$

in which $P(E_0|E_0')$ is the conditional probability that the particle leaving the barrier region with energy E_0' in the unstable mode y_0 returns to the barrier with an energy E_0. The conditional probability satisfies detailed balance,

$$P(E_0|E_0') \, e^{-\beta E_0'} \; = \; P(E_0'|E_0) \, e^{-\beta E_0} \,. \tag{14.20}$$

As a consequence of Eq. (14.20), the distribution $p(E_0)$ approaches $p_{eq}(E_0)$ for E_0 a few $k_B T$ below V_b.

The key quantity now is the energy loss from the unstable mode to the reservoir as the particle moves from the barrier to the well and then back to the barrier. The equations of motion for the stable mode α resulting from Eq. (14.2) with Eq. (14.3) reads

$$\ddot{y}_\alpha(t) + \mu_\alpha^{(b)\,2} y_\alpha(t) \; = \; g_\alpha F(t) \,, \qquad \alpha \neq 0 \,, \tag{14.21}$$

where

$$g_\alpha \; = \; u_{\alpha 0}/u_{00} \,. \tag{14.22}$$

The force pulse $F(t)$ is due to the anharmonic potential contribution $V^{(nl)}(q)$. The coupling of the unstable mode to the reservoir is weak under the condition

$$\varepsilon \equiv \sum_{\alpha=1}^{N} g_\alpha^2 \; = \; \frac{1}{u_{00}^2} - 1 \ll 1 \,. \tag{14.23}$$

For $\varepsilon \ll 1$, we may disregard the back-action of the stable modes on the force pulse so that the nonlinear force on the bath mode depends only on the time dependence of the trajectory of the unstable mode. Then we have

$$F(t) \; = \; -u_{00} \frac{1}{M} \left. \frac{\partial V^{(nl)}(q)}{\partial q} \right|_{q \, = \, q_b + u_{00} y_0(t)/\sqrt{M}} \,. \tag{14.24}$$

Equation (14.21) describes a forced harmonic oscillator and can be solved for arbitrary initial values of y_α and $\dot{y}_\alpha$ at $t = 0$. The energy transferred on average per cycle from the unstable mode to the stable mode y_α is

$$\begin{aligned}
\Delta E_{cycle} \; = \; & \frac{1}{2} g_\alpha^2 \int_{-\infty}^{\infty} dt \int_{-\infty}^{\infty} dt' \, \cos\left[\mu_\alpha^{(b)}(t - t')\right] F(t) F(t') \\
& + g_\alpha \int_{-\infty}^{\infty} dt \, \left[\dot{y}_\alpha(0) \cos\left(\mu_\alpha^{(b)} t\right) - y_\alpha \mu_\alpha^{(b)} \sin\left(\mu_\alpha^{(b)} t\right)\right] F(t) \,.
\end{aligned} \tag{14.25}$$

The time integrations are over the history of the trajectory of the unstable mode which starts at time $t = -\infty$ at the barrier, moves through the well, and returns to the barrier at time $t = +\infty$. To leading order, the equation of motion of the unstable mode is decoupled from the bath modes and takes the form

$$\ddot{y}_0(t) - \omega_R^2 y_0 + u_{00} \frac{1}{M} \left. \frac{\partial V^{(nl)}(q)}{\partial q} \right|_{q = q_b + u_{00} y_0(t)/\sqrt{M}} \; = \; 0 \,. \tag{14.26}$$

Next, we choose for the initial values the thermal distribution

$$\mu_\alpha^{(b)\,2}\langle y_\alpha^2(0)\rangle = \langle \dot{y}_\alpha^2(0)\rangle = k_B T \,, \quad \text{and} \quad \langle y_\alpha(0)\rangle = \langle \dot{y}_\alpha(0)\rangle = \langle y_\alpha(0)\dot{y}_\alpha(0)\rangle = 0 \,.$$

Then the energy transferred on average from the particle to the reservoir per cycle is

$$\Delta E \equiv \langle \Delta E_{\text{cycle}}\rangle = \frac{1}{2}\int_{-\infty}^{\infty} dt \int_{-\infty}^{\infty} dt' \,\Phi(t-t')F(t)F(t') \tag{14.27}$$

with the dissipation kernel

$$\Phi(t-t') = \sum_{\alpha=1}^{N} g_\alpha^2 \cos\left[\mu_\alpha^{(b)}(t-t')\right] \,. \tag{14.28}$$

The Laplace transform of the kernel $\Phi(t-t')$ may be written as

$$\widehat{\Phi}(z) = \sum_{\alpha=1}^{N} g_\alpha^2 \frac{z}{z^2 + \mu_\alpha^{(b)\,2}} = \frac{1}{u_{00}}\frac{z}{z^2 + z\widehat{\gamma}(z) - \omega_b^2} - \frac{z}{z^2 - \omega_R^2} \,. \tag{14.29}$$

The first expression is the direct Laplace transform of (14.28). The second form is expressed in terms of the spectral damping function $\widehat{\gamma}(z)$ defined in Eq. (3.34). The derivation of this form is less direct and involves the inversion of the force constant matrix $U^{(b)}$ [311].

The uncertainty of the mean square energy loss for the thermal distribution of the initial values $\dot{y}_\alpha(0)$ and $y_\alpha(0)$ is expressed by

$$\delta E^2 \equiv \langle \Delta E_{\text{cycle}}^2\rangle - \langle \Delta E_{\text{cycle}}\rangle^2 = 2k_B T\,\Delta E \,. \tag{14.30}$$

With Eq. (14.27) and Eq. (14.30), the conditional probability $P(E_0|E_0')$ is given by the Gaussian distribution

$$P(E_0|E_0') = \left(4\pi k_B T\Delta E\right)^{-1/2} \exp\left[-\left(E_0 - E_0' + \Delta E\right)^2 \Big/ 4k_B T\Delta E\right] \,. \tag{14.31}$$

Since the solution of Eq. (14.19) for $p(E_0)$ approaches $p_{\text{eq}}(E_0)$ for $V_b - E_0 \gg k_B T$, it is convenient to make the ansatz

$$p(E_0) = p_{\text{eq}}(E_0)\, e^{\beta(E_0 - V_b)/2}\, \phi[\,\beta(E_0 - V_b)] \,, \tag{14.32}$$

which transforms Eq. (14.19) into a Wiener-Hopf equation with a symmetric kernel that can be solved by standard techniques [312]. Insertion of the resulting solution for $p(E_0)$ into Eq. (14.14) gives the classical escape rate k_{cl} in the form (11.16) with the attempt frequency (11.17), in which the transmission factor ϱ is given by

$$\varrho = \frac{\omega_R}{\omega_b} \exp\left(\frac{1}{\pi}\int_{-\infty}^{\infty} \frac{dx}{1+x^2}\ln\left[1 - \exp\left(-\frac{1+x^2}{4}\frac{\Delta E}{k_B T}\right)\right]\right) \,. \tag{14.33}$$

For $\Delta E \gg k_B T$, the transmission factor approaches the Grote-Hynes value $\varrho = \omega_R/\omega_b$ exponentially fast. The corrections are of order $\exp(-\Delta E/4k_B T)$. Nonequilibrium effects in $p(E_0)$ are only important for ΔE of the order of $k_B T$ or smaller. In the opposite limit $\Delta E \ll k_B T$, we find from Eq. (14.33) linear dependence on ΔE,

$$\varrho = \frac{\omega_R}{\omega_b} \frac{\Delta E}{k_B T} . \qquad (14.34)$$

For weak damping, ΔE coincides with the energy loss given in Eq. (11.31) and the expression (14.34) agrees with the weak-damping result by Kramers given in Eq. (11.39).

A deeper analytical and numerical study of the energy loss formula (14.27) with (14.28) is given in Ref. [311]. A multidimensional generalization of the turnover theory for activated rate processes is reported in Ref. [313]. This concludes our discussion of the Kramers turnover problem.

14.3   The crossover temperature

At zero temperature, the amplitude of the bounce is maximal and the period is infinity. With increasing temperature, the amplitude and the period of the bounce decreases. Eventually, the bounce has shrinked to a small oscillation around the barrier point q_b. In the $N+1$-dimensional configuration space, this is a periodic harmonic oscillation of the unstable mode y_0 with frequency ω_R. The bounce $q_B(\tau)$ is the projection of the periodic orbit in $\{q, x\}$ space onto the original q-axis. There is no periodic solution in the well of the upside-down potential $-V(q, x)$ with a frequency smaller than ω_R. The temperature

$$T_0 \equiv \hbar\omega_R/2\pi k_B , \qquad (14.35)$$

where ω_R is the positive root of Eq. (14.13), is the crossover temperature [314, 306] at which the action of the bounce $q_B(\tau)$ coincides with the action of the constant solution $\bar{q} = q_b$. Roughly speaking, at temperature $T = T_0$, there is a crossover between thermal hopping and quantum tunneling. The crossover regime will be considered more closely in Chapter 16.

The relation (14.35) with (14.13) is valid for any linear damping mechanism. Since $\hat{\gamma}(\omega_R)$ in Eq. (14.13) is positive, the crossover temperature T_0 is always lowered by dissipation. Further, T_0 increases monotonously towards the value $\hbar\omega_b/2\pi k_B$ of the undamped case as the memory-friction relaxation time is increased while $\hat{\gamma}(\omega = 0)$ is held fixed. In the particular case of Ohmic damping, $\hat{\gamma}(z) = \gamma$, and a cubic potential of the form (11.40), for which $\omega_b = \omega_0$, we obtain from Eq. (14.13)

$$\omega_R = \omega_b \left(\sqrt{1 + \alpha^2} - \alpha \right) , \qquad (14.36)$$

where $\alpha = \gamma/2\omega_0 = \gamma/2\omega_b$ is the usual dimensionless damping parameter. From this we find the monotonous behaviour

$$T_0 = (\hbar\omega_b/2\pi k_B) \left(\sqrt{1 + \alpha^2} - \alpha \right) . \qquad (14.37)$$

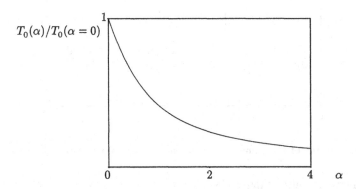

Figure 14.1: The normalized crossover temperature $T_0(\alpha)/T_0(\alpha = 0)$ is shown for Ohmic damping as a function of the parameter $\alpha = \gamma/2\omega_{\mathrm{b}}$.

In the strong damping limit, Eq. (14.37) simplifies to the form

$$T_0 = \frac{\hbar}{2\pi k_{\mathrm{B}}} \frac{\omega_{\mathrm{b}}^2}{\gamma} . \tag{14.38}$$

The monotonous decrease of the crossover temperature with increasing damping strength is sketched in Fig. 14.1.

With the effective barrier frequency ω_{R}, the general relation (12.57) between the rate and the imaginary part of the free energy takes the form

$$k = -\kappa \, (2/\hbar) \operatorname{Im} F , \tag{14.39}$$

where

$$\begin{aligned}
\kappa &= 1 ; & \hbar\beta\omega_{\mathrm{R}} &\geq 2\pi , \\
\kappa &= \hbar\beta\omega_{\mathrm{R}}/2\pi ; & \hbar\beta\omega_{\mathrm{R}} &< 2\pi .
\end{aligned} \tag{14.40}$$

Additional reasons for the validity of the expressions (14.39) with (14.40) are presented in Section 15.4. The unified formula (14.39) with (14.40) has been also derived from a study of the reactive flux through a parabolic barrier [299].

The various temperature regions in quantum statistical decay are shown in Fig. 14.2. Below T_0, the functional integral is dominated by the bounce trajectory $q_{\mathrm{B}}(\tau)$, and the predominant escape mechanism is quantum mechanical tunneling. Above T_0, the functional integral is dominated by the constant trajectory $\bar{q}(\tau) = q_{\mathrm{b}}$ [resulting in the Arrhenius factor $\mathrm{e}^{-\beta V_{\mathrm{b}}}$], and the relevant decay process is thermally activated escape over the barrier. The fluctuation modes about $\bar{q}(\tau) = q_{\mathrm{b}}$ yield quantum corrections in the prefactor of the rate formula thereby enhancing the rate above the classical expression. They become increasingly important as the temperature is lowered. Special care is required in the evaluation of the functional integral in a narrow crossover region about T_0 (cf. Chapter 16). We have now prepared the ground to work out the rate formula in the particular temperature regimes.

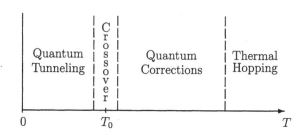

Figure 14.2: Sketch of the dominant escape mechanism as a function of temperature.

15. Thermally activated decay

15.1   Rate formula above the crossover regime

Consider the decay of the metastable state at temperatures fairly above T_0 so that thermal activation prevails. The rate may be written in the form

$$k = f_{cl} c_{qm} \, e^{-V_b/k_B T} , \qquad (15.1)$$

where f_{cl} is the attempt frequency in the classical limit and c_{qm} is a (dimensionless) factor describing enhancement of the classical rate by quantum effects. We now compute f_{cl} and c_{qm} for arbitrary frequency-dependent damping.

We proceed by splitting the partition function (4.210) for the damped system into the contributions arising from the Gaussian fluctuations about the constant paths $\bar{q}(\tau) = 0$ and $\bar{q}(\tau) = q_b$ and write

$$Z = Z_0 + Z_b = Z_0 \left(1 + Z_b/Z_0 \right) . \qquad (15.2)$$

Now, we recall the treatment of the fluctuation modes given for the partition function in Section 4.3. A path with $\hbar\beta$-periodic fluctuations about the constant path $\bar{q}(\tau) = 0$ may be written as

$$q(\tau) \longrightarrow q^{(0)}(\tau) = \sum_{n=-\infty}^{\infty} Y_n \, e^{i\nu_n \tau} , \qquad (15.3)$$

where again $\nu_n = 2\pi n/\hbar\beta$. Inserting (15.3) into the expression (4.62) for the "dissipative" action and disregarding terms of third and higher order in Y_n, we find[1]

$$S[q^{(0)}(\cdot)] = \frac{1}{2} M\hbar\beta \sum_{n=-\infty}^{\infty} \Lambda_n^{(0)} Y_n Y_{-n} , \qquad (15.4)$$

where the $\{\Lambda_n^{(0)}\}$ are the eigenvalues of the fluctuation operator $\mathbf{\Lambda}_0$ defined in Eq. (4.219). As discussed already in Subsection 4.3.3 [cf. Eq. (4.220)], we have

---

[1] Throughout the following we shall use the simpler notation $S[q]$ for $S_{eff}^E[q]$.

$$\Lambda_n^{(0)} = \nu_n^2 + \omega_0^2 + |\nu_n| \widehat{\gamma}(|\nu_n|) . \tag{15.5}$$

Similarly, the path $q^{(b)}(\tau)$ with $\hbar\beta$-periodic fluctuations about q_b

$$q(\tau) \longrightarrow q^{(b)}(\tau) = q_b + \sum_{n=-\infty}^{\infty} X_n \, e^{i\nu_n \tau} \tag{15.6}$$

has the second order action

$$S[q^{(b)}(\cdot)] = \hbar\beta V_b + \frac{1}{2} M \hbar\beta \sum_{n=-\infty}^{\infty} \Lambda_n^{(b)} X_n X_{-n} , \tag{15.7}$$

where

$$\Lambda_n^{(b)} = \nu_n^2 - \omega_b^2 + |\nu_n| \widehat{\gamma}(|\nu_n|) . \tag{15.8}$$

All modes $\{Y_n\}$ and $\{X_n\}$ except of X_0 have positive eigenvalues. The contributions of these modes to the partition function may be calculated in the usual way by carrying out the Gaussian integrals over the amplitude sets $\{Y_n\}$ and $\{X_n\}$ using the functional measure (4.227). The mode X_0 needs special treatment. Since the corresponding eigenvalue $\Lambda_0^{(b)} = -\omega_b^2$ is negative, the integral over X_0 is divergent. The divergence is related to the fact that the action for the constant path $\bar{q}(\tau) = q_b$ is a saddle point in function space with the *unstable* direction along X_0. In fact, this should not be a surprise since we are trying to evaluate the free energy of an unstable system. Langer [288] was the first who explained that the functional integral can still be defined by deforming the integration contour of the variable X_0 into the upper half of the complex plane along the direction of steepest descent. This leads to a positive imaginary part for the quantity Z_b. Clearly, it would be inconsistent to keep a real contribution to Z_b/Z_0 in the expression (15.2) since the ratio Z_b/Z_0 contains the exponentially small factor $e^{-\beta V_b}$, while non-Gaussian corrections to Z_0 are disregarded. On the other hand, it is consistent to keep the exponentially small imaginary part of Z_b as it is the leading imaginary part. The resulting imaginary part of the free energy

$$\mathrm{Im}\, F = -(\beta Z_0)^{-1} \mathrm{Im}\, Z_b \tag{15.9}$$

is then given by

$$\mathrm{Im}\, F = -(1/2\beta)\,(D_0/|D_b|)^{1/2}\, e^{-\beta V_b} , \tag{15.10}$$

where D_0 and D_b are the determinants related to the second order action functionals (15.4) and (15.7) (cf. Subsections 4.3.2 and 4.3.3),

$$D_0 = \omega_0^2 \prod_{n=1}^{\infty} \left(\Lambda_n^{(0)}\right)^2 , \qquad D_b = -\omega_b^2 \prod_{n=1}^{\infty} \left(\Lambda_n^{(b)}\right)^2 . \tag{15.11}$$

As in Eq. (12.55) or Eq. (12.60), the factor $1/2$ in Eq. (15.10) is related to the fact that the integral over X_0 along the direction of steepest descent extends only over one wing of the Gaussian function $\exp(-M\beta|\Lambda_0^{(b)}|X_0^2/2)$.

Using Eqs. (12.56), (15.10) and (14.35), we obtain the rate above T_0 in the form

$$
\begin{aligned}
k &= \frac{1}{\hbar\beta_0}\left(\frac{D_0}{|D_{\mathrm{b}}|}\right)^{1/2} e^{-\beta V_{\mathrm{b}}} = \frac{1}{\hbar\beta_0}\frac{\omega_0}{\omega_{\mathrm{b}}}\prod_{n=1}^{\infty}\frac{\Lambda_n^{(0)}}{\Lambda_n^{(\mathrm{b})}}\, e^{-\beta V_{\mathrm{b}}} \\
&= \frac{\omega_0}{2\pi}\frac{\omega_{\mathrm{R}}}{\omega_{\mathrm{b}}}\prod_{n=1}^{\infty}\frac{\nu_n^2+\omega_0^2+\nu_n\widehat{\gamma}(\nu_n)}{\nu_n^2-\omega_{\mathrm{b}}^2+\nu_n\widehat{\gamma}(\nu_n)}\, e^{-\beta V_{\mathrm{b}}} \,.
\end{aligned}
\tag{15.12}
$$

Upon comparing Eq. (15.12) with Eq. (15.1), we find for the classical attempt frequency the previous form (14.18), and for the quantum mechanical enhancement factor the infinite product form [306]

$$
c_{\mathrm{qm}} = \prod_{n=1}^{\infty}\frac{\Lambda_n^{(0)}}{\Lambda_n^{(\mathrm{b})}} = \prod_{n=1}^{\infty}\frac{\nu_n^2+\omega_0^2+\nu_n\widehat{\gamma}(\nu_n)}{\nu_n^2-\omega_{\mathrm{b}}^2+\nu_n\widehat{\gamma}(\nu_n)}\,.
\tag{15.13}
$$

The resulting form (15.13) for c_{qm} is substantiated by an independent dynamical approach [304]. In the classical limit $T \gg T_0$, the factor c_{qm} approaches unity, and hence the rate expression (15.12) leads to the proper classical form [cf. Eq. (14.17) with (14.18)]

$$
k_{\mathrm{cl}} = \frac{\omega_0}{2\pi}\frac{\omega_{\mathrm{R}}}{\omega_{\mathrm{b}}}\, e^{-\beta V_{\mathrm{b}}} \,.
\tag{15.14}
$$

Compared with the TST formula (11.2), the classical rate is reduced by the transmission factor $\varrho = \omega_{\mathrm{R}}/\omega_{\mathrm{b}}$, which describes the effect of *recrossings* of the barrier top by the particle in the moderate to large damping region. The result agrees with the rate expression obtained from the extension of Kramers' approach to the case of memory friction [303].[2] The expression (15.14) also agrees with the classical rate expression obtained below in Section 15.4 from multidimensional transition state theory. There, however, the interpretation of the reduction factor ϱ will be different.

The effect of memory friction on the rate is reflected by the renormalization of the barrier frequency. Interestingly, the renormalized frequency ω_{R} is the same which enters the definition of the crossover temperature T_0. For frequency-independent damping, $\widehat{\gamma}(z) = \gamma$, the attempt frequency (14.18) simplifies to Kramers' celebrated result, Eq. (11.17) with Eq. (11.12).

In summary, the crucial assumptions underlying the generalized Kramers formula (15.14) are as follows [264]:

i. The coupling to the environment is so strong that the rate is controlled by the diffusive dynamics near to the barrier top.

ii. The effective potential has a parabolic shape around the barrier top.

iii. The stochastic dynamics is represented by the *linear* Langevin equation (2.4) with a Gaussian random force with classical power spectrum (2.8).

---

[2]In the chemical literature, the factor $\varrho = \omega_{\mathrm{R}}/\omega_{\mathrm{b}}$, where ω_{R} is determined by Eq. (14.13), is known as the Grote-Hynes correction to classical TST.

15.2   Quantum corrections in the preexponential factor

At temperatures well below the classical regime but still above the crossover temperature, the rate is enhanced by quantum corrections,

$$k = c_{\text{qm}} k_{\text{cl}} , \qquad (15.15)$$

where c_{qm} is given in Eq. (15.13).

The leading quantum correction at high T is found upon rewriting the expression (15.13) as an exponential of a sum of logarithms and expanding each logarithm in powers of $1/\hbar\beta$. One then finds in leading order in $1/\hbar\beta$ a surprisingly simple expression which is independent of damping for arbitray $\tilde{\gamma}(\omega)$ [314],

$$c_{\text{qm}} = \exp\left(\frac{\hbar^2 (\omega_0^2 + \omega_{\text{b}}^2)}{4\pi^2} \sum_{n=1}^{\infty} \frac{1}{n^2}\right) = \exp\left(\frac{\hbar^2 (\omega_0^2 + \omega_{\text{b}}^2)}{24 (k_{\text{B}}T)^2}\right) . \qquad (15.16)$$

We see that the escape is enhanced by two different quantum effects. First, quantum fluctuations increase the mean energy in the well. Second, when a particle is thermally excited almost to the barrier top, quantum fluctuations allow for tunneling through the remaining small barrier. Both effects lead to an effective reduction of the barrier.

For frequency-independent damping, the infinite product in Eq. (15.13) can be expressed exactly in terms of gamma functions [304]

$$c_{\text{qm}} = \frac{\Gamma(1 + \lambda_b^+/\nu)\,\Gamma(1 + \lambda_b^-/\nu)}{\Gamma(1 + \lambda_0^+/\nu)\,\Gamma(1 + \lambda_0^-/\nu)} , \qquad (15.17)$$

where $\nu = \nu_1 = 2\pi k_{\text{B}}T/\hbar$ is the smallest Matsubara frequency, and where

$$\lambda_{\text{b}}^{\pm} = \frac{\gamma}{2} \pm \left(\frac{\gamma^2}{4} + \omega_{\text{b}}^2\right)^{1/2} , \qquad \lambda_0^{\pm} = \frac{\gamma}{2} \pm \left(\frac{\gamma^2}{4} - \omega_0^2\right)^{1/2} . \qquad (15.18)$$

Plots of the enhancement factor c_{qm} versus temperature are shown in Fig. 15.1.

At strong friction $\gamma \gg \omega_0, \omega_{\text{b}}$ and $T \gg T_0$, Eq. (15.17) simplifies to

$$c_{\text{qm}} = \exp\left\{\frac{\hbar\beta(\omega_{\text{b}}^2 + \omega_0^2)}{2\pi\gamma}\left[\psi\left(1 + \frac{\hbar\beta\gamma}{2\pi}\right) - \psi(1)\right]\right\} , \qquad (15.19)$$

where $\psi(z)$ is the digamma function. This form reduces in the high temperature regime $T \gg \hbar\gamma/2\pi k_{\text{B}}$ to the leading quantum correction given above in Eq. (15.16).

In the temperature regime

$$T_0 \ll T \ll \hbar\gamma/2\pi k_{\text{B}} , \qquad (15.20)$$

where T_0 is given in Eq. (14.38), the expression (15.17) takes the form

$$c_{\text{qm}} = \exp\left\{\frac{\hbar\beta(\omega_{\text{b}}^2 + \omega_0^2)}{2\pi\gamma} \ln\left(\frac{\hbar\beta\gamma}{2\pi}\right)\right\} = \left(\frac{\gamma^2 T_0}{\omega_{\text{b}}^2 T}\right)^{(1+\omega_0^2/\omega_{\text{b}}^2)T_0/T} . \qquad (15.21)$$

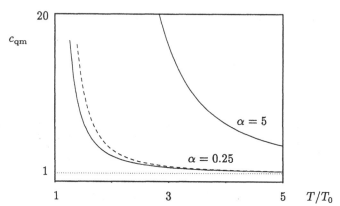

Figure 15.1: The quantum correction factor c_{qm} is shown as a function of the scaled temperature T/T_0 for a system with $\omega_0 = \omega_{\mathrm{b}}$ and frequency-independent damping for $\alpha = 0.25$ and $\alpha = 5$. The dashed curve is a plot of the formula (15.16) for $\alpha = 0.25$.

This factor can enhance the rate quite significantly, even well above the crossover temperature. For instance, for $T = 4T_0$ and $\omega_0 = \omega_{\mathrm{b}}$, one gets $c_{\mathrm{qm}} = \gamma/2\omega_{\mathrm{b}} \gg 1$.

As T approaches the crossover temperature T_0 from above, the eigenvalue $\Lambda_1^{(\mathrm{b})}$ goes to zero. As a result, the quantum correction factor c_{qm} becomes singular. This indicates that the Gaussian treatment of the fluctuation modes $Y_{\pm 1}$ is not a valid approximation in this regime. The proper treatment of this mode in the crossover regime is carried out in Chapter 16.

15.3   The quantum Smoluchowski equation approach

At strong friction $\gamma \gg \omega_0$, the evolution of the probability density in coordinate space $p(q, t)$ of a classical Brownian particle is described by the Smoluchowski diffusion equation (11.19). In the temperature regime (15.20), the dynamics of the Brownian particle $p(q, t)$ can still be described by a diffusion equation of the form (11.19), but the diffusion coefficient D is above the classical value $D_{\mathrm{cl}} = k_{\mathrm{B}}T$ because of quantum fluctuations. When the drift potential is harmonic, $V(q) = \frac{1}{2}M\omega_0^2 q^2$, it is quite natural to assume that the diffusion coefficient is increased in accordance with the enhancement of the coordinate spread by quantum fluctuations,

$$D_{\mathrm{cl}} = k_{\mathrm{B}}T = M\omega_0^2 \langle q^2 \rangle_{\mathrm{cl}} \longrightarrow \widetilde{D}_{\mathrm{qm}} = M\omega_0^2 [\langle q^2 \rangle_{\mathrm{cl}} + \lambda], \qquad (15.22)$$

where $\lambda = \langle q^2 \rangle - \langle q^2 \rangle_{\mathrm{cl}}$ is the quantum mechanical part of the coordinate spread at strong friction. Explicit expressions for λ are given in Eqs. (6.103) and (6.104).

For a nonlinear potential $V(q)$, it is now an obvious thing to generalize the expression (15.22) to a position-dependent diffusion coefficient $D_{\mathrm{qm}}(q) = \widetilde{D}_{\mathrm{qm}}(q)$, where

$\tilde{D}_{\rm qm}(q) = k_{\rm B}T + \lambda V''(q)$. We then arrive at the Smoluchowski diffusion equation

$$\frac{\partial p(q,t)}{\partial t} = \frac{1}{M\gamma}\frac{\partial}{\partial q}\hat{L}_{\rm qm}(q)p(q,t) \qquad (15.23)$$

with the position-dependent quantum Smoluchowski flux operator

$$\hat{L}_{\rm qm}(q) = \frac{\partial V(q)}{\partial q} + \frac{\partial}{\partial q}D_{\rm qm}(q) . \qquad (15.24)$$

Here, the quantum Smoluchowski equation (QSE) has been deduced heuristically. The quantum correction in the diffusion coefficient $D_{\rm qm}(q)$ has been substantiated by a path integral evaluation of the propagating function J introduced in Section 5.5 [315], and by different reasoning in Ref. [316].

Unfortunately, the QSE with the diffusion coefficient $D_{\rm qm}(q) = \tilde{D}_{\rm qm}(q)$ has a weakness which shows up in order λ^2 and in higher orders of λ. Namely, for a periodic potential $V(q) = V(q+nL)$ the diffusion coefficient $\tilde{D}_{\rm qm}(q)$ invalidates corresponding periodicity of the equilibrium potential $\psi(q) = \int_0^q dx\, V'(x)/\tilde{D}_{\rm qm}(x)$ [cf. Eq. (11.25)] in order λ^2 and in higher orders of λ. As a result, for a periodic drift potential $V(q)$, there would be a nonzero stationary equilibrium current, which would violate the second law of thermodynamics in order λ^2. The appropriate modification of the diffusion coefficient, which coincides with $\tilde{D}_{\rm qm}(q)$ in first order in λ, is [317]

$$D_{\rm qm}(q) = k_{\rm B}T/[1 - \lambda V''(q)/k_{\rm B}T] . \qquad (15.25)$$

With this form, the QSE does not yield a stationary current in any orders of λ for a periodic drift potential, so that the QSE is fully reconciled with the second law of thermodynamics. The equilibrium position distribution resulting from the QSE is

$$p_{\rm eq}(q) = \frac{N_0}{D_{\rm qm}(q)}\,{\rm e}^{-\psi(q)} , \qquad (15.26)$$

where the equilibrium thermodynamic potential $\psi(q)$ is given by

$$\psi(q) = \beta\left\{V(q) - \tfrac{1}{2}\lambda\beta[V'(q)]^2\right\} . \qquad (15.27)$$

For a harmonic well, the potential $\psi(q)$ agrees in order λ with the exponent $\frac{1}{2}q^2/\langle q^2\rangle$ of the equilibrium position distribution $< q|\hat{\rho}_\beta|q >$ given in Eq. (6.116). Since $\langle q^2\rangle$ gets smaller as γ is increased, the equilibrium distribution for a particle in a well is squeezed by friction quite substantially.

We are now ready to study the escape of the quantum Brownian particle from a metastable well following the lines given in Section 11.3. Upon using the expressions (15.27) and (15.25), we immediately get from the expression (11.28) the result

$$k = \frac{\sqrt{V''(0)|V''(q_{\rm b})|}}{2\pi M\gamma}\,{\rm e}^{-\beta V_{\rm b}}\,{\rm e}^{\beta\lambda\{V''(0)+|V''(q_{\rm b})|\}/2} . \qquad (15.28)$$

Here, the last exponential factor describes enhancement of the rate by quantum fluctuations above the classical rate in the Smoluchowski regime. With the explicit forms (6.100), (6.103) or (6.104) applicable in the respective temperature regimes, the enhancement factor in the expression (15.28) coincides with the earlier expressions (15.16), (15.19) and (15.21), respectively, which have been calculated from the fluctuation determinant of the path integral in the free energy method.

Finally, we mention that, in difference to the original work [315, 317], the QSE (15.23) with the flux operator (15.24) and the diffusion coefficient (15.25) has no quantum correction in the drift potential. If we would add the term $\Delta V_{qm} = \frac{1}{2}\lambda V''(q)$ to the drift potential, as done in Refs. [315, 317], the quantum mechanical enhancement factor in Eq. (15.28) would be doubled. The absence of quantum fluctuations in the QSE (15.23) with (15.24) has been also noticed in Refs. [265] and [316].

15.4 Multidimensional quantum transition state theory

With the above considerations, the escape problem for temperatures above T_0 is basically solved. However, the presence of the factor κ in the formula (12.57) has invoked some criticism [318, 262]. Therefore, we find it useful to confirm the results of the Im F method by the periodic orbit approach.

An independent unique approach to the quantum statistical decay problem is based on the multidimensional quantum transition state theory (MQTST) put forward by Miller in Ref. [276]. In the semiclassical limit, the "thermal propagator" $< q\,|\,e^{-\hat{H}\tau/\hbar}\,|\,q' >$ is dominated by *periodic orbit* trajectories in the upside-down potential landscape $-V(q, \boldsymbol{x})$ of the $(N+1)$-dimensional configuration space. The periodic orbit is a solution of the equations of motion (4.34) analytically continued to imaginary time with total energy conserved, Eq. (13.1). The qualitative behaviour of the periodic orbit has been discussed already in Chapter 13.

Following Gutzwiller [301], we now introduce a coordinate z_0 which measures distance along the curved periodic trajectory and N other coordinates $\{z_j\}$, $j = 1, \cdots, N$, being locally orthogonal displacements away from it. We shall call z_0 the *escape coordinate*, and $z_0^{(PO)}(\tau)$ the periodic orbit trajectory evoluting along the escape coordinate. It is convenient to choose the phase of the periodic orbit such that

$$z_0^{(PO)}(\pm\hbar\beta/2) = z_0^{(m)} ; \qquad z_0^{(PO)}(0) = z_0^{(ex)} , \qquad (15.29)$$

where $z_0^{(m)}$ and $z_0^{(ex)}$ are the turning points on the surfaces Σ_m and Σ_{ex} introduced in Chapter 13, respectively.

We choose a flat dividing surface Σ_b which delimits the domain of attraction about the metastable minimum from the exterior region. The escape path crosses the dividing plane Σ_b perpendicularly at $z_0^{(b)}$, where $z_0^{(b)}$ is the point on the escape coordinate for which $V(q, \boldsymbol{x})$ is a maximum. The transition state approximation implies that the escaping particle crosses the dividing surface Σ_b only once. With this choice of coordinates, we thus have similar to Eq. (12.10)

$$\hat{F}\hat{P} \xrightarrow{\text{TST}} \delta\left(\hat{z}_0 - z_0^{(b)}\right) \frac{\hat{p}_0}{M} \Theta(\hat{p}_0) = \frac{1}{2M} \delta\left(\hat{z}_0 - z_0^{(b)}\right) \left(|\hat{p}_0| + \hat{p}_0\right). \tag{15.30}$$

Again, the second term in the bracket does not contribute. Eq. (12.11) with (12.12) is generalized to

$$k = \frac{1}{Z_0} \frac{1}{2\pi\hbar} \int_0^\infty dE\, p(E)\, e^{-\beta E}, \tag{15.31}$$

where Z_0 is the partition function of the global system, and

$$p(E) = \lim_{\epsilon \to 0^+} \text{Im} \left\{ \int_0^\infty d\tau\, e^{(E+i\epsilon)\tau/\hbar} \right. \\ \left. \times \int d\boldsymbol{z}\, \delta\left(z_0 - z_0^{(b)}\right) |\dot{z}_0| < \boldsymbol{z}|\, e^{-\hat{H}\tau/\hbar} |\boldsymbol{z}> \right\}. \tag{15.32}$$

The quantity $p(E)$ represents the inclusive transmission probability at the total energy E. As the quantum statistical decay rate (12.11) is essentially the Boltzmann-averaged transmission probability, it is clear that $p(E)$ should include all possible partitions of the total energy E into the energy left in the escape coordinate and the individual energies in the transverse degrees of freedom. This, indeed, is the case as we shall see shortly [cf. Eq. (15.40) below].

According to Gutzwiller's semiclassical analysis [301], a single periodic orbit in the upside-down potential gives the contribution

$$p_1(E) = \text{Im} \frac{-\mathcal{J}_1}{\sqrt{2\pi\hbar}} \int_0^\infty d\tau\, \left| \frac{\partial^2 S_{\text{cl}}}{\partial \tau^2} \right|^{1/2} \prod_{\alpha=1}^N \frac{1}{2\sinh[\tau\mu_\alpha(z)/2]}\, e^{-[S_{\text{cl}}(\tau) - E\tau]/\hbar}. \tag{15.33}$$

The Euclidean action for a single periodic orbit $z_0^{(\text{PO})}(\tau')$ with period $\bar{\tau}$ is

$$S_{\text{cl}}(\bar{\tau}) = W(E) + E\bar{\tau}, \tag{15.34}$$

where E is the conserved total energy (13.1) of this particular path, and

$$W(E) \equiv \int_0^{\bar{\tau}} d\tau' \left(M\dot{q}^2(\tau') + \sum_{\alpha=1}^N m_\alpha \dot{x}_\alpha^2(\tau') \right) \tag{15.35}$$

is the abbreviated action for the global system. The factor $\mathcal{J}_1$ keeps track of the phase factors picked up by the periodic orbit at the two conjugate points. Finally, the N-fold product in Eq. (15.33) results from the Gaussian fluctuations perpendicular to the reaction coordinate. Here, the set $\{\mu_\alpha(z)\}$ ($\alpha = 1, 2, \cdots, N$) represents the dynamical stability (angular) frequencies of the periodic orbit. Computation of the integral over τ in Eq. (15.33) by steepest descent, as in Section 12.3, gives

$$p_1(E) = -\mathcal{J}_1 \left(\prod_{\alpha=1}^N \frac{1}{2\sinh[\bar{\tau}(E)\mu_\alpha(E)/2]} \right) e^{-W(E)/\hbar}. \tag{15.36}$$

Next, consider the contribution of n cycles at total energy E. Evidently, the abbreviated action is $nW(E)$, the total time spent is $\tau_n = n\bar{\tau}$, and the overall phase factor is $\mathcal{J}_n = (-1)^n$. Summing over all possibilities, we find

$$p(E) = \sum_{n=1}^{\infty} (-1)^{n-1} e^{-nW(E)/\hbar} \prod_{\alpha=1}^{N} \frac{1}{2\sinh[n\bar{\tau}(E)\mu_\alpha(E)/2]} \,. \qquad (15.37)$$

Upon expanding the $\sinh^{-1}$ functions into a geometrical series of exponentials, Eq. (15.37) can be rewritten in the form

$$p(E) = \sum_{n_1=0}^{\infty} \sum_{n_2=0}^{\infty} \cdots \sum_{n_N=0}^{\infty} \left\{ 1 + \exp\left[\left(W(E) - W'(E) \sum_{\alpha=1}^{N} \left(n_\alpha + \tfrac{1}{2}\right) \hbar\mu_\alpha(E) \right)/\hbar \right] \right\}^{-1}.$$
$$(15.38)$$

It is convenient to introduce the energy E_{esc} left in the escape coordinate while the transverse degrees of freedom are excited to the modes $\{n_\alpha\}$,

$$E_{\text{esc}} = E - \sum_{\alpha=1}^{N} \left(n_\alpha + 1/2\right) \hbar\mu_\alpha(E) \,. \qquad (15.39)$$

Next we observe that the argument of the exponential function in Eq. (15.38) represents the leading terms of a Taylor expansion of $W(E_{\text{esc}})$ around $W(E)$. Thus finally, the microcanonical cumulative transmission probability can be written as [319, 264]

$$p(E) = \sum_{n_1=0}^{\infty} \sum_{n_2=0}^{\infty} \cdots \sum_{n_N=0}^{\infty} \frac{1}{1 + e^{W(E_{\text{esc}})/\hbar}} \,. \qquad (15.40)$$

For $T > T_0$, the integral in Eq. (12.11) receives the main contributions from energies $E_{\text{esc}} \gtrsim V_{\text{b}}$. In this regime, we may approximate the potential $V(q)$ by the parabolic barrier form (14.1) and use the normal mode analysis discussed in Sect. 14.1. Near the barrier top, where $V(q) = V_{\text{b}} - \frac{1}{2}M\omega_{\text{b}}^2(q - q_{\text{b}})^2$, the escape coordinate z_0 coincides with the normal coordinate associated with the eigenvalue $\mu_0^{(\text{b})}$, and the stability frequencies of the periodic orbit are represented by the eigenfrequencies $\mu_\alpha^{(\text{b})}$ ($\alpha = 1, 2, \cdots, N$). In the parabolic barrier approximation, the periodic orbit is oscillating along $z_0 = y_0$ with a constant period $2\pi/\omega_{\text{R}}$, and the abbreviated action simply is

$$W(E_{\text{esc}}) \approx W_{\text{h}}(E_{\text{esc}}) = 2\pi(V_{\text{b}} - E_{\text{esc}})/\omega_{\text{R}} \,. \qquad (15.41)$$

Upon interchanging the integration over E with the summations, we obtain [319]

$$k = \frac{1}{Z_0} \frac{1}{2\pi\hbar} \sum_{n_1=0}^{\infty} \sum_{n_2=0}^{\infty} \cdots \sum_{n_N=0}^{\infty} e^{-\beta(E_\perp + V_{\text{b}})} \int_0^{\infty} dE \, \frac{e^{-\beta(E - E_\perp - V_{\text{b}})}}{1 + e^{-\beta_0(E - E_\perp - V_{\text{b}})}} \,, \qquad (15.42)$$

where $\beta_0 = 2\pi/\hbar\omega_R$, and where $E_\perp$ is the energy in the transverse modes,

$$E_\perp = \sum_{\alpha=1}^{N} \left(n_\alpha + \tfrac{1}{2}\right) \hbar\mu_\alpha^{(b)} . \tag{15.43}$$

The partition function of the well regime is conveniently expressed in terms of the eigenfrequencies $\mu_\alpha^{(0)}$, as discussed in Subsection 4.3.5 and specified in Eq. (4.234). Extending the integration interval in Eq. (15.42) from $-\infty$ to ∞, the integral yields the function $\pi/[\beta_0 \sin(\pi\beta/\beta_0)]$. Corrections due to the actually finite lower bound of the integral in Eq. (15.42) are exponentially small for $\beta V_b \gg 1$. It is now straightforward to perform the summations. We then find

$$k = \frac{\omega_R}{2\pi} \frac{\sinh(\hbar\beta\mu_0^{(0)}/2)}{\sin(\hbar\beta\omega_R/2)} \prod_{\alpha=1}^{N} \frac{\sinh(\hbar\beta\mu_\alpha^{(0)}/2)}{\sinh(\hbar\beta\mu_\alpha^{(b)}/2)} \, e^{-\beta V_b} . \tag{15.44}$$

Next, the product of the sinh factors can be transformed to read

$$\prod_{\alpha=0}^{N} \frac{\sinh(\hbar\beta\mu_\alpha^{(0)}/2)}{|\sinh(\hbar\beta\mu_\alpha^{(b)}/2)|} = \prod_{\alpha=0}^{N} \left(\frac{\mu_\alpha^{(0)}}{|\mu_\alpha^{(b)}|} \prod_{n=1}^{\infty} \frac{\mu_\alpha^{(0)\,2} + \nu_n^2}{\mu_\alpha^{(b)\,2} + \nu_n^2} \right) = \frac{\omega_0}{\omega_b} \prod_{n=1}^{\infty} \frac{\nu_n^2 + \omega_0^2 + \nu_n\widehat{\gamma}(\nu_n)}{\nu_n^2 - \omega_b^2 + \nu_n\widehat{\gamma}(\nu_n)} . \tag{15.45}$$

In the first equality we have used for the sinh functions the infinite product representation (4.234). The second equality follows with Eqs. (4.237), (4.239) and (4.241), and with the corresponding relations in which $-\omega_b^2$ is substituted for ω_0^2. Thus we find

$$k = \frac{\omega_R}{2\pi} \frac{\omega_0}{\omega_b} \prod_{n=1}^{\infty} \frac{\nu_n^2 + \omega_0^2 + \nu_n\widehat{\gamma}(\nu_n)}{\nu_n^2 - \omega_b^2 + \nu_n\widehat{\gamma}(\nu_n)} \, e^{-\beta V_b} . \tag{15.46}$$

The expression (15.46) agrees with the Im F result (15.12). Thus we have verified this formula by an independent method.

The periodic orbit approach applies also to the quantum tunneling regime $T < T_0$. In this case, the tunneling rate is again given by Eq. (15.31) with Eq. (15.40). However, the abbreviated action $W(E_{esc})$ must take into account the actual barrier shape.

In the MQTST ansatz (15.30), recrossings of the dividing surface Σ_b are excluded. In this method, the reduction factor $\varrho = \omega_R/\omega_b$ appears because the barrier frequency ω_R of the potential $V(q, \boldsymbol{x})$ along the escape coordinate y_0 is smaller than the barrier frequency of the bare potential $V(q)$. In contrast to MQTST, in the standard treatment of the decay by Langevin or Fokker-Planck equation methods, the reduction factor $\varrho = \omega_R/\omega_b$ originates from diffusive recrossings of the barrier top along the particle's coordinate q [264]. Thus, although the two physical pictures are different, the results are in correspondence.

The form (15.44) reveals the multi-dimensional character of the barrier crossing process. These features are somewhat hidden in the formula (15.46). The latter form is especially convenient for systems in which we eventually take the limit $N \to \infty$.

16. The crossover region

We have seen already in Section 15.2 that the quantum correction factor $c_{\rm qm}$ increases with decreasing temperature, and even diverges as $T \to T_0$ because the eigenvalue $\Lambda_1^{(\rm b)} = \Lambda_{-1}^{(\rm b)} = \nu_1^2 - \omega_{\rm b}^2 + |\nu_1|\widehat{\gamma}(|\nu_1|)$ in Eq. (15.7) vanishes at this temperature. Near $T = T_0$, the eigenvalue $\Lambda_1^{(\rm b)}$ may be expanded in powers of the parameter

$$\varepsilon \equiv (T_0 - T)/T_0 \,, \tag{16.1}$$

which is chosen negative above T_0 for later purposes. In leading order in ε we have from Eq. (15.8)

$$\Lambda_1^{(\rm b)} = -\varepsilon\,\Omega^2 \,, \tag{16.2}$$

where

$$\Omega^2 = \omega_{\rm b}^2 + \omega_{\rm R}^2[\,1 + \partial\widehat{\gamma}(\omega_{\rm R})/\partial\omega_{\rm R}\,] \,. \tag{16.3}$$

The rate expression (15.12) can be written in the form

$$k = \frac{\omega_{\rm R}}{2\pi}\frac{\Omega^2}{\Lambda_1^{(\rm b)}}\,A\,{\rm e}^{-\beta V_{\rm b}} \,, \tag{16.4}$$

where the smooth temperature dependence of the prefactor near $T = T_0$ is allocated to the dimensionless prefactor

$$A \equiv \frac{\omega_0}{\omega_{\rm b}}\frac{\nu_1^2 + \omega_0^2 + \nu_1\widehat{\gamma}(\nu_1)}{\Omega^2}\prod_{n=2}^{\infty}\frac{\nu_n^2 + \omega_0^2 + \nu_n\widehat{\gamma}(\nu_n)}{\nu_n^2 - \omega_{\rm b}^2 + \nu_n\widehat{\gamma}(\nu_n)} \,. \tag{16.5}$$

Substituting the expression (16.2) into Eq. (16.4), we obtain for T slightly above T_0

$$k = \frac{\omega_{\rm R}}{2\pi}\frac{A}{-\varepsilon}\,{\rm e}^{-\beta V_{\rm b}} \,. \tag{16.6}$$

The prefactor A can equivalently be expressed in terms of the partition function of the global system $Z_0^{(\rm tot)}$ and the eigenfrequencies $\mu_\alpha^{(\rm b)}$ for $\alpha \geq 1$. Upon extracting the almost-zero factor $\sinh(\hbar\beta\omega_{\rm R}/2) \approx -\pi\varepsilon$, and using the form (4.234), we find from Eq. (15.44)

$$A = \frac{1}{2\pi}\frac{1}{Z_0^{(\rm tot)}}\prod_{\alpha=1}^{N}\frac{1}{2\sinh(\hbar\beta\mu_\alpha^{(\rm b)}/2)} \,. \tag{16.7}$$

In the Ohmic case, the infinite product in Eq. (16.5) can be expressed again in terms of gamma functions [cf. Eq. (15.17) with (15.18)]. We find

$$A = \frac{\omega_0}{\omega_{\rm b}}\frac{\nu^2}{\Omega^2}\frac{\Gamma(2 + \lambda_{\rm b}^+/\nu)\Gamma(2 + \lambda_{\rm b}^-/\nu)}{\Gamma(1 + \lambda_0^+/\nu)\Gamma(1 + \lambda_0^-/\nu)} \,. \tag{16.8}$$

It is obvious from the derivation that the singularity in Eq. (16.6) at $T = T_0$ is unphysical. Rather it is an artefact of the Gaussian approximation for the fluctuation modes $X_{\pm 1}$ in Eq. (15.7). The singularity is removed by evaluating these modes beyond the Gaussian approximation. The vanishing eigenvalue points to the fact that at $T = T_0$ the action of the constant path $q(\tau) = q_b$ is degenerate with the action of the periodic bounce path $q(\tau) = q_B(\tau)$. The accurate calculation of the integral over the modes $X_{\pm 1}$ slightly above T_0 is different from the calculation slightly below T_0. The cases $T > T_0$ and $T < T_0$ were discussed in Refs. [308] and [306], respectively. The reader is also referred to the discussion in Ref. [263].

16.1 Beyond steepest descent above T_0

To regularize the disturbing integrals, the second-order action (15.7) must be supplemented by terms of higher order in the amplitudes X_1 and X_{-1}. Upon expanding the potential $V(q)$ about the barrier top beyond the parabolic barrier form,

$$V(q) = V_b - \frac{1}{2} M\omega_b^2 (q - q_b)^2 + \sum_{k=3}^{\infty} \frac{1}{k} Mc_k (q - q_b)^k , \qquad (16.9)$$

the extension of the second order action (15.7) is found as

$$S[q^{(b)}(\cdot)] = \hbar\beta V_b + \frac{1}{2} M\hbar\beta \left(\sum_{n=-\infty}^{\infty} \Lambda_n^{(b)} X_n X_{-n} \right.$$
$$\left. + 2c_3 \left(X_{-2}X_1^2 + X_2 X_{-1}^2 + 2X_0 X_1 X_{-1} \right) + 3c_4 X_1^2 X_{-1}^2 \right) ,$$

where we have kept terms up to the fourth order in $X_{\pm 1}$. After integrating out the amplitudes X_0 and $X_{\pm n}$ ($n \geq 2$), the action of the residual modes $X_{\pm 1}$ becomes

$$\Delta S_1^{(b)} = \frac{1}{2} M\hbar\beta \left(2\Lambda_1^{(b)} X_1 X_{-1} + B_4 X_1^2 X_{-1}^2 \right) . \qquad (16.10)$$

The coefficient B_4 measures the strength of the leading anharmonic contribution,

$$B_4 = 4c_3^2/\omega_b^2 - 2c_3^2/\Lambda_2^{(b)} + 3c_4 , \qquad (16.11)$$

and is positive for the usual case of a barrier which is wider than a parabolic barrier.[1]

We now withdraw the Gaussian approximation for the fluctuation modes X_1 and X_{-1} which has lead us to the spurious form (16.1) near $T = T_0$. The removal amounts to the substitution [cf. the functional measure (4.227)]

$$\frac{1}{\Lambda_1^{(b)}} \longrightarrow \frac{1}{\widetilde{\Lambda}_1^{(b)}} \equiv i \frac{M\beta}{2\pi} \int_{-\infty}^{\infty} dX_1 \int_{-\infty}^{\infty} dX_{-1} \exp(-\Delta S_1^{(b)}/\hbar) . \qquad (16.12)$$

Upon introducing polar coordinates (r, φ) defined by $X_{\pm 1} \equiv (r/\sqrt{2})\, e^{\pm i\varphi}$, we obtain

$$\frac{1}{\widetilde{\Lambda}_1^{(b)}} = M\beta \int_0^{\infty} dr\, r \exp\left\{ -\tfrac{1}{2} M\beta\Lambda_1^{(b)} r^2 - \tfrac{1}{8} M\beta B_4 r^4 \right\} , \qquad (16.13)$$

---

[1] A brief discussion of the role of the coefficient B_4 is given at the end of Section 16.3.

which may be written in terms of the function erfc(z) given in Eq. (12.31) as

$$\frac{1}{\widetilde{\Lambda}_1^{(b)}} = \left(\frac{\pi M\beta}{2B_4}\right)^{1/2} \operatorname{erfc}\left(\Lambda_1^{(b)}\left(\frac{M\beta}{2B_4}\right)^{1/2}\right) \exp\left(\Lambda_1^{(b)\,2}\frac{M\beta}{2B_4}\right). \tag{16.14}$$

With the form (16.2) for $\Lambda_1^{(b)}$, we then get the compact expression

$$\Omega^2/\widetilde{\Lambda}_1^{(b)} = \sqrt{\pi}\,\kappa\,\operatorname{erfc}(-\kappa\varepsilon)\,e^{\kappa^2\varepsilon^2}, \tag{16.15}$$

in which the dimensionless parameter κ is given by

$$\kappa = \Omega^2\sqrt{M\beta/2B_4}. \tag{16.16}$$

Finally, upon substituting $\Omega^2/\widetilde{\Lambda}_1^{(b)}$ for $\Omega^2/\Lambda_1^{(b)}$ in Eq. (16.4), the thermal rate is found in the form

$$k = \frac{\omega_R}{2\pi}A\sqrt{\pi}\kappa\,\operatorname{erfc}(-\kappa\varepsilon)\,e^{\kappa^2\varepsilon^2-\beta V_b}. \tag{16.17}$$

To summarize, the problematic term $1/\Lambda_1^{(b)}$ has been regularized by taking into account non-Gaussian fluctuations of the corresponding mode. The improved rate formula (16.17) is valid from very high T, where pure thermal activation prevails, down to $T = T_0$. Before entering into the discussion of the result achieved, let us consider the crossing rate for T slightly below T_0.

16.2   Beyond steepest descent below T_0

Below T_0, the bounce trajectory $q_B(\tau)$ exists as a third extremal action path besides the constant paths $q(\tau) = 0$ and $q(\tau) = q_b$. Since the bounce action $S_B \equiv S_{\text{eff}}^E[q_B]$ is smaller than the action $\hbar\beta V_b$ of the trivial saddle point $q(\tau) = q_b$, the bounce gives the leading contribution to the barrier partition function Z_b and to $\operatorname{Im} F$. Because the bounce is a periodic path with period $\hbar\beta$, it may be written as Fourier series,

$$q_B(\tau) = q_b + \sum_{n=-\infty}^{\infty} Q_n\,e^{i\nu_n\tau}. \tag{16.18}$$

A change of the phase of the bounce $q_B(\tau) \to q_B(\tau + \tau_0)$ has no influence on the numerical value of the action. Because of the time translation invariance of the action, we encounter a *zero mode* in the quantum fluctuations about the bounce. It is convenient to choose $q_B(\tau) = q_B(-\tau)$ or, equivalently, $Q_n = Q_{-n}$. Near T_0, the amplitudes Q_n are small and they can be calculated from the equation of motion (13.2) as a power series in $\sqrt{\varepsilon}$, where again $\varepsilon \equiv (T_0 - T)/T_0$. The leading terms are

$$Q_1 = \left(\frac{\varepsilon\Omega^2}{B_4}\right)^{1/2}, \qquad Q_0 = -\frac{2c_3}{\omega_b^2}Q_1^2, \qquad Q_2 = \frac{c_3}{\Lambda_2^{(b)}}Q_1^2, \tag{16.19}$$

with which we find for the bounce action upon using Eq. (16.16)

$$S_B = \hbar\beta V_b - \hbar\kappa^2\varepsilon^2 + \mathcal{O}(\varepsilon^3) \,. \tag{16.20}$$

In the next step, we add quantum fluctuations to the bounce trajectory,

$$q(\tau) = q_B(\tau) + \sum_{n=-\infty}^{\infty} F_n \, e^{i\nu_n\tau} \,, \tag{16.21}$$

and calculate the change of the action by the fluctuation modes. The result expression is conveniently written in the form

$$S[q] = S_B + S^{(1)} + S^{(2)} \,, \tag{16.22}$$

where $S^{(1)}$ is the second order action, and the relevant anharmonic contributions are in $S^{(2)}$. For T slightly below T_0, we may use the expressions (16.19) for the Fourier coefficients of the bounce. Then the action $S^{(1)}$ may be written as

$$S^{(1)} = \frac{M\hbar\beta}{2}\Big(-\omega_b^2 G_0^2 + 2\Lambda_2^{(b)} G_2 G_{-2} + \varepsilon\Omega^2(F_1+F_{-1})^2 + 2\sum_{n=3}^{\infty}\Lambda_n^{(b)} F_n F_{-n}\Big) \,. \tag{16.23}$$

Here we have kept only the terms of leading order in ε, and we have introduced

$$G_0 \equiv F_0 - (2c_3/\omega_b^2)(\varepsilon\Omega^2/B_4)^{1/2}(F_1+F_{-1}) \,,$$
$$G_{\pm 2} \equiv F_{\pm 2} + (2c_3/\Lambda_2^{(b)})(\varepsilon\Omega^2/B_4)^{1/2}F_{\pm 1} \,. \tag{16.24}$$

The fluctuation mode G_0 reduces the action which verifies that the bounce is indeed a saddle point. The integral over the amplitude G_0 must again be carried out by deforming the integration contour as described in Section 12.5. This leads to an imaginary contribution to the partition function from the barrier region with the already familiar factor $1/2$ [cf. Eq. (12.53)].

Next, observe that the twofold degenerate eigenvalue $\Lambda_1^{(b)} = \Lambda_{-1}^{(b)}$ defined in Eq. (15.8), which would become negative below T_0, is split into the eigenvalues $\Lambda_{1,-}^{(B)}$ and $\Lambda_{1,+}^{(B)}$ of the second order action (16.23) as T passes through T_0 from above. The degeneracy is removed because the constant path bifurcates into the bounce path at the crossover temperature. Substituting $F_{\pm 1} \equiv 2^{-1/2}(F_{1,+}\pm F_{1,-})$ into Eq. (16.23), we see that the eigenvalue $\Lambda_{1,-}^{(B)}$ of the mode $F_{1,-}$ is exactly zero,[2] while the eigenvalue of the mode $F_{1,+}$ is $\Lambda_{1,+}^{(B)} = 2\varepsilon\Omega^2$. Further, we see from Eq. (16.23) that the eigenvalues with $n \geq 2$ are in leading order independent of ε. Thus we find for the determinant D_B' connected with the second-order action functional (16.23),

$$|D_B'| = 2\varepsilon\Omega^2\,|D_B''| \quad \text{with} \quad |D_B''| = |D_b''| \equiv \omega_b^2\Big(\prod_{n=2}^{\infty}\Lambda_n^{(b)}\Big)^2 \,, \tag{16.25}$$

where the single prime is to indicate that the zero eigenvalue is omitted in the determinant, and where terms of order ε^2 are disregarded.

---

[2]The zero eigenvalue is due to the time-translation symmetry of the bounce mentioned before.

Because of the smallness of the eigenvalues $\Lambda_{1,\pm}^{(B)}$, the Gaussian approximation for the modes $F_{1,\pm}$ breaks down. Thus, the action (16.23) must be supplemented by terms of the third and fourth order in the amplitudes F_1 and F_{-1}. These higher-order terms contain nonlinear couplings between the amplitudes F_1 and F_{-1} and other amplitudes. The relevant contribution has the form

$$
\begin{aligned}
S^{(2)} \;=\; & -M\hbar\beta \left[\, c_3 \left(F_2 F_{-1}^2 + F_{-2} F_1^2 + 2F_0 F_1 F_{-1}\right) \right. \\
& \left. + \; 3c_4 Q_1 \left(F_1^2 F_{-1} + F_1 F_{-1}^2\right) + 3c_4 F_1^2 F_{-1}^2/2 \,\right] .
\end{aligned}
\tag{16.26}
$$

The integrations over the amplitudes F_0 and $F_{\pm 2}$ are again Gaussian and can be peformed by completing the square. Then we are left with the integral over the amplitudes F_1 and F_{-1} weighted with the action

$$
\Delta S_1^{(B)} \;=\; \tfrac{1}{2} M B_4 \hbar\beta \left[Q_1^2 (F_1 + F_{-1})^2 + 2Q_1 (F_1 + F_{-1}) F_1 F_{-1} + F_1^2 F_{-1}^2 \right] .
\tag{16.27}
$$

Here the terms of third and fourth order are due to the action $S^{(2)}$. Upon introducing polar cordinates (ρ, φ) with origin at $-Q_1$,

$$
F_{\pm 1} \;=\; (\rho/\sqrt{2}) \, e^{\pm i\varphi} - Q_1 ,
\tag{16.28}
$$

the action turns out to be independent of φ. Thus, the φ mode is the zero mode connected with changes of the phase of the bounce, while the ρ mode describes amplitude fluctuations. These fluctuations can be as large as the bounce amplitude Q_1. Therefore, we have to keep all terms up to the order ρ^4.

Abandonment of the Gaussian approximation amounts to the substitution

$$
\begin{aligned}
\frac{1}{\sqrt{\Lambda_{1,+}^{(B)}\Lambda_{1,-}^{(B)}}} \;\longrightarrow\; \frac{1}{\Lambda_1^{(B)}} \;&\equiv\; i\frac{M\beta}{2\pi} \int_{-\infty}^{\infty} dF_1 \int_{-\infty}^{\infty} dF_{-1} \, \exp(-\Delta S_1^{(B)}/\hbar) \\
&= M\beta \int_0^{\infty} d\rho \, \rho \exp\left\{ -\tfrac{1}{8}\beta M B_4 (\rho^2 - 2Q_1^2)^2 \right\} .
\end{aligned}
\tag{16.29}
$$

As in the case $T > T_0$, we thus reach an expression with an error function,

$$
1/\Lambda_1^{(B)} \;=\; (\sqrt{\pi}\kappa/\Omega^2)\, \text{erfc}\,(-\kappa\varepsilon) .
\tag{16.30}
$$

Altogether, the expression (15.10) is replaced by

$$
\text{Im}\, F = -\frac{1}{2\beta\Lambda_1^{(B)}} \left(\frac{D_0}{|D_B''|}\right)^{1/2} e^{-S_B/\hbar} ,
\tag{16.31}
$$

where the determinant $|D_B''|$ is given in Eq. (16.25). Upon using the relation (12.57), and the expressions (15.11), (16.5) and (16.20), the decay rate is found in the form

$$
k \;=\; \frac{1}{\hbar\beta}\, A\, \sqrt{\pi}\kappa\, \text{erfc}\,(-\kappa\varepsilon)\, e^{\kappa^2\varepsilon^2 - \beta V_b} .
\tag{16.32}
$$

At the crossover temperature $T = T_0$, we have $1/\hbar\beta = \omega_R/2\pi$. Thus, the formula (16.32) exactly coincides at this temperature with the previous result (16.17) evaluated for temperatures slightly above T_0. From this we infer that the expression (16.32) is valid in the crossover region above and below T_0.

Below T_0, the coefficient B_4 defined in Eq. (16.11) can be related to the change of the bounce period per unit energy. This is shown as follows. First, we write down the formal expansion of the action about $\hbar\beta_0$,

$$S(\hbar\beta) = \hbar\beta V_b + \frac{1}{2} S''(\hbar\beta_0)\left(\hbar\beta - \hbar\beta_0\right)^2 + \mathcal{O}\left((\hbar\beta - \hbar\beta_0)^3\right). \tag{16.33}$$

In the second step, we identify Eq. (16.33) with Eq. (16.20) and use the relations[3]

$$S''(\tau) = -\frac{1}{W''(z)} = -\frac{1}{|\tau'(z)|}, \tag{16.34}$$

where $W(z)$ is the abbreviated action for one periodic orbit at total energy z and where $\tau(z)$ is the respective period [cf. Eq.(15.34)]. We then find

$$|\tau'| \equiv |\tau'(z = V_b)| = \frac{B_4}{M\Omega^4}\hbar\beta_0 = \frac{\hbar}{2}\frac{\beta_0^2}{\kappa^2}. \tag{16.35}$$

Upon solving Eq. (16.35) for κ and inserting into Eq. (16.32), we obtain [319]

$$k = \frac{A}{2}\left(\frac{2\pi}{\hbar|\tau'|}\right)^{1/2} \mathrm{erfc}\left[\left(\frac{\hbar}{2|\tau'|}\right)^{1/2}(\beta_0 - \beta)\right] \exp\left(-\beta V_b + \frac{\hbar(\beta - \beta_0)^2}{2|\tau'|}\right), \tag{16.36}$$

which is exactly in the form (12.30). This rate expression applies again for T slightly above and below T_0.

16.3 The scaling region

The energy $M\Omega^4/B_4$ is usually of the order of the barrier height. Then the parameter κ is of the order of $(V_b/\hbar\omega_R)^{1/2}$. Thus we have $\kappa \gg 1$ in the semiclassical limit. The improved formula (16.17) [or (16.32)] is only required in the region $|\kappa\varepsilon| \lesssim 1$, or

$$|T - T_0| \lesssim T_0/\kappa, \tag{16.37}$$

in which the argument of the erfc function is of order one or smaller. Thus, for $\kappa \gg 1$, the crossover region is narrow on the temperature scale T_0.

For T well above this regime, i.e., for $\kappa\varepsilon < -1$, we may use the asymptotic form

$$\sqrt{\pi}\,\kappa\, e^{\kappa^2\varepsilon^2}\, \mathrm{erfc}(-\kappa\varepsilon) \approx -1/\varepsilon. \tag{16.38}$$

Substituting this form, Eq. (16.17) reduces to the expression (16.6).

For T well below the crossover region ($\kappa\varepsilon > 1$), we have $\mathrm{erfc}(-\kappa\varepsilon) \approx 2$. The zero mode normalization factor (cf. Section 17.1)

$$S_0 \equiv M\int_0^{\hbar\beta} d\tau\, \dot{q}_B^2(\tau) \tag{16.39}$$

---

[3]The prime (double prime) denotes the first (second) derivative with respect to the argument.

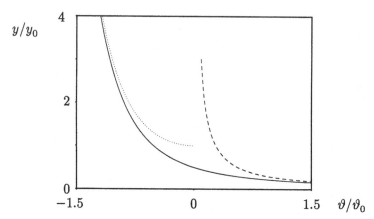

Figure 16.1: The scaled rate y/y_0 is shown as a function of the scaled temperature ϑ/ϑ_0. The high temperature formula (16.6) is represented by a dashed line and the low temperature formula (16.41) as a dotted line. The crossover function (16.46) smoothly matches onto these functions below and above the crossover region.

takes for the periodic bounce path (16.18) with (16.19) to leading order in ε the form

$$S_0 = 2M\hbar\beta\nu_1^2 Q_1^2 = 8\pi^2 \frac{M\Omega^2}{B_4\hbar\beta_0}\varepsilon \overset{(16.16)}{=} 16\pi^2 \left(\frac{\kappa}{\hbar\beta_0\Omega}\right)^2 \hbar\varepsilon . \qquad (16.40)$$

Employing the second form, the rate expression (16.32) can be rewritten as

$$k = \left(\frac{S_0}{2\pi\hbar}\right)^{1/2} \left(\frac{D_0}{|D'_B|}\right)^{1/2} e^{-S_B/\hbar} , \qquad (16.41)$$

where $|D'_B|$ is given in Eq. (16.25). Another useful expression for the rate follows from Eq. (16.36) by substituting the form (16.7) for the quantity A. The result is

$$k = \frac{1}{Z_0}\left(\prod_{\alpha=1}^{N} \frac{1}{2\sinh(\hbar\beta\mu_\alpha^{(b)}/2)}\right) \frac{1}{(2\pi\hbar|\tau'|)^{1/2}} e^{-S_B/\hbar} . \qquad (16.42)$$

The expressions (16.41) and (16.42) hold in the regime $\kappa^{-1} < \varepsilon \ll 1$. They smoothly match well below T_0 on the semiclassical quantum tunneling rate formula expressed in terms of properties of the bounce, which is discussed subsequently in Section 17.1.

The quantum-statistical decay rate exhibits in the crossover region a universal scaling behaviour [263]. Defining a temperature scale ϑ_0 and a frequency scale y_0,

$$\vartheta_0 \equiv T_0/\kappa , \qquad y_0 = (\omega_R/2\pi)\sqrt{\pi}\,\kappa A , \qquad (16.43)$$

there follows from Eq. (16.32) that the quantity

$$y/y_0 \equiv (\gamma/y_0)\, e^{\beta V_b} \qquad (16.44)$$

is represented by a *universal* function of $\vartheta/\vartheta_0 \equiv \kappa(T - T_0)/T_0$,

$$y/y_0 = U(\vartheta/\vartheta_0) . \tag{16.45}$$

The scaling function $U(z)$ is given by the integral representation

$$U(z) = \frac{2}{\sqrt{\pi}} \int_0^\infty dt \, e^{-t^2 - 2zt} = \text{erfc}(z) \, e^{z^2} . \tag{16.46}$$

The universal behaviour (16.45) is independent of the specific form of the metastable potential and also independent of the dissipative mechanism. Only the scale factors y_0 and ϑ_0 depend on the particular system under consideration. In Fig. 16.1, the scaled rate y/y_0 is shown as a function of the scaled temperature ϑ/ϑ_0 in comparison with the high and low temperature formulas (16.6) and (16.41), respectively.

Let us finally briefly discuss the significance of the anharmonicity factor B_4 defined in Eq. (16.11). When $B_4 > 0$, the bounce amplitude and action increases monotonously with the bounce period, i.e., when the temperature is decreased. Thus, for $B_4 > 0$, the transition from the constant path to the periodic path at $T = T_0$ is a second order phase transition [307, 308]. In the limit $B_4 \to 0$, the parameter κ given in Eq. (16.16) diverges. Thus, the scaling function (16.46) becomes meaningless in this limit. To regularize the interpolating function in the case $B_4 = 0$, we have to include in the action $\Delta S_1^{(b)}$ in Eq. (16.10) the term of sixth order,

$$\Delta S_1^{(b)} = \frac{1}{2} M\hbar\beta \Big(2\Lambda_1^{(b)} X_1 X_{-1} + B_6 X_1^3 X_{-1}^3 \Big) . \tag{16.47}$$

The coefficient B_6 is found for the potential (16.9) as

$$B_6 = 20c_6/3 - 2c_4^2/\Lambda_3^{(b)} . \tag{16.48}$$

The integrand in Eq. (16.12) is now defined in terms of the action (16.47).

Eventually, we find again the universal scaling form (16.45) for the thermal-to-quantum crossover. However, the scaling function $U(z)$ and the scale factor κ are different. The dimensionless parameter κ reads

$$\kappa = \Omega^2 \left(M^2\beta^2/4B_6 \right)^{1/3} , \tag{16.49}$$

and the scaling function $U(z)$ has the integral representation

$$U(z) = \frac{2}{\sqrt{\pi}} \int_0^\infty dt \, e^{-t^3 - 2zt} . \tag{16.50}$$

When the barrier is made steeper, the coefficient B_4 is decreased. In the parameter regime $B_4 < 0$, the barrier is even indented compared with a parabolic one. In this case, the bounce action is not any more a monotonous function of the bounce period, and the classical-to-quantum transition has the characteristics of a phase transition of first order. Physically, this is due to the fact that tunneling near to the barrier top is less favorable than at a lower level. Above the critical temperature of the first order transition, we find again universal behaviour. The scaling form is obtained by analytical continuation of the expression (16.45) with (16.46) for complex B_4 from the regime $\mathrm{Re}\, B_4 > 0$ to the regime $\mathrm{Re}\, B_4 < 0$. Since the case $B_4 < 0$ is exotic and of minor physical importance, it will not be discussed further.

17. Dissipative quantum tunneling

We now come to consider the thermodynamic rate expression for temperatures well below the crossover region, where the decay is dominated by quantum tunneling.

17.1 The quantum rate formula

At temperatures below T_0, the relevant stationary point describing the barrier region is the bounce trajectory $q_{\mathrm{B}}(\tau)$ which obeys the equation of motion (13.2), where the dissipative kernel $k(\tau)$ is defined in Eq. (4.45). To second order in the fluctuations about the extremal path, $q(\tau) = q_{\mathrm{B}}(\tau) + \xi(\tau)$, the action takes the form

$$S[q] \;=\; S_{\mathrm{B}} + \frac{1}{2}M \int_0^{\hbar\beta} d\tau\, \xi(\tau)\, \Lambda[q_{\mathrm{B}}(\tau)]\, \xi(\tau)\,, \qquad (17.1)$$

where the fluctuation operator $\Lambda[q_{\mathrm{B}}(\tau)]$ is defined in Eq. (4.214). Expanding $\xi(\tau)$ into the normalized eigenmodes $\chi(\tau)$ according to

$$\xi(\tau) \;=\; \sum_n c_n \chi_n(\tau)\,, \qquad (17.2)$$

where $\chi_n(\tau)$ is normalized on the interval $(0, \hbar\beta)$, the action becomes

$$S[q] \;=\; S_{\mathrm{B}} + \frac{1}{2}M \sum_n \Lambda_n[q_{\mathrm{B}}]\, c_n^2\,. \qquad (17.3)$$

The $\Lambda_n[q_{\mathrm{B}}]$ are the eigenvalues of $\Lambda[q_{\mathrm{B}}]$ for periodic boundary conditions. The determinant of the operator $\Lambda[q_{\mathrm{B}}]$ in the diagonal representation reads

$$D[q_{\mathrm{B}}] \;\equiv\; \det\left(\Lambda[q_{\mathrm{B}}]\right) \;=\; \prod_n \Lambda_n[q_{\mathrm{B}}]\,. \qquad (17.4)$$

Differentiation of the equation of motion (13.20 with respect to time shows that $\dot{q}_{\mathrm{B}}(\tau)$ is an $\hbar\beta$ periodic eigenmode of $\Lambda[q_{\mathrm{B}}(\tau)]$ with eigenvalue zero,

$$\Lambda[q_{\mathrm{B}}]\, \dot{q}_{\mathrm{B}}(\tau) \;=\; 0\,. \qquad (17.5)$$

The normalized zero mode reads

$$\chi_1(\tau) \;=\; \sqrt{M/S_0}\, \dot{q}_{\mathrm{B}}(\tau)\,, \qquad (17.6)$$

where S_0 is defined in Eq. (16.39). The mode $\chi_1(\tau)$ describes a time translation of the bounce. For an infinitesimal phase shift $\delta\tau$, we have

$$q_{\rm B}(\tau + \delta\tau) = q_{\rm B}(\tau) + \sqrt{S_0/M}\,\chi_1(\tau)\,\delta\tau \,, \qquad (17.7)$$

and the eigenvalue is zero because the action is invariant under changes of the phase.

With Eqs. (17.2) and (17.7), the formally divergent integral over the zero mode amplitude c_1 can be rewritten as an integral over the phase time τ_0 of the bounce which varies over the finite interval from zero to $\hbar\beta$ [276, 320, 321]. Upon using Eq. (17.3) and comparing Eq. (17.2) with Eq. (17.7), we find the substitution rule

$$\frac{1}{\sqrt{\Lambda_1[q_{\rm B}]}} \;\rightarrow\; \sqrt{\frac{M}{2\pi\hbar}} \int dc_1 \exp\left(-M\Lambda_1[q_{\rm B}]c_1^2/2\hbar\right) \;\rightarrow\; \sqrt{\frac{S_0}{2\pi\hbar}} \int_0^{\hbar\beta} d\tau_0 = \sqrt{\frac{S_0}{2\pi\hbar}}\hbar\beta \,.$$

While the bounce $q_{\rm B}(\tau)$ is nodeless, the zero mode (17.6) has one node. By the node-counting theorem, there exists a nodeless eigenmode of $\Lambda[q_{\rm B}]$ which must have a lower, i.e., a *negative* eigenvalue. The negative eigenvalue is again related to the fact that we are dealing with an unstable system. All the other eigenvalues of $\Lambda[q_{\rm B}]$ are positive. For T well below T_0, even the smallest positive eigenvalue, which has developed out of the quasi-zero eigenvalue $\Lambda_{1,+}^{\rm B} = 2\varepsilon\Omega^2$ near T_0, is sufficiently large that the respective mode can be treated in Gaussian approximation. In the next step, the integration contour for the *unstable* mode with the negative eigenvalue is deformed as described above. By the analytic continuation, the partition function gets an imaginary contribution. Finally, we use the relation $k = -(2/\hbar\beta)\,{\rm Im}\,(Z_{\rm b}/Z_0)$.

Collecting the various terms, the quantum decay rate at temperature T emerges in the concise form

$$k = f_{\rm qm}\,e^{-S_{\rm B}/\hbar} \,. \qquad (17.8)$$

The exponent is determined by the bounce action, which is the effective action (4.62) of a single bounce, $S_{\rm B} = S_{\rm eff}^{\rm E}[q_{\rm B}(\cdot)]$. The bounce path $q_{\rm B}(\tau)$ is a stationary point of this action. It is a periodic path with period $\hbar\beta$ in the upside-down potential obeying the equation of motion (13.2).

The prefactor or quantum mechanical attempt frequency is given by

$$f_{\rm qm} = \left(\frac{S_0}{2\pi\hbar}\right)^{1/2} \left(\frac{D_0}{|D'[q_{\rm B}]|}\right)^{1/2} \,. \qquad (17.9)$$

Here, S_0 is twice the kinetic part of the bounce action [see Eq. (16.39)], and $D'[q_{\rm B}]$ is the determinant (17.4) connected with the Gaussian fluctuations about the bounce path, and the prime indicates that the zero eigenvalue is omitted.

Slightly below T_0, the rate (17.8) with (17.9) matches the previous expression (16.41) with the actions $S_{\rm B}$ and S_0 given in Eq. (16.20) and Eq. (16.40), respectively. Further, we have $|D'[q_{\rm B}]| = |D_{\rm B}'|$, where $|D_{\rm B}'|$ is given in Eq. (16.25).

As in the case $T > T_0$ discussed in Section 15.4, it is possible to rewrite Eq. (17.8) with Eq. (17.9) in a form which directly displays the multi-dimensional character of the tunneling process. In the "many-dimensional WKB" approach the system-plus-reservoir complex is visualized as tunneling entity which moves along the (curved) most probable escape path (periodic orbit) $z_0^{(PO)}(\tau)$ in the $(N+1)$-dimensional configuration space. In this illustration, the bounce trajectory $q_B(\tau)$ is the projection of the periodic orbit[1] onto the reaction coordinate q. In the multi-dimensional picture, the exponent in the tunneling formula is given by the expression [cf. Eqs. (15.34) and (15.35)]

$$S_B/\hbar = \frac{2}{\hbar} \int_{z_0^{(m)}}^{z_0^{(ex)}} dz_0 \sqrt{2M[V(q,x) - E]} + E\beta , \qquad (17.10)$$

where dz_0 measures distance along the curved orbit with period $\hbar\beta$, and where E is the conserved energy of this path. The turning points are as in Eq. (15.29). At zero temperature, where $E = 0$ and $E\hbar\beta = 0$, the bounce action S_B coincides with S_0.

Consider next the preexponential factor of the tunneling rate. It is convenient to split the determinant of the fluctuations about the periodic orbit into the contribution from *longitudinal* fluctuations along the escape coordinate

$$\xi_\parallel(\tau) \equiv y_0(\tau) - y_0^{(PO)}(\tau) , \qquad (17.11)$$

and into the *transverse* determinant describing fluctuations in the coordinates $y_n(\tau)$ ($n = 1, 2, \cdots, N$) which are locally perpendicular to the escape coordinate. The latter determinant is conveniently expressed in terms of the dynamical stability frequencies $\mu_\alpha^{(B)}$ ($\alpha = 1, 2, \cdots, N$) of the periodic orbit. We then have [264, 319]

$$k = \frac{1}{Z_0} \left(\prod_{\alpha=1}^{N} \frac{1}{2\sinh(\hbar\beta\mu_\alpha^{(B)}/2)} \right) \frac{1}{\sqrt{2\pi\hbar|\tau'(E_{\hbar\beta})|}} e^{-S_B/\hbar} , \qquad (17.12)$$

where $E_{\hbar\beta}$ is the total energy of the periodic orbit with period $\hbar\beta$. This form for the quantum statistical tunneling rate is valid down to zero temperature and smoothly matches on the previous formula (16.42) at temperatures slightly below T_0.

An alternative derivation of the formula (17.8) with (17.9) has been given for zero temperature by Schmid [322] using a many-dimensional WKB approach for the quasi-stationary ground-state wave function. The generalization to finite temperatures is presented in Ref. [323].

For weak Ohmic dissipation, the bounce action and the attempt frequency may be calculated perturbatively in the Ohmic coupling $\alpha = \gamma/2\omega_0$. For the cubic metastable potential (11.41), one obtains as extension of the WKB result (12.36) with (12.39) at $T = 0$ the bounce action

$$S_B(T = 0) = \frac{36}{5} \frac{V_b}{\omega_0} \left(1 + \frac{45\zeta(3)}{\pi^3} \alpha + \mathcal{O}(\alpha^2) \right) , \qquad (17.13)$$

where $\zeta(3)$ is a Riemann number, and the quantum mechanical prefactor [325]

---

[1]As a result of the projection, the energy of the path $q_B(\tau)$ is not conserved, and the equation of motion becomes nonlocal in time.

$$f_{\rm qm}(T = 0) = 12\sqrt{6\pi}\,\frac{\omega_0}{2\pi}\sqrt{\frac{V_{\rm b}}{\hbar\omega_0}}\Big(1 + 2.86\,\alpha + \mathcal{O}(\alpha^2)\Big)\,. \qquad (17.14)$$

Dissipative quantum tunneling from the ground state in a metastable well was studied first by Caldeira and Leggett [77]. The qualitative conclusion of their study was that damping suppresses quantum tunneling by an exponential factor, which depends linearly on α for weak damping, as given in Eq. (17.13) .

17.2 Thermal enhancement of macroscopic quantum tunneling

At nonzero temperature, the tunneling rate is enhanced by thermal assistance since now there is a finite probability that the particle tunnels from an excited state in the well. The leading thermal enhancement at low T arises from the temperature dependence of the bounce action. It is convenient to write

$$k(T) = k(0)\,e^{\mathcal{A}(T)} \qquad \text{with} \qquad \mathcal{A}(T) = [\,S_{\rm B}(0) - S_{\rm B}(T)\,]/\hbar\,. \qquad (17.15)$$

The asymptotic expansion of the bounce action about zero temperature is discussed in some detail in Ref. [326] for a general metastable potential and arbitrary frequency-dependent damping. By applying the asymptotic Euler-Maclaurin expansion [88], which associates the Fourier coefficients Q_n of the bounce at low T with the Fourier representation $Q(\nu)$ of the zero temperature bounce trajectory, it is found that the leading contribution to $\mathcal{A}(T)$ is due to the temperature dependence of the kernel $K(\tau)$ in the influence action $S_{\rm infl}^{\rm E}[q_{\rm B}]$ given in Eq. (4.52). We obtain

$$\begin{aligned}
\mathcal{A}(T) &= \frac{1}{2\hbar}\,[\,K_T(0) - K_{T=0}(0)\,]\int_{-\infty}^{\infty} d\tau\, q_{\rm B}^{(0)}(\tau)\int_{-\infty}^{\infty} d\tau'\, q_{\rm B}^{(0)}(\tau') \\
&= \frac{q_{\rm b}^2\tau_{\rm B}^2}{2\pi\hbar}\int_0^{\infty} d\omega\, J(\omega)\Big(\coth(\omega\hbar\beta/2) - 1\Big)\,,
\end{aligned} \qquad (17.16)$$

where $q_{\rm B}^{(0)}(\tau)$ is the zero temperature bounce. In the second form, we have used Eq. (4.55) with Eq. (4.57), and we have introduced the width or "length" of the zero temperature bounce,

$$\tau_{\rm B} \equiv \frac{1}{q_{\rm b}}\int_{-\infty}^{\infty} d\tau\, q_{\rm B}^{(0)}(\tau)\,. \qquad (17.17)$$

Since the metastable minimum is chosen at $q = 0$, we have $q_{\rm B}^{(0)}(\tau \to \pm\infty) \to 0$. For a spectral density $J(\omega) \propto \omega^s$, the "dissipative" kernel $k(\tau)$ in the equation of motion decays for $T = 0$ as $|\tau|^{-(1+s)}$. Therefore, we have asymptotically $q_{\rm B}(\tau) \propto |\tau|^{-(1+s)}$. This verifies that the bounce width (17.17) is finite for $s > 0$.

For the spectral density $J(\omega) = M\gamma_s\omega_{\rm ph}^{1-s}\omega^s$ [cf. Eq. (3.58)], the enhancement function (17.16) takes the form

$$\mathcal{A}(T) = 2\Gamma(1 + s)\zeta(1 + s)\frac{M\gamma_s q_{\rm b}^2}{2\pi\hbar}\omega_{\rm ph}^2\tau_{\rm B}^2\left(\frac{T}{T_{\rm ph}}\right)^{1+s}\,, \qquad (17.18)$$

where $\Gamma(z)$ is Euler's gamma function and $\zeta(z)$ is Riemann's zeta function. Thus, the thermal enhancement is sensitive to the spectral form of the dissipative mechanism, and it qualitatively differs from the undamped case. For an undamped system, the enhancement is exponentially weak, $\mathcal{A}(T) \propto e^{-\hbar\omega_0/k_BT}$ as we can see from the expression (12.41). In contrast, for a damped system, we have algebraic enhancement, $\mathcal{A}(T) \propto T^{1+s}$. The exponent of the algebraic law is independent of the particular form of the metastable potential and is therefore a distinctive feature of the particular damping [305, 326]. Hence a precise measurement of the power of the algebraic enhancement of the tunneling probability at low T gives direct information about the spectral properties of the environmental coupling. The properties of the barrier and additional dependence on friction enter only through the square of the factor $\tau_B q_b$.

For Ohmic damping, $J(\omega) = M\gamma\omega$, we have

$$\mathcal{A}(T) = \frac{\pi^2}{3} \frac{M\gamma q_B^2}{2\pi\hbar} \left(\frac{\tau_B k_B T}{\hbar} \right)^2 , \qquad (17.19)$$

and the next-to-leading order term varies with temperature as T^4.

Consider next the bounce width. In the limit $\gamma \to 0$, we find for a cubic potential from Eq. (17.17) with Eq. (12.38) for the width of the bounce $\tau_B = 6/\omega_0$. For strong damping, $\alpha \equiv \gamma/2\omega_0 \gg 1$, the bounce is given below in Eq. (17.27) yielding the width $\tau_B = 4\pi\alpha/\omega_0$. Thus, the qualitative conclusion is that the bounce width increases with damping. This is again a signature of the increasing tunneling distance in the $\{q, x\}$ configuration space discussed in Chapter 13.

The universal enhancement (17.19) is due to thermally excited states of the environment, and not to thermally excited states in the well. A discussion of the formula (17.19) from the view of thermal quantum noise-theory is given in Ref. [327]. Finally, we remark that a similar universal low temperature behaviour $\propto T^{s+1}$ occurs in the free energy of damped quantum systems [208] (cf. Sections 6.5 and 21.9), and, e.g., in transport properties [328] (cf. Subsection 24.5.1).

17.3 Quantum decay in a cubic potential for Ohmic friction

In this section, we restrict the attention to a cubic potential and Ohmic friction. At $T = 0$, there are two relevant dimensionless damping parameters. One parameter is $\alpha = \gamma/2\omega_0 = \eta/2M\omega_0$, which measures the damping strength in units of the well frequency. The other is

$$K = c\alpha V_b/\hbar\omega_0 \propto \eta V_b/V''(0) , \qquad (17.20)$$

where c is a numerical constant of order one.[2] The parameter K is independent of the mass, wheras $\alpha \propto 1/\sqrt{M}$. Therefore, α and K are independent parameters. In the overdamped regime $\alpha \gg 1$, the inertia term in the equation of motion (13.2) and

---

[2]The definition of the parameter K is given in Subsection 17.4.

in the bounce action is small compared to the friction term and may therefore be disregarded. The ratio K/α is large in the semiclassical limit.

For general α, the quantum rate expression (17.8) with (17.9) must be computed numerically. In the regime $\alpha \gg 1$, $K \gg 1$ and $T < T_0$, the tunneling rate can be calculated in analytic form for the cubic metastable potential. We now turn to the discussion of these particular cases.

17.3.1 Bounce action and quantum prefactor

For the cubic metastable potential (11.40), the bounce obeys the equation of motion

$$-\ddot{q}_B(\tau) + \omega_0^2 q_B(\tau) - \frac{\omega_0^2}{q_b} q_B^2(\tau) + \frac{1}{M} \int_0^{\hbar\beta} d\tau' \, k(\tau - \tau') q_B(\tau') = 0 , \qquad (17.21)$$

or equivalently in Fourier space upon writing $q_B(\tau) = \sum_\ell Q_\ell \, e^{i\nu_\ell \tau}$,

$$\left(\nu_\ell^2 + |\nu_\ell|\hat{\gamma}(|\nu_\ell|) + \omega_0^2 \right) Q_\ell = \frac{\omega_0^2}{q_b} \sum_m Q_{\ell+m} Q_m . \qquad (17.22)$$

With use of the equation of motion (17.22), the bounce action may be written as

$$S_B = \frac{M\omega_0^2}{6q_b} \hbar\beta \sum_{\ell,m} Q_\ell Q_m Q_{\ell+m} , \qquad (17.23)$$

and the zero mode normalization factor (16.39) is $S_0 = M\hbar\beta \sum_m \nu_m^2 Q_m^2$. The equation (17.22) may be solved numerically by successive iteration, starting for instance with the zero order ansatz $Q_m \propto e^{-m}$.

Consider next the fluctuations $y(\tau)$ about the bounce. Writing $y(\tau) = \sum_\ell Y_\ell \, e^{i\nu_\ell \tau}$, the eigenvalue equation for the fluctuation operator $\Lambda[q_B(\tau)]$ reads

$$[\nu_\ell^2 + \gamma|\nu_\ell| + \omega_0^2] Y_\ell^{(n)} - \frac{2\omega_0^2}{q_b} \sum_m Q_{\ell+m} Y_m^{(n)} = \Lambda_n Y_\ell^{(n)} . \qquad (17.24)$$

The lowest odd mode of this equation is the translational mode which has one node and eigenvalue zero, $y_{\text{zero}}(\tau) = \dot{q}_B(\tau) = i \sum_\ell \nu_\ell Q_\ell \, e^{i\nu_\ell \tau}$. Since the Fourier coefficients Q_m are small for large m, the eigenvalues Λ_n for large n can be calculated perturbatively about $\Lambda_n^{(b)}$. For small n, the eigenvalue Λ_n can be calculated numerically by diagonalization of truncated $N \times N$ matrices. In the numerical implementation, low eigenvalues require large N up to $N = 250$.

A numerical caculation of tunneling rates at $T = 0$ was given first by Chang and Chakravarty [332]. Accurate numerical calculations at finite temperatures in the range $0.1\, T_0 \leq T \leq T_0$ and $0.1 \leq \alpha \leq 1$ are presented in Ref. [263]. The work gives tables both for the exponent and the prefactor of the quantum rate formula (17.8) with (17.9) in this range of parameters. The numerical data are in remarkable agreement with the measured lifetime of the metastable zero-voltage state in the current-biased Josephson junction [266, 333].

The characteristic features of the quantum decay rate as a function of friction strength and temperature are summarized below in Section 17.5.

17.3.2 Analytic results for strong Ohmic dissipation

For large Ohmic damping, $\alpha \gg 1$, the inertia term $\nu_\ell^2 Q_\ell$ in Eq. (17.22) can be disregarded. In the absence of the inertia term, the equation of motion for the Fourier coefficients Q_ℓ is solved in analytic form with the ansatz

$$Q_\ell = a\, e^{-b|\ell|} . \tag{17.25}$$

The coefficients are found to read

$$a = q_{\rm b} T/T_0 , \qquad \text{and} \qquad b = \operatorname{artanh}(T/T_0) , \tag{17.26}$$

where $T_0 \equiv \hbar\omega_0/(4\pi\alpha k_{\rm B})$ is the crossover temperature in the strong damping limit, Eq. (14.38). With the expressions (17.25) and (17.26), the bounce trajectory in the temperature regime $0 \le T \le T_0$ takes the analytic form [308]

$$q_{\rm B}(\tau) = q_{\rm b}\frac{T}{T_0}\sum_\ell e^{-b|\ell|}\, e^{i\nu_\ell \tau} = q_{\rm b}\frac{(T/T_0)^2}{1 - \sqrt{1 - (T/T_0)^2}\cos(\nu\tau)} , \tag{17.27}$$

where $\nu = \nu_1 = 2\pi/\hbar\beta$. At $T = T_0$, the bounce path (17.27) coalesces with the constant path $q = q_{\rm b}$. For $T > T_0$, the solution (17.27) is not real anymore. The action (17.23) pertaining to the trajectory (17.27) is found in the analytic form

$$S_{\rm B}(T) = 6\pi\alpha\frac{V_{\rm b}}{\omega_0}\left[1 - \frac{1}{3}\left(\frac{T}{T_0}\right)^2\right] , \qquad T \le T_0 . \tag{17.28}$$

In the limit $T \to 0$, the bounce (17.27) reduces to the Lorentzian form [307]

$$q_{\rm B}^{(0)}(\tau) = \frac{2\alpha}{\omega_0}q_{\rm b}\int_{-\infty}^{+\infty} d\nu\, e^{-2\alpha|\nu|/\omega_0}\, e^{i\nu\tau} = \frac{2q_{\rm b}}{1 + (\omega_0\tau/2\alpha)^2} . \tag{17.29}$$

For this path, the exit point on the other side of the barrier is at $q_{\rm ex} = 2q_{\rm b}$, while $q_{\rm ex} = 3q_{\rm b}/2$ for $\alpha = 0$. The qualitative conclusion is that $q_{\rm ex}$ is a monotonous function of α varying in the range $3q_{\rm b}/2 \le q_{\rm ex} \le 2q_{\rm b}$. We remark that $q_{\rm ex} > 3q_{\rm b}/2$ is possible because the bounce does not conserve energy when α is nonzero.[3]

The action at zero temperature is [263]

$$S_{\rm B}(T = 0) = 6\pi\alpha\frac{V_{\rm b}}{\omega_0}\left[1 + \frac{1}{4\alpha^2} + \mathcal{O}(\alpha^{-4})\right] , \tag{17.30}$$

where we have also given the leading correction $\propto 1/\alpha^2$. The width (17.17) of the bounce path (17.29) is obtained as $\tau_{\rm B} = 4\pi\alpha/\omega_0$. With this result, the thermal enhancement expression (17.19) is in agreement with the T^2 contribution in the action (17.28). Interestingly enough, we see from the expression (17.28) that the T^2 law strictly holds up to the crossover temperature, and there are no other temperature corrections in the exponential factor of the rate for $\alpha \gg 1$.

---

[3]The periodic orbit in the full $\{q, x\}$ space has conserved total energy. The point $q_{\rm ex}$ is the projection of the exit point $P_{\rm ex}$ of the periodic orbit onto the q-axis (cf. Fig. 13.1 in Chapter 13).

Consider next the fluctuations $y(\tau)$ about the bounce. For the low-energy modes we may disregard the inertia term $\nu_\ell^2 Y_\ell$ in Eq. (17.24). The truncated eigenvalue equation for the even modes with zero and with two nodes can be solved in analytic form with the ansatz

$$Y_\ell = (C + |\ell|)\, e^{-b|\ell|} \, . \tag{17.31}$$

The resulting eigenvalues are

$$\Lambda_{0/1} = -\omega_0^2 \left(1 \pm \sqrt{1 + 4[1 - (T/T_0)^2]} \right) /2 \, . \tag{17.32}$$

As a consequence of metastability, the lowest eigenvalue Λ_0 turns out negative. Observe also that the lowest odd mode is the zero mode.

Turning to the higher eigenvalues Λ_n ($|n| \geq 2$), we first observe that at the crossover temperature $T = T_0$, at which the periodic bounce is degenerated into a point, $q_{\rm B}(\tau) = q_{\rm b}$, the eigenvalues Λ_n ($|n| \geq 2$) are given by [cf. Eq. (15.8)]

$$\Lambda_n = \nu_n^2 + \gamma |\nu_n| + \omega_0^2 - 2\omega_{\rm b}^2 = \nu_n^2 + \gamma |\nu_n| + \omega_0^2 - 4\pi\gamma/\hbar\beta_0 \, , \tag{17.33}$$

where $\beta_0 = 1/k_{\rm B}T_0$. It is tempting to replace for $T < T_0$, the inverse crossover temperature β_0 in Eq. (17.33) by the actual inverse temperature β. This yields

$$\Lambda_n = \nu_n^2 + \gamma |\nu_n| + \omega_0^2 - 2\gamma\nu \, , \qquad (|n| \geq 2) \, , \tag{17.34}$$

where $\nu = \nu_1 = 2\pi/\hbar\beta$. The expression (17.34) found by this intuitive argument is indeed the correct result for the eigenvalues Λ_n ($|n| > 2$) in the regime $0 < T \leq T_0$, as was shown in Ref. [307]. We remark that in the expression (17.34), the kinetic term ν_n^2 is needed for $|n| \gg 1$, while it is irrelevant for small n. Readily, the action factor related to the zero mode normalization is calculated as

$$S_0 = M \int_0^{\hbar\beta} d\tau\, \dot{q}_{\rm B}^2(\tau) = 6\pi \frac{V_{\rm b}}{\gamma} \left(1 - \frac{T^2}{T_0^2} \right) \, . \tag{17.35}$$

To obtain the pre-exponential factor (17.9), we gather up the expression (15.11) for the determinant D_0 and the expressions (17.32), (17.33) and (17.35). We then find the pre-exponential factor of the rate expression in the form

$$f_{\rm qm} = A_1 A_2 \, , \tag{17.36}$$

$$A_1 = \sqrt{\frac{S_0}{2\pi\hbar}} \frac{\omega_0}{\sqrt{|\Lambda_0|\Lambda_1}} \, , \qquad A_2 = \frac{\prod_{n=1}^{\infty}[\nu_n^2 + \gamma\nu_n + \omega_0^2]}{\prod_{n=2}^{\infty}[\nu_n^2 + \gamma\nu_n + \omega_0^2 - 2\gamma\nu]} \, . \tag{17.37}$$

Inserting Eqs. (17.32) and (17.35) into the expression for A_1, the temperature dependence cancels out and we obtain

$$A_1 = \sqrt{\frac{3V_{\rm b}}{\hbar\gamma}} \frac{1}{\omega_0} \, . \tag{17.38}$$

The infinite products in the expression for A_2 can be expressed in terms of gamma functions. We find [cf. Eq. (15.17)]

$$A_2 = \left(\omega_0^2 - \gamma\nu\right) \frac{\Gamma(1+\lambda_B^+/\nu)\,\Gamma(1+\lambda_B^-/\nu)}{\Gamma(1+\lambda_0^+/\nu)\,\Gamma(1+\lambda_0^-/\nu)} \,, \tag{17.39}$$

where

$$\lambda_0^\pm = \gamma/2 \pm \sqrt{\gamma^2/4 - \omega_0^2}\,,$$
$$\lambda_B^\pm = \gamma/2 \pm \sqrt{\gamma^2/4 - \omega_0^2 + 2\gamma\nu}\,. \tag{17.40}$$

For strong damping, $\alpha \gg 1$, we obtain from Eq. (17.40) the relations

$$\lambda_B^+ = \lambda_0^+ + 2\kappa\,, \quad \text{and} \quad \lambda_B^- = \lambda_0^- - 2\nu\,, \tag{17.41}$$

with which the partial prefactor A_2 is readily evaluated as

$$A_2 = \gamma^4/\omega_0^2\,. \tag{17.42}$$

Thus, the pre-exponential factor of the rate turns out to be temperature-independent. Collecting the factors A_1 and A_2, and the bounce action (17.28), we reach the following final expression for the crossing rate (17.8) in the regime $0 \le T \le T_0$,

$$k = 8\sqrt{6}\,\alpha^{7/2}\omega_0 \sqrt{\frac{V_b}{\hbar\omega_0}} \exp\left[-6\pi\alpha\frac{V_b}{\hbar\omega_0}\left(1 - \frac{T^2}{3T_0^2}\right)\right]\,. \tag{17.43}$$

Subleading contributions to the rate expression are given in Ref. [263]. This concludes our discussion of the quantum-statistical escape of an overdamped particle from the metastable well of a cubic potential in the semiclassical limit.

17.4 Quantum decay in a tilted cosine washboard potential

The phenomenon of incoherent tunneling of a macroscopic variable has become most clearly visible in superconducting quantum interference devices, where the phase difference ψ of the Cooper pair wave function across the Josephson junction plays the role of the tunneling coordinate (cf. Subsections 3.2.2 and 4.2.10). The measured lifetimes of the zero voltage state in current-biased Josephson systems were found to be in excellent agreement [329] with the theoretical predictions.

The standard phenomenological model for a Josephson junction is the resistively shunted junction (RSJ) model [101] (cf. Subsection 3.2.2). Within the RSJ model, the deterministic equation of motion for the phase difference ψ of the pair wave function across the junction with capacitance C and effective shunt resistance R reads[4]

$$C\left(\frac{\phi_0}{2\pi}\right)^2\frac{d^2\psi}{dt^2} + \frac{1}{R}\left(\frac{\phi_0}{2\pi}\right)^2\frac{d\psi}{dt} + E_J\sin\psi - I_{\text{ext}}\frac{\phi_0}{2\pi} = 0\,, \tag{17.44}$$

where $\phi_0 = h/2e$ is the flux quantum, I_{ext} is the externally applied bias current, and $E_J = I_c\phi_0/2\pi$ is the Josephson coupling energy. The current I_c is the maximum supercurrent that the junction will support. Equation (17.44) describes classical damped motion of the phase in the potential (4.195).

---

[4]The RSJ model is Markovian and corresponds to the substitution $Y^*(\omega) = 1/R$ in Eq. (3.155).

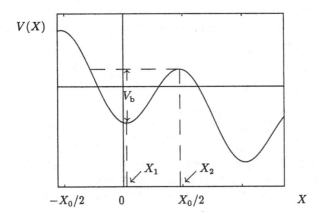

Figure 17.1: Sketch of the tilted washboard potential.

The dynamical equation (17.44) is formally equivalent to the equation of motion for a Brownian particle of mass M and position X in the absence of fluctuations,

$$M\ddot{X}(t) + M\gamma\dot{X}(t) + \frac{\partial V(X)}{\partial X} = 0 \,. \tag{17.45}$$

The potential $V(X)$ is of trigonometric form with a sloping background resulting in a potential drop $2\pi V_{\text{tilt}}$ per length X_0 as sketched in Fig. 17.1,

$$V(X) = -V_0 \cos\left(\frac{2\pi X}{X_0}\right) - V_{\text{tilt}}\frac{2\pi X}{X_0} \,. \tag{17.46}$$

In the mapping of the two models, the equivalence relations are

$$\frac{X}{X_0} \triangleq \frac{\psi}{2\pi} \,, \qquad MX_0^2 \triangleq C\phi_0^2 \,, \qquad \gamma \triangleq \frac{1}{RC} \,,$$
$$V_0 \triangleq E_{\text{J}} \,, \qquad V_{\text{tilt}} \triangleq I_{\text{ext}}\frac{\phi_0}{2\pi} \,. \tag{17.47}$$

The model (17.45) with (17.46) is archetypal for many diffusion problems in condensed matter physics.

The two macroscopically distinguishable states in the current-biased Josephson junction are the zero voltage state and the voltage state. According to the relation (3.93), the first state corresponds to the particle being trapped in a well, and the second to the particle sliding down the cascade of wells. The particle in the well oscillates with the "plasma" frequency

$$\omega_0 = \frac{2\pi}{X_0}\frac{(V_0^2 - V_{\text{tilt}}^2)^{1/4}}{\sqrt{M}} \qquad \text{for} \qquad V_0 > V_{\text{tilt}} \,. \tag{17.48}$$

For the potential $V(X)$ given in Eq. (17.46), the barrier frequency ω_b coincides again with the well frequency ω_0. The minima and maxima of the tilted washboard potential are located in the principal interval $0 \leq X < X_0$ at

$$X_1 = (X_0/2\pi) \arcsin (V_{\text{tilt}}/V_0) , \qquad \text{and} \qquad X_2 = X_0/2 - X_1 , \qquad (17.49)$$

and the barrier height is given by

$$V_b = V(X_2) - V(X_1) = 2\sqrt{V_0^2 - V_{\text{tilt}}^2} - 2V_{\text{tilt}} \arccos(V_{\text{tilt}}/V_0) . \qquad (17.50)$$

For $V_{\text{tilt}} = V_0$, the cascade of barriers ceases to exist. In the regime $V_{\text{tilt}} > V_0$, the slope of the potential is negative everywhere, so that the particle slides all the way down and cannot be trapped even in the absence of inertia.

In the regime $V_{\text{tilt}} < V_0$, the usual dimensionless damping parameter is

$$\alpha \equiv \frac{\gamma}{2\omega_0} = \frac{\gamma\sqrt{M}X_0}{4\pi(V_0^2 - V_{\text{tilt}}^2)^{1/4}} . \qquad (17.51)$$

For the potential (17.46), it is convenient to define the second dimensionless friction parameter K as

$$K \equiv \frac{M\gamma X_0^2}{2\pi\hbar} = 4\pi\alpha\frac{\sqrt{V_0^2 - V_{\text{tilt}}^2}}{\hbar\omega_0} . \qquad (17.52)$$

The definition of K is in correspondence with the definition of the Kondo parameter in Eq. (4.154).

Consider now the crossing rate from the higher to the lower well which we denote by k^+ henceforth. We are interested in the regime $V_{\text{tilt}} < V_0$, $T < T_0$ in the semiclassical limit $K \gg 1$. In the tunneling regime, the crossing rate is dominated by the one-bounce contribution. The bounce trajectory in imaginary time obeys the equation of motion

$$-M\ddot{X}_B(\tau)+V_0\frac{2\pi}{X_0} \sin \left(\frac{2\pi}{X_0}X_B(\tau)\right) - \frac{2\pi V_{\text{tilt}}}{X_0} + \int_0^{\hbar\beta} d\tau' \, k(\tau-\tau')X_B(\tau') = 0 . \quad (17.53)$$

In the high-friction regime $\alpha \gg 1$, the inertia term $M\ddot{X}_B(\tau)$ in Eq. (17.53) is small compared to the friction term and can therefore be dropped. The truncated equation of motion for the bounce can be solved in analytic form, as discovered by Korshunov [330]. The solution is conveniently parametrized in terms of the temperature scales

$$k_BT_0 \equiv \frac{\sqrt{V_0^2 - V_{\text{tilt}}^2}}{K} = \frac{\hbar\omega_b}{4\pi\alpha} , \qquad \text{and} \qquad k_BT_1 \equiv \frac{V_{\text{tilt}}}{K} . \qquad (17.54)$$

Here, T_0 is the usual crossover temperature between quantum tunneling and thermal hopping in the strong damping limit, Eq. (14.38), while the energy scale k_BT_1 characterizes the effective strength of the bias. In the regime $T \ll T_1$, the thermal energy is negligibly small compared to the bias energy, while for $T \gg T_1$ the bias may be disregarded, as we shall see shortly. The bounce trajectory is found in the form

$$X_B(\tau) = X_1 + \frac{X_0}{\pi} \arctan\left(\frac{T^2/T_0 T_1}{1 - \sqrt{1 - (T/T_0)^2}\cos(2\pi\tau/\hbar\beta)}\right). \tag{17.55}$$

In the zero temperature limit, we obtain from Eq. (17.55) the bounce as

$$X_B(\tau) = X_1 + \frac{X_0}{\pi} \arctan\left(\frac{2T_0/T_1}{1 + (\omega_0\tau/2\alpha)^2}\right). \tag{17.56}$$

In the limit $T \to T_0$, the bounce is contracted to the constant path $X_B(\tau) = X_2$.
The action of the bounce path (17.55) takes the analytic form

$$S_B/\hbar = K \ln\left(\frac{V_0^2}{V_{tilt}^2[1 + (T/T_1)^2]}\right) + 2K\left(1 - \frac{T_1}{T}\arctan\frac{T}{T_1}\right). \tag{17.57}$$

Consider next the determinant of the fluctuation operator. The two lowest eigenvalues for symmetric modes with zero and two nodes are found as

$$\Lambda_{0/1} = \omega_0^2\left[\frac{1}{2} - \frac{V_{tilt}^2}{V_0^2}\left(1 + \frac{T^2}{T_1^2}\right) \mp \sqrt{\frac{1}{4} + \frac{V_{tilt}^4}{V_0^4}\left(1 + \frac{T^2}{T_1^2}\right)\left(1 - \frac{T^2}{T_0^2}\right)}\right]. \tag{17.58}$$

From this we get for the product

$$|\Lambda_0|\Lambda_1 = \omega_0^4 \frac{V_{tilt}^2}{V_0^2}\left(1 + \frac{T^2}{T_1^2}\right)\left(1 - \frac{T^2}{T_0^2}\right). \tag{17.59}$$

The eigenvalues for $n \geq 2$ are in the form (17.34) with ω_0 given in Eq. (17.48). The action determining the normalization of the zero mode is

$$S_0 = M \int_0^{\hbar\beta} d\tau\, \dot{X}_B^2(\tau) = 4\pi\frac{(V_0^2 - V_{tilt}^2)^{3/2}}{\gamma V_0^2}\left(1 - \frac{T^2}{T_0^2}\right). \tag{17.60}$$

Writing the pre-exponential factor of the rate as in Eq. (17.36) with (17.37), we find

$$A_1 = \frac{1}{\sqrt{\alpha\hbar\omega_0}}\frac{M^{3/2}\omega_0^2}{V_{tilt}}\left(\frac{X_0}{2\pi}\right)^3\frac{1}{\sqrt{1 + (T/T_1)^2}}, \tag{17.61}$$

$$A_2 = \gamma^4/\omega_0^2 = 16\alpha^4\omega_0^2.$$

Thus we obtain, using Eq. (17.61), the quantum mechanical attempt frequency as

$$f_{qm} = A_1 A_2 = \frac{16\alpha^{7/2}M^{3/2}\omega_0^4}{\sqrt{\hbar\omega_0}V_{tilt}}\left(\frac{X_0}{2\pi}\right)^3\frac{1}{\sqrt{1 + (T/T_1)^2}}. \tag{17.62}$$

Substituting the bounce action (17.57) and the attempt frequency (17.62) into

$$k^+ = f_{qm}\, e^{-S_B/\hbar}, \tag{17.63}$$

we obtain the forward tunneling rate to the next lower well [cf. Ref. [330]].
In the limit $X_0, V_0, V_{tilt} \to \infty$ with the following quantities constrained as

$$\lim_{X_0, V_0, V_{\text{tilt}} \to \infty} \left(\frac{2\pi}{X_0}\right)^3 (V_0 + V_{\text{tilt}}) = 24 \frac{V_{\text{b}}}{X_{\text{b}}^3} , \qquad \lim_{X_0, V_0, V_{\text{tilt}} \to \infty} \frac{2\pi}{X_0} (V_0 - V_{\text{tilt}}) = \frac{3}{2} \frac{V_{\text{b}}}{X_{\text{b}}} ,$$

the tilted washboard potential mutates into the cubic potential given by Eq. (11.40) with Eq. (11.41) (up to a constant shift). It is straightforward to see that in this limit the results (17.55) – (17.62) coincide with the corresponding results for the cubic metastable potential given in Subsection 17.3.2. Thus, the rate expression (17.63) with (17.57) and (17.62) reduces to the previous result, Eq. (17.43).

In the opposite limit of a small bias, the bounce consists of a weakly bound, widely spaced instanton–anti-instanton pair. The distance τ_s between the instanton and the anti-instanton is large compared to the width α/ω_{b} of an instanton so that the action of the bounce changes only weakly when the distance τ is moved away from the stationary point $\tau = \tau_s$. To study this case, we write the bounce action (17.57) as the action for two instantons at a distance τ,

$$S(\tau)/\hbar = 2S_{\text{inst}}/\hbar + W(\tau) - \epsilon\tau , \tag{17.64}$$

$$S_{\text{inst}}/\hbar = K\left[1 + \ln(\pi V_0/K\hbar\omega_c)\right] , \tag{17.65}$$

$$W(\tau) = 2K \ln\left[(\hbar\beta\omega_c/\pi) \sin(\pi\tau/\hbar\beta)\right] . \tag{17.66}$$

The first term in Eq. (17.64) is twice the action of an instanton, the second term $W(\tau)$ represents the interaction between the instantons at a distance τ, and the last term is the bias action for a potential drop $\hbar\epsilon = 2\pi V_{\text{tilt}}$ on the distance X_0. In the decomposition, we have introduced a suitable reference frequency ω_c. The role of the reference frequency is discussed below in Section 18.1.

The action (17.64) is extremal for $\tau = \tau_s$, where

$$\tau_s = (\hbar\beta/\pi) \operatorname{arccot} (\hbar\beta\epsilon/2\pi K) . \tag{17.67}$$

The action $S(\tau_s)$ coincides with the bounce action S_{B} given in Eq. (17.57).

The forward tunneling rate (17.63) with Eq. (17.62) can be written as

$$k^+ = \frac{K\gamma^2}{\pi} \left(\frac{\pi K}{\epsilon^2 + (2\pi K/\hbar\beta)^2}\right)^{1/2} e^{-S(\tau_s)/\hbar} , \tag{17.68}$$

where we have introduced the bias frequency

$$\epsilon = 2\pi V_{\text{tilt}}/\hbar . \tag{17.69}$$

When the instanton–anti-instanton pair is only weakly bound, the negative eigenvalue Λ_0 of the fluctuation operator is close to zero and therefore the Gaussian approximation for the breathing mode of the bounce breaks down. We find from Eq. (17.58)

$$\Lambda_0 \approx - \omega_0^2(V_{\text{tilt}}^2/V_0^2) \left(1 + T^2/T_1^2\right) . \tag{17.70}$$

Thus it should be expected that the formula (17.68) is not appropriate for weak bias, $V_{\text{tilt}} \ll V_0$, and low temperatures, $T, T_1 \ll T_0$. To deal with this case, we take one step back by withdrawing the execution of the breathing mode integral for the bounce by steepest descent. At this earlier stage, the rate expression (17.68) takes the form

$$k^+ = \frac{K\gamma^2}{\pi} \operatorname{Im} \int_{\mathcal{C}} d\tau \, e^{-S(\tau)/\hbar} , \qquad (17.71)$$

where $S(\tau)$ is the bounce action (17.64) of the instanton–anti-instanton pair as a function of the distance τ and $\mathcal{C}$ is a contour which may be deformed to pass through the stationary point of the action. The location of this point is given by Eq. (17.67). Since we have $\partial^2 W/\partial\tau^2|_{\tau=\tau_s} < 0$, the point τ_s is a saddle point, and the direction of steepest descent is perpendicular to the real axis of the complex τ-plane. Deforming the integration path at the saddle point along the direction of steepest descent, the expression (17.71) takes the form

$$k^+ = \frac{\Delta^2}{2} \operatorname{Im} \int_{\tau_s}^{\tau_s+i\infty} d\tau \, e^{\epsilon\tau - W(\tau)} . \qquad (17.72)$$

Here, Δ is the usual tight-binding tunnel matrix element (cf. Sections 18.1 and 20.2) which is determined by the instanton trajectory,

$$\Delta/2 = f_{\text{inst}} \, e^{-S_{\text{inst}}/\hbar} . \qquad (17.73)$$

Comparing Eq. (17.71) with Eqs. (17.72), (17.73), and substituing the instanton action, Eq. (17.65), we obtain $f_{\text{inst}} = \sqrt{K/2\pi}\,\gamma$, and the tunneling matrix element

$$\Delta/2 = \sqrt{K/2\pi}\,\gamma\, e^{-K} \left(K\hbar\omega_c/\pi V_0 \right)^K . \qquad (17.74)$$

The relation (17.74) expresses the transfer matrix element Δ in terms of the parameters of the original extended system. We shall rederive this important result quite differently in Section 25.4 by use of a self-duality symmetry of the model.

If we had evaluated Eq. (17.72) by steepest descent, we would recover the previous result (17.68). Fortunately, it is possible to perform the integration exactly. Observing that the analytically continued function $Q(z) = W(\tau = iz)$ is an analytic function in the strip $0 > \operatorname{Im} z > -\hbar\beta$ of the complex z-plane, we may write

$$k^+ = \frac{\Delta^2}{4} \int_{-\infty-ic}^{\infty-ic} dz \, e^{i\epsilon z - Q(z)} , \qquad (17.75)$$

where c is an arbitrary constant in the interval $0 < c < \hbar\beta$. For $K < \frac{1}{2}$, we may put $c = 0^+$. We then have

$$k^+ = \frac{\Delta^2}{2} \left(\frac{\pi}{\hbar\beta\omega_c} \right)^{2K} \int_0^\infty dt \, \frac{\cos(\epsilon t - \pi K)}{\sinh^{2K}(\pi t/\hbar\beta)} . \qquad (17.76)$$

The integration can be carried out in analytic form without approximation [88]. We then obtain the one-bounce contribution to the forward rate in the form

$$k^+(T,\epsilon) = \frac{K}{4\pi^2} \hbar\gamma^2\beta \left(\frac{2K}{\beta V_0} \right)^{2K} e^{-2K} \frac{|\Gamma(K + i\hbar\beta\epsilon/2\pi)|^2}{\Gamma(2K)} e^{\hbar\beta\epsilon/2} . \qquad (17.77)$$

The expression (17.75) is a contour integral which encircles the singularity at $t = 0$. Thus, the final expression (17.77) is not limited to the regime $K < \frac{1}{2}$ but also holds for $K \geq \frac{1}{2}$.[5] Further discussion of this point is given below Eq. (20.78).

In the absence of the bias, $\epsilon = 0$, the crossing rate takes the form

$$k^+(T, 0) = \frac{K^2 \Gamma(K)}{2\pi^{3/2}\Gamma(K + 1/2)} \frac{\hbar\gamma^2}{V_0} e^{-2K} \left(\frac{K k_{\mathrm{B}} T}{V_0}\right)^{2K-1} . \qquad (17.78)$$

In the opposite limit $T = 0$ and nonzero bias, we find

$$k^+(0, \epsilon) = \frac{K^2}{\pi \Gamma(2K)} \frac{\hbar\gamma^2}{V_0} e^{-2K} \left(\frac{K \hbar \epsilon}{\pi V_0}\right)^{2K-1} . \qquad (17.79)$$

The temperature T_1, where $k_{\mathrm{B}} T_1 = \hbar|\epsilon|/2\pi K$, is a sort of a crossover temperature between the regimes where either Eq. (17.78) or Eq. (17.79) applies. For $T \ll T_1$, the expression (17.79) is appropriate, while for $T \gg T_1$ the formula (17.78) is valid.

Up to now, we have considered the crossing rate from the higher to the lower well, $k = k^+$. The backward rate from the lower to the higher well obeys detailed balance,

$$k^-(T, \epsilon) = e^{-\hbar\epsilon/k_{\mathrm{B}}T} k^+(T, \epsilon) . \qquad (17.80)$$

Within steepest descent for the breathing mode, the backward rate k^- is determined by a different branch of the arctan function in the action (17.57). A general discussion of the detailed balance property is given below in Subsection 20.2.1.

We remark that the leading thermal enhancement calculated from the expression (17.77) is in agreement with the formula (17.19).

These results can be applied directly to macroscopic quantum tunneling in a voltage-biased Josephson junction. The corresponding model is introduced in Subsection 3.4.3 [see also Eq. (17.44)]. We have the parameter identifications $K = 1/\rho$, $\hbar\epsilon = 2eV_{\mathrm{x}}/\rho$, and $\gamma = E_{\mathrm{c}}/\pi\hbar\rho$, where $\rho = R/R_{\mathrm{Q}}$ [$R_{\mathrm{Q}} = 2\pi\hbar/4e^2$ and $E_{\mathrm{c}} = 2e^2/C$ are the resistance quantum and charging energy for Cooper pairs]. The overdamped limit corresponds to $1/4\alpha^2 = 2\pi^2\rho^2 E_{\mathrm{J}}/E_{\mathrm{c}} \ll 1$, as follows with the correspondence relations (17.47). The dc current through the Ohmic impedance R in the circuit sketched in Fig. 3.3 (see Subsection 3.4.3) is $I = (V_{\mathrm{x}} - V_{\mathrm{a}})/R$, where $V_{\mathrm{a}} = \hbar\langle\dot\psi\rangle/2e$. For $\rho \ll 1$ and small V_{x}, the phase slip is determined by the single-bounce contribution (17.63), $\langle\dot\psi\rangle = 2\pi k^+$. At $T = 0$ and in the overdamped limit, the rate k^+ is given in Eq. (17.79). Thus we find

$$I(V_{\mathrm{x}}) = \frac{V_{\mathrm{x}}}{R} \left(1 - \frac{\pi}{2}(\pi\rho)^{-2/\rho - 5/2}\left(\frac{eV_{\mathrm{x}}}{E_{\mathrm{J}}}\right)^{2/\rho}\left(\frac{E_{\mathrm{c}}}{eV_{\mathrm{x}}}\right)^2 + \mathcal{O}\left[\left(\frac{E_{\mathrm{c}}}{eV_{\mathrm{x}}}\right)^4\right]\right) . \qquad (17.81)$$

When the junction is in the zero voltage state, the voltage drop occurs at the resistor, yielding Ohm's law. Macroscopic quantum tunneling into a finite voltage state contributes the *nonlinear* term in the $I(V_{\mathrm{x}})$ characteristics. Further discussion of the voltage-biased Josephson junction is given in Subsections 20.3.4, 25.3, and 25.5.1.

---

[5]The contour integral (17.75) with the singular integrand $e^{-W(iz)}$, where $W(\tau)$ is given in Eq. (17.66), is analogous to Hankel's contour integral of the reciprocal gamma function [331].

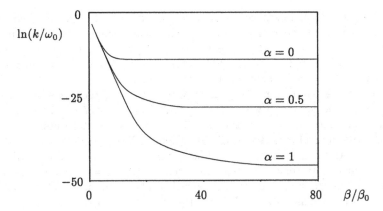

Figure 17.2: Arrhenius plot of the decay rate for a system with a cubic plus quadratic potential ($V_{\rm b} = 5\hbar\omega_0$) and Ohmic damping ($\alpha = \gamma/2\omega_0$) for various values of α.

17.5 Concluding remarks

The imaginary-time functional integral approach has provided an almost complete description of the quantum statistical decay of a metastable state extending from thermally activated decay at high temperatures down to very low temperatures where the system tunnels out of the ground state. The main features of the decay are summarized in the Arrhenius plot shown in Fig. 17.2. In this diagram, the classical rate is represented by a falling straight line. The rate flattens out towards a finite value at $T = 0$ due to quantum tunneling. For an undamped system ($\alpha = 0$), the transition between the classical and quantum regime is rather sharp. In the presence of damping, the classical rate is reduced only slightly because the attempt frequency gets smaller by the factor $\omega_{\rm R}/\omega_{\rm b}$. On the other hand, damping causes an exponentially strong suppression of the zero-temperature tunneling rate. Furthermore, the crossover temperature is lowered and the transition between thermally activated decay and tunneling becomes more gradual when damping is increased. For strong Ohmic dissipation, there is a large region in which thermal and quantum fluctuations interplay and thermal enhancement is governed by the power law given in Eq. (17.19).

The qualitative features of the tunneling rate can be seen rather directly by using the "many-dimensional WKB" approach. The smallest value of the global potential $V(q, x)$ [cf. the definition of $V(q, x)$ in Eq. (3.13)] on the hyperplane of constant q is independent of the bath coupling, but the minimum is located away from the q-axis at $x_\alpha = (c_\alpha/m_\alpha\omega_\alpha^2)\, q$. The reduction of the tunneling rate by dissipation can therefore be viewed as being due to the fact that the path under the barrier of the global potential $V(q, x)$, i.e. the path between the turning points in Eq. (17.10), is longer, both for zero and finite temperatures, compared to the tunneling length of the undamped system in the potential $V(q)$.

The phenomenon of macroscopic quantum tunneling has been observed experimentally in a large number of physical and chemical systems [264]. An especially attractive physical system is the current-biased Josephson junction or rf superconducting quantum interference device (SQUID) system since all relevant junction parameters can be measured independently. This provided accurate tests of the theory without adjustable parameters. On a $\log k$ versus T^2 plot, the experimentally observed slope can be compared with the predicted behaviour (17.19). Indeed, experiments have confirmed most of the theoretical predictions very accurately [329, 266]. Surveys of experimental results in MQT are given in Refs. [333, 334].

Evidence both for non-Ohmic and Ohmic dissipation has been discovered experimentally in diverse systems. Thermal enhancement with $s = 5$ in Eq. (17.18) was found in experiments on proton tunneling in hydrated protein powders [335]. The data analysis of tunneling rates for Li^+ impurities in diluted perovskite $K_{1-x}Li_xTaO_3$ was found to be consistent with $s = 3$ [336].

Dielectric relaxation of orientational defects in polycrystalline H_2O and D_2O ices in the crossover regime for different samples was shown to be in perfect agreement with the universal scaling law (16.45) with (16.46), and the thermal enhancement was found in the Ohmic form, Eq. (17.15) with (17.19) [337].

The discovery of single-molecule magnets (SMM) like $Mn_{12}ac$ opened about 15 years ago the new research field of nanomagnetism. SMM proved to be an ideal test ground for theories which are applied at the borderline between classical and quantum mechanical behaviour. They provided clear evidence of quantum tunneling of the magnetization [148, 150, 153].

PART IV

THE DISSIPATIVE TWO–STATE SYSTEM

One of the most interesting aspects of quantum theory is the phenomenon of constructive and destructive interference. The phase coherence between different quantum mechanical states may lead to clockwise motion. The simplest model involving quantum coherence is a two-state system. The problem of a quantum system whose state is effectively confined to a two-dimensional Hilbert space is often encountered in physics and chemistry. For instance, imagine a quantum mechanical particle tunneling clockwise forth and back between two different localized states. In reality, such a system is strongly affected by the surroundings. We will see that the coupling can lead to qualitative changes in the behaviours: the environment-induced fluctuations can destroy quantum coherence and can even lead to a phase transition to a state in which quantum tunneling is quenched.

18. Introduction

Anderson *et al.* [110] and, independently, Phillips [111] postulated in 1972 the existence of two-level systems in glasses to explain low temperature anomalies of the specific heat in these amorphous materials. Although the true microscopic nature of the tunneling entities in glasses is unclear, they can be visualized by a particle tunneling in a double well along an (unknown) reaction coordinate. It turned out in ultrasonic experiments as an important difference between dielectric and metallic glasses that the lifetime of the tunneling eigenstates is drastically shorter in the case of metals. Golding *et al.* [338] explained this unexpected behaviour by a nonadiabatic coupling of the tunneling entity to the conduction electrons in metals. Shortly later, Black and Fulde [339] extended this idea to describe relaxation processes in a superconducting environment by following the lines sketched above in Subsections 3.3.3 and 4.2.8. The theoretical predictions were confirmed by G. Weiss *et al.* [340] in ultrasonic experiments on superconducting amorphous metals. They showed that the lifetime is reduced when the environment is switched from superconducting to normalconducting, thus demonstrating the significance of the electronic coupling. A survey of the early developments and the perturbative treatment of this coupling is given in Ref. [112]. For a nonperturbative treatment see Ref. [341]. The nonlinear acoustic response of amorphous metals has been studied by Stockburger *et al.* [342], and the results for the dynamical susceptibility were found in good agreement with experiments [343]. A comprehensive review of the physics of tunneling systems in amorphous and crystalline solids with emphasis on the thermodynamic, acoustic, dielectric and optical properties has been given in Ref. [344].

The tunneling of light particles like small polarons, hydrogen isotopes or muons in solids has been thoroughly studied since several decades. While earlier work was mainly concerned with the significance of polaron effects [119] – [125], more recent attention has focused on the singular transient response of the fermionic environment in metals at low temperatures. The nonadiabatic influence of conduction electrons on the motion of interstitials in metals was proposed by Kondo [347] to explain the anomalous temperature dependence of muon diffusion in host metals at low temperatures. This mechanism has been confirmed experimentally for incoherent tunneling of μ^+ and p in Cu, Al, and Sc [92], for defects in mesoscopic wires [96, 95], and for tunneling of H and D in metals and superconductors. Comprehensive reviews of experiment [345] and theory [346] are available. The particle tunnels either randomly or coherently between particular interstitial sites. Experimentally, one passes from the incoherent to the coherent dynamics by lowering the temperature. The transition becomes apparent, e.g., in a change from quasielastic to inelastic neutron scattering.

Electron transfer reactions are ubiquitous in chemical and biological systems. In its simplest form, an electron localized at a donor site is tunneling to the acceptor site. Marcus' theory [97, 98] provides the appropriate scenario to describe such processes. Often the interaction between the charge and the polarization cloud of the environment is so strong that tunneling is possible only with the assistance of favorable equilibrium fluctuations in the environmental modes. Because of the small mass, electron transfer is strongly influenced by quantum effects even at room temperatures.

Examples of quantum coherence in an effective two-state system are the inversion resonance of the NH_3 molecule, strangeness oscillations of a neutral K-meson beam, and coherent tunneling of light interstitials in tunneling centers. Another important, but apparently quite different example of a two-state system is a rf SQUID ring threaded by an external flux near half a flux quantum (cf. Subsection 3.2.2). Such a system might be the appropriate vehicle for the observation of "macroscopic quantum coherence" (MQC) on a true macroscopic scale [262, 348].

The behaviour of the two-state system (TSS) is strongly influenced by the dissipative coupling to the heat bath's dynamical degrees of freedom. We shall consider linear couplings to the heat bath that are sensitive to the value of σ_z. For instance, a dipole-local-field coupling provides a simple physical model for this type of coupling. To be definite, we choose $H_I = -q \sum_\alpha c_\alpha x_\alpha$, where $q = \sigma_z q_0/2$ with q_0 being the spatial distance of the two localized states. It should be adequate in many cases of interest that the response of the environment to a perturbation can be considered as linear. Then, a bath which is represented by a set of harmonic oscillators with a coupling linear in the coordinates captures the essential physics we wish to describe. Since the TSS is like a spin, the relevant Hamiltonian has become known in the literature as the "spin-boson" Hamiltonian. We have introduced the spin-boson model already in Section 3.2. The relevant Hamiltonian is given in Eq. (3.86) or Eq. (3.88).

Despite its apparent simplicity, the spin-boson model cannot be solved exactly by any known method (apart from some limited regimes of the parameter space). Not only is the spin-boson model nontrivial mathematically, it is also nontrivial physically.

The environment acts on the TSS by a fluctuating force $\xi(t) = \sum_\alpha c_\alpha x_\alpha(t)$. For a bath with linear response, the modes $x_\alpha(t)$ obey Gaussian statistics. Therefore the dynamics of the bath is fully characterized by the force autocorrelation function in thermal equilibrium $\langle \xi(t)\xi(0)\rangle_\beta$, which is simply a superposition of harmonic oscillator correlation functions [cf. Eq. (5.33)]. In the formal path integral expression for the reduced density matrix, the environment reveals itself through an influence functional $\mathcal{F}$. In Chapters 4 and 5 we have given several useful forms for $\mathcal{F}$ applicable to thermodynamics and dynamics, respectively.

18.1 Truncation of the double-well to the two-state system

18.1.1 Shifted oscillators and orthogonality catastrophe

In Section 3.2 we have already briefly addressed the reduction of a double well system to a two-state system. Before resuming the discussion of the reduction for the dissipative case let us first consider the adiabatic limit, in which the environmental modes instantaneously adapt themselfes to the particle's position. The oscillator part of the Hamiltonian (3.86) for the two positions $\sigma_z = \pm 1$ of the particle reads

$$
\begin{aligned}
H_\pm &= \frac{1}{2}\sum_\alpha \left(\frac{p_\alpha^2}{m_\alpha} + m_\alpha \omega_\alpha^2 x_\alpha^2 \mp q_0 c_\alpha x_\alpha \right) \\
&= \sum_\alpha \left[\hbar\omega_\alpha b_\alpha^\dagger b_\alpha \mp \tfrac{1}{2}\hbar\lambda_\alpha \left(b_\alpha + b_\alpha^\dagger \right) \right] ,
\end{aligned}
\tag{18.1}
$$

where in the second line we have introduced the language of creation and annihilation operators, and where $c_\alpha = \sqrt{2\hbar m_\alpha \omega_\alpha}\,\lambda_\alpha/q_0$. Upon introducing the shifted operator

$$
b_{\pm,\alpha} = b_\alpha \mp \frac{1}{2}\frac{\lambda_\alpha}{\omega_\alpha} ,
\tag{18.2}
$$

the terms linear in b_α, $b_\alpha^\dagger$ cancel out, and we get the Hamiltonian $H_\pm$ in normal form

$$
H_\pm = \sum_\alpha H_{\pm,\alpha} \quad \text{with} \quad H_{\pm,\alpha} = \hbar\left(\omega_\alpha b_{\pm,\alpha}^\dagger b_{\pm,\alpha} - \frac{\lambda_\alpha^2}{4\omega_\alpha} \right) .
\tag{18.3}
$$

Since the shifted operator obeys the same commutation relations as the original one, the two different vacuum states for the boson α are defined by $b_{\pm,\alpha}|0_{\pm,\alpha}\rangle = 0$. The normalized many-boson states $\{|n_{\pm,\alpha}\rangle\}$ can be gained from the vacuum state $|0_{\pm,\alpha}\rangle$ in the usual way, $|n_{\pm,\alpha}\rangle = (b_{\pm,\alpha}^\dagger)^{n_{\pm,\alpha}}|0_{\pm,\alpha}\rangle/\sqrt{n_{\pm,\alpha}!}$. Action of the one annihilation operator on the vacuum state of the other yields

$$b_{\mp,\alpha}|0_{\pm,\alpha}\rangle \;=\; \pm\frac{\lambda_\alpha}{\omega_\alpha}|0_{\pm,\alpha}\rangle \;. \tag{18.4}$$

The Hamiltonians $H_{\pm,\alpha}$ describe two harmonic oscillators with eigenfrequency ω_α of which the centers have spatial distance

$$s_\alpha \;=\; \frac{c_\alpha}{m_\alpha\omega_\alpha^2}q_0 \;=\; \sqrt{\frac{2\hbar}{m_\alpha\omega_\alpha}}\frac{\lambda_\alpha}{\omega_\alpha} \;. \tag{18.5}$$

Hence the vacuum or ground states of $H_{-,\alpha}$ and $H_{+,\alpha}$ can mutually be transformed into each other with the displacement operation

$$|0_{\pm,\alpha}\rangle \;=\; e^{i\Omega_{\mp,\alpha}}|0_{\mp,\alpha}\rangle \quad \text{with} \quad \Omega_{\mp,\alpha} \;=\; \pm\frac{s_\alpha p_\alpha}{\hbar} \;=\; \pm i\frac{\lambda_\alpha}{\omega_\alpha}[\,b_{\mp,\alpha}^\dagger - b_{\mp,\alpha}\,] \;. \tag{18.6}$$

With use of the commutation relation of $b_{\pm,\alpha}$ with $b_{\pm,\alpha}^\dagger$ we find that the vacuum state of the one oscillator can be expressed in terms of a coherent state of the displaced other oscillator,

$$|0_{\pm,\alpha}\rangle \;=\; \exp\left(-\frac{\lambda_\alpha^2}{2\omega_\alpha^2}\right)\sum_{n=0}^{\infty}\frac{(\pm 1)^n}{\sqrt{n!}}\left(\frac{\lambda_\alpha}{\omega_\alpha}\right)^n|n_{\mp,\alpha}\rangle \;. \tag{18.7}$$

Thus, the probability to excite n bosons with energy $\hbar\omega_\alpha$ in a sudden transition from the ground state at $\sigma_z = -1$ to the state $\sigma_z = 1$ is a Poissonian distribution,

$$p_{n,\alpha} \;=\; |\langle 0_{-,\alpha}|n_{+,\alpha}\rangle|^2 \;=\; \exp\left(-\frac{\lambda_\alpha^2}{\omega_\alpha^2}\right)\frac{1}{n!}\left(\frac{\lambda_\alpha}{\omega_\alpha}\right)^{2n} \tag{18.8}$$

with mean particle number $\bar{n}_\alpha = \sum_\alpha n\,p_{n,\alpha} = \lambda_\alpha^2/\omega_\alpha^2$. The probability for transition without excitation of bosons with frequency ω_α is

$$p_{0,\alpha} \;=\; |\langle 0_{-,\alpha}|0_{+,\alpha}\rangle|^2 \;=\; \exp\left(-\bar{n}_\alpha\right) \;. \tag{18.9}$$

We can now use these findings to calculate the dressed amplitude for transitions between the states $\sigma_z = \pm 1$. Considering the boson drag in the adiabatic limit, the bare amplitude $\langle -|+\rangle = \Delta_0$ is renormalized by an exponential dressing factor, which is known as Franck-Condon factor,

$$\Delta \;=\; \Delta_0\prod_\alpha\langle 0_{-,\alpha}|0_{+,\alpha}\rangle \;=\; \Delta_0\exp\left(-\frac{1}{2}\sum_\alpha\frac{\lambda_\alpha^2}{\omega_\alpha^2}\right) \;. \tag{18.10}$$

The bare amplitude is reduced because of the displacement of the bosonic cloud in the transition. For any finite number of boson modes the dressed tunneling amplitude stays finite. In the continuum limit, we obtain upon introducing the smooth spectral density (3.89)

$$\Delta = \Delta_0 \exp\left(-\frac{1}{2}\int d\omega \, \frac{G(\omega)}{\omega^2}\right).$$ (18.11)

When the integral diverges, the two displaced ground states of the bosons become orthogonal. In this case, they represent two different worlds which do not mutually interfere. This situation is dubbed *orthogonality catastrophe*. Ultraviolet-divergence of the integral is irrelevant, since there is always some physical cutoff in the spectral function $G(\omega)$, as we have discussed in Subsection 3.1.5.

In the super-Ohmic case $G(\omega) \propto \omega^s$ with $s > 1$, the integral is infrared-convergent and hence there is no orthogonality catastrophe. This is an indication for the possibility of elastic tunneling processes without dynamical involvement of the reservoir. The integral is infrared-divergent, however, when the spectral density is Ohmic or sub-Ohmic, $s \leq 1$. This case is more subtle since tunneling is quenched in the adiabatic limit discussed hitherto. The appropriate treatment is given in Section 20.2.

18.1.2 Adiabatic renormalization

Consider a quantum particle moving in a double-well potential $V(q)$ as described in Sect. 3.2. Assume that the particle interacts with an environment as given in Eq. (3.11). We now wish to see under which conditions the continuous system can be reduced to a discrete two-state system. Roughly speaking, the reduction is possible if there is a wide separation of the relevant energy scales. In the first stage of the truncation procedure, the system is coupled to only high-frequency bosons with $\omega > \omega_c$, where ω_c is chosen to simultaneously satisfy the conditions

$$\hbar|\epsilon|, \, k_B T \ll \hbar\omega_c \ll \hbar\omega_R(\omega_c), \quad \text{and} \quad \Delta(\omega_c) \ll \omega_c.$$ (18.12)

Here, ω_R is the dressed well frequency [cf. Eq. (14.13) with $\omega_b \to \omega_0$], and $\hbar\Delta(\omega_c)$ is the dressed tunnel splitting. The frequencies ω_R and Δ are renormalized by the modes with $\omega > \omega_c$. Under conditions (18.12), the excited states in each well will not be significantly populated. Therefore, they can be ignored so that we can replace the Hamiltonian of the continuous double-well system by an effective two-state Hamiltonian with a renormalized tunnel splitting. In the second stage of the reduction scheme, the remaining low-frequency bosons with $\omega < \omega_c$ are coupled to the discrete system. As we shall see, these modes will severely influence transitions between the two low-lying states. Thus finally, we have reached the dissipative two-state or spin-boson Hamiltonian [cf. Eq. (3.88)]

$$H = -\frac{\hbar\Delta}{2}\sigma_x - \frac{\hbar\epsilon}{2}\sigma_z - \frac{1}{2}\sigma_z\sum_{\alpha \, \epsilon \, \text{lf}} \hbar\lambda_\alpha\left(b_\alpha + b_\alpha^\dagger\right) + \sum_{\alpha \, \epsilon \, \text{lf}} \hbar\omega_\alpha b_\alpha^\dagger b_\alpha,$$ (18.13)

in which the spectral density $G_{\text{lf}}(\omega)$ subsumes the coupling to modes with $\omega \lesssim \omega_c$,

$$G_{\mathrm{lf}}(\omega) = \sum_{\alpha \in \mathrm{lf}} \lambda_\alpha^2 \, \delta(\omega - \omega_\alpha) \, . \tag{18.14}$$

The spectral density $G_{\mathrm{lf}}(\omega)$ includes all bath modes that are slow on the time scale Δ^{-1} and hence cause long-ranged memory effects.

In contrast, fast modes are dynamically important only on short time scales. Since fast modes can quickly adjust themselves to the slow tunneling motion, they can be treated adiabatically as discussed in the preceding subsection. For $\Delta = 0$, the two lowest energy eigenfunctions of the *symmetric* global system at $T = 0$ are given by

$$
\begin{aligned}
|\psi_\pm\rangle &= \tfrac{1}{\sqrt{2}} \left(|+\rangle \prod_{\alpha \in \mathrm{hf}} |0_{+,\alpha}\rangle \pm |-\rangle \prod_{\alpha \in \mathrm{hf}} |0_{-,\alpha}\rangle \right) \\
&= \tfrac{1}{\sqrt{2}} \left(|+\rangle \prod_{\alpha \in \mathrm{hf}} e^{i\Omega_\alpha/2} |0_\alpha\rangle \pm |-\rangle \prod_{\alpha \in \mathrm{hf}} e^{-i\Omega_\alpha/2} |0_\alpha\rangle \right) ,
\end{aligned}
\tag{18.15}
$$

where $|\pm\rangle$ denotes the state $\sigma_z = \pm 1$ of the TSS, and $|0_{\pm,\alpha}\rangle$ are the shifted ground states of the α'th oscillator for these TSS states. The translation operator Ω_α is defined as in Eq. (18.6), but depends on the original cration and annihilation operators,

$$\Omega_\alpha = s_\alpha p_\alpha / \hbar = i(\lambda_\alpha / \omega_\alpha)[b_\alpha^\dagger - b_\alpha] \, . \tag{18.16}$$

The level splitting at $T = 0$ dressed by the high frequency modes is given by

$$
\begin{aligned}
\Delta &= \Delta_0 \prod_{\alpha \in \mathrm{hf}} \langle 0_\alpha| \, e^{i\Omega_\alpha} \, |0_\alpha\rangle \\
&= \Delta_0 \exp\left(-\frac{1}{2} \sum_{\alpha \in \mathrm{hf}} \frac{\lambda_\alpha^2}{\omega_\alpha^2} \right) = \Delta_0 \exp\left(-\frac{1}{2} \int_0^\infty d\omega \, \frac{G_{\mathrm{hf}}(\omega)}{\omega^2} \right) .
\end{aligned}
\tag{18.17}
$$

The Franck-Condon factor represents the polarization cloud of phonons with frequencies above ω_c in the adiabatic limit. In the approximation (18.17) for the high-frequency modes of the bath, it is not distinguished whether the Franck-Condon factor originates from a single bath mode, or whether there is in reality a continuum of bath states in the regime $\omega > \omega_c$ [349].

The reduction procedure relying on the expression (18.17) has been studied by Sethna [160] in the context of tunneling centers in solids which are described by a spectral density $G(\omega) \propto \omega^3$ at low frequencies (cf. Subsection 4.2.4). In this case, also low-frequency modes may be included in the Franck-Condon factor. The case of Ohmic dissipation is more subtle due to an inherent infrared divergence, as adressed already in the preceeding subsection (see also below).

18.1.3 Renormalized tunnel matrix element

Let us now examine the dressed tunnel matrix element Δ in the Ohmic case beyond the crude treatment (18.17). The spectral density is [cf. Eq. (4.153)]

$$G(\omega) = 2K\omega \tag{18.18}$$

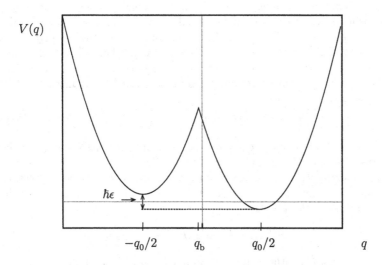

Figure 18.1: The asymmetric double well described in Eq. (18.20).

with the Kondo coupling parameter [cf. Eqs. (17.20) and (17.52)]

$$K \equiv \frac{\eta q_0^2}{2\pi\hbar} = \frac{M\gamma q_0^2}{2\pi\hbar} = \frac{v\alpha}{\pi}, \qquad (18.19)$$

where η is the Ohmic viscosity. In the last form, we have introduced the usual dimensionless damping parameter $\alpha = \gamma/2\omega_0$ and the dimensionless "barrier height" $v \equiv M\omega_0^2 q_0^2/\hbar\omega_0$. In the semiclassical limit, we have $v \gg 1$.

For the slightly asymmetric double well potential sketched in Fig. 18.1,

$$V(q) = \begin{cases} M\omega_0^2(q + q_0/2)^2/2 + \hbar\epsilon/2, & q < q_{\mathrm{b}}, \\ M\omega_0^2(q - q_0/2)^2/2 - \hbar\epsilon/2, & q > q_{\mathrm{b}}, \end{cases} \qquad (18.20)$$

the renormalized tunnel matrix element [cf. Eq. (17.73)]

$$\Delta/2 = f_{\mathrm{inst}}\, e^{-S_{\mathrm{inst}}/\hbar} \qquad (18.21)$$

can be calculated in analytic form [350] for arbitrary damping strength α. The potential (18.20) has a cusp at the barrier top which is located at $q_{\mathrm{b}} = -\hbar\epsilon/M\omega_0^2 q_0$. The bounce is the extremal path in the upside-down potential $-V(q)$ which starts near $+q_0/2$ and then moves forth and back through the valley, and finally returns to the starting point after time $\hbar\beta$. Because of the weak bias, the bounce is a weakly-bound instanton–anti-instanton pair, as explained in Subsection 17.4. The interval τ betweeen the centers of the instanton and the anti-instanton is a collective variable which parametrizes the breathing mode. The bounce obeys a linear equation of motion of the form (6.122) which is supplemented by jump conditions in the velocity and in the acceleration. The discontinuities are situated at the times where the path passes the cusp of the potential on the way forth and back. The calculation is similar to the proceeding explained in Subsection 6.7 and is reported in Ref. [350]. It is found that the bounce action can be decomposed as in Eq. (17.64). The instanton part of the action is obtained in the form

$$S_{\text{inst}}/\hbar \;=\; [\,v^2/2 - \pi^2 K^2\,]\langle q^2\rangle/q_0^2 + K[\,C_{\text{E}} + \ln(\omega_0/\omega_c)\,]\,, \qquad (18.22)$$

where ω_c is a reference frequency, $C_{\text{E}} = 0.5772\ldots$ is Euler's constant, and $\langle q^2\rangle$ is the coordinate dispersion of the damped harmonic oscillator. In the parameter regime (18.12), it is consistent to take $\langle q^2\rangle$ at zero temperature. The relevant expression for the dispersion $\langle q^2\rangle$ is given in Eqs. (6.101) – (6.114). In the weak damping limit, $\alpha \ll 1$, we obtain from Eq. (18.22)

$$S_{\text{inst}}/\hbar \;=\; S_{\text{inst}}^{(0)}/\hbar + K[\,\ln(\omega_0/\omega_c) - c_0\,] + \mathcal{O}(K^2)\,, \qquad (18.23)$$

where $S_{\text{inst}}^{(0)} = M\omega_0 q_0^2/4$ is the instanton action for zero damping. The numerical constant is $c_0 = 1/2 - C_{\text{E}} \approx -0.0772$.

In the opposite strong-damping limit, $\alpha \gg 1$, we find from Eq. (18.22)

$$S_{\text{inst}}/\hbar \;=\; K[\,\ln(\omega_0/\omega_c) + \ln v - \ln K - c_\infty\,] + \mathcal{O}[\,\ln(K/v)\,]\,, \qquad (18.24)$$

where $c_\infty = \ln(2\pi) - C_{\text{E}} \approx 1.2606$. For strong damping, the form (18.24) is generally valid. Only the numerical factor c_∞ depends on the particular shape of the potential. The terms $K \ln v$ and $K \ln K$ are independent of the shape of the double-well potential for fixed distance q_0 between the wells. This is because the instanton width in the limit $\alpha \gg 1$ is $\omega_R^{-1} = \alpha\omega_0^{-1} = (\pi K/v)\omega_0^{-1}$ [cf. Eq. (14.36) for $\alpha \gg 1$]. The inverse width ω_R provides the effective high-frequency cutoff of the problem.

The instanton action has been estimated for the quartic double-well potential, Eq. (3.75), in Refs. [351, 352]. The numerical constant for weak damping was obtained in Ref. [352], using a variational method, as $c_0 \approx -0.2392$, while in Ref. [351] it was found $c_0 \approx 0.2006$. In the strong damping limit, Dorsey et al. [352] found the form (18.24) with $c_\infty = 1.24$, while Chakravarty and Kivelson [351] gave $c_\infty = 1.594$.

For the cosine potential (17.46), the instanton action in the regime $\alpha \gg 1$ has been given already in Eq. (17.65). This expression is again in the form (18.24) with the constant $c_\infty = \ln(4\pi) \approx 2.5310$. For this particular potential, also the prefactor f_{inst} is known in analytic form for $\alpha \gg 1$ [cf. Eq. (17.74)],

$$f_{\text{inst}} \;=\; \sqrt{K/2\pi}\,\gamma \;=\; \sqrt{2\pi}\,K^{3/2}\omega_0/v\,. \qquad (18.25)$$

On the other hand, for the double well with the cusp at the barrier top, Eq. (18.20), one finds for $\alpha \gg 1$ [350]

$$f_{\text{inst}} \;=\; \sqrt{2.042\ln(\pi K/v)/\pi K}\;\, v\omega_0/2\pi\,. \qquad (18.26)$$

The quite different dependence of the prefactor f_{inst} on the parameters K and v in Eqs. (18.25) and (18.26) originates from the denominator of the determinantal factor [cf. Eq. (17.36)]

$$A_2 = \frac{\prod_{n=1}^{\infty}[\nu_n^2 + \gamma \nu_n + \omega_0^2]}{\prod_{n=2}^{\infty} \Lambda_n}. \tag{18.27}$$

The quantities Λ_n ($n \geq 2$) are the positive eigenvalues of the second-order fluctuation operator about the instanton–anti-instanton path.

For the cosine potential, we have $A_2 = 16\alpha^4\omega_0^2 \propto (K/v)^4$ [cf. Eq. (17.61)], whereas for the cusp potential $A_2 = 2.042 \ln(\pi K/v)\omega_0^2$, as discussed in Ref. [350]. This explains why Eqs. (18.25) and (18.26) differ by a factor $\propto (K/v)^2/\sqrt{\ln(K/v)}$.

It is plausible to assume that the behaviour $f_{\text{inst}} \propto K^{3/2}\omega_0/v$ for $\alpha \gg 1$ [cf. Eq. (18.25)] is generally found for any smooth double-well potential for which the barrier frequency ω_b is of the order of the well frequency ω_0. The prefactor (18.26) is different because the barrier frequency ω_b is singular for a cusp-shaped barrier.

It remains to demonstrate that the unphysical cutoff frequency ω_c cancels out in the calculation of physical quantities using the truncated model (18.13) with (18.14). In the Ohmic case, we have $\Delta \propto \exp[-K \ln(\omega_0/\omega_c)]$, as follows from Eq. (18.21) with (18.22). Furthermore, in physical quantities, Δ and the instanton–anti-instanton interaction $W(\tau)$ are combined in the form $\Delta^2 \exp[-W(\tau)]$ [cf. the rate expression (17.72)]. Substituting the form (17.66) for $W(\tau)$, we see that ω_c cancels out. However, when we consider the pair interaction $W(\tau)$ in the core regime, $\omega_c\tau \lesssim 1$, the cancellation is only partial. From this we infer that the cancellation is only in leading order of ϵ/ω_c, ω_c/ω_b, ω_c/ω_0, and $1/\hbar\beta\omega_c$ [352].

Let us finally resume the adiabatic renormalization scheme in which we successively integrate out high frequencies. To extend the adiabatic renormalization (18.17) to lower frequencies, we define the dressed tunneling matrix element

$$\Delta_d(\omega_\ell) \equiv \Delta \exp\left(-\frac{1}{2}\int_{\omega_\ell}^{\infty} d\omega \, \frac{G_{\text{lf}}(\omega)}{\omega^2}\right). \tag{18.28}$$

Since $\Delta_d < \Delta$, we may iteratively integrate out oscillators with frequencies in the range $\omega_\ell(\Delta) > \omega > \omega_\ell(\Delta_d) \equiv p\Delta_d$, where p is unspecified, but generally large, $p \gg 1$. The procedure converges to a dressed matrix element Δ_r which is self-consistently determined by the relation [89]

$$\Delta_r = \Delta \exp\left(-\frac{1}{2}\int_{p\Delta_r}^{\infty} d\omega \, \frac{G_{\text{lf}}(\omega)}{\omega^2}\right). \tag{18.29}$$

The treatment of the slow modes can be improved by employing the flow equation formalism by Wegner [354] in which the Hamiltonian is diagonalized by continuous unitary tranformations. This method leads to the self-consistent relation[1] [356, 357]

$$\Delta_r = \Delta \exp\left(-\frac{1}{2}\int_0^{\infty} d\omega \, \frac{G_{\text{lf}}(\omega)}{\omega^2 - \Delta_r^2}\right), \tag{18.30}$$

---

[1] A similar integral is also found for the finite-T modification, Eq. (19.42) with (19.43).

where the principal value of the improper integral is taken. The expression (18.29) disregards frequencies below $p\Delta_r$. On the other hand, in the form (18.30), slow modes $\omega < \Delta_r$ lead to an increase of the renormalized tunneling matrix element. The difference of Eq. (18.30) from (18.29) is negligible in the Ohmic and super-Ohmic case.

Consider now a spectral density of the algebraic form[2] [cf. Eq. (3.58)]

$$G_{\text{lf}}(\omega) = 2\delta_s\omega_{\text{ph}}^{1-s}\omega^s f(\omega/\omega_c), \qquad f(\omega/\omega_c) = e^{-\omega/\omega_c}. \qquad (18.31)$$

In the super-Ohmic case, $s > 1$, the integral in Eq. (18.28) remains integrable in the limit $\omega_\ell \to 0$. It is then natural to introduce a Franck-Condon factor which includes *all* modes of the reservoir. Thus we define

$$\Delta_{\text{eff}} \equiv \Delta_d(\omega_\ell = 0) = \Delta e^{-B_s} \qquad \text{for} \qquad s > 1, \qquad (18.32)$$

$$B_s = \delta_s\Gamma(s-1)(\omega_D/\omega_{\text{ph}})^{s-1}, \qquad (18.33)$$

where we have used an exponential cutoff with the Debye frequency, $\omega_c = \omega_D$.

For Ohmic damping, we have $d\ln[\Delta_d(\omega_\ell)]/d\ln(\omega_\ell) = K$. Therefore, the effects of the low-frequency modes at $T = 0$ crucially depend on the value of K. For $K > 1$, the renormalized tunneling matrix element Δ_r is iterated to zero, $\Delta_r = \Delta_d(\omega_\ell = 0) = 0$. This is the localization phenomenon obtained by Chakravarty [133] and by Bray and Moore [134] using renormalization group methods. For $K < 1$, we find from Eq. (18.29) or from Eq. (18.30), disregarding an unspecified numerical constant,

$$\Delta_r = (\Delta/\omega_c)^{K/(1-K)}\Delta \qquad \text{for} \qquad K < 1. \qquad (18.34)$$

This form was also found with a variational treatment of the sluggish modes [353].

For subsequent convenience, we also introduce a slightly modified frequency scale in which Δ_r is multiplied by the factor $[\Gamma(1-2K)\cos(\pi K)]^{1/2(1-K)}$,

$$\Delta_{\text{eff}} = [\Gamma(1-2K)\cos(\pi K)]^{1/2(1-K)}(\Delta/\omega_c)^{K/(1-K)}\Delta, \qquad K < 1. \qquad (18.35)$$

The additional factor is of order one for all $K \lesssim \frac{1}{2}$ and is singular at $K = 1$. The effective frequency Δ_{eff} is equal to Δ for $K = 0$ and equal to $\pi\Delta^2/2\omega_c$ for $K = \frac{1}{2}$. We shall see in Subsection 21.3.1 that Δ_{eff} is the proper inverse time scale for $0 \le K < 1$.

In the sub-Ohmic case, $s < 1$, we may put $\omega_c \to \infty$ since the adiabatic dressing integral is ultraviolet convergent. In the adiabatic scheme, Eq. (18.29), Δ_r is iterated to zero with an essential singularity. Thus we expect that tunneling is quenched at zero temperature, i.e., the particle is effectively trapped in one of the two localized states. On the other hand, the improved relation (18.30) predicts different behaviour for very weak sub-Ohmic coupling [358] (c is a numerical constant of order unity)

$$\delta_s^{1/(1-s)}\omega_{\text{ph}} < c\Delta. \qquad (18.36)$$

---

[2]Here we distinguish the frequency scale ω_{ph} from the cutoff ω_c. In the super-Ohmic case, we shall sometimes identify ω_c with the Debye frequency ω_D. For $s = 1$, the parameter δ_1 coincides with Kondo's parameter K and with the parameter α introduced by Leggett *et al.* [89].

In this regime, the relation (18.30) gives a nonzero renormalized tunneling matrix element. The findings are in accord with results from a study of the related Ising model [cf. Section 19.5] using a variational method [349]. The regime (18.36) is, however, of little practical interest. At finite temperatures, the iteration procedure stops at $\omega_\ell = 1/\hbar\beta$. We then obtain $\Delta_r = \Delta \exp[-\delta_s(\hbar\beta\omega_{ph})^{1-s}/(1-s)]$, which is qualitatively correct [89], as we shall see in Subsection 20.2.7.

18.1.4 Polaron transformation

In particular cases it is useful to introduce a basis of dressed states in which the oscillator α is shifted by a displacement $-\frac{1}{2}s_\alpha\sigma_z$, where $s_\alpha = q_0 c_\alpha/m_\alpha\omega_\alpha^2$, as discussed in Subsection 18.1.1. The unitary operator which transforms to the basis of the displaced harmonic oscillator states is

$$U = \exp[-\tfrac{1}{2}i\sigma_z\Omega], \quad \Omega = \sum_\alpha s_\alpha p_\alpha/\hbar = i\sum_\alpha (\lambda_\alpha/\omega_\alpha)[b_\alpha^\dagger - b_\alpha]. \quad (18.37)$$

The polaron transformation diagonalizes the last three terms in the Hamiltonian (18.13). The transformed Hamiltonian $\widetilde{H} = U^{-1}HU$ takes the exact form

$$\widetilde{H} = -\frac{\hbar\Delta}{2}\Big(|R\rangle\langle L|\,e^{i\Omega} + |L\rangle\langle R|\,e^{-i\Omega} \Big) - \frac{\hbar\epsilon}{2}\sigma_z + \sum_\alpha \hbar\omega_\alpha b_\alpha^\dagger b_\alpha, \quad (18.38)$$

where we have dropped an irrelevant constant.

Alternatively, instead of transforming the Hamiltonian, we may transform the tunneling operator as $\widetilde{\sigma}_x = U\sigma_x U^{-1}$. The polaron-transformed tunneling operator $\widetilde{\sigma}_x$ acts in the full system-plus-reservoir space and takes the form

$$\widetilde{\sigma}_x = |R\rangle\langle L|\,e^{-i\Omega} + \text{h.c.} = |R\rangle\langle L|\prod_\alpha \int dx_\alpha |x_\alpha\rangle\langle x_\alpha - s_\alpha| + \text{h.c.} \quad (18.39)$$

The operation of $\widetilde{\sigma}_x$ transfers the particle from one localized state to the other and simultaneously shifts the set of bath oscillators $\{\alpha\}$ by the set of displacements $\{s_\alpha\}$. In the dressed basis, pictorially, the particle drags behind it a polaronic cloud. The equilibrium autocorrelation function of the dressed tunneling operator evolving under the full untransformed Hamiltonian, $\widetilde{\sigma}_x(t) = e^{iHt/\hbar}\widetilde{\sigma}_x(0)\,e^{-iHt/\hbar}$, differs from the autocorrelation function of the bare σ_x. This will be discussed in Subsection 21.6.4.

18.2 Pair interaction in the charge picture

18.2.1 Analytic expression for any s and arbitrary cutoff ω_c

We have introduced a charge picture for the spin-boson model in Subsection 4.2.3. The charges are lined up with alternating sign, and the imaginary-time charge interaction $W(\tau)$ is the second integral of the kernel $\mathcal{K}(\tau)$ [cf. Eqs. (4.68) and (4.70)],

$$W(\tau) = \int_0^\infty d\omega \, \frac{G_{\text{lf}}(\omega)}{\omega^2} \, \frac{\cosh(\omega\hbar\beta/2) - \cosh[\omega(\hbar\beta/2 - \tau)]}{\sinh(\omega\hbar\beta/2)} , \qquad (18.40)$$

where the spectral density $G_{\text{lf}}(\omega)$ represents the coupling to modes with frequencies below ω_c. The function $W(\tau)$ is real for real τ and satisfies the reflection property

$$W(\tau) = W(\hbar\beta - \tau), \qquad \text{for} \qquad 0 \le \tau < \hbar\beta . \qquad (18.41)$$

In dynamical expressions, the function $W(\tau)$ is analytically continued to real time,

$$Q(t) \equiv W(\tau = it) = \int_0^\infty d\omega \, \frac{G_{\text{lf}}(\omega)}{\omega^2} \left\{ \coth\left(\frac{\hbar\beta\omega}{2}\right)\left(1 - \cos(\omega t)\right) + i\sin(\omega t) \right\} . \qquad (18.42)$$

For complex time $z = t - i\tau$, the pair interaction $Q(z)$ has the symmetry

$$Q(-z - i\hbar\beta) = Q(z) . \qquad (18.43)$$

The function $Q(z)$ is analytic in the strip $0 \ge \text{Im}\, z > -\hbar\beta$.

Consider now the pair correlation function $Q(z)$ for the spectral density $G_{\text{lf}}(\omega)$, Eq. (3.90) with Eq. (3.58) [the coupling constant $\delta_s = M\gamma_s q_0^2/2\pi\hbar$ is dimensionless],

$$G_{\text{lf}}(\omega) \doteq 2\delta_s\omega_{\text{ph}}^{1-s}\omega^s \, e^{-\omega/\omega_c} . \qquad (18.44)$$

For the form (18.44), the integration in Eq. (18.42) can be performed exactly. We find for the pair interaction for complex time z and general s the analytic form [359]

$$\begin{aligned}
Q(z) = {} & 2\delta_s\Gamma(s-1)\left(\frac{\omega_c}{\omega_{\text{ph}}}\right)^{s-1}\left\{ \left(1 - (1 + i\omega_c z)^{1-s}\right) + 2(\hbar\beta\omega_c)^{1-s}\zeta(s-1, 1+\kappa) \right. \\
& \left. - (\hbar\beta\omega_c)^{1-s}\left[\zeta\left(s-1, 1+\kappa+i\frac{z}{\hbar\beta}\right) + \zeta\left(s-1, 1+\kappa-i\frac{z}{\hbar\beta}\right)\right] \right\} . \qquad (18.45)
\end{aligned}$$

Here, $\zeta(z, q)$ is Riemann's generalized zeta function [87], $\Gamma(z)$ is Euler's gamma function, and

$$\kappa = 1/\hbar\beta\omega_c . \qquad (18.46)$$

The expression (18.45) satisfies the symmetry relation (18.43). For integer $s > 1$ ($s = 2, 3, \cdots$) and $\kappa \ll 1$, the temperature-dependent part of the pair interaction can be expressed in terms of polygamma functions. Explicit forms for $W(\tau)$ are easily obtained from Eq. (18.45) by analytic continuation. For later convenience, we also introduce the function $X(t) \equiv Q(t - i\hbar\beta/2)$,

$$X(t) \equiv X_1 - X_2(t) = \int_0^\infty d\omega \, \frac{G_{\text{lf}}(\omega)}{\omega^2}\left[\coth(\tfrac{1}{2}\hbar\beta\omega) - \frac{\cos(\omega t)}{\sinh(\tfrac{1}{2}\hbar\beta\omega)}\right] . \qquad (18.47)$$

The integral representation for $X(t)$ follows from Eq. (18.42). Using the functional relation $\zeta(x, q+1) = \zeta(x, q) - q^{-x}$, we find from Eq. (18.45) the explicit form

$$\begin{aligned}
X(z) = {} & 2\delta_s\Gamma(s-1)\left(\frac{\omega_c}{\omega_{\text{ph}}}\right)^{s-1}\left\{ 1 + 2(\hbar\beta\omega_c)^{1-s}\zeta(s-1, 1+\kappa) \right. \qquad (18.48) \\
& \left. - (\hbar\beta\omega_c)^{1-s}\left[\zeta\left(s-1, \frac{1}{2}+\kappa+i\frac{z}{\hbar\beta}\right) + \zeta\left(s-1, \frac{1}{2}+\kappa-i\frac{z}{\hbar\beta}\right)\right] \right\} .
\end{aligned}$$

The expressions (18.45) and (18.48) hold for general s and arbitrary temperature.

We see from Eq. (18.42) that for a bosonic bath $Q'(t) = \mathrm{Re}\, Q(t)$ is temperature-dependent, whereas $Q''(t) = \mathrm{Im}\, Q(t)$ is temperature-independent. A spin-$\frac{1}{2}$ bath described by the spectral density Eq. (3.187) shows opposite behaviour, namely $Q'(t)$ is temperature-independent, and the temperature dependence is captured by $Q''(t)$. The explicit calculation of the spin-bath correlation function for the various spectral densities and the study of the implications with respect to thermodynamics and dynamics is left to the reader.

18.2.2 Ohmic dissipation and universality limit

Taking in Eqs. (18.45) and (18.48) the limit $s \to 1$ we obtain the pair interaction for the Ohmic spectral density [cf. footnote to Eq. (18.31)]

$$Q(z) = 2K \ln\left(\frac{\hbar\beta\omega_c\Gamma^2(1+\kappa)}{\Gamma(\kappa + iz/\hbar\beta)\Gamma(1 + \kappa - iz/\hbar\beta)}\right), \tag{18.49}$$

$$X(z) = 2K \ln\left(\frac{\hbar\beta\omega_c\Gamma^2(1+\kappa)}{\Gamma(\frac{1}{2} + \kappa + iz/\hbar\beta)\Gamma(\frac{1}{2} + \kappa - iz/\hbar\beta)}\right). \tag{18.50}$$

We shall refer to the situation in which $\hbar\omega_c$ is the largest energy scale of the open system as the strict Ohmic case. Formally, this case corresponds to the limit $\kappa \to 0$ in Eqs. (18.49) and (18.50). We then arrive at the so-called scaling form[3]

$$Q(t) \equiv Q'(t) + iQ''(t) = 2K \ln\left(\frac{\hbar\beta\omega_c}{\pi}\sinh\frac{\pi|t|}{\hbar\beta}\right) + i\pi K \,\mathrm{sgn}\,(t), \tag{18.51}$$

$$X(t) = 2K \ln\left(\frac{\hbar\beta\omega_c}{\pi}\cosh\frac{\pi t}{\hbar\beta}\right). \tag{18.52}$$

Correspondingly, the charge interaction in imaginary time has the scaling form

$$W(\tau) = 2K \ln\left[(\hbar\beta\omega_c/\pi)\sin(\pi\tau/\hbar\beta)\right]. \tag{18.53}$$

We shall see in Subsection 19.1.1 that in the Coulomb gas representation for the partition function the charge interactions can be divided into dipole self-interactions and interdipole interactions. Each factor Δ^2 is associated with a dipole self-interaction, which for dipole length τ is $e^{-W(\tau)} \propto \omega_c^{-2K}$. Correspondingly, in the propagating function from a diagonal to a diagonal state of the RDM, each factor Δ^2 comes with a factor $e^{-Q'(t)}$, as can be seen from the exact formal expressions given below in Subsection 21.2.2. Thus, with the forms (18.51) and (18.53), Δ and ω_c occur only in the combination $\Delta^2/\omega_c^{2K} = \Delta_r^{2-2K}$, where Δ_r is defined in Eq. (18.34).

We call an observable which is a function of Δ_r without other dependence on ω_c as universal. Vice versa, any extra dependence on the cutoff ω_c is called non-universal. The universality or scaling limit is

---

[3]In the scaling limit, effects of the core regime of the pair interaction, $\tau \lesssim 1/\omega_c$, are disregarded.

$$\omega_c \to \infty \quad \text{with} \quad \Delta_r \quad \text{fixed} \, . \tag{18.54}$$

In the scaling limit, the Ohmic TSS is equivalent to the anisotropic Kondo model, and the energy scale $\hbar\Delta_r$ corresponds to the Kondo energy, up to a numerical factor which is of order 1 for $0 < K \lesssim \frac{1}{2}$. The correspondence is discussed in Section 19.4.

For $K \geq \frac{1}{2}$, the form (18.53) requires a short-time regularization. It is convenient to employ a hard-sphere repulsion at a distance $\tau_c = 1/\omega_c$ between the charges [cf. Eqs. (19.101) and (19.109) given below]. We shall see that the partition function explicitly depends on the cutoff for $K \geq \frac{1}{2}$, whereas the specific heat depends on the cutoff for $K \geq \frac{3}{2}$ and is universal for $K < \frac{3}{2}$ [cf. the discussion in Subsection 19.2.1].

19. Thermodynamics

In this chapter, we consider the effects of a dissipative environment on thermodynamic properties. First, we investigate the equilibrium properties of the open two-state system. Then we discuss, based on exact formal expressions for the partition function, the relationship of the Ohmic spin-boson model with variants of the Kondo model and with the $1/r^2$ Ising model. Finally, we calculate thermal tunneling rates by using the method of analytic continuation of the free energy. Emphasis is put on electron transfer in a solvent and on interstitial tunneling in solids.

19.1   Partition function and specific heat

We now derive an exact formal expression for the partition function in the form of a power series in Δ^2. We discuss various limits and introduce suitable approximations which then are applied to Ohmic and non-Ohmic spectral densities.

19.1.1   Exact formal expression for the partition function

In the semiclassical limit, the partition function is dominated by the classical paths in the upside-down potential. In an upside-down double well, the only classical paths available at low temperatures [apart from the trivial paths $q(\tau) = \pm q_0/2$] are sequences of well-separated instantons or kinks[1]. In the tight-binding two-state case, the path $\sigma(\tau) = (2/q_0)q(\tau)$ flips suddenly, say at times s_j, between the eigenvalues ± 1 of the Pauli matrix σ_z, as sketched in Fig. 4.1. Since each kink is followed by an anti-kink and vice versa, it is useful to group a given path consisting of $2m$ transitions into m kink–anti-kink pairs (bounces). We shall denote the lengths of the bounces by $\{\tau_j\}$ and the intervals between bounces by $\{\rho_j\}$,

$$\tau_j = s_{2j} - s_{2j-1} \, ; \qquad \rho_j = s_{2j+1} - s_{2j} \, , \qquad (j = 1, \dots, m) \, , \tag{19.1}$$

[1]Instantons or kinks are paths that interpolate at times $-\infty$ and $+\infty$ between the two distinct ground states at $-q_0/2$ and $q_0/2$, respectively. The center of the instanton is the time where the transition occurs. In the tight-binding limit, the instanton has zero width and its shape is step-like.

where $s_0 = 0$ and $s_{2m+1} = \hbar\beta$. In the equivalent charge picture, the bounces are dipoles, and the intervals τ_j and ρ_j are intra-dipole and inter-dipole lengths, respectively. Each kink–anti-kink pair or dipole contributes to the partition function a term $\Delta^2/4$ times a factor depending on the bias. Then we can write the partition function in the form of a power series in Δ^2. The contribution from the right/left well, which is the lower/higher well for $\epsilon > 0$, is [the label (E) refers to Euclidean quantities]

$$Z_{\mathrm{R/L}} = \sum_{m=0}^{\infty} \left(\frac{\Delta}{2}\right)^{2m} \int_0^{\hbar\beta} ds_{2m} \int_0^{s_{2m}} ds_{2m-1} \cdots \int_0^{s_2} ds_1\, B_{\mathrm{R/L},m}^{(\mathrm{E})}(\{s_j\}) \mathcal{F}_m^{(\mathrm{E})}(\{s_j\}) . \quad (19.2)$$

The kink centers (charge positions) s_j are the collective coordinates of the problem. The function $B_{\mathrm{R/L},m}^{(\mathrm{E})}$ subsumes all dependence on the bias,

$$B_{\mathrm{R/L},m}^{(\mathrm{E})} = \exp\left(\pm \tfrac{1}{2}\hbar\beta\epsilon \mp \epsilon \sum_{j=1}^{2m} (-1)^j s_j\right) = \exp\left(\pm \tfrac{1}{2}\hbar\beta\epsilon \mp \epsilon \sum_{j=1}^{m} \tau_j\right) . \quad (19.3)$$

The first form expresses the bias factor in terms of the flip times $\{s_j\}$, whereas the second form is given in terms of the periods $\{\tau_j\}$ spent in the left/right well.

The factor $\mathcal{F}_m^{(\mathrm{E})}$ carries the environmental influences in the form of kink or charge interactions. The influence function $\mathcal{F}_m^{(\mathrm{E})}$ has been given explicitly in Eq. (4.70),

$$\mathcal{F}_m^{(\mathrm{E})} = \exp\left\{ \sum_{j=2}^{2m} \sum_{i=1}^{j-1} (-1)^{i+j} W(s_j - s_i) \right\} . \quad (19.4)$$

The expression (19.2) with Eqs. (19.3) and (19.4) is the "Coulomb gas" representation of the partition function. It is in the form of a grand-canonical sum of unit charges ± 1 stringed with alternating sign. The influences of the environment are in the charge interaction $W(\tau)$ which is defined in Eq. (18.40).

Alternatively, the sequence of alternating charges may be viewed as a sequence of dipoles. Thus we may rewrite Eq. (19.4) as

$$\mathcal{F}_m^{(\mathrm{E})} = \exp\left(-\sum_{j=1}^{m} W(\tau_j) \right) \exp\left(-\sum_{n=1}^{m-1} \sum_{j=1}^{m-n} \Lambda_{j+n,j}^{(\mathrm{E})} \right) . \quad (19.5)$$

Here, the consecutively numbered dipoles are arranged in chronological order. The first exponential factor represents the self-interactions of the m dipoles, and the second one describes the interactions between these units. The terms $\Lambda_{j+1,j}^{(\mathrm{E})}$ are the nearest-neighbour terms, the terms $\Lambda_{j+2,j}^{(\mathrm{E})}$ are the next-nearest-neighbour terms, etc. The interaction of dipole j with a chronologically later dipole k is given by

$$\Lambda_{k,j}^{(\mathrm{E})} = W(s_{2k} - s_{2j-1}) + W(s_{2k-1} - s_{2j}) - W(s_{2k} - s_{2j}) - W(s_{2k-1} - s_{2j-1}) . \quad (19.6)$$

We remark again that Δ is already adiabatically renormalized for the modes with $\omega > \omega_c$. Therefore, if the two-state system was actually obtained from an extended system as described in Section 18.1, the physically relevant quantity Δ_{eff} becomes independent of ω_c to leading order. Just as well, the spin-boson model may be considered as an independent model in which Δ is a free parameter which does not depend on ω_c at all. The latter view is taken when the spin-boson model is connected with the Kondo model and with the Ising model [cf. Sections 19.4 and 19.5].

It is useful to write the time-ordered expression (19.2) as a Laplace integral,

$$Z_{\text{R/L}} = \frac{1}{2\pi i} \int_{\mathcal{C}} d\lambda \; e^{\lambda \hbar \beta} \, z_{\text{R/L}}(\lambda) \,, \tag{19.7}$$

where $\mathcal{C}$ is the standard Bromwich contour, i. e., any contour from $-i\infty$ to $+i\infty$ lying entirely to the right of all singularities of $z_{\text{R/L}}(\lambda)$. We then obtain

$$z_{\text{R/L}}(\lambda) = \frac{1}{\lambda \mp \epsilon/2} \sum_{m=0}^{\infty} \left(\frac{\Delta}{2}\right)^{2m} \left(\prod_{j=1}^{m} \int_0^\infty d\tau_j \, e^{-(\lambda \pm \epsilon/2)\tau_j} \int_0^\infty d\rho_j \, e^{-(\lambda \mp \epsilon/2)\rho_j}\right) \mathcal{F}_m^{(\text{E})} \,. \tag{19.8}$$

The partition function of the two-state system is

$$Z = Z_{\text{R}}(\epsilon) + Z_{\text{L}}(\epsilon) = Z_{\text{R}}(\epsilon) + Z_{\text{R}}(-\epsilon) = Z_{\text{L}}(\epsilon) + Z_{\text{L}}(-\epsilon) \,. \tag{19.9}$$

The expressions (19.7) – (19.9) represent the exact formal solution for the partition function of the spin-boson model. The result is valid for linear friction with arbitrary frequency dependence.

19.1.2 Static susceptibility and specific heat

The occupation probability of the right/left well in thermal equilibrium is given by

$$P_{\text{R/L}}^{(\text{eq})} = Z_{\text{R/L}}/Z \,, \tag{19.10}$$

yielding for the thermal expectation value of σ_z

$$\langle \sigma_z \rangle^{(\text{eq})} \equiv \frac{1}{Z} \text{tr} \left\{ \sigma_z \, e^{-\beta H} \right\} = P_{\text{R}}^{(\text{eq})} - P_{\text{L}}^{(\text{eq})} = \frac{Z_{\text{R}} - Z_{\text{L}}}{Z} \,. \tag{19.11}$$

Alternatively, we may express $\langle \sigma_z \rangle^{(\text{eq})}$ directly in terms of the free energy F,

$$\langle \sigma_z \rangle^{(\text{eq})} = \frac{2}{\hbar \beta} \frac{1}{Z} \frac{\partial Z}{\partial \epsilon} = \frac{2}{\hbar \beta} \frac{\partial \ln Z}{\partial \epsilon} = -\frac{2}{\hbar} \frac{\partial F}{\partial \epsilon} \,. \tag{19.12}$$

The response of the TSS to an external static bias is described by the static nonlinear susceptibility [2]

$$\overline{\chi}_z(T, \epsilon) = \frac{2}{\hbar} \frac{\partial \langle \sigma_z \rangle^{(\text{eq})}}{\partial \epsilon} = \frac{4}{\beta \hbar^2} \frac{\partial^2 \ln Z}{\partial \epsilon^2} = -\frac{4}{\hbar^2} \frac{\partial^2 F}{\partial \epsilon^2} \,. \tag{19.13}$$

---

[2]We mark the static susceptibility by a bar, and the dynamical susceptibility by a tilde. Throughout part IV, we choose a normalization which differs by a factor of 4 compared to the 1993 edition in order to have agreement with the standard form (21.232) of the fluctuation-dissipation theorem for the σ_z autocorrelation function. There is also a factor 4 difference with respect to the literature on the s-d model, since there the corresponding susceptibility χ_{sd} is normalized according to the correlation function for the spin $S_z = \sigma_z/2$. The equivalence relation is $\overline{\chi}_z = 4\chi_{sd}/(g\mu_{\text{B}})^2$.

In context with the Kondo problem, $\bar{\chi}_z$ describes the response of the impurity to a magnetic field perturbation $-g\mu_B\hbar\sigma_z/2$. In the two-state system, the static susceptibility $\bar{\chi}_z$ is the response to a static strain field or bias. We shall see in Subsection 21.2.4 that the static susceptibility can be also calculated in a dynamical approach.

Finally, the specific heat is readily obtained from the thermodynamic relation

$$c \equiv c_V/k_B = \beta^2 \partial^2 \ln Z/\partial\beta^2 . \tag{19.14}$$

The thermodynamic method presented in this chapter gives the partition function in the form of a power series in Δ^2. We see from the defining relations (19.11) – (19.14) that we have to perform resummations of the resulting series in order to obtain the thermodynamic response functions in the form of power series representations in Δ^2. In Chapter 21, we shall consider the thermodynamic response functions by a different approach. There we derive within a *dynamical approach* directly exact formal series expressions in Δ^2 for $\langle\sigma_z\rangle^{(eq)}$ and for $\ln Z$. As an advantage over the series for Z, they give upon differentiation directly the respective series expansions for the thermodynamic response functions.

19.1.3 The self-energy method

In the absence of dissipation, we have $\mathcal{F}_m^{(E)} = 1$. Then the series (19.8) for the Laplace transform of the partition function of the right well is a geometrical series which can be summed to the form[3]

$$z_{R/L}^{(0)}(\lambda) = \frac{1}{\left(\lambda \mp \epsilon/2\right)} \frac{1}{\left(1 - \dfrac{\Delta^2}{4}\dfrac{1}{(\lambda^2 - \epsilon^2/4)}\right)} = \frac{\lambda \pm \epsilon/2}{\lambda^2 - (\Delta^2 + \epsilon^2)/4} . \tag{19.15}$$

The expression (19.15) has simple poles at $\lambda = \pm\Delta_b/2$, where

$$\Delta_b \equiv (\Delta^2 + \epsilon^2)^{1/2} . \tag{19.16}$$

The transform $z_{R/L}^{(0)}(\lambda)$ is easily inverted to obtain $Z_{R/L}^{(0)}(\hbar\beta)$,

$$Z_{R/L}^{(0)}(\hbar\beta) = \cosh(\hbar\beta\Delta_b/2) \pm (\epsilon/\Delta_b)\sinh(\hbar\beta\Delta_b/2) . \tag{19.17}$$

We remark that the same form for $Z_{R/L}^{(0)}(\hbar\beta)$ is found if we substitute into the defining expression $Z_{R/L}^{(0)}(\hbar\beta) = < R/L| \, e^{-\beta H_{TSS}} \, |R/L >$ the state relation (3.82).

The specific heat resulting from $Z^{(0)}(\hbar\beta)$ shows the familiar Schottky anomaly

$$c = (\hbar\beta\Delta_b/2)^2 \operatorname{sech}^2(\hbar\beta\Delta_b/2) . \tag{19.18}$$

From this we find that the specific heat of the undamped system is exponentially small at low temperatures $k_B T \ll \hbar\Delta_b$, $c \approx (\hbar\Delta_b/k_B T)^2 \exp(-\hbar\Delta_b/k_B T)$, while it depends algebraically on temperature, $c \approx (\hbar\Delta_b/2k_B T)^2$ at high temperatures $k_B T \gg \hbar\Delta_b$.

---

[3]For later convenience, we write Δ instead of Δ_0.

For the open system, we may write $z_{\mathrm{R/L}}(\lambda)$ in the form

$$z_{\mathrm{R/L}}(\lambda) \;=\; \frac{1}{[\,z_{\mathrm{R/L}}^{(0)}(\lambda)\,]^{-1} - \Sigma_{\mathrm{R/L}}(\lambda)}\,, \qquad (19.19)$$

where $\Sigma_{\mathrm{R/L}}(\lambda)$ captures the effects of the interaction with the environment. Following the Green function terminology, we shall refer to the function $\hbar\Sigma_{\mathrm{R/L}}(\lambda)$ as the self-energy. Formally, the analog of the self-energy is an isobaric ensemble of interacting kink–anti-kink pairs or dipoles. In graphical terms, the self-energy diagrams are *irreducible*, i.e., they cannot be divided into disconnected subdiagrams without cutting inter-dipole interaction lines. To proceed, we define a modified influence functional which is irreducible by construction. Subtracting all reducible parts of $\mathcal{F}_n^{(\mathrm{E})}$, the irreducible influence functional reads

$$\widetilde{\mathcal{F}}_n^{(\mathrm{E})} \;\equiv\; \mathcal{F}_n^{(\mathrm{E})} - \sum_{j=2}^{n} (-1)^j \sum_{m_1,\cdots,m_j \geq 1} \mathcal{F}_{m_1}^{(\mathrm{E})}\,\mathcal{F}_{m_2}^{(\mathrm{E})}\cdots\mathcal{F}_{m_j}^{(\mathrm{E})}\,\delta_{m_1+\cdots+m_j,n}\,. \qquad (19.20)$$

Here again we choose the convention that in each term the dipoles are consecutively numbered and arranged in chronological order. In the subtracted terms, the bath correlations are only inside of the individual factors $\mathcal{F}_{m_j}^{(\mathrm{E})}$. To illustrate the general expression (19.20), we explicitly give the case $n = 3$,

$$\widetilde{\mathcal{F}}_3^{(\mathrm{E})} \;\equiv\; \mathcal{F}_3^{(\mathrm{E})} - \mathcal{F}_2^{(\mathrm{E})}\,\mathcal{F}_1^{(\mathrm{E})} - \mathcal{F}_1^{(\mathrm{E})}\,\mathcal{F}_2^{(\mathrm{E})} + \mathcal{F}_1^{(\mathrm{E})}\,\mathcal{F}_1^{(\mathrm{E})}\,\mathcal{F}_1^{(\mathrm{E})}$$

$$= \; \mathrm{e}^{-W(\tau_1)-W(\tau_2)-W(\tau_3)} \qquad\qquad\qquad\qquad (19.21)$$

$$\times \left\{ \left(\mathrm{e}^{-\Lambda_{3,1}^{(\mathrm{E})}} - 1\right)\mathrm{e}^{-\Lambda_{3,2}^{(\mathrm{E})}-\Lambda_{2,1}^{(\mathrm{E})}} + \left(\mathrm{e}^{-\Lambda_{3,2}^{(\mathrm{E})}} - 1\right)\left(\mathrm{e}^{-\Lambda_{2,1}^{(\mathrm{E})}} - 1\right)\right\}\,.$$

From this form, we can directly see the irreducibility of $\widetilde{\mathcal{F}}_3^{(\mathrm{E})}$.

With the irreducible influence function (19.20), the series in Δ^2 for $\Sigma_{\mathrm{R/L}}(\lambda)$ reads

$$\Sigma_{\mathrm{R/L}}(\lambda) \;=\; \sum_{m=1}^{\infty} \Sigma_{\mathrm{R/L}}^{(m)}(\lambda)\,, \qquad (19.22)$$

where

$$\Sigma_{\mathrm{R/L}}^{(1)}(\lambda) \;=\; \frac{\Delta^2}{4}\int_0^\infty d\tau\, \mathrm{e}^{-(\lambda\pm\epsilon/2)\tau}\left(\mathrm{e}^{-W(\tau)} - 1\right)\,, \qquad (19.23)$$

and for $m \geq 2$,

$$\Sigma_{\mathrm{R/L}}^{(m)}(\lambda) \;=\; \left(\frac{\Delta^2}{4}\right)^{m}\left(\prod_{j=1}^{m}\int_0^\infty d\tau_j\, \mathrm{e}^{-(\lambda\pm\epsilon/2)\tau_j}\right)\left(\prod_{k=1}^{m-1}\int_0^\infty d\rho_k\, \mathrm{e}^{-(\lambda\mp\epsilon/2)\rho_k}\right)\widetilde{\mathcal{F}}_m^{(\mathrm{E})}\,.$$

$$(19.24)$$

Upon expanding the expression (19.19) in powers of $\Sigma_{\mathrm{R/L}}(\lambda)$ and substituting the series expansion (19.22) with the definitions (19.23) and (19.24), the original series

(19.8) is recovered. This verifies that the expression (19.19) with Eqs. (19.22) – (19.24) is a formally exact rearrangement of the original series (19.8).

In practical computations, one may truncate the series (19.22) for the self-energy at a given order, and then invert the Laplace transform (19.19) to obtain $Z_{R/L}(\hbar\beta)$. In cases where explicit inversion is difficult, much can nevertheless be learned from a study of the singularities of $z_{R/L}(\lambda)$.

19.1.4 The limit of high temperatures

For temperatures $T \gg \hbar\Delta_{\text{eff}}/k_B$, the leading contributions to the partition function are the terms with $m = 0$ and $m = 1$ in Eq. (19.2). We find from Eq. (19.2) or (19.8)

$$Z_{R/L} = e^{\pm\hbar\beta\epsilon/2}\left(1 + \frac{\Delta^2}{4}\int_0^{\hbar\beta}d\tau\,[\hbar\beta - \tau]\,e^{\mp\epsilon\tau - W(\tau)} + \mathcal{O}(\Delta^4)\right). \qquad (19.25)$$

The symmetry property (18.41) suggests to map the interval $\hbar\beta/2 \leq \tau \leq \hbar\beta$ onto the interval $0 \leq \tau \leq \hbar\beta/2$. We then obtain for $Z = Z_R + Z_L$ the expression

$$Z(\beta) = 2\cosh(\tfrac{1}{2}\hbar\beta\epsilon) + \frac{\hbar\beta\Delta^2}{2}\int_0^{\hbar\beta/2}d\tau\,\cosh[\epsilon(\tfrac{1}{2}\hbar\beta - \tau)]\,e^{-W(\tau)}. \qquad (19.26)$$

Thus, the leading high-temperature dependence of the partition function of the TSS is determined by the one-dipole contribution.

19.1.5 Noninteracting–kink-pair approximation

With decreasing temperature, the terms with $m > 1$ of the series expression (19.2) become relevant. Let us now assume for the moment that the width of a typical kink–anti-kink pair is very much smaller than the typical interval between such pairs. This seems to be intuitively obvious since the average width $\langle\tau\rangle$ tends to be suppressed relative to the average interval $\langle\rho\rangle$ because of the self-interactions in Eq. (19.5). The picture we then have is that the kink pairs or dipoles form a dilute "gas" in the fixed "volume" $\hbar\beta$, and we may argue that the inter-dipole interactions $\Lambda_{j,k}^{(E)}$ may safely be ignored relative to the intra-dipole or self-interactions. Within the noninteracting–kink-pair approximation, we have $\mathcal{F}_{2m}^{(E)} = \exp[-\sum_{j=1}^m W(\tau_j)]$. With this form, the terms in the series (19.2) are convolutions, and the series (19.8) for the Laplace transform becomes a geometrical series which can be summed to the form

$$z_{R/L}^{(1)}(\lambda) = \frac{1}{\left(\lambda \mp \epsilon/2\right)}\frac{1}{\left(1 - K_{R/L}(\lambda)/[\lambda \mp \epsilon/2]\right)}, \qquad (19.27)$$

$$K_{R/L}(\lambda) = \frac{\Delta^2}{4}e^{i\varphi}\int_0^\infty du\,e^{-[\lambda \pm \epsilon/2]\tau(u)}e^{-W[\tau(u)]}, \qquad (19.28)$$

where $\tau(u) = u\,e^{i\varphi}$, and $0 \le \varphi \le \frac{1}{2}\pi$. Here we have used the analytic properties of the function $W(\tau)$. The original contour is $\varphi = 0$. In numerical computations, it is favourable to choose $\varphi = \pi/2$. The expression (19.27) can be rewritten in the form (19.19) in which the self-energy is the expression $\Sigma_{\mathrm{R/L}}^{(1)}(\lambda)$ given in Eq. (19.23). Thus, the noninteracting–kink-pair approximation corresponds to considering the self-energy in order Δ^2. We then obtain

$$Z_{\mathrm{R/L}}^{(1)}(\hbar\beta) = \frac{1}{2\pi i} \int_{\mathcal{C}} d\lambda \, \frac{e^{\lambda\hbar\beta}}{[\,z_{\mathrm{R/L}}^{(0)}(\lambda)\,]^{-1} - \Sigma_{\mathrm{R/L}}^{(1)}(\lambda)} \ . \qquad (19.29)$$

Unfortunately, the kink–anti-kink pairs or dipoles do *not* actually form a dilute gas except for fairly strong damping. In this regime, the attractive self-interaction of the pair is strong enough at short distances that the breathing-mode integral $K_{\mathrm{R/L}}(\lambda)$ can be expanded about $\lambda = \pm\epsilon$,

$$K_{\mathrm{R/L}}(\lambda) = K_{\mathrm{R/L}}(\pm\epsilon) + (\lambda \mp \epsilon)\, K_{\mathrm{R/L}}'(\pm\epsilon) + \cdots . \qquad (19.30)$$

Now, the average dipole length $\langle \tau \rangle$, which can be estimated as $|K_{\mathrm{R/L}}'(\pm\epsilon)|/K_{\mathrm{R/L}}(\pm\epsilon)$, is small compared with the average interval $\langle \rho \rangle$ which is of the order of $1/K_{\mathrm{R/L}}(\pm\epsilon)$ provided that

$$|K_{\mathrm{R/L}}'(\pm\epsilon)| \equiv \frac{\Delta^2}{4} \int_0^\infty d\tau \, \tau \, e^{\mp\,\epsilon\tau - W(\tau)} \ll 1 \ . \qquad (19.31)$$

When this condition is met, the dipoles form a dilute gas. If we truncate the series (19.30) after the first term, then $z_{\mathrm{R/L}}(\lambda)$ has a simple pole at $\lambda = \pm\,\epsilon/2 + K_{\mathrm{R/L}}(\pm\epsilon)$. The transform $z_{\mathrm{R/L}}^{(1)}(\lambda)$ is easily inverted. The free energy associated with $Z_{\mathrm{R/L}}^{(1)}$ is

$$F_{\mathrm{R/L}}^{(1)} = \mp\frac{\hbar\epsilon}{2} - \frac{\hbar\Delta^2}{4} \int_0^\infty d\tau \, e^{\mp\,\epsilon\tau - W(\tau)} \ . \qquad (19.32)$$

If the condition of strong damping and/or high temperature is not met, the noninteracting–kink-pair approximation fails.[4] The approximation is inconsistent in particular regimes since it does not properly take into account the reflection property (18.41) of the interactions. Moreover, in the weak-coupling limit, the interactions $\Lambda_{j,k}^{(E)}$ render contributions to the first order in δ_s even for zero bias, as we can see from the expression (19.2) with (19.5). Thus, it is *not* consistent to make the noninteracting–kink-pair approximation in the regime of weak to moderate damping and/or low to moderate temperatures.

---

[4]Compared to the noninteracting-blip approximation (NIBA) which is useful in the calculation of dynamical quantities [cf. Section 21.3], this approximation is valid in a smaller region of the parameter space. In the NIBA, there are partial cancellations among interblip correlations. Corresponding cancellations are absent in the present case.

19.1.6 Weak-damping limit

In the weak-damping limit, the coupling to the environment can be treated as a perturbation. It is clear that a perturbative expansion of the partition function itself is not very meaningful. Rather we aim at a perturbative calculation of the self-energy to first order in the coupling parameter δ_s. In the second stage, we then calculate the shift of the poles of $z_{R/L}(\lambda)$ to linear order in δ_s.

It is convenient to employ the renormalized frequency scale Δ_{eff}, as defined in Eqs. (18.32) and (18.35) for the super-Ohmic and Ohmic case, respectively. To proceed, we follow Ref. [360]. We expand $\Sigma_{R/L}(\lambda)$ as in Eq. (19.22) and consider each term of the series to linear order in δ_s. Thus we may write for weak coupling (wc)

$$\Sigma_{R/L}^{(\text{wc})}(\lambda) = \Sigma_{R/L}'(\lambda) + \Sigma_{R/L}''(\lambda) , \qquad (19.33)$$

where $\Sigma_{R/L}'(\lambda)$ is the contribution from the intra-dipole interaction,

$$\Sigma_{R/L}'(\lambda) = \frac{\Delta_{\text{eff}}^2}{4} \int_0^\infty d\tau \left(2q - W_{T=0}(\tau) \right) e^{-(\lambda \pm \epsilon/2)\tau} , \qquad (19.34)$$

which is of order Δ_{eff}^2. The term $2q$ is a counter term. It will be chosen such that the effective level spacing at $T = 0$ comes out correctly as $\hbar\Delta_b$, where

$$\Delta_b = (\Delta_{\text{eff}}^2 + \epsilon^2)^{1/2} . \qquad (19.35)$$

The other term $\Sigma_{R/L}''(\lambda)$ in Eq. (19.33) captures the contributions from the inter-dipole interactions. We find from Eq. (19.24) the series

$$\Sigma_{R/L}''(\lambda) = -\sum_{m=2}^\infty \left(\frac{\Delta_{\text{eff}}^2}{4} \right)^m \left(\prod_{j=1}^m \int_0^\infty d\tau_j \, e^{-(\lambda \pm \epsilon/2)\tau_j} \right) \left(\prod_{k=1}^{m-1} \int_0^\infty d\rho_k \, e^{-(\lambda \mp \epsilon/2)\rho_k} \right) \Lambda_{m,1}^{(E)} , \qquad (19.36)$$

where $\Lambda_{k,j}^{(E)}$ is defined in Eq. (19.6). The series (19.36) describes sequences of dipoles in which the first dipole interacts with the last one. The intermediate dipoles are noninteracting and form a grand-canonical ensemble in the "volume" ρ. For the term of order Δ_{eff}^{2m} we have

$$\rho = \sum_{j=2}^{m-1} \tau_j + \sum_{k=1}^{m-1} \rho_k . \qquad (19.37)$$

The expression (19.36) is irreducible in the sense specified in Subsection 19.1.3. Note that the term $\Sigma_{R/L}''(\lambda)$ is disregarded in the noninteracting–kink-pair approximation. Physically, $\Sigma_{R/L}''(\lambda)$ describes one-phonon thermal processes in which the system may make any number of tunneling transitions in the interval between emission and absorption of the phonon. The grand-canonical ensemble of non-interacting kinks in the interval ρ can be summed in analytic form yielding the imaginary-time propagator with initial and final state $\sigma = +1$ (right well) [cf. Eq. (19.17)],

$$Z_{R/L}^{(0)}(\rho) = \cosh(\rho\Delta_b/2) \pm (\epsilon/\Delta_b)\sinh(\rho\Delta_b/2) . \qquad (19.38)$$

With this summation carried out, the quantity $\Sigma''_{R/L}(\lambda)$ takes the form

$$\Sigma''_{R/L}(\lambda) = -\frac{\Delta^4_{\text{eff}}}{16}\int_0^\infty d\tau_1\, d\rho\, d\tau_2\, e^{-(\lambda\pm\epsilon/2)(\tau_1+\tau_2)}\, e^{-(\lambda\mp\epsilon/2)\rho}\, Z^{(0)}_{R/L}(\rho)\,\Lambda^{(E)}_{2,1}\,, \quad (19.39)$$

where

$$\Lambda^{(E)}_{2,1} = W(\tau_1+\rho+\tau_2) + W(\rho) - W(\tau_1+\rho) - W(\rho+\tau_2)\,. \quad (19.40)$$

The self-energy correction $\hbar\Sigma^{(wc)}_{R/L}(\lambda)$ shifts the poles of $z_{R/L}(\lambda)$ in Eq. (19.19) from $\lambda = \pm\Delta_b/2$ to $\lambda = \pm\Omega/2$. To linear order in δ_s, the frequency Ω is given by

$$\Omega^2 = \Delta^2_b - 2\left(\Delta_b \pm \epsilon\right)\Sigma_{R/L}(\lambda = \Delta_b/2)\,. \quad (19.41)$$

The expressions (19.34) and (19.39) are conveniently evaluated by using the spectral representation (18.40) for the pair interaction $W(\tau)$ and by commuting the frequency integral with the time integrals. We readily obtain

$$\Omega^2 = \Delta^2_{\text{eff}} + \epsilon^2 - 2\Delta^2_{\text{eff}}\,\mathrm{Re}\,v\!\left((\Delta^2_{\text{eff}}+\epsilon^2)^{1/2}\right)\,, \quad (19.42)$$

where Δ_{eff} is the dressed tunneling matrix element at $T = 0$ introduced for the particular cases in Eqs. (18.32) and (18.35), and

$$v(z) \equiv \int_0^\infty d\omega\,\frac{G_{1f}(\omega)}{\omega^2 - z^2}\,\frac{1}{e^{\omega\hbar\beta} - 1}\,. \quad (19.43)$$

Thus, the frequency Ω turns out symmetric in ϵ. The last term in Eq. (19.42) vanishes at zero temperature. Picking up the two poles by the contour integral (19.7), we find $Z_{R/L}$ exactly in the form (19.17) in which Δ_b is replaced by Ω,

$$Z_{R/L}(\hbar\beta) = \cosh(\hbar\beta\Omega/2) \pm (\epsilon/\Omega)\sinh(\hbar\beta\Omega/2)\,. \quad (19.44)$$

To summarize, the form of the partition function is left unchanged for weak damping, but the level splitting becomes renormalized and depends on the spectral density of the coupling to the reservoir and on temperature.

19.1.7   The self-energy method revisited: partial resummation

In Subsection 19.1.3, we have explained the systematic series expansion in Δ^2 for the self-energy. In this section, we give an iterative scheme where in each step the self-energy is calculated with insertions of the imaginary-time propagator which has been obtained in the preceding step.

Assume that we have calculated the self-energy in the nth step of the iteration scheme up to terms with n irreducible dipoles. We then have the form

$$\Sigma_{R/L,n}(\lambda) = \Sigma^{(1)}_{R/L}(\lambda) + \sum_{m=2}^n \Sigma^{(m)}_{R/L,n}(\lambda)\,. \quad (19.45)$$

The single-dipole contribution $\Sigma_{\mathrm{R/L}}^{(1)}(\lambda)$ is the same in each step and is defined in Eq. (19.23). Further assume that in all terms with $m \geq 2$ we have made the insertions of the propagator, which has been calculated in the previous step, into the intervals the system spends in the right well. The self-energy (19.45) defines the imaginary-time propagator with boundary conditions $\sigma(0) = \sigma(\rho) = \pm 1$ (right/left state) according to the integral expression

$$Z_{\mathrm{R/L}}^{(n)}(\rho) \;=\; \frac{1}{2\pi i}\int_C d\lambda\, \frac{e^{\lambda\rho}}{[\,z_{\mathrm{R/L}}^{(0)}(\lambda)\,]^{-1} - \Sigma_{\mathrm{R/L},n}(\lambda)}\,. \qquad (19.46)$$

In the next step of the iteration, we calculate the self-energy as

$$\Sigma_{\mathrm{R/L},n+1}(\lambda) \;=\; \Sigma_{\mathrm{R/L}}^{(1)}(\lambda) + \sum_{m=2}^{n+1} \Sigma_{\mathrm{R/L},n+1}^{(m)}(\lambda)\,, \qquad (19.47)$$

in which $\Sigma_{\mathrm{R/L},n+1}^{(m)}(\lambda)$ for $m \geq 2$ is given by

$$\Sigma_{\mathrm{R/L},n+1}^{(m)}(\lambda) \;=\; \left(\frac{\Delta^2}{4}\right)^m \left(\prod_{j=1}^{m}\int_0^\infty d\tau_j\, e^{-(\lambda \pm \epsilon/2)\tau_j}\right)\left(\prod_{k=1}^{m-1}\int_0^\infty d\rho_k\, e^{-(\lambda \mp \epsilon/2)\rho_k}\, Z_{\mathrm{R/L}}^{(n)}(\rho_k)\right)$$

$$\times \exp\left(-\sum_{j=1}^{m} W(\tau_j) - \sum_{n=2}^{m-1}\sum_{j=1}^{m-n}\Lambda_{j+n,j}^{(\mathrm{E})}\right)\prod_{j=1}^{m-1}\left(\exp\left(-\Lambda_{j+1,j}^{(\mathrm{E})}\right) - 1\right). \qquad (19.48)$$

Here, m is the number of dipoles in the "bare" self-energy term. The $\{\tau_j\}$ are the dipole lengths (periods spent in the left well), and the $\{\rho_k\}$ are the intervals between the dipoles. In Eq. (19.48), the bare self-energy term is dressed by inserting in each interval ρ_k the Euclidean propagator obtained in the previous step, Eq. (19.46). The first term in the second line contains the self-interactions of the dipoles and all inter-dipole interactions, except of those between nearest neighbours. The nearest-neighbour interactions are contained in the last term and we have subtracted in each interaction factor a one in order that the expression (19.48) is irreducible.

The procedure starts with the computation of the imaginary-time propagator $Z_{\mathrm{R/L}}^{(1)}(\rho)$ in the noninteracting–kink-pair approximation, Eq. (19.29). We then insert this form into the interdipole interval of two irreducible dipoles. Thus we obtain the improved self-energy $\Sigma_{\mathrm{R/L},2}(\lambda) = \Sigma_{\mathrm{R/L}}^{(1)}(\lambda) + \Sigma_{\mathrm{R/L},2}^{(2)}(\lambda)$ which gives the propagator $Z_{\mathrm{R/L}}^{(2)}(\rho)$ according to Eq. (19.46), and so on.

19.2 Ohmic dissipation

So far, the treatment has been quite general since we did not specify the spectral density of the coupling. Consider now the Ohmic case. In the most interesting regime $k_{\mathrm{B}}T \ll \hbar\omega_c$, the pair interaction is given by the scaling form (18.53). We now give results in various limits. Then we turn to the exactly solvable case $K = 1/2$.

19.2.1   General results

Substituting the form (18.53) for $W(\tau)$ into the high temperature formula (19.25), the integral must be supplemented by a hard-sphere cutoff at $\tau = 1/\omega_c$ for $K \geq \frac{1}{2}$. The cutoff is irrelevant for the specific heat in the range $K < \frac{3}{2}$, whereas it is relevant for $K > \frac{3}{2}$ (see below). Using Eqs. (19.9) and (19.14), we find for the specific heat in the regime $K < 1$ and zero bias ($\epsilon = 0$) the analytic expression

$$c = \frac{(1-K)(1-2K)}{\cos(\pi K)} \left(\frac{(2\pi)^K}{2\Gamma(1-K)} \right)^2 \theta^{2K-2} + \mathcal{O}\left(\theta^{4K-4}\right), \qquad (19.49)$$

where θ is a dimensionless temperature, and the scale Δ_{eff} is defined in Eq. (18.35),

$$\theta = k_B T / \hbar \Delta_{\text{eff}}. \qquad (19.50)$$

The high temperature formula is valid for temperatures where $\theta \gg 1$.

For weak-damping $K \ll 1$ and general ϵ, the level splitting $\hbar\Omega$ has been calculated in Subsection 19.1.6. Substituting the Ohmic form of the spectral density $G_{\text{lf}}(\omega)$ [the case $s = 1$ in Eq. (18.44)] into Eq. (19.42) with Eq. (19.43), we find for all T

$$\Omega^2 = \Delta_{\text{eff}}^2 + \epsilon^2 + 2K\Delta_{\text{eff}}^2 \left\{ \text{Re}\,\psi\left(\frac{i\hbar\sqrt{\Delta_{\text{eff}}^2 + \epsilon^2}}{2\pi k_B T} \right) - \ln\left(\frac{\hbar\sqrt{\Delta_{\text{eff}}^2 + \epsilon^2}}{2\pi k_B T} \right) \right\}, \quad (19.51)$$

where $\psi(z)$ is the digamma function. Inserting this form into Eq. (19.44) and using Eqs. (19.9) and (19.14), we obtain the specific heat in the regime $\theta \ll 1$ in the form

$$c = K\varrho^2 \sum_{n=1}^{\infty} (2n-1)\pi^{2n} |B_{2n}| (2\varrho\theta)^{2n-1}, \qquad \text{for} \quad K \ll 1, \qquad (19.52)$$

where B_m is a Bernoulli number and $\varrho = \Delta_{\text{eff}}/\sqrt{\Delta_{\text{eff}}^2 + \epsilon^2}$. The series (19.52) is the exact asymptotic expansion in the weak-damping limit. We note that the noninteracting–kink-pair approximation does not describe correctly the low-temperature regime. Namely, it predicts also even powers of θ in the low-temperature expansion for the specific heat. In reality, these terms are absent.

When the specific heat $c(T)$ varies linear with temperature and the static susceptibility $\overline{\chi}_z(T, \epsilon)$ defined in Eq. (19.13) approaches a constant as $T \to 0$, a useful quantity is the so-called Wilson ratio

$$R \equiv \lim_{T \to 0} 4\, c(T, \epsilon) \Big/ k_B T\, \overline{\chi}_z(T, \epsilon), \qquad (19.53)$$

where the factor of 4 is chosen for convenience [see footnote to Eq. (19.13)].

For weak damping and zero bias, we find from Eq. (19.44) the form $\overline{\chi}_z(0,0) = 2/(\hbar\Delta_{\text{eff}})$. Combining this expression with the weak-damping result (19.52), we obtain for $K \ll 1$ the Wilson ratio

$$R = 2K\pi^2/3. \qquad (19.54)$$

In the language of the equivalent Kondo problem [see Section 19.4], the system shows Fermi liquid behaviour at low temperatures. Surprisingly, the weak-damping result (19.54) turns out to be exact for all K in the regime $0 < K < 1$ [208]. Moreover, it also holds for nonzero bias, and, in this case, it is even valid for $K > 1$. The corresponding proof is given in Section 21.9, following a different line of arguments.

The damped system behaves Schottky-like in the regime $0 < K < 1$ (cf. the discussion in Ref. [359]). The specific heat increases algebraically as the temperature is raised above zero, instead of the exponentially small enhancement in the zero-damping limit. It then approaches a maximum near $T = \hbar\Delta_{\rm eff}/k_{\rm B}$. Well above this value, c drops again to zero algebraically, and the exponents depend on the coupling constant K [cf. Eq. (19.49)].

The regime $K > 1$ corresponds to the ferromagnetic sector of the Kondo problem. This case is mathematically easier since the perturbation expansion in Δ^2 is valid down to zero temperature. The formula (19.49) can be analytically continued to the regime $3/2 > K > 1$ using standard functional relations of the gamma functions. We find

$$c = (K-1)^2 \frac{\pi^{3/2}}{2} \frac{\Gamma(3/2 - K)}{\Gamma(2 - K)} \frac{\Delta^2}{\omega_{\rm c}^2} \left(\frac{\pi k_{\rm B} T}{\hbar \omega_{\rm c}} \right)^{2K-2} . \tag{19.55}$$

This term is the leading one for all T in the regime $\hbar\beta\omega_{\rm c} \gg 1$. Thus, the Schottky peak has faded away and the specific heat increases monotonously with T.

For strong damping $K > 3/2$, the leading contribution to the specific heat originates from the hard-sphere cutoff at $\tau = 1/\omega_{\rm c}$ mentioned above and after Eq.(18.53). This term describes Fermi liquid behaviour for all T below $\hbar\omega_{\rm c}/k_{\rm B}$ [cf. Ref. [359]],

$$c = \frac{\pi^2}{6} \frac{K}{(2K - 3)} \frac{\Delta^2}{\omega_{\rm c}^2} \frac{k_{\rm B} T}{\hbar \omega_{\rm c}} \qquad \text{for} \qquad K > 3/2 . \tag{19.56}$$

For the particular value $K = 3/2$, the specific heat is given by the sum of the expressions (19.55) and (19.56). Each term is singular in the limit $K \to 3/2$. However, in the sum the singularities cancel each other. We get

$$c = \frac{\pi^2}{12} \frac{\Delta^2}{\omega_{\rm c}^2} \frac{k_{\rm B} T}{\hbar \omega_{\rm c}} \qquad \text{for} \qquad K = 3/2 . \tag{19.57}$$

Thus, the specific heat is regular for the Kondo parameter $K = \frac{3}{2}$.

19.2.2   The special case $K = \frac{1}{2}$

For the special value $\frac{1}{2}$ of K, the influence interaction factor (19.4) or (19.5) with the pair interaction (18.53) can be decomposed into a sum of $m(m-1)$ terms which represent the combinatorical possibilities of dividing up m kinks and m anti-kinks into kink–anti-kink pairs. In each term, the given sequence of kinks and anti-kinks (cf. Fig. 4.1) is left unchanged and only the self-interactions of each pair or dipole are taken into account. For instance, the influence factor for two kink pairs or dipoles with dipole lengths τ_1 and τ_2 and distance ρ,

$$\mathcal{F}_2^{(\mathrm{E})} \equiv e^{-W(\tau_1)-W(\tau_2)-W(\rho)-W(\tau_1+\rho+\tau_2)+W(\tau_1+\rho)+W(\rho+\tau_2)} , \qquad (19.58)$$

can be decomposed as

$$\mathcal{F}_2^{(\mathrm{E})} = e^{-W(\tau_1)-W(\tau_2)} + e^{-W(\rho)-W(\tau_1+\rho+\tau_2)} . \qquad (19.59)$$

The equivalence of Eq. (19.59) with Eq. (19.58) holds for the Ohmic form (18.53) with $K = \frac{1}{2}$. Substituting the respective decomposition for $\mathcal{F}_m^{(\mathrm{E})}$ into the series (19.2), the integrals over the flip times can be carried out in analytic form (cf. Ref. [359] for details). The coefficients of the resulting series expression for Z are expressed in terms of the integral

$$K_m = \int_{x_0}^{\infty} dx \, \frac{x^m}{\sinh x} , \qquad (19.60)$$

where $x_0 = \pi/\hbar\beta\omega_c$ is a hard-sphere cutoff. In reality, the cutoff is only needed for the $m = 0$ moment. The partition function for the symmetric system is found as

$$Z = 2\exp\left(\sum_{m=1}^{\infty} \frac{(-1)^{m-1}}{m!} \frac{K_{m-1}}{(2\pi\theta)^m} \right) = 2\exp\left(\int_{x_0}^{\infty} dx \, \frac{1-e^{-x/2\pi\theta}}{x \, \sinh x} \right) , \qquad (19.61)$$

where θ is given in Eq. (19.50). For $K = \frac{1}{2}$, we have

$$\theta = \frac{k_{\mathrm{B}}T}{\hbar\gamma} , \qquad \text{where} \qquad \gamma \equiv \Delta_{\mathrm{eff}}(K = \tfrac{1}{2}) = \frac{\pi}{2} \frac{\Delta^2}{\omega_c} . \qquad (19.62)$$

Upon inserting the form (19.61) into Eq. (19.14), we obtain for the specific heat

$$c = \frac{1}{2\pi\theta} - \frac{1}{(2\pi\theta)^2} \int_0^{\infty} dx \, \frac{x \, e^{-x/2\pi\theta}}{\sinh x} , \qquad (19.63)$$

which is free of a short-distance singularity. The integral can be expressed in terms of the trigamma function [87] as

$$c = \frac{1}{2\pi\theta} - \frac{1}{2(2\pi\theta)^2} \, \psi'\left(\frac{1}{2} + \frac{1}{4\pi\theta} \right) . \qquad (19.64)$$

At this point, we anticipate that the case $K = \frac{1}{2}$ corresponds to the "Toulouse limit" of the anisotropic Kondo model [132], which in turn is equivalent to the resonance level model [361, 363] with zero Coulomb interaction.[5] The respective model describes a d-level near the Fermi surface which is coupled to a bath of spinless fermions with a band width $2\hbar\omega_c$ around the Fermi level. The coupling to the fermions

---

[5]The equivalence of the Ohmic two-state model to these models is discussed in Section 19.4.

is assumed constant within the band. We put $\hbar\omega_c \ll E_F$. Then the dispersion relation can be linearized about the Fermi wave vector. Measuring the momentum from the reference value $p_F = mv_F$, we have

$$E(k) - E_F \equiv \hbar\epsilon_k = \hbar v_F k . \qquad (19.65)$$

The Toulouse Hamiltonian reads (L is a normalization length)

$$H_T = \sum_k \hbar\epsilon_k c_k^\dagger c_k + \hbar\epsilon_d d^\dagger d + \frac{V}{\sqrt{L}} \sum_k \left(c_k^\dagger d + d^\dagger c_k \right) . \qquad (19.66)$$

To show the equivalence, we first consider the partition function of this model. For simplicity, we place the d-level at the Fermi energy, $\epsilon_d = 0$, for the moment. The perturbation series in the interaction picture reads

$$Z_T = 2 \sum_{m=0}^{\infty} \left(\frac{V^2}{L\hbar^2} \right)^m \int_0^{\hbar\beta - \tau_c} ds_{2m} \int_0^{s_{2m} - \tau_c} ds_{2m-1} \cdots \int_0^{s_2 - \tau_c} ds_1 \, \mathcal{G}_m(\{s_j\}) , \qquad (19.67)$$

$$\mathcal{G}_m(\{s_j\}) \equiv \sum_{k_1, \cdots, k_{2m}} \exp\left(\sum_{j=1}^m \left(\epsilon_{k_{2j-1}} s_{2j-1} - \epsilon_{k_{2j}} s_{2j} \right) \right) \langle c_{k_{2m}} c_{k_{2m-1}}^\dagger \cdots c_{k_2} c_{k_1}^\dagger \rangle_\beta , \qquad (19.68)$$

where $\langle \cdots \rangle_\beta$ means thermal average with respect to the fermionic reservoir, and where we have already extracted the time-dependent exponential factors. Next, we aim at writing the thermal average in terms of electron-hole excitations. These are expressed in terms of the Fermi function $f(\epsilon)$ as [cf. Eq. (4.131)]

$$\langle c_k^\dagger c_{k'} \rangle_\beta = \delta_{kk'} f(\epsilon_k) , \quad \text{and} \quad \langle c_k c_{k'}^\dagger \rangle_\beta = \delta_{kk'} [1 - f(\epsilon_k)] . \qquad (19.69)$$

The thermal fermion propagator is given by

$$D(\tau) \equiv \sum_k \langle c_k^\dagger c_k \rangle_\beta \, e^{\epsilon_k \tau} = \sum_k \langle c_k c_k^\dagger \rangle_\beta \, e^{-\epsilon_k \tau} . \qquad (19.70)$$

Taking the continuum limit and observing that the density of states per unit length

$$\rho \equiv dk / 2\pi d\epsilon_k = (2\pi v_F)^{-1} \qquad (19.71)$$

is constant, we have

$$D(\tau) = \rho L \int_{-\omega_c}^{\omega_c} d\epsilon \, \frac{e^{-\epsilon\tau}}{1 + e^{-\hbar\beta\epsilon}} . \qquad (19.72)$$

Since the integral in Eq. (19.72) is well-defined for infinite limits, we may remove the ultraviolet cutoff, and instead of this regularize the propagator by introducing a hard-sphere repulsion at $\tau = \tau_c \equiv 1/\omega_c$. Thus we find

$$D(\tau) = \Theta(\tau - \tau_c) \frac{\pi\rho L}{\hbar\beta} \frac{1}{\sin(\pi\tau/\hbar\beta)} = \Theta(\tau - \tau_c) \rho L\omega_c \, e^{-W_{K=1/2}(\tau)} . \qquad (19.73)$$

In the second form, we have expressed the fermionic thermal propagator in terms of the kink pair interaction (18.53) for $K = \frac{1}{2}$.

Using Wick's theorem, the thermal average in Eq. (19.68) can be decomposed into a series which represents all possibilities of building products of electron-hole contractions. Using the correspondence relation (19.73), one finds that the Wick decomposition of $\mathcal{G}_m(\{s_j\})$ coincides with the before-mentioned decomposition of the influence function $\mathcal{F}_m^{(\mathrm{E})}(\{s_j\})$. We have

$$\mathcal{G}_m(\{s_j\}) = (\rho L \omega_c)^m \mathcal{F}_m^{(\mathrm{E})}(\{s_j\}) . \qquad (19.74)$$

This is the key relation uncovering the correspondence of the resonance level model (19.66) with the Ohmic spin-boson model for the special damping parameter $K = \frac{1}{2}$. Substituting Eq. (19.74) into Eq. (19.67), the partition function takes the form

$$Z_{\mathrm{T}} = 2 \sum_{m=0}^{\infty} \left(\frac{V^2 \rho \omega_c}{\hbar^2} \right)^m \int_0^{\hbar\beta - \tau_c} ds_{2m} \int_0^{s_{2m} - \tau_c} ds_{2m-1} \cdots \int_0^{s_2 - \tau_c} ds_1 \, \mathcal{F}_m^{(\mathrm{E})}(\{s_j\}) . \quad (19.75)$$

Evidently, this series is identical in form with the corresponding series for the partition function of the Ohmic two-state system for $K = \frac{1}{2}$. The mapping is completed by the parameter identifications (for completeness we generalize to the case $\epsilon_{\mathrm{d}} \neq 0$)

$$\gamma \equiv \pi \Delta^2 / 2\omega_c = 2\pi \rho V^2 / \hbar^2 , \qquad \text{and} \qquad \epsilon = \epsilon_{\mathrm{d}} . \qquad (19.76)$$

The latter relation connects the bias of the TSS with the energy of the d-level.

The equations of motion for the single-particle imaginary-time Green's function can easily be solved in Fourier space, yielding the fermionic Matsubara sums

$$G^{(\mathrm{E})}(\tau) \equiv \langle T_\tau d^\dagger(\tau) d(0) \rangle_\beta = \frac{1}{\hbar\beta} \sum_{n \, \mathrm{odd}} \frac{e^{-i\nu_n \tau}}{[-i\nu_n - \epsilon_{\mathrm{d}} - i\gamma \, \mathrm{sgn}(\nu_n)/2]} , \qquad (19.77)$$

$$g_k^{(\mathrm{E})}(\tau) \equiv \langle T_\tau c_k^\dagger(\tau) d(0) \rangle_\beta = \frac{1}{\hbar\beta} \sum_{n \, \mathrm{odd}} \frac{V/\hbar\sqrt{L}}{(-i\nu_n - \epsilon_k)} \frac{e^{-i\nu_n \tau}}{[-i\nu_n - \epsilon_{\mathrm{d}} - i\gamma \, \mathrm{sgn}(\nu_n)/2]} .$$

We are now ready to calculate the partition function. Putting $H(\lambda) = H_0 + \lambda H_1$, where H_1 is the interaction term in Eq. (19.66), the change of the free energy by the interaction is given by [159]

$$\Delta F = \left(V/\sqrt{L} \right) \int_0^1 d\lambda \sum_k \left\langle c_k^\dagger d + d^\dagger c_k \right\rangle_\lambda , \qquad (19.78)$$

where $\langle \cdots \rangle_\lambda$ denotes the *thermal average* $\langle H_1 \rangle_\lambda = \mathrm{tr} \{ e^{-\beta H(\lambda)} H_1 \} / \mathrm{tr} \, e^{-\beta H(\lambda)}$. Substituting the expression (19.77) for $g_k^{(\mathrm{E})}(0)$, the thermal expectation value and finally the free energy (19.78) is easily calculated. Thus, we find for the partition function the expression [the hard core in Eq. (19.75) is replaced by a high-frequency cutoff]

$$Z_T = 2\cosh\left(\frac{\hbar\beta\epsilon_d}{2}\right)\exp\left\{\frac{\hbar\beta}{2\pi}\int_{-\omega_c}^{\omega_c}d\omega\,\tanh\left(\frac{\hbar\beta\omega}{2}\right)\arctan\left(\frac{\gamma}{2(\omega-\epsilon_d)}\right)\right\}, \quad (19.79)$$

where the first factor is due to the term $\hbar\epsilon_d d^\dagger d$ of the Hamiltonian, and the second factor results from the coupling term. To transform the "fermionic" expression (19.79) into the bosonic form (19.61), we employ the integral representations

$$\tanh\left(\frac{\hbar\beta\omega}{2}\right) = \frac{2}{\pi}\int_0^\infty dx\,\frac{\sin\left(\hbar\beta\omega x/\pi\right)}{\sinh x},$$

$$\qquad\qquad (19.80)$$

$$1 - e^{-x/2\pi\theta} = \frac{\hbar\beta}{\pi^2}\int_{-\infty}^\infty d\omega\,\sin\left(\frac{\hbar\beta\omega x}{\pi}\right)\arctan\left(\frac{\gamma}{2\omega}\right),$$

We then obtain from Eq. (19.79)

$$Z_T = 2\cosh\left(\frac{\hbar\beta\epsilon_d}{2}\right)\exp\left\{\int_{x_0}^\infty dx\,\frac{[1 - e^{-x/2\pi\theta}]\cos(\mu x/2\pi\theta)}{x\sinh x}\right\}, \quad (19.81)$$

where $\mu \equiv 2\epsilon_d/\gamma$. Finally, by virtue of the integral representation

$$\cosh\left(\frac{\hbar\beta\epsilon_d}{2}\right) = \lim_{x_0\to 0}\exp\left\{\int_{x_0}^\infty dx\,\frac{1 - \cos(\mu x/2\pi\theta)}{x\sinh x}\right\}, \quad (19.82)$$

and the relation $\epsilon_d = \epsilon$, the expression (19.81) takes the form

$$Z_T = 2\exp\left\{\int_{x_0}^\infty dx\,\frac{1 - e^{-x/2\pi\theta}\cos(\epsilon x/\pi\gamma\theta)}{x\sinh x}\right\}. \quad (19.83)$$

The final expression (19.83) for the partition function is exactly the generalization of the former result (19.61) to nonzero bias. It is now straightforward to generalize also the specific heat expression (19.64) to nonzero bias. We find for all T

$$c = \frac{1}{2\pi\theta} - \frac{1}{8\pi^2\theta^2}\,\mathrm{Re}\left\{\left(1 + 2i\epsilon/\gamma\right)^2\psi'\left(\frac{1}{2} + \frac{1 + 2i\epsilon/\gamma}{4\pi\theta}\right)\right\}, \quad (19.84)$$

where $\psi'(z) = d\psi(z)/dz$ is the trigamma function. From Eq. (19.84), we obtain for the specific heat the leading low-temperature behaviour

$$c = \frac{2\pi}{3}\,\frac{\gamma^2}{\gamma^2 + 4\epsilon^2}\,\frac{k_B T}{\hbar\gamma}. \quad (19.85)$$

Substituting the expression (19.83) into Eq. (19.13) and performing the derivations yields the static susceptibility $\overline{\chi}_z(T,\epsilon)$. We find that the linear susceptibility

$\overline{\chi}_z^{(\ell)}(\theta) = \overline{\chi}_z(\theta, \epsilon = 0)$ is connected with the specific heat for zero bias by the relation $\overline{\chi}_z^{(\ell)}(\theta) = 8[1 - 2\pi\theta c(\theta)]/\pi\hbar\gamma$. Thus, the Wilson ratio reads for all T

$$R_\ell(\theta) \equiv 4\frac{c(\theta)}{\theta\hbar\gamma\overline{\chi}_z^{(\ell)}(\theta)} = \frac{\pi c(\theta)}{2\theta[1 - 2\pi\theta c(\theta)]}. \tag{19.86}$$

This function has the limiting value $R_\ell(0) = \pi^2/3$, which is in agreement with the previous result (19.54). Finally, substituting Eq. (19.83) into Eq. (19.13), we obtain for the static *nonlinear* susceptibility the analytic form

$$\overline{\chi}_z(T, \epsilon) = (2\beta/\pi^2) \operatorname{Re}\psi'[\tfrac{1}{2} + \hbar\beta(\gamma + 2i\epsilon)/4\pi]. \tag{19.87}$$

We shall take up again the case $K = \tfrac{1}{2}$ in Section 21.6.

19.3 Non-Ohmic spectral densities

In this section, we extend the discussion of the thermodynamics of the open TSS to non-Ohmic spectral densities of the form (18.44) with $s \neq 1$. Here we restrict the discussion to the unbiased case ($\epsilon = 0$).

19.3.1 The sub-Ohmic case

The important point that distinguishes the sub-Ohmic case from the super-Ohmic case is that the pair correlation function at zero temperature following from the analytic continuation of the expression (18.45),

$$W_{T=0}(\tau) = 2\delta_s\Gamma(s - 1)\Big(\frac{\omega_c}{\omega_{\rm ph}}\Big)^{s-1}\Big(1 - (1 + \omega_c\tau)^{1-s}\Big), \tag{19.88}$$

increases monotonously with τ in the sub-Ohmic case ($s < 1$), while it approaches a finite constant in the super-Ohmic case ($s > 1$). Therefore, the damping is usually large even at zero temperature for $s < 1$. We have for $\omega_c\tau \gg 1$

$$W_{T=0}(\tau) = 2\delta_s|\Gamma(s - 1)|(\omega_{\rm ph}\tau)^{1-s} - 2|B_s|, \qquad |B_s| = \delta_s|\Gamma(s - 1)|(\omega_{\rm ph}/\omega_c)^{1-s}.$$

Evaluating $K'_{\rm R}(0)$ [cf. Eq. (19.31)] with this form for $\epsilon = 0$, we see that the dilute gas condition $|K'_{\rm R}(0)| \ll 1$ corresponds to the condition

$$\rho_s^{1/(1-s)} \equiv \delta_s^{1/(1-s)}\omega_{\rm ph}/\Delta_{\rm eff} \gg 1, \tag{19.89}$$

where $\Delta_{\rm eff} = \Delta\, {\rm e}^{|B_s|}$. In the temperature regime

$$\delta_s^{(1+s)/(1-s)}\Big(\hbar\omega_{\rm ph}/k_{\rm B}T\Big)^{1+s} = \delta_s^{(1+s)/(1-s)}\Big(T_{\rm ph}/T\Big)^{1+s} \gg 1, \tag{19.90}$$

we may linearize $\exp[-W(\tau)]$ in $W(\tau)$, and we may expand this term about $\tau = 0$,

$$e^{-W(\tau)} = e^{-W_{T=0}(\tau)}\left\{1 + 2\delta_s\Gamma(1+s)\zeta(1+s)(T/T_{\rm ph})^{s-1}(\tau/\hbar\beta)^2\right\}, \qquad (19.91)$$

where terms of order $(\tau/\hbar\beta)^4$ are disregarded. We then find from Eq. (19.26) with Eq. (19.91) for the specific heat, Eq. (19.14), the contribution of order Δ^2,

$$c = q_s\, e^{p_s}\Gamma[3/(1-s),\, p_s](\Delta_{\rm eff}/\omega_{\rm ph})^2\,(T/T_{\rm ph})^s, \qquad (19.92)$$

where $\Gamma(z, a)$ is the incomplete gamma function [87], and where

$$\begin{aligned} q_s &= s^2(s+1)\zeta(s+1)\Big(2\delta_s\Gamma(s)/(1-s)\Big)^{(s+2)/(s-1)}, \\ p_s &= 2\delta_s(\omega_{\rm ph}/\omega_c)^{1-s}\Gamma(s)/(1-s). \end{aligned} \qquad (19.93)$$

In the regime (19.89), the condition (19.90) is satisfied not only for $k_{\rm B}T \ll \hbar\Delta$ but also for $k_{\rm B}T_{\rm ph} \gtrsim k_{\rm B}T \gg \hbar\Delta$. Thus, the Schottky peak is absent below $T_{\rm ph}$, and the specific heat steadily increases like in the Ohmic strong-damping case.

We may readily perform the limit $\omega_c \to \infty$ in Eq. (19.92), yielding

$$c = q_s\Gamma[3/(1-s)]\,(\Delta_{\rm eff}/\omega_{\rm ph})^2\,(T/T_{\rm ph})^s. \qquad (19.94)$$

We conclude this subsection with the remark that we recover the Ohmic result (19.56) for $K \gg 1$ if we perform the limit $s \to 1^-$ in Eq. (19.92) keeping ω_c finite.

19.3.2 The super-Ohmic case

Because of the low spectral density of low energy excitations in the super-Ohmic case, we are usually in the weak-coupling regime at low temperatures. For weak damping, we may expand the self-energy in a series in δ_s. Employing a dimensional analysis, it is straightforward to see that the term which is linear in δ_s, which is the "one-phonon process", is the dominant one in the parameter regime $\delta_s(\omega_c/\omega_{\rm ph})^{s-1} \ll 1$. Then we can go back to the formulas (19.42) – (19.44). From these we obtain the asymptotic expansion for a symmetric system at low T in the form

$$c = \delta_s\left(\frac{\Delta_{\rm eff}}{\omega_{\rm ph}}\right)^{s-1}\sum_{n=0}^{\infty}(2n+s)\Gamma(2n+s+2)\zeta(2n+s+1)\,\theta^{2n+s}, \qquad (19.95)$$

where $\zeta(z)$ is Riemann's zeta function, and where $\theta = k_{\rm B}T/\hbar\Delta_{\rm eff}$ with $\Delta_{\rm eff}$ given in Eq. (18.32). Hence the specific heat behaves as T^s as $T \to 0$. In the Ohmic limit $s \to 1$, the expression simplifies to the expansion (19.52) with $\varrho = 1$. In the high-temperature limit $k_{\rm B}T \gg \hbar\Delta_{\rm eff}$ we find from Eqs. (19.42) – (19.44)

$$c = \frac{1}{4\theta^2}\left\{1 - 2\delta_s(3-s)(2-s)\Gamma(s-1)\zeta(s-1)\left(k_{\rm B}T/\hbar\omega_{\rm ph}\right)^{s-1}\right\}. \qquad (19.96)$$

For a discussion of other parameter regimes, we refer to Ref. [359].

19.4   Relation between the Ohmic TSS and the Kondo model

In this section, we summarize the relation between the spin-boson system with Ohmic dissipation and the Kondo problem. In its simplest form, the Kondo problem deals with a single magnetic impurity of spin 1/2 which interacts via an exchange scattering potential with a band of conduction electrons. The essence of the model has its origin in the constant density of electron-hole excitations in the vicinity of the Fermi surface. In the adiabatic scheme, the constant density of states gives rise to Anderson's orthogonality catastrophe [364], and to logarithmic infrared divergences in a variety of physical quantities, e. g., in the static magnetic susceptibility, or in the soft X-ray absorption and emission of metals [177, 365, 132]. These phenomena have been popularized by Kondo as the Fermi surface effects [127].

The relationship between the spin-boson model and the Kondo model is due to the fact that the low-energy electron-hole excitations have bosonic character and can be interpreted in terms of density fluctuations [178]. In one dimension, there is an exact mapping between fermions and bosons [366]. Here we restrict our attention to the anisotropic Kondo model and to the resonance level model (see Ref. [132]). Generalizations to multi-channel Kondo impurity models which show non-Fermi liquid behaviour, and the possible experimental relevance are reviewed in Ref. [367]. In Subsection 19.2.2 we have already discussed a particular case of the correspondence. There we have shown that the Ohmic spin-boson model with the Kondo parameter $K = \frac{1}{2}$ maps on the Toulouse model which is a special case of the Kondo model.

19.4.1   Anisotropic Kondo model

The Kondo model is, at first glance, a simple model in which one assumes that a magnetic impurity interacts with the conduction electrons via a pointlike exchange interaction, so that only s-wave scattering occurs. This reduces the problem to an essentially one-dimensional one. Further, it is assumed that, out of the many different bands which may exist in a solid, only a narrow band of half-width $\hbar \omega_c = \hbar / \tau_c$ which is chosen symmetric about the Fermi energy E_F interacts with the impurity. Linearizing the dispersion relation as in Eq. (19.65), the Kondo Hamiltonian is given by

$$H_K = \hbar v_F \sum_{k,\sigma} k c_{k,\sigma}^\dagger c_{k,\sigma} + J\boldsymbol{S} \cdot \boldsymbol{s}(0)/\hbar \,, \qquad (19.97)$$

which is also referred to as the s-d Hamiltonian. The operator $c_{k,\sigma}^\dagger$ creates a conduction electron with momentum $\hbar k$ and spin polarization $\sigma = \pm 1$. Further, $S_i = \frac{1}{2}\hbar\tau^i$ is the spin operator of the impurity where τ^i ($i = 1, 2, 3$) are the Pauli matrices (to make a distinction from the TSS operators, we choose τ^i). The effective spin operator due to the conduction electrons at the impurity site $\boldsymbol{r} = 0$ is conveniently expressed in terms of Wannier operators for electrons localized at the origin with spin polarization σ,

$$c_\sigma^\dagger(0) = L^{-1/2} \sum_k c_{k,\sigma}^\dagger \,, \qquad (19.98)$$

where $k = 2\pi n/L$ ($n \in \{0, \pm 1, \ldots\}$), and L is a normalization length. We then have

$$s_i(0) = \frac{\hbar}{2} \sum_{\sigma,\sigma'} c_\sigma^\dagger(0) \tau_{\sigma,\sigma'}^i c_{\sigma'}(0) . \qquad (19.99)$$

In the original Kondo problem, the exchange interaction satisfies rotational invariance, $J_\parallel = J_\perp$. However, in order to be able to relate the Kondo problem to the spin-boson problem, it is essential to generalize the isotropic coupling in the original model (19.97) to the case where the exchange constants $J_\parallel$ for the $S_z s_z$ term and $J_\perp$ for $S_x s_x + S_y s_y$ are independent parameters and arbitrarily large.

The anisotropic Kondo Hamiltonian takes the form

$$H_K = \hbar v_F \sum_{k,\sigma} k c_{k,\sigma}^\dagger c_{k,\sigma} + \frac{\hbar J_\parallel}{4} \tau_z \sum_\sigma \sigma c_\sigma^\dagger c_\sigma + \frac{\hbar J_\perp}{2} \left(\tau_+ c_\downarrow^\dagger c_\uparrow + \tau_- c_\uparrow^\dagger c_\downarrow \right) \qquad (19.100)$$

with $\tau_\pm = (\tau_x \pm i\tau_y)/2$. The $J_\parallel$ term describes scattering of the fermions at the impurity in which the spin polarization is conserved while the $J_\perp$ term describes spin-flip scattering. The parameters $J_\parallel$ and $J_\perp$ have dimension velocity. The dimensionless coupling constants are $\rho J_\parallel$ and $\rho J_\perp$, and the constant density of states ρ is as in Eq. (19.71), $\rho = (2\pi v_F)^{-1}$.

To show the equivalence between the anisotropic Kondo model, Eq. (19.100), and the spin-boson model, I follow the steps of Yuval and Anderson [368], who have cast the partition function for the impurity into the "Coulomb gas" form

$$Z_K = \sum_{m=0}^\infty \left(\frac{\rho J_\perp \cos^2 \delta_K}{2\tau_c} \right)^{2m} \int_0^{\hbar\beta - \tau_c} ds_{2m} \int_0^{s_{2m} - \tau_c} ds_{2m-1} \cdots \int_0^{s_2 - \tau_c} ds_1 \qquad (19.101)$$

$$\times \exp \left\{ 2 \left(1 - \frac{2}{\pi} \delta_K \right)^2 \sum_{j>k=1}^{2m} (-1)^{j+k} \ln \left[\frac{\hbar\beta}{\pi\tau_c} \sin \left(\frac{\pi(s_j - s_k)}{\hbar\beta} \right) \right] \right\} .$$

Here, $\tau_c = 1/\omega_c$ is the hard-core length of the Coulomb charges, and δ_K is the scattering phase shift at the singular non-spin-flip potential $J_\parallel \delta(x)/4$. The series (19.101) corresponds to the Coulomb gas representation (19.2) with Eq. (19.4) and the Ohmic form (18.53) for $W(\tau)$, and $\epsilon = 0$. The correspondence of the parameters is

$$\Delta = \rho J_\perp \cos^2(\delta_K)/\tau_c ; \qquad K = (1 - 2\delta_K/\pi)^2 . \qquad (19.102)$$

The scattering phase shift $\delta_K(J_\parallel)$ depends on the regularization prescription for the singular scattering potential [132]. If we choose a separable form, we obtain

$$\delta_K(J_\parallel) = \arctan(\pi\rho J_\parallel/4) , \qquad (19.103)$$

while the smoothing prescription $\delta(x) \to (2\pi)^{-1} \int_{-k_c}^{k_c} dk\, e^{ikx}$, where k_c is large, gives $\delta_K = \pi\rho J_\parallel/4$. In Eq. (19.103), only the contribution which is linear in $J_\parallel$ is universal. This implies that the correspondence between the anisotropic Kondo model and the Ohmic TSS is universal in the regime $\rho J_\perp = \Delta/\omega_c \ll 1$ and $\rho|J_\parallel| = |1 - K| \ll 1$.

A local magnetic field h_m in z-direction which couples to the impurity spin, and not to the conduction electrons, gives to H_K the additional contribution $-\frac{1}{2}g\mu_\mathrm{B}h_\mathrm{m}\tau_z$. Thus, the bias energy $\hbar\epsilon$ in the TSS corresponds to $g\mu_\mathrm{B}h_\mathrm{m}$. Introducing the static susceptibility χ_{sd} with zero Landé factor for the electron band, the Wilson ratio is found as [cf. Eq. (57) in Ref. [361], and footnote to Eq. (19.13)]

$$R = \lim_{T\to 0}\frac{(g\mu_\mathrm{B})^2 c(T)}{T\chi_{sd}} = \frac{2\pi^2}{3}(1 - 2\delta_\mathrm{K}/\pi)^2, \qquad \chi_{sd}/(g\mu_\mathrm{B})^2 = \overline{\chi}_z/4. \quad (19.104)$$

This result is in agreement with the Wilson ratio for the Ohmic TSS (see the discussion in Section 21.9). We see from Eq. (19.103) that the regime $-\infty < \rho J_\parallel < \infty$ maps on the regime $4 > K > 0$. The critical coupling $K = 1$ separates the ferromagnetic Kondo regime $\rho J_\parallel < 0$, which corresponds to the regime $K > 1$, from the more intriguing antiferromagnetic regime $\rho J_\parallel > 0$, i. e., $K < 1$. The special choice $\delta_\mathrm{K}(J_\parallel) = (2 - \sqrt{2})\pi/4$, which corresponds to the case $K = \frac{1}{2}$, is known in the literature as the Toulouse limit of the anisotropic Kondo model. This limiting case is exactly solvable in closed form [132].

In the antiferromagnetic regime, the Kondo temperature $T_\mathrm{K} \propto \Delta_\mathrm{r}$ separates the perturbative regime $T > T_\mathrm{K}$, in which physical quantities, e.g., the magnetic susceptibility, can be expanded in powers of $J_\perp$, from the nonperturbative regime. In the regime $\rho J_\perp \ll \rho J_\parallel \ll 1$, which is equivalent to $\Delta/\omega_c \ll 1 - K \ll 1$, the Kondo temperature is given by [132]

$$k_\mathrm{B}T_\mathrm{K} = \frac{2}{\pi}\left(\frac{J_\perp}{2J_\parallel}\right)^{1/\rho J_\parallel}\hbar\omega_c = \frac{2}{\pi}\left(\frac{1}{2(1-K)}\right)^{1/(1-K)}\hbar\Delta_\mathrm{r}. \quad (19.105)$$

The second equality follows with the equivalence relations (19.102), and with (18.34). In the antiferromagnetic sector, the properties for $T \ll T_\mathrm{K}$ can be described phenomenologically in the spirit of the Landau theory of a Fermi liquid [362].

19.4.2 Resonance level model

The other model of interest is the *resonance level model*. This model has been introduced as a generalization of the Toulouse model (19.66) by Schlottmann [363], and independenly by Vigmann and Finkels'teĭn [361] (for a review of the resonance level model, see Ref. [132]). The corresponding Hamiltonian describes a d–level near the Fermi surface interacting with a band of spinless fermions. In generalization of the model (19.66), the fermions interact with each other through a repulsive contact potential which represents the (screened) Coulomb interaction. Linearizing again the dispersion relation about the Fermi momentum, the Hamiltonian takes the form

$$\begin{aligned} H_\mathrm{RL} = {} & \hbar v_\mathrm{F}\sum_k k c_k^\dagger c_k + \hbar\epsilon_\mathrm{d} d^\dagger d + \frac{V}{\sqrt{L}}\sum_k\left(c_k^\dagger d + d^\dagger c_k\right) \\ & + \frac{U}{L}\sum_{k,k'}\left(c_k^\dagger c_{k'} - c_{k'}c_k^\dagger\right)\left(d^\dagger d - \tfrac{1}{2}\right). \end{aligned} \quad (19.106)$$

The partition function of this model is equivalent to Eq. (19.2) if we identify

$$\Delta^2 = 4\rho\omega_c \frac{V^2}{\hbar^2}\cos^2(\delta_{\rm RL}) ; \quad K = \frac{1}{2}\left(1 - \frac{2}{\pi}\delta_{\rm RL}\right)^2 ; \quad \epsilon = \epsilon_{\rm d} , \qquad (19.107)$$

where $\delta_{\rm RL} = \pi\rho U/\hbar + \mathcal{O}(U^2)$ is the scattering phase for the contact potential $U\delta(x)$, and where again $\rho = (2\pi v_{\rm F})^{-1}$. In the absence of the repulsive interaction, $U = 0$, we have $K = \frac{1}{2}$. This case corresponds to the Toulouse limit of the antiferromagnetic Kondo Hamiltonian. Perturbation expansion in ρU in the resonant level model corresponds to perturbation expansion about $K = \frac{1}{2}$ in the two-state model.

Here we have discussed the similarity between the Ohmic two-state system and the Kondo problem on the level of the partition function. The comparison of these models on the basis of their Hamiltonians is set out in Ref. [89]. We should like to remark that the mapping of the Ohmic two-state system neither to the anisotropic Kondo model nor to the resonant level model is exact. However, the low-energy excitations of these models are very similar, and therefore also the physical phenomena of these models are very similar at low temperatures.

19.5 Equivalence of the Ohmic TSS with the $1/r^2$ Ising model

It has been shown by Anderson and Yuval [369] that the Ising model with ferromagnetic $1/r^2$ pair interaction can be mapped on the Kondo model. The inverse square ferromagnetic Ising model is another specific realization of the Coulomb gas model. The Ising spins represent the imaginary time history of the single impurity spin of the Kondo model. The short-time regularization $\tau_c = 1/\omega_c$ is provided by the lattice, $N\tau_c = \hbar\beta$. With the correspondence of the Kondo model to the TSS explained in the previous section, the inverse-square Ising model is also equivalent to the Ohmic TSS.

Consider the one-dimensional N-site Ising model for spins $\{S_j = \pm 1\}$ with ferromagnetic long-range interactions, a magnetic field term, and periodic boundary conditions. In the state representation, the partition function reads

$$Z_{\rm Ising} = \sum_{S_1,\cdots,S_N} \exp\left\{ -\sum_{j>i} V(j-i) S_j S_i - \frac{h_{\rm m}}{2}\sum_j S_j\right\} . \qquad (19.108)$$

The first term in the exponent represents the spin interactions, and the second term is a magnetic field contribution. Next, we go over from the state representation (19.108) to the charge representation. Concerning the interaction term, this corresponds to switching over in the influence functional from the form (4.69) to the form (4.70). The Coulomb gas representation of the partition function reads [370]

$$Z_{\rm Ising} = \sum_{n=0}^{\infty} y^{2n} \int_0^{\hbar\beta-\tau_c} \frac{ds_{2n}}{\tau_c} \int_0^{s_{2n}-\tau_c} \frac{ds_{2n-1}}{\tau_c} \cdots \int_0^{s_2-\tau_c} \frac{ds_1}{\tau_c} \qquad (19.109)$$

$$\times \exp\left\{ 4\sum_{j=2}^{2n}\sum_{i=1}^{j-1}(-1)^{j+i}U[(s_j - s_i)/\tau_c] + h_{\rm m}\sum_{j=1}^{2n}(-1)^j s_j/\tau_c\right\} ,$$

where $V(k)$ is the second derivative of $U(k)$ in discretized form,

$$V(k) = U(k+1) + U(k-1) - 2U(k), \qquad (19.110)$$

and where $y = \exp[2U(0)]$ is the fugacity of the Coulomb gas. Upon requiring correspondence of Eq. (19.109) with the partition function of the Ohmic TSS, Eq. (19.2) with Eqs. (19.3), (19.4), and (18.53), we obtain

$$U(n) = \frac{K}{2} \ln \left[\left(\frac{N}{\pi} \right) \sin \left(\frac{\pi n}{N} \right) \right], \qquad n \geq 1, \qquad (19.111)$$

and

$$y = \Delta/2\omega_c, \qquad N = \hbar\beta/\tau_c, \qquad \epsilon = -h_m/\tau_c. \qquad (19.112)$$

Substituting the form (19.111) into Eq. (19.110), we find

$$V(n) = -\frac{K}{2} \frac{(\pi/N)^2}{\sin^2(\pi n/N)} + \mathcal{O}[(\pi/N)^4], \qquad n \geq 2. \qquad (19.113)$$

In the limit $N \to \infty$, we obtain $2V(1) = 2U(0) + KC$, where $C = \ln 2$.[6] In this limit, the interaction between the spin states depends inversely on the square of the distance. The Hamiltonian of the Ising model is found in the usual notation as

$$H_{\text{Ising}} = \frac{\hbar}{N\tau_c} \left(-\frac{J_{NN}}{2} \sum_j S_{j+1} S_j - \frac{J_{LR}}{2} \sum_{j>i} \frac{(\pi/N)^2 S_j S_i}{\sin^2[\pi(j-i)/N]} + \frac{h_m}{2} \sum_j S_j \right), \qquad (19.114)$$

where $J_{LR} = K$, and $J_{NN} + J_{LR} = -2V(1)$. The correspondence relations with the Ohmic TSS are

$$\frac{\Delta}{2\omega_c} = \exp[-J_{NN} - (1+C)J_{LR}], \qquad K = J_{LR}, \qquad (19.115)$$

and the relations (19.112). The equivalence between the Ohmic TSS and the $1/r^2$ Ising model has been utilized recently [371, 372]. In these studies, the imaginary-time correlation function is calculated by Monte Carlo simulations on the Ising system and then continued to real-time by employing a Padé approximation.

The equivalence of the TSS with a long-range ferromagnetic Ising model can also be established for non-Ohmic dissipation. For $G_{\text{lf}}(\omega) \propto \omega^s$, the spin interaction $V(j - i)$ in the Ising model at zero temperature falls off as $(j - i)^{-(1+s)}$. Using the correspondence, a number of properties proven rigorously for the Ising model [349, 373] can directly be transferred to the dissipative two-state system.

20. Electron transfer and incoherent tunneling

Before turning to the full dynamics of the TSS in Chapter 21, we now first consider tunneling rates using the thermodynamic method discussed already in Part III. The results shall be confirmed by the dynamical approach given in Chapter 21.

---

[6]The constant C depends on the particular hard-core regularization chosen.

20.1 Electron transfer

Electron transfer plays an important role in many processes in chemistry and biology. A striking example is the ultrafast primary electron tranfer step found in photosynthetic reaction centers which is responsible for the high efficiency of the photosynthetic mechanism. Significant progress with the understanding of the essence of electron transfer (ET) processes in condensed matter has been made by Marcus [97, 98]. He was the first who recognized the importance of the solvent environment and employed linear response theory to describe the effects of the solvent coupled to the electronic degree of freedom. In the ordinary ET reaction, there is a free energy barrier separating reactants and products. The barrier is because the donor and acceptor states are strongly solvated, and the transfer of the electron then requires a reorganization of the environment. At ordinary temperatures, the electron has to tunnel through the barrier from the reactant to the product state. The tunneling is only effective if suitable bath fluctuations bring reactant and product energy levels into resonance. In the classical limit, the transfer rate is determined by two factors. First, a Boltzmann factor with the activation energy required for bath fluctuations. Second, an attempt frequency prefactor describing the probability for tunneling once the levels are in resonance. The latter factor is mainly determined by the overlap between the electronic wave functions localized on different redox sites. For large electronic coupling, the reaction is adiabatic whereas, for weak electronic coupling, the reaction is nonadiabatic.

It is implied by the success of the classical Marcus theory that the electron can be described in terms of a discrete variable at all temperatures of interest. Here we restrict the attention to the important case of a two-state system: a *donor* state D located at $\sigma = -1$ and an *acceptor* state A located at $\sigma = +1$, and we describe the solvent in terms of a linearly responding heat bath. Then the proper Hamiltonian is the spin-boson Hamiltonian, as given in Eq. (3.88) with (3.80) or in Eq. (18.13),

$$H = -\tfrac{1}{2}\hbar\Delta\,\sigma_x - \tfrac{1}{2}\hbar\epsilon\,\sigma_z + \tfrac{1}{2}\mu\mathcal{E}(t)\,\sigma_z + H_{\mathrm{R}}\,. \qquad (20.1)$$

Here, H_{TSS} describes the tunneling of the electron between the donor and the acceptor state, and H_{R} represents the solvent as a Gaussian reservoir. The collective bath mode $\mu\mathcal{E}(t) = -\tfrac{1}{2}q_0\sum_\alpha c_\alpha x_\alpha(t)$ is coupled to the spin operator σ_z. It can be thought of as a fluctuating dynamical polarization energy. The polarization field $\mathcal{E}(t)$ vanishes on average, and μ is the difference of the dipole moments of the two electronic states.

All effects of the solvent on the electron transfer are contained in the properties of the bath correlation function [cf. Eqs. (4.68) and (18.42)]

$$\left(\frac{\mu}{\hbar}\right)^2 \Big\langle \mathcal{E}(z)\mathcal{E}(0) \Big\rangle_\beta = \frac{d^2 Q(z)}{dz^2} = \int_0^\infty d\omega\, G(\omega)\,\frac{\cosh[\omega(\hbar\beta/2 - iz)]}{\sinh[\omega\hbar\beta/2]}, \qquad (20.2)$$

where $z = t - i\tau$ is a complex time.

20.1.1 Adiabatic bath

An adiabatic bath is a sluggish bath. It is characterized by only zero frequency fluctuations and is therefore not dynamical. In practice, electron transfer (ET) systems are adiabatic when $\omega_c \ll \Delta$ and $\omega_c \ll k_B T/\hbar$, where ω_c is a typical bath frequency, and when frequencies significantly higher than ω_c are absent in the bath. The case of an adiabatic bath can be solved in analytical form [374].

In the adiabatic regime, Eq. (20.2) reduces to

$$\mu^2 \langle \mathcal{E}(z)\mathcal{E}(0) \rangle_\beta \;\longrightarrow\; \mu^2 \langle |\mathcal{E}|^2 \rangle_\beta \;=\; 2\Lambda_{cl}/\beta \,, \tag{20.3}$$

where

$$\Lambda_{cl} \;=\; \hbar \int_0^\infty d\omega \, \frac{G(\omega)}{\omega} \tag{20.4}$$

is the classical *reorganization energy*, or *solvation energy*, or *coincidence energy*.

For $G(\omega) = 2\delta_s \omega_{ph}^{1-s} \omega^s \, e^{-\omega/\omega_c}$, the reorganization energy is given by

$$\Lambda_{cl} \;=\; 2\delta_s \Gamma(s) \, (\omega_c/\omega_{ph})^{s-1} \hbar\omega_c \,. \tag{20.5}$$

In particular, the Ohmic case $s = 1$ gives with $\delta_1 = K$

$$\Lambda_{cl} \;=\; 2K\hbar\omega_c \,. \tag{20.6}$$

For a linearly responding bath, the polarization energy $\mu\mathcal{E}$ has the normalized Gaussian distribution with mean square (20.3),

$$\rho(\mathcal{E}) \;=\; \frac{\mu}{2} \sqrt{\frac{\beta}{\pi\Lambda_{cl}}} \, \exp\left(-\beta\mu^2\mathcal{E}^2/4\Lambda_{cl} \right) . \tag{20.7}$$

In the adiabatic limit, the polarization is slow, and the Born-Oppenheimer method gives for the Hamiltonian $H_{TSS} + \frac{1}{2}\mu\mathcal{E}\sigma_z$ two electronic eigenstates with energies

$$E_\pm(\mathcal{E}) \;=\; \pm\hbar\Omega(\mathcal{E})/2 \,, \qquad \Omega(\mathcal{E}) \;\equiv\; \sqrt{\Delta^2 + (\epsilon - \mu\mathcal{E}/\hbar)^2} \,. \tag{20.8}$$

The adiabatic partition function is an average with the Gaussian probability distribution (20.7) of Boltzmann factors, where the activation energies are given by Eq. (20.8),

$$Z_{ad} \;=\; \int_{-\infty}^\infty d\mathcal{E} \, \rho(\mathcal{E}) \left(e^{-\beta E_+(\mathcal{E})} + e^{-\beta E_-(\mathcal{E})} \right) = 2\int_{-\infty}^\infty d\mathcal{E} \, \rho(\mathcal{E}) \cosh[\hbar\beta\Omega(\mathcal{E})/2] \,. \tag{20.9}$$

The expression (20.9) is also the partition function of a 1D Ising model with *infinite range* interactions. To discuss the qualitative features, we rewrite Eq. (20.9) as

$$Z_{ad} \;=\; \frac{\mu}{2} \sqrt{\frac{\beta}{\pi\Lambda_{cl}}} \int_{-\infty}^\infty d\mathcal{E} \left(e^{-\beta F_+(\mathcal{E})} + e^{-\beta F_-(\mathcal{E})} \right) , \tag{20.10}$$

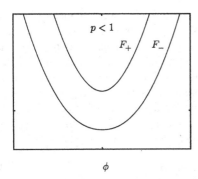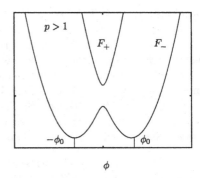

Figure 20.1: The adiabatic Born-Oppenheimer surfaces $F_+(\phi)$ and $F_-(\phi)$ for the symmetric spin-boson model, Eq.(20.13). The parameter p is chosen as $p = 1/2$ (left) and $p = 8$ (right). The distance between the surfaces at $\phi = 0$ is $\hbar\Delta$.

where

$$F_\pm(\mathcal{E}) = \frac{\mu^2 \mathcal{E}^2}{4\Lambda_{\rm cl}} \pm \frac{\hbar}{2}\sqrt{\Delta^2 + (\epsilon - \mu\mathcal{E}/\hbar)^2} \qquad (20.11)$$

are the adiabatic potential energies for the polarization energy $\mu\mathcal{E}$.

Consider first the symmetric case $\epsilon = 0$, where the two redox states are solvated with the same energy. The adiabatic surfaces are qualitatively different depending on the size of the parameter

$$p = \Lambda_{\rm cl}/\hbar\Delta . \qquad (20.12)$$

Measuring the polarization energy in units of $\hbar\Delta$, $\phi = \mu\mathcal{E}/\hbar\Delta$, we then have

$$F_\pm(\phi) = \frac{\hbar\Delta}{2}\left(\frac{\phi^2}{2p} \pm \sqrt{1 + \phi^2}\right) . \qquad (20.13)$$

The potential $F_+(\phi)$ is monostable with its minimum at $\phi = 0$ for all p, while $F_-(\phi)$ is monostable with minimum at $\phi = 0$ for $p \leq 1$ only. If $p > 1$, the potential $F_-(\phi)$ is bistable with minima $F_-^{\rm (min)} = -[p + 1/p]\hbar\Delta/4$ at $\phi = \pm\phi_0$, where $\phi_0 = \sqrt{p^2 - 1}$. This behaviour is illustrated in Fig. 20.1. Thus, if $p > 1$, the average spin $\langle\sigma_z\rangle$ is nonzero as $\beta \to \infty$. Hence, the *adiabatic* spin-boson model exhibits broken symmetry in the ground state which is a type of fluctuation-induced *self-trapping*. Whether trapping occurs for an adiabatic bath is decided by the competition between the reorganization energy $\Lambda_{\rm cl}$ and the energy for resonant tunneling $\hbar\Delta$. Slowly fluctuating fields may impede tunneling and may even prevent transitions. When spontaneous fluctuations are absent, i. e., when the reorganization energy is zero, then $p = 0$ and the particle tunnels clockwise.

The localization transition for an adiabatic bath, $\omega_c \ll k_{\rm B}T/\hbar$, is not sensitive to whether the spectral density is sub-Ohmic, Ohmic or super-Ohmic, since only the value of the integral in Eq. (20.4) decisively matters. The situation is different for a *dynamical* bath, $\omega_c \gg k_B T/\hbar$ and $\omega_c \gg \Delta$. The localization phenomenon is conveniently studied either by considering the flow of the parameters in the renormalization group equations [89] or by employing a variational method [349]. One finds that the ground state exhibits symmetry breaking in the sub-Ohmic case when the inequality (18.36) is reversed [349, 358]. In the Ohmic case, the localization transition occurs at the Kondo parameter $K = 1$ [133, 134, 89]. Here again the slow modes are responsible for the symmetry breaking. The phase diagram as a function of Δ/ω_c for $s \leq 1$ is discussed in Ref. [349]. It is shown by using an energy-entropy argument that there is *no* phase transition to localization when

$$\int_0^\infty d\omega \, G(\omega)/\omega^2 \; < \; \infty \, . \qquad (20.14)$$

Thus, the symmetry of the ground state is not spontaneously broken for super-Ohmic dissipation of any strength.

For an adiabatic bath, also the dynamics can be solved in analytic form. Consider as an example the thermal equilibrium correlation function $C^+(t) \equiv \langle \sigma_z(t)\sigma_z(0)\rangle_\beta$. The procedure essentially consists of two stages. In the first stage, the real-time dynamics of the TSS is calculated for a constant polarization energy $\mu\mathcal{E}$. We find[1]

$$C_0^+(t;\mathcal{E}) \; = \; \frac{(\epsilon - \mu\mathcal{E}/\hbar)^2}{\Omega^2} + \frac{\Delta^2}{\Omega^2}\cos(\Omega t) - i\tanh\left(\frac{\hbar\beta\Omega}{2}\right)\frac{\Delta^2}{\Omega^2}\sin(\Omega t) \, , \qquad (20.15)$$

where the tunneling frequency $\Omega = \Omega(\mathcal{E})$ is defined in Eq. (20.8). In the second stage, the correlation function (20.15) is averaged with the adiabatic weight functional as in Eq. (20.9). This yields

$$C_{ad}^+(t) \; = \; \frac{2}{Z_{ad}} \int_{-\infty}^\infty d\mathcal{E} \, \rho(\mathcal{E}) \cosh[\hbar\beta\Omega(\mathcal{E})/2]\, C_0^+(t;\mathcal{E}) \, . \qquad (20.16)$$

Thus, the dynamical problem is reduced to quadrature. The two-state correlation function of the free system is inhomogeneously broadened by the adiabatic bath fluctuations. Due to the time-independent part in Eq. (20.15), the "adiabatic" expression (20.16) does not decay to zero at large times. To describe the decay of the correlations, we have to add dynamical fluctuations [see Chapter 21].

20.1.2 Marcus theory for electron transfer

In the absence of nuclear (solvent) tunneling, the electron transfer dynamics is well understood in two different limits. In the *adiabatic limit*, the electronic coupling is so large that the parameter p is of order one or smaller. When p is above one, but not extremely large, the rate is controlled by the motion of the solvent on the lower electronic surface [cf. Fig. 20.1 (right)] and well described by the classical activated rate theory [98]. It is expected to depend on the electronic coupling Δ like

---

[1]Here we assume that the two eigenstates with energies $E_\pm(\mathcal{E})$ are thermally occupied. The real part of Eq. (20.15) is the result of a standard quantum mechanical calculation for a biased TSS, and the imaginary part follows with the fluctuation-dissipation theorem [cf. Eqs. (6.53) and (6.54)].

$$k \propto \exp[-(F_-^* - \hbar\Delta/2)/k_B T] \,, \tag{20.17}$$

where F_-^* is the free energy barrier for $\Delta = 0$, $F_-^* = \Lambda_{cl}/4$.

In the opposite *nonadiabatic limit*, the electronic coupling is weak, $p \gg 1$. To lowest order in the coupling Δ, the transfer rate is given by the Golden Rule formula

$$k \propto \Delta^2 \exp(-\Lambda_{cl}/4k_B T) \,. \tag{20.18}$$

In the two limits, one assumes a separation of time scales. The adiabatic limit is characterized by slow fluctuations of the solvent (nuclear degree of freedom) such that the Born-Oppenheimer approximation for the electron can be used. In the nonadiabatic limit, the solvent fluctuations are assumed to be fast. Then the ET is determined by thermal excitations of the solvent to the vicinity of the crossing region which are followed by a fast transversal of this region.

The nonadiabatic regime is conveniently discussed with the adiabatic potential surfaces (20.11). For $\Delta = 0$, the free energy curves become two intersecting parabola,

$$F_\pm(\mathcal{E}) = \pm \frac{\hbar\epsilon}{2} + \frac{1}{4\Lambda_{cl}} (\mu\mathcal{E} \mp \Lambda_{cl})^2 - \frac{\Lambda_{cl}}{4} \,. \tag{20.19}$$

which are known in the literature as the *Marcus parabola*. The parabola are the adiabatic energy functions for *diabatic* states ("diabatic" states are the eigenstates for $\Delta = 0$). In Fig. 20.2, we have given the Marcus parabola schematically for three different values of the bias parameter ϵ. Here we choose ϵ positive. The higher local minimum in Figs. 20.2 (b) and (c) which is situated at $\mathcal{E} = \Lambda_{cl}/\mu$ is the donor (D) state, and the lower local minimum situated at $\mathcal{E} = -\Lambda_{cl}/\mu$ is the acceptor (A) state. Pictorially, the reorganization energy is the difference of the free energy $F_-(\mathcal{E})$ between the donor state and the acceptor state,

$$\Lambda_{cl} = F_-(\mathcal{E} = \Lambda_{cl}/\mu) - F_-(\mathcal{E} = -\Lambda_{cl}/\mu) \,. \tag{20.20}$$

Since the nuclear degree of freedom (polarization) is slow compared to the electronic degree of freedom, the transition is determined by the *Franck-Condon principle*. This principle states that the transitions between the electronic states occur vertically in Fig. 20.2; that is, the nuclear reaction coordinate $\mathcal{E}$ does not change during the electronic transition process. The vertical process, however, is energetically unfavorable unless the bath configuration $\mathcal{E}$ is in the *Landau-Zener* regime, $\mathcal{E} \approx \mathcal{E}^*$, where $\mathcal{E}^* = \hbar\epsilon/\mu$ is the point of intersection of the two curves. The activation (free) energies for this particular bath fluctuation relative to the D and A state are

$$\begin{aligned} F_D^* &\equiv F_+(\mathcal{E}^*) - F_+(\mathcal{E} = +\Lambda_{cl}/\mu) = (\hbar\epsilon - \Lambda_{cl})^2/4\Lambda_{cl} \,, \\ F_A^* &\equiv F_-(\mathcal{E}^*) - F_-(\mathcal{E} = -\Lambda_{cl}/\mu) = (\hbar\epsilon + \Lambda_{cl})^2/4\Lambda_{cl} \,. \end{aligned} \tag{20.21}$$

In the symmetric case $\epsilon = 0$, we have $F_D^* = F_A^* = \Lambda_{cl}/4$. With increasing bias, the activation energy relative to the donor state becomes smaller, and therefore the $D \to A$ (forward) rate becomes larger. When the condition $\hbar\epsilon = \Lambda_{cl}$ is reached,

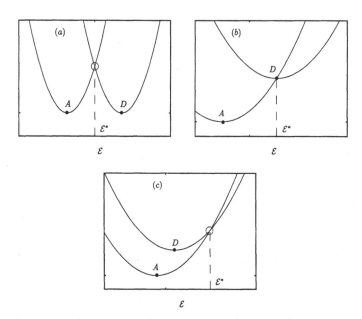

Figure 20.2: Free energy of the diabatic ($\Delta = 0$) electronic states as a function of the nuclear reaction coordinate $\mathcal{E}$ ("Marcus parabola"). The figures show the schematic dependence on the bias ϵ. (a) Symmetric case $\epsilon = 0$. (b) Activationless case $\epsilon = \Lambda_{\mathrm{cl}}/\hbar$. (c) Inverted regime $\epsilon > \Lambda_{\mathrm{cl}}/\hbar$. The intersection point of the two parabola is at $\mathcal{E}^* = \hbar\epsilon/\mu$.

the free energy barrier has completely disappeared. Since $F_D^* = 0$ at this point, we have activationless transfer which depends only weakly on temperature. As the bias is increased further, the activation energy grows again. For this reason, the regime $\hbar\epsilon > \Lambda_{\mathrm{cl}}$ is called the *inverted regime*. Altogether, the Marcus theory predicts that the transfer rate is a non-monotonous function of the bias. The sweeping success of Marcus' theory is founded on the discovery of this characteristic behaviour in ET measurements.

For low temperatures, nuclear tunneling in the bath modes becomes relevant [98]. This collective tunneling phenomenon is generally thought to be prevailing in the inverted regime. However, at low T, nuclear tunneling may even dominate the transfer dynamics for a symmetric system. The quantum effects depend on the low-frequency behaviour of the spectral density, the coupling strength, and the cutoff frequency. This will be discussed in some detail for the nonadiabatic regime in Section 20.2.

In a *classical* description of activated barrier crossing, the rate expression consists of the Boltzmann factor $\exp(-\beta F_{D/A}^*)$ and of the pre-exponential factor or attempt frequency [cf. Eq. (10.4)]. The Boltzmann factor represents the probability for a fluctuation so that $\mathcal{E} = \mathcal{E}^*$. It has been attempted to integrate higher-order effects in the electronic coupling by using a Landau-Zener pre-exponential factor which describes the probability for tunneling on resonance $\mathcal{E} = \mathcal{E}^*$ and depends on the details of the dynamics in the Landau-Zener transition region [98, 375, 376]. Another possibility is to include terms of higher order in the electronic coupling Δ as a geometrical series [377, 378]. When the collective bath motion is slow compared with the motion of the TSS, we may substitute into the formally exact series given below in Eq. (21.261) for the bath correlation function (18.42) the approximate form $Q(t) = i[\Lambda_{cl}t - \Lambda_{cl}\omega_R t^2]/\hbar + \Lambda_{cl}t^2/\hbar^2\beta$, where ω_R is a characteristic bath frequency. Then the integrations over $s_1, \ldots, s_{n-1}$ in Eq. (21.261) lead to a product of $n-1$ delta functions which render the $\tau_2, \ldots, \tau_n$ integrals trivial, and the τ_1 integral Gaussian. Thus we obtain in the classical regime for the rate from the D state to the A state (forward rate) the Zusman-type form

$$k_{cl}^+(T, \epsilon) = \frac{\hbar\Delta^2}{4 + 2\pi\hbar\Delta^2/\Lambda_{cl}\omega_R} \left(\frac{\pi}{\Lambda_{cl}k_B T}\right)^{1/2} \exp\left(-\frac{(\hbar\epsilon - \Lambda_{cl})^2}{4\Lambda_{cl}k_B T}\right) . \qquad (20.22)$$

The prefactor describes dynamical recrossings of the dividing surface. The backward rate from the A state to the D state obeys detailed balance,

$$k_{cl}^-(T, \epsilon) = k_{cl}^+(T, -\epsilon) = k_{cl}^+(T, \epsilon) \exp(-\hbar\beta\epsilon) . \qquad (20.23)$$

The frequency scale ω_R in Eq. (20.22) depends on details of the solvent model [377]. It may be identified, e.g., with the renormalized frequency defined in Eq. (14.13). The classical forward rate has a maximum at $\hbar\epsilon = \Lambda_{cl}$, and it is *symmetric* around this maximum. At very high temperatures, the Arrhenius factor is saturated, and the rate shows a universal $T^{-1/2}$ temperature dependence. In the adiabatic ("large" Δ) limit, the Marcus parabola are split up and the transfer is described by Transition State Theory (TST) on the lower energy surface. The rate expression becomes independent of the electronic coupling. We obtain from Eq. (20.22)

$$k_{cl,TST}^+(T, \epsilon) = \frac{\omega_R}{2\pi} \left(\frac{\pi\Lambda_{cl}}{k_B T}\right)^{1/2} \exp\left(-\frac{(\hbar\epsilon - \Lambda_{cl})^2}{4\Lambda_{cl}k_B T}\right) . \qquad (20.24)$$

In the opposite nonadiabatic ("small" Δ) limit, the expression (20.22) reduces to the classical *Golden Rule* rate

$$k_{cl,GR}^+(T, \epsilon) = \frac{\hbar\Delta^2}{4} \left(\frac{\pi}{\Lambda_{cl}k_B T}\right)^{1/2} \exp\left(-\frac{(\hbar\epsilon - \Lambda_{cl})^2}{4\Lambda_{cl}k_B T}\right) . \qquad (20.25)$$

Whether the expression (20.24) or the formula (20.25) applies depends on whether the "adiabaticity parameter"

$$g = \pi\hbar\Delta^2/2\omega_R\Lambda_{cl} \qquad (20.26)$$

is large or small. For $g \ll 1$, we have the nonadiabatic rate expression (20.25), while for $g \gg 1$, we obtain the Δ independent adiabatic rate (20.24). The rate in the adiabatic limit is generally much larger than in the nonadiabatic limit.

Attempts for a unified treatment of the crossover region between the adiabatic and nonadiabatic limits, in which the assumed separation of time scales is violated, have been largely founded on an imaginary-time formulation and on an assumed relationship between the ET rate and an analytic continuation of the free energy [cf. Section 12.5]. Among these approaches are treatments which are based on the centroid free energy method [cf. Section 12.6] put forward, e.g., by Gehlen et al. [379] and extended later on to complex-valued centroid coordinates by Stuchebrukhov and Song [380], and on numerical instanton methods [381].

A more complete theoretical treatment of the crossover region which is free of questionable assumptions is based on the real-time path integral formulation. The respective discussion is given in Section 21.8.

20.2 Incoherent tunneling in the nonadiabatic regime

When the global system is ergodic, the reduced system approaches thermal equilibrium as time passes by. For sufficiently high temperatures and/or strong coupling, the system exhibits overdamped exponential relaxation towards the equilibrium state. This regime can be investigated by computation of the evolution of occupation probabilities. We shall postpone the corresponding real-time study to the next chapter, though. In this section, our concern is to calculate the transition rates of the spin-boson model using the free energy (imaginary-time) method. Later on, we shall confirm that the results obtained agree with those of the real-time approach.

In Section 12.1, we have given formally exact quantum mechanical rate expressions which form the basis of numerical computation. Here we confine ourselves to the so-called nonadiabatic transfer limit. This is the limit in which the tunneling matrix element Δ is small compared to characteristic vibrational frequencies and to the reorganization frequency $\Lambda_{cl}/\hbar$. Then we may employ perturbation theory in the tunneling matrix element Δ. The so-called "Golden Rule" approach to the spin-boson tunneling rate, which is widely used in chemical physics and in condensed matter physics, treats the system-environment coupling to all orders and the transfer matrix in second order. In the terminology of the extended system, the nonadiabatic rate corresponds to the one-bounce rate expression. The nonadiabatic regime is relevant, e.g., for electron transfer and for defect tunneling in solids (cf. the recent review in Ref. [344]). Systematic corrections to the nonadiabatic rate are discussed within a dynamical approach in Subsection 21.8.

20.2.1 General expressions for the nonadiabatic rate

In the nonadiabatic regime, the expression (19.32) describes the dominant tunneling contribution to the free energy connected with the right/left state. We have discussed already in Subsection 17.4 that the stationary point $\bar{\tau}$ of the integrand in Eq. (19.32) is a saddle point with the direction of steepest descent along real time. Thus, by the corresponding deformation of the integration path, the free energy (19.32) acquires an imaginary part which is

$$\text{Im}\, F_{\text{L/R}} = -\frac{\hbar\Delta^2}{8i} \int_{\bar{\tau}-i\infty}^{\bar{\tau}+i\infty} d\tau\, e^{\pm\,\epsilon\tau - W(\tau)} = -\frac{\hbar\Delta^2}{8} \int_{-\infty-i\bar{\tau}}^{\infty-i\bar{\tau}} dz\, e^{\pm\,i\epsilon z - Q(z)} . \qquad (20.27)$$

The analytically continued function $Q(z) = W(\tau = iz)$ is the complex pair interaction defined in Eq. (18.42). Now observe that $\text{Im}\,F_{\text{L/R}}$ is related to the forward/backward tunneling rate $k^{\pm}$ from the energetically higher/lower state $\sigma = \mp1$ to the lower/higher state $\sigma = \pm1$ by [cf. Eq (14.39) with $\kappa = 1$]

$$k^{\pm} = -(2/\hbar)\,\text{Im}\, F_{\text{L/R}} . \qquad (20.28)$$

Putting $z = t - i\bar{\tau}$, we obtain for the forward/backward rate the integral expression

$$k^{\pm}(\epsilon) = \frac{\Delta^2}{4}\, e^{\pm\epsilon\bar{\tau}} \int_{-\infty}^{\infty} dt\, e^{\pm\,i\epsilon t}\, e^{-Q(t-i\bar{\tau})} . \qquad (20.29)$$

Because $Q(z)$ is analytic in the strip $0 \geq \text{Im}\, z > -\hbar\beta$, we may choose for $\bar{\tau}$ an arbitrary value in the interval $0 \leq \bar{\tau} < \hbar\beta$. Setting $\bar{\tau} = 0$, Eq. (20.29) takes the form

$$k^{\pm}(\epsilon) = \frac{\Delta^2}{4} \int_{-\infty}^{\infty} dt\, e^{\pm\,i\epsilon t}\, e^{-Q(t)} = \frac{\Delta^2}{2} \int_{0}^{\infty} dt\, \cos\left(\pm\,\epsilon t - Q''(t)\right) e^{-Q'(t)} . \qquad (20.30)$$

Another convenient choice is $\bar{\tau} = \hbar\beta/2$ in Eq. (20.29) [395]. We then have

$$k^{\pm}(\epsilon) = \frac{\Delta^2}{4}\, e^{\hbar\beta\epsilon/2} \int_{-\infty}^{\infty} dt\, e^{\pm\,i\epsilon t}\, e^{-X(t)} = \frac{\Delta^2}{2}\, e^{\pm\,\hbar\beta\epsilon/2} \int_{0}^{\infty} dt\, \cos(\epsilon t)\, e^{-X(t)} , \qquad (20.31)$$

where the function $X(t)$ is an even function of t and is defined in Eq. (18.47).

The backward rate $k^-(\epsilon)$ is related to the forward rate $k^+(\epsilon)$ simply by

$$k^-(\epsilon) = k^+(-\epsilon) = e^{-\hbar\beta\epsilon}\, k^+(\epsilon) . \qquad (20.32)$$

The first form is the definition of the backward rate in terms of the forward rate, and the second form is obvious from Eq. (20.31). Thus, the forward and backward tunneling rates satisfy the *principle of detailed balance*. It is evident from the derivation that detailed balance formally is a consequence of the analytic properties of the function $Q(z)$ in the strip $0 \geq \text{Im}\, z > -\hbar\beta$, and of the reflection property (18.43).

The above rate expressions are nonperturbative in the bath coupling. They correspond to "Golden Rule" in the electronic coupling and are exact in the nonadiabatic limit. The same forms are also found in the dynamical approach given below [cf. Sec. 21.3 and Subsec. 21.8]. Finally, we remark that equivalent expressions have been obtained already in the sixties [382, 120], and were later rederived repeatedly.

20.2.2   Probability for energy exchange: general results

Further insight into the understanding of the quantum rate expression (20.29) is gained by rewriting it in the form

$$k^+(\epsilon) = \tfrac{1}{2}\pi\hbar\Delta^2 P(\hbar\epsilon) , \qquad (20.33)$$

where $P(E)$ is the Fourier transform of the correlation function $e^{-Q(z)}$,

$$P(E) = \frac{1}{2\pi\hbar} \int_{-\infty}^{\infty} dz\, e^{iEz/\hbar - Q(z)} = \frac{e^{\beta E/2}}{2\pi\hbar} \int_{-\infty}^{\infty} dt\, e^{iEt/\hbar - X(t)} . \qquad (20.34)$$

In electron transfer processes, the function $P(E)$ represents the nuclear contribution to the rate. It is nothing but a compact expression for the thermally averaged nuclear Franck-Condon overlap integrals [375, 98]. The function $P(E)$ is equally useful to describe the influence of an electromagnetic environment in single charge tunneling [383], as discussed below in Section 20.3. It is the essence of the quantum rate expression (20.33), and satisfies a number of general properties holding without specification of the spectral density $G(\omega)$. First of all, $P(E)$ is normalized as

$$\int_{-\infty}^{\infty} dE\, P(E) = e^{-Q(0)} = 1 . \qquad (20.35)$$

Secondly, we see from Eq. (20.32) that $P(E)$ obeys the detailed balance relation

$$P(-E) = e^{-\beta E} P(E) . \qquad (20.36)$$

The properties (20.35) and (20.36) imply that $P(E)$ is a positive function of E. It is therefore natural to regard $P(E)$ for $E > 0$ as the spectral probability density function that the reservoir absorbs the energy E from the system during the tunneling event. Correspondingly, $P(E)$ with $E < 0$ is the probability density for emission of energy from the reservoir. Thirdly, using the defining expression (20.34) for $P(E)$, we find the sum rule

$$\int_{-\infty}^{\infty} dE\, EP(E) = -i\hbar \left.\frac{dQ(z)}{dz}\right|_{z=0} = \Lambda_{\mathrm{cl}} . \qquad (20.37)$$

To obtain the second form, we have used the spectral representation (18.42) for $Q(z)$ and we have observed that the resulting expression is just the classical bath reorganization energy defined in Eq. (20.4). The l.h.s. is the balance for the mean energy absorbed and emitted by the environment, $\langle E \rangle_{\mathrm{abs}} - \langle E \rangle_{\mathrm{em}}$. Thus, the sum rule (20.37) expresses that the net absorbed energy is equal to the reorganization energy. A related discussion of the role of the function $P(E)$ is given in Ref. [384].

To consolidate the meaning of $P(E)$, we invoke an old technique originally introduced by Lax [382] for the study of optical line shape functions, and employed in the ET theory by Jortner [385] (see also Refs. [375, 383, 146]).

Consider first the case where the thermal reservoir consists of a single oscillator with eigenfrequency ω_0. The respective spectral density is

$$G(\omega) = \lambda_0^2 \delta(\omega - \omega_0) . \qquad (20.38)$$

Assume that the oscillator is in the ground state. Then the correlation function $Q(z)$ takes the form $Q(z) = \rho[1 - e^{-i\omega_0 z}]$ where $\rho = (\lambda_0/\omega_0)^2$. Upon expanding $e^{-Q(z)}$ in powers of $e^{-i\omega_0 z}$, the function $P(E)$ takes the form of an infinite series of delta functions,

$$P(E) = \sum_{m=0}^{\infty} p_m \, \delta(E - m\hbar\omega_0) \,, \qquad p_m = \frac{\rho^m}{m!} e^{-\rho} \,. \qquad (20.39)$$

The quantities $p_1, p_2, \cdots, p_m$ are the probabilities that the oscillator absorbs $1, 2, \cdots, m$ quanta of energy $\hbar\omega_0$. The probabilities $\{p_m\}$ form a normalized Poisson distribution with mean number ρ of absorbed quanta. Thus, $P(E)$ is the probability function for absorption of energy E resulting from the individual uncorrelated absorption processes .

At finite temperature, the oscillator does not only absorb, but does also emit quanta. The respective mean numbers of bosons are $\rho_a = (1 + n)\rho$, and $\rho_e = n\rho$, where $n = 1/[\exp(\beta\hbar\omega_0) - 1]$ is the Bose distribution function in thermal equilibrium. With the Poissonian probabilites $p_{m,a} = \rho_a^m \, e^{-\rho_a}/m!$ for absorption of m quanta and $p_{\ell,e} = \rho_e^\ell e^{-\rho_e}/\ell!$ for emission of ℓ quanta, the probability for exchange of energy E of the quantum oscillator in thermal equilibrium with the TSS takes in generalization of Eq. (20.39) the form

$$P(E) = e^{-(\rho_e + \rho_a)} \sum_{\ell,m} \frac{\rho_e^\ell \rho_a^m}{\ell! \, m!} \delta\Big(E - (m - \ell)\hbar\omega_0\Big) \,. \qquad (20.40)$$

Putting $k = m - \ell$ as summation variable the sum over the variable ℓ takes the form of the ascending series of the modified Bessel functions $I_k(z)$ [88]. Putting $\mu = \hbar\beta\omega_0/2$ we readily obtain

$$P(E) = e^{-\rho \coth \mu} \sum_{k=-\infty}^{\infty} I_k(\rho/\sinh \mu) \, e^{k\mu} \, \delta(E - k\hbar\omega_0) \,. \qquad (20.41)$$

We now show that the same expression follows from the second form for $P(E)$ in Eq. (20.34). Upon using the expression (18.47) for $X(t)$ and the form (20.38) for $G(\omega)$ we get

$$e^{-X(t)} = e^{-\rho \coth \mu} \exp[\,(\rho/\sinh \mu)\cos(\omega_0 t)\,]$$

$$= e^{-\rho \coth \mu} \sum_{k=-\infty}^{\infty} I_k(\rho/\sinh \mu) \, e^{-ik\omega_0 t} \,. \qquad (20.42)$$

Here the second line follows upon identifying the second exponential factor in the first line with the generating functional of the modified Bessel functions [88]. With the series expression (20.42) the time integration in Eq. (20.34) can be done easily, and the result is just the expression (20.41).

It is straightforward to generalize the discussion to two and three, and finally to infinitely many reservoir oscillators. This confirms that $P(E)$ is the probability for exchange of energy with the environment resulting from uncorrelated exchange of bosons between the reservoir and the TSS in compliance with detailed balance.

Often in nature, a system is influenced by a number of different dissipative mechanisms connected with different bath correlation functions $Q_1(z), \cdots, Q_n(z)$. Then, the correlation factor $e^{-Q(z)}$ can be factorized into the separate contributions. The function $P(E)$, which involves the totality of energy exchange processes, can now be written as a convolution of the partial probability densities $P_j(E)$ connected with the individual correlation functions $Q_j(z)$ [385],

$$P(E) = \int_{-\infty}^{\infty} dE_1 \cdots dE_{n-1} \, P_1(E - E_1) P_2(E_1 - E_2) \cdots P_n(E_{n-1}) \,. \qquad (20.43)$$

Correspondingly, also the total forward/backward rate can be written as a convolution of the partial contributions modulo a factor.

In numerical computations of $P(E)$ for a given spectral density $G_{\mathrm{lf}}(\omega)$, the most difficult task is to calculate the probability function at $T = 0$, $P_0(E)$, for E near to zero since $Q_0(z) \equiv Q_{T=0}(z)$ depends algebraically on z in the regime $\omega_c|z| \gg 1$ for $s \neq 1$, and logarithmically on z for $s = 1$. Fortunately, the computation is facilitated by employing an integral equation for $P_0(E)$. Following Ref. [386], we start out from a relation for the bath correlation function $Q_0(z)$ at zero temperature,

$$\frac{d}{dz} e^{-Q_0(z)} = -i \int_0^{\infty} d\omega \, \frac{G_{\mathrm{lf}}(\omega)}{\omega} e^{-i\omega z} \, e^{-Q_0(z)} \,. \qquad (20.44)$$

Taking the Fourier transform and integrating by parts the resulting integral on the l.h.s., we then obtain for $P_0(E)$ the integral equation [375, b] (see also Refs. [387, 146]),

$$E \, P_0(E) = \int_0^E dE' \, \frac{G_{\mathrm{lf}}[(E - E')/\hbar]}{(E - E')/\hbar} \, P_0(E') \,. \qquad (20.45)$$

Thus we have avoided to calculate $Q_0(t)$. Numerical integration of Eq. (20.45) yields $P_0(E)$ up to a multiplicative constant, the starting value $P_0(0)$. Subsequently, the constant can be fixed by the normalization condition (20.35).

At finite temperature, we write $Q_1(z) = Q(z) - Q_0(z)$. The direct integration of Eq. (20.34) for $P_1(E)$ is without difficulties, since $Q_1(z)$ is exponentially cut off. Finally, $P(E)$ is obtained as a convolution of $P_0(E)$ and $P_1(E)$ [cf. Eq. (20.43)].

We conclude this subsection with calculating the integral (20.34) by the method of steepest descent. Putting

$$R(z) = Q(z) - iEz/\hbar \,, \qquad (20.46)$$

the saddle point z^* obeys the relation $R'(z^*) = 0$ (the prime denotes differentiation with respect to the argument). We find $z^* = -i\tau_{\mathrm{B}}$ with $0 < \tau_{\mathrm{B}} \leq \hbar\beta/2$, and

$$P(E) = \frac{1}{\hbar} \left(\frac{1}{2\pi R''(z^*)} \right)^{1/2} e^{-\beta F_{\mathrm{qm}}} \qquad \text{with} \qquad F_{\mathrm{qm}} = R(z^*)/\beta \,. \qquad (20.47)$$

The quantities F_{qm} and $R''(z^*)$ are positive real. Here, F_{qm} is the "quantum activation free energy" [388] of the donor state, and $R''(z^*)$ introduces quantum effects in the pre-exponential factor. The expression (20.47) is still formal. In general, it is not possible to invert the condition $R'(z^*) = 0$ in analytic form to find the saddle point z^*. The various cases where the steepest-descent formula (20.47) for $P(E)$ is justified and explicit expressions can be given are discussed in the following subsections.

In the classical limit, $\kappa \to \infty$, F_{qm} coincides with the classical activation energy F_D^* for the donor state introduced in Eq. (20.21). We then obtain

$$P(E) = \sqrt{\beta/4\pi\Lambda_{cl}}\, e^{-\beta F_D^*} \qquad \text{with} \qquad F_D^* = (E - \Lambda_{cl})^2/4\Lambda_{cl} . \qquad (20.48)$$

This expression is in correspondence with the Marcus form (20.25).

20.2.3 The spectral probability density for absorption at $T = 0$

Consider the probability density for absorption of energy by the reservoir ($E > 0$) at zero temperature. When $T = 0$, the pair interaction (18.45) reduces to the form

$$Q_{T=0}(z) = 2\delta_s[\Gamma(s)/(s-1)](\omega_c/\omega_{ph})^{s-1}[1 - (1 + i\omega_c z)^{1-s}] . \qquad (20.49)$$

Substituting Eq. (20.49) into Eq. (20.46), the saddle point condition $R'(z^*) = 0$ gives $z^* = -i\tau_B$ with $\tau_B = [(\Lambda_{cl}/E)^{1/s}-1]/\omega_c$, where Λ_{cl} is the solvation energy, Eq. (20.5). The length τ_B is the width of the bounce (kink–anti-kink pair) in imaginary time. In steepest-descent approximation, we find from Eq. (20.47)

$$P_{T=0}(E) = \frac{1}{\sqrt{2\pi s\Lambda_{cl}\hbar\omega_c}}\left(\frac{\Lambda_{cl}}{E}\right)^{\frac{1+s}{2s}} \exp\left\{ -\frac{E}{\hbar\omega_c} + \frac{1}{s-1}\frac{\Lambda_{cl}}{\hbar\omega_c}\left[s\left(\frac{E}{\Lambda_{cl}}\right)^{\frac{s-1}{s}} - 1\right]\right\} . \qquad (20.50)$$

The expression (20.50) is asymptotically exact in the regime

$$[2\delta_s\Gamma(s)]^{1/s}(E/\hbar\omega_{ph})^{(s-1)/s} = (E/\Lambda_{cl})^{(s-1)/s}\Lambda_{cl}/\hbar\omega_c \gg (s+2)(s+1)/8s . \qquad (20.51)$$

Physically, this is the regime of multi-phonon emission processes. It roughly corresponds to the regime $E \lesssim \hbar\omega_{ph}$ in the sub-Ohmic case, $E \gtrsim \hbar\omega_{ph}$ in the super-Ohmic case, and to $\delta_1 \equiv K \gg 1$ in the Ohmic case.

The probability for the transfer of energy to the environment has a maximum at $E = E_{max}$, where E_{max} obeys the transcendental equation

$$[1 - (\Lambda_{cl}/E_{max})^{1/s}]E_{max} + [(s+1)/2s]\hbar\omega_c = 0 . \qquad (20.52)$$

The expressions (20.50) – (20.52) hold for general s. The form of the ascending flank for E well below E_{max} depends on the particular value of s.

In the sub-Ohmic case, $s < 1$, the formula (20.50) is valid down to $E = 0$. The function $P(E)$ has a pronounced maximum at $E = E_{max}$. In the limit $\omega_c \to \infty$, the position of the maximum is selfconsistently determined from Eq. (20.52) as

$$E_{max} = [2s/(1+s)]^{s/(1-s)}[2\delta_s\Gamma(s)]^{1/(1-s)}\hbar\omega_{ph} . \qquad (20.53)$$

As a result of the high density of low-energy excitations, the probability of the TSS to lose only a small amount of energy is exponentially small and therefore elastic tunneling processes are absent,

$$P_{T=0}(E \to 0) \propto E^{-(1+s)/2s} \exp[-a_s(\Lambda_{\mathrm{cl}}/E)^{\frac{1}{s}-1}] . \qquad (20.54)$$

Hence, barrier crossing is completely quenched for a symmetric system at zero temperature [with the exception of the special case (18.36)].

The descending flank of $P_{T=0}(E)$ for E above $E_{\max}$ is formed by multi-phonon emission. These processes gradually die out as E is increased further, so that the steepest descent formula (20.50) ceases to be valid. For $E \gg E_{\max}$, the influence of the environment is only weak and the probability for absorption of energy by the reservoir is determined by the one-phonon process. Putting $\mathrm{e}^{-Q(t)} \approx 1 - Q(t)$ and substituting the representation (18.42) for $Q(t)$, we obtain[2] for $E \gg E_{\max}$ and $s < 1$

$$P_{T=0}(E) = \frac{1}{E} \frac{G_{\mathrm{lf}}(E/\hbar)}{E/\hbar} = \frac{1}{\Gamma(s)} \frac{\Lambda_{\mathrm{cl}}}{(\hbar\omega_c)^2} \left(\frac{\hbar\omega_c}{E} \right)^{2-s} \mathrm{e}^{-E/\hbar\omega_c} . \qquad (20.55)$$

From this we see that the one-phonon contribution to $P(E)$ directly reflects the spectral properties of the reservoir coupling modulo a factor E^{-2}.

The *super-Ohmic* case with $s > 2$ shows opposite behaviour. Now, the low-energy tail of $P_{T=0}(E)$ ($E \ll E_{\max}$) is determined by the one-phonon process, Eq. (20.55). With increasing E, multi-phonon processes become active and form the ascending flank of $P_{T=0}(E)$. For $E \gtrsim E_{\max}$, the asymptotic multi-phonon expression (20.50) is valid. The maximum of $P_{T=0}(E)$ is situated at $E_{\max} = \Lambda_{\mathrm{cl}} - \frac{1}{2}(1 + s)\hbar\omega_c$. In the limit $E \to \infty$, $P_{T=0}(E)$ drops to zero faster than algebraically. Physically, this is a consequence of the exponential cutoff in the spectral density $G_{\mathrm{lf}}(\omega)$.

Finally, consider the Ohmic case. Taking the limit $s \to 1$ in Eq. (20.50), we find

$$P_{T=0}(E) = \sqrt{\frac{K}{\pi}} \frac{\mathrm{e}^{2K}}{(2K)^{2K}} \frac{\mathrm{e}^{-E/\hbar\omega_c}}{E} \left(\frac{E}{\hbar\omega_c} \right)^{2K} . \qquad (20.56)$$

The term $\propto E^{2K-1}$ in $P(E)$ holds for all K, whereas the prefactor is only valid for $K \gg 1$ [cf. Eq. (20.81) below]. The maximum of the probability for absorption of energy by the reservoir is at $E_{\max} = (2K - 1)\hbar\omega_c = \Lambda_{\mathrm{cl}} - \hbar\omega_c$. Well below the maximum ,$P(E)$ has a power law form with an exponent which depends on the coupling strength and can be positive or negative.

20.2.4 Crossover from quantum-mechanical to classical behaviour

We now generalize the discussion to finite temperature. Consider first the case $E = 0$ in the steepest descent result (20.47). For $E = 0$, the saddle point is at $z^* = -i\hbar\beta/2$.

---

[2]At finite T, the one-phonon emission process has an additional factor $1/[1 - \exp(-\beta E)]$.

It is therefore natural to switch to the representation (20.31) in which the integrand is stationary at $t = 0$. Expanding the integral (18.47) for $X(t)$ about $t = 0$, we find

$$X(t) = \Lambda_1 \beta/4 + (\Lambda_2/\hbar^2\beta)\, t^2 + \mathcal{O}(t^4)\,, \tag{20.57}$$

$$\Lambda_1 = \frac{4}{\beta} \int_0^\infty d\omega\, \frac{G_{\mathrm{lf}}(\omega)}{\omega^2} \frac{[\cosh(\frac{1}{2}\hbar\beta\omega) - 1]}{\sinh(\frac{1}{2}\hbar\beta\omega)}\,, \tag{20.58}$$

$$\Lambda_2 = \frac{\hbar^2\beta}{2} \int_0^\infty d\omega\, G_{\mathrm{lf}}(\omega)\, \frac{1}{\sinh(\frac{1}{2}\hbar\beta\omega)}\,. \tag{20.59}$$

The quantities Λ_1 and Λ_2 are "quantum-corrected" reorganization energies. For the form (18.44) for $G_{\mathrm{lf}}(\omega)$ we find from Eq. (18.48) the analytical expressions

$$\Lambda_1 = 8k_{\mathrm{B}}TB_s + b_s(\kappa)\, k_{\mathrm{B}}T\,(T/T_{\mathrm{ph}})^{s-1}\,, \tag{20.60}$$

$$\Lambda_2 = a_s(\kappa)\, k_{\mathrm{B}}T\,(T/T_{\mathrm{ph}})^{s-1}\,, \tag{20.61}$$

$$b_s(\kappa) = 16\delta_s\Gamma(s-1)\Big\{\zeta(s-1, 1+\kappa) - \zeta(s-1, \tfrac{1}{2}+\kappa)\Big\}\,, \tag{20.62}$$

$$a_s(\kappa) = 2\delta_s\Gamma(s+1)\zeta(s+1, \tfrac{1}{2}+\kappa)\,, \tag{20.63}$$

where $T_{\mathrm{ph}} = \hbar\omega_{\mathrm{ph}}/k_{\mathrm{B}}$. The first term in Eq. (20.60) gives rise to the adiabatic dressing factor introduced in Eqs. (18.32). We see from the integral representations (20.58) and (20.59) that $\Lambda_1 \geq \Lambda_2$. In the limit $\kappa \equiv k_{\mathrm{B}}T/\hbar\omega_c \to \infty$, the quantum reorganization energies Λ_1 and Λ_2 approach the classical reorganization energy,

$$\lim_{\kappa\to\infty} \Lambda_1 = \Lambda_{\mathrm{cl}}\,, \qquad \lim_{\kappa\to\infty} \Lambda_2 = \Lambda_{\mathrm{cl}}\,. \tag{20.64}$$

These limiting cases can be seen either from the integral representations or by substituing the asymptotic expansion of the zeta function into Eqs. (20.60) and (20.61). In the quantum regime, $\kappa \ll 1$, the coefficient functions take the form

$$b_s(0) = 16\delta_s\Gamma(s-1)[\,2 - 2^{s-1}]\zeta(s-1)\,, \tag{20.65}$$

$$a_s(0) = 2\delta_s\Gamma(s+1)[\,2^{s+1} - 1]\zeta(s+1)\,. \tag{20.66}$$

Substituting Eq. (20.57) into Eq. (20.31), we obtain the probability for exchange of zero energy as (cf. Eq. (4.4) in Ref. [389])

$$P(0) = \sqrt{\beta/4\pi\Lambda_2}\, e^{-\beta\Lambda_1/4}\,. \tag{20.67}$$

Within the bounce picture discussed in Subsection 17.4, the formula (20.67) has the following interpretation. The bounce consists of an instanton–anti-instanton pair with (imaginary-time) distance $\hbar\beta/2$, which is half of the bounce period $\hbar\beta$. The exponent $\Lambda_1/4k_{\mathrm{B}}T$ is the interaction between the instantons,

$$\Lambda_1/4k_{\mathrm{B}}T \;=\; W(\hbar\beta/2) \;=\; Q(-i\hbar\beta/2) \;=\; X(0)\,, \qquad (20.68)$$

and $\Lambda_2/k_{\mathrm{B}}T$ is directly connected with the negative eigenvalue of the breathing mode of the bounce. Hence it is a measure for the stiffness of the bounce width.

The leading correction to the formula (20.67) arises from the $\mathcal{O}(t^4)$-term in Eq. (20.57). This contribution is small when the condition

$$32\delta_s\left(\frac{T}{T_{\mathrm{ph}}}\right)^{s-1} \frac{[\,\Gamma(s+1)\,\zeta(s+1,\tfrac{1}{2}+\kappa)\,]^2}{\Gamma(s+3)\,\zeta(s+3,\tfrac{1}{2}+\kappa)} \;\gg\; 1 \qquad (20.69)$$

is satisfied. Physically, this condition holds in the regime of multi-phonon processes. In the Ohmic case, Eq. (20.69) corresponds to the damping regime $K \gg 1$. In the super-Ohmic case, the condition (20.69) is met for temperatures well above T_{ph}. For a sub-Ohmic bath, the condition (20.69) is fulfilled when T is fairly below T_{ph}.

For an exchanged energy E in the range

$$E \;\ll\; (1+2\kappa)\Lambda_2\,, \qquad (20.70)$$

the saddle point z^* is still near to $-i\hbar\beta/2$. Thus we may use again the short-time expansion (20.57) for $X(t)$. We then find an explicit expression for the form (20.47),

$$P(E) \;=\; \sqrt{\beta/4\pi\Lambda_2}\,\mathrm{e}^{-\beta F_{\mathrm{qm}}} \qquad \text{with} \qquad F_{\mathrm{qm}}(E) \;=\; \Lambda_1/4 - E/2 + E^2/4\Lambda_2\,. \quad (20.71)$$

This expression includes quantum effects and satisfies the detailed balance relation (20.36). Corrections to the formula (20.71) are found to be small if we meet the conditions (20.69) and (20.70). The energy range allowed by Eq. (20.70) sensitively depends on temperature. It shrinks to zero $\propto T^s$ as $T \to 0$, as follows with Eq. (20.61). In the classical limit, $\kappa \to \infty$, the influence of *every* bath is characterized by only one parameter, the reorganization energy Λ_{cl}. In this limit, the expression (20.71) reduces to the classical probability function (20.48), and the condition (20.70) is also valid when the exchanged energy E is of the order of Λ_{cl} or larger.

As the temperature is lowered to the quantum regime, the particular spectral properties of the spectral density $G_{\mathrm{lf}}(\omega)$ become relevant which indicates that nuclear tunneling becomes effective. The quantum effects are captured by the deviations of the activation energies Λ_1 and Λ_2 from the classical solvation energy Λ_{cl}. We should like to remark that the above expressions are valid for general s. However, the conditions (20.69) and (20.70) depend in the quantum regime crucially on the parameter s.

As E is increased, the stationary point z^* moves from $z^* = -i\hbar\beta/2$ towards $z^* = 0$. For energies where the condition (20.70) is violated, the short-time expansion about $z = -i\hbar\beta/2$, Eq. (20.57), is not any more applicable, and the expression (20.71) is incorrect. When z^* is distant from $-i\hbar\beta/2$ the expression (20.47) must be evaluated numerically. However, when z^* is near to zero, which is the case when E and/or T is very large, we can analyze again the expression (20.47) in analytic form.

We now study the expression (20.47) near to the maximum of $P(E)$ in the Ohmic and super-Ohmic case. For E near to E_{max} and $s \geq 1$, the stationary point is found to be near $z = 0$ for all T. A consistent analysis can thus be performed upon expanding $Q(z)$ in Eq. (20.34) about $z = 0$ up to fourth order in z. We find

$$Q(z) = i\frac{\Lambda_{cl}}{\hbar}z + \frac{A_2}{2!}(\omega_c z)^2 - i\frac{A_3}{3!}(\omega_c z)^3 - \frac{A_4}{4!}(\omega_c z)^4 + \mathcal{O}(\omega_c z)^5 . \qquad (20.72)$$

For the spectral density (18.31), the dimensionless coefficient functions read

$$A_2(\kappa) = s\left[1 + 2\kappa^{s+1}\zeta(s+1, 1+\kappa)\right]\Lambda_{cl}/\hbar\omega_c ,$$
$$A_3(\kappa) = s(s+1)\Lambda_{cl}/\hbar\omega_c , \qquad (20.73)$$
$$A_4(\kappa) = s(s+1)(s+2)\left[1 + 2\kappa^{s+3}\zeta(s+3, 1+\kappa)\right]\Lambda_{cl}/\hbar\omega_c .$$

Then we find the normalized Gaussian distribution

$$P(E) = \frac{1}{\sqrt{2\pi\langle E^2\rangle}}\exp\left(-\frac{(E - E_{max})^2}{2\langle E^2\rangle}\right) . \qquad (20.74)$$

The position of the maximum and the width of the distribution are given by

$$E_{max} = \Lambda_{cl} - \frac{A_3}{2A_2}\hbar\omega_c ,$$
$$\langle E^2\rangle \equiv 2k_B T_{eff}\Lambda_{cl} = \left(1 + \frac{A_3^2}{2A_2^3} - \frac{A_4}{2A_2^2}\right)A_2(\hbar\omega_c)^2 . \qquad (20.75)$$

The expression (20.74) with (20.75) and (20.73) correctly describes $P(E)$ near to the maximum for all T. Observe that the transfer of energy is activationless at the maximum for all temperatures. In the zero temperature limit, we find from Eq. (20.75)

$$E_{max} = \Lambda_{cl} - \tfrac{1}{2}(1+s)\hbar\omega_c , \qquad \langle E^2\rangle = s\hbar\omega_c\Lambda_{cl} - \tfrac{1}{2}(s+1)(\hbar\omega_c)^2 . \qquad (20.76)$$

The resulting expression for $P(E)$ is consistent with the previous result (20.50) around the maximum. In the classical limit, $\kappa \gg 1$, E_{max} and $\langle E^2\rangle$ are in Marcus form,

$$E_{max} = \Lambda_{cl} , \qquad \langle E^2\rangle = 2k_B T\Lambda_{cl} . \qquad (20.77)$$

Thus, the position of the maximum is shifted upwards in energy to the reorganization energy as the temperature is raised from the quantum regime ($\kappa \ll 1$) to the classical regime ($\kappa \gg 1$). Simultaneously, the width of the Gaussian distribution is substantially broadened as temperature is increased. The effective temperature T_{eff} in Eq. (20.75) approaches T at high temperatures and becomes temperature-independent as the actual temperature decreases. The expression (20.74) with Eq. (20.75) takes consistently into account all quantum effects for E around E_{max}. It is the systematic generalization of a nonadiabatic rate formula discussed by Garg et al. [377].[3] These authors take into account the quantum effects included in the

---

[3]The model in Ref. [377] differs in the description of the medium. The reaction coordinate is modeled by a damped harmonic oscillator leading to a spectral density of the form (3.173).

function $A_2(\kappa)$, and they disregard those described by $A_3(\kappa)$ and $A_4(\kappa)$. The functions $A_3(\kappa)$ and $A_4(\kappa)$ vanish in the classical limit, but may become relevant in the quantum regime. The treatment in Ref. [377] has several precursors [390] and has sometimes been termed "semiclassical" [390, 388].

In the classical limit, the probability $P(E)$ is symmetric about the maximum. In the quantum regime, the function $P(E)$ is asymmetric. The function $P(E)$ is enhanced above $E_{\max}$ because of the smaller width of the barrier in the inverted regime which leads to a higher tunneling probability. This behaviour is confirmed, e.g., by the $T = 0$ expressions (20.50) and (20.56). The asymmetry of $P(E)$ around the maximum is reduced as the temperature and/or the damping strength is increased.

20.2.5 The Ohmic case

Consider first the Ohmic case, $s = 1$, in the scaling regime $k_B T$, $\hbar|\epsilon| \ll \hbar\omega_c$, in which the rate expression (20.29) can be calculated in analytic form. Substituting the scaling form (18.51) into Eq. (20.29), the Golden Rule rate is found to lowest order in $k_B T/\hbar\omega_c$ and in ϵ/ω_c as [391, 392, 393]

$$k^+(T, \epsilon) \;=\; \frac{1}{4}\frac{\Delta^2}{\omega_c}\left(\frac{\hbar\omega_c}{2\pi k_B T}\right)^{1-2K}\frac{|\Gamma(K + i\hbar\epsilon/2\pi k_B T)|^2}{\Gamma(2K)}\, e^{\hbar\epsilon/2k_B T}\,. \tag{20.78}$$

With the form (18.51), the integrand in Eq. (20.29) is singular at the origin. Naively, one would therefore expect that the result (20.78) is only valid in the regime $K < \frac{1}{2}$. However, observe that we have calculated a contour integral which actually bypasses the singularity. Hence the expression (20.78) can be analytically continued to the regime $K \geq \frac{1}{2}$ and therefore holds without restriction on the parameter K. This is easily substantiated by use of the representation (20.31) with the form (18.52) for $X(t)$. The resulting integrand is nonsingular on the real axis for $K \geq 0$, and the respective integral is tabulated in Ref. [87]. In the end, we find the previous form (20.78).

For $K < 1$, it is convenient to absorb the ω_c dependence into the effective tunneling matrix element $\Delta_{\rm eff}$ defined in Eq. (18.35). Then Eq. (20.78) is rewritten as

$$k^+(T, \epsilon) \;=\; \Delta_{\rm eff}\,\frac{\sin(\pi K)}{2\pi}\left(\frac{\hbar\Delta_{\rm eff}}{2\pi k_B T}\right)^{1-2K}|\Gamma(K + i\hbar\epsilon/2\pi k_B T)|^2\, e^{\hbar\epsilon/2k_B T}\,. \tag{20.79}$$

For weak damping $K \ll 1$, this expression reduces to the form

$$k^+(T, \epsilon) \;=\; \left(\frac{2\pi k_B T}{\hbar\Delta_{\rm eff}}\right)^{2K}\frac{\pi K\Delta_{\rm eff}^2}{[\,(2\pi K k_B T/\hbar)^2 + \epsilon^2\,]}\frac{\epsilon}{[\,1 - \exp(-\hbar\epsilon/k_B T)\,]}\,. \tag{20.80}$$

At zero temperature, we can calculate the rate exactly for arbitrary ω_c. Substituting the form $Q_{T=0}(z) = 2K\ln(1 + i\omega_c z)$ into the integral (20.34), we obtain

$$k^+(0, \epsilon) \;=\; \frac{\pi}{2\Gamma(2K)}\frac{\Delta^2}{\omega_c}\left(\frac{\epsilon}{\omega_c}\right)^{2K-1}e^{-\epsilon/\omega_c}\,, \tag{20.81}$$

whereas $k^-(0, \epsilon) = 0$, as follows with Eq. (20.32). Hence at $T = 0$, there are only transitions from the upper to the lower well, as expected. We can also calculate the asymptotic enhancement at low T for arbitrary ω_c. Expanding $Q(z)$ up to terms of order T^2, we formally obtain $k^+(T, \epsilon) = \exp[\,K(\pi^2/3)\partial^2/(\partial\hbar\beta\epsilon)^2\,]\,k^+(0, \epsilon)$, yielding

$$k^+(T, \epsilon) = k^+(0, \epsilon) \exp\left\{ \frac{K}{3} \left(\frac{\pi}{\hbar\beta} \right)^2 \left[\left(\frac{2K-1}{|\epsilon|} - \frac{1}{\omega_c} \right)^2 - \frac{2K-1}{\epsilon^2} \right] \right\}. \quad (20.82)$$

The finite-temperature enhancement of the rate varies as T^2. This in agreement with the corresponding result for an extended biased potential, Eq. (17.19). The width of the instanton-anti-instanton pair in steepest descent is $\bar{\tau} = 2K/\epsilon$, so that for $K \gg 1$ also the prefactor of the T^2 law corresponds with Eq. (17.19). When $\hbar\epsilon$ is chosen near to the maximum of the rate, the result (20.82) is consistent with the low-temperature expansion of the rate formula (20.74) with (20.75) for the Ohmic case.

On the other hand, in the absence of a bias ($\epsilon = 0$), Eqs. (20.78) and (20.79) give

$$k^\pm(T, 0) = \frac{\sqrt{\pi}\Gamma(K)}{4\Gamma(K + \frac{1}{2})} \frac{\Delta^2}{\omega_c} \left(\frac{\pi k_B T}{\hbar\omega_c} \right)^{2K-1} = \frac{\Delta_{\text{eff}}}{2} \frac{\Gamma(K)}{\Gamma(1-K)} \left(\frac{\hbar\Delta_{\text{eff}}}{2\pi k_B T} \right)^{1-2K}. \quad (20.83)$$

The forms containing Δ_{eff} apply in the regime $K < 1$, while the forms with the explicit ω_c dependence are also valid in the regime $K > 1$. We remark that, for the particular potentials discussed in Subsection 17.4 and Section 18.1, the transfer matrix element may be expressed in terms of the original potential parameters.

In the regime $0 \le K < 1$, the above Golden Rule results are valid if the temperature or the bias exceeds a certain value [cf. the discussion in Subsection 21.3.1]. For $K < \frac{1}{2}$, low temperatures and weak bias, the dynamics may actually be coherent and therefore cannot be simply described in terms of a single rate. For $\frac{1}{2} < K < 1$, the Golden rule approximation may break down at low T. On the other hand, for $K > 1$, the above Golden Rule rate expressions are valid down to zero temperature. Further discussions of this point and of adiabatic corrections are given in Subsection 21.3.2.

Next, consider the crossover from the quantum to the classical regime in the bias range given in Eq. (20.70). The quantum activation energies Λ_1 and Λ_2 take the form

$$\Lambda_1 = 8K k_B T \ln[\,\hbar\beta\omega_c\Gamma^2(1 + \kappa)/\Gamma^2(\tfrac{1}{2} + \kappa)\,], \quad (20.84)$$

$$\Lambda_2 = 2K k_B T \psi'(\tfrac{1}{2} + \kappa), \quad (20.85)$$

where $\kappa = 1/\beta\hbar\omega_c$. With these forms, the rate formula (20.33) with (20.71) reads

$$k^+(T, \epsilon) = \frac{\hbar\beta\Delta^2}{4} \left(\frac{\pi\, e^{\hbar\beta\epsilon}}{2K\psi'(\tfrac{1}{2} + \kappa)} \right)^{1/2} \left(\frac{\Gamma^2(\tfrac{1}{2} + \kappa)}{\hbar\beta\omega_c\Gamma^2(1 + \kappa)} \right)^{2K} \exp\left(-\frac{(\hbar\beta\epsilon)^2}{8K\psi'(\tfrac{1}{2} + \kappa)} \right). \quad (20.86)$$

This reduces in the quantum regime $k_B T \ll \hbar\omega_c$ to the form

$$k^+(T,\epsilon) = k^+(T,0) \exp\left\{\frac{\hbar\beta\epsilon}{2} - \frac{1}{K}\left(\frac{\hbar\beta\epsilon}{2\pi}\right)^2\right\}, \qquad (20.87)$$

$$k^+(T,0) = \frac{\sqrt{\pi}}{4\sqrt{K}}\frac{\Delta^2}{\omega_c}\left(\frac{\pi}{\hbar\beta\omega_c}\right)^{2K-1}. \qquad (20.88)$$

The bias dependence is correctly described by Eq. (20.87) for $\epsilon \ll K\pi k_B T/\hbar$. In the regime $K \gtrsim 1$, we have $\Gamma(K)/\Gamma(K+\frac{1}{2}) \approx 1/\sqrt{K}$. Thus, in this damping regime, the steepest-descent expression (20.88) coincides with the exact formula (20.83).

In the classical regime, $k_B T \gg \hbar\omega_c$, Eq. (20.86) reduces to the form

$$k^+(T,\epsilon) = \frac{\hbar\Delta^2}{4}\left(\frac{\pi}{2K\hbar\omega_c k_B T}\right)^{1/2} \exp\left(-\frac{\hbar(\epsilon - 2K\omega_c)^2}{8K\omega_c k_B T}\right). \qquad (20.89)$$

The general expression (20.86) describes the transition from the quantum rate expression, Eq. (20.87) with (20.88), to the classical rate, Eq. (20.89).

Finally, the rate expression near to the maximum is given by Eqs. (20.33), (20.74), and (20.75) with the coefficient functions $A_2 = 2K[1 + 2\kappa^2\psi'(1+\kappa)]$, $A_3 = 4K$, and $A_4 = 4K[3 + \kappa^4\psi^{(3)}(1+\kappa)]$. As T is increased from $\kappa \ll 1$ to $\kappa \gg 1$, $\epsilon_{\max}$ moves from $(2K-1)\omega_c$ to $2K\omega_c$, and the width $\langle\epsilon^2\rangle$ expands from $(2K-1)\omega_c^2$ to $4K\omega_c k_B T/\hbar$.

The above rate expressions are directly relevant for a large variety of tunneling problems involving an Ohmic heat bath, e.g., conduction electron reservoirs in metals, and resistive electromagnetic environments.

We remark that the theoretical interest in Ohmic dissipation stems, to a large part, from the power-law behaviours of the transfer rate discussed in this subsection.

20.2.6 Exact nonadiabatic rates for $K = 1/2$ and $K = 1$

For the special value $K = \frac{1}{2}$, the nonadiabatic rate (20.29) can be calculated for the full pair interaction (18.49) by contour integration. Thus we find the rate in analytic form for arbitrary value of $\hbar\beta\omega_c$. Closing the contour in the half-plane $\mathrm{Im}\, z > 0$, we pick up the contributions from the simple poles of $\Gamma(\kappa + iz/\hbar\beta)$. The poles are at $z = i(n+\kappa)\hbar\beta$, where $n = 0, 1, 2, \cdots$ and $\kappa = 1/\hbar\beta\omega_c$. The residua are proportional to $\Gamma(n+2\kappa+1)$. Upon using Euler's integral representation for the function $\Gamma(n+2\kappa+1)$ [87], the contributions from the sequence of poles can be summed. Eventually, the remaining integration can be performed. We readily find [394]

$$k^+(T,\epsilon; K = \tfrac{1}{2}) = \frac{\Gamma(\frac{1}{2}+\kappa)}{\sqrt{\pi}\Gamma(1+\kappa)} \frac{e^{\hbar\beta\epsilon/2}}{[\cosh(\hbar\beta\epsilon/2)]^{1+2\kappa}}\frac{\gamma}{2}, \qquad (20.90)$$

where $\gamma \equiv \Delta_{\text{eff}}(K = \frac{1}{2}) = \pi\Delta^2/2\omega_c$. Note that the expression (20.90) satisfies the detailed balance condition (20.32). For the symmetric system, Eq. (20.90) becomes

$$k^+(T,0; K = \tfrac{1}{2}) = k^-(T,0; K = \tfrac{1}{2}) = \frac{\Gamma(\frac{1}{2}+\kappa)}{\sqrt{\pi}\Gamma(1+\kappa)}\frac{\gamma}{2}. \qquad (20.91)$$

In the low-temperature regime $\kappa \equiv \hbar\beta\omega_c \gg 1$, the rate (20.91) becomes temperature-independent, $k_+ = k_- = \gamma/2$, which is in agreement with the expression (20.83) for $K = \frac{1}{2}$. In the opposite classical limit $\hbar\beta\omega_c \ll 1$, we find

$$k^+(T, 0; K = \tfrac{1}{2}) = \frac{\Delta^2}{4}\left(\frac{\pi\hbar}{\omega_c k_B T}\right)^{1/2} \propto 1/\sqrt{T} . \tag{20.92}$$

In the Ohmic case with damping strength $K = \frac{1}{2}$, the solvation energy Λ_{cl} is equal to $\hbar\omega_c$, as we see from Eq. (20.6). Thus we recover the high-temperature limit of the formula (20.25).

As a function of the bias, the rate (20.90) has a maximum at $\epsilon = \epsilon_{max}$, where

$$\hbar\epsilon_{max} = k_B T \ln(1 + \hbar\omega_c/k_B T) . \tag{20.93}$$

In the classical limit, this reduces to the Marcus result $\epsilon_{max} = \omega_c$. As well, the expression (20.90) reduces to the Marcus form (20.25). As T is lowered, the maximum of the rate shifts to smaller values for ϵ_{max} and finally approaches $\epsilon_{max} = 0$ as $T \to 0$.

Making use of the convolution property

$$k^+(\epsilon; 2K) = \frac{2}{\pi\Delta^2} \int_{-\infty}^{\infty} d\epsilon' \, k^+(\epsilon'; K) k^+(\epsilon - \epsilon'; K) , \tag{20.94}$$

which follows from Eqs. (20.33) and (20.43), we can obtain the exact rate in analytic form also for $K = 1$. Substituting Eq. (20.90) in Eq. (20.94), we find

$$k^+(T, \epsilon; K = 1) = \frac{\Delta^2}{2\hbar\beta\omega_c^2} \left(\frac{\Gamma(\frac{1}{2} + \kappa)}{\Gamma(1 + \kappa)}\right)^2 \frac{e^{\hbar\beta\epsilon/2} \mathcal{Q}_{2\kappa}[\coth(\hbar\beta\epsilon/2)]}{[\sinh(\hbar\beta\epsilon/2)]^{1+2\kappa}} , \tag{20.95}$$

where $\mathcal{Q}_\nu(z)$ is the Legendre function of the second kind [87]. Upon using the property $\mathcal{Q}_\nu(z) = z^{-\nu-1} F_\nu(z^2)$, where $F_\nu(z^2)$ is a function of z^2, it is straightforward to see that the expression (20.95) satisfies detailed balance. For zero bias, we obtain

$$k^+(T, 0; K = 1) = \frac{\pi\Delta^2}{2\hbar\beta\omega_c^2} \frac{\Gamma^4(1 + 2\kappa)}{\Gamma^4(1 + \kappa)\Gamma(2 + 4\kappa)} , \tag{20.96}$$

which agrees at low temperatures with the temperature dependence in Eq. (20.83) (where $K = 1$). In the high-temperature limit, we recover again the $T^{-1/2}$ Marcus law. At high temperatures, the rate expression (20.95) as a function of the bias is symmetric around the maximum which is at $\epsilon = \epsilon_{max}$, where $\epsilon_{max} = \Lambda_{cl}/\hbar = 2\omega_c$. With decreasing temperature, the maximum is shifted to $\epsilon_{max} = \Lambda_{cl}/2\hbar$, and the rate becomes asymmetric around this value. In summary, the special cases $K = \frac{1}{2}$ and $K = 1$ show in the various limits the general characteristic features discussed above.

20.2.7 The sub-Ohmic case $(0 < s < 1)$

The sub-Ohmic case is distinguished by the high density of low-energy excitations. We therefore expect that the dissipative effects become most noticeable at low T and weak bias. The forward rate for a biased system at zero temperature is given by

$$k^+(0, \epsilon) = (\pi\hbar/2) \Delta^2 P_{T=0}(\hbar\epsilon) . \tag{20.97}$$

The behaviour of $P_{T=0}(\hbar\epsilon)$ has been discussed in Subsection 20.2.3.

Consider next the low-temperature corrections in the regime $k_B T \ll \hbar\epsilon$. We should expect that the ascending shoulder of $P(E)$ is shifted to lower energies as the temperature is raised above zero since the particle can then also absorb energy from the reservoir. This effect can easily be studied asymptotically. Since the form (20.49) cuts off the Fourier integral (20.34) for $P(E)$ at $|z| \gtrsim 1/(\omega_{\rm ph}\delta_s^{1/(1-s)})$, the temperature-dependent term of $Q(z)$ in Eq. (18.45) may be expanded in powers of z^2,

$$\Delta Q(z) = 2\delta_s \Gamma(s+1)\zeta(s+1) (\hbar\beta\omega_{\rm ph})^{-(s+1)}(\omega_{\rm ph}z)^2 + \mathcal{O}[(\omega_{\rm ph}z)^4] . \qquad (20.98)$$

The leading temperature correction for $k_B T \ll E_{\max}$ is given by the factor $\exp[-\Delta Q(z^*)]$ where $z^* = -i\tau_B$ is the stationary point for $T = 0$. Since $\Delta Q(z^*)$ is negative, this term actually leads to an enhancement of the probability function and hence of the crossing rate. We readily find

$$k^+(T,\epsilon) = k^+(0,\epsilon) \exp\left\{ 2\delta_s\Gamma(s+1)\zeta(s+1)(\omega_{\rm ph}\tau_B)^2\left(\frac{T}{T_{\rm ph}}\right)^{1+s}\right\}, \qquad (20.99)$$

where we have put $\hbar\omega_{\rm ph} = k_B T_{\rm ph}$. Note that, with Eqs. (3.58), (3.90), and (17.17), the thermal enhancement factor in Eq. (20.99) is exactly in the form discussed in Section 17.2 for a general extended metastable potential [cf. Eq. (17.15) with Eq. (17.18)]. The enhancement function $\mathcal{A}(T)$ [the curly bracket in Eq. (20.99)] varies as T^{1+s}. This is in agreement with the general conclusions reported in Section 17.2.

In the limit $s \to 1^-$, the condition (20.51) reduces to the condition $K \gg 1$. Performing this limit in Eq. (20.99) with Eq. (20.50), the resulting expression is consistent with the Ohmic result (20.82) with Eq. (20.81) for large K and $\hbar\epsilon \ll \Lambda_{\rm cl}$.

Finally, we turn to the discussion of the rate for a symmetric system. In the low-temperature regime determined by the condition (20.69) we may use the rate expression (20.67) with the activation energies (20.60) and (20.61). Thus we find

$$k^+(T,0) = \frac{\Delta^2 e^{2|B_s|}}{4\omega_{\rm ph}} \left(\frac{\pi}{a_s(\kappa)}\right)^{1/2} \left(\frac{T_{\rm ph}}{T}\right)^{(1+s)/2} \exp\left\{ -\frac{b_s(\kappa)}{4}\left(\frac{T_{\rm ph}}{T}\right)^{1-s}\right\}, $$
$$(20.100)$$

where $a_s(\kappa)$ and $b_s(\kappa)$ are defined in Eqs. (20.62) and (20.63).

In the regime $k_B T \ll \hbar\omega_c$, the coefficient functions $a_s(0)$ and $b_s(0)$ are given in Eqs. (20.65) and (20.66).[4] As T goes to zero, the rate vanishes with an essential singularity for the reasons mentioned above.

By carefully taking the limit $s \to 1^-$ in Eq. (20.100), we obtain the Ohmic results for $\epsilon = 0$, Eq. (20.86) and Eq. (20.88), respectively. The generalization to the biased case can be taken from Subsection 20.2.4.

---

[4]We remark that the corresponding factors in Eq. (6.12) of Ref. [89] are incorrect.

20.2.8 The super-Ohmic case ($s > 1$)

We now study the tunneling rate for super-Ohmic spectral densities and we identify the cutoff frequency ω_c with the Debye frequency. We write $X(t)$ as in Eq. (18.47),

$$X(t) = X_1 - X_2(t) , \qquad (20.101)$$

where $X_1 = \int_0^\infty d\omega\, G_{1f}(\omega) \coth(\tfrac{1}{2}\hbar\beta\omega)/\omega^2$ is the Huang-Rhys factor [382] describing adiabatic phonon dressing of the tunneling matrix element. For $s > 2$, the functions X_1 and $X_2(t)$ are individually nonsingular, whereas for $s \leq 2$ only $X(t)$ is nonsingular.

Consider first the regime well below the Debye temperature, $T \ll T_D \equiv \hbar\omega_D/k_B$. Then we may put $\kappa = 0$ in Eq. (18.48). It is natural to absorb the adiabatic term X_1 into the definition of a renormalized tunneling matrix element. We write

$$\widetilde{\Delta} = \Delta\, e^{-X_1/2} = \Delta_{\text{eff}}\, e^{-D_s} , \qquad \Delta_{\text{eff}} = \Delta\, e^{-B_s} . \qquad (20.102)$$

Here, Δ_{eff} is the polaron-dressed tunneling matrix element at $T = 0$ defined in Eq. (18.32) with Eq. (18.33). The Franck-Condon factor e^{-D_s} describes the reduction of the tunneling amplitude by thermal excitations of the polaron cloud. We have

$$D_s = d_s (T/T'_{\text{ph}})^{s-1} \quad \text{with} \quad d_s = \Gamma(s-1)\zeta(s-1)/\pi^2 , \qquad (20.103)$$

where we have introduced for convenience the temperature scale

$$T'_{\text{ph}} = T_{\text{ph}}/(2\delta_s\pi^2)^{1/(s-1)} . \qquad (20.104)$$

For $s = 3$, T'_{ph} coincides with the temperature scale used in the recent literature [395, 396]. Here we restrict the attention to the most relevant regime

$$\hbar\Delta_{\text{eff}}/k_B \ll T'_{\text{ph}} \ll T_D . \qquad (20.105)$$

For $T \ll T'_{\text{ph}}$, the dressed matrix element $\widetilde{\Delta}$ is practically temperature-independent. Putting $\kappa = 0$ in Eq. (18.45), we have

$$X_2(t) = 2[\Gamma(s-1)/\pi^2]\,(T/T'_{\text{ph}})^{s-1} \operatorname{Re} \zeta(s-1, \tfrac{1}{2} + it/\hbar\beta) . \qquad (20.106)$$

The forward tunneling rate expression (20.31) takes the form

$$k^+(T,\epsilon) = \frac{\widetilde{\Delta}^2}{4}\, e^{\hbar\beta\epsilon/2} \int_{-\infty}^{\infty} dt\, e^{i\epsilon t}\, e^{X_2(t)} . \qquad (20.107)$$

Expansion of the integrand in powers of $X_2(t)$ gives the multi-phonon series for the rate. The term linear in $X_2(t)$ is the one-phonon process. It takes the form

$$k^+(T,\epsilon) = \frac{\pi}{2}\, \frac{\widetilde{\Delta}^2}{\epsilon^2}\, \frac{G_{1f}(\epsilon)}{[1 - \exp(-\hbar\beta\epsilon)]} = \frac{1}{2\pi}\, \frac{\widetilde{\Delta}^2}{\epsilon}\, \frac{(\hbar\epsilon/k_B T'_{\text{ph}})^{s-1}}{[1 - \exp(-\hbar\beta\epsilon)]} , \qquad (20.108)$$

yielding the one-phonon relaxation rate

$$\gamma_{\rm r} \equiv k^+ + k^- = (\pi\tilde{\Delta}^2/2\epsilon^2)\, G_{\rm lf}(\epsilon)\coth(\hbar\beta\epsilon/2)\,. \tag{20.109}$$

Thus, for nonzero bias and T above $\hbar\epsilon/k_{\rm B}$, the one-phonon rate is proportional to T.

For $s < 3$, the one-phonon rate diverges in the limit $\epsilon \to 0$. For $s = 3$, the forward rate remains finite in this limit,

$$k^+(T,0) = (\hbar\tilde{\Delta}^2/2\pi k_{\rm B}T'_{\rm ph})\, T/T'_{\rm ph}\,. \tag{20.110}$$

Thus, the rate varies in the regime $T \ll T'_{\rm ph}$, in which $\tilde{\Delta} \approx \Delta_{\rm eff}$, *linearly* with T [125].

The one-phonon process vanishes for $s > 3$ as $\epsilon \to 0$. Thus for a symmetric system, the leading contribution for $T \ll T'_{\rm ph}$ is the "two-phonon assisted" process which is

$$k^+(T,0) = \frac{\pi}{2}\frac{\tilde{\Delta}^2}{4}\int_0^\infty \frac{d\omega}{\omega^4}\frac{G_{\rm lf}^2(\omega)}{\sinh^2(\hbar\beta\omega/2)} = \frac{\Gamma(2s-3)\zeta(2s-4)}{2\pi^3}\frac{\hbar\tilde{\Delta}^2}{k_{\rm B}T_{\rm ph}}\left(\frac{T}{T_{\rm ph}}\right)^{2s-3}\,. \tag{20.111}$$

Thus in the regime $T \ll T'_{\rm ph}$, the rate varies as T^{2s-3}. This gives for the diffusion constant[5] in the case $s = 5$ the familiar T^7 law [123] in the small-polaron model [120].

At this point, we should like to remark that the present treatment relies on the assumption that the dynamics is incoherent. However, for $T \ll T'_{\rm ph}$, the dynamics is actually coherent and therefore the Golden Rule approach is inadequate. The relevant discussion within a dynamical approach is given in Section 21.4. There it turns out that the rate (20.109) is the rate describing relaxation of the incoherent part of $\langle\sigma_z\rangle_t$ towards the equilibrium value $\langle\sigma_z\rangle_\infty$ for $\epsilon \gg \tilde{\Delta}$ [cf. Eqs. (21.178), (21.179), and (21.181)]. Similar conclusions hold for the two- and multi-phonon rate. The discussion of the dynamics beyond the one-phonon process is given in Subsection 21.3.3.

For T of the order of $T'_{\rm ph}$ or larger, phonon processes of higher order contribute, and for $T \gg T'_{\rm ph}$ the full multi-phonon series has to be taken into account. In this regime, the formula (20.71) applies. It is convenient to write the activation energies Λ_1 and Λ_2 given in Eqs. (20.60) and (20.61) in the form

$$\Lambda_1 = 4[2B_s - c_s(\kappa)(T/T'_{\rm ph})^{s-1}]k_{\rm B}T\,, \qquad \Lambda_2 = d_s(\kappa)(T/T'_{\rm ph})^{s-1}k_{\rm B}T/4\,, \tag{20.112}$$

$$c_s(\kappa) = 2\Gamma(s-1)\{\zeta(s-1,\tfrac{1}{2}+\kappa) - \zeta(s-1;1+\kappa)\}/\pi^2\,, \tag{20.113}$$

$$d_s = 4\Gamma(s+1)\zeta(s+1,\tfrac{1}{2}+\kappa)/\pi^2\,. \tag{20.114}$$

We then obtain

$$k^+(T,\epsilon) = \frac{\hbar\Delta^2\,e^{-2B_s}}{2k_{\rm B}T'_{\rm ph}}\sqrt{\frac{\pi}{d_s(\kappa)}}\left(\frac{T'_{\rm ph}}{T}\right)^{(s+1)/2} \tag{20.115}$$

$$\times \exp\left[\frac{\hbar\epsilon}{2k_{\rm B}T} + c_s(\kappa)\left(\frac{T}{T'_{\rm ph}}\right)^{s-1} - \frac{1}{d_s(\kappa)}\left(\frac{\hbar\epsilon}{k_{\rm B}T'_{\rm ph}}\right)^2\left(\frac{T'_{\rm ph}}{T}\right)^{s+1}\right]\,.$$

---

[5]In the incoherent tunneling regime, the diffusion constant is proportional to the tunneling rate [cf. Subsection 24.2.1].

The rate expression (20.115) is the result of the integral expression (20.107) in steepest descent approximation. It practically holds in the entire regime $T \gtrsim T'_{\text{ph}}$.[6] In the regime $T'_{\text{ph}} \lesssim T \ll T_{\text{D}}$, we have $\kappa \approx 0$, and hence the coefficients $c_s(0)$ and $d_s(0)$ are temperature independent. For zero bias, the rate varies as $k \propto T^{-(s+1)/2} \exp[\, c_s(0)(T/T'_{\text{ph}})^{s-1} \,]$. We have $c_s(0) = (2^s - 4)\Gamma(s-1)\zeta(s-1)/\pi^2 > 0$ for $s > 1$. Thus the thermal polaronic effects lead to an exponential enhancement of the rate. Observe that $c_s(0)$ is regular at $s = 2$. Further, the singularity at $s = 1$ cancels a singularity in the term B_s. Thus, Λ_1 is regular at $s = 1$. At higher temperatures, the dependence on κ becomes relevant, and in the classical regime $\kappa \gg 1$, the expression (20.115) matches with the classical rate (20.25). In this limit, the adiabatic dressing factor e^{-2B_s} is fully compensated by thermal effects and thus is absent. In the context of the small-polaron problem, the regime $T \gg T_{\text{D}}$ has been studied in Refs. [120, b] and [397].

In conclusion, the rate formula (20.115) covers the entire domain of multi-phonon processes which extends from the quantum regime up to the classical regime.

20.2.9   Incoherent defect tunneling in metals

For defect tunneling in metals, the major dissipative influences at low temperatures are due to the interaction with conduction electrons. At higher temperatures, the coupling to acoustic phonons becomes important. The former coupling leads to a decrease of the rate with increasing temperature for $K < 1/2$ and $\epsilon = 0$, as can be seen from Eq. (20.83). The latter coupling causes phonon-assisted exponential enhancement of the rate at higher temperatures [cf. Eq. (20.115) for $\epsilon = 0$]. Thus, the concerted influences lead to a minimum of the rate as a function of temperature.

To study incoherent tunneling near the minimum of the rate, it is necessary to consider the combined spectral density

$$G_{\text{lf}}(\omega) = 2K\omega\, e^{-\omega/\omega_c} + 2\delta_3 \omega_{\text{ph}}^{-2} \omega^3\, e^{-\omega/\omega_{\text{D}}} . \qquad (20.116)$$

Interestingly, for the spectral density (20.116), the nonadiabatic tunneling rate can be found in closed analytical form in the entire temperature regime.

Consider first the regime $T \ll T_{\text{D}}$, $\hbar\omega_c/k_{\text{B}}$. For $\kappa_{\text{el}} = \kappa_{\text{ph}} \approx 0$, we obtain from the general expression (18.48) for the pair interaction $X(t)$ the form

$$X(t) = X_{\text{el}}(t) + X_{\text{ph}}(t) ,$$
$$X_{\text{el}}(t) = 2K \ln\left((\hbar\beta\omega_c/\pi) \cosh(\pi t/\hbar\beta)\right) , \qquad (20.117)$$
$$X_{\text{ph}}(t) = 2B_3 + \phi/3 - \phi/\cosh^2(\pi t/\hbar\beta) ,$$

where

$$\phi \equiv (T/T'_{\text{ph}})^2 = 2\pi^2 \delta_3 (k_{\text{B}} T/\hbar\omega_{\text{ph}})^2 . \qquad (20.118)$$

---

[6]The precise conditions for the validity of the form (20.115) are given in Eqs. (20.69) and (20.70).

Substituting Eq. (20.117) into Eq. (20.31), the tunneling rate may be written as

$$k^+(T, \epsilon; K) = k_{el}^+(T, \epsilon; K)\, \mathcal{A}_{ph}(T, \epsilon)\,, \qquad (20.119)$$

where k_{el}^+ is the forward rate in the presence of the conduction electrons alone, Eq. (20.78) or Eq. (20.79), and where $\mathcal{A}_{ph}(T, \epsilon)$ represents the activation factor due to the contact with the phonon bath in the presence of the fermionic excitations. Expanding the factor $\exp[\phi/\cosh^2(\pi t/\hbar\beta)]$ in the integrand of Eq. (20.31) in a power series in $\phi/\cosh^2(\pi t/\hbar\beta)$, and integrating each term, the resulting series turns out as a generalized hypergeometric series $_2F_2(a, b; c, d; z)$ [88]. The phonon activation factor takes the form [395]

$$\mathcal{A}_{ph}(T, \epsilon) = e^{-2B_3}\, e^{-\phi/3}\,_2F_2(K + i\hbar\beta\epsilon/2\pi,\ K - i\hbar\beta\epsilon/2\pi;\ K,\ K + \tfrac{1}{2};\ \phi)\,. \qquad (20.120)$$

Equation (20.119) with Eq. (20.120) and Eq. (20.78) or (20.79) constitutes the exact nonadiabatic rate expression in the entire regime $T \ll T_D$, $\hbar\omega_c/k_B$. If we substitute the hypergeometric series into Eq. (20.119), we obtain the expansion of the tunneling rate in terms of the multi-phonon processes in the presence of the electronic influences.

It is instructive to consider this result more closely in various limits. For zero bias, $k_{el}(T, 0)$ is given in Eq. (20.83), and the phonon factor (20.120) reduces to a degenerate hypergeometric function (Kummer function) [88]. We then have

$$k^\pm(T, 0; K) = \frac{\sqrt{\pi}\,\Gamma(K)}{4\Gamma(K + \tfrac{1}{2})}\left(\frac{\hbar\omega_c}{\pi k_B T}\right)^{1-2K}\frac{\Delta^2}{\omega_c}\, e^{-(2B_3+\phi/3)}\,_1F_1(K;\ K + \tfrac{1}{2};\ \phi)\,. \qquad (20.121)$$

In the temperature regime $T \ll T_{ph}'$, we may expand the Kummer function in a power series in ϕ. This corresponds to a categorization of the rate in terms of the one-phonon process, two-phonon process, etc., but in which the Ohmic contribution is fully taken into account [398].

In the opposite regime $T \gg T_{ph}'$ (but still well below the Debye temperature), we may substitute the asymptotic representation of the Kummer function for $\phi \gg 1$. Then the expression (20.121) reduces to the form

$$k^+(T, 0; K) = \frac{1}{4\sqrt{\pi}}\frac{\hbar\Delta^2\, e^{-2B_3}}{k_B T_{ph}'}\left(\frac{\pi k_B T}{\hbar\omega_c}\right)^{2K}\left(\frac{T_{ph}'}{T}\right)^2 \exp\left(\frac{2T^2}{3T_{ph}'^2}\right)\,, \qquad (20.122)$$

where the term $e^{-\phi/3}$ is absorbed into the last exponential factor. This expression is nonperturbative in the phonon coupling. For $K = 0$, Eq. (20.122) coincides with an early result by Holstein [120, b] and by Pirc and Gosar [399].

On the other hand, for $K = 0$ and $\epsilon \neq 0$, the expression (20.119) with (20.120) and with $\widetilde{\Delta}^2 = \Delta^2\, e^{-(2B_3+\phi/3)}$ reduces to the form

$$k^+(T, \epsilon; 0) = \frac{1}{2\pi}\frac{\epsilon}{1 - e^{-\hbar\beta\epsilon}}\left(\frac{\hbar\widetilde{\Delta}}{k_B T_{ph}'}\right)^2\,_2F_2\left(1 + i\frac{\hbar\beta\epsilon}{2\pi},\ 1 - i\frac{\hbar\beta\epsilon}{2\pi};\ 2,\ \frac{3}{2};\ \phi\right)\,. \qquad (20.123)$$

Next, consider a symmetric system in the absence of the coupling to conduction electrons, $K = 0$. We obtain from Eq. (20.123) the multi-phonon series expansion

$$k^+(T, 0; 0) = \frac{\widetilde{\Delta}}{2\pi} \frac{\hbar\widetilde{\Delta}}{k_B T'_{\text{ph}}} \sum_{n=1}^{\infty} \frac{\Gamma(\frac{3}{2})}{n\Gamma(n + \frac{1}{2})} \left(\frac{T}{T'_{\text{ph}}}\right)^{2n-1}. \qquad (20.124)$$

The terms $n = 1$ and $n = 2$ correspond to the one-phonon process, Eq. (20.110), and two-phonon process, Eq. (20.111). Alternatively, we can derive the form (20.124) from Eq. (20.121). However, to circumvent a divergence in the limit $K \to 0$, the diagonal (zero-phonon) process has to be subtracted [120, 89]. The asymptotic representation of the series (20.124) in the limit $T \gg T'_{\text{ph}}$ is given in Eq. (20.122) for $K = 0$.

As the temperature is increased further, the parameter $\kappa_{\text{ph}} = 1/\hbar\beta\omega_D$ moves away from zero. Since we have $\Lambda_2^{(\text{ph})} \gg \Lambda_2^{(\text{el})}$ for $T \gg T'_{\text{ph}}$, we find for a symmetric system

$$k^+(T, 0; K) = \frac{\hbar\Delta^2\, e^{-2B_3}}{2k_B T'_{\text{ph}}} \sqrt{\frac{\pi}{d_3(\kappa_{\text{ph}})}} \left(\frac{\pi k_B T}{\hbar\omega_c}\right)^{2K} \left(\frac{T'_{\text{ph}}}{T}\right)^2 \exp\left\{c_3(\kappa_{\text{ph}})\left(\frac{T}{T'_{\text{ph}}}\right)^2\right\},$$

where
$$\qquad (20.125)$$

$$c_3(\kappa) = 2[\zeta(2, \tfrac{1}{2} + \kappa) - \zeta(2, 1 + \kappa)]/\pi^2, \qquad d_3(\kappa) = 24\zeta(4, \tfrac{1}{2} + \kappa)/\pi^2. \quad (20.126)$$

The expression (20.125) describes the crossover between the multi-phonon quantum rate expression (20.122) and the classical Marcus form[7]

$$k^+(T, 0; K) = \frac{\hbar\Delta^2}{4} \left(\frac{\pi k_B T}{\hbar\omega_c}\right)^{2K} \left(\frac{\pi}{\Lambda_{\text{cl}}^{(\text{ph})} k_B T}\right)^{1/2} \exp\left(-\frac{\Lambda_{\text{cl}}^{(\text{ph})}}{4k_B T}\right), \qquad (20.127)$$

where $\Lambda_{\text{cl}}^{(\text{ph})} = 2\delta_3(\omega_D/\omega_{\text{ph}})^2\hbar\omega_D$. Here we tacitly have assumed that we are still in the scaling limit with regard to the electron bath, $k_B T \ll \hbar\omega_c$. The appropriate generalizations to the biased case and to the crossover to the classical limit with respect to the electron bath are straightforward and left to the reader.

The theory of tunneling and diffusion of light interstitials in metals has been reviewed in Ref. [346]. The cooperation of the electron bath with the phonon bath leads to a minimum of the tunneling rate as a function of temperature for $K < \frac{1}{2}$. The minimum has been observed for muon diffusion in Al and in Cu [92], for incoherent hydrogen tunneling in Niobium [91, 93], and for defect tunneling in mesoscopic Bi wires [96, 95, 396]. Recently, the data for jump rates of defects in mesoscopic Bi wires [95] have been analyzed by using a roughly guessed formula interpolating between the one-phonon rate and the asymptotic expression (20.122) [400].

For tunneling of hydrogen in Niobium, the minimum of the crossing rate is found for T near T_D. The corresponding rate has been studied numerically in this regime [401]. It would now be interesting to apply the formula (20.125) to this problem. One final remark is appropriate. Because of disorder in real systems, the above rate expressions should be averaged with a distribution of bias energies if a quantitative comparison of theory with experiment is attempted.

---

[7]Note that the effects of the electrons are still quantum mechanical.

20.3   Single charge tunneling

A tunnel junction embedded in an electrical circuit forms a quantum system violating Ohm's law. The tunneling processes of electrons through the barrier excite electromagnetic modes, rendering inelastic scattering processes. Thereby the current is reduced at low voltage, an effect called dynamical Coulomb blockade (DCB). The Coulomb blockade regime is most developed in the weak-tunneling limit, $R_T \gg R_K$, where R_T is the tunneling resistance, and $R_K = 2\pi\hbar/e^2$ is the resistance quantum.

Control of single electron tunneling processes by tuning of gate voltages may be achieved in the DCB regime. The common principle of single electron devices, such as turnstiles, pumps, and transistors, is to transfer electrons one by one in systems of small tunnel junctions (cf. the contributions by D. Esteve, and by D. V. Averin and K. H. Likharev in Ref. [145]). As an indispensable ground work, it is necessary to understand the DCB effect in charge tunneling through a single junction.

20.3.1   Weak-tunneling regime

For weak tunneling, the wave function of an excess electron is localized near to the barrier. Then the discrete charge representation (3.168) applies and tunneling through the barrier can be treated in Golden Rule approximation which formally corresponds to incoherent tunneling in a two-state system in the nonadiabatic limit. The effects of the electromagnetic environment are in the phase-phase equilibrium correlation function $Q_\varphi(t)$ given in Eq. (3.160). We rely on the global model (3.167) introduced in Section 3.4, and we consider for simplicity the case of a constant tunneling amplitude, $T_{k,k'} = T_T$, and we assume for a normal junction a constant density of states in the electron band around the Fermi energy E_F. It is convenient to combine the relevant junction properties in the tunneling resistance R_T,

$$\frac{1}{R_T} = \frac{4\pi e^2}{\hbar} N_R(0) N_L(0) \Omega_R \Omega_L |T_T|^2 , \qquad (20.128)$$

where $N_{R/L}(0)$ and $\Omega_{R/L}$ are the density of states and the volume of the right/left electrode, respectively. We assume that the tunneling resistance R_T is large compared to the resistance quantum $R_K = 2\pi\hbar/e^2$ so that thermal equilibrium in each electrode is maintained and the tunneling term H_T, Eq. (3.165), can be treated as a perturbation. The tunneling current may be written as (we put $v = eV_a/\hbar$)

$$I(V_a) = e\,[\,k^+(v) - k^-(v)\,] = e\,\big[\,1 - e^{-\hbar\beta v}\,\big]\,k^+(v) , \qquad (20.129)$$

where $k^+(v)$ is the (forward) rate for tunneling from the left electrode to the right electrode, and $k^-(v)$ is the backward tunneling rate. In the second form, we have used the detailed balance relation (20.32). The calculation of the rate to second order in H_T is straightforward. The thermal average of the quasiparticle modes in the electrodes introduces the Fermi distribution function $f(\omega)$, as in Eq. (4.131) with Eq. (4.132). Switching to the frequency variables $\omega' = E_{k'}/\hbar$ and $\omega'' = E_{k''}/\hbar$, we obtain

$$k^+(v) = \frac{\hbar}{2\pi} \frac{R_K}{R_T} \int_{-\omega_c}^{\omega_c} d\omega' \int_{-\omega_c}^{\omega_c} d\omega'' \, f(\omega') \, f(-\omega'') \, P_{em}(\hbar\omega' + \hbar v - \hbar\omega'') . \qquad (20.130)$$

Since the cutoff ω_c in the electron band is, in general, the largest frequency of the problem, we eventually take the limit $\omega_c \to \infty$.

The formula (20.130) has a neat interpretation. The integrand is proportional to the spectral probability density for an occupied state at energy $\hbar\omega'$ in the left electrode and an empty state at energy $\hbar\omega''$ in the right electrode times the probability density $P_{em}(\hbar\omega' + \hbar v - \hbar\omega'')$ that the electromagnetic environment absorbs the energy $\hbar(\omega' + v - \omega'')$. The bias energy $\hbar v = eV_a$ expresses the difference of the Fermi energies in the two electrodes due to the ideal voltage source V_a. The probability density of the electromagnetic environment to absorb the energy E is given by

$$P_{em}(E) = \frac{1}{2\pi\hbar} \int_{-\infty}^{\infty} dt \, e^{iEt/\hbar - Q_{em}(t)} . \qquad (20.131)$$

For the electromagnetic environment described by the Hamiltonian (3.154) the function $Q_{em}(t)$ corresponds to the phase correlation function $Q_\varphi(t)$ which describes the time correlations of the fluctuations of the phase jump $\varphi(t)$ at the junction due to the electromagnetic environment, $Q_{em}(t) = Q_\varphi(t)$,

$$e^{-Q_{em}(t)} \equiv \left\langle e^{i\varphi(t)} e^{-i\varphi(0)} \right\rangle_\beta = e^{-\langle [\varphi(0) - \varphi(t)] \varphi(0) \rangle_\beta} . \qquad (20.132)$$

To obtain the second form, we have employed the Gaussian statistical properties of the fluctuating phase $\varphi(t)$. The phase correlation function $Q_{em}(t) \equiv \langle [\varphi(0) - \varphi(t)] \varphi(0) \rangle_\beta$ can be written as [cf. Eqs. (3.160) and (3.161)]

$$Q_{em}(t) = \int_0^\infty d\omega \, \frac{G_{em}(\omega)}{\omega^2} \left\{ \coth\left(\frac{\hbar\beta\omega}{2}\right)\left(1 - \cos(\omega t)\right) + i\sin(\omega t) \right\} \qquad (20.133)$$

with

$$G_{em}(\omega) = (e^2/\pi\hbar)\, \omega^2 \tilde{\chi}''(\omega) = 2\omega \, \mathrm{Re} \, Z_t^*(\omega)/R_K . \qquad (20.134)$$

The general properties of the probability function $P_{em}(E)$ have been discussed already in Subsection 20.2.2. In particular, the relation (20.36) for $P_{em}(E)$ ensures detailed balance for the rate, which we have employed already in Eq. (20.129).

A resistive environment with Ohmic impedance is described by the spectral density (3.170),

$$G_{em}(\omega) = \frac{2\alpha\,\omega}{1 + (\omega/\omega_R)^2} , \qquad (20.135)$$

where $\alpha = R/R_K$, and $\omega_R = 1/RC = E_c/\pi\hbar\alpha$, and where $E_c = e^2 = /2C$ is the charging energy. For such environment, the absorbed net energy (20.37), i.e. the reorganization energy Λ_{cl}, coincides with the charging energy,

$$\Lambda_{cl} = E_c . \qquad (20.136)$$

Thus, for a high-impedance environment, the junction behaves classically down to fairly low temperatures, and the Gaussian form corresponding to expression (20.25),

$$P_{\text{em}}(E) = \left(\frac{1}{4\pi E_c k_{\text{B}} T}\right)^{1/2} \exp\left(-\frac{(E - E_c)^2}{4E_c k_{\text{B}} T}\right) \qquad (20.137)$$

applies in this regime. With increasing quality factor Q_{qual} in the spectral density (3.173), the probability function $P_{\text{em}}(E)$ changes from the smooth form (20.137) to the resonant characteristics given in Eq. (20.41), where ω_0 corresponds to ω_{L}.

Next, we return to the rate expression (20.130. With the substitution $\omega'' = \omega + \omega'$, the ω'-integration in Eq. (20.130) can be done, yielding

$$\int_{-\infty}^{\infty} d\omega' f(\omega') f(-\omega' - \omega) = \int_{-\infty}^{\infty} d\omega' \frac{f(\omega') - f(\omega' + \omega)}{1 - e^{-\beta \hbar \omega}} = \frac{\omega}{1 - e^{-\beta \hbar \omega}}. \qquad (20.138)$$

With this relation, the rate expression (20.130) is reduced to the single integral

$$k^+(v) = \hbar \int_{-\infty}^{\infty} d\omega \, k_0^+(\omega) P_{\text{em}}(\hbar v - \hbar \omega), \qquad (20.139)$$

$$k_0^+(\omega) = \frac{1}{2\pi} \frac{R_{\text{K}}}{R_{\text{T}}} \frac{\omega}{1 - e^{-\hbar \beta \omega}}. \qquad (20.140)$$

For "elastic" tunneling we have $P_{\text{em}}(\hbar v - \hbar \omega) = \delta(\hbar v - \hbar \omega)$ and hence $k^+(v) = k_0^+(v)$. Thus, the rate $k_0^+(v)$ describes tunneling of the fermionic quasiparticle through the barrier in the absence of coupling to the electromagnetic environment.

It is instructive to see that the tunneling rate expression (20.130) for the fermionic entity can be transformed into an expression of the form (20.30) which describes tunneling of a bosonic entity. To prepare the ground, we first observe that the detailed balance relation (20.36) is met if we write the function $P(\hbar \omega)$, Eq. (20.34), which describes the reservoir's probability density for absorption of energy $\hbar \omega$, as [402]

$$P(\hbar \omega) = D(\omega) f(-\omega) / \hbar \omega_c, \qquad (20.141)$$

where $f(\omega)$ is the Fermi function. The function $D(\omega)$ is symmetric about the Fermi energy, $D(\omega) = D(-\omega)$, and the normalization is $\int_0^{\infty} d\omega \, D(\omega) = \omega_c$, as follows from Eq. (20.35). As we shall see, the function $D(\omega)$ can be imagined as an effective density of states for fermionic quasiparticles.

Consider next the Fourier transform of the correlator $\exp[-Q_{\text{ohm}}(t, K)]$, where $Q_{\text{ohm}}(t; K)$ is the bath correlation function in the Ohmic scaling limit, Eq. (18.51),

$$Q_{\text{ohm}}(t; K) = 2K \ln[(\hbar \beta \omega_c / \pi) \sinh(\pi t / \hbar \beta)] + i\pi K \, \text{sgn}(t). \qquad (20.142)$$

The Fourier integral

$$\frac{1}{2\pi \hbar} \int_{-\infty}^{\infty} dt \, e^{-i\omega t} \exp\left[-Q_{\text{ohm}}(t; K)\right] = \frac{1}{\hbar \omega_c} D_{\text{ohm}}(\omega; K) f(\omega) \qquad (20.143)$$

can be done in analytic form [cf. Eq. (20.29) with Eq. (20.78)], yielding ($|\omega| < \omega_c$)

$$D_{\text{ohm}}(\omega; K) = \frac{1}{\Gamma(2K)} \left(\frac{\hbar \beta \omega_c}{2\pi}\right)^{1-2K} \frac{|\Gamma(K + i\hbar \beta \omega / 2\pi)|^2}{|\Gamma(1/2 + i\hbar \beta \omega / 2\pi)|^2}. \qquad (20.144)$$

Observing that the density of states is constant in the fermionic band $|\omega| < \omega_c$ for the particular value $K = \frac{1}{2}$, $D_{\text{ohm}}(\omega, \frac{1}{2}) = 1$, we can directly relate for $K = \frac{1}{2}$ the Fermi function with the $P(E)$–function discussed in Subsection 20.2.2,

$$f(\omega) \;=\; \frac{\omega_c}{2\pi} \int_{-\infty}^{\infty} dt\, e^{-i\omega t}\, \exp[\,-Q_{\text{ohm}}(t; K = \tfrac{1}{2})\,] \,. \qquad (20.145)$$

Next, we substitute the Fourier representation (20.145) both for $f(\omega')$ and for $f(-\omega'')$ and the Fourier representation (20.131) for $P_{\text{em}}(t)$ into the expression (20.130). Since the expression (20.130) is in the form of a convolution, we immediately get with use of the obvious relation $2Q_{\text{ohm}}(t, \frac{1}{2}) = Q_{\text{ohm}}(t, 1)$

$$k^{+}(v) \;=\; \left(\frac{\omega_c}{2\pi}\right)^2 \frac{R_{\text{K}}}{R_{\text{T}}} \int_{-\infty}^{\infty} dt\, e^{ivt}\, \exp[\,-Q_{\text{ohm}}(t; K = 1) - Q_{\text{em}}(t)\,] \,. \qquad (20.146)$$

Thus we have found a surprising result: tunneling of a fermionic particle through a barrier embedded in a conductor is like tunneling of a boson which undergoes Ohmic dissipation with Kondo parameter $K = 1$. In the presence of an electromagnetic environment, the respective correlation function $Q_{\text{em}}(t)$ is simply added to the the Ohmic correlation function.

For an Ohmic impedance, the spectral density $G_{\text{em}}(\omega)$ is given by the Drude form (20.135). The correlation function $Q_{\text{em}}(t; \alpha)$ for the algebraic cutoff at frequency ω_{R} differs in essence from the form (18.51), being appropriate for exponential cutoff at frequency ω_c, by an adiabatic correction

$$Q_{\text{em}}(t; \alpha) \;=\; Q_{\text{ohm}}(t; \alpha) + \Delta Q_{\text{em}}(\alpha) \,, \qquad (20.147)$$

$$\Delta Q_{\text{em}}(\alpha) \;=\; 2\alpha \ln(\omega_{\text{R}}/\omega_c) + 2\alpha \zeta_{\text{D}} \,, \qquad (20.148)$$

$$\zeta_{\text{D}} \;=\; \psi\!\left(1 + \frac{\hbar\beta\omega_{\text{R}}}{2\pi}\right) - \psi(1) - \ln\left(\frac{\hbar\beta\omega_{\text{R}}}{2\pi}\right) \xrightarrow{\hbar\beta\omega_{\text{R}} \gg 1} C_{\text{E}} \,. \qquad (20.149)$$

The relation $\zeta_{\text{D}} = C_{\text{E}}$ holds in the scaling regime, where $C_{\text{E}} = 0.577\ldots$ is Euler's constant. With this, the forward tunneling rate (20.146) takes the form

$$k^{+}(v) \;=\; \left(\frac{\omega_c}{2\pi}\right)^2 \frac{R_{\text{K}}}{R_{\text{T}}}\, e^{-\Delta Q_{\text{em}}(\alpha)} \int_{-\infty}^{\infty} dt\, e^{ivt}\, \exp[\,-Q_{\text{ohm}}(t; K = 1 + \alpha)\,] \,. \qquad (20.150)$$

Thus, there is an entire low-energy regime where weak tunneling of electrons coupled to a resistive electromagnetic environment behaves exactly like tunneling of bosons embedded in an Ohmic environment with Ohmic damping parameter

$$K = 1 + \alpha \,. \qquad (20.151)$$

One remark is in order. The expression (20.150) is the leading contribution to the rate in the weak-tunneling limit. Generalization to the full weak-tunneling power series in $R_{\text{K}}/R_{\text{T}}$, and to the related strong-tunneling series will be given in Chapter 26.

Alternatively, we can transfer the effects of the electromagnetic environment into an effective frequency- and temperature-dependent tunneling density. Writing

$$\frac{1}{2\pi\hbar}\int_{-\infty}^{\infty}dt\,e^{-i\omega t}\,e^{-[Q_{\text{ohm}}(t;K=1)+Q_{\text{em}}(t)]/2}\;=\;\frac{1}{\hbar\omega_c}D_{\text{eff}}(\omega)f(\omega)\,,\tag{20.152}$$

the forward rate (20.146) is transformed into

$$k^+(v)\;=\;\frac{1}{2\pi}\frac{R_{\text{K}}}{R_{\text{T}}}\int_{-\omega_c}^{\omega_c}d\omega\,D_{\text{eff}}(\omega)D_{\text{eff}}(\omega+v)f(\omega)\,f(-\omega-v)\,.\tag{20.153}$$

In this formulation, the effects of the environment are described in terms of an effective frequency-dependent tunneling density of states of the fermionic charge carrier. In the absence of the environmental coupling, $Q_{\text{em}}(\omega)=0$, we have $D_{\text{eff}}(\omega)=1$, so that $k^+(v)$ takes the form (20.140).

The expressions (20.146) and (20.153) are bosonic and fermionic representations, respectively, of the same physical tunneling process.

20.3.2 The current–voltage characteristics

Substituting the expression (20.139) into Eq. (20.129), the current-voltage characteristics of a single junction embedded in an electromagnetic environment is found as

$$I(V_a)\;=\;\int_{-\infty}^{\infty}dE\,\frac{1-e^{-\beta eV_a}}{1-e^{-\beta E}}\,P_{\text{em}}(eV_a-E)\,I_0(E/e)\,,\tag{20.154}$$

where

$$I_0(V)\;=\;V/R_{\text{T}}\tag{20.155}$$

is the Ohmic tunneling current. The expression (20.154) satisfies the plausible relation $I(-V_a)=-I(V_a)$, as follows with use of the detailed balance relation (20.36).

When the impedance $Z(\omega=0)=R$ of the environment is very small compared to the resistance quantum R_{K}, the transfer of the charge through the junction is predominantly elastic tunneling, $P_{\text{em}}(E)\approx\delta(E)$. This leads us to the form (20.140) for the forward tunneling rate, and finally with Eq. (20.129) or directly from Eq. (20.154) to the Ohmic law $I(V_a)=V_a/R_{\text{T}}$, warranting the interpretation of R_{T} as a tunneling resistance.

In the opposite limit of a high-impedance environment, the Gaussian distribution (20.137) for $P_{\text{em}}(E)$ is very narrow at low T about $E=E_c$ and can well be approximated by $P_{\text{em}}(E)=\delta(E-E_c)$. In this limit, the energy absorbed by the environment equals the charging energy, and the current is given by

$$I(V_a)\;=\;\Theta(V_a-E_c/e)\,[V_a-E_c/e]/R_{\text{T}}\,.\tag{20.156}$$

The Coulomb gap $eV_a>E_c$ is in correspondence with the energy balance (3.151).

At zero temperature, we have $P_{\text{em}}(E)=0$ for $E<0$ since the bath cannot hand in energy anymore. In this limit, the expression (20.154) reduces to

$$I(V_a)\;=\;\Theta(V_a)\frac{1}{eR_{\text{T}}}\int_0^{eV_a}dE\,(eV_a-E)\,P_{\text{em}}(E)\,.\tag{20.157}$$

Direct information about the distribution of energy absorbed by the environment is gained from the second derivative of the current with respect to the applied voltage,

$$\frac{d^2 I(V_a)}{dV_a^2} = \frac{e}{R_T} P_{em}(eV_a) . \tag{20.158}$$

When $eV_a \gg k_B T$ and $P_{em}(eV_a) \ll P_{em}(0)$, we find from Eq. (20.154)

$$I(V_a) = \frac{1}{eR_T} \int_{-\infty}^{\infty} dE \, (eV_a - E) \, P_{em}(E) . \tag{20.159}$$

Using the sum rules (20.35) and (20.37), and assuming a resistive environment for which Λ_{cl} equals the charging energy, Eq. (20.136), we find the linear current-voltage characteristics at low temperatures in the classical regime $k_B T \ll e^2/2C < eV_a$ as

$$I(V_a) = [V_a - e/2C]/R_T . \tag{20.160}$$

The shift in voltage by $e/2C$ is a manifestation of the Coulomb blockade in the classical regime.

Consider now the current–voltage characteristics in the low temperature quantum regime $k_B T \ll eV_a \ll e^2/2C$. From the nonadiabatic rate formula (20.150) we can immediately read off that for an Ohmic impedance described by the spectral density (20.135) the analytic rate expressions (20.78) – (20.83) can directly be applied to single-electron tunneling in a resistive environment if we employ the correspondences

$$K \leftrightarrow 1 + \alpha , \quad \text{and} \quad \frac{\Delta^2}{4} \leftrightarrow \left(\frac{\omega_c}{2\pi}\right)^2 \frac{R_K}{R_T} e^{-\Delta Q_{em}(\alpha)} . \tag{20.161}$$

Upon using these correspondences and the rate expression (20.81), the current-voltage characteristics in the regime $k_B T \ll eV_a \ll \hbar\omega_R$ is found to read

$$I(V_a) = \frac{e^{-2\alpha C_E}}{\Gamma(2 + 2\alpha)} \frac{V_a}{R_T} \left(\frac{\pi\alpha eV_a}{E_c}\right)^{2\alpha} , \quad \alpha = R/R_K . \tag{20.162}$$

We see that the sharp Coulomb gap in Eq. (20.156) is rounded by the quantum fluctuations. The dynamical Coulomb-blockade shows itself in a superlinear behaviour $I(V_a) \propto V_a^{1+2\alpha}$ in the regime $k_B T \ll eV_a$, or equivalently in the so-called zero-bias anomaly $dI(V_a)/dV_a \propto V_a^{2\alpha}$, instead of the voltage-independent conductance without the electromagnetic influences. In view of the substitution rule $K \to 1 + \alpha$ in the correspondence, it is not surprising that the current shows the characteristic power laws which we have already encountered in the discussion of the Ohmic case in Subsection 20.2.5. The tunneling density of states associated with the Ohmic impedance is obtained from Eq. (20.144) with Eq. (20.161) at zero temperature for $\omega \ll \omega_R$ as

$$D_{em}(\omega) = e^{-\alpha C_E} (|\omega|/\omega_R)^\alpha /\Gamma(1 + \alpha) . \tag{20.163}$$

The tunneling density of states is nonanalytic at the Fermi energy and it is diluted near the Fermi level for $\alpha > 0$ compared with the constant tunneling density of states for $\alpha = 0$. In the expression (20.153), the zero-bias anomaly originates from the dilute density of states near the Fermi level. The power law in Eq. (20.162) directly reflects the nonanalytic behaviour $D_{em}(\omega \to 0) \propto |\omega|^\alpha$.

We should like to remark that we have the same power-law form as in Eq. (20.162) for any other environment with a finite impedance at $\omega = 0$, where $\alpha = Z(0)/R_K$. Only the prefactor depends on the spectral form of the impedance at finite frequencies.

Consider next the linear conductance at finite temperature. Using the correspondences (20.161), we obtain from Eq. (20.129) with Eq. (20.83) the expression

$$\frac{dI}{dV_a}\bigg|_{V_a=0} = \frac{e^{-2\alpha C_E}}{R_T} \frac{\sqrt{\pi}\Gamma(1+\alpha)}{2\Gamma(\frac{3}{2}+\alpha)} \left(\frac{\pi^2 \alpha k_B T}{E_c}\right)^{2\alpha}. \tag{20.164}$$

In the high voltage regime $eV_a \gg E_c$, we may use in Eq. (20.157) for $P_{em}(E)$ the form (20.55). Upon substituting for the spectral density $G_{em}(\omega)$ the expression (20.135), we obtain the leading correction to the strict Coulomb gap form (20.156),

$$I(V_a) = \frac{1}{R_T}\left(V_a - \frac{e}{2C} + \frac{\alpha}{\pi^2}\frac{e^2}{4C^2}\frac{1}{V_a}\right), \quad \text{for} \quad eV_a \gg E_c. \tag{20.165}$$

Thus, the actual offset is smaller than the offset in Eq. (20.156).

Let us finally consider an LC transmission line described by the sub-Ohmic spectral density (3.175) with (3.176). Putting $s = \frac{1}{2}$ in the former result (20.50) and taking the limit $\omega_c \to \infty$, we find for the probability function $P_{em}(E)$ the expression[8]

$$P_{em}(E) = \sqrt{\frac{eV_c}{4\pi E^3}} \exp\left(-\frac{eV_c}{4E}\right) \quad \text{with} \quad V_c = \frac{4eR_0}{C_0 R_K}. \tag{20.166}$$

The function $P_{em}(E)$ has a maximum at $E = eV_c/6$. Upon inserting this form into the current (20.157), we find exponential suppression of the current $\propto \exp(-V_c/4V_a)$ for $V_a \ll V_c$ instead of the power law suppression in the Ohmic case, Eq. (20.162). The strong suppression of the charge transfer is due to the much higher density of low-frequency excitations of the LC transmission line. Finally, we remark that at higher voltages the current characteristics is similar to the case of an Ohmic resistive environment.

20.3.3   Weak tunneling of 1D interacting electrons

In 1D quantum wires, the electron-electron interaction leads to a break-down of the Fermi liquid model. The usual quasiparticle description does not apply any more. Instead, the low-energy excitations of the correlated system are collective bosonic excitations, which can be viewed as collective density fluctuations of a harmonic fluid. The corresponding Luttinger liquid model in bosonic representation is described in Chapter 26. Now assume that the transport through the quantum wire is impeded by a barrier or impurity. The corresponding transport problem is treated

---

[8]The expression for V_c is misprinted in Ref. [146].

below in Section 26.1. Already at this point, however, it is appropriate to demonstrate the formal similarity of correlated electron-tunneling through a strong barrier with weak single-electron tunneling in the presence of a resistive electromagnetic environment. To this aim, we only have to anticipate the particular form of the equilibrium correlation function of the phase $\phi(0,t)$ of the fermion field (26.2), $4\pi\langle [\phi(0,0) - \phi(0,t)]\phi(0,0)\rangle_\beta \equiv Q(t; 1/g)$. Here, g is a dimensionless interaction constant in the Luttinger model. The case $g < 1$ corresponds to repulsive electron-electron interaction, and $g = 1$ is the Fermi liquid limit. We find from Eq. (26.22) that the phase correlation function $Q(t; 1/g)$ is formally identical with the Ohmic bath correlation function $Q_{\mathrm{ohm}}(t; K)$ given in Eq. (20.142), with $K = 1/g$,

$$Q_{\mathrm{ohm}}(t; 1/g) \equiv \frac{2}{g}\ln\left[\frac{\hbar\beta\omega_c}{\pi}\sinh\left(\frac{\pi t}{\hbar\beta}\right)\right] + i\frac{\pi}{g}\,\mathrm{sgn}(t)\,. \qquad (20.167)$$

In the bosonic representation, the forward tunneling rate of the unit charge through the barrier is given, in analogy with Eq. (20.146), by

$$k^+(v) = \left(\frac{\omega_c}{2\pi}\right)^2\frac{R_K}{R_T}\int_{-\infty}^{\infty}dt\,e^{ivt}\exp\left[-Q_{\mathrm{ohm}}(t; 1/g)\right]\,, \qquad (20.168)$$

where $V_a = \hbar v/e$ is the applied voltage at the impurity. Thus we find a direct formal correspondence between weak tunneling of 1D interacting electrons and weak tunneling of electrons coupled to an electrical circuit with Ohmic impedance. In the correspondence we have

$$1 + \alpha = 1/g\,. \qquad (20.169)$$

We shall discuss below in Chapter 26, in particular in Subsection 26.1.4, that the correspondence als holds for joint tunneling of two, three and many charges, and that it is also valid in the weak-barrier, or equivalently strong-tunneling regime.

Obviously, we may reverse the previous arguments and introduce an effective tunneling density of states of the fermionic entity, $D_{\mathrm{ohm}}(\omega; 1/g)$, as in Eq. (20.144). Thus we find in analogy with Eq. (20.153) the expression [402]

$$k^+(v) = \frac{1}{2\pi}\frac{R_K}{R_T}\int_{-\omega_c}^{\omega_c}d\omega\,D_{\mathrm{ohm}}(\omega; 1/2g)D_{\mathrm{ohm}}(\omega + v; 1/2g)f(\omega)f(-\omega - v)\,. \qquad (20.170)$$

It is also easy to see from Eq. (20.129) with the form (20.168), and with use of the expression (20.143) for $Q_{\mathrm{ohm}}(t; 1/2g) = \frac{1}{2}Q_{\mathrm{ohm}}(t; 1/g)$, that the current may be written as

$$I(V_a) = \frac{\hbar\omega_c}{eR_T}\left(1 - e^{-\hbar\beta v}\right)f(-v)D_{\mathrm{ohm}}(v; 1/g)\,, \qquad (20.171)$$

where $D_{\mathrm{ohm}}(\omega; K)$ is given in Eq. (20.144).

We have $D_{\mathrm{ohm}}(\omega; 1) = (\omega/\omega_c)\coth(\beta\hbar\omega/2)$, so that Ohm's law $I(V_a) = V_a/R_T$ is recovered from Eq. (20.171) in the limit $g \to 1$.

Here we have restricted the attention to charge transport through a single barrier. In a double junction system, besides sequential tunneling, higher order tunneling processes may occur in which the Coulomb barrier is bypassed by only virtually occupying the island. A theoretical description of resonant tunneling for correlated electrons using the above concept of effective tunneling densities is reported in Ref. [403]. A recent comprehensive review of mesoscopic electron transport is given in Ref. [404], and a nonperturbative formalism in the presence of strong Coulomb interactions has been developed and applied to resonant tunneling in Ref. [405].

20.3.4   Tunneling of Cooper pairs

Tunneling of Cooper pairs through a Josephson junction is also affected by the electromagnetic environment. For weak Josephson coupling energy, $E_J \ll E_c = 2e^2/C$, we may calculate the crossing rate for a Cooper pair in Golden Rule approximation with respect to the Josephson coupling energy. Since the tunneling entities are bosons, the rate can be written in a form analogous to Eq. (20.29) [cf. Fig. 3.3 and Subsection 3.4.3]. With the externally applied voltage V_x, the rate expression is

$$k^+ = \frac{E_J^2}{4\hbar^2} \int_{-\infty}^{\infty} dt \, e^{i2eV_x t/\hbar} \, e^{-Q_\psi(t)} \, . \tag{20.172}$$

The bias energy is $2eV_x$ because the charge transferred by a Cooper pair across the junction is $2e$, and the phase correlation function for Cooper pairs $Q_\psi(t)$ is given in Eq. (3.179). With the correspondences

$$\hbar\Delta \,\hat{=}\, E_J = (\hbar/2e)I_c \, , \qquad \hbar\epsilon \,\hat{=}\, 2eV_x \, , \qquad K \,\hat{=}\, \rho = R/R_Q \, , \tag{20.173}$$

where $R_Q = 2\pi\hbar/4e^2$, and with

$$Q_\psi(t) \equiv Q_{\text{ohm}}(t; K = \rho) + 2\rho \ln(\omega_R/\omega_c) + 2\rho\zeta \, , \tag{20.174}$$

we can directly apply the results for incoherent tunneling of a boson with Ohmic friction presented in Section 20.2 to Cooper pair tunneling. The logarithmic term in Eq. (20.174) takes into account that the cut-off frequency is

$$\omega_R = 1/Z(0)C = E_c/\pi\rho\hbar \, , \tag{20.175}$$

instead of ω_c. The quantity ζ accounts for the particular high-frequency dependence of the total impedance $Z_t^*(\omega)$ in adiabatic approximation,

$$\zeta = \zeta_D + \int_0^{\infty} \frac{d\omega}{\omega} \left[\frac{\text{Re}\, Z_t^*(\omega)}{\rho R_Q} - \frac{1}{1 + (\pi\rho\hbar\omega/E_c)^2} \right] \, . \tag{20.176}$$

The term ζ_D, given in Eq. (20.149), arises from the Drude form (20.135), and the integral accounts for deviation of the actual $Z_t^*(\omega)$ from this behaviour.

Introducing the function $P_\psi(E)$ as the Fourier transform of $e^{-Q_\psi(t)}$ [see Subsection 20.2.2 for the physical meaning and the properties of $P_\psi(E)$], the current-voltage characteristic takes the form

$$I(V_x) = 2e(k^+ - k^-) = \frac{\pi e(E_J e^{-\rho\zeta})^2}{\hbar} \left[1 - e^{-2\beta eV_x} \right] P_\psi(2eV_x) \, . \tag{20.177}$$

Using the rate expression (20.78) and the above correspondences, the current-voltage characteristics is found to read [406]

$$I(V_x) = \frac{\pi e \rho}{\hbar} \frac{(E_J \, e^{-\rho \zeta})^2}{E_c} \left(\frac{\beta E_c}{2\pi^2 \rho}\right)^{1-2\rho} \frac{|\Gamma(\rho + i\beta e V_x/\pi)|^2}{\Gamma(2\rho)} \sinh(\beta e V_x) \, . \qquad (20.178)$$

This expression holds for a wide range of temperatures in the weak tunneling regime $E_J \ll E_c$. For small ρ, the supercurrent-voltage characteristics (20.178) shows a peak at voltage $V = \pi \rho / e \beta$ which becomes increasingly marked as temperature is lowered. The current at zero temperature is found from (20.178) as

$$I(V_x) = \frac{\pi^{5/2} \rho}{2 \, \Gamma(\rho) \Gamma(\rho + \frac{1}{2})} \left(\frac{E_J \, e^{-\rho \zeta}}{e V_x}\right)^2 \left(\frac{\pi \rho e V_x}{E_c}\right)^{2\rho} \frac{V_x}{R} \, . \qquad (20.179)$$

Hence the supercurrent exhibits the zero bias anomaly $I(V_x) \propto V_x^{2\rho-1}$ [387]. It describes suppression of the current by the Coulomb blockade effect for $\rho > 1$. Also the zero bias conductance at finite temperature exhibits power law behaviour,

$$\frac{dI}{dV_x}\bigg|_{V_x=0} = \frac{1}{R_Q} \frac{\sqrt{\pi}}{2} \frac{\Gamma(\rho)}{\Gamma(\rho + \frac{1}{2})} \left(\frac{E_J \, e^{-\rho \zeta}}{E_c/\rho \pi^2}\right)^2 \left(\frac{\beta E_c}{\rho \pi^2}\right)^{2-2\rho} \, . \qquad (20.180)$$

For small impedance $Z(0) \ll R_Q$, we get from Eq. (20.178)

$$I(V_x) = \frac{\pi e}{\hbar} \frac{E_J^2 \, e^{-2\rho \zeta}}{E_c} \rho^{2\rho} \left(\frac{\beta E_c}{2\pi^2}\right)^{1-2\rho} \frac{2\pi^2 \rho \beta e V_x}{(\beta e V_x)^2 + (\pi \rho)^2} \, , \qquad (20.181)$$

which reduces further in the classical limit $k_B T \gg \hbar \omega_R$ to the form

$$I(V_x) = \frac{I_c^2 \, e^{-2\rho \zeta}}{2} \frac{Z(0) \, V_x}{V_x^2 + [2 e Z(0) k_B T / \hbar]^2} \, . \qquad (20.182)$$

At $\rho < 1$, which is the usual case, the conductance found from Eq. (20.179) diverges in the zero bias limit. Also the conductance (20.180) diverges with decreasing temperature as $T^{2\rho-2}$. This indicates onset of strong tunneling in the low-energy regime. The singularity in the weak-tunneling conductance is smoothed out by taking into account terms of higher order in E_J^2 (see Section 25). At this point, a preliminary final remark is appropriate. The case of weak Josephson coupling is related to the case of large Josephson coupling [Eq. (17.81)] by a duality symmetry. This issue is discussed in Subsection 25.3.

20.3.5 Tunneling of quasiparticles

Quasiparticle tunneling in a superconducting junction is similar to quasiparticle tunneling in a normal junction. The key difference is that the density of states of the BCS quasiparticles strongly depends on frequency, in particular in the region slightly above the gap frequency. The normalized density of states $\mathcal{N}_{qp}(\omega)$ of the BCS quasiparticle excitations is given in Eq. (4.166). Since the tunneling rate obeys detailed

balance, Eq. (20.32), we have again the form (20.129) for the current. In alteration
to Eq. (20.130), the forward rate for quasiparticle tunneling is given by

$$k^+(v) = \frac{\hbar}{2\pi} \frac{R_K}{R_T} \int_{-\omega_c}^{\omega_c} d\omega \int_{-\omega_c}^{\omega_c} d\omega' \, \mathcal{N}_{qp}(\omega)\mathcal{N}_{qp}(\omega') \, f(\omega)f(-\omega') \, P_{em}(\hbar\omega + \hbar v - \hbar\omega') \, .$$

Other forms analogous to those given for quasiparticle tunneling in a normal junction
are easily found. For instance, in analogy with Eq. (20.154) the quasiparticle current
through the junction in the presence of the environment, $I_{qp,em}$, is related to the
quasiparticle current in the absence of it, $I_{qp,0}$, by the integral relation [387]

$$I_{qp,em}(V_a, T) = \int_{-\infty}^{\infty} dE \, \frac{1 - e^{-\beta eV_a}}{1 - e^{-\beta E}} \, P_{em}(eV_a - E) \, I_{qp,0}(E/e, T) \, . \qquad (20.183)$$

The quasiparticle current in the absence of the environment reads

$$I_{qp,0}(V_a, T) = \frac{\hbar}{eR_T} \int_{-\infty}^{\infty} d\omega \, \mathcal{N}_{qp}(\omega)\mathcal{N}_{qp}(\omega + eV_a/\hbar) \, [\, f(\omega) - f(\omega + eV_a/\hbar)] \, . \qquad (20.184)$$

At zero temperature, the ω-integral can be done in analytic form. The resulting
expression is given in terms of the hypergeometric function $_2F_1(z)$,

$$I_{qp,0}(V_a, 0) = \Theta(V_a - V_g) \frac{V_a}{R_T} \frac{V_a}{2(V_a + V_g)} \qquad (20.185)$$

$$\times \left\{ B(\tfrac{1}{2}, \tfrac{1}{2}) \, _2F_1(\tfrac{1}{2}, \tfrac{1}{2}; 1; z) - \frac{(V_a - V_g)^2}{V_a^2} B(\tfrac{3}{2}, \tfrac{1}{2}) \, _2F_1(\tfrac{1}{2}, \tfrac{3}{2}; 2; z) \right\} ,$$

where the gap voltage is $V_g = 2\hbar\Delta_g/e$ and $z = (V_a - V_g)^2/(V_a + V_g)^2$.

The quasiparticle current $I_{qp,0}(V_a, 0)$ is zero in the interval $0 \le V_a < V_g$. At the
gap voltage, $V_a = V_g$, the quasiparticle current jumps from zero to the finite value

$$I_{qp,0}(V_g, 0) = \frac{\pi}{4} \frac{V_g}{R_T} \, . \qquad (20.186)$$

For voltage $V_a \gg V_g$, the quasiparticle current approaches the strict Ohmic behaviour

$$I_{qp,0}(V_a \gg V_g, 0) = \frac{V_a}{R_T} \, . \qquad (20.187)$$

For an environment described by an Ohmic impedance R, we find from Eq.
(20.183) upon using findings from Subsection 20.2.5, in particular Eq. (20.81), the
anomalous threshold behaviour (we put $\alpha = R/R_K$)

$$I_{qp,em}(V_a, 0) \propto \Theta(V_a - V_g) \, (V_a - V_g)^{2\alpha} \, . \qquad (20.188)$$

Hence the jump at the threshold is dissolved due to quantum fluctuations of the
Ohmic electromagnetic environment.

At voltage $V_a \gg V_g$, the quasiparticle current varies as

$$I_{qp,em}(V_a, 0) \propto V_a^{2\alpha+1} \, . \qquad (20.189)$$

The deviation of $I_{qp,em}(V_a, 0)$ from the Ohmic law $I(V_a) \propto V_a$ is again a signature of
dynamical Coulomb blockade by an Ohmic electromagnetic environment.

21. Two-state dynamics

Up to now, we have mainly discussed thermodynamic properties of the open two-state system. As far as dynamics is concerned, we have been limited to the study of nonadiabatic tunneling rates in the incoherent regime. We now present the real-time approach based on the method given in Chapter 5. The approach will cover the full dynamics in a unified manner for different kinds of the initial preparation and for arbitrary linear dissipation, both in the incoherent and oscillatory regime. We shall provide explicit expressions in most regions of the parameter space. Emphasis is put on the regime in which the system shows quantum coherent oscillations.

21.1 Initial preparation, expectation values, and correlations

We have discussed already in Sections 5.1 – 5.4 initial conditions, preparation functions and propagating functions for a general global system. We now come to the specification for the damped two-state system. In the sequel, we put emphasis on answering questions which are of experimental relevance.

21.1.1 Product initial state

Consider a global system for which the initial state of the density matrix is factorizing,

$$W_{\mathrm{fc}}(t=0,\bar{\sigma}) = |R><R| \otimes \exp\{-\beta[H_{\mathrm{Res}} - \bar{\sigma}\mathfrak{E}(t=0)]\}/Z_{\mathrm{Res}} . \qquad (21.1)$$

The TSS is prepared in the eigenstate $\sigma = +1$ of σ_z (right well), and the bath is in a shifted canonical distribution. Here, H_{Res} is the bare bath Hamiltonian (3.3), $\mathfrak{E}(t)$ is the collective bath mode defined in Eq. (3.85), and $\bar{\sigma}$ is a control parameter for the shift of the bath in the initial state at time zero.

The influence functional tailored to the initial state (21.1) is conveniently expressed in terms of an influence functional of the Feynman-Vernon form in which the system-reservoir coupling is switched on at time t_0 where $t_0 \leq 0$. Substituting the spin path (4.65) into the influence functional (5.21) with Eq. (5.22) and observing that the last term in Eq. (5.22) does not contribute since $\sigma^2(t) = \sigma'^2(t)$, we find

$$\mathcal{F}[\sigma, \sigma'; t_0] = \exp\left\{ -\frac{1}{4}\int_{t_0}^{t}dt'\int_{t_0}^{t'}dt'' \right. \qquad (21.2)$$

$$\left. \times \left(\sigma(t') - \sigma'(t')\right)\left(\mathfrak{L}(t'-t'')\,\sigma(t'') - \mathfrak{L}^*(t'-t'')\,\sigma'(t'')\right)\right\} .$$

We have substituted $L(t) = (\hbar/q_0^2)\mathfrak{L}(t)$, and $\mathfrak{L}(t)$ is related to $Q(t)$ by $\ddot{Q}(t) = \mathfrak{L}(t)$.

Imagine that the bath is in canonical equilibrium of the bare Hamiltonian (3.3) and the system is suddenly prepared at time zero in the state $\sigma = +1$. Then, the choice $\bar{\sigma} = 0$ in Eq. (21.1) corresponds to the situation in which the system evolves out of this state before the bath has relaxed to the shifted thermal equilibrium distribution. Mathematically, this is achieved by switching on the system-bath coupling at time $t_0 = 0$. We shall refer to this case as the preparation class A henceforth. Preparation according to class A might be relevant in electron transfer reactions where a particular electronic donor state is suddenly prepared by photoinjection.[1] The corresponding influence functional is

$$\mathcal{F}_{\overline{\sigma}=0}[\sigma, \sigma'] = \mathcal{F}[\sigma, \sigma'; t_0 = 0] . \tag{21.3}$$

The other important case is when $\overline{\sigma} = +1$ in Eq. (21.1), referred to as class B in the sequel. This product initial state is prepared by holding the system for some large time in the state $\sigma = +1$, so that the environment could have come into thermal equilibrium with it. The product initial state can be imposed, for example, by applying a strong negative bias $-\hbar\epsilon_0\Theta(-t)$ with $\epsilon_0 \gg \Delta$ for all times $t < 0$, so that the system is trapped in the state $\sigma = +1$. At time zero the constraint is released, and for $t > 0$ the dynamics is governed by the spin-boson Hamiltonian (18.13). Such initial preparation is achieved, e.g., in the rf SQUID device by a suitable choice of the applied magnetic field (cf. Subsection 3.2.2). The relevant influence functional is

$$\mathcal{F}_{\overline{\sigma}}[\sigma, \sigma'] = \mathcal{F}[\sigma, \sigma'; t_0 \to -\infty] , \tag{21.4}$$

in which the double path $\sigma(t')$, $\sigma'(t')$ is constrained for all times $t' < 0$ to the state $\sigma = \sigma' = \overline{\sigma}$. Upon using Eqs. (21.3) and (21.4), we find the relation

$$\mathcal{F}_{\overline{\sigma}}[\sigma, \sigma'] = \mathcal{F}_{\overline{\sigma}=0}[\sigma, \sigma'] \exp\left\{ i\frac{\overline{\sigma}}{2} \int_0^t dt' \left[\sigma(t') - \sigma'(t') \right] \dot{Q}''(t') \right\} , \tag{21.5}$$

where $\dot{Q}''(t)$ is the time derivative of the imaginary part of the complex bath correlation function $Q(t)$ defined in Eq. (18.42). Thus, the effects of preparation class A can be described for the Hamiltonian (3.86) in terms of a particular time-dependent bias, $\epsilon(t) = \overline{\sigma}\dot{Q}''(t)$.

The expression (21.5) establishes the connection between the two different kinds of preparation of the thermal bath. For a strict Ohmic spectral density ($\omega_c \to \infty$), we have $\dot{Q}''(t) \propto \delta(t)$. Hence the integral in Eq. (21.5) vanishes, and the different preparation has no effect. In the general case, the effects of different preparation (class A or class B) vanishes on a time scale of order $1/\omega_c$ and therefore are relevant only in the adiabatic limit, in which ω_c is of the order of Δ or smaller.

All information on the two-state system at a later time $t > 0$ is contained in the reduced density matrix. The diagonal elements $\rho_{1,1}$ and $\rho_{-1,-1}$ are the population probabilities of the two states, and the off-diagonal elements $\rho_{-1,1}$ and $\rho_{1,-1}$ are the coherences. The reduced density matrix can be written as a linear combination of the Pauli matrices and of the unit matrix, $\rho(t) = \frac{1}{2}[\mathbf{1} + \sum_{j=x,y,z}\langle\sigma_j\rangle_t\sigma_j]$, where

$$\langle\sigma_j\rangle_t \equiv \text{tr}_{\text{Res}}\left\{ \exp[-\beta H_{\text{Res}} + \overline{\sigma}\beta\mathfrak{E}(0)] < R|\sigma_j(t)|R > \right\}/Z_{\text{Res}} \tag{21.6}$$

---
[1]The observability of electronic coherence in ET reactions for preparation A is studied in Refs. [407, 408], and references therein.

is the expectation value of σ_j, and $\sigma_j(t)$ is taken in the Heisenberg representation with respect to the full Hamiltonian H,

$$\sigma_j(t) = e^{iHt/\hbar} \sigma_j e^{-iHt/\hbar} . \tag{21.7}$$

The specifications for the preparation classes A and B are given in Subsection 21.2.3. The expectation values $\langle \sigma_j \rangle_t$ are related to the reduced density matrix by

$$
\begin{aligned}
\langle \sigma_z \rangle_t &= \rho_{1,1}(t) - \rho_{-1,-1}(t) , \\
\langle \sigma_x \rangle_t &= \rho_{1,-1}(t) + \rho_{-1,1}(t) , \\
\langle \sigma_y \rangle_t &= i\rho_{1,-1}(t) - i\rho_{-1,1}(t) .
\end{aligned}
\tag{21.8}
$$

In all our subsequent studies we choose the initial condition $\rho_{\sigma,\sigma'}(t = 0) = \delta_{\sigma,1}\delta_{\sigma',1}$. The quantity $\langle \sigma_z \rangle_t$ describes the difference of the populations of the two localized states for initial population of the right state. It gives immediate information about the tunneling dynamics and is most directly relevant in studies of "macroscopic quantum coherence" (MQC). The understanding of the TSS dynamics is completed by the knowledge of the coherences $\langle \sigma_x \rangle_t$ and $\langle \sigma_y \rangle_t$ [cf. Section 4.1]. We obtain from Eq. (4.6) the bound

$$\langle \sigma_x \rangle_t^2 + \langle \sigma_y \rangle_t^2 + \langle \sigma_z \rangle_t^2 \leq 1 . \tag{21.9}$$

The equals sign holds in the absence of damping. Using Eq. (21.7) and the commutation relation $(H, \sigma_z) = i\hbar\Delta\,\sigma_x$, one finds

$$\langle \sigma_y \rangle_t = -\frac{1}{\Delta} \frac{d\langle \sigma_z \rangle_t}{dt} . \tag{21.10}$$

Hence the coherence $\langle \sigma_y \rangle_t$ is proportional to the tunneling current.

In the absence of the system-bath coupling, it is straightforward to calculate the transition amplitudes and to join them together to construct the density matrix.[2] Alternatively, we may evaluate directly for zero damping the expressions for the expectation values $\langle \sigma_j \rangle_t$ given below in Subsection 21.2.3. In any event, the result is

$$
\begin{aligned}
\langle \sigma_z \rangle_t^{(0)} &= \epsilon^2/\Delta_b^2 + (\Delta^2/\Delta_b^2) \cos(\Delta_b t) , \\
\langle \sigma_x \rangle_t^{(0)} &= (\epsilon\Delta/\Delta_b^2)[1 - \cos(\Delta_b t)] , \\
\langle \sigma_y \rangle_t^{(0)} &= (\Delta/\Delta_b) \sin(\Delta_b t) .
\end{aligned}
\tag{21.11}
$$

Observe that these expressions saturate the bound in Eq. (21.9). The expression for $\langle \sigma_z \rangle_t^{(0)}$ is sometimes called Rabi's formula. The RDM displays oscillations with the transition frequency $\Delta_b = (\Delta^2 + \epsilon^2)^{1/2}$, which reveals the phase-coherent dynamics. For $\epsilon = 0$, we have $(H, \sigma_x) = 0$, and hence $\sigma_x(t) = \sigma_x(0)$. Therefore, $\langle \sigma_x \rangle_t^{(0)}$ is frozen up at the initial value, which is zero.

---

[2]For a discussion of the bare TSS and the fictitious spin $\frac{1}{2}$ system, we refer the reader to Ref. [409].

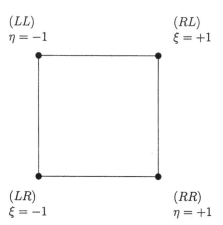

Figure 21.1: Graphical representation of the four states of the reduced density matrix. The diagonal states are labelled by $\eta = \pm 1$, and the off-diagonal states by $\xi = \pm 1$.

The expectation values $\langle \sigma_j \rangle_t$ $(j = x, y, z)$ of the dissipative two-state system are conveniently expressed in terms of the Feynman-Vernon two-time conditional propagating function $J(\zeta, t; \zeta_0, t_0)$ introduced in Eqs. (5.11), (5.12). Here, ζ denotes one of the four states of the reduced density matrix [see Fig. 21.1]. The RDM has two diagonal states (populations), denoted by $\zeta = (\eta, 0) = \eta$, and two off-diagonal states (coherences), denoted by $\zeta = (0, \xi) = \xi$ (in the sequel, we drop the redundant zero). We now assume that the TSS starts out at time zero from the diagonal state $\eta_0 = 1$, and the reservoir's initial state may be either class A or class B, as discussed above. Propagation of the RDM under the full Hamiltonian is then given by

$$\rho_\zeta(t) = J(\zeta, t; \eta_0 = 1, 0) , \qquad (21.12)$$

and the expectation values of the populations and coherences take the form

$$\langle \sigma_z \rangle_t = \sum_{\eta = \pm 1} \eta \, J(\eta, t; \eta_0 = 1, 0) , \qquad (21.13)$$

$$\langle \sigma_x \rangle_t = \sum_{\xi = \pm 1} J(\xi, t; \eta_0 = 1, 0) , \qquad (21.14)$$

$$\langle \sigma_y \rangle_t = i \sum_{\eta = \pm 1} \xi \, J(\xi, t; \eta_0 = 1, 0) . \qquad (21.15)$$

We shall take up these expressions below in Subsection 21.2.3.

21.1.2 Thermal initial state

In many cases of interest, the system under consideration, e.g. a tunneling system in a solid, can not be prepared in a particular pure state. In the usual experimentally reproducible situation, before measurements take place, the system has relaxed to the thermal equilibrium state which is entangled with the environment. After that, it is prepared by a measurement of a certain observable, say at time zero. The measurement of the same observable at a later time t gives then direct information about the equilibrium autocorrelation function of this observable.

For the TSS, the occurring equilibrium autocorrelation functions are ($j = x, y, z$)

$$C_j(t) \equiv \langle \sigma_j(t)\sigma_j(0)\rangle_\beta - \langle\sigma_j\rangle_\infty^2 = \mathrm{tr}\left\{e^{-\beta H}\,\sigma_j(t)\sigma_j(0)\right\}/Z - \langle\sigma_j\rangle_\infty^2, \quad (21.16)$$

where $\sigma_j(t)$ is given in Eq. (21.7). As in Eq. (6.11), we have subtracted the equilibrium average, so that $\lim_{t\to\infty} C_j(t) \to 0$. This is necessary that the Fourier transform of $C_j(t)$ is well-defined. The subtraction term is absent for $C_y(t)$, since $\langle\sigma_y\rangle_\infty = 0$, as follows from Eq. (21.10). Actually, $C_y(t)$ can be found from $C_z(t)$ by differentiation, $C_y(t-t') = \partial^2/(\partial t\,\partial t')\,C_z(t-t')/\Delta^2$, as follows with use of the commutation relation $(H, \sigma_z) = i\hbar\Delta\,\sigma_y$. The real part of $C_j(t)$ is the symmetric autocorrelation function in thermal equilibrium of the observable σ_j, while the imaginary part of $C_j(t)$ is connected with the linear response $\chi_j(t)$ to an external force coupled to σ_j. We have

$$S_j(t) \equiv \mathrm{Re}\,C_j(t) = \tfrac{1}{2}\langle\,[\,\sigma_j(t)\sigma_j(0) + \sigma_j(0)\sigma_j(t)\,]\,\rangle_\beta - \langle\sigma_j\rangle_\infty^2 \quad (21.17)$$

$$\hbar\chi_j(t) \equiv -2\,\Theta(t)\,\mathrm{Im}\,C_j(t) = i\,\Theta(t)\,\langle\,[\,\sigma_j(t)\sigma_j(0) - \sigma_j(0)\sigma_j(t)\,]\,\rangle_\beta\,. \quad (21.18)$$

If we restrict the attention to the dynamics of the TSS, and disregard for a moment the average of the environmental modes, we get upon using the representations (3.81)

$$C_z(t) = \frac{1}{Z}\Big\{ <R|\,e^{-\beta H}|R><R|\sigma_z(t)|R> - <L|\,e^{-\beta H}|L><L|\sigma_z(t)|L>$$

$$+ <R|\,e^{-\beta H}|L><L|\sigma_z(t)|R> - <L|\,e^{-\beta H}|R><R|\sigma_z(t)|L> \Big\} - \langle\sigma_z\rangle_\infty^2\,,$$

and

$$C_x(t) = \frac{1}{Z}\Big\{ <R|\,e^{-\beta H}|R><R|\sigma_x(t)|L> + <L|\,e^{-\beta H}|L><L|\sigma_x(t)|R>$$

$$+ <R|\,e^{-\beta H}|L><L|\sigma_x(t)|L> + <L|\,e^{-\beta H}|R><R|\sigma_x(t)|R> \Big\} - \langle\sigma_x\rangle_\infty^2\,.$$

Next, we take into account that the TSS is actually entangled with the environment when the first measurement at time zero takes place. To deal with this case, we proceed as follows. Suppose that at time t_p, where $t_p < 0$, the system is released from the RDM state ζ_p, and then propagates under the full Hamiltonian. At time $t_0 = 0$, the system is measured to be in the state ζ_0, and eventually at later time t a second measurement is performed in which it is found in the state ζ. The corresponding three-time conditional propagating function for this sequence of events is denoted by $J(\zeta, t; \zeta_0, 0; \zeta_p, t_p)$. For an ergodic system, the correlations of a dynamical variable at time t with the same (or a different) dynamical variable at time zero do not depend on the particular initial state of the reduced system chosen at time t_p, if the time of preparation t_p is displaced to the infinite past. Thus, for convenience, we may choose at time t_p in the infinite past a product initial state, in which the reservoir is in thermal equilibrium and the system in a diagonal state, say in the state $\eta_p = +1$.[3] Then, before the first measurement of a TSS observable at time zero is performed, the TSS has already equilibrated with the environment and is in an entangled canonical state of the system-plus-reservoir complex. With these preliminary thoughts, we are now in the position to study equilibrium correlation functions within the standard real-time influence functional approach introduced in Section 5.5.

Consider first the autocorrelation function of the population, in which the system finally ends up in a diagonal state of the RDM. Regarding the symmetric part, the system dwells at time zero in a diagonal state, whereas for the antisymmetric part it occupies at this time an off-diagonal state. Taking into account the respective weight factors of these states, we get

$$S_z(t) = \lim_{t_p \to -\infty} \sum_{\eta=\pm 1} \sum_{\eta_0=\pm 1} \eta \eta_0 \, J(\eta, t; \eta_0, 0; \eta_p, t_p) - \langle \sigma_z \rangle_\infty^2 \,, \qquad (21.19)$$

$$\hbar \chi_z(t) = \lim_{t_p \to -\infty} 2i \, \Theta(t) \sum_{\eta=\pm 1} \sum_{\xi_0=\pm 1} \eta \xi_0 \, J(\eta, t; \xi_0, 0; \eta_p, t_p) \,. \qquad (21.20)$$

The symmetric population correlations are relevant, e.g., for the description of inelastic neutron scattering from interstitials or defects in a lattice. The differential cross section of this process is given by [201]

$$\frac{\partial^2 \Sigma_{\text{inel}}}{\partial \Omega \, \partial \omega} = \frac{\Sigma_{\text{inc}}}{4\pi} \left(\frac{k_f}{k_i} \right) S_{\text{inel}}(\boldsymbol{k}, \omega) \,, \qquad (21.21)$$

where $\hbar \boldsymbol{k}$ and $\hbar \omega$ are the momentum and energy transfer for the neutrons, and Σ_{inc} is the incoherent cross section of the scatterer. The scattering function or dynamic structure factor for a scatterer with position $\boldsymbol{q}(t)$ at time t is given by

$$S_{\text{inel}}(\boldsymbol{k}, \omega) = \frac{1}{2\pi} \int_{-\infty}^{\infty} dt \, e^{i\omega t} \left(\left\langle e^{-i\boldsymbol{k} \cdot \boldsymbol{q}(0)} \, e^{i\boldsymbol{k} \cdot \boldsymbol{q}(t)} \right\rangle_\beta - \left| \left\langle e^{i\boldsymbol{k} \cdot \boldsymbol{q}} \right\rangle_\beta \right|^2 \right) \,. \qquad (21.22)$$

For a defect tunneling between positions $\frac{1}{2}\boldsymbol{q}_0$ and $-\frac{1}{2}\boldsymbol{q}_0$, the operator $\boldsymbol{q}(t)$ takes the form $\boldsymbol{q}(t) = \frac{1}{2}\boldsymbol{q}_0 \sigma_z(t)$. We then find

$$S_{\text{inel}}(\boldsymbol{k}, \omega) = \sin^2 \left(\tfrac{1}{2}\boldsymbol{k} \cdot \boldsymbol{q}_0 \right) \widetilde{C}_z^-(\omega) / 2\pi \,. \qquad (21.23)$$

Here,

$$\widetilde{C}_z^-(\omega) = \int_{-\infty}^{\infty} dt \, e^{i\omega t} \left(\langle \sigma_z(0) \sigma_z(t) \rangle_\beta - \langle \sigma_z \rangle_\infty^2 \right) \qquad (21.24)$$

---

[3]This product state may be prepared, e.g., according to the preparation class A, or according to the preparation class B, which have been introduced in Subsection 21.1.1.

is the Fourier transform of the pseudo-spin correlation function $C_z(-t)$. Using the spectral relation (6.21), we may express $\widetilde{C}_z^-(\omega)$ in terms of the symmetric correlation function $\widetilde{S}_z(\omega)$

$$\widetilde{C}_z^-(\omega) = \frac{2\widetilde{S}_z(\omega)}{1 + e^{\beta \hbar \omega}} . \tag{21.25}$$

The differential cross section for *inelastic* neutron scattering takes the form

$$\frac{\partial^2 \Sigma_{\text{inel}}}{\partial \Omega \, \partial \omega} = \frac{\Sigma_{\text{inc}}}{4\pi^2} \left(\frac{k_{\text{f}}}{k_{\text{i}}} \right) \sin^2 \left(\frac{\boldsymbol{k} \cdot \boldsymbol{q}_0}{2} \right) \frac{\widetilde{S}_z(\omega)}{1 + e^{\beta \hbar \omega}} . \tag{21.26}$$

Hence, inelastic neutron scattering gives data directly for the spectral function $\widetilde{S}_z(\omega)$.

The other important equilibrium correlation function is that of the tunneling or coherence operator σ_x. Since the polaron unitary operator (18.37) does not commute with σ_x, the correlation function of the bare coherence operator,

$$C_x(t) = \text{tr} \{ e^{-\beta H} e^{iHt/\hbar} \sigma_x e^{-iHt/\hbar} \sigma_x \}/Z - [\text{tr} \{ e^{-\beta H} \sigma_x \}/Z]^2 , \tag{21.27}$$

differs from that of the dressed coherence operator (18.39),

$$C_x^{(p)}(t) = \text{tr} \{ e^{-\beta H} e^{iHt/\hbar} \widetilde{\sigma}_x e^{-iHt/\hbar} \widetilde{\sigma}_x \}/Z - [\text{tr} \{ e^{-\beta H} \widetilde{\sigma}_x \}/Z]^2 , \tag{21.28}$$

$$= \text{tr} \{ e^{-\beta \widetilde{H}} e^{i\widetilde{H}t/\hbar} \sigma_x e^{-i\widetilde{H}t/\hbar} \sigma_x \}/Z - [\text{tr} \{ e^{-\beta \widetilde{H}} \sigma_x \}/Z]^2 . \tag{21.29}$$

With use of Eq. (18.38) we see that the forms (21.28) and (21.29) are equivalent. The undressed and dressed coherence correlations are subject to different bath correlations, as we shall see in Subsection 21.2.5. Due to the operation of σ_x at time zero, the TSS makes at this time a transition from an off-diagonal to a diagonal state or vice versa. Hence the coherence correlations are described in terms of a four-time propagating function $J(\zeta, t; \zeta_+, 0_+; \zeta_-, 0_-, \zeta_{\text{p}}, t_{\text{p}})$. The system finally ends up in an off-diagonal state. Assuming that the system is again prepared in the infinite past in a diagonal state, we get

$$S_x(t) = \lim_{t_{\text{p}} \to -\infty} \frac{1}{2} \left\{ \sum_{\{\xi, \xi_+, \eta_- = \pm 1\}} J(\xi, t; \xi_+, 0_+; \eta_-, 0_-; \eta_{\text{p}}, t_{\text{p}}) \right.$$
$$\left. + \sum_{\{\xi, \eta_+, \xi_- = \pm 1\}} J(\xi, t; \eta_+, 0_+; \xi_-, 0_-; \eta_{\text{p}}, t_{\text{p}}) \right\} - \langle \sigma_x \rangle_\infty^2 , \tag{21.30}$$

and

$$\hbar \chi_x(t) = \lim_{t_{\text{p}} \to -\infty} i \Theta(t) \left\{ \sum_{\{\xi, \xi_+, \eta_- = \pm 1\}} \xi_+ \eta_- J(\xi, t; \xi_+, 0_+; \eta_-, 0_-; \eta_{\text{p}}, t_{\text{p}}) \right.$$
$$\left. + \sum_{\{\xi, \eta_+, \xi_- = \pm 1\}} \eta_+ \xi_- J(\xi, t; \eta_+, 0_+; \xi_-, 0_-; \eta_{\text{p}}, t_{\text{p}}) \right\} . \tag{21.31}$$

We see that both $S_x(t)$ and $\chi_x(t)$ capture two types of contributions. In type A (given in the respective first line), the system hops at time zero from a diagonal to an off-diagonal state, whereas in type B (the respective second line), it hops at this time from an off-diagonal to a diagonal state. The coherence correlations (21.30) and (21.31) are more intricate than the population correlations (21.19) and (21.20). We shall continue the discussion of the coherence correlations in Subsection 21.2.5.

21.2 Exact formal expressions for the system dynamics

21.2.1 Sojourns and blips

According to the discussion in Chapter 5, dynamical quantities are expressed in terms of double path integrals of the form

$$\int \mathcal{D}\sigma(\cdot) \int \mathcal{D}\sigma'(\cdot)\, \mathcal{A}[\sigma(\cdot)]\mathcal{A}^*[\sigma'(\cdot)]\mathcal{F}[\sigma(\cdot),\sigma'(\cdot)] \qquad (21.32)$$

with appropriately chosen boundary conditions for the spin paths $\sigma(t')$ and $\sigma'(t')$. The functional $\mathcal{A}[\sigma(\cdot)]$ is the probability amplitude for the free (undamped) TSS to follow the path $\sigma(t')$, and $\mathcal{F}[\sigma(\cdot),\sigma'(\cdot)]$ is the real-time influence functional discussed in Section 5.5. For the two-state system, the paths $\sigma(t')$ and $\sigma'(t')$ jump between the two discrete values $+1$ and -1, as depicted in Fig. 4.1. According to Eq. (5.28), it is convenient to define antisymmetric and symmetric spin paths $\xi(t')$ and $\eta(t')$,

$$\xi(t') \equiv \tfrac{1}{2}\left[\sigma(t') - \sigma'(t')\right]; \qquad \eta(t') \equiv \tfrac{1}{2}\left[\sigma(t') + \sigma'(t')\right], \qquad (21.33)$$

satisfying the relation $\xi(t')\eta(t') = 0$.

Since the spin paths $\xi(t')$ and $\eta(t')$ are piecewise constant with sudden jumps in between, it is convenient to perform integrations by parts in the expression (21.2). Then the influence functional takes the form (apart from possible boundary terms)

$$\mathcal{F}[\sigma,\sigma';t_0] \;=\; \exp\left(\int_{t_0}^{t}dt' \int_{t_0}^{t'}dt''[\,\dot{\xi}(t')Q'(t'-t'')\dot{\xi}(t'') + i\dot{\xi}(t')Q''(t'-t'')\dot{\eta}(t'')\,]\right),$$

where $Q(t) = Q'(t) + iQ''(t)$ is the complex bath correlation function defined in Eq. (18.42). This form is especially convenient for the piecewise constant spin path with discontinuous jumps [cf. Fig. 4.1, and Eq. (21.35) below].

The double path sum can be visualized as a path sum for a single path that visits the four states of the reduced density. The combinatorial problem facing us is the sum of paths a walker can go along the edges of the square sketched in Fig. 21.1. A period the path spends in a diagonal state has been dubbed *sojourn* by Leggett et al. [89], and a period the path dwells in an off-diagonal state has been termed *blip*. During a sojourn, the function $\xi(\tau)$ is zero, whereas during a blip interval the function $\eta(\tau)$ is zero. There are two sojourn states, labelled by $\eta_j = +1$ [state (RR)] and $\eta_j = -1$ [state (LL)]. Similarly, there are two kinds of blips, and we assign the label $\xi_j = +1$ to the off-diagonal state (RL) and the label $\xi_j = -1$ to the off-diagonal state (LR). For later convenience we introduce blip lengths τ_j and sojourn lengths s_j, respectively. Assuming that the path starts out from a sojourn state, we have

$$\tau_j = t_{2j} - t_{2j-1}, \qquad s_j = t_{2j+1} - t_{2j}. \tag{21.34}$$

A general sojourn-to-sojourn path along the edges of the square in Fig. 21.1 making $2n$ transitions at intermediate times t_j $(j = 1, 2, \ldots, 2n)$ is parametrized by

$$
\begin{aligned}
\eta^{(n)}(t') &= \sum_{j=0}^{n} \eta_j [\Theta(t' - t_{2j}) - \Theta(t' - t_{2j+1})], \\
\xi^{(n)}(t') &= \sum_{j=1}^{n} \xi_j [\Theta(t' - t_{2j-1}) - \Theta(t' - t_{2j})].
\end{aligned}
\tag{21.35}
$$

Introducing the notation $Q_{j,k} = Q(t_j - t_k)$, we may define the bath correlations $\Lambda_{j,k}$ between the blip-pair $\{j, k\}$ and the blip-sojourn correlations $X_{j,k}$ between the sojourn k and a later blip j in the compact form

$$
\begin{aligned}
\Lambda_{j,k} &= Q'_{2j,2k-1} + Q'_{2j-1,2k} - Q'_{2j,2k} - Q'_{2j-1,2k-1}, \\
X_{j,k} &= Q''_{2j,2k+1} + Q''_{2j-1,2k} - Q''_{2j,2k} - Q''_{2j-1,2k+1}.
\end{aligned}
\tag{21.36}
$$

With these expressions, the influence functional for the path (21.35) reads

$$
\begin{aligned}
\mathcal{F}^{(n)} &= G_n H_n, \\
G_n &= \exp\left[-\sum_{j=1}^{n} Q'_{2j,2j-1} \right] \exp\left[-\sum_{j=2}^{n} \sum_{k=1}^{j-1} \xi_j \xi_k \Lambda_{j,k} \right], \\
H_n &= \exp\left[i \sum_{j=1}^{n} \sum_{k=0}^{j-1} \xi_j X_{j,k} \eta_k \right] = \exp\left[i \sum_{k=0}^{n-1} \sum_{j=k+1}^{n} \xi_j X_{j,k} \eta_k \right].
\end{aligned}
\tag{21.37}
$$

It is natural to interpret the expression (21.37) by means of a charge picture. The paths (21.35) represent a sequence of blip and sojourn dipoles labelled by ξ_j and η_j, respectively. This is because the system is after every second transition again in a blip or in a sojourn state. The sojourn dipole η_j has length $t_{2j+1} - t_{2j}$ and the blip dipole ξ_j has length $t_{2j} - t_{2j-1}$. The values $\eta_j = +1$ and $\xi_j = +1$ correspond to $(+, -)$ dipoles, and the values $\eta_j = -1$ and $\xi_j = -1$ correspond to $(-, +)$ dipoles. The functions $Q'(\tau)$ and $Q''(\tau)$ represent the interaction between two blip charges and between a sojourn and a blip charge, respectively, of distance τ, and each pair of equal sign. The function $\Lambda_{j,k}$ and $X_{j,k}$ describe the dipole-dipole interaction between two blip dipoles and between a sojourn dipole and a succeeding blip dipole, respectively, each of them of the same type.

The blip interactions are bundled up in the real-valued function G_n. The first exponential factor contains the intrablip or intradipole interactions, and the second exponential factor represents the interblip correlations. The interactions between the sojourns and all subsequent blip states are in the phase factor H_n.

The function G_n is a filtering function which suppresses long blips. Hence, dwelling of the system in a sojourn state is favoured. Physically, this is because the environment is continuously measuring σ_z and therefore suppresses occupation of off-diagonal states, and quantum interference between the eigenstates of σ_z.

It is straightforward to write down also the various weight factors resulting from the amplitude product $\mathcal{A}[\sigma]\,\mathcal{A}^\star[\sigma']$ for the TSS or spin Hamiltonian (3.80). The weight to switch per unit time from a diagonal state η to an off-diagonal state ξ (or vice versa) is

$$-i\,\xi\eta\,\Delta/2 \ . \tag{21.38}$$

Thus we have the weight factor $-i\Delta/2$ for transitions $(RL) \Longleftrightarrow (RR)$ and $(LL) \Longleftrightarrow (LR)$, and $i\Delta/2$ for transitions $(RL) \Longleftrightarrow (LL)$ and $(RR) \Longleftrightarrow (LR)$.

The weight to stay in a sojourn is unity, while the weight factor to stay in the jth blip with label ξ_j and length τ_j is $\exp(i\epsilon\xi_j\tau_j)$. The blip factors for n blip states are accumulated in the overall bias factor

$$B_n \ = \ \exp\left(i\epsilon \sum_{j=1}^{n} \xi_j \tau_j \right) \ . \tag{21.39}$$

The prescription to sum over all paths along the edges of the square sketched in Fig. 21.1 now means (i) to sum over the possible intermediate sojourn and blip states the paths with a given number of transitions can visit, (ii) to integrate over the time-ordered jump times of these paths, and (III) to sum over the possible numbers of transitions the system may take,

$$\sum_{\text{all paths}} \cdots \rightarrow \sum_{n=0}^{\infty} \sum_{\{\eta_j=\pm1\}} \sum_{\{\xi_j=\pm1\}} \int_{t_i}^{t_f} dt_n \int_{t_i}^{t_n} dt_{n-1} \cdots \int_{t_i}^{t_2} dt_1 \cdots \ . \tag{21.40}$$

We have explained in Section 5.5 that for correlated system-reservoir initial states at time zero, e.g. in equilibrium correlation functions, it is essential to consider the evolution of the system since a fictitious time t_p at which the global system is prepared in some factorizing initial state, where t_p is eventually sent to the infinite past (cf. the contours $\mathcal{C}_{\mathrm{III}}$ or $\mathcal{C}_{\mathrm{IV}}$ specified in Fig. 5.4. With this aim in view, we divide the integrations over the time-ordered flip times $\{t_j\}$ into a negative and a positive time branch. It is convenient to introduce the compact notation

$$\int_{t_\mathrm{p}}^{t} \mathcal{D}_{k,\ell}\{t_j\} \ \times \ \cdots \ \equiv \ \int_0^t dt_{\ell+k} \cdots \int_0^{t_{\ell+2}} dt_{\ell+1} \int_{t_\mathrm{p}}^{0} dt_\ell \cdots \int_{t_\mathrm{p}}^{t_2} dt_1 \, \Delta^{k+\ell} \ \times \ \cdots \ . \tag{21.41}$$

Here, ℓ is the number of steps in the negative time branch $t_\mathrm{p} < t' < 0$ and k is the number of steps in the positive time branch, $0 < t' < t$. For saving of writing, we have included in the definition the tunneling matrix factor. We are now ready to discuss the conditional propagating functions for the various boundary conditions of interest.

21.2.2 Conditional propagating functions

We have seen in Section 21.1 that the relevant expectation values and equilibrium autocorrelation functions can be expressed in terms of two-time, three-time, and four-time conditional propagating functions.

If the expectation values $\langle \sigma_j \rangle_t$ ($j = x$, y, z) imply a factorizing initial state of the global system at time zero, they can be expressed in terms of the two-time conditional propagating function $J(\zeta, t; \eta_0, 0)$. Here, the initial state of the system is the sojourn state η_0, and the preparation of the reservoir may be performed either according to class A or according to class B. When the final state is again a sojourn, the number of transitions the system makes is even, while it is odd, when the final state is a blip.

Upon collecting the various weight factors discussed in the previous subsection we obtain for the two-time correlation function in the case of a sojourn as final state the series expression

$$J(\eta, t; \eta_0, 0) = \delta_{\eta,\eta_0} + \eta\,\eta_0 \sum_{m=1}^{\infty} \frac{(-1)^m}{2^{2m}} \int_0^t \mathcal{D}_{2m,0}\{t_j\} \sum_{\{\xi_j=\pm1\}} G_m B_m \sum_{\{\eta_j=\pm1\}'} H_m \,, \quad (21.42)$$

while in the case of a blip as final state, we have

$$J(\xi, t; \eta_0, 0) = -i\,\xi\,\eta_0 \sum_{m=1}^{\infty} \frac{(-1)^{m-1}}{2^{2m-1}} \int_0^t \mathcal{D}_{2m-1,0}\{t_j\} \sum_{\{\xi_j=\pm1\}'} G_m B_m \sum_{\{\eta_j=\pm1\}'} H_m \,. \quad (21.43)$$

The prime in $\{\eta_j = \pm1\}'$ and $\{\xi_j = \pm1\}'$ indicates that the initial sojourn and the final sojourn/blip state are fixed as indicated in the argument of the propagating function.

The three-time conditional propagating function for being in a sojourn state at time zero and time t reads

$$J(\eta, t; \eta_0, 0; \eta_p, t_p) = \delta_{\eta,\eta_0}\delta_{\eta_0,\eta_p} \quad (21.44)$$

$$+ \delta_{\eta,\eta_0}\eta_0\eta_p \sum_{n=1}^{\infty} \left(-\frac{1}{4}\right)^n \int_{t_p}^t \mathcal{D}_{0,2n}\{t_j\} \sum_{\{\xi_j=\pm1\}} G_n B_n \sum_{\{\eta_j=\pm1\}'} H_n$$

$$+ \delta_{\eta_0,\eta_p}\eta\eta_0 \sum_{m=1}^{\infty} \left(-\frac{1}{4}\right)^m \int_{t_p}^t \mathcal{D}_{2m,0}\{t_j\} \sum_{\{\xi_j=\pm1\}} G_m B_m \sum_{\{\eta_j=\pm1\}'} H_m$$

$$+ \eta\eta_p \sum_{m=1}^{\infty}\sum_{n=1}^{\infty} \left(-\frac{1}{4}\right)^{m+n} \int_{t_p}^t \mathcal{D}_{2m,2n}\{t_j\} \sum_{\{\xi_j=\pm1\}} G_{m+n} B_{m+n} \sum_{\{\eta_j=\pm1\}'} H_{m+n} \,.$$

The sum over arrangements $\{\xi_j\}$ and $\{\eta_j\}$ extends over the possible values ± 1 of all the intermediate states the system may take for $2n$ transitions in the interval $t_{\mathrm{p}} < t' < 0$ and $2m$ transitions in the interval $0 < t' < t$. The prime in $\{\eta_j\}'$ is to indicate that the sojourns at times t_{p}, 0, and t are fixed as indicated in the arguments of the propagating function. Correspondingly, we find for occupation of a blip state at time zero

$$
\begin{aligned}
J(\eta, t; \xi_0, 0; \eta_{\mathrm{p}}, t_{\mathrm{p}}) \;=\; & \eta\eta_{\mathrm{p}} \sum_{m=1}^{\infty}\sum_{n=1}^{\infty} \left(-\frac{1}{4}\right)^{n+m-1} \int_{t_{\mathrm{p}}}^{t} \mathcal{D}_{2m-1,2n-1}\{t_j\} \\
& \times \sum_{\{\xi_j=\pm 1\}'} G_{m+n-1} B_{m+n-1} \sum_{\{\eta_j=\pm 1\}'} H_{m+n-1} \, ,
\end{aligned}
\tag{21.45}
$$

where the sum is over all sequences of blips and sojourns which are in accordance with the constraints indicated by the arguments of the propagating function.

The discussion of the coherence correlations is postponed to Subsection 21.2.5.

21.2.3   The expectation values $\langle \sigma_j \rangle_t$ ($j = x, y, z$)

Consider first the population $\langle \sigma_z \rangle_t$. Upon substituting the series expression (21.42) for the two-time conditional propagating function into into Eq. (21.13), and performing the summation over the intermediate sojourn states, $\{\eta_j = \pm 1\}'$, we find the exact formal series expression [89]

$$
\langle \sigma_z \rangle_t \;=\; 1 + \sum_{m=1}^{\infty} (-1)^m \int_0^t \mathcal{D}_{2m,0}\{t_j\} \, \frac{1}{2^m} \sum_{\{\xi_j=\pm 1\}} \left(F_m^{(+)} B_m^{(\mathrm{s})} - F_m^{(-)} B_m^{(\mathrm{a})} \right) .
\tag{21.46}
$$

The bias dependence is divided up into the symmetric and antisymmetric functions

$$
B_m^{(\mathrm{s})} \;=\; \cos\!\left(\epsilon \sum_{j=1}^{m} \xi_j \tau_j\right) ; \qquad B_m^{(\mathrm{a})} \;=\; \sin\!\left(\epsilon \sum_{j=1}^{m} \xi_j \tau_j\right) ,
\tag{21.47}
$$

and the effects of the environment are in the influence functions

$$
F_m^{(+)} \;=\; G_m \prod_{k=0}^{m-1} \cos\left(\phi_{k,m}\right) , \qquad F_m^{(-)} \;=\; G_m \sin\left(\phi_{0,m}\right) \prod_{k=1}^{m-1} \cos\left(\phi_{k,m}\right) .
\tag{21.48}
$$

The term G_m describes the intra- and inter blip correlations, and the residual factors contain the phase correlations. The bath correlations between the kth sojourn and the $m - k$ succeeding blips are combined in the phase

$$
\phi_{k,m} \;=\; \sum_{j=k+1}^{m} \xi_j X_{j,k} \, .
\tag{21.49}
$$

Finally, the sum over the labels $\{\xi_j\}$ runs over all intermediate blip states, $\{\xi_j = \pm 1\}$.

For subsequent purposes, we introduce the notation

$$\langle \sigma_z \rangle_t \equiv P(t) = P_R(t) - P_L(t) = 2P_R(t) - 1 \,, \tag{21.50}$$

where $P_{R/L}(t)$ is the occupation probability of the right/left well, respectively. According to our initial condition, we have $P(0) = 1$. The function $P(t)$ may be split into the components which are symmetric (s) and antisymmetric (a) under inversion of the bias $(\epsilon \to -\epsilon)$,

$$P(t) = P_s(t) + P_a(t) \,. \tag{21.51}$$

When the damping of the system persists until time infinity, the system is ergodic and it relaxes to the thermal equilibrium state[4]. Then the part $P_s(t)$, which is even in the bias, goes to zero as $t \to \infty$, whereas the part $P_a(t)$, which is odd in the bias, approaches the equilibrium value $P(t \to \infty) = P_\infty = \langle \sigma_z \rangle_\infty = \langle \sigma_z \rangle^{(eq)}$,

$$\langle \sigma_z \rangle_\infty = P_\infty = \lim_{t \to \infty} \sum_{m=1}^{\infty} (-1)^{m-1} \int_0^t \mathcal{D}_{2m,0}\{t_j\} \frac{1}{2^m} \sum_{\{\xi_j = \pm 1\}} F_m^{(-)} B_m^{(a)} \,. \tag{21.52}$$

This expression determines the equilibrium distribution via computation of a dynamical quantity. On the other hand, the equilibrium distribution P_∞ may be expressed in terms of pure thermodynamic quantities, namely the partition functions Z_R and Z_L of the left and right state, respectively. We have [cf. Eqs. (19.10) and (19.11)]

$$\langle \sigma_z \rangle_\infty = \langle \sigma_z \rangle^{(eq)} = (Z_R - Z_L)/(Z_R + Z_L) \,. \tag{21.53}$$

When the system is ergodic, the two different expressions (21.52) and (21.53) coincide.

It is also straightforward to derive the series expressions for the coherences $\langle \sigma_x \rangle_t$ and $\langle \sigma_y \rangle_t$. Upon employing the series expression (21.43), the relations (21.14) and (21.15) yield [410]

$$\langle \sigma_x \rangle_t = \sum_{m=1}^{\infty} (-1)^{m-1} \int_0^t \mathcal{D}_{2m-1,0}\{t_j\} \frac{1}{2^m} \sum_{\{\xi_j = \pm 1\}} \xi_m \left(F_m^{(+)} B_m^{(a)} + F_m^{(-)} B_m^{(s)} \right) , \tag{21.54}$$

$$\langle \sigma_y \rangle_t = \sum_{m=1}^{\infty} (-1)^{m-1} \int_0^t \mathcal{D}_{2m-1,0}\{t_j\} \frac{1}{2^m} \sum_{\{\xi_j = \pm 1\}} \left(F_m^{(+)} B_m^{(s)} - F_m^{(-)} B_m^{(a)} \right) . \tag{21.55}$$

The expressions (21.46), (21.54) and (21.55) are exact formal expressions for the time evolution of the reduced density matrix. Observe that the expressions (21.46) and (21.55) satisfy the relation (21.10).

Let us finally comment on the different kinds of preparation discussed in Subsection 21.1.1. In the expressions (21.46), (21.54), and (21.55) the differences between

---

[4]This is the case, when the spectral density $G(\omega)$ has at low frequencies the power-law form $\propto \omega^s$ with $s < 2$ [cf. Subsection 3.1.5 and Section 7.4]

class A and class B are captured by the correlations of the blips with the initial sojourn, as follows from Eq. (21.5) with Eq. (21.33). These correlations are described by the functions $X_{j,0}$ given in Eq. (21.36). For class A, the initial sojourn begins at time zero. Thus we have $X_{j,0}^{(A)} = Q''_{2j,1} + Q''_{2j-1,0} - Q''_{2j,0} - Q''_{2j-1,1}$. On the other hand, for class B, the initial sojourn has length infinity. In this limit, the second and third term in $X_{j,0}$ cancel out, yielding $X_{j,0}^{(B)} = Q''_{2j,1} - Q''_{2j-1,1}$. For an Ohmic spectral density and the scaling limit (18.54), the bath correlation function is given by Eq. (18.51). In this limit, the preparation classes A and B coincide, $X_{j,0}^{(A)} = X_{j,0}^{(B)}$. Substituting $Q''(t) = \pi K \, \mathrm{sgn}\,(t)$ into Eqs. (21.48), the influence functions simplify to the forms

$$F_m^{(+)} = [\cos(\pi K)]^m \, G_m \, ; \qquad F_m^{(-)} = \xi_1 \sin(\pi K)[\cos(\pi K)]^{m-1} \, G_m \, . \qquad (21.56)$$

We shall use these forms below in Section 21.6.

21.2.4 Correlation and response function of the populations

Upon inserting the series expression (21.44) into Eq. (21.19) and summing again over the intermediate sojourns, we obtain for the symmetric correlation function [208]

$$S_z(t) \;=\; S_z^{(\mathrm{unc})}(t) + R(t) \, , \qquad (21.57)$$

$$S_z^{(\mathrm{unc})}(t) \;=\; P_{\mathrm{s}}(t) + P_\infty \, [\, P_{\mathrm{a}}(t) - P_\infty \,] \, . \qquad (21.58)$$

Here, $P_{\mathrm{s/a}}(t)$ is the part of $\langle \sigma_z \rangle_t$ in Eq. (21.46) which is symmetric (antisymmetric) in the bias, and $P_\infty = P_{\mathrm{a}}(t \to \infty) = \langle \sigma_z \rangle_\infty$ is the equilibrium value of σ_z. The term $S_z^{(\mathrm{unc})}(t)$ is the contribution to $S_z(t)$ in which the entanglement of the system with the reservoir at time zero is disregarded, which formally corresponds to the neglect of all bath correlations $Q(\tau)$ between the negative-time and the positive-time branch of the full spin path. If we had chosen a product initial state for the system-plus-reservoir complex at time zero (just before the first measurement takes place), then Eq. (21.58) would be the resulting expression for the symmetric σ_z autocorrelation function. The residual term $R(t)$ in Eq. (21.57) describes all dynamical effects which result from the system-bath correlations in the initial state at time zero. We find

$$R(t) \;=\; \lim_{t_{\mathrm{p}} \to -\infty} \sum_{m=1}^{\infty} \sum_{n=1}^{\infty} (-1)^{m+n-1} \int_{t_{\mathrm{p}}}^{t} \mathcal{D}_{2m,2n}\{t_j\}$$
$$\times \; \frac{1}{2^{m+n}} \sum_{\{\xi_j = \pm 1\}} \left[\, F_{m,n} B_m^{(\mathrm{s})} B_n^{(\mathrm{s})} - \left(F_{m,n} - F_m^{(-)} F_n^{(-)} \right) B_m^{(\mathrm{a})} B_n^{(\mathrm{a})} \, \right] , \qquad (21.59)$$

where

$$F_{m,n} \;=\; G_{m+n} \sin\left(\phi_{n,m+n}\right) \sin\left(\phi_{0,m+n}\right) \prod_{\substack{k=1 \, (k \neq n)}}^{m+n-1} \cos\left(\phi_{k,m+n}\right) \, . \qquad (21.60)$$

The second term in the round bracket in Eq. (21.59) $\propto F_m^{(-)} F_n^{(-)}$ is a subtraction term in which the bath correlations are only inside of the negative and positive time branch. It is easily seen from Eqs. (21.46) and (21.52) that this term is just the counter term of the contribution $P_\infty P_a(t)$ in Eq.(21.58). We should expect that at low temperatures and long times the correlation term $R(t)$ gives significant contributions. Indeed, this we shall find below. For later convenience, we also give the simplified form of Eq. (21.60) holding in the scaling limit (18.54) with (18.51),

$$F_{m,n} = \xi_1 \xi_{n+1} \left[\sin(\pi K) \right]^2 \left[\cos(\pi K) \right]^{m+n-2} G_{m+n} . \tag{21.61}$$

Likewise, the response function (21.20) is found with use of the expression (21.45) as

$$\chi_z(t) = \lim_{t_p \to -\infty} \frac{4}{\hbar} \sum_{m=1}^{\infty} \sum_{n=1}^{\infty} \frac{(-1)^{m+n}}{2^{m+n}} \int_{t_p}^{t} \mathcal{D}_{2m-1,2n-1}\{t_j\} \sum_{\{\xi_j = \pm 1\}} \xi_n F_{m+n-1}^{(-)} B_{m+n-1}^{(s)} , \tag{21.62}$$

where $t > 0$ and where $F_m^{(-)}$ is defined in Eq. (21.48). The peculiarity of this expression is that the TSS is in a blip state at time zero.

The static nonlinear susceptibility is defined by the relations

$$\overline{\chi}_z \equiv \widetilde{\chi}_z(\omega = 0) = \int_0^\infty dt\, \chi_z(t) . \tag{21.63}$$

We now substitute the form (21.62) into Eq. (21.63) and use that the integrals over the last interval at negative times ($\tau_n^- = -t_{2n-1}$) and the first interval at positive times ($\tau_n^+ = t_{2n}$) are of the form ($\tau_n = \tau_n^- + \tau_n^+$ is the full blip length)

$$\int_0^\infty d\tau_n^-\, d\tau_n^+\, f(\tau_n^- + \tau_n^+) = \int_0^\infty d\tau_n\, \tau_n f(\tau_n) .$$

Upon rearrangement of the double series, we then obtain the series expression

$$\overline{\chi}_z = \lim_{t \to \infty} \frac{2}{\hbar} \sum_{m=1}^{\infty} (-1)^{m-1} \int_0^t \mathcal{D}_{2m,0}\{t_j\} \frac{1}{2^m} \sum_{\{\xi_j \pm 1\}} F_m^{(-)} B_m^{(s)} \left(\sum_{n=1}^{m} \xi_n \tau_n \right) , \tag{21.64}$$

which expresses the static susceptibility as the asymptotic long-time limit of a dynamical quantity. Now, upon comparing the series (21.64) with the series (21.52), we see that they are related by

$$\overline{\chi}_z = \frac{2}{\hbar} \frac{\partial \langle \sigma_z \rangle_\infty}{\partial \epsilon} , \tag{21.65}$$

which agrees with the thermodynamic relation (19.13). Thus we have confirmed this relation by a dynamical approach. Our subsequent study of the dynamics of the dissipative two-state system is based upon the above exact formal series expressions.

21.2.5   Correlation and response function of the coherences

In this subsection, we restrict for simplicity the attention to the case of Ohmic dissipation in the scaling limit (18.54) with (18.51). The case of the coherence correlation function is more complicated than the case of the population correlations for two reasons. First, the system makes an additional step at time zero. This step, however, is not dynamical, but is just enforced by the operation of σ_x. Secondly, the system ends up in an off-diagonal state at time t. The final off-diagonal state gives rise to a boundary term in the influence function which exactly appears as if there would be an extra step at time t back to a diagonal state. For these two reasons, the influence function introduces additional bath correlation terms, resulting in an overall factor $(\Delta_{\rm r}/\Delta)^2 = (\Delta_{\rm r}/\omega_{\rm c})^{2K}$ in the series expression for $C_x(t)$. Hence $C_x(t)$ is non-universal and therefore vanishes in the scaling limit, as observed first by Guinea [411].

Universality is restored, however, by taking into account the adiabatic dynamics of the bath modes when the particle tunnels from the one localized state to the other. These effects are included in the correlation function $C_x^{(\rm pd)}(t)$ of the *polaron-dressed* tunneling operator, Eq. (21.28). Since the dressed operator $\tilde{\sigma}_x$ acts in the full system-plus-reservoir space and the correlated initial state involves a particular preparation of the reservoir, the procedure of the elimination of the bath modes has to be reconsidered. The relevant analysis has been performed in Ref. [412]. In the end, one finds a modified influence functional. This, however, can be cast into the standard form at the expense of introducing modified system paths. The findings can be put in simple terms. First, the terms in Eqs. (21.30) and (21.31) with $\xi = -\xi_+ = \eta_-$ for group A, and $\xi = -\eta_+ = \xi_-$ for group B, vanish again in the scaling limit. In the residual terms, which have $\xi = \xi_+$ for group A and $\xi = \eta_+$ for group B, the system's jumps at time zero and time t actually do not give rise to bath correlations in the influence functional. In the equivalent charge picture, the corresponding charges are removed by the polaron transformation. For this reason, these terms turn out to be universal. At this point, we anticipate that the equilibrium expectation value $\langle \sigma_x \rangle_\infty$ contributing to the expression (21.30) is non-universal and therefore vanishes in the scaling limit [see the discussion after Eq.(21.144)].

Because of the additional step the system makes (which is without a factor of Δ), it is convenient to introduce the modified integration symbol [cf. Eq.(21.41)]

$$\int_{t_p}^t \tilde{\mathcal{D}}_{k,\ell}\{t_j\} \times \cdots \equiv \int_0^t dt_{\ell+k+1} \cdots \int_0^{t_{\ell+3}} dt_{\ell+2} \int_{t_p}^0 dt_\ell \cdots \int_{t_p}^{t_2} dt_1 \, \Delta^{k+\ell} \times \cdots , \quad (21.66)$$

and the extra step is exactly at time zero, $t_{\ell+1} = 0$. For both groups, the system is finally in a blip state. The symmetric correlation function $S_x(t) = \operatorname{Re} C_x^{(\rm pd)}(t)$ is found as $S_x(t) = S_x^{\rm A}(t) + S_x^{\rm B}(t)$, where

$$S_x^{\rm A}(t) = \frac{1}{2} \sum_{m=1}^\infty \left(-\frac{\cos(\pi K)}{2} \right)^{m-1} \int_{-\infty}^t \tilde{\mathcal{D}}_{2m-2,0}\{t_j\} \sum_{\{\xi_j = \pm 1\}_{\rm A}} G_m^{\rm A} B_m^{(\rm s)} , \quad (21.67)$$

$$S_x^B(t) = -\sum_{m=2}^{\infty}\sum_{n=1}^{\infty}\left(-\frac{\cos(\pi K)}{2}\right)^{n+m-1}\sin^2(\pi K)\int_{-\infty}^{t}\widetilde{\mathcal{D}}_{2m-1,2n-1}\{t_j\} \qquad (21.68)$$

$$\times \sum_{\substack{\{\xi_j=\pm1\} \\ \xi_n=\xi_{n+1}=-\xi_{m+n}}} \xi_{m+n}\,\xi_1\,G_{m+n}^B B_{m+n}^{(s)}\,.$$

Correspondingly, the response function $\chi_x(t) = -(2/\hbar)\,\Theta(t)\,\mathrm{Im}\,C_x^{(\mathrm{pd})}(t)$ is found as

$$\chi_x^A(t) = \frac{1}{\hbar}\sum_{m=1}^{\infty}\sum_{n=1}^{\infty}\left(-\frac{\cos(\pi K)}{2}\right)^{m+n-1}\tan(\pi K)\int_{-\infty}^{t}\widetilde{\mathcal{D}}_{2m-2,2n}\{t_j\} \qquad (21.69)$$

$$\times \sum_{\{\xi_j=\pm1\}_A} \xi_{m+n}\,\xi_1\,G_{m+n}^A B_{m+n}^{(s)}\,,$$

$$\chi_x^B(t) = \frac{1}{\hbar}\sum_{m=1}^{\infty}\sum_{n=1}^{\infty}\left(-\frac{\cos(\pi K)}{2}\right)^{m+n-1}\tan(\pi K)\int_{-\infty}^{t}\widetilde{\mathcal{D}}_{2m-1,2n-1}\{t_j\} \qquad (21.70)$$

$$\times \sum_{\{\xi_j=\pm1\}_B} [\,\sin^2(\pi K)\,\xi_{n+1} + \cos^2(\pi K)\,\xi_{n+m}\,]\,\xi_1\,G_{m+n}^B B_{m+n}^{(s)}\,.$$

The summations over the sojourn states are already performed. The subscripts $\{\ldots\}_A$ and $\{\ldots\}_B$ indicate that the blip labels are subject to the constraints

$$\xi_{m+n} = \xi_{n+1} \quad \text{(group A)}, \qquad \xi_{m+n} = -\xi_n \quad \text{(group B)}\,. \qquad (21.71)$$

The charge interaction $Q'(t)$ is given in Eq. (18.51). The interaction factors G_{m+n}^A and G_{m+n}^B differ from the form (21.37) for G_{m+n} by the removal of the charges situated at the origin, $t' = 0$, and at the end point, $t' = t$. Just for this reason, the remaining $2m + 2n - 2$ blip charges come with the universal factor $\Delta_r^{(2-2K)(m+n-1)}$, and there is no extra dependence on Δ and ω_c.

Equations (21.67) – (21.70) are exact formal expressions for the polaron-dressed coherence correlation function in the scaling limit.

21.2.6 Generalized exact master equation and integral relations

We have seen from the exact formal expressions for the expectation values $\langle\sigma_j\rangle_t$ that the system's transitions are correlated with each other through the influence functions $F_m^{(\pm)}(\{t_j\})$. Notwithstanding these formidable intricacies, it is possible to describe the dynamics of the conditional populations $P(i,t;j,0)$ [i and j label diagonal states] for any N-state system in terms of a set of exact generalized master equations (GME)

$$\dot{P}(i,t;j,0) = -\sum_{k=-N/2}^{N/2}\int_0^t dt'\,K(i,t;k,t')P(k,t';j,0)\,, \qquad t > 0\,. \qquad (21.72)$$

Conservation of probability provides the sum rule $\sum_i K(i, t; j, t') = 0$.

Let us now consider the case of a two-state system, $P(\pm 1, t; 1, 0) = \frac{1}{2}[1 \pm \langle \sigma_z \rangle_t]$. Upon using the sum rule for the kernels, and introducing linear combinations which are even and odd under bias inversion, $K_z^{(s/a)}(t, t') = K(-1, t; -1, t') \pm K(1, t; 1, t')$, the GME for $\langle \sigma_z \rangle_t$ with product initial state at time zero reads [413]

$$\frac{d\langle \sigma_z \rangle_t}{dt} = \int_0^t dt' \, [\, K_z^{(a)}(t, t') - K_z^{(s)}(t, t') \langle \sigma_z \rangle_{t'} \,] \,. \qquad (21.73)$$

Irreducible kernels and self-energies

The kernels are, by definition, the irreducible components in the exact formal series (21.46). Irreducibility of a kernel means that it cannot be cut into two uncorrelated pieces at an intermediate sojourn without removing bath correlations across this sojourn. We now define *irreducible* influence clusters $\widetilde{F}_n^{(\pm)}$ by subtraction of the reducible components in $F_n^{(\pm)}$, which are in product form. Since the bias factor factorizes correspondingly in the subtractions, as we can see from the cluster-wise $\{\xi\}$-summations, we consider the product $F_n^{(\pm)} B_n^{(s/a)}$. For a path with n blips and time growing from right to left in each term, we find

$$\widetilde{F}_n^{(\pm)} B_n^{(s/a)} \equiv F_n^{(\pm)} B_n^{(s/a)} \qquad (21.74)$$

$$- \sum_{j=2}^n (-1)^j \sum_{m_1, \cdots, m_j} F_{m_1}^{(+)} B_{m_1}^{(s)} F_{m_2}^{(+)} B_{m_2}^{(s)} \cdots F_{m_j}^{(\pm)} B_{m_j}^{(s/a)} \delta_{m_1 + \cdots + m_j, n} \,.$$

The inner sum is over positive integers m_j. The subtractions represent all possibilities of factorizing the influence functions into clusters. For instance, the $n = 3$ term reads

$$\widetilde{F}_3^{(\pm)} B_3^{(s/a)} = F_3^{(\pm)} B_3^{(s/a)} - F_2^{(+)} B_2^{(s)} F_1^{(\pm)} B_1^{(s/a)} \qquad (21.75)$$

$$- F_1^{(+)} B_1^{(s)} F_2^{(\pm)} B_2^{(s/a)} + F_1^{(+)} B_1^{(s)} F_1^{(+)} B_1^{(s)} F_1^{(\pm)} B_1^{(s/a)} \,.$$

In the subtractions, the bath correlations are only inside the individual cluster factors $F_{m_j}^{(\pm)}$, and there are no bath correlations between the clusters. We are now ready to give the exact formal series expressions for the kernels $K_z^{(s/a)}(t, t')$ of the GME. The kernels are constructed by matching the iterative solution of Eq. (21.73) with the exact formal series expression (21.46). Eventually, we find

$$K_z^{(s/a)}(t, t') = \Delta^2 F_1^{(\pm)}(t, t') B_1^{(s/a)}(t, t') \qquad (21.76)$$

$$+ \sum_{n=2}^{\infty} (-1)^{n-1} \frac{\Delta^{2n}}{2^n} \int_{t'}^t dt_{2n-1} \cdots \int_{t'}^{t_3} dt_2 \sum_{\{\xi_j = \pm 1\}} \widetilde{F}_n^{(\pm)} B_n^{(s/a)} \,.$$

The product function $\widetilde{F}_n^{(\pm)} B_n^{(s/a)}$ depends on $2n$ flip times, where the first transition or flip just occurs at $t_1 = t'$, and the last one at $t_{2n} = t$. The intermediate flip times are integrated out.

The GME (21.73) with the kernels (21.76) is the exact dynamical equation for general (state-independent) dissipation.

The coherence $\langle\sigma_x\rangle_t$ is connected with $\langle\sigma_z\rangle_t$ by the exact integral relation

$$\langle\sigma_x\rangle_t = \int_0^t dt' [K_x^{(s)}(t,t') + K_x^{(a)}(t,t')\langle\sigma_z\rangle_{t'}] . \tag{21.77}$$

The kernels $K_x^{(s/a)}(t,t')$ are given again in the form of series expressions involving the modified influence functions. We find upon matching Eq. (21.77) with the exact formal series expression (21.54)

$$K_x^{(s/a)}(t,t') = \Delta F_1^{(\mp)}(t,t') B_1^{(s/a)}(t,t') \tag{21.78}$$

$$+ \sum_{n=2}^{\infty} (-1)^{n-1} \frac{\Delta^{2n-1}}{2^n} \int_{t'}^t dt_{2n-1} \cdots \int_{t'}^{t_3} dt_2 \sum_{\{\xi_j=\pm1\}} \xi_n \widetilde{F}_n^{(\mp)} B_n^{(s/a)} .$$

It is appropriate to remark that the GME (21.73) and the integral relation (21.77) are also valid when the bias and the tunneling matrix element are externally driven. The only modification would be to employ the substitution (22.5), when the tunneling amplitude is time-dependent, and to use the modified bias factors (22.3) with (22.2) when the bias is time-dependent.

In the absence of explicit time-dependence, the kernels $K_z^{(s/a)}(t,t')$ and $K_x^{(s/a)}(t,t')$ depend only on the relative time $t - t'$. Then, the expressions (21.73) and (21.77) are convolutions which can solved by switching to the Laplace transforms.[5] We obtain

$$\langle\sigma_z(\lambda)\rangle = \frac{1 + \widehat{K}_z^{(a)}(\lambda)/\lambda}{\lambda + \widehat{K}_z^{(s)}(\lambda)} = \frac{1 + \Sigma_z^{(a)}(\lambda)/\lambda}{\lambda + \lambda\Delta^2/(\lambda^2+\epsilon^2) + \Sigma_z^{(s)}(\lambda)} , \tag{21.79}$$

$$\langle\sigma_x(\lambda)\rangle = \frac{1}{\lambda}\Sigma_x^{(s)}(\lambda) + \left(\epsilon\Delta/(\lambda^2+\epsilon^2) + \Sigma_x^{(a)}(\lambda)\right)\langle\sigma_z(\lambda)\rangle . \tag{21.80}$$

For subsequent convenience we have introduced the self-energies $\hbar\Sigma_z^{(s/a)}(\lambda)$ and $\hbar\Delta\Sigma_x^{(s/a)}(\lambda)$. The self-energies are the Laplace transforms of the kernels (21.76) and (21.78) in which the respective kernels of the isolated system are subtracted,[6]

$$\Sigma_z^{(s)}(\lambda) \equiv \widehat{K}_z^{(s)}(\lambda) - \lambda\Delta^2/(\lambda^2+\epsilon^2) , \qquad \Sigma_z^{(a)}(\lambda) \equiv \widehat{K}_z^{(a)}(\lambda) ,$$

$$\Sigma_x^{(a)}(\lambda) \equiv \widehat{K}_x^{(a)}(\lambda) - \epsilon\Delta/(\lambda^2+\epsilon^2) , \qquad \Sigma_x^{(s)}(\lambda) \equiv \widehat{K}_x^{(s)}(\lambda) . \tag{21.81}$$

By definition, the self-energies $\hbar\Sigma_z^{(s/a)}(\lambda)$ and $\hbar\Delta\Sigma_x^{(s/a)}(\lambda)$ are zero for vanishing coupling to the reservoir. Once the self-energies are known, the expressions (21.79) and (21.80) can then be inverted to obtain $\langle\sigma_z\rangle_t$ and $\langle\sigma_x\rangle_t$.

---

[5]We use for the Laplace transform the convention as given in Eqs. (6.32) and (6.33).

[6]The quantity $\Sigma_z^{(s/a)}(\lambda)$ has dimension frequency, whereas $\Sigma_x^{(s/a)}(\lambda)$ is dimensionless.

When the system is ergodic, the expectation values relax to the respective thermal equilibrium state. The residua of the pole at $\lambda = 0$ of $\langle\sigma_z(\lambda)\rangle$ and $\langle\sigma_x(\lambda)\rangle$ give us directly $\langle\sigma_z\rangle_{t\to\infty}$ and $\langle\sigma_x\rangle_{t\to\infty}$, as the contributions of all other singularities in the complex λ-plane fade away in the course of time. Since the self-energies $\Sigma_z^{(s/a)}(\lambda)$ and $\Sigma_x^{(s/a)}(\lambda)$ have well-defined limits for $\lambda \to 0$, we find for the equilibrium values the exact formal expressions

$$\langle\sigma_z\rangle_\infty = \frac{\Sigma_z^{(a)}(0)}{\Sigma_z^{(s)}(0)}, \qquad \langle\sigma_x\rangle_\infty = \Sigma_x^{(s)}(0) + \left(\frac{\Delta}{\epsilon} + \Sigma_x^{(a)}(0)\right)\langle\sigma_z\rangle_\infty. \qquad (21.82)$$

When explicit inversion of the transforms (21.79) and (21.80) is not possible, much about the dynamics can nevertheless be learned from a study of the singularities.

Despite the progress achieved by the exact results (21.79) and (21.80), the formal expressions for the kernels as well as the series expression for $C_z^{(\pm)}(t)$ obtained in the preceding section, while exact, are extremely cumbersome. Nevertheless, they can be evaluated in certain limits by analytical methods. This is discussed in the sequel. Some exact results in the regime of incoherent exponential relaxation at $T = 0$ are discussed in Subsection 21.8.1.

21.3   The noninteracting–blip approximation (NIBA)

We continue the discussion of the dynamics with describing the noninteracting-blip approximation (NIBA) which turns out to be exact in various limits [89].

The simple assumption underlying the NIBA is that the average time $\langle s\rangle$ spent by the system in a diagonal state is very large compared to the average time $\langle\tau\rangle$ spent in an off-diagonal state. This assumption leads to two simple prescriptions:

(1) Set the sojourn-blip correlations $X_{j,k}$ equal to zero when $j \neq k + 1$, and put $X_{k+1,k} = Q''(t_{2k+2} - t_{2k+1})$. Thus, the correlations between a sojourn k and the subsequent blips reduce to the intrablip phase correlation $\phi_{k,m} = \xi_{k+1}Q''(\tau_{k+1})$.

(2) Set all interblip interactions $\Lambda_{j,k}$ in the factor G_n in Eq. (21.37) equal to zero.

In the NIBA, the influence functional (21.37) reduces to a factorized form of intra-blip bath correlations in which only the signs of the blip phase terms depend on the labels of the respective preceding sojourns,

$$\mathcal{F}_{\text{NIBA}}^{(n)} = \prod_{j=1}^{n}\exp\left\{-Q'(\tau_j) + i\xi_j\eta_{j-1}Q''(\tau_j)\right\}. \qquad (21.83)$$

For a strict Ohmic spectral density, i.e., in the scaling limit (18.54) with (18.51), the assumption (1) is exact. Thus in this important case, the NIBA consists in the neglect of the interblip interactions $\Lambda_{j,k}$.

Generally, the NIBA can be justified in at least three different limits [89]:

(a) Weak-coupling and zero bias: because of the $\{\xi_j\}$-summations in Eq. (21.76), those contributions to the kernel $K_z^{(s)}(\tau)$ which are *linear* in the interblip correlations $\Lambda_{j,k}$ cancel each other. Thus, the interblip correlations of the kernel $K_z^{(s)}(\tau)$ are of second order in the coupling δ_s, whereas the intrablip correlations are of first order in δ_s. Hence the NIBA for $\langle\sigma_z\rangle_t$ is exact for zero bias in the weak damping (one-phonon process) limit. In all other cases, i.e., in the kernel $K_z^{(s/a)}(\tau)$ for nonzero bias, and in the kernel $K_x^{(s/a)}(\tau)$ for zero and nonzero bias, the cancellations among first-order interblip correlations are incomplete.

(b) For $s > 1$ at zero temperature and $s > 2$ at finite temperatures, $Q'(t \approx \Delta^{-1})$ differs only little from the asymptotic value $Q'(t \to \infty)$. Thus, since the average length of a sojourn is of the order of Δ^{-1} or larger, the interblip interactions $\Lambda_{j,k}$ tend to be small compared with the intrablip interactions because of the partial cancellations in Eq. (21.36).

(c) Long blips are suppressed when the function $Q'(t)$ increases with t at long times. This occurs at $T = 0$ for sub-Ohmic damping, $s < 1$, and at finite temperatures for $s < 2$.[7] Thus, the NIBA is justified (1) in the sub-Ohmic case at all T, (2) in the Ohmic case for large damping and/or high temperature, and (3) in the super-Ohmic case with $s < 2$ for high temperatures. Long blips are also suppressed, independently of $Q(t)$, when the bias is very large.

Thus, the NIBA gives consistent results in three physically quite different regimes.

The picture we now have is that of a noninteracting gas of blips in a one-dimensional volume of length t. All influences of the environment are contained in the intrablip interactions which are fully taken into account.[8] Because of the absence of interblip correlations, the modified influence functions (21.74) are zero for all $n > 1$. Thus, within the NIBA, the kernels given in Subsection 21.2.6 are determined by the one-blip contribution in Eq. (21.76) and in Eq. (21.78). Switching to the self-energies (21.81), we have in the NIBA[9]

$$\Sigma_z^{(s)}(\lambda) = \Delta^2 \int_0^\infty d\tau \, e^{-\lambda\tau} \cos(\epsilon\tau)\left(e^{-Q'(\tau)}\cos[Q''(\tau)] - 1\right),$$

$$\Sigma_z^{(a)}(\lambda) = \Delta^2 \int_0^\infty d\tau \, e^{-\lambda\tau} \sin(\epsilon\tau) \, e^{-Q'(\tau)} \sin[Q''(\tau)],$$

$$\Sigma_x^{(s)}(\lambda) = \Delta \int_0^\infty d\tau \, e^{-\lambda\tau} \cos(\epsilon\tau) \, e^{-Q'(\tau)} \sin[Q''(\tau)],$$

$$\Sigma_x^{(a)}(\lambda) = \Delta \int_0^\infty d\tau \, e^{-\lambda\tau} \sin(\epsilon\tau)\left(e^{-Q'(\tau)}\cos[Q''(\tau)] - 1\right). \qquad (21.84)$$

---

[7]At $T = 0$, we have $Q'(t \to \infty) \propto t^{1-s}$, whereas for finite temperatures $Q'(t \to \infty) \propto t^{2-s}$, as follows from Eq. (18.45).

[8]A simple derivation of the NIBA in the Heisenberg picture is given in Ref. [414].

[9]The generalization for time-dependent driving is studied in Section 22.1.

Substituting these forms for the self-energies into Eqs. (21.79) and (21.80) and Laplace inverting the resulting expression gives the evolution of the damped TSS in the NIBA. The NIBA gives a good qualitative (and often quantitative) account of the dynamics over the most interesting time regime of several Δ_r^{-1}. However, it does not describe the long-time behaviour correctly for $\langle\sigma_j\rangle_t$ and $C_j^{\pm}(t)$ $(j=x,y,z)$ at low T.

Before computing $\langle\sigma_j\rangle_t$ and $C_j^{\pm}(t)$ for particular spectral densities, we briefly discuss several flaws of the NIBA occurring in the long-time limit at low temperatures. First, since blip-sojourn and blip-blip interactions are absent, the system-bath correlations in the initial state are disregarded in the NIBA. As a result, the function $S_z(t)$ coincides with the function $S_z^{(\mathrm{unc})}(t)$ given in Eq. (21.58), in which the components $P_{\mathrm{s/a}}(t)$ and $P_\infty = \langle\sigma_z\rangle_\infty$ are the NIBA expressions. Thus, the algebraic long-time tails of $S_z(t)$ at $T=0$ (cf. Subsection 21.7) are disregarded in the NIBA. Secondly, the equilibrium values $\langle\sigma_z\rangle_\infty$ and $\langle\sigma_x\rangle_\infty$ for $\epsilon \neq 0$ are qualitatively incorrect at low temperatures. In the NIBA, we have

$$\Sigma_z^{(\mathrm{s})}(0) = k^+ + k^- , \qquad \Sigma_z^{(\mathrm{a})}(0) = k^+ - k^- , \tag{21.85}$$

where $k^{\pm}$ are the nonadiabatic rate expressions discussed in Subsection 20.2.1. We find from Eq. (21.82) upon using the detailed balance relation (20.32)

$$\langle\sigma_z\rangle_\infty = \tanh\left(\frac{\hbar\epsilon}{2k_{\mathrm{B}}T}\right) , \tag{21.86}$$

$$\langle\sigma_x\rangle_\infty = \frac{\Delta}{\epsilon}\tanh\left(\frac{\hbar\epsilon}{2k_{\mathrm{B}}T}\right) . \tag{21.87}$$

Hence the NIBA predicts symmetry breaking and strict localization in the lower well at zero temperature, even when the bias is infinitesimal, $\langle\sigma_z\rangle_{\infty|T=0} = \mathrm{sgn}\,(\epsilon)$. Furthermore, we get from Eq. (21.87) $\langle\sigma_x\rangle_{\infty|T=0} = \Delta/|\epsilon|$. This expression violates the bound $|\langle\sigma_x\rangle_t| \leq 1$ in the regime $|\epsilon| < \Delta$.

To get the correct weak-damping forms for $\langle\sigma_z\rangle_\infty$ and $\langle\sigma_x\rangle_\infty$ we follow a simple thought. First, assuming that the eigenstates of H_{TSS} are thermally occupied, we get

$$\langle\sigma_z\rangle_\infty = \frac{1}{Z}\left[<R|e^{-\beta H_{\mathrm{TSS}}}|R> - <L|e^{-\beta H_{\mathrm{TSS}}}|L>\right] ,$$

$$\langle\sigma_x\rangle_\infty = \frac{1}{Z}\left[<R|e^{-\beta H_{\mathrm{TSS}}}|L> + <L|e^{-\beta H_{\mathrm{TSS}}}|R>\right] .$$

Next, we write the localized states $|R>$ and $|L>$ in terms of the eigenstates $|g>$ and $|e>$ of the Hamiltonian H_{TSS}. With use of the relations (3.82) we then obtain

$$\langle\sigma_z\rangle_\infty = \frac{\epsilon}{\Delta_{\mathrm{b}}}\tanh\left(\frac{\hbar\Delta_{\mathrm{b}}}{2k_{\mathrm{B}}T}\right) , \tag{21.88}$$

$$\langle\sigma_x\rangle_\infty = \frac{\Delta}{\Delta_{\mathrm{b}}}\tanh\left(\frac{\hbar\Delta_{\mathrm{b}}}{2k_{\mathrm{B}}T}\right) . \tag{21.89}$$

The first expression shows that the particle is not confined in reality to one of the wells at zero temperature, $\langle\sigma_z\rangle_{\infty|T=0} = \epsilon/\Delta_b$, and the second expression satisfies the bound $|\langle\sigma_x\rangle_{\infty|T=0}| \leq 1$.

This observation shows that the NIBA fails for a biased system in the regime $k_BT \lesssim \hbar\Delta_b$. The appropriate treatment of this regime within the dynamical approach is given in Section 21.4.

The expressions (21.86) and (21.87) match the exact weak-damping expressions (21.88) and (21.89), respectively, in the temperature regime $T \gtrsim \hbar\Delta_b/k_B$. This points to the fact that the NIBA gives the behaviour of both the expectation values $\langle\sigma_j\rangle_t$ and the equilibrium correlation function $C_z^\pm(t)$ correctly for all times at temperatures above $\hbar\Delta_b/k_B$.[10] In the remainder of this section we describe the dynamics of the TSS within the NIBA for particular cases.

21.3.1 Symmetric Ohmic system in the scaling limit

We begin the discussion of the Ohmic scaling limit (18.54) by considering the symmetric case $\epsilon = 0$. Substituting the form (18.51) for the bath correlation function $Q(\tau)$, the resulting integral over the blip length τ in Eq. (21.84) can be evaluated in analytic form. The kernel $\widehat{K}_z^{(s)}(\lambda)$ is found to read

$$\widehat{K}_z^{(s)}(\lambda) = g(\lambda), \qquad (21.90)$$

$$g(\lambda) = \Delta_{\text{eff}}\left(\frac{\hbar\beta\Delta_{\text{eff}}}{2\pi}\right)^{1-2K}\frac{h(\lambda)}{K+\hbar\beta\lambda/2\pi}; \quad h(\lambda) = \frac{\Gamma(1+K+\hbar\beta\lambda/2\pi)}{\Gamma(1-K+\hbar\beta\lambda/2\pi)}, \quad (21.91)$$

where Δ_{eff} is the effective tunnel matrix element defined in Eq. (18.35). The Laplace transform of the population $\langle\sigma_z(\lambda)\rangle$ and the spectral function $\widetilde{S}_z(\omega)$ of the symmetric equilibrium correlation function [cf. Eq. (21.26)] read

$$\langle\sigma_z(\lambda)\rangle = \frac{1}{\lambda+g(\lambda)}, \qquad \widetilde{S}_z(\omega) = \text{Re}\,\frac{2}{-i\omega+g(\lambda=-i\omega)}. \qquad (21.92)$$

Consider now first the dynamics of the symmetric TSS at zero temperature. Taking in Eq. (21.91) the limit $T\to 0$, we get $g(\lambda) = \Delta_{\text{eff}}(\Delta_{\text{eff}}/\lambda)^{1-2K}$, and thus

$$\langle\sigma_z(\lambda)\rangle = \frac{1}{\lambda+\Delta_{\text{eff}}(\Delta_{\text{eff}}/\lambda)^{1-2K}}. \qquad (21.93)$$

This is the Laplace transform of a known special function, the Mittag-Leffler function $E_\nu(z)$ [392, a], of which the integral representation is given in Eq. (7.44) [229]. We

---

[10]The attentive reader may have noticed that the argument has been given for a weakly damped system. However, since any system behaves more classically as the damping is increased, this condition should be valid for any damping strength.

have

$$\langle \sigma_z \rangle_t \equiv P(t) = E_{2-2K}[-(\Delta_{\text{eff}} t)^{2-2K}]$$

$$= \sum_{m=0}^{\infty} \frac{(-1)^m}{\Gamma[1 + (2 - 2K)m]} (\Delta_{\text{eff}} t)^{(2-2K)m} . \qquad (21.94)$$

As $K \to 0$, we recover from Eq. (21.94) the persistent oscillation $\langle \sigma_z \rangle_t = \cos(\Delta t)$, whereas for $K = 1/2$ we get pure relaxation $\langle \sigma_z \rangle_t = \exp[-(\pi \Delta^2 / 2\omega_c)t]$. Here we have employed the relation $\Delta_{\text{eff}}(K = \frac{1}{2}) = \pi \Delta^2 / 2\omega_c$, which follows from Eq. (18.35).

Since the NIBA is exact in order Δ^2, the expression (21.94) correctly describes the leading time dependence at short times,

$$\langle \sigma_z \rangle_t = 1 - (\Delta_{\text{eff}} t)^{2-2K} / \Gamma(3 - 2K) + \mathcal{O}[(\Delta_{\text{eff}} t)^{4-4K}] , \qquad \Delta_{\text{eff}} t \ll 1 . \qquad (21.95)$$

The transform (21.93) is very useful in understanding the behaviour of $\langle \sigma_z \rangle_t$. For $0 < K < \frac{1}{2}$, the expression (21.93) has the following singularities:

(1) A branch point at $\lambda = 0$. The complex λ-plane is cut along the negative real axis, and in the cut plane the integrand is single-valued.

(2) A complex conjugate pair of simple poles on the principal sheet at

$$\lambda_{1/2} \equiv -\gamma \pm i\Omega = \Delta_{\text{eff}} \exp[\pm i \tfrac{\pi}{2(1-K)}] . \qquad (21.96)$$

Thus we may write

$$P(t) = P_{\text{coh}}(t) + P_{\text{inc}}(t) . \qquad (21.97)$$

The cut gives a negative contribution $P_{\text{inc}}(0) = -K/(1 - K)$ to the initial value $P(0) = 1$, and it yields the asymptotic behaviour as $\Delta_{\text{eff}} t \to \infty$,

$$P_{\text{inc}}(t) = \sum_{n=1}^{\infty} \frac{(-1)^{n-1}}{\Gamma[1 - (2 - 2K)n]} \frac{1}{(\Delta_{\text{eff}} t)^{(2-2K)n}} . \qquad (21.98)$$

The leading contribution is negative and decays as $1/(\Delta_{\text{eff}} t)^{2-2K}$. The branch point is absent and hence $P_{\text{inc}}(t) = 0$ for $K = 0$ and for $K = \frac{1}{2}$.

The residues of the pair of complex conjugate poles give the coherent part

$$P_{\text{coh}}(t) = \frac{1}{1 - K} \cos(\Omega t) e^{-\gamma t} . \qquad (21.99)$$

The oscillation frequency Ω and the dephasing rate γ are given by

$$\Omega = \Delta_{\text{eff}} \cos[\tfrac{\pi K}{2(1-K)}] , \quad \text{and} \quad \gamma = \Delta_{\text{eff}} \sin[\tfrac{\pi K}{2(1-K)}] . \qquad (21.100)$$

The quality factor of the oscillation Q at $T = 0$ is a function of K and it is independent of the frequency scale Δ_{eff}. The expressions (21.100) give

$$Q \equiv \Omega/\gamma = \cot\left[\frac{\pi K}{2(1-K)}\right]. \tag{21.101}$$

Recently, it was argued by Lesage and Saleur [415] by using a generalization of the boundary conditions changing operators of conformal field theory [416] that the dominant long-time behaviour of $P(t)$ is a damped oscillation of the form (21.99). They found in their approach that the quality factor of the oscillation in the regime $0 < K < \frac{1}{2}$ is precisely given by Eq. (21.101). Thus, the arguments given in Ref. [415] support that the quality factor found in in the NIBA is exact. Only the frequency scale turns out to be slightly modified. In our notation, the results for Ω and γ by Lesage and Saleur read

$$\Omega = \Delta_{\text{LS}} \cos\left[\frac{\pi K}{2(1-K)}\right], \quad \text{and} \quad \gamma = \Delta_{\text{LS}} \sin\left[\frac{\pi K}{2(1-K)}\right], \tag{21.102}$$

where Δ_{LS} is related to the effective frequency Δ_{eff} given in Eq. (18.35) by

$$\Delta_{\text{LS}} = \sin\left[\frac{\pi K}{2(1-K)}\right] a(K) \Delta_{\text{eff}}, \tag{21.103}$$

where

$$a(K) = \frac{\Gamma(\frac{K}{2(1-K)})}{\sqrt{\pi}\,\Gamma(\frac{1}{2(1-K)})} \left(\frac{\Gamma(\frac{1}{2}+K)\Gamma(1-K)}{\sqrt{\pi}}\right)^{\frac{1}{2(1-K)}}. \tag{21.104}$$

The limiting behaviour of (21.103), as $K \to 0$, is $\Delta_{\text{LS}} = \Delta_{\text{eff}}[1 + \mathcal{O}(K^2)]$. Similarly, as $K \to \frac{1}{2}$, we have $\Delta_{\text{LS}} = \Delta_{\text{eff}}[1 + \mathcal{O}(1-2K)]$. Thus, the NIBA frequency scale Δ_{eff} in Eq. (21.100) coincides with the scale Δ_{LS} in these limits. For intermediate values of K, the scale Δ_{LS} differs from Δ_{eff} by only a few percent; e.g., for $K = \frac{1}{4}$ we find from Eq. (21.103) $\Delta_{\text{LS}} \approx 1.038\,\Delta_{\text{eff}}$.

For $\frac{1}{2} < K < 1$, the poles of the Laplace transform (21.93) are not anymore on the principal sheet, so that $P(t)$ in the NIBA is fully determined by the branch-cut contribution[11] $P_{\text{inc}}(t)$, which is a completely monotonous function of t. However, the power law form (21.98) is qualitatively incorrect. As opposed to this sluggish decay, the nonequilibrium correlation function $P(t)$ decays in reality exponentially fast [cf. the discussion in Subsection 21.7.1].

Let us now turn to finite temperatures. Now, the branch point is resolved into an infinite sequence of poles lying on the negative real λ-axis at $\lambda_n = -(2\pi/\hbar\beta)n - v_n$ $(n = 1, 2, \ldots)$, where $-K \leq (\hbar\beta/2\pi)v_n \leq K$. Hence $P_{\text{inc}}(t) = \sum_{n=1}^{\infty} A_n \exp(\lambda_n t)$. The amplitude A_n vanishes linearly with K as $K \to 0$, so that $P_{\text{inc}}(t)$ can be disregarded for weak damping. The coherent dynamics is again determined by a pair of complex conjugate simple poles, and $P_{\text{coh}}(t)$ is in the form (21.99). For $K \ll 1$ and $0 \leq$

---

[11]Since there is no coherent contribution for $K > 1/2$ (in the limit $\Delta_{\text{eff}}/\omega_c \to 0$) at $T = 0$, it is unlikely that there would be any one for $T > 0$ in the damping regime $K > 1/2$.

$k_BT \lesssim \hbar\Delta_{\mathrm{eff}}$, we can determine the position of the poles iteratively from Eq. (21.92). We find for the frequency Ω and the dephasing rate γ the forms[12]

$$\Omega(T) = \Delta_{\mathrm{eff}} \{1 + K[\operatorname{Re}\psi(i\hbar\beta\Delta_{\mathrm{eff}}/2\pi) - \ln(\hbar\beta\Delta_{\mathrm{eff}}/2\pi)]\} ,$$
$$\gamma(T) = \tfrac{1}{2}\pi K\Delta_{\mathrm{eff}} \coth(\hbar\beta\Delta_{\mathrm{eff}}/2) . \tag{21.105}$$

The dephasing rate is in Korringa form. At $T = 0$, it is consistent with Eq. (21.100) for $K \ll 1$. The same temperature dependence of the level splitting and the relaxation rate has been observed in a variety of physical systems involving electron-hole excitations, such as interstitials in metals [93] and rare impurities in metals [417]. The generalization of Eq. (21.105) to the biased case is given in Section 21.4.

For intermediate to high temperatures, $k_BT \gtrsim \hbar\Delta_{\mathrm{eff}}$, and general K, we may expand the function $h(\lambda)$ given in Eq. (21.91) about $\lambda = 0$ up to terms of order λ^2. Then the condition $\lambda + g(\lambda) = 0$ yields a quadratic equation for the poles of $\langle\sigma_z(\lambda)\rangle$. Introducing the dimensionless quantities

$$x = (\hbar\beta/2\pi)\lambda , \qquad u = (\hbar\beta\Delta_{\mathrm{eff}}/2\pi)^{2K-2}h(0) , \tag{21.106}$$

the quadratic equation for x reads

$$(1 - g_2 u)x^2 + (K + g_1 u)x + u = 0 , \tag{21.107}$$

$$g_1 = K^{-1} - \pi\cot(\pi K) , \qquad g_2 = \tfrac{1}{2}[\psi'(1-K) - \psi'(1+K) - g_1^2] . \tag{21.108}$$

The roots of Eq. (21.107) are complex conjugate in the coherent and real in the incoherent regime. The temperature $T^*(K)$ at which the transition between the two "phases" occurs is found from Eq. (21.107) as [418]

$$T^*(K) = \frac{\hbar\Delta_{\mathrm{eff}}}{k_B} \left\{ \frac{\Gamma(K)}{K\Gamma(1-K)} \left[1 + \pi K\cot(\pi K) + 2\sqrt{W(K)} \right] \right\}^{1/2(1-K)} , \tag{21.109}$$

where $W(K) = \pi K\cot(\pi K) - K^2 g_2(K)$. In the entire regime $0 \le K \le \tfrac{1}{2}$, the formula (21.109) has a relative error of less than 0.4% as compared with the bound calculated numerically from Eq. (21.92) by Garg [419, 89]. The bound for the coherence regime monotonously increases when K is decreased. For $K \ll 1$, we find from Eq. (21.109)

$$T^*(K) = [(2\pi)^K/\pi K]^{1/(1-K)}\hbar\Delta_r/k_B \approx \hbar\Delta_r/k_B\pi K . \tag{21.110}$$

The transition temperature varies inversely with K for weak damping. The formula (21.109) gives also the exact result $T^* = \hbar\Delta_{\mathrm{eff}}/k_B\pi$ for $K = \tfrac{1}{2}$.

---

[12]We remark again that the NIBA is exact for a symmetric system in the weak-damping limit. The expression for $\Omega(T)$ coincides with the former result (19.51) obtained from a rather sophisticated imaginary-time calculation ($\epsilon = 0$).

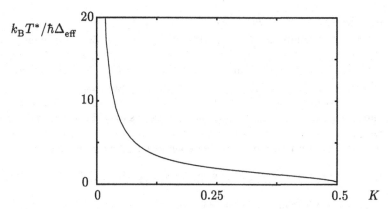

Figure 21.2: The temperature $T^*(K)$ for crossover from the coherent to the incoherent phase is shown as a function of K.

21.3.2 Weak Ohmic damping and moderate-to-high temperature

In this subsection, we study the case of weak Ohmic dissipation, $K \ll 1$, in the regime

$$T_b \lesssim T \ll \hbar\omega_c/k_B , \qquad T_b \equiv \hbar(\Delta_{\rm eff}^2 + \epsilon^2)^{1/2} . \qquad (21.111)$$

The regime $0 \leq T \lesssim T_b$ is discussed in Section 21.4. The case $K \ll 1$ applies, e.g., to interstitial tunneling in metals. In addition, the weak-damping case is of fundamental importance with regard to "Macroscopic Quantum Coherence" (MQC) experiments since the dynamics might be coherent in the accessible temperature regime.

The argument of this subsection is based on the observation that in the temperature regime (21.111) the characteristic fluctuations of the collective bath mode $\mathfrak{E}(t)$ [cf. Eq. (5.34)] are memory-less,

$$\mathrm{Re}\, \langle \delta\mathfrak{E}(t)\, \delta\mathfrak{E}(0) \rangle_T = K\pi\hbar k_B T\, \delta(t) . \qquad (21.112)$$

The Markovian form (21.112) is consistent with the long-time or high-temperature limit of the Ohmic bath correlation function (18.51),

$$Q(t) = 2K\left\{\frac{\pi|t|}{\hbar\beta} + \ln\left(\frac{\hbar\beta\omega_c}{2\pi}\right)\right\} + i\pi K\,\mathrm{sgn}(t) . \qquad (21.113)$$

The real part $Q'(t)$ is $4/\hbar^2$ times the second integral of the force autocorrelation (21.112), as follows with Eqs. (4.68) and (18.42). Because in Eq. (21.112) the correlator of the collective bath mode $\mathfrak{E}(t)$ is memory-less, the form (21.113) is the Markov limit for $Q(t)$. The expression (21.113) emerges directly from Eq. (18.51) by taking the high-temperature limit, $t \gg \hbar\beta/2\pi$. Physically, the logarithmic term in Eq. (21.113) is an adiabatic Franck-Condon factor of the form (18.27) which is made up by all modes in the frequency range $\omega_\ell = 2\pi/\hbar\beta \leq \omega \leq \omega_c$. The dynamical (first) term is due to the modes with $\omega \ll 2\pi/\hbar\beta$. Use of the form (21.113) instead of the full expression (18.51) in the breathing mode integral for $\Sigma_z^{(s)}(\lambda)$ in Eq. (21.84) is equivalent to the substitution $h(\lambda) \rightarrow h(0)$ in Eq. (21.91). This is a consistent approximation when the relevant poles $\lambda_{1/2}$ of $\langle\sigma_z(\lambda)\rangle$ satisfy the condition $\hbar\beta|\lambda_{1/2}| \ll 1$. This condition is fulfilled in the regime (21.111), which concludes the argument.

For the Markov form (21.113) the interblip interactions cancel out exactly. Thus, $\Lambda_{jk} = 0$, and hence the noninteracting-blip assumption is *exact* in the regime (21.111).

It is convenient to absorb the adiabatic Franck-Condon term $2K\ln(\hbar\beta\omega_c/2\pi)$ in Eq. (21.113) into an effective temperature-dependent tunneling matrix element,

$$\Delta_T \equiv \Delta_{\text{eff}} \left(2\pi k_B T/\hbar\Delta_{\text{eff}}\right)^K \sqrt{h(0)} = \Delta_r \left(2\pi k_B T/\hbar\Delta_r\right)^K , \qquad (21.114)$$

where we have used $\Delta_{\text{eff}}^{2-2K} h(0) = \Delta_r^{2-2K}[1+\mathcal{O}(K^2)]$, which follows from Eq. (18.35). With these preliminaries, it is now obvious that the kernel (21.90) reduces to the simple form

$$\widehat{K}_z^{(s)}(\lambda) = \Delta_T^2/[\lambda + 2\pi K/\hbar\beta] . \qquad (21.115)$$

Consider now first the *symmetric* system. Substituting Eq. (21.115) into the expression (21.79) and setting $\epsilon = 0$, we obtain[13]

$$\langle\sigma_z(\lambda)\rangle = \frac{\lambda + 2\pi K/\hbar\beta}{\lambda[\lambda + 2\pi K/\hbar\beta] + \Delta_T^2} = \frac{\lambda - \lambda_1 - \lambda_2}{(\lambda - \lambda_1)(\lambda - \lambda_2)} . \qquad (21.116)$$

The poles $\lambda_{1/2}$ of $\langle\sigma_z(\lambda)\rangle$ are again complex-conjugate for $T < T^*$, and real-valued for $T > T^*$. The transition temperature $T^*(K)$ separating the coherent from the incoherent "phase" is given in Eq. (21.110). For small K, it varies with $1/K$, $T^*(K) = \hbar\Delta_r/k_B\pi K$. In the coherent regime $\hbar\Delta_{\text{eff}}/k_B \lesssim T \leq T^*(K)$, we have $\lambda_{1/2} = -\gamma \pm i\Omega$. The population $\langle\sigma_z\rangle_\beta$ shows damped oscillations,

$$\langle\sigma_z\rangle_t = \frac{\cos(\Omega t - \phi)}{\cos(\phi)} e^{-\gamma t} , \qquad \phi = \arctan(\gamma/\Omega) . \qquad (21.117)$$

The corresponding spectral function $\widetilde{S}_z(\omega)$ is a superposition of two inelastic lines of Lorentzian shape centered at $\omega = \pm\Omega(T)$ with line width $2\gamma(T)$,

$$\widetilde{S}_z(\omega) = \frac{\gamma + (\Omega - \omega)\tan\phi}{(\omega - \Omega)^2 + \gamma^2} + \frac{\gamma + (\Omega + \omega)\tan\phi}{(\omega + \Omega)^2 + \gamma^2} . \qquad (21.118)$$

The parameters of the equivalent damped oscillator are

$$\Omega(T) = \Delta_T\sqrt{1 - (T/T^*)^{2-2K}} ; \qquad \gamma(T) = \pi K k_B T/\hbar . \qquad (21.119)$$

---

[13]The pole condition is the weak-damping limit of Eq. (21.107). The generalization to larger values of K is analogous to Eq. (21.107). Only the coefficients of the algebraic equation are modified.

The expressions (21.119) for Ω and γ smoothly match Eq. (21.105) near $T = \hbar\Delta_r/k_B$. The oscillation frequency increases $\propto T^2$ at low temperatures,[14] as we can see from Eq. (21.105). For T above $\hbar\Delta_r/k_B$ the form (21.119) applies, and the oscillation frequency has a maximum at $T = K^{1/(2-2K)}T^*$. Above this temperature, $\Omega(T)$ decreases monotonously and approaches zero for $T = T^*$. The coherent quantum oscillations "dephase" on the time scale $1/\gamma(T)$ which decreases inversely with T.

For T above T^*, the two poles of $\langle\sigma_z(\lambda)\rangle$ in Eq. (21.116) are situated on the real negative λ-axis at $\lambda = -\gamma_1$ and $\lambda = -\gamma_2$. This yields in the time domain

$$\langle\sigma_z\rangle_t = \frac{\gamma_1}{\gamma_1 - \gamma_2} e^{-\gamma_2 t} - \frac{\gamma_2}{\gamma_1 - \gamma_2} e^{-\gamma_1 t}, \tag{21.120}$$

where

$$\gamma_{1/2} = \pi K/\hbar\beta \pm \sqrt{(\pi K/\hbar\beta)^2 - \Delta_T^2}. \tag{21.121}$$

As T is increased further, the poles move in opposite direction along the negative real λ-axis. The rate γ_1 increases with temperature and attains linear dependence on T in the regime $T \gg T^*$, $\gamma_1 = 2\pi K k_B T/\hbar$. In contrast, the rate γ_2 decreases with increasing temperature and takes for $T \gg T^*(K)$ the asymptotic form

$$\gamma_2 \equiv \gamma_r(T) = \frac{1}{K}\Delta_r\left(\frac{\hbar\Delta_r}{2\pi k_B T}\right)^{1-2K}. \tag{21.122}$$

Fairly above T^*, the weight of the pole λ_1 is very small, so that only the relaxation pole moving toward the origin, $\lambda = -\gamma_r$, is relevant. Then, the spectral function $\widetilde{S}_z(\omega)$ is described by a quasi-elastic Lorentzian peak, and $\langle\sigma_z\rangle_t$ decays exponentially,

$$\widetilde{S}_z(\omega) = \frac{2\gamma_r}{\omega^2 + \gamma_r^2}, \qquad \langle\sigma_z\rangle_t = e^{-\gamma_r t}. \tag{21.123}$$

The decay rate γ_r coincides with twice the forward "Golden Rule" tunneling rate given in Eq. (20.83), $\gamma_r = k^+ + k^-$ ($K \ll 1$). Since the tunneling relaxation rate varies with temperature as T^{2K-1}, the width of the basaltic peak shrinks with increasing temperature when $K < 1/2$. The anomalous T^{2K-1} law has been predicted first by Kondo [347]. This temperature regime is often referred to as the Kondo regime.

The analysis is easily extended to the *biased* system. We have with Eq. (21.91)

$$\widehat{K}_z^{(s)}(\lambda) = \tfrac{1}{2}[g(\lambda + i\epsilon) + g(\lambda - i\epsilon)],$$
$$\widehat{K}_z^{(a)}(\lambda) = i\tan(\pi K)\tfrac{1}{2}[g(\lambda + i\epsilon) - g(\lambda - i\epsilon)], \tag{21.124}$$

and similar forms hold in the NIBA for $\widehat{K}_x^{(s/a)}(\lambda)$.

---

[14]The T^2-law for $\Omega(T)$ gives a linear contribution to the specific heat (cf. Subsection 19.2.1).

In the Markov limit, we use the form (21.113) in Eq. (21.84) which amounts to substituting $h(0)$ for $h(\lambda)$ in Eq. (21.91). Eventually, we find [410]

$$
\begin{aligned}
\widehat{K}_z^{(\mathrm{s})}(\lambda) &= \Delta_T^2[\lambda + 2\pi K/\hbar\beta]/[\,(\lambda + 2\pi K/\hbar\beta)^2 + \epsilon^2\,]\,, \\
\widehat{K}_z^{(\mathrm{a})}(\lambda) &= \Delta_T^2\pi K\epsilon/[\,(\lambda + 2\pi K/\hbar\beta)^2 + \epsilon^2\,]\,, \\
\widehat{K}_x^{(\mathrm{s})}(\lambda) &= (\pi K/\Delta)\widehat{K}_z^{(\mathrm{s})}(\lambda)\,, \\
\widehat{K}_x^{(\mathrm{a})}(\lambda) &= \widehat{K}_z^{(\mathrm{a})}(\lambda)/(\pi K\Delta)\,.
\end{aligned}
\tag{21.125}
$$

With these expressions, the expectation values (21.79) and (21.80) are found as

$$
\langle \sigma_z(\lambda) \rangle = \frac{1}{\lambda}\left(1 + \frac{(\pi K\epsilon - \lambda - 2\pi K/\hbar\beta)\Delta_T^2}{N(\lambda)} \right)\,,
\tag{21.126}
$$

$$
\langle \sigma_x(\lambda) \rangle = \widehat{K}_x^{(\mathrm{s})}(\lambda)/\lambda + \widehat{K}_x^{(\mathrm{a})}(\lambda)\,\langle \sigma_z(\lambda) \rangle\,,
\tag{21.127}
$$

where

$$
N(\lambda) = \lambda[\,(\lambda + 2\pi K/\hbar\beta)^2 + \epsilon^2\,] + \Delta_T^2[\,\lambda + 2\pi K/\hbar\beta\,]\,.
\tag{21.128}
$$

The dynamics is determined by four singularities in the complex λ-plane: a simple pole at $\lambda = 0$, the residuum of which represents the equilibrium value,

$$
\langle \sigma_z \rangle_\infty = \frac{\hbar\epsilon}{2k_\mathrm{B}T}\,;\qquad \langle \sigma_x \rangle_\infty = \frac{\Delta_T}{\Delta}\frac{\hbar\Delta_T}{2k_\mathrm{B}T}\,,
\tag{21.129}
$$

and three simple poles which are located at the zeros of the cubic equation $N(\lambda) = 0$. At quick glance, it seems that there occur additional singularities in $\langle \sigma_x(\lambda) \rangle$, which are in the kernels $\widehat{K}_x^{(\mathrm{s/a})}(\lambda)$ and are located at the zeros of the equation $(\lambda + 2\pi K/\hbar\beta)^2 + \epsilon^2 = 0$. However, the residua come in pairs which cancel each other.

The characteristics of the cubic equation $N(\lambda) = 0$ is as follows [418]. For a bias in the range $|\epsilon| < \Delta_T/8$, we may distinguish three temperature regimes with qualitatively different behaviours. For $T < T_1$ and $T > T_2$, where $T_2 > T_1$, one of the zeros of $N(\lambda)$ is negative real, and the other two are complex conjugate. The former gives rise to a relaxation term and the latter to damped oscillations,

$$
\lambda_1 = -\gamma_\mathrm{r}\,,\qquad \lambda_{2/3} = -\gamma \pm i\Omega\,.
\tag{21.130}
$$

In the regime $T_1 < T < T_2$, all zeros are negative real. Thus in the intermediate regime, the system shows over-damped relaxation. The transition temperatures are

$$
T_{2/1} = \frac{\hbar\Delta_T}{4\sqrt{2}\pi K k_\mathrm{B}}\left\{ \frac{1 + 20\epsilon^2/\Delta_T^2 - 8\epsilon^4/\Delta_T^4 \pm (1 - 8\epsilon^2/\Delta_T^2)^{3/2}}{\epsilon^2/\Delta_T^2} \right\}^{1/2}\,.
\tag{21.131}
$$

At the critical bias strength $|\epsilon| = \epsilon_\mathrm{c}$, where $\epsilon_\mathrm{c} = \Delta_T/\sqrt{8}$, the transition temperatures T_1 and T_2 coincide, $T_1(\epsilon_\mathrm{c}) = T_2(\epsilon_\mathrm{c}) = (3/2)^{1/3}\hbar\Delta_T/(4\pi K k_\mathrm{B})$.

For $|\epsilon| \ll \epsilon_c$, we have $T_1 = [\hbar(\Delta_T - \epsilon^2/2\Delta_T)/\pi K + \mathcal{O}(\epsilon^4)]/k_B$, and the upper transition temperature is $T_2 = [\hbar\Delta_T^2/4\pi K|\epsilon| + \hbar|\epsilon|/2\pi K + \mathcal{O}(|\epsilon|^3)]/k_B$.

For $|\epsilon| > \epsilon_c$, the pole condition $N(\lambda) = 0$ yields a real root and two complex conjugate ones, as given in Eq. (21.130), for all temperatures.

The Vieta relations for γ_r, γ and Ω read $(T \gtrsim T_b)$

$$\gamma_r + 2\gamma = 2\vartheta\,, \tag{21.132}$$

$$\gamma_r(\gamma^2 + \Omega^2) = \Delta_T^2\vartheta\,, \tag{21.133}$$

$$\gamma^2 + 2\gamma\gamma_r + \Omega^2 = \Delta_T^2 + \epsilon^2 + \vartheta^2 \tag{21.134}$$

where we have introduced the "thermal" frequency being scaled by a factor $2\pi K$

$$\vartheta = 2\pi K k_B T/\hbar\,. \tag{21.135}$$

For temperatures well below $T_0 \equiv \hbar(\Delta_T^2 + \epsilon^2)^{1/2}/2\pi K k_B$, we find

$$
\begin{aligned}
\gamma_r &= \frac{\Delta_T^2}{\Delta_T^2 + \epsilon^2}\vartheta - \frac{\Delta_T^2\epsilon^4}{(\Delta_T^2 + \epsilon^2)^2}\vartheta^3 + \mathcal{O}(\vartheta^5)\,, \\
\gamma &= \frac{\Delta_T^2 + 2\epsilon^2}{2(\Delta_T^2 + \epsilon^2)}\vartheta + \frac{\Delta_T^2\epsilon^4}{2(\Delta_T^2 + \epsilon^2)^2}\vartheta^3 + \mathcal{O}(\vartheta^5)\,, \\
\Omega^2 &= \Delta_T^2 + \epsilon^2 - \frac{\Delta_T^2(\Delta_T^2 + 4\epsilon^2)}{4(\Delta_T^2 + \epsilon^2)^2}\vartheta^2 + \mathcal{O}(\vartheta^4)\,,
\end{aligned}
\tag{21.136}
$$

whereas for $T \gg T_0$

$$
\begin{aligned}
\gamma_r &= \frac{\Delta_T^2}{\vartheta} + \Delta_T^2(\Delta_T^2 - \epsilon^2)\frac{1}{\vartheta^3} + \mathcal{O}(1/\vartheta^5)\,, \\
\gamma &= \vartheta - \frac{\Delta_T^2}{2\vartheta} - \frac{\Delta_T^2(\Delta_T^2 - \epsilon^2)}{2}\frac{1}{\vartheta^3} + \mathcal{O}(1/\vartheta^5)\,, \\
\Omega^2 &= \epsilon^2 + \frac{\Delta_T^2(4\epsilon^2 - \Delta_T^2)}{4\epsilon^2}\frac{1}{\vartheta^2} + \mathcal{O}(1/\vartheta^4)\,.
\end{aligned}
\tag{21.137}
$$

For $T \ll T_0$, the relaxation rate γ_r and the dephasing rate γ grow linearly with T to leading order, whereas the oscillation frequency decreases with growing T. For $T \gg T_0$, the relaxation rate shows Kondo-like behaviour $\gamma_r \propto T^{2K-1}$ to leading order. The oscillation frequency approaches the bias frequency at high T, whereas the dephasing rate grows linearly with T.

Laplace inversion of the expression (21.126) for $\langle\sigma_z(\lambda)\rangle$ yields in the time regime

$$\langle\sigma_z\rangle_t = a_1 e^{-\gamma_r t} + [(1 - a_1 - \langle\sigma_z\rangle_\infty)\cos\Omega t + a_2\sin\Omega t]e^{-\gamma t} + \langle\sigma_z\rangle_\infty\,, \tag{21.138}$$

with the initial condition $\langle \sigma_z \rangle_{t=0} = 1$ and $\langle \dot{\sigma}_z \rangle_{t=0} = 0$. The amplitudes are given by

$$
\begin{aligned}
a_1 &= [\Omega^2 + \gamma^2 - \Delta_T^2 - (\Omega^2 + \gamma^2)\langle \sigma_z \rangle_\infty]/D \ , \\
a_2 &= [(\gamma_r - \gamma)a_1 + \gamma(1 - \langle \sigma_z \rangle_\infty)]/\Omega \ ,
\end{aligned}
\tag{21.139}
$$

where $D = \Omega^2 + (\gamma - \gamma_r)^2$. Similarly, we obtain from Eq. (21.127)

$$
\langle \sigma_x \rangle_t = b_1 e^{-\gamma_r t} + [-(b_1 + \langle \sigma_x \rangle_\infty) \cos \Omega t + b_2 \sin \Omega t] e^{-\gamma t} + \langle \sigma_x \rangle_\infty \ , \tag{21.140}
$$

with the initial conditions $\langle \sigma_x \rangle_{t=0} = 0$ and $\langle \dot{\sigma}_x \rangle_{t=0} = \pi K \Delta_T^2 / \Delta$. The amplitudes b_1 and b_2 are found to read

$$
\begin{aligned}
b_1 &= \left(\frac{\epsilon \Delta_T^2}{\Delta} - \frac{\epsilon^2(\gamma^2 + \Omega^2)}{\gamma^2 + \Omega^2 - \Delta_T^2} \langle \sigma_x \rangle_\infty \right) \frac{1}{D} \ , \\
b_2 &= \left(\pi K \frac{\Delta_T^2}{\Delta} + (\gamma_r - \gamma)b_1 - \gamma \langle \sigma_x \rangle_\infty \right) \frac{1}{\Omega} \ .
\end{aligned}
\tag{21.141}
$$

The expressions (21.136) – (21.141) describe the dynamics of the expectation values $\langle \sigma_z \rangle_t$ and $\langle \sigma_x \rangle_t$ for $K \ll 1$ in the Markov regime $T \gtrsim T_b$. One remark is appropriate. Since the damping strength K is multiplied by temperature in the pole condition (21.128), perturbative treatment of the damping breaks down even for small K when T is sufficiently large. This is immediately obvious also from the rich characteristics of the cubic equation (21.128).

Consider next the spin correlation function. In the NIBA, the term $R(t)$ in Eq. (21.57) vanishes, and we find from Eq. (21.58) with Eqs. (21.79) and (21.81)

$$
\widetilde{S}_z(\omega) = 2 \operatorname{Re} \frac{-i\omega - \langle \sigma_z \rangle_\infty \widehat{K}_z^{(a)}(-i\omega)}{-i\omega \left[-i\omega + \widehat{K}_z^{(s)}(-i\omega) \right]} \ , \tag{21.142}
$$

where the kernels are given in Eq. (21.124), and $\langle \sigma_z \rangle_\infty$ in Eq. (21.86). The same form has been found in Ref. [420] using a perturbative Liouville relaxation method. The dynamical susceptibility $\widetilde{\chi}_z(\omega)$ in the NIBA is given below in Eq. (22.18). We have already discussed after Eq. (21.84) that the NIBA gives inconsistent results for a biased system at low T and small frequencies (long times). For instance, the weight of the quasi-elastic line at the origin described by Eq. (21.142) is $(\epsilon/\Omega)^2 - \langle \sigma_z \rangle_\infty^2$. With the NIBA form (21.86) for $\langle \sigma_z \rangle_\infty$, the weight becomes negative at low T, which is unphysical.

For temperatures in the Markov regime (21.111), the NIBA is correct for all frequencies. Substituting the NIBA kernels (21.125) into Eq. (21.142) and decomposing the resulting expression into partial fractions, we obtain the spectral spin correlation function as a superposition of three Lorentzians [we use $\langle \sigma_z \rangle_\infty = (\epsilon/\Omega) \tanh(\beta \hbar \Omega/2)$],

$$
\widetilde{S}_z(\omega) = \sin^2 \theta \sum_{\zeta = \pm 1} \frac{\gamma + (\Omega - \zeta \omega) \tan \phi}{(\omega - \zeta \Omega)^2 + \gamma^2} + \frac{\cos^2 \theta}{\cosh^2(\beta \hbar \Omega/2)} \frac{2\gamma_r}{\omega^2 + \gamma_r^2} \ , \tag{21.143}
$$

where $\cos\theta = \epsilon/\Omega$, and $\tan\phi = \gamma/\Omega + (\gamma_r/\Omega)(\epsilon^2 - \langle\sigma_z\rangle_\infty^2\Omega^2)/(\Omega^2 - \epsilon^2)$. The inelastic peaks centered at $\omega = \pm\Omega$ are signatures of coherent tunneling at frequency Ω with dephasing time $1/\gamma$. The quasi-elastic peak at $\omega = 0$ represents incoherent relaxation to the thermal equilibrium state on the time scale $1/\gamma_r$. As the temperature approaches the crossover temperature T_0 from below, the peaks of $\widetilde{S}_z(\omega)$ merge into a single quasi-elastic line. Well above this temperature $\widetilde{S}_z(\omega)$ is given by [392]

$$\widetilde{S}_z(\omega) = \frac{2\gamma_r}{\omega^2 + \gamma_r^2}\,, \tag{21.144}$$

with the line width γ_r given in Eq. (21.137).

Finally, one remark concerning the scaling limit (18.54) is appropriate. We see from the above results that the expectation value $\langle\sigma_z\rangle_t$ is universal in the sense specified in Subsection 18.2.2. In contrast, the expectation values $\langle\sigma_x\rangle_t$ and $\langle\sigma_y\rangle_t$ are non-universal since they have an overall factor $\Delta_r/\Delta \propto (\Delta_r/\omega_c)^K$. Hence, the off-diagonal elements of the reduced density matrix vanish in the scaling limit (18.54).

21.3.3 The super-Ohmic case

We now study the behaviour of $\langle\sigma_z\rangle_t$ for a symmetric system with a super-Ohmic spectral density, Eq. (18.44) with $s > 2$. Our analysis is based on the expression

$$\langle\sigma_z(\lambda)\rangle = \frac{\lambda}{\lambda^2 + \widetilde{\Delta}^2 + \lambda\Sigma_z(\lambda)}\,, \tag{21.145}$$

where we have introduced the polaron dressed tunneling matrix element $\widetilde{\Delta}$ defined in Eq. (20.102). The self-energy in the NIBA, Eq. (21.84), takes the form

$$\Sigma_z^{(B)}(\lambda) = \widetilde{\Delta}^2 \int_0^\infty d\tau\, e^{-\lambda\tau} \left(e^{-[Q'(\tau) - X_1]} \cos[Q''(\tau)] - 1 \right). \tag{21.146}$$

The label (B) refers to "blip". For low-to-moderate temperatures, we expect to find the poles of Eq. (21.145) near $\pm i\widetilde{\Delta}$. Upon iteration, we find the pole condition

$$\lambda^2 + \lambda\Sigma_z^{(B)}(\pm i\widetilde{\Delta}) + \widetilde{\Delta}^2 = 0\,. \tag{21.147}$$

The imaginary part of the self-energy leads to a shift of the oscillation frequency.[15] However, since generally $\mathrm{Im}\,\Sigma_z^{(B)}(\pm i\widetilde{\Delta}) \ll \mathrm{Re}\,\Sigma_z^{(B)}(\pm i\widetilde{\Delta})$ for $s > 2$ [422], here we disregard the imaginary part of the self-energy. We define $\mathrm{Re}\,\Sigma_z^{(B)}(\pm i\widetilde{\Delta}) \equiv \Upsilon$.

Employing a manoeuvre similar to that which has led us to Eq. (20.31), we find

$$\Upsilon = \frac{\widetilde{\Delta}^2}{2} \cosh\left(\tfrac{1}{2}\hbar\beta\widetilde{\Delta}\right) \int_{-\infty}^\infty dt\, e^{i\widetilde{\Delta}t} [e^{X_2(t)} - 1]\,. \tag{21.148}$$

---

[15]The shift due to the one-phonon process is given in Eq. (21.178) [cf. also Eq. (19.42)].

where $X_2(t)$ is given in Eqs. (20.101) and (20.106). Expansion of the integrand in powers of $X_2(t)$ gives the multi-phonon series for the dephasing rate. For $s = 3$, we obtain a hypergeometric series [cf. Eq. (20.123)],

$$\Upsilon = \frac{\widetilde{\Delta}}{2\pi} \coth\left(\tfrac{1}{2}\hbar\beta\widetilde{\Delta}\right) \left(\frac{\hbar\widetilde{\Delta}}{k_{\mathrm{B}} T'_{\mathrm{ph}}}\right)^2 {}_2\mathrm{F}_2\left(1 + i\frac{\hbar\beta\widetilde{\Delta}}{2\pi}, \, 1 - i\frac{\hbar\beta\widetilde{\Delta}}{2\pi}; \, 2, \frac{3}{2}; \, \phi\right), \quad (21.149)$$

where ϕ and T'_{ph} are given in Eq. (20.118). It is convenient to write $\Upsilon = \Upsilon_+ + \Upsilon_-$, where Υ_+ captures the even and Υ_- the odd multi-phonon processes. At temperatures $T \lesssim \hbar\widetilde{\Delta}/k_{\mathrm{B}}$, only the one-phonon process is relevant because of the condition (20.105),

$$\Upsilon_- = (\widetilde{\Delta}/2\pi)(\hbar\widetilde{\Delta}/k_{\mathrm{B}} T'_{\mathrm{ph}})^2 \coth(\tfrac{1}{2}\hbar\beta\widetilde{\Delta}), \qquad \Upsilon_+ \ll \Upsilon_-. \quad (21.150)$$

At higher temperatures, $\hbar\widetilde{\Delta}/k_{\mathrm{B}} \lesssim T \ll T_{\mathrm{D}}$, we find from Eq. (21.149) the expansions in terms of even and odd multi-phonon processes,

$$\Upsilon_+ = \frac{\widetilde{\Delta}}{\pi} \frac{\hbar\widetilde{\Delta}}{k_{\mathrm{B}} T} \sum_{n=1}^{\infty} \frac{\Gamma(\tfrac{3}{2})}{2n\Gamma(2n + \tfrac{1}{2})} \left(\frac{T}{T'_{\mathrm{ph}}}\right)^{4n}, \quad (21.151)$$

$$\Upsilon_- = \frac{\widetilde{\Delta}}{\pi} \frac{\hbar\widetilde{\Delta}}{k_{\mathrm{B}} T'_{\mathrm{ph}}} \frac{T}{T'_{\mathrm{ph}}} \sum_{n=1}^{\infty} \frac{\Gamma(\tfrac{3}{2})}{(2n-1)\Gamma(2n - \tfrac{1}{2})} \left(\frac{T}{T'_{\mathrm{ph}}}\right)^{4n-4}. \quad (21.152)$$

With growing temperature, higher multi-phonon processes become increasingly relevant, and for T above T'_{ph}, but still well below the Debye temperature, we may use the asymptotic expression for the multi-phonon series. We find from Eq. (21.148) by steepest descent for general s [cf. Eq. (20.115) for $\kappa = 0$ and $\epsilon = 0$]

$$\Upsilon_\pm = \tfrac{1}{2}\Upsilon = \frac{\hbar\Delta^2 \, \mathrm{e}^{-2B_s}}{2k_{\mathrm{B}} T'_{\mathrm{ph}}} \sqrt{\frac{\pi}{d_s(0)}} \left(\frac{T'_{\mathrm{ph}}}{T}\right)^{(s+1)/2} \exp\left\{c_s(0)\left(\frac{T}{T'_{\mathrm{ph}}}\right)^{s-1}\right\}. \quad (21.153)$$

For $s = 3$, this form reduces to

$$\Upsilon_\pm = \tfrac{1}{2}\Upsilon = \frac{1}{4\sqrt{\pi}} \frac{\hbar\Delta^2 \, \mathrm{e}^{-2B_3}}{k_{\mathrm{B}} T'_{\mathrm{ph}}} \left(\frac{T'_{\mathrm{ph}}}{T}\right)^2 \exp\left(\frac{2T^2}{3T'^2_{\mathrm{ph}}}\right). \quad (21.154)$$

In the under-damped regime, the poles of Eq. (21.145) are at $\lambda_{1/2} = -\gamma \pm i\Omega$ with

$$\gamma = \tfrac{1}{2}\Upsilon, \qquad \Omega = \sqrt{\widetilde{\Delta}^2 - \tfrac{1}{4}\Upsilon^2}, \quad (21.155)$$

and the populations undergo damped oscillations

$$\langle \sigma_z \rangle_t = \cos(\Omega t - \phi)\, \mathrm{e}^{-\gamma t} / \cos(\phi), \qquad \phi = \arctan(\gamma/\Omega). \quad (21.156)$$

The temperature T^*, at which the transition from under-damped to overdamped behaviour occurs, is determined by the equality $\Upsilon = 2\widetilde{\Delta}$, as follows from Eq. (21.155). With Eqs. (20.102) and (21.153), we obtain for T^* the transcendental equation

$$T^* = T'_{\text{ph}} \left[\frac{1}{[c_s(0) + d_s]} \ln \left(\frac{\Delta\, e^{-B_s}}{A_s(T^*)} \right) \right]^{1/(s-1)}, \qquad (21.157)$$

where $A_s(T)$ is the preexponential factor in the expression (21.153).

For T above T^*, we have [observe the subtle difference to Eq. (21.120)],

$$\langle\sigma_z\rangle_t = \frac{1}{(\gamma_1 - \gamma_2)} \left(\gamma_1\, e^{-\gamma_1 t} - \gamma_2\, e^{-\gamma_2 t} \right), \qquad \gamma_{1/2} = \tfrac{1}{2}\Upsilon \pm \sqrt{\tfrac{1}{4}\Upsilon^2 - \widetilde{\Delta}^2}. \quad (21.158)$$

For T far above T^*, we have $\Upsilon \gg 2\widetilde{\Delta}$. Hence the residuum of the relaxation pole at $\lambda = -\gamma_2$ becomes very small and therefore the contribution of this pole may be disregarded. Thus we find the simple relaxation behaviour

$$\langle\sigma_z\rangle_t = e^{-\Upsilon t}. \qquad (21.159)$$

Until now, our study of the super-Ohmic case has been restricted to the NIBA. Recently, an approximate summation of the self-energy $\hbar\Sigma_z(\lambda)$ in all orders of $\widetilde{\Delta}^2$ has been put forward [422]. The approximation consists in disregarding all bath correlations except for the intra-sojourn correlations in the terms $n \geq 2$ in the series (21.76). This converts the series into a convolution. Switching to the Laplace transform, each blip-sojourn sequence in the terms $n \geq 2$ contributes a factor $-\Sigma_z^{(S)}(\lambda)/\lambda$, where the factor $1/\lambda$ originates from the blip interval, and $\hbar\Sigma_z^{(S)}(\lambda)$ is the self-energy contribution from a single sojourn. The resulting expression is a geometrical series which is summed to

$$\Sigma_z(\lambda) = \Sigma_z^{(B)}(\lambda) - \frac{\widetilde{\Delta}^2}{\lambda} \frac{\Sigma_z^{(S)}(\lambda)/\lambda}{1 + \Sigma_z^{(S)}(\lambda)/\lambda}. \qquad (21.160)$$

Because of the $\{\xi_j\}$-summation in Eq. (21.76), the odd multi-phonon contributions to $\Sigma_z^{(S)}(\lambda)$ cancel out, and we obtain

$$\Sigma_z^{(S)}(\lambda) = \widetilde{\Delta}^2 \int_0^\infty ds\, e^{-\lambda s} \Big(\cosh[Q'(s) - X_1] \cos[Q''(s)] - 1 \Big). \qquad (21.161)$$

Substituting the expression (21.160) into Eq. (21.145), we find

$$\langle\sigma_z(\lambda)\rangle_t = \frac{\lambda + \Sigma_z^{(S)}(\lambda)}{[\lambda + \Sigma_z^{(S)}(\lambda)][\lambda + \Sigma_z^{(B)}(\lambda)] + \widetilde{\Delta}^2}. \qquad (21.162)$$

The iteration of the pole condition in Eq. (21.162) leads to the substitutions $\Sigma_z^{(S)}(\lambda) \to \Upsilon_+$, $\Sigma_z^{(B)}(\lambda) \to \Upsilon_+ + \Upsilon_-$, yielding

$$\langle\sigma_z(\lambda)\rangle = \frac{\lambda + \Upsilon_+}{[\lambda + \Upsilon_+][\lambda + \Upsilon_+ + \Upsilon_-] + \widetilde{\Delta}^2}, \qquad (21.163)$$

where Υ_+ and Υ_- are given in Eq. (21.150) for $k_{\mathrm{B}}T \lesssim \hbar\widetilde{\Delta}$, and for higher temperatures in Eqs. (21.151) and (21.152).[16] This expression describes again underdamped oscillations for $T < T^*$,

$$\langle \sigma_z \rangle_t = \cos(\Omega t - \phi)\, e^{-\gamma t}/\cos(\phi)\,, \qquad \phi = \arctan(\Upsilon_-/2\Omega)\,, \qquad (21.164)$$

with

$$\gamma = \Upsilon_+ + \tfrac{1}{2}\Upsilon_-\,, \qquad \Omega = \widetilde{\Delta}\sqrt{1 - (\Upsilon_-/2\widetilde{\Delta})^2}\,. \qquad (21.165)$$

Observe the deviations from the NIBA result (21.155), (21.156), in particular the additional factor 2 in the *even* multi-phonon contributions to the dephasing rate γ.

For $T_{\mathrm{D}} \gg T \gg T'_{\mathrm{ph}}$, the asymptotic expression both of the series (21.151) and (21.152) is given for general s in Eq. (21.153), and for $s = 3$ in Eq. (21.154). The oscillation ceases to exist when the condition $\Upsilon_- = 2\widetilde{\Delta}$ is reached. This results in a transcendental equation for the transition temperature T^* of the form (21.157) with an extra factor of 2 in the argument of the logarithm. For $T > T^*$, we have

$$\langle \sigma_z \rangle_t = \frac{\gamma_1 - \tfrac{1}{2}\Upsilon}{\gamma_1 - \gamma_2} e^{-\gamma_1 t} + \frac{\tfrac{1}{2}\Upsilon - \gamma_2}{\gamma_1 - \gamma_2} e^{-\gamma_2 t}\,, \qquad \gamma_{1/2} = \tfrac{3}{4}\Upsilon \pm \sqrt{(\tfrac{1}{4}\Upsilon)^2 - \widetilde{\Delta}^2}\,. \quad (21.166)$$

For T well above T^*, the rate γ_1 is near Υ, whereas γ_2 approaches $\tfrac{1}{2}\Upsilon$. Hence the residuum of the second relaxation contribution becomes negligibly small, and we obtain the NIBA result (21.159). As the temperature is increased further, the crossover to the classical Marcus regime takes place. This has been discussed already in some detail in Subsection 20.2.8.

In Eq. (21.160) a special interblip correlation term is summed in all orders in $\widetilde{\Delta}^2$. The treatment of interblip correlations is not systematic. Nevertheless, the characteristic features of the resulting dynamics are found to be in qualitative agreement with the NIBA results. There is only a marginal difference in the crossover temperature T^*. The major difference is an additional factor of 2 in the even multi-phonon contribution Υ_+ to the rate around the transition temperature.

21.4 Weak-coupling theory beyond the NIBA for a biased system

For weak damping and temperature in the range of $T_{\mathrm{b}} = \hbar(\Delta_{\mathrm{eff}}^2 + \epsilon^2)^{1/2}/k_{\mathrm{B}}$ or smaller, the Markov assumption (21.112) and the corresponding form (21.113) for $Q(\tau)$ is not valid, and the justification of the noninteracting-blip approximation is questionable.

---

[16]For $k_{\mathrm{B}}T \lesssim \hbar\widetilde{\Delta}$, a possible generalization is to insert in the blip interval in Eq. (21.146) and in the sojourn interval in Eq. (21.161) the corresponding propagators of the free system [cf. the discussion in Subsections 19.1.6, 19.1.7, and 21.4.1]. The iteration of the pole condition results in the substitutions $\Sigma_z^{(\mathrm{S})}(\lambda) \to \tfrac{1}{2}(1+p)\Upsilon_+$, $\Sigma_z^{(\mathrm{B})}(\lambda) \to \tfrac{1}{2}(1+p)\Upsilon_+ + \Upsilon_-$, where $p = \hbar\beta\widetilde{\Delta}/\sinh(\hbar\beta\widetilde{\Delta})$ [422]. Since the quantity p is different from 1 only for $T \lesssim \hbar\widetilde{\Delta}/k_{\mathrm{B}}$, we then have $T \ll T'_{\mathrm{ph}}$ [cf. Eq. (20.105)], and hence $\Upsilon_+ \ll \Upsilon_-$. This shows that the expression (21.163) is also valid in the regime $T \lesssim \hbar\widetilde{\Delta}/k_{\mathrm{B}}$.

For a symmetric system, the NIBA is found to be inconsistent for the coherence $\langle \sigma_x \rangle_t$, as we have remarked already in item (a) in Section 21.3, and after Eq. (21.87). For nonzero bias, the NIBA breaks down for the same reason at temperatures below T_b for the population $\langle \sigma_z \rangle_t$ and the coherence $\langle \sigma_x \rangle_t$, as well as for the correlation functions. In all these cases, the interblip correlations are already relevant for the one-phonon process. This is confirmed by the subsequent study. The bias energy is important for tunneling systems in crystals with higher defect concentration because of strain fields and in amorphous materials, e.g., in dielectric and metallic glasses.

We now deal with the interblip correlations systematically to first order in the coupling strength. To be general, we assume for the spectral density $G_{1f}(\omega)$ the power-law form (18.31). We consider the regime $T \lesssim T_b = \hbar (\Delta_{\text{eff}}^2 + \epsilon^2)^{1/2}/k_B$ for weak damping, where Δ_{eff} is the effective tunneling coupling given in Eq. (18.32) for a super-Ohmic, and in Eq. (18.35) for an Ohmic bath, and $\Delta_{\text{eff}} = \Delta$ in the sub-Ohmic case.

21.4.1 The one-boson self-energy

Interestingly enough, the series (21.76) and (21.78) for the kernels $K_z^{(s/a)}(t, t')$ and $K_x^{(s/a)}(t, t')$ can be summed in analytic form for weak damping with a strategy similar to that presented in Subsection 19.1.6[421, 413, 410]. We proceed by considering the blip correlations to linear order in $\Lambda_{j,k}$ in the series expression (21.76). Switching to the Laplace transform the self energy $\Sigma_z^{(s)}(\lambda)$ is found to read

$$\Sigma_z^{(s)}(\lambda) = \Delta_{\text{eff}}^2 \int_0^\infty d\tau \, e^{-\lambda \tau} \cos(\epsilon \tau) \Big(q - Q'(\tau) \Big) \tag{21.167}$$

$$- \lambda \sum_{n=2}^\infty (-\Delta_{\text{eff}}^2)^n \Big(\prod_{j=1}^n \int_0^\infty d\tau_j \, ds_j \, e^{-\lambda(s_j + \tau_j)} \Big) \Lambda_{n1} \sin(\epsilon \tau_1) \sin(\epsilon \tau_n) \prod_{k=2}^{n-1} \cos(\epsilon \tau_k) \, .$$

Here we have included the full (nonperturbative) adiabatic dressing of the tunneling coupling. The first term is the self-energy in the NIBA, and q is a counterterm so that the oscillation frequency at $T = 0$ will emerge as $\Delta_b = \sqrt{\Delta_{\text{eff}}^2 + \epsilon^2}$. The residual term represents the contribution of the one-boson interblip correlations. In diagrammatic language, the series is the sum of all *irreducible* multi-blip diagrams in which the first blip interacts with the last blip via one-boson exchange. The insertion of noninteracting blips in between accounts for all intermediate uncorrelated tunneling events. The series of the insertions can be summed and yields [cf. Eq. (21.11)]

$$P_0(s) = \langle \sigma_z \rangle_s^{(0)} = \frac{\epsilon^2}{\Delta_b^2} + \frac{\Delta_{\text{eff}}^2}{\Delta_b^2} \cos(\Delta_b s) \, , \qquad \Delta_b = (\Delta_{\text{eff}}^2 + \epsilon^2)^{1/2} \, , \tag{21.168}$$

where s is the interval between the first and the last blip. Thus we get

$$\Sigma_z^{(s)}(\lambda) = \Delta_{\text{eff}}^2 \int_0^\infty d\tau \, e^{-\lambda \tau} \cos(\epsilon \tau) \Big(q - Q'(\tau) \Big)$$

$$- \Delta_{\mathrm{eff}}^4 \int_0^\infty d\tau_2 \, ds \, d\tau_1 \, \mathrm{e}^{-\lambda(\tau_2 + s + \tau_1)} \sin(\epsilon\tau_2) \, P_0(s) \sin(\epsilon\tau_1) \qquad (21.169)$$

$$\times \left[Q'(\tau_1 + \tau_2 + s) + Q'(s) - Q'(\tau_1 + s) - Q'(\tau_2 + s) \right].$$

Upon using the spectral representation (18.42) and interchanging the frequency integral and the time integrals [421], we get

$$\Sigma_z^{(\mathrm{s})}(\lambda) \;=\; -\frac{2\Delta_{\mathrm{eff}}^2}{\Delta_{\mathrm{b}}^2} \frac{\Delta_{\mathrm{eff}}^2 \lambda^3 u(\lambda) + \epsilon^2(\lambda^2 + \Delta_{\mathrm{b}}^2) \, \mathrm{Re}\left\{(\lambda - i\Delta_{\mathrm{b}})u(\lambda + i\Delta_{\mathrm{b}})\right\}}{(\lambda^2 + \epsilon^2)^2}. \qquad (21.170)$$

where the function $u(z)$ is given by

$$u(z) \;=\; \frac{1}{2} \int_0^\infty d\omega \frac{G_{\mathrm{lf}}(\omega)}{\omega^2 + z^2} \left(\coth(\tfrac{1}{2}\hbar\beta\omega) - 1 \right). \qquad (21.171)$$

It is related to the function $v(y)$ occurring in the imaginary-time approach [cf. Eq. (19.43)] by analytic continuation, $u(z) = v(y = -iz)$.

Similarly, the selfenergy $\Sigma_z^{(\mathrm{a})}(\lambda)$ represents the correlations between the first sojourn and the last blip with insertion of all possible tunneling events in between. The corresponding expression is

$$\Sigma_z^{(\mathrm{a})}(\lambda) \;=\; \Delta_{\mathrm{eff}}^2 \int_0^\infty d\tau \, \mathrm{e}^{-\lambda\tau} \sin(\epsilon\tau) \, Q''(\tau)$$

$$- \Delta_{\mathrm{eff}}^4 \int_0^\infty d\tau_2 \, ds \, d\tau_1 \, \mathrm{e}^{-\lambda(\tau_2 + s + \tau_1)} \sin(\epsilon\tau_2) \, P_0(s) \cos(\epsilon\tau_1) \qquad (21.172)$$

$$\times \left[Q''(\tau_1 + \tau_2 + s) - Q''(\tau_1 + s) \right].$$

Use of the spectral representation (18.42) for $Q''(\tau)$ gives

$$\Sigma_z^{(\mathrm{a})}(\lambda) \;=\; i\epsilon \frac{\Delta_{\mathrm{eff}}^2(\lambda^2 + \Delta_{\mathrm{b}}^2)}{\Delta_{\mathrm{b}}(\lambda^2 + \epsilon^2)} [w(\lambda + i\Delta_{\mathrm{b}}) - w(\lambda - i\Delta_{\mathrm{b}})], \qquad (21.173)$$

where

$$w(z) \;=\; \frac{1}{2} \int_0^\infty d\omega \frac{G_{\mathrm{lf}}(\omega)}{\omega(\omega^2 + z^2)}. \qquad (21.174)$$

Corresponding expressions are found from Eq. (21.78) for the selfenergies $\Sigma_x^{(\mathrm{s/a})}(\lambda)$,

$$\Sigma_x^{(\mathrm{s})}(\lambda) \;=\; \frac{\Delta_{\mathrm{eff}}^2}{\Delta} \int_0^\infty d\tau \, \mathrm{e}^{-\lambda\tau} \cos(\epsilon\tau) \, Q''(\tau)$$

$$- \frac{\Delta_{\mathrm{eff}}^4}{\Delta} \int_0^\infty d\tau_2 \, ds \, d\tau_1 \, \mathrm{e}^{-\lambda(\tau_2 + s + \tau_1)} \cos(\epsilon\tau_2) \, P_0(s) \cos(\epsilon\tau_1) \qquad (21.175)$$

$$\times \left[Q''(\tau_1 + \tau_2 + s) - Q''(\tau_1 + s) \right],$$

$$\Sigma_x^{(a)}(\lambda) = \frac{\Delta_{\text{eff}}^2}{\Delta} \int_0^\infty d\tau \, e^{-\lambda\tau} \sin(\epsilon\tau)\left(q - Q'(\tau)\right)$$

$$+ \frac{\Delta_{\text{eff}}^4}{\Delta} \int_0^\infty d\tau_2 \, ds \, d\tau_1 \, e^{-\lambda(\tau_2+s+\tau_1)} \cos(\epsilon\tau_2) \, P_0(s) \sin(\epsilon\tau_1) \quad (21.176)$$

$$\times \left[Q'(\tau_1+\tau_2+s) + Q'(s) - Q'(\tau_1+s) - Q'(\tau_2+s) \right].$$

We conclude with the remark that these expressions for the self-energies $\Sigma_x^{(s/a)}(\lambda)$ can be cast in forms whch are similar to the expressions (21.173) and (21.170).

21.4.2 Populations and coherences (super-Ohmic and Ohmic)

To determine the dynamics, we substitute the expressions (21.170) and (21.173) into Eq. (21.79) and study the singularities of $\langle\sigma_z(\lambda)\rangle$. First, we see that the self-energy term shifts the two complex-conjugate poles away from the imaginary axis into the left half-plane to $\lambda = -\gamma \pm i\Omega$. These poles render damped oscillations in the time domain, and inelastic spectral lines in frequency regime. Furthermore, there occur additional poles at $\lambda = -\gamma_r$ and at $\lambda = 0$. The former is the relaxation pole. It contributes in the time regime a term describing exponential relaxation with the rate γ_r, and in frequency space a quasi-elastic peak of width γ_r centered at $\omega = 0$. The residuum of the latter pole gives the equilibrium value $\langle\sigma_z\rangle_\infty$ and turns out as

$$\langle\sigma_z\rangle_\infty = \frac{\Sigma_z^{(a)}(0)}{\Sigma_z^{(s)}(0)} = \frac{\epsilon}{\Delta_b} \tanh\left(\frac{\hbar\Delta_b}{2k_BT}\right) \approx \frac{\epsilon}{\Omega} \tanh\left(\frac{\hbar\Omega}{2k_BT}\right), \quad (21.177)$$

which is the weak-damping form anticipated already in Eq. (21.88).

Computation of the shifts of the other three poles induced by the one-boson contribution to the self-energy $\Sigma_z^{(s)}(\lambda)$ yields

$$\Omega^2 = \Delta_{\text{eff}}^2[1 - 2\,\text{Re}\,u(i\Delta_b)] + \epsilon^2,$$

$$\gamma_r = \frac{\pi}{2}\frac{\Delta_{\text{eff}}^2}{\Delta_b^2} S(\Delta_b) = \frac{\pi}{2} \sin^2(\varphi)\, S(\Delta_b), \quad (21.178)$$

$$\gamma = \frac{1}{2}\gamma_r + \frac{\pi}{2}\frac{\epsilon^2}{\Delta_b^2} S(0) = \frac{1}{2}\gamma_r + \frac{\pi}{2}\cos^2(\varphi)\, S(0),$$

valid for general spectral density $G_{\text{lf}}(\omega)$. The noise power $S(\omega)$ originates from emission and absorption of a single boson in thermal equilibrium (see Subsection 3.1.4),

$$S(\omega) = G_{\text{lf}}(\omega) \coth(\hbar\omega/2k_BT). \quad (21.179)$$

The relaxation rate γ_r represents the inverse time scale for relaxation of the TSS to the ground state. It is proportional to the spectral power $S(\omega)$ at the level splitting frequency. The decoherence rate γ_r is the inverse time scale for loss of phase coherence between the two states. It is a combination of the relaxation contribution $\frac{1}{2}\gamma_{\rm r}$ and the so-called "pure-dephasing" contribution due to longitudinal noise at zero frequency. This expression is only meaningful if the noise power is regular around $\omega = 0$. The noise power $S(0)$ is zero in the super-Ohmic case, and $\propto T$ in the Ohmic case. In addition, the level splitting $\hbar\Omega$ is renormalized due to transverse noise. This effect is the analogue of the Lamb shift. For Ohmic dissipation, we obtain the expressions

$$\Omega^2 = \Delta_{\rm b}^2 + 2K[\,{\rm Re}\,\psi(i\hbar\Delta_{\rm b}/2\pi k_{\rm B}T) - \ln(\hbar\Delta_{\rm b}/2\pi k_{\rm B}T)\,]\,,$$
$$\gamma_{\rm r} = \pi K \coth(\hbar\Delta_{\rm b}/2k_{\rm B}T)\Delta_{\rm eff}^2/\Delta_{\rm b}\,, \qquad (21.180)$$
$$\gamma = \gamma_{\rm r}/2 + 2\pi K(\epsilon^2/\Delta_{\rm b}^2)k_{\rm B}T/\hbar\,.$$

In the temperature range $k_{\rm B}T \gtrsim \hbar\Delta_{\rm b}$, these expressions for γ and $\gamma_{\rm r}$ smoothly map on the "Markov" results given in Eq. (21.136). At higher temperature, $T \gtrsim \hbar(\Delta_T^2 + \epsilon^2)^{1/2}/2\pi K k_{\rm B}$, the "one-phonon" approximation breaks down, and the rates are determined by the cubic equation (21.128), which yields in the high temperature limit the expressions given in Eq. (21.137).

For $s < 1$ in the spectral density (18.31), we have $\lim_{\omega \to 0} S(\omega) \to \infty$, which indicates that the dephasing rate γ becomes infinitely large by the one-boson process. It is most unlikely that this feature is spoilt by multi-phonon processes. This leads us to the conclusion that the dynamics is fully incoherent for sub-Ohmic damping even if the parameter δ_s is infinitesimal. It is difficult to assess the domain of parameters in which the one-phonon process is the leading one. Generally, we expect that the parameter domain is substantially smaller than in the Ohmic or super-Ohmic case.[17]

The transform (21.79) with (21.173) is easily inverted to obtain $\langle\sigma_z\rangle_t$. We find[18]

$$\langle\sigma_z\rangle_t = [\epsilon^2/\Omega^2 - \langle\sigma_z\rangle_\infty]\,{\rm e}^{-\gamma_{\rm r}t} + \langle\sigma_z\rangle_\infty \qquad (21.181)$$
$$+ \{(\Delta_{\rm eff}^2/\Omega^2)\cos\Omega t + [\,(\gamma_{\rm r}\epsilon^2 + \gamma\Delta_{\rm eff}^2)/\Omega^3 - \gamma_{\rm r}\langle\sigma_z\rangle_\infty/\Omega\,]\sin\Omega t\}\,{\rm e}^{-\gamma t}\,.$$

The spectral spin correlation function $\widetilde{S}_z(\omega)$ is found in the form (21.143) in which $\langle\sigma_z\rangle_\infty$ is given by Eq. (21.177). With the expression (21.177), the weight of the quasielastic peak is $(\epsilon^2/\Omega^2)[\,1 - \tanh^2(\hbar\Omega/2k_{\rm B}T)\,]$, which is positive down to zero temperature. Thus, the NIBA flaw that the weight of the quasielastic peak becomes negative at low T is dissolved by the interblip correlations.

Following similar lines, it is also straightforward to calculate $\langle\sigma_x\rangle_t$. The resulting expression is

$$\langle\sigma_x\rangle_t = \left[\frac{\epsilon\Delta_{\rm eff}^2}{\Delta\Omega^2} - \langle\sigma_x\rangle_\infty\right]{\rm e}^{-\gamma_{\rm r}t} + \left[-\frac{\epsilon\Delta_{\rm eff}^2}{\Delta\Omega^2}\cos\Omega t + b_2\sin\Omega t\right]{\rm e}^{-\gamma t} + \langle\sigma_x\rangle_\infty\,, \qquad (21.182)$$

where

$$b_2 = (\Delta_{\rm eff}^2/\Delta\Omega)[\,\pi K + \epsilon\,(\gamma_{\rm r} - \gamma)/\Omega^2\,] - \gamma_{\rm r}\langle\sigma_x\rangle_\infty/\Omega\,. \qquad (21.183)$$

---

[17]The sub-Ohmic incoherent tunneling rate is discussed in Subsection 20.2.7.
[18]We deliberately substitute Ω for $\Delta_{\rm b}$ in the amplitudes.

The equilibrium value is

$$\langle \sigma_x \rangle_\infty = \frac{\Delta_{\text{eff}}^2}{\Delta \Delta_b} \tanh \left(\frac{\hbar \Delta_b}{2k_B T} \right) \approx \frac{\Delta_{\text{eff}}^2}{\Delta \Omega} \tanh \left(\frac{\hbar \Omega}{2k_B T} \right). \tag{21.184}$$

The above expressions for $\langle \sigma_z \rangle_t$ and $\langle \sigma_x \rangle_t$ match in the Ohmic case with the expressions (21.136) – (21.141) at temperatures $T \gtrsim T_b$. Thus, we have found in the Ohmic case for $K \ll 1$ analytic expressions for the dynamics in the entire temperature range of interest. It is interesting to note that dynamical approach presented here gives the correct equilibrium vales for $\langle \sigma_z \rangle_\infty$ and $\langle \sigma_x \rangle_\infty$ which we have derived above in Eqs. (21.88) and (21.89) by a simple thermostatic consideration.

In the weak-damping limit, a useful quantity is the difference of the population of the ground state and the excited state, which is

$$N(t) = (\epsilon/\Omega)\langle \sigma_z \rangle_t + (\Delta/\Omega)\langle \sigma_x \rangle_t. \tag{21.185}$$

With use of the above results for the expectation values, we get

$$N(t) = \tanh(\hbar \Omega / 2k_B T)[1 - e^{-\gamma_r t}] + (\epsilon/\Omega)[e^{-\gamma_r t} + (\gamma_r/\Omega) \sin(\Omega t) e^{-\gamma t}], \tag{21.186}$$

which describes relaxation from the initial value ϵ/Ω to the equilibrium value $\tanh(\hbar \Omega / 2k_B T)$. There is also a minor coherent contribution which is of order K.

The analysis can be generalized to the case of time-dependent external field coupled to σ_z and to σ_x by employing the substitutions given in Subsection 22.1.1. The dynamics of $\langle \sigma_z \rangle_t$ under influence of a monochromatic high-frequency field coupled to σ_z has been studied in Ref. [413]. Alternative approaches based on second-order perturbation in the TSS-bath coupling have been frequently employed in this parameter regime. In these treatments, the adiabatic renormalization of the bare tunneling matrix element is usually disregarded. For a discussion of $\langle \sigma_z \rangle_t$ and $N(t)$ under monochromatic low-frequency driving, we refer to Refs. [423, 424].

21.5 The interacting-blip chain approximation

A systematic improvement on the NIBA consists in taking into account, besides the intra-blip correlations, the nearest-neighbour correlations between blips and the full phase correlations between neighbouring sojourn-blip pairs. Diagrammatically, we then have a chain of blips in which the nearest-neighbour interblip correlations are fully included. A pictorial description illustrating the contribution of three blips to $\langle \sigma_z \rangle_t$ is sketched in Fig. 21.3. Because of the chain-like structure, this approximation has been dubbed "interacting-blip chain approximation" (IBCA) [425].

In the IBCA, the bath influence function $\mathcal{F}^{(n)}$ in Eq. (21.37) is approximated as

$$\mathcal{F}_{\text{IBCA}}^{(n)} = \exp \left\{ -\sum_{j=1}^{n} \left(Q_{2j,2j-1}' - i\xi_j \eta_{j-1} X_{j,j-1} \right) - \sum_{j=2}^{n} \xi_j \xi_{j-1} \Lambda_{j,j-1} \right\}. \tag{21.187}$$

The nearest-neighbour blip-blip and sojourn-blip interactions read

$$
\begin{aligned}
\Lambda_{j,j-1} &= Q'_{2j,2j-3} + Q'_{2j-1,2j-2} - Q'_{2j,2j-2} - Q'_{2j-1,2j-3}\,, \\
X_{j,j-1} &= Q''_{2j,2j-1} + Q''_{2j-1,2j-2} - Q''_{2j,2j-2}\,.
\end{aligned}
\tag{21.188}
$$

Next, our task is to establish the dynamical equations for the reduced density matrix in the IBCA. To this end, we introduce the conditional probabilities per unit time $R_+(t;\tau)$ and $R_-(t;\tau)$ for the particle to be released from the sojourn state $\eta_0 = +1$ at time zero and to hop at time $t-\tau$ into the final blip state $\xi_f = +1$ and $\xi_f = -1$, respectively, and afterwards to remain there until time t. In addition, we define a kernel matrix $Y_{\xi,\xi'}(\tau_2, s_1, \tau_1)$ representing the possible elementary blip-sojourn-blip processes. The kernel describes a two-step transition from the blip state ξ' which has been visited for a period τ_1 via an intermediate sojourn state to the blip state ξ which is visited for a period τ_2. The time spent in the one or the other intermediate sojourn state is s_1. The kernel contains the intrablip interaction of the last blip and the interactions of this blip with the preceding blip and with the intermediate sojourn. Summation over the two intermediate sojourn states yields

$$
Y_{\xi,\xi'}(\tau_2, s_1, \tau_1) = -\tfrac{1}{2}\,\xi\xi'\Delta^2 \exp[\,-Q'(\tau_2) - \xi\xi'\Lambda_{2,1}\,]\cos X_{21}\,.
\tag{21.189}
$$

A further element of the dynamical equations is the bias phase factor which takes into account the influences of the deterministic forces. For a stay in the blip state $\xi = \pm1$, say lasting from time $t'-\tau$ until time t', one has to write a factor $B_\pm(\tau) = \exp(\pm i\epsilon\tau)$, and in the case of a time dependent bias [cf. Chapter 22]

$$
B_\pm(t',\tau) = \exp\left(\pm i \int_{t'-\tau}^{t'} dt''\, \epsilon(t'') \right).
\tag{21.190}
$$

In numerical computation, a major difficulty arises from the fact that the bias factor (21.190) depends on absolute times, the initial time $t'-\tau$ and the final time t' of the blip. When the preparation of the initial state belongs to class B [cf. Subsection 21.1.1], the initial sojourn has length infinity. Hence in this case, the first sojourn-blip

Figure 21.3: Three-blip contribution to $\langle\sigma_z\rangle_t$ in the IBCA. The solid line represents a sojourn and the curly line a blip interval. The blip-blip correlations $\Lambda_{j,j-1}$ are symbolically sketched by a dashed curve and the sojourn-blip correlations by dotted curve. The intrablip interactions and the individual interactions in $\Lambda_{j,j-1}$ are not displayed.

pair plays a special role. The corresponding amplitude depends on the blip length τ alone and is represented by the factor

$$A_\pm(\tau) = \mp i\,(\Delta/2)\exp[-Q'(\tau) \pm iQ''(\tau)] \,. \tag{21.191}$$

The iteration of the elementary sojourn-blip sequence (21.189) generates all paths the system can go. The sum over all possibilities of stringing together sojourn-blip sequences can be combined in a set of integral equations. Piecing together the elements (21.189) – (21.191), we find

$$\begin{aligned} R_\pm(t';\tau) &= B_\pm(t',\tau)\Big[A_\pm(\tau) + \sum_{k=\pm}\int_0^{t'-\tau} ds \int_0^{t'-\tau-s} d\tau' \\ &\quad \times Y_{\pm,k}(\tau,s,\tau')\, R_k(t'-\tau-s;\tau')\Big]\,. \end{aligned} \tag{21.192}$$

The coupled integral equations (21.192) are the dynamical equations in the IBCA. As a consequence of the bath correlations, they are not in the form of convolutions, even not when the bias is static.

Integration of the conditional probabilities $R_\pm(t;\tau)$ over the length τ of the final blip state gives the off-diagonal elements of the RDM at time t. We have

$$\langle \sigma_x \rangle_t = \int_0^t d\tau\,[R_+(t;\tau) + R_-(t;\tau)]\,, \tag{21.193}$$

$$\langle \sigma_y \rangle_t = i\int_0^t d\tau\,[R_+(t;\tau) - R_-(t;\tau)]\,. \tag{21.194}$$

Finally, the population $\langle \sigma_z \rangle_t$ results from integration of Eq. (21.194),

$$\langle \sigma_z \rangle_t = 1 - \Delta \int_0^t dt'\,\langle \sigma_y \rangle_{t'}\,, \tag{21.195}$$

as follows from Eq. (21.10). The integration over t' takes into account that the final step back to the diagonal at time t' is in the interval $0 \le t' \le t$. In conclusion, the dynamical problem of finding the reduced density matrix is solved up to quadratures once the conditional quantities $R_\pm(t';\tau)$ are known in the interval $0 \le \tau \le t' \le t$.

An efficient numerical algorithm consists in solving (21.192) by iteration on an equidistant grid in time [425].

The method presented here differs from the iterative solution of the GME (21.73) (cf. Ref. [413]). To include nearest-neighbour blip correlations in the GME, the respective kernel has to be considered at least in order Δ^4. However, iteration of the GME does not lead to linked blip clusters of higher order than those included in the kernel. Furthermore, the GME (21.73) is in the form of a convolution in the absence of time-dependent deterministic forces, while the dynamical equation (21.192) is generally in nonconvolutive form.

As the nearest-neighbour blip correlations constitute the most relevant corrections to the NIBA, the IBCA applies for longer propagation times than the NIBA. The IBCA is most suitable for moderate-to-strong damping.

Systematic improvement of the IBCA is possible along two lines of development. For weak-to-moderate damping, we may insert, in analogy with the proceeding in Subsection 21.4, all possible tunneling events of the undamped system into the intervals of the chain-links. For higher damping, the first step to do would be to include all next-to-nearest-neighbour interblip correlations. The relevant kernels $Y_{\xi,\xi',\xi''}$ would then depend on three blip labels and on five time intervals, namely the lengths of three blips and of two sojourns in between. Upon book-keeping more and more time intervals in the kernels, the range in which the bath correlations are taken into account exactly is systematically enlarged. The corresponding generalization of the numerical algorithm is clear, but the numerical costs increase drastically with each step.

21.6 Ohmic dissipation with K at and near $\frac{1}{2}$: exact results

For the special case $K = \frac{1}{2}$, the Ohmic TSS can be mapped on the Toulouse model, as we have discussed already in Subsection 19.2.2. The equations of motion for the imaginary-time Green's function of the d-level are solved without difficulties and yield the expression (19.77). The relevant equilibrium correlation functions can be expressed in terms of this function, as we shall see below in Subsection 21.6.6). Alternatively, we can directly sum the exact series expressions given in Subsections 21.2.3 and 21.2.4 [207]. As an advantage over the fermionic approach, the latter method facilitates the calculation of nonequilibrium conditional probabilities.

21.6.1 Grand-canonical sums of collapsed blips and sojourns

The reason that the path summation is possible in analytic form in the particular case $K = \frac{1}{2}$ can most easily be understood using the concept of *collapsed* blips and *collapsed* sojourns. In the corresponding charge picture, neighbouring charges of opposite sign form *collapsed* dipoles which have vanishing dipole moment and hence are not interacting with other charges.

The computation is done by putting $K = \frac{1}{2} - \kappa$ and eventually taking the limit $\kappa \to 0$. In the scaling limit (18.54) with (18.51), the bath influence factors $F_m^{(\pm)}$ and $F_{m,n}$ are given in Eqs. (21.56) and (21.61). The central point now is that every factor $\cos(\pi K) = \sin(\pi \kappa)$ occurring in the series expressions given in Subsection 21.2.3 is zero as $\kappa \to 0$. To obtain a nonvanishing contribution, every $\cos(\pi K)$ factor must be compensated by a simple pole at $K = \frac{1}{2}$. The singularity required is provided by the short-distance singular behaviour of the intradipole interaction in the breathing mode integral of a dipole, $\lim_{\tau \to 0} e^{-Q'(\tau)} \approx (\omega_c \tau)^{-1+2\kappa}$. The dipole length may represent either a sojourn or a blip interval. The integral over the dipole length is singular when the bias factor is even in the bias, and it is regular when the bias factor is odd in the bias. Thus, the nonvanishing contribution has the form

$$I(K = \tfrac{1}{2}) \;\equiv\; \lim_{\kappa \to 0} \Delta^2 \cos(\tfrac{1}{2}\pi - \kappa\pi) \int_0^{\infty} d\tau \, e^{-Q'(\tau)} \cos(\epsilon\tau) f(\tau)$$

$$= \lim_{\kappa \to 0} \Delta^2 \sin(\pi\kappa) \int_0^{\infty} d\tau \, \frac{1}{(\omega_c\tau)^{1-2\kappa}} \cos(\epsilon\tau)\, f(\tau) \qquad (21.196)$$

$$= \lim_{\kappa \to 0} \frac{\Delta^2}{\omega_c} f(0) \sin(\pi\kappa)\Gamma(2\kappa) \;=\; \frac{\pi}{2} \frac{\Delta^2}{\omega_c} \;=\; \gamma\,.$$

The function $f(\tau)$ represents the exponential interaction factor of the dipole with all the other charges. In the third line, we have utilized that a dipole with zero dipole length does not interact with other charges, $f(0) = 1$. From this we see that a collapsed dipole depends neither on the bias strength nor on temperature, and that the expression $I(K = \frac{1}{2})$ coincides with the frequency scale γ introduced in Eq. (19.63). Furthermore, collapsed dipoles are noninteracting, and the grand-canonical sum of these entities can be performed without difficulty.

During an extended sojourn of length s, the system may make visits of duration zero to either of the two blip states. We refer to these lightning visits of blip states as collapsed blips (CB). All the possibilities of these short visits within a sojourn interval s form a grand-canonical ensemble of these entities. Taking into account the multiplicity of the possible intermediate blip and sojourn states, and the minus sign associated with each factor Δ^2, the grand-canonical sum of noninteracting collapsed blips in the sojourn interval s adds up to the exponential factor

$$U_{\mathrm{CB}}(s) \;=\; e^{-\gamma s}\,, \qquad (21.197)$$

which we refer to as CB form factor. Equivalently, during an extended blip state of length τ, the system may make any number of visits of duration zero to a sojourn state. Altogether, they represent a noninteracting gas of collapsed sojourns (CS). The grand-canonical sum of these entities in the blip interval τ yields the CS form factor

$$U_{\mathrm{CS}}(\tau) \;=\; e^{-\gamma\tau/2}\,, \qquad (21.198)$$

where we have taken into acount the different multiplicity factor for collapsed sojourns.[19] With these preliminaries, the evaluation is considerably facilitated.

21.6.2 The expectation value $\langle \sigma_z \rangle_t$ for $K = \frac{1}{2}$

Consider now first the symmetric system at $T = 0$. For $K = \frac{1}{2}$, only collapsed blips contribute, as follows from the form (21.56) for $F_m^{(+)}$. Since collapsed blips are noninteracting, the noninteracting-blip assumption [cf. Section 21.3] becomes exact. The Laplace-transformed kernel $\widehat{K}_z^{(s)}(\lambda)$ takes the form

---

[19] After a short visit to a sojourn state, the system must return to the same blip state, whereas after a short visit to a blip state, the system can return to either sojourn state. This is the reason for the multiplicity factor 2 in the exponent of the CB form factor.

$$0 \qquad\qquad\qquad\qquad t \qquad\quad 0 \qquad\qquad\qquad\qquad\qquad t$$

Figure 21.4: The diagrams for $P_{\mathrm{s}}(t)$ (left) and $P_{\mathrm{a}}(t)$ (right). The full (dashed) lines represent sojourns (blips). A full box represents the insertion of a CB form factor U_{CB} within a sojourn interval, and the empty box stands for a CS form factor U_{CS} inserted in a blip interval. The downward and upward spikes symbolize the remaining solitary flips or charges.

$$\widehat{K}_z^{(\mathrm{s})}(\lambda) \;=\; \lim_{K \to 1/2} \Delta^2 \cos(\pi K) \int_0^\infty d\tau \, \frac{\mathrm{e}^{-\lambda\tau}}{(\omega_{\mathrm{c}}\tau)^{2K}} \;=\; \gamma \,, \qquad (21.199)$$

and the Laplace transform of the population $\langle\sigma_z\rangle_t$ is $\langle\sigma_z(\lambda)\rangle = 1/[\lambda+\gamma]$. Thus, in the time domain, the system's populations show incoherent exponential relaxation

$$\langle\sigma_z\rangle_t \;=\; \mathrm{e}^{-\gamma t} \qquad\qquad\qquad\qquad (21.200)$$

with the relaxation rate $\gamma = \frac{\pi}{2}\frac{\Delta^2}{\omega_{\mathrm{c}}}$. The expression (21.200) is immediately comprehensible: it is represented by a single sojourn of length t which is dressed by a CB form factor, as sketched diagrammatically in Fig. 21.4 (left). Observing that the CB form factor is insensitive to bias and to temperature, it is evident that the expression (21.200) represents the symmetric contribution $P_{\mathrm{s}}(t)$ to $\langle\sigma_z\rangle_t$ for any bias strength and any temperature.

Consider next the antisymmetric contribution $P_{\mathrm{a}}(t)$ to $\langle\sigma_z\rangle_t$. We see from Eqs. (21.46) – (21.48) that the occuring influence function $F_m^{(-)}$ has one $\cos(\pi K)$ factor less than $F_m^{(+)}$ and that it is combined with the odd bias factor $B_m^{(\mathrm{a})}$. Consequently, each sequence of blips and sojourns contributing to $P_{\mathrm{a}}(t)$ arranges as follows. Because of the factor ξ_1 in $F_m^{(-)}$, the first blip is an extended blip, say of finite length τ, and it comes with a bias factor $\sin(\epsilon\tau)$. During this blip state, the system may make any number of lightning visits of duration zero to a sojourn state. Summation of all these events yields a CS form factor $U_{\mathrm{CS}}(\tau)$. The extended blip is followed in the remaining sojourn interval $s = t - \tau$ by a grand-canonical sum of collapsed blips, which gives a CB form factor $U_{\mathrm{CB}}(t-\tau)$. The resulting expression for $P_{\mathrm{a}}(t)$ is diagrammatically sketched in Fig. 21.4 (right). Joining the various components together, and taking into account the interaction between the two remaining solitary charges, we obtain

$$P_{\mathrm{a}}(t) \;=\; \Delta^2 \int_0^\infty d\tau \int_0^\infty ds \, \Theta(t-\tau-s) \sin(\epsilon\tau) \, \mathrm{e}^{-Q'(\tau)} \, \mathrm{e}^{-\gamma\tau/2} \, \mathrm{e}^{-\gamma s} \,. \qquad (21.201)$$

Using Eqs. (21.200), (21.201), and (18.51), we finally get

$$\langle\sigma_z\rangle_t \;=\; \mathrm{e}^{-\gamma t} + 2 \int_0^t \frac{d\tau}{\hbar\beta} \, \frac{\sin(\epsilon\tau)}{\sinh(\pi\tau/\hbar\beta)} \left[\, \mathrm{e}^{-\gamma\tau/2} - \mathrm{e}^{-\gamma t} \, \mathrm{e}^{\gamma\tau/2} \,\right]. \qquad (21.202)$$

Consider now first the case $T = 0$. At long times $t \gg 1/\gamma$, we find from Eq. (21.202) damped oscillations towards the equilibrium value with a quality factor $Q = 2\epsilon/\gamma$,

$$\langle \sigma_z \rangle_t \approx \langle \sigma_z \rangle_\infty + \frac{8}{\pi} \frac{\gamma^2}{\gamma^2 + 4\epsilon^2} \frac{\sin(\epsilon t)\,e^{-\gamma t/2}}{\gamma t} , \qquad (21.203)$$

and the expectation value in thermal equilibrium is

$$\langle \sigma_z \rangle_\infty = (2/\pi) \arctan(2\epsilon/\gamma) . \qquad (21.204)$$

Thus, there is nonzero occupation of the higher (left) state even at zero temperature.

At finite temperature, the leading asymptotic behaviour is

$$\langle \sigma_z \rangle_t \approx \langle \sigma_z \rangle_\infty + \frac{16}{\hbar\beta\gamma} \frac{\gamma^2}{\gamma^2 + 4\epsilon^2} \sin(\epsilon t)\,e^{-(\gamma+\nu)t/2} , \qquad (21.205)$$

where $\nu = 2\pi/\hbar\beta$. The equilibrium value for all T and ϵ is given by

$$\langle \sigma_z \rangle_\infty = 2 \int_0^\infty \frac{d\tau}{\hbar\beta} \frac{\sin(\epsilon\tau)\,e^{-\gamma\tau/2}}{\sinh(\pi\tau/\hbar\beta)} = \frac{2}{\pi} \operatorname{Im} \psi\left(\frac{1}{2} + \frac{\hbar\gamma}{4\pi k_B T} + i\frac{\hbar\epsilon}{2\pi k_B T}\right) , \qquad (21.206)$$

where $\psi(z)$ is the digamma function. From these results for $K = \frac{1}{2}$, we can draw two conclusions. First, a bias leads to an increase of the quality factor of the oscillation, whereas the amplitude is decreased. Second, with increasing temperature, the quality factor of the oscillation decreases, whereas the amplitude is increased. Similar behaviour has been found in numerical studies also for $K \neq \frac{1}{2}$ [372].

As we have dicussed at the end of Subsection 21.3.2, the expectation values $\langle \sigma_x \rangle_t$ and $\langle \sigma_y \rangle_t$ are non-universal. Hence they are zero in the universality limit (18.54).

21.6.3   The case $K = \frac{1}{2} - \kappa$; coherent–incoherent crossover

For a symmetric system at zero temperature, the NIBA expression (21.94) gives for $K = \frac{1}{2} - \kappa$ with $|\kappa| \ll 1$ a coherent contribution $P_{coh}(t)$ of the form (21.99), where

$$\Omega = 2\pi\kappa\,\Delta_{eff} , \qquad \gamma = \Delta_{eff} , \qquad (21.207)$$

and an incoherent part

$$P_{inc}(t) = -2\kappa/(\Delta_{eff}t)^{1+2\kappa} . \qquad (21.208)$$

Thus, NIBA predicts the coherent-incoherent transition to occur at $K = \frac{1}{2}$.

To investigate the transition beyond NIBA, we now follow Ref. [426] and perform a systematic expansion of K about the exactly solvable case $K = \frac{1}{2}$. For $K = \frac{1}{2} - \kappa$, the blip length is finite, and within every blip one has to take into account all sequences of collapsed sojourns. This crucially modifies the kernel $K_z^{(s)}(\lambda)$ in the limit $\lambda \to 0$, as compared with the NIBA. The grand-canonical sum of collapsed sojourns within a blip of length τ simply gives a factor $\exp(-\Delta_{eff}\tau/2)$, as explained in the preceding subsection. Thus we have

$$K_z^{(s)}(\lambda) = \Delta^2 \pi \kappa \int_0^\infty d\tau \, \frac{e^{-(\lambda+\Delta_{\rm eff}/2)\tau}}{(\omega_c \tau)^{1-2\kappa}} = \Delta_{\rm eff} \left[\frac{\lambda}{\Delta_{\rm eff}} + \frac{1}{2} \right]^{-2\kappa}. \tag{21.209}$$

With the regularized kernel (21.209), the pole condition reads

$$\lambda [\, 1/2 + \lambda/\Delta_{\rm eff}\,]^{2\kappa} = -\Delta_{\rm eff}\,. \tag{21.210}$$

This gives for the oscillation frequency and the decay rate the expressions

$$\Omega = 2\pi\kappa\,\Delta_{\rm eff}[\,1+\mathcal{O}(\kappa)\,]\,, \qquad \gamma = \Delta_{\rm eff}\,[\,1+\mathcal{O}(\kappa)\,]\,, \tag{21.211}$$

which coincide in the explicitly given orders with the NIBA expressions (21.207). From this we conclude that the value $K = \frac{1}{2}$ is exactly the critical damping strength where the coherent-incoherent transition occurs at zero temperature in the limit $\Delta/\omega_c \to 0$, as already predicted by the NIBA.

The kernel (21.209) shifts the branch point of $\langle \sigma_z(\lambda) \rangle$ from $\lambda = 0$ to $\lambda = -\Delta_{\rm eff}/2$ and thereby removes the spurious algebraic long-time tail (21.208). For $\kappa > 0$, the leading cut contribution at times $t \gg \Delta_{\rm eff}^{-1}$ is given by

$$P_{\rm inc}(t) = -2\kappa \, \frac{\exp(-\Delta_{\rm eff}t/2)}{(\Delta_{\rm eff}t)^{1+2\kappa}}\,, \tag{21.212}$$

while for $\kappa < 0$ we find

$$P_{\rm inc}(t) = 8|\kappa| \, \frac{\exp(-\Delta_{\rm eff}t/2)}{(\Delta_{\rm eff}t)^{1+2|\kappa|}}\,. \tag{21.213}$$

Thus the unphysical algebraic long-time tail (21.208) predicted by the NIBA is actually suppressed by an exponential decay factor. It is straightforward to see that for $\kappa > 0$ $(K < \frac{1}{2})$ the cut contribution is negative for all times, while it becomes positive for $\kappa < 0$ $(K > \frac{1}{2})$. For $K \geq \frac{1}{2}$, the expectation value $\langle \sigma_z \rangle_t$ is decaying monotonously from the initial value $+1$ to zero, which means that the dynamics is fully incoherent. In marked contrast to the NIBA result (21.208), the power of the algebraic decay factor does not depend on the sign of κ.

21.6.4 Equilibrium σ_z autocorrelation function

We begin the discussion of the equilibrium correlation functions by considering the symmetric σ_z autocorrelation function. Starting out from the expression (21.57) with Eq. (21.59), and gathering up the results achieved in Subsection 21.6.2, we find

$$S_z^{(\rm unc)}(t) = e^{-\gamma t} + \langle \sigma_z \rangle_\infty [\, P_{\rm a}(t) - \langle \sigma_z \rangle_\infty \,]\,, \tag{21.214}$$

where $P_{\rm a}(t)$ and $\langle \sigma_z \rangle_\infty$ are given in Eqs. (21.201) and (21.206), respectively.

Because of the factor $\xi_1 \xi_{n+1} \sin^2(\pi K)$ in the expression (21.61) for $F_{m,n}$, the correlation term $R(t)$ basically consists of two extended blips, the one being located in the negative-time and the other in the positive-time branch. The interval of each of the two blips is filled with a gas of collapsed sojourns, yielding form factors $U_{\rm CS}$ as

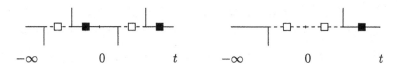

Figure 21.5: The left diagram represents the correlation contribution $R(t)$ to $S_z(t)$, and the right diagram pictorially gives the response function $\chi_z(t)$. The symbols are as in Fig. 21.4.

explained in Subsection 21.6.1. The two extended blips are followed in the respective remaining interval of the negative- and positive-time branches by a grand-canonical sum of collapsed noninteracting blips, giving form factors U_{CB}. The correlation term $R(t)$ is sketched in Fig. 21.5 (left).

For $K = \frac{1}{2}$, the blip interaction factor for two blips of type ξ and $\pm\xi$ with de-markation times t_1, t_2 ($t_2 > t_1$) and t_3, t_4 ($t_4 > t_3$), respectively, can be decomposed as [cf. the similar decomposition (19.59) for imaginary times]

$$
\begin{aligned}
G_2(\xi,\xi) &= e^{-Q'_{21}} e^{-Q'_{43}} + e^{-Q'_{32}} e^{-Q'_{41}} , \\
G_2(\xi,-\xi) &= e^{-Q'_{21}} e^{-Q'_{43}} - e^{-Q'_{31}} e^{-Q'_{42}} .
\end{aligned}
\tag{21.215}
$$

Performing the ξ summation, we find that the contribution from the first terms for $G_2(\xi,\xi)$ and $G_2(\xi,-\xi)$ cancels the subtraction term in Eq. (21.59). Piecing together these observations, we find for the correlation term $R(t)$ the form

$$
\begin{aligned}
R(t) &= -\frac{\Delta^4}{2} \int_0^\infty d\tau_1 \, d\tau_2 \, ds_1 \, ds_2 \, \Theta(t - \tau_2 - s_2)\, e^{-\gamma(\tau_1+\tau_2)/2}\, e^{-\gamma s_1}\, e^{-\gamma(t-\tau_2-s_2)} \\
&\quad \times \Big\{ \cos[\,\epsilon(\tau_1 + \tau_2)\,]\, e^{-Q'(s_1+s_2)}\, e^{-Q'(\tau_1+s_1+s_2+\tau_2)} \\
&\quad + \cos[\,\epsilon(\tau_1 - \tau_2)\,]\, e^{-Q'(\tau_1+s_1+s_2)}\, e^{-Q'(s_1+s_2+\tau_2)} \Big\} .
\end{aligned}
\tag{21.216}
$$

Here, τ_1 and τ_2 are the lengths of the two extended blips. The length of the interme-diate sojourn is $s_1 + s_2$, where s_1 is the interval until time zero and s_2 is the residual interval at positive times. The pair interactions $Q'(\tau)$ convey correlations between the negative-time and the positive-time branch.

The expression (21.216) is readily evaluated by introducing the arguments of the function $Q'(\tau)$ as new variables and by performing the other integrations. Combining the resulting form for $R(t)$ with the expression (21.214), we find in the end for the symmetrized correlation function the analytic form

$$
S_z(t) = e^{-\gamma t} - F_1^2(t) - F_2^2(t) ,
\tag{21.217}
$$

$$
\begin{aligned}
F_1(t) &= \frac{\Delta^2}{2\gamma} \int_0^\infty d\tau \, \sin(\epsilon\tau)\, e^{-Q'(\tau)} \left[e^{-\gamma|t-\tau|/2} + e^{-\gamma(t+\tau)/2} \right] , \\
F_2(t) &= \frac{\Delta^2}{2\gamma} \int_0^\infty d\tau \, \cos(\epsilon\tau)\, e^{-Q'(\tau)} \left[e^{-\gamma|t-\tau|/2} - e^{-\gamma(t+\tau)/2} \right] .
\end{aligned}
\tag{21.218}
$$

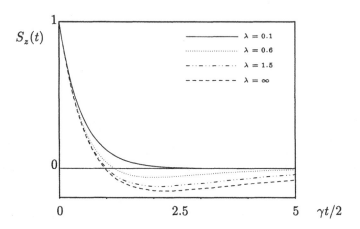

Figure 21.6: The formula (21.217) for $S_z(t)$ is plotted for zero bias as a function of $\gamma t/2$ for different values of the scaled inverse temperature $\lambda = \hbar\gamma/2\pi k_{\rm B}T$. In the high temperature case $\lambda = 0.1$, the curve for $S_z(t)$ cannot be resolved from the function $\langle\sigma_z\rangle_t = {\rm e}^{-\gamma t}$.

The result (21.217) with Eq. (21.218) is exact for all T and ϵ, and it holds for all t. For $\gamma t \gg 1$ and $T = 0$, the functions $F_1(t)$ and $F_2(t)$ are asymptotically given by

$$F_1(t) \approx \frac{4}{\pi}\frac{\gamma^2}{\gamma^2+4\epsilon^2}\frac{\sin(\epsilon t)}{\gamma t}\,, \qquad F_2(t) \approx \frac{4}{\pi}\frac{\gamma^2}{\gamma^2+4\epsilon^2}\frac{\cos(\epsilon t)}{\gamma t}\,. \qquad (21.219)$$

Substituting Eq. (21.219) in Eq. (21.217), we find for the symmetric system at $T = 0$ the algebraic long-time behaviour

$$\lim_{t\to\infty} S_z(t) \approx -\left(\frac{4}{\pi\gamma}\right)^2\frac{1}{t^2}\,. \qquad (21.220)$$

At very low T, the algebraic law (21.220) holds in the intermediate time region $\gamma^{-1} \ll t \ll \hbar\beta$, while in the asymptotic limit $t \gg \hbar\beta$ one has exponential decay with a rate given by the lowest Matsubara frequency $\nu_1 = 2\pi/\hbar\beta$ [207],

$$\lim_{t\to\infty} S_z(t) \approx -(8k_{\rm B}T/\hbar\gamma)^2{\rm e}^{-\nu_1 t}\,. \qquad (21.221)$$

Note that $S_z(t)$ and $S_z^{({\rm unc})}(t)$ approach zero from opposite sides as $t \to \infty$. The behaviour of $S_z(t)$ is shown for different temperatures in Fig. 21.6.

For nonzero bias and short to intermediate times $t \lesssim 1/\gamma$, the correlation function $S_z(t)$ is given by Eq. (21.214). At $T = 0$, we find for t near $1/\gamma$ damped oscillations,

$$S_z(t) \approx \frac{8}{\pi}\frac{\gamma^2}{\gamma^2+4\epsilon^2}\langle\sigma_z\rangle_\infty\frac{\sin(\epsilon t)\,{\rm e}^{-\gamma t/2}}{\gamma t}\,. \qquad (21.222)$$

From this we draw two conclusions. First, a bias can induce a transition from incoherent relaxation to damped coherent oscillation. Numerical simulation results lead

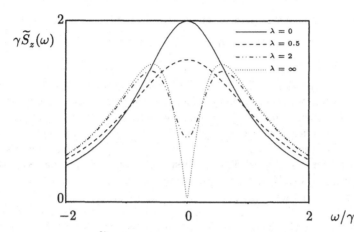

Figure 21.7: The function $\gamma \widetilde{S}_z(\omega)$ is shown for $\epsilon = 0$ for a set of values of the scaled inverse temperature $\lambda = \hbar\gamma/2\pi k_{\mathrm{B}}T$.

one to assume that the transition generally occurs in the regime $\frac{1}{2} < K < 1$ when the bias exceeds a critical value [372]. Secondly, the quality factors for the bias-induced oscillations of $\langle \sigma_z \rangle_t$ and $S_z(t)$ are the same. This observation holds generally, as we can infer from the form (21.214) which dominates the behaviour of $S_z(t)$ in the time domain $t \lesssim 1/\Delta_{\mathrm{eff}}$.

At times $t \gg 1/\gamma$ and zero temperature, the decay of the correlations is dominated by the residual term $R(t)$. We find from Eq. (21.217) with Eq. (21.218) that $S_z(t)$ drops algebraically,

$$\lim_{t\to\infty} S_z(t) = -\left(\frac{\hbar\overline{\chi}_z}{2}\right)^2 \frac{1}{t^2}, \qquad \overline{\chi}_z = \frac{8}{\pi\hbar\gamma} \frac{\gamma^2}{\gamma^2 + 4\epsilon^2}. \qquad (21.223)$$

Here we have introduced the static susceptibility $\overline{\chi}_z$ at $T = 0$, which has been calculated from Eq. (21.65) with Eq. (21.204), or alternatively from Eq. (19.87).

The effects of the correlations in the initial state also show up clearly in frequency space. Upon Fourier transforming the expression (21.217), we obtain

$$\widetilde{S}_z(\omega) = \coth\left(\frac{\hbar\omega}{2k_{\mathrm{B}}T}\right) \frac{2}{\pi} \frac{\gamma}{\omega[\omega^2 + \gamma^2]}\left(\omega\Phi''(\omega) - \gamma\Phi'(\omega)\right), \qquad (21.224)$$

where $\Phi'(\omega)$ and $\Phi''(\omega)$ are the real and imaginary parts of the complex function

$$\Phi(\omega) = \psi(x_+) + \psi(x_-) - \psi(x_+ - i\hbar\beta\omega/2\pi) - \psi(x_- - i\hbar\beta\omega/2\pi), \qquad (21.225)$$

where $\psi(x)$ is the digamma function, and $x_\pm = \frac{1}{2} + (\frac{1}{2}\gamma \pm i\epsilon)\hbar\beta/2\pi$. At $T = 0$, we have

$$\begin{aligned} \Phi'(\omega) &= -\tfrac{1}{2}\ln\left\{[\gamma^2 + 4(\omega+\epsilon)^2][\gamma^2 + 4(\omega-\epsilon)^2]/[\gamma^2 + 4\epsilon^2]^2\right\}, \\ \Phi''(\omega) &= \arctan[2(\omega+\epsilon)/\gamma] + \arctan[2(\omega-\epsilon)/\gamma]. \end{aligned} \qquad (21.226)$$

The low-frequency behaviour at zero temperature is nonanalytic and is given by

$$\widetilde{S}_z(\omega \to 0) = \pi(\hbar\overline{\chi}_z/2)^2 |\omega| . \tag{21.227}$$

While the spectral function $\widetilde{S}_z^{(\mathrm{unc})}(\omega)$ is temperature-independent for zero bias and is given by the Lorentzian form $2\gamma/(\omega^2 + \gamma^2)$, the expression (21.224) is drastically different at $T = 0$ in the low-frequency regime, as shown by the curve with $\lambda = \infty$ in Fig. 21.7. As T is increased, the trough of $\widetilde{S}_z(\omega)$ is levelled off. At high temperatures where $\lambda \ll 1$, the difference between $\widetilde{S}_z(\omega)$ and $\widetilde{S}_z^{(\mathrm{unc})}(\omega)$ becomes negligibly small.

We now study the response function $\chi_z(t)$. The analysis again proceeds by employing the concept of collapsed dipoles. First, we recognize from the formally exact expression for $\chi_z(t)$, Eq. (21.62), that the system is in a blip state at time zero. Secondly, we see from the form (21.56) for the influence factor $F_m^{(-)}$ that there is only one extended blip for $K = \frac{1}{2}$. It is then clear that the system hops into the extended blip state in the negative time branch and finishes off this state only at a positive time. Within the extended blip, there are again insertions of collapsed sojourns, resulting in a CS form factor U_{CS}. The extended blip is followed by an extended sojourn with insertions of collapsed blips. This is described by a CB form factor U_{CB}. The respective diagram is shown in Fig. 21.5 (right). In mathematical terms, we then have

$$\chi_z(t) = \frac{2}{\hbar}\,\Theta(t)\,\Delta^2 \int_0^\infty d\tau_1\, d\tau_2\, ds_2\, \delta(t - \tau_2 - s_2)$$
$$\times\, \cos[\,\epsilon(\tau_1 + \tau_2)\,]\, e^{-Q'(\tau_1 + \tau_2)}\, e^{-\gamma(\tau_1 + \tau_2)/2}\, e^{-\gamma s_2} , \tag{21.228}$$

where $\tau_1 + \tau_2$ is the overall length of the blip and s_2 is the remaining sojourn length. Introducing $\tau = \tau_1 + \tau_2$ as a new intergation variable and performing the other integrals, we find for the response function the exact analytic expression

$$\chi_z(t) = (4/\hbar)\,\Theta(t)\, F_2(t)\, e^{-\gamma t/2} , \tag{21.229}$$

where the function $F_2(t)$ is defined in Eq. (21.218).

Substituting the form (21.219) for $F_2(t)$, we find that the response function at $T = 0$ decays asymptotically as

$$\chi_z(t) \approx \frac{16}{\pi\hbar} \frac{\gamma^2}{\gamma^2 + 4\epsilon^2} \frac{\cos(\epsilon t)\, e^{-\gamma t/2}}{\gamma t} , \qquad \gamma t \gg 1 . \tag{21.230}$$

Upon Fourier transforming Eq. (21.229), we find for the dynamical susceptibility for all values of T and ϵ the exact result

$$\widetilde{\chi}_z(\omega) = \frac{2}{\pi} \frac{1}{\hbar\omega} \frac{\gamma}{\omega + i\gamma}\, \Phi(\omega) , \tag{21.231}$$

where $\Phi(\omega)$ is the complex function defined in Eq. (21.225).

Two remarks are appropriate. First, we see that the independently derived expressions (21.224) and (21.231) satisfy the fluctuation-dissipation theorem

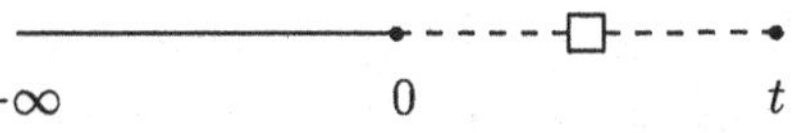

Figure 21.8: The diagram describing $S_x(t)$ with symbols as in Fig. 21.4. The bullets mark transitions which are free of bath correlations because of the modified influence functional.

$$\widetilde{S}_z(\omega) \;=\; \hbar \coth(\hbar\beta\omega/2)\, \widetilde{\chi}_z''(\omega) \;. \tag{21.232}$$

Secondly, the static susceptibility $\overline{\chi}_z = \widetilde{\chi}_z(\omega = 0)$ found from Eq. (21.231) coincides with the result (19.87) found from a thermodynamic approach, and normalized as in Eq. (19.13). This form is also in agreement with the form found from the dispersion relation

$$\overline{\chi}_z \;=\; \frac{2}{\pi} \int_0^\infty d\omega\, \frac{\widetilde{\chi}_z''(\omega)}{\omega} \;. \tag{21.233}$$

21.6.5 Equilibrium σ_x autocorrelation function

With the concept of collapsed blips and collapsed sojourns explained in Subsection 21.6.1, it is possible to evaluate also the series expressions (21.67) – (21.70) for the dressed coherence correlation function in the scaling limit for $K = \frac{1}{2}$.

Consider first the symmetrized correlation function. Again we assign a factor $\cos(\pi K)$ to each collapsing dipole. We then find that in each term of the series (21.68) there is a surplus of one $\cos(\pi K)$ factor. As a result, the contribution $S_x^{\mathrm{B}}(t)$ vanishes as $K \to \frac{1}{2}$. In the contribution $S_x^{\mathrm{A}}(t)$, the system dwells in the initial sojourn state η throughout the negative-time branch. At time zero it then hops into the blip state $\xi = -\eta$. Afterwards, it stays there until time t except for lightning visits of sojourn states, which altogether sum up to a CS form factor $\mathrm{e}^{-\gamma t/2}$. Piecing together the bias factor of the blip state with the CS form factor, we find damped oscillations behaviour

$$S_x(t) \;=\; \cos(\epsilon t)\, \mathrm{e}^{-\gamma t/2} \;. \tag{21.234}$$

The contribution $S_x^{\mathrm{A}}(t)$ is sketched diagrammatically in Fig. 21.8. Since all charges merge into collapsed dipoles, there are no excess charges, and hence $S_x(t)$ does not depend on temperature.

Next, we study the response function $\chi_x(t)$. The contribution of group A is sketched in Fig. 21.9 (left). In the negative-time branch, there is an extended blip state which is followed by an extended sojourn state, both being dressed with CS and CB form factor, respectively. At time zero, the system hops back into a blip state and stays there until time t. The extended blip state is again dressed by a CS form factor. In mathematical terms we have

$$\chi_x^{\mathrm{A}}(t) \;=\; \Theta(t)\, \frac{2}{\hbar}\, \sin(\epsilon t)\, \mathrm{e}^{-\gamma t/2} \Delta^2 \int_0^\infty d\tau \int_0^\infty ds\, \sin(\epsilon\tau)\, \mathrm{e}^{-Q'(\tau)}\, \mathrm{e}^{-\gamma\tau/2}\, \mathrm{e}^{-\gamma s} \;. \tag{21.235}$$

Now, the double integral times the factor Δ^2 is just $\langle \sigma_z \rangle_\infty = P_\mathrm{a}(t \to \infty)$ given in Eq. (21.206), as follows with Eq. (21.201). In the end, we find [207]

Figure 21.9: The contribution of group A (left) and group B (right) to $\chi_x(t)$. The symbols are as in Fig. 21.4. Each diagram has only one extended dipole.

$$\chi_x^A(t) = (2/\hbar)\,\langle\sigma_z\rangle_\infty\,\sin(\epsilon t)\,e^{-\gamma t/2}\,. \qquad (21.236)$$

Thus we have again exponential decay. This is, because each extended interval, except for the first sojourn, is cut off exponentially by the respecetve form factor.

Now we turn to the contributions of group B. Because of the $\cos^2(\pi K)$-factor in the second term in the square bracket of Eq. (21.70), the respective contribution vanishes $\propto (K - \frac{1}{2})^2$ as $K \to \frac{1}{2}$, whereas the first term in the square bracket is nonzero in this limit. In this contribution, the system hops from a sojourn into a blip at negative time $-\tau$ and stays there until time zero, where it returns to a sojourn state. At time s, it hops again into a blip state and dwells in this state until time t. Again, each extended blip interval is dressed with a CS form factor, as discussed above. However, because of the factor ξ_{n+1} in the first term of the square bracket in Eq. (21.70) and the summation over the values ± 1, the extended sojourn in the positive-time branch is free of collapsed blips. The extended dipole of length $\tau + s$ introduces correlations between the negative- and positive-time branches as depicted in Fig. 21.9 (right). In terms of a formula we have

$$\chi_x^B(t) = \frac{\Delta^2}{\hbar} \int_0^\infty d\tau \int_0^t ds\, e^{-\gamma(t+\tau-s)/2}\, e^{-S(\tau+s)} \cos[\epsilon(t - \tau - s)]\,. \qquad (21.237)$$

Introducing the dipole length $\tau + s$ as a new integration variable, performing the other integrations, and combining the resulting expression with Eq. (21.236), we find

$$\chi_x(t) = (2/\hbar)[\sin(\epsilon t)F_1(t) + \cos(\epsilon t)F_2(t)]\,, \qquad (21.238)$$

where the functions $F_1(t)$ and $F_2(t)$ have been given already in Eq. (21.218). The function $\chi_x(t)$ describes the linear response of the system to a variation of the tunneling splitting Δ. Using Eq. (21.219), we find asymptotically algebraic decay,

$$\chi_x(t) \approx \frac{8}{\pi\hbar}\frac{\gamma^2}{\gamma^2 + 4\epsilon^2}\frac{1}{\gamma t} \qquad \text{for} \qquad t \gg 1/\gamma\,. \qquad (21.239)$$

Next, consider the spectral properties. Taking the Fourier transform of $S_x(t)$ given in Eq. (21.234), we obtain

$$\widetilde{S}_x(\omega) = \gamma\,\frac{\omega^2 + \epsilon^2 + \gamma^2/4}{(\omega^2 + \epsilon^2 + \gamma^2/4)^2 - 4\epsilon^2\omega^2}\,. \qquad (21.240)$$

On the other hand, we may calculate $\widetilde{\chi}_x''(\omega)$ from the expression (21.238). We then find that the resulting expression is in accordance with the fluctuation-dissipation theorem

$$\hbar\widetilde{\chi}_x''(\omega) = \tanh(\hbar\beta\omega/2)\,\widetilde{S}_x(\omega)\,. \tag{21.241}$$

Finally, the real part of the dynamical susceptibility is found in the unbased case as

$$\widetilde{\chi}_x'(\omega) = \frac{8\gamma}{\pi\hbar}\frac{1}{\omega^2+\gamma^2/4}\,\mathrm{Re}\left\{\psi\left(\frac{1}{2}+\frac{\hbar\gamma}{4\pi k_{\mathrm{B}}T}\right)-\psi\left(\frac{1}{2}+i\frac{\hbar\omega}{2\pi k_{\mathrm{B}}T}\right)\right\}\,. \tag{21.242}$$

From this we find that the static susceptibility $\overline{\chi}_x = \widetilde{\chi}_x(\omega=0)$ diverges logarithmically for zero bias as temperature goes to zero.

21.6.6 Correlation functions in the Toulouse model

Additional insights are gained by calculating the equilibrium tunneling and coherence correlation functions in the fermionic model (19.66). The equivalence relations of the TSS operators with the fermionic operators of the d level are

$$\sigma_z = 2d^\dagger d - 1\,, \qquad \sigma_x = d^\dagger + d\,, \qquad \sigma_y = i(d-d^\dagger)\,. \tag{21.243}$$

To proceed, we rewrite the Matsubara sum (19.77) for $G^{(\mathrm{E})}(\tau)$ as a contour integral. Analytic continuation to real time gives for the Green function $G(t) \equiv \langle T_t d^\dagger(t)d(0)\rangle_\beta$ of the d-level the expression

$$G(\pm t) = G^{(\mathrm{E})}(\tau=\pm it) = \pm\tfrac{1}{2}\,e^{\mp i\epsilon t}\,e^{-\gamma t/2} + \tfrac{1}{2}\,e^{\mp i\epsilon t}[\,F_1(t)\mp iF_2(t)\,]\,, \tag{21.244}$$

where the functions $F_1(t)$ and $F_2(t)$ are defined by the integral representations

$$F_1(t) = \frac{\gamma}{2\pi}\int_{-\infty}^{\infty}d\omega\,\frac{\cos(\omega t)}{\omega^2+\gamma^2/4}\,\tanh\left(\frac{\hbar(\omega+\epsilon)}{2k_{\mathrm{B}}T}\right)\,,$$

$$F_2(t) = \frac{\gamma}{2\pi}\int_{-\infty}^{\infty}d\omega\,\frac{\sin(\omega t)}{\omega^2+\gamma^2/4}\,\tanh\left(\frac{\hbar(\omega+\epsilon)}{2k_{\mathrm{B}}T}\right)\,. \tag{21.245}$$

The key relation linking the fermionic representation to the bosonic one is

$$\tanh(\hbar\omega/2k_{\mathrm{B}}T) = (2\omega_c/\pi)\int_0^\infty d\tau\,\sin(\omega\tau)\,e^{-Q'_{K=1/2}(\tau)}\,, \tag{21.246}$$

where $Q'_{K=1/2}(\tau)$ is the function (18.51) for $K=\frac{1}{2}$. Upon substituting Eq. (21.246), the "fermionic" expressions (21.245) are transformed into the bosonic representations (21.218). Thus, Eqs. (21.218) and (21.245) are different integral representations of the same functions. Next, we substitute the form (21.243) for σ_z into Eq. (21.16), use Eq. (21.18) for $i=z$, and observe that the two-particle Green function factorizes into products of one-particle Green functions. Thus we find

$$S_z(t) = 1+4\,\mathrm{Re}\left\{G^2(0^+)-G(0^+)-G(t)G(-t)\right\}\,,$$

$$\chi_z(t) = (8/\hbar)\Theta(t)\,\mathrm{Im}\left\{G(t)G(-t)\right\}\,. \tag{21.247}$$

Substituting the forms (21.244) for $G(\pm t)$, we recover the earlier results (21.217) and (21.229).

In the fermionic representation, it is straightforward to calculate also the symmetric σ_j autocorrelation function $S_j(t) = \mathrm{Re}\,\langle\sigma_j(t)\sigma_j(0)\rangle_\beta$ and the response function $\chi_j(t) = -(2/\hbar)\Theta(t)\,\mathrm{Im}\,\langle\sigma_j(t)\sigma_j(0)\rangle_\beta$ for $j = x,\,y$. By virtue of the relations (21.243) we find

$$S_y(t) = S_x(t)\,, \quad \text{and} \quad \chi_y(t) = \chi_x(t)\,. \tag{21.248}$$

The resulting expressions are

$$S_x(t) = \mathrm{Re}\,\{\,G(t) - G(-t)\,\} = \cos(\epsilon t)\,\mathrm{e}^{-\gamma t/2}\,, \tag{21.249}$$

$$\chi_x(t) = -\frac{2}{\hbar}\Theta(t)\mathrm{Im}\,\{\,G(t) - G(-t)\,\} = \frac{2}{\hbar}\Theta(t)[\,\sin(\epsilon t)F_1(t) + \cos(\epsilon t)F_2(t)\,]\,.$$

These expressions coincide with the analogous correlation functions of the spin-boson model, Eqs. (21.234) and (21.238). Because σ_x in Eq. (21.243) is the bare tunneling operator, it is evident that we have worked in the representation (21.29). Thus, the Toulouse Hamiltonian (19.66), and the resonant-level Hamiltonian (19.106), directly correspond to the polaron-transformed Hamiltonian $\widetilde{H}$ given in Eq.(18.38). This concludes the discussion of the exactly solvable case $K = \frac{1}{2}$.

21.7   Long–time behaviour at $T = 0$ for $K < 1$: general discussion

We have seen in the previous section that the grand-canonical sum of collapsed non-interacting dipoles gives rise to a sojourn or blip form factor. For K different from $\frac{1}{2}$ the dipoles do not collaps anymore and therefore are interacting with each other. For this reason, the grand-canonical sum of extended dipoles can not be performed in analytic form any more. We argue that in the limit of long times the alternating series of these entities nevertheless add up to an effective form factor which cuts off the respective time interval at a length of order $1/\Delta_\mathrm{r}$. With these preliminaries, the asymptotic dynamics can be understood in terms of two rules:

(I) Every time interval which is free of a form factor for $K = \frac{1}{2}$ is free of a form factor also for $K \neq \frac{1}{2}$. We shall refer to such intervals as bare intervals.

(II) The totality of charge arrangement placed between bare intervals form a grand-canonical ensemble (charge cluster) which effectively acts as a form factor. The form factor restricts the respective interval to a length of order $1/\Delta_\mathrm{r}$.

Rule (I) applies to blip states in the series expressions given in Subsections 21.2.4 and 21.2.5 which have weight factors ξ_j. For instance in the expression (21.61), these special blips are the first blip in the negative- and the first blip in the positive-time branch. For these blips, the $\{\xi_j\}$ summations lead to cancellations among the interactions stretching over the respective preceding sojourn interval. As a result, these sojourns stay bare. We now apply these rules to correlation functions of the TSS.

21.7.1   The populations

The noninteracting-blip approximation does not describe the long-time dynamics of the population difference $\langle \sigma_z \rangle_t$ correctly. In the Ohmic case for $K < 1$ and in the absence of a bias, the expectation value $\langle \sigma_z \rangle_t$ is found to decay asymptotically as $\propto (\Delta_r t)^{-(2-2K)}$, as follows from Eq. (21.98). The algebraic decay in the NIBA originates from a branch point of the Laplace-transformed kernel $\widehat{K}_z^{(s)}(\lambda) \propto \lambda^{2K-1}$ at $\lambda = 0$ [cf. Eq. (21.93)]. The asymptotic behaviour is changed qualitatively by the interblip correlations, as we have shown for the particular case $K = \frac{1}{2} - \kappa$ in Subsection 21.6.3. It is difficult to work out the effects of the interblip correlations quantitatively for general K since many different frequency scales act in combination. This phenomenon is well-known from the closely related Kondo problem in the antiferromagnetic sector. The resummation of the alternating series (21.76) of blip sequences results in kernels $\widehat{K}_z^{(s/a)}(\lambda)$ which are regular at $\lambda = 0$ (see Subsection 21.8.1). Thus, $\langle \sigma_z \rangle_t$, or the envelope function of it, approaches the equilibrium value exponentially fast, as follows by Laplace inversion of Eq. (21.79). Further discussion of relaxation and decoherence, and several exact analytical results at $T = 0$ are given in Section 21.8.

The exponential decay towards the equilibrium distribution $\langle \sigma_z \rangle_\infty$ also follows immediately from the above two rules. As in the diagrams in Fig. 21.4, there is (at long time) only a single cluster both for $P_s(t)$ and $P_a(t)$, and hence exponential decay.

21.7.2   The population correlations and generalized Shiba relation

With rules (I) and (II), we see from Fig 21.5 (right) that the response function $\chi_z(t)$ is diagrammatically represented by a single neutral cluster surrounding the origin of the time axis. Hence, $\chi_z(t)$ decays exponentially at long times.

Similarly, the contribution $S_z^{(\mathrm{unc})}(t)$ to $S_z(t)$ given in Eq. (21.58) decays exponentially, as follows with the results obtained in Subsection 21.7.1. Employing rules (I, II) to the diagram in Fig. 21.5 (left), we find that the correlation term $R(t)$ is represented by two neutral clusters, the one in the negative time branch near the origin, the other in the positive time branch near t. Therefore, at asymptotic times, the clusters are separated by an interval of length t. Since the interval is bare, the clusters are interacting with the unscreened dipole-dipole interaction which in the Ohmic case is $\ddot{Q}'(t) = -2K/t^2$. Hence we should expect that $S_z(t)$ decays as $1/t^2$ for Ohmic dissipation.

To put this argument in concrete form, we expand the blip interaction factor G_{n+m} in Eq. (21.61) under the assumption that the effective length of the two clusters is small compared to the interval t between the clusters. We then have

$$F_{m,n} \;\to\; F_m^{(-)}(\{\tau_k\})\, F_n^{(-)}(\{\tau_j\}) \left(1 - \ddot{Q}'(t) \sum_{j=1}^{n} \sum_{k=n+1}^{n+m} \xi_j \xi_k \tau_j \tau_k \right), \qquad (21.250)$$

where $F_n^{(-)}$ and $F_m^{(-)}$ are the influence functions (21.56) for the clusters with n blips in the negative and m blips in the positive time branch. Next, we substitute Eq. (21.250) into Eq. (21.59). Because of the $\{\xi_j\}$ summations, those terms which are odd in the blip labels ξ_j for the individual time branches cancel out. The remaining contribution can be written in the symmetric form

$$
\begin{aligned}
R(t) &= \ddot{Q}'(t) \lim_{t_q \to \infty} \sum_{m=1}^{\infty} (-1)^m \int_0^{t_q} \mathcal{D}_{2m,0}\{t_j\} \frac{1}{2^m} \sum_{\{\xi_j = \pm 1\}} B_m^{(s)} F_m^{(-)} \left(\sum_{k=1}^m \xi_k \tau_k \right) \\
&\times \lim_{t_p \to -\infty} \sum_{n=1}^{\infty} (-1)^n \int_{t_p}^0 \mathcal{D}_{0,2n}\{t_j\} \frac{1}{2^n} \sum_{\{\xi_j = \pm 1\}} B_n^{(s)} F_n^{(-)} \left(\sum_{\ell=1}^n \xi_\ell \tau_\ell \right) .
\end{aligned}
$$

This expression describes two neutral clusters which are interacting with each other through the dipole-dipole interaction $\ddot{Q}'(t)$. Next, we observe upon inspection of the expression (21.64) that the series expansion for each cluster can be identified with the static susceptibility $\overline{\chi}_z$ at $T = 0$ (apart from a missing factor $2/\hbar$). With this identification we obtain for $S_z(t)$ the exact asymptotic behaviour

$$
S_z(t) = \left(\frac{\hbar \overline{\chi}_z}{2} \right)^2 \ddot{Q}'(t) = -2K \left(\frac{\hbar \overline{\chi}_z}{2} \right)^2 \frac{1}{t^2} ; \qquad t \gg \hbar \overline{\chi}_z . \tag{21.251}
$$

This relation holds for any bias and for any K smaller than one. Interestingly, the bias and the damping parameter K enter into (21.251) only implicitly through the static zero temperature susceptibility, except for the extra factor K. For $K > 1$, the Kondo temperature $T_K \equiv 2/(\pi k_B \overline{\chi}_z)$ is zero in the scaling limit, and the asymptotic expansion discussed here does not exist. In this damping regime, the high-temperature expansion is valid down to $T = 0$. In the absence of a bias, one obtains from Eq. (21.177) in the zero damping limit $\overline{\chi}_z \to 2/\hbar\Delta$, while for $K = 1/2$ one has $\overline{\chi}_z = 8/(\pi\hbar\gamma)$ [cf. Eq.(21.223)], where $\gamma = \pi\Delta^2/2\omega_c$. From the correspondence with the Kondo model [132, 89], we find in the damping regime $1 - K \ll 1$ with Eq. (19.105) $\overline{\chi}_z = [2(1 - K)]^{K/(1-K)}/(\hbar\Delta_r)$, where $\Delta_r = (\Delta/\omega_c)^{K/(1-K)}\Delta$.

We remark that the zero temperature behaviour (21.251) is also valid at finite temperatures in the time region $\hbar\overline{\chi}_z \ll t \ll \hbar\beta$. At times $t \gg \hbar\beta$, the correlation function $C(t)$ approaches the equilibrium value exponentially fast with a rate given by the smallest Matsubara frequency $\nu_1 = 2\pi/\hbar\beta$.

From Eq. (21.251) we may also infer the behaviour in frequency space near $\omega = 0$, known as the *generalized Shiba relation*, [20]

$$
\lim_{\omega \to 0} \widetilde{S}_z(\omega \to 0)/|\omega| = 2\pi K(\hbar\overline{\chi}_z/2)^2 , \tag{21.252}
$$

or equivalently

$$
\lim_{\omega \to 0} \hbar\widetilde{\chi}_z''(\omega)/\omega = 2\pi K(\hbar\overline{\chi}_z/2)^2 . \tag{21.253}
$$

These relations are analogous to a relation that has been proven by Shiba [427] for the Anderson model. While Shiba's derivation is essentially based upon a particle number conservation law, the proof given here is based upon the exact formal solution.

---

[20]For the normalization of $\overline{\chi}_z$, see Eq. (19.13) and footnote, and Eq. (19.104).

Finally, we should like to relate the asymptotic behaviour of the damped spin to that of the damped linear oscillator discussed in Subsection 6.4.1 using the correspondences

$$S_{\mathrm{osc}}(t) \;\hat{=}\; \tfrac{1}{4}q_0^2\, S_{\mathrm{spin}}(t) \,, \qquad \chi_{\mathrm{osc}}(t) \;\hat{=}\; \tfrac{1}{4}q_0^2\, \chi_{\mathrm{spin}}(t) \,, \qquad M\gamma \;\hat{=}\; 2\pi\hbar K/q_0^2 \,,$$

where $\tfrac{1}{2}q_0$ is identified with the spread of the equivalent undamped oscillator in the ground state, $\tfrac{1}{2}q_0 = \sqrt{\langle q^2 \rangle} = \sqrt{\hbar/2m\omega_0}$ [cf. also text below Eq. (3.185)]. With these correspondences, the asymptotic behaviour (6.63) maps exactly on the form (21.251). This supports the conjecture that a generalized Shiba relation holds for any system with Ohmic damping and nonzero static susceptibility at $T = 0$ [see also the generalized Wilson ratio for non-Ohmic spectral densities discussed in Section 21.9]. With the correspondences, it is easy to find from Eq. (6.69) the generalization of the Shiba relation for the spin-boson model in the super-Ohmic case with $s < 2$,

$$\lim_{\omega \to 0^+} \hbar\widetilde{\chi}_z''(\omega)/|\omega|^s \;=\; 2\pi\delta_s\omega_{\mathrm{ph}}^{1-s}(\hbar\overline{\chi}_z/2)^2 \,. \tag{21.254}$$

Alternatively, we may substitute in the first form in Eq. (21.251) the super-Ohmic pair interaction at $T = 0$ as found from Eq. (18.45), and then switch to the Fourier transform. Eventually, we find again the result (21.254). Thus the spin correlation function $S_z(t)$ decays asymptotically as t^{-1-s}.

In the sub-Ohmic weak-damping regime (18.36), the static susceptibility is finite. Then, the asymptotic behaviour of $S_z(t)$ is again described by the form (21.251) in which $Q(t)$ is the sub-Ohmic kernel, and hence $S_z(t) \propto t^{-1-s}$. Then there holds again a Shiba relation of the form (21.254).

The generalized Shiba relation (21.252), or Eq. (21.253), has been derived first in Ref. [208]. Recently, this relation has been corroborated by diverse other methods. It has been verified numerically by employing the correspondence with the anisotropic Kondo model [428, 429], and by numerical integration of the flow equations resulting from a continuous sequence of infinitesimal unitary transformations [356, 357]. The relation has been confirmed also analytically by using nonperturbative methods derived from a Bethe ansatz [430].

21.7.3   The coherence correlation function

Finally, consider the asymptotic dynamics of the coherence correlation function at $T = 0$. Using rules (I) and (II), one finds that the contribution of group A has a single neutral cluster surrounding the origin of the time axis. Hence, both $S_x^A(t)$ and $\chi_x^A(t)$ decay exponentially fast. In group B, we have two clusters with opposite charges ± 1, one in each time branch. Since in each branch the initial sojourn is free of insertions, the two clusters are situated near the origin and near t. Since the intermediate sojourn interval is bare, they interact with the unscreened charge-charge interaction $e^{-Q'(t)} \propto t^{-2K}$. This interaction directly determines the long-time behaviour of $S_x(t)$ and $\chi_x(t)$ for $0 < K < 1$,

$$S_x(t) \propto e^{-Q'(t)} \propto t^{-2K} , \qquad \chi_x(t) \propto e^{-Q'(t)} \propto t^{-2K} . \qquad (21.255)$$

Thus, the coherence correlations decay with a power law in which the exponent depends on the damping strength. This is

The asymptotic behaviours given in Eq. (21.255) are consistent with the fluctuation-dissipation theorem. Upon matching the Fourier transforms of the expressions (21.255) with the fluctuation-dissipation theorem (21.241) and taking the inverse transforms we find

$$S_x(t) = (\hbar/2) \cot(\pi K) \chi_x(t) \quad \text{for} \quad t \gg 1/\Delta_r . \qquad (21.256)$$

For $K = \frac{1}{2}$, the function $\chi_x(t)$ decays asymptotically as $1/t$ whereas $S_x(t)$ decays exponentially, as we can see from Eqs. (21.239) and (21.234). This is consistent with the FDT-relation (21.241) since the $\cot(\pi K)$-factor vanishes at $K = \frac{1}{2}$.

The law (21.255) leads to a different behaviours of the linear static susceptibility for K below and above $\frac{1}{2}$. For $K < \frac{1}{2}$, the slow decay of $\chi_x(t)$ implies that the linear static susceptibility diverges algebraically, $\overline{\chi}_x^{(\ell)} \propto T^{2K-1}$, as $T \to 0$, which indicates that the response at zero temperature is actually nonlinear. On the other hand, for $K > \frac{1}{2}$, the decay of $\chi_x(t)$ is sufficiently fast, so that the linear static susceptibility is finite at $T = 0$. Recently, these properties have been confirmed numerically [431].

In summary, the decay of the population correlation function $S_z(t)$ at long times is determined by the unscreened dipole interaction, whereas the decay of the coherence correlation function $S_x(t)$ is determined by the unscreened charge interaction.

21.8   From weak to strong tunneling: relaxation and decoherence

21.8.1   Incoherent tunneling beyond the nonadiabatic limit

Upon taking in the GME (21.73) the Markovian limit, we obtain the rate equation

$$d\langle \sigma_z \rangle_t / dt = k^+ - k^- - (k^+ + k^-)\langle \sigma_z \rangle_t , \qquad (21.257)$$

where the forward/backward rate from the state $\sigma = \pm 1$ to the state $\sigma = \mp 1$ is

$$k^\pm = \frac{1}{2} \int_0^\infty d\tau \left[K_z^{(s)}(\tau) \pm K_z^{(a)}(\tau) \right] = \frac{1}{2} \left[\widehat{K}_z^{(s)}(\lambda = 0) \pm \widehat{K}_z^{(a)}(\lambda = 0) \right] . \qquad (21.258)$$

Eq. (21.257) describes the population dynamics in the incoherent regime. The solution is[21]

---

[21]Strictly speaking, the total transfer rate $k = k^+ + k^-$ is given by the smallest real solution of the equation $k - \widehat{K}_z^{(s)}(\lambda = -k) = 0$. When k is small compared to the other frequency scales of the system, this relation reduces to the form (21.258). In the coherent regime, Eq. (21.257) is not valid (see Eq. (21.181)). In this regime, the rate $k \stackrel{\frown}{=} \gamma_r$ describes the decay of the incoherent part of $\langle \sigma_z \rangle_t$.

$$\langle \sigma_z \rangle_t = \langle \sigma_z \rangle_\infty + [1 - \langle \sigma_z \rangle_\infty] e^{-(k^+ + k^-)t} . \tag{21.259}$$

The equilibrium value

$$\langle \sigma_z \rangle_\infty = \frac{k^+ - k^-}{k^+ + k^-} \tag{21.260}$$

is in agreement with the exact formal expression given in Eq. (21.82).

Upon introducing the blip and sojourn intervals (21.34) in the series expression (21.76) and switching to the Laplace transform, we obtain the series expression

$$k^\pm = \sum_{n=1}^\infty (-1)^{n-1} \frac{\Delta^{2n}}{2^{n+1}} \int_0^\infty d\tau_n \left[\prod_{j=1}^{n-1} ds_j \, d\tau_j \right] \sum_{\{\xi_j = \pm 1\}} \left(\widetilde{F}_n^{(+)} B_n^{(s)} \pm \widetilde{F}_n^{(-)} B_n^{(a)} \right) . \tag{21.261}$$

The rate expression is in the form of a series in the number of bounces (blips). The $n = 1$ term is the nonadabatic rate, which has been discussed in Sec. 20.2. The terms with $n > 1$ represent the adiabatic corrections. They depend significantly on the interblip correlations and would vanish if the interblip interactions were disregarded.

A graph-theoretic approach to the computation of adiabatic corrections in the real-time path integral formulation has been given by Stockburger and Mak [432]. The linked cluster sum considered by these authors is the graphical equivalent of the series expression (21.261) with the irreducible influence functions $\widetilde{F}_n^{(\pm)}$. They presented an approximate resummation of the multi-blip contributions under the assumption that the separation of blips is larger than $1/\omega_c$. The numerical studies show two competing adiabatic effects. At high temperatures, the rate is reduced (compared to the nonadiabatic case) because of correlated re-crossings of the Landau-Zener region. For low enough temperatures, the rate is enhanced over the NIBA rate because of an effective reduction of the reaction barrier by the electronic coupling and by nuclear tunneling. The exact resummation of the series expressions (21.76) and (21.261) in the weak-damping limit is given above in Section 21.4.

Exact solution in analytic form at $T = 0$

Interestingly, in the Ohmic scaling limit at $T = 0$, the incoherent rate can be determined in analytic form in all orders in Δ. The analysis of the Coulomb gas representation (21.261) performed in Ref. [433] reveals a close relationship between the rate contribution of order Δ^{2n} in the TSS with the rate k_n^+ from site 0 to site n in the Schmid model, which is of the same order in Δ (see Section 25.6). With the explicit form of the rate k_n^+ given below in Eq. (25.126), one then obtains the weak-tunneling series of the TSS (forward) rate in the analytic form

$$k^+(\epsilon) = \frac{\epsilon}{2\sqrt{\pi}} \sum_{m=1}^\infty \frac{1}{m!} \frac{\Gamma(Km)\left[1 - \cos(2\pi Km)\right]}{\Gamma[\frac{3}{2} + (K-1)m]} \left(\frac{\epsilon}{\epsilon_{sb}} \right)^{(2K-2)m} . \tag{21.262}$$

The Kondo-like frequency scale ϵ_{sb} is related to the parameters of the TSS by

$$\epsilon_{sb}^{2-2K} = \frac{\Gamma^2(1-K)}{2^{2K}} \frac{\Delta^2}{\omega_c^{2K}} . \tag{21.263}$$

For rational K, the series (21.262) can be written as linear combination of hypergeometric functions. In the regime $K \gtrsim 1$, the weak-tunneling series (21.262) is absolutely converging when the bias is small/large enough. The leading term is the nonadiabatic or golden rule rate (20.81). Expressed in terms of ϵ_{sb}, it reads

$$k_{\text{gr}}^{+}(\epsilon) = \sin^2(\pi K) \frac{\Gamma(K)}{\Gamma(\frac{1}{2} + K)} \frac{\epsilon}{\sqrt{\pi}} \left(\frac{\epsilon}{\epsilon_{\text{sb}}} \right)^{2K-2}. \qquad (21.264)$$

At $K = \frac{1}{2}$, the golden rule rate $k_{\text{gr}}^{+}(\epsilon)$ coincides with the full rate $k^{+}(\epsilon)$.

Upon performing transformations analogously to those executed below in Subsection 25.5.3, the series (21.262) can be transformed into the integral representation

$$k^{+}(\epsilon) = \text{Re}\, \frac{\epsilon}{2\pi i} \int_{\mathcal{C}} \frac{dz}{z} \left\{ \sqrt{z - 1 - z^K u_2} - \sqrt{z - 1 - z^K u_1} \right\}, \qquad (21.265)$$

where $u_1 = (\epsilon/\epsilon_{\text{sb}})^{2K-2}$ and $u_2 = e^{i2\pi K} u_1$. The contour $\mathcal{C}$ starts at the origin, encircles the branch point in counter-clockwise sense, and returns to the origin. The representation (21.265) converges not only in the regime where the series (21.262) does, but for all values of K and of ϵ/ϵ_0, . This allows us to determine the asymptotic series.

Consider first the regime $K < 1$. Upon changing variable z to $y = z^{1-K}/u_1$ and $y = z^{1-K}/u_2$, and expanding the ensuing integrands in powers of $\epsilon/\epsilon_{\text{sb}}$, we get [433]

$$k^{+}(\epsilon) = \frac{\epsilon_{\text{sb}}}{2\sqrt{\pi}} \sum_{n=0}^{\infty} b_n(K) \frac{1}{n!} \frac{\Gamma[(\frac{1}{2} - n)\frac{K}{1-K}]}{(\frac{1}{2} - n)\Gamma[(\frac{1}{2} - n)\frac{1}{1-K}]} \left(\frac{\epsilon}{\epsilon_{\text{sb}}} \right)^{2n}. \qquad (21.266)$$

The function $b_n(K)$ is given by

$$b_n(K) = \begin{cases} 2\sin^2[\frac{\pi K}{1-K}(\frac{1}{2} - n)], & \text{for} \qquad K < \frac{1}{3}, \\ 1, & \text{for} \qquad \frac{1}{3} < K < 1. \end{cases} \qquad (21.267)$$

For very weak damping $K \ll 1$, both the Taylor series (21.262) and the asymptotic series (21.266) can be summed to the simple form ($\epsilon_{\text{sb}} = \Delta_{\text{eff}}$)

$$k^{+}(\epsilon) = \pi K \epsilon_{\text{sb}}^2 / \sqrt{\epsilon_{\text{sb}}^2 + \epsilon^2}, \qquad (21.268)$$

which agrees with the earlier result (21.180), $k^{+} = \gamma_{\text{r}}(T = 0)$.

In the zero bias limit, the asymptotic series (21.266) reduces to the $n = 0$-term. This yields for the forward rate of the unbiased TSS the expression

$$k^{+}(0) = \frac{b_0(K)}{\sqrt{\pi}} \frac{\Gamma[\frac{K}{2(1-K)}]}{\Gamma[\frac{1}{2(1-K)}]} \epsilon_{\text{sb}}. \qquad (21.269)$$

In the narrow regime $1 - K \ll 1$, we get from Eq. (21.269)

$$k^{+}(0) = \sqrt{\frac{2}{\pi}(1 - K)}\, \epsilon_{\text{sb}}, \quad \text{with} \quad \epsilon_{\text{sb}} = \left(\frac{1}{2(1-K)} \right)^{\frac{1}{1-K}} \Delta_{\text{r}} = \frac{2}{\pi} \frac{k_{\text{B}} T_K}{\hbar}. \qquad (21.270)$$

The second form relates ϵ_{sb} to the Kondo temperature T_K of the anisotropic Kondo model in the corresponding regime $\rho J_\perp \ll \rho J_\parallel$ [see Eq. (19.105)].

Consider next the asymptotic expansion for $K > 1$ and large bias. This may be derived from (21.265) upon changing from variable z to $t = e^{-i\pi} u_{1/2} z^K$. We then find in the section $K = p + \kappa$, where $p = 1, 2, \cdots$, and $0 \leq \kappa < 1$,

$$k^+(\epsilon) = \frac{\epsilon}{\sqrt{\pi}} \sum_{m=1}^{\infty} \frac{(-1)^m}{m!} \frac{\Gamma(\frac{m}{K}) \sin[\frac{1+p}{K}m\pi] \sin(\frac{p}{K}m\pi)}{K \Gamma[\frac{3}{2} + (\frac{1}{K} - 1)m]} \left(\frac{\epsilon}{\epsilon_{sb}} \right)^{(2/K-2)m}, \qquad (21.271)$$

From this we see that the rate is zero for a symmetric system. This confirms that there is a transition to self-trapping at $K = 1$ in the Ohmic scaling limit (18.54), as is known also from the anisotropic Kondo model. The localization transition has been discussed by Chakravarty [133], Bray and Moore [134], and by Hakim $et\ al.$ [135].

In Fig. 21.10 (left) the normalized rate k^+/k_{gr}^+ is plotted for different $K < 1$. The horizontal line represents the particular case $K = \frac{1}{2}$. For $K < \frac{1}{4}$, the full rate is always lower than the golden rule rate. Hence the numerous multi-bounce contributions interfere destructively in this regime. For $\frac{1}{2} < K < 1$, the multi-bounce contributions interfere constructively for all x so that the full rate is always above the golden rule rate. In the regime $\frac{1}{4} < K < \frac{1}{2}$, the normalized rate goes through a maximum as tunneling is increased, and finally falls below one. This reflects constructive interference at small and intermediate x, and destructive interference at large x.

Fig. 21.10 (right) shows plots of the normalized rate k^+/k_{gr}^+ for $K > 1$. The normalized rate goes through a maximum which is shifted to higher x when K is increased. At large enough x, the rate k^+ falls below the golden rule rate. Hence there is constructive interference of the tunneling terms at small and intermediate x, and destructive interference in the strong-tunneling regime.

21.8.2   Decoherence at zero temperature: analytic results

One might guess from the known special cases that there is in the spin-boson model a close relationship between the relaxation rate $k^+(\epsilon)$ and the decoherence rate $\gamma(\epsilon)$. The former describes relaxation to the ground state of the TSS, and the latter loss of phase coherence, which reveals itself in the damping of the oscillatory dynamics as given, e.g., in the expression (21.181). In the weak damping limit, we simply have $\gamma(\epsilon) = k^+(\epsilon)/2$, as we can see from the last relation in Eq. (21.180). In Ref. [433], there has been posed the conjecture that this relation holds in the regime $0 < K \leq \frac{1}{3}$, and that a similar relation holds in the regime $\frac{1}{3} < K \leq \frac{1}{2}$, but with the coefficient $b_n(K) = 2\sin^2[\frac{\pi K}{1-K}(\frac{1}{2} - n)]$ in Eq. (21.266) instead of $b_n(K) = 1$. With this conjecture, which is in agreement with the relations known in special cases, the strong-tunneling series of the decoherence rate at $T = 0$ in the regime $0 < K \leq \frac{1}{2}$ is found to read

$$\gamma(\epsilon) = \frac{\epsilon_{sb}}{2\sqrt{\pi}} \sum_{n=0}^{\infty} \sin^2[\pi(\frac{1}{2} - n)\frac{K}{1-K}] \frac{1}{n!} \frac{\Gamma[(\frac{1}{2} - n)\frac{K}{1-K}]}{(\frac{1}{2} - n)\Gamma[(\frac{1}{2} - n)\frac{1}{1-K}]} \left(\frac{\epsilon}{\epsilon_{sb}} \right)^{2n}. \qquad (21.272)$$

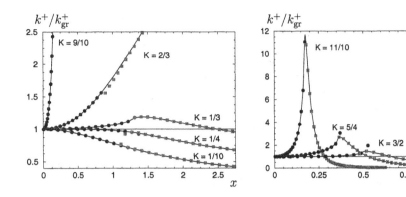

Figure 21.10: The scaled rate k^+/k_{gr}^+ is plotted versus $x = (\Delta_{\mathrm{r}}/\epsilon)^{1-K} = (\epsilon/\omega_{\mathrm{c}})^K \Delta/\epsilon$ for various values of K, $K < 1$ (left) and $K > 1$ (right). The circles are calculated from the weak-tunneling series, and the squares from the asymptotic strong-tunneling series. The full curve is the respective hypergeometric function expression.

A significant cheque of the expression (21.272) is the asymptotic strong-tunneling limit $\epsilon \to 0$, in which we get

$$\gamma(0) = \frac{1}{\sqrt{\pi}}\sin^2\left[\frac{\pi K}{2(1-K)}\right]\frac{\Gamma\left[\frac{K}{2(1-K)}\right]}{\Gamma\left[\frac{1}{2(1-K)}\right]}\,\epsilon_0 \,. \tag{21.273}$$

The decoherence rate of the unbiased Ohmic TSS has been calculated within the framework of integrable QFT [415]. The respective result, which is given above in Eqs. (21.102)-(21.104) is in full agreement with the expression (21.273) for $K \leq \frac{1}{2}$. The expression (21.272) matches also the known expressions in the biased case in the regimes $K \ll 1$ and K close to $\frac{1}{2}$. These congruences may be regarded as clear hints that the strong tunneling series (21.272) is indeed the actual analytic expression for the decoherence rate. Assuming that the conjecture is correct, the strong-tunneling series (21.272) represents the exact lower bound for decoherence in the SB model in the scaling limit in the regime $0 < K \leq \frac{1}{2}$. The bound is saturated at $T = 0$.

21.9   Thermodynamics from dynamics

In numerical computations of dynamical correlation functions, often the imaginary-time correlation function is calculated by standard Monte Carlo simulations and then continued to real time by using, e.g., Padé approximant methods or image reconstruction techniques like Max-Ent [434]. However, the analytic continuation of a function which is blurred with numerical noise is ill-defined: small errors in the input data can lead to exponentially enhanced errors in the output data.

Here we wish to disseminate that sometimes the opposite proceeding, namely computation of thermodynamics from the asymptotic dynamics may have considerable advantages.

In the standard thermodynamic approach, first of all the partition function is computed, which then provides the basis to calculate the static susceptibility, specific heat, etc. This proceeding has two major disadvantages. First, it is difficult to perform approximations in the Coulomb gas representation for the partition function which preserve the symmetry of the imaginary-time interaction, $W(\tau) = W(\hbar\beta - \tau)$. Secondly, the fugacity expansion for Z is often found to converge badly. Remarkably, the fugacity expansion for $\ln Z$ or the free energy usually converges much faster [cf. the discussion in Ref. [435] for an unordered Coulomb gas].

Instead of using the imaginary-time approach for the partition function Z, we can find the thermodynamic properties in a dynamical approach by studying the equilibrium state which the system takes at asymptotic times, if it is ergodic. To proceed, we first note that the equilibrium distribution $\langle\sigma_z\rangle_\infty$ is related to the partition function Z by $\langle\sigma_z\rangle_\infty = (2/\hbar\beta)\partial \ln Z/\partial\epsilon = -(2/\hbar)\partial F/\partial\epsilon$ [cf. Eq. (19.12)]. Thus, we directly obtain the free energy if we integrate the formal expression for $\langle\sigma_z\rangle_\infty$ with respect to the bias. Since the integration constant does not depend on Δ and β, it may be set equal to zero without any restriction. It is just in this way that we directly obtain an exact formal series expression for $\ln Z$ or the free energy.[22] We now briefly sketch this approach, and then use it to show that the damped system has a surprising universal behaviour at low temperatures (see also [208]).

Starting with the formal solution for $\langle\sigma_z\rangle_\infty$ given by Eq. (21.52), we find for the free energy the expression

$$F(T,\epsilon) = \lim_{t\to\infty} \frac{\hbar}{2} \sum_{m=1}^{\infty}(-1)^{m-1}\int_0^t \mathcal{D}_{2m,0}\{t_j\}\, \frac{1}{2^m}\sum_{\{\xi_j\}} F_m^{(-)} A_m^{(s)}\,, \qquad (21.274)$$

where

$$A_m^{(s)} = \cos\left(\epsilon\sum_{j=1}^{m}\xi_j\tau_j\right)\Big/\sum_{j=1}^{m}\xi_j\tau_j\,. \qquad (21.275)$$

We are not allowed to interchange in Eq. (21.274) $\lim_{t\to\infty}$ with the summation of the series since the limit would give a divergent result for each term.

Secondly, we use the fact that the static (nonlinear) susceptibility $\overline{\chi}_z$ is related to $\langle\sigma_z\rangle_\infty$ by $\overline{\chi}_z = (2/\hbar)\partial\langle\sigma_z\rangle_\infty/\partial\epsilon$ [cf. Eq. (19.13)]. Using this relation, we obtain from the series expression (21.52) for $\langle\sigma_z\rangle_\infty$ the exact formal solution for $\overline{\chi}_z$ in the form

$$\overline{\chi}_z(T,\epsilon) = \lim_{t\to\infty} \frac{2}{\hbar} \sum_{m=1}^{\infty}(-1)^{m-1}\int_0^t \mathcal{D}_{2m,0}\{t_j\}\, \frac{1}{2^m}\sum_{\{\xi_j\}} F_m^{(-)} C_m^{(s)}\,, \qquad (21.276)$$

where

$$C_m^{(s)} = \cos\left(\epsilon\sum_{j=1}^{m}\xi_j\tau_j\right)\sum_{j=1}^{m}\xi_j\tau_j\,. \qquad (21.277)$$

---

[22]We recall that the imaginary-time approach directly leads to a series expression for the partition function Z (cf. Chapter 19).

Let us now determine the asymptotic low temperature expansion of F and $\overline{\chi}_z$. First, we observe that the temperature enters into the expressions (21.274) and (21.276) only through the pair interaction $Q'(\tau)$. Secondly, we write $Q'(\tau) = Q'_0(\tau) + Q'_1(\tau)$, where $Q'_0(\tau)$ is the pair interaction for zero temperature, and $Q'_1(\tau)$ is the finite temperature contribution. The leading correction at low T is easily computed for the form (18.44) of $G_{\mathrm{lf}}(\omega)$ from Eq. (18.45), and is

$$Q'_1(\tau) = \gamma_s (\hbar\beta\widetilde{\omega})^{1-s}(\tau/\hbar\beta)^2\left\{1 + \mathcal{O}[(\tau/\hbar\beta)^2]\right\}, \qquad (21.278)$$

where

$$\gamma_s = 2\delta_s\Gamma(1+s)\zeta(1+s), \qquad (21.279)$$

and where $\Gamma(z)$ and $\zeta(z)$ are the gamma and Riemann zeta function. With this expression the blip correlation factor G_m defined in Eq. (21.37) takes the low temperature expansion

$$G_m = G_m^{(0)}\left\{1 - \gamma_s\widetilde{\omega}^{1-s}\left(\frac{k_B T}{\hbar}\right)^{1+s}\left(\sum_{j=1}^m \xi_j\tau_j\right)^2 + \mathcal{O}(T^{3+s})\right\}, \qquad (21.280)$$

where $G_m^{(0)}$ is the blip interaction factor at zero temperature. Inserting now (15.96) into the expressions (15.92) and (15.93), and observing that the resulting powers of $\sum_j \xi_j\tau_j$ can be generated by differentiations with respect to the bias, we immediately find for F and $\overline{\chi}_z$ the low temperature expansions

$$F(T,\epsilon) = F(0,\epsilon) - \frac{1}{4}\gamma_s\overline{\chi}_z(0,\epsilon)\,(\hbar\widetilde{\omega})^{1-s}\,(k_B T)^{1+s} + \mathcal{O}(T^{3+s}), \qquad (21.281)$$

$$\overline{\chi}_z(T,\epsilon) = \overline{\chi}_z(0,\epsilon) + (\gamma_s/\hbar^2)\overline{\chi}''_z(0,\epsilon)\,(\hbar\widetilde{\omega})^{1-s}\,(k_B T)^{1+s} + \mathcal{O}(T^{3+s}). \qquad (21.282)$$

Here, $F(0,\epsilon$ and $\overline{\chi}_z(0,\epsilon)$ are the free energy and static nonlinear susceptibility at zero temperature and $\overline{\chi}''_z = \partial^2\overline{\chi}_z^{(0)}/(\partial\epsilon)^2$. The expressions (21.281) and (21.282) are the asymptotic expansions of F and $\overline{\chi}_z$ for a two-state system described by a spectral density of the form (18.44) under the limitations discussed below. The leading temperature dependence is given by the power law T^{1+s}. The prefactor depends on the factor γ_s and on $\overline{\chi}_z^{(0)}$. Notice that the bias enters only through the nonlinear susceptibility at zero temperature. Observing now that the specific heat is defined by

$$c(T) = -k_B^{-1}T\partial^2 F/(\partial T)^2, \qquad (21.283)$$

we immediately see that the leading dependence of $c(T)$ on temperature is T^s. Thus, it is natural to adjust the definition of the Wilson ratio, Eq. (19.53), to non-Ohmic spectral densities ($s \neq 1$). We define the generalized Wilson ratio

$$R_s \equiv \lim_{T\to 0}\frac{4c(T)}{\overline{\chi}_z\,(\hbar\widetilde{\omega})^{1-s}(k_B T)^s}, \qquad (21.284)$$

where $\overline{\chi}_z$ is the nonlinear static susceptibility. We then obtain from Eq. (21.281)

$$R_s = 2s\Gamma(2+s)\zeta(1+s)\delta_s . \qquad (21.285)$$

These results for the free energy, the specific heat, and the Wilson ratio are generally valid provided that the static susceptibility at zero temperature exists. This is the case for general s when the system is *biased* (nonlinear static susceptibility). The linear susceptibility is finite in the super-Ohmic case for all δ_s, and in the Ohmic case for K in the regime $0 < K < 1$. Thus, in these cases the relation (21.285) is also valid. For $s = 1$ and $K > 1$, and in the sub-Ohmic case $s < 1$ and any $\delta_s > 0$, the limit $\lim_{\epsilon \to 0} \overline{\chi}_z(T = 0, \epsilon)$ is divergent. Hence in these parameter regimes, the above analysis is not applicable. Instead, the high temperature expansions for $\overline{\chi}(T, \epsilon = 0)$ and $F(T)$, which are the expansions in powers of Δ^2, Eq. (21.274) and Eq. (21.276), are valid down to $T = 0$. For an Ohmic bath, the Wilson ratio simplifies to

$$R_1 = 2K\pi^2/3 . \qquad (21.286)$$

Using the correspondence relation (19.102) for K, we see that the Wilson ratio (21.286) of the Ohmic TSS coincides with the Wilson ratio (19.104) of the s-d model. Recently, the Wilson ratio has been studied numerically for the s-d model, and excellent agreement with the formula (21.286) has been found [429].

22. The driven two-state system

With the advance of laser technology and the possibility for experimental time resolution in the sub-picosecond regime, there has been growing interest in the study of the dynamics of quantum systems that are driven by strong time-dependent external fields. Quantum dynamics of explicitly time-dependent Hamiltonians shows a variety of novel effects, such as the phenomenon of coherent destruction of tunneling [436], stabilization of localized states which would otherwise decay [437], and the possibility of controlling the tunneling dynamics with pulsed monochromatic light [438, a] [439] and sinusoidal fields [438, b] [439]. For a review, see Refs. [440, 441]. Here we restrict the attention to a two-state system which is simultaneously exposed to a fluctuating force by the surroundings and to deterministic time-dependent forces.

22.1 Time-dependent external fields

We generalize the Hamiltonian (18.13) of the global system by taking into account the coupling with external time-dependent fields. We assume that the external fields couple to the system's operators σ_z and σ_x only, and have no effect on the bath.

22.1.1 Diagonal and off-diagonal driving

The coupling of an external field to σ_z describes, e.g., an electric field coupled to the dipole moment of the TSS, or, in the rf SQUID device discussed in Subsection

3.2.2, a time-dependent magnet flux threading the ring. This coupling leads to a temporal modulation of the bias which is superimposed on the static bias. This entails substituting in the Hamiltonian (18.13)

$$\epsilon \quad \Longrightarrow \quad \epsilon(t) = \epsilon_0 + \epsilon_1(t) . \tag{22.1}$$

Here, ϵ_0 is the bias frequency related to the intrinsic static strain field, and $\epsilon_1(t)$ is a bias modulation due to the externally applied time-dependent force. Regarding electron transfer in a solvent, it is conceivable to control charge tunneling by application of strong continuous laser fields. For charge transfer in nanostructured devices, we may think of regulating the dynamics by turning on microwave irradiation or a high-frequency voltage. In pump-probe set-ups, the relevant system is subject to a pulse-shaped driving force. The modulation of the bias energy leads to modified bias phases. The accumulated phases for a path with a sequence of m blips is given by

$$\varphi_m = \sum_{j=1}^{m} \xi_j \vartheta(t_{2j}, t_{2j-1}) , \quad \text{where} \quad \vartheta(t_2, t_1) = \int_{t_1}^{t_2} dt' \, \epsilon(t') . \tag{22.2}$$

The bias factors replacing the forms (21.47) are expressed in terms of φ_m as

$$B_m^{(s)} = \cos(\varphi_m) , \qquad B_m^{(a)} = \sin(\varphi_m) . \tag{22.3}$$

Secondly, a "quadrupole"-like time-dependent coupling induces a temporal variation of the barrier opacity. A barrier modulation can be realized, e.g., in a superconducting loop with two Josephson junctions [442]. The quadrupole coupling leads to a *multiplicative* modification of the tunneling coupling. For harmonic pulsation of the barrier, we have in the Hamiltonian (18.13) the substitution [443]

$$\Delta \quad \Longrightarrow \quad \Delta(t) = \Delta \exp[\mu \cos(\nu t)] , \tag{22.4}$$

where ν is the angular frequency and μ a suitable dimensionless amplitude.

22.1.2 Exact formal solution

In the exact formal series expressions obtained in Chapter 21, the time-dependent tunneling matrix element is accounted for by the substitution [e.g., in Eq. (21.41)]

$$\Delta^m \quad \longrightarrow \quad \Delta_m(\{t_j\}) = \prod_{j=1}^{m} \Delta(t_j) . \tag{22.5}$$

In the nonconvolutive case, the Laplace transformed master equation (21.73) for $P(t) = \langle \sigma_z \rangle_t$ takes with the kernel $\widehat{K}_\lambda^{(s/a)}(t) = \int_0^\infty d\tau \, e^{-\lambda \tau} K_z^{(s/a)}(t+\tau, t)$ the form[1]

$$\lambda \widehat{P}(\lambda) = 1 + \int_0^\infty dt \, e^{-\lambda t} \left[\widehat{K}_\lambda^{(a)}(t) - \widehat{K}_\lambda^{(s)}(t) P(t) \right] . \tag{22.6}$$

---

[1]We restrict the attention to the discussion of $\langle \sigma_z \rangle_t$. The corresponding study of $\langle \sigma_x \rangle_t$ is without difficulty and left to the reader.

Consider now first the case of a periodic bias modulation with period $\mathcal{T} = 2\pi/\omega$,

$$\epsilon(t) = \epsilon(t + 2\pi n/\omega) . \tag{22.7}$$

Then the kernels $\widehat{K}_\lambda^{(s/a)}(t)$ have the periodicity of the driving field and can be written as a Fourier series,

$$\widehat{K}_\lambda^{(s/a)}(t) = \sum_{m=-\infty}^{\infty} k_m^{(s/a)}(\lambda)\, e^{-im\omega t} . \tag{22.8}$$

Substituting Eq. (22.8) into Eq. (22.6), one finds the solution of Eq. (22.6) in recursive form [444]. The population $P(t)$ consists of a transient and a persistent oscillatory part. The transient behaviour is determined by the zeros of the characteristic equation

$$\lambda + im\omega + k_0^{(s)}(\lambda + im\omega) = 0 . \tag{22.9}$$

The primary solutions λ_0 are those for $m = 0$, and the time scale for the decay of the corresponding transients is set by the smallest real negative part. All the other solutions of Eq. (22.9) for $m = \pm 1, \pm 2, \ldots$, are found by multiple shifts by the driving frequency in imaginary direction, $\lambda_m = \lambda_0 - im\omega$. The corresponding transient contributions decay on the same time scale as the primary transients.

After the transient behaviour has died, $P(t)$ oscillates with the period $\mathcal{T}$. This asymptotic regime is described by the Fourier series [444] (see also Ref. [445])

$$\lim_{t \to \infty} P(t) = P^{(\mathrm{as})}(t) = \sum_{m=-\infty}^{\infty} p_m\, e^{-im\omega t} . \tag{22.10}$$

The coefficient p_m is found from Eq. (22.6) as the residuum of the pole of $\widehat{P}(\lambda)$ at $\lambda = -im\omega$. We obtain the recursive relations

$$
\begin{aligned}
p_0 &= \frac{k_0^{(a)}(0)}{k_0^{(s)}(0)} - \sum_{m\neq 0} \frac{k_m^{(s)}(0)}{k_0^{(s)}(0)} p_m , \\[2mm]
p_m &= \frac{1}{-im\omega}\left(k_m^{(a)}(-im\omega) - \sum_n k_{m-n}^{(s)}(-im\omega)\, p_n \right) .
\end{aligned}
\tag{22.11}
$$

For monochromatic modulation

$$\epsilon(t) = \epsilon_0 + \epsilon_1 \cos(\omega t) , \tag{22.12}$$

the kernel coefficients $k_{2m}^{(a)}(\lambda)$ and $k_{2m+1}^{(s)}(\lambda)$ are zero for $\epsilon_0 = 0$. With this property, we find from Eq. (22.11) for zero static bias the selection rule[2]

$$p_{2m} = 0 . \tag{22.13}$$

Thus only odd multiples of the fundamental frequency are contained in $P^{(\mathrm{as})}(t)$.

___

[2] $\langle\sigma_y\rangle_t^{(\mathrm{as})}$ obeys the same selection rule [cf. Eq. (21.10)], and $\langle\sigma_x\rangle_t^{(\mathrm{as})}$ has only even harmonics.

In the general case, the kernels $k_m^{(s/a)}(\lambda)$ are power series expressions in Δ^2. In the NIBA, they take the simple form

$$
\begin{aligned}
k_m^{(s)}(\lambda) &= \Delta^2 \int_0^\infty d\tau \, e^{-\lambda\tau} e^{-Q'(\tau)} \cos[Q''(\tau)] A_m^{(s)}(\tau) \,, \\
k_m^{(a)}(\lambda) &= \Delta^2 \int_0^\infty d\tau \, e^{-\lambda\tau} e^{-Q'(\tau)} \sin[Q''(\tau)] A_m^{(a)}(\tau) \,.
\end{aligned}
\tag{22.14}
$$

with

$$
\begin{aligned}
A_{2m}^{(s)}(\tau) &= (-1)^m \, e^{-im\omega\tau} \cos(\epsilon_0\tau) J_{2m}\left(\frac{2\epsilon_1}{\omega}\sin\frac{\omega\tau}{2}\right) \,, \\
A_{2m}^{(a)}(\tau) &= (-1)^m \, e^{-im\omega\tau} \sin(\epsilon_0\tau) J_{2m}\left(\frac{2\epsilon_1}{\omega}\sin\frac{\omega\tau}{2}\right) \,,
\end{aligned}
\tag{22.15}
$$

$$
\begin{aligned}
A_{2m+1}^{(s)}(\tau) &= (-1)^{m+1} \, e^{-i(m+1/2)\omega\tau} \sin(\epsilon_0\tau) J_{|2m+1|}\left(\frac{2\epsilon_1}{\omega}\sin\frac{\omega\tau}{2}\right) \,, \\
A_{2m+1}^{(a)}(\tau) &= (-1)^m \, e^{-i(m+1/2)\omega\tau} \cos(\epsilon_0\tau) J_{|2m+1|}\left(\frac{2\epsilon_1}{\omega}\sin\frac{\omega\tau}{2}\right) \,,
\end{aligned}
\tag{22.16}
$$

where $J_n(z)$ is a Bessel function of the first kind. We see from these forms that the oscillating bias suppresses long blips. Thus, whenever the NIBA is appropriate in the absence of the oscillating field, it is justified even better in the presence of time-dependent driving. One can now use these forms to calculate the dynamics of $P(t)$ numerically in the NIBA [441].

22.1.3 Linear response

When the driving amplitude ϵ_1 is small compared to the driving frequency ω, we may expand the kernels (22.14) in a power series in ϵ_1. Within linear response, the relevant kernels are of order unity and of order ϵ_1. We then find in the Fourier series (22.10) the limitation $|m| \leq 1$. The term $k_0^{(s/a)}(\lambda)$ is independent of ϵ_1, and $k_{\pm1}^{(s/a)}(\lambda) = \mathcal{O}(\epsilon_1)$. As a result, $P^{(as)}(t)$ behaves as

$$
P^{(as)}(t) = P_\infty + \hbar\epsilon_1 \left[\widetilde{\chi}(\omega) \, e^{-i\omega t} + \widetilde{\chi}(-\omega) \, e^{i\omega t} \right] / 4 \,,
\tag{22.17}
$$

where $P_\infty = \lim_{\epsilon_1 \to 0} k_0^{(a)}(0) / k_0^{(s)}(0) = \tanh(\frac{1}{2}\hbar\beta\epsilon_0)$ [cf. Eq. (21.86)]. The evaluation of the recursive relations (22.11) for $|m| \leq 1$ gives

$$
\widetilde{\chi}(\omega) = \lim_{\epsilon_1 \to 0} \frac{4}{\hbar\epsilon_1 \left[-i\omega + k_0^{(s)}(-i\omega)\right]} \left(k_1^{(a)}(-i\omega) - P_\infty k_1^{(s)}(-i\omega) \right) \,.
\tag{22.18}
$$

This is the dynamical susceptibility in the NIBA. The reader may easily convince himself that the expression (22.18) is directly connected with the spectral correlation function (21.142) by the fluctuation-dissipation theorem (21.232).

22.1.4   The Ohmic case with Kondo parameter $K = \frac{1}{2}$

For an Ohmic heat bath with damping strength $K = \frac{1}{2}$ and $\hbar\beta\omega_c \gg 1$, the dynamics can be solved exactly, up to quadratures, in the presence of diagonal and off-diagonal driving. In generalization of the expression (21.202), we obtain [445, 443]

$$
P(t) = \exp\left(-\int_0^t d\tau\,\gamma(\tau)\right) + P_a(t), \qquad \gamma(\tau) = \frac{\pi}{2}\frac{\Delta^2(\tau)}{\omega_c},
$$

$$
P_a(t) = \int_0^t dt_2\,\Delta(t_2)\exp\left(-\int_{t_2}^t d\tau\,\gamma(\tau)\right)
\tag{22.19}
$$

$$
\times \int_0^{t_2} dt_1\,\Delta(t_1)\sin[\vartheta(t_2,t_1)]\,e^{-Q'(t_2-t_1)}\exp\left(-\frac{1}{2}\int_{t_1}^{t_2} d\tau'\,\gamma(\tau')\right),
$$

where $\vartheta(t_2, t_1)$ is given in Eq. (22.2). The asymptotic dynamics is described by $P_a(t)$.

Consider now the case of time-independent tunneling coupling in some more detail. We find from Eq. (22.19) for the Fourier amplitudes p_m the exact expression

$$
p_m(\omega, \epsilon_1) = \frac{\Delta^2}{-im\omega + \gamma}\int_0^\infty d\tau\,e^{+im\omega\tau - \gamma\tau/2 - Q'(\tau)}A_m^{(a)}(\tau).
\tag{22.20}
$$

In Fig. 22.1, the spectral amplitude of the fundamental frequency, $\eta_1(\omega, \epsilon_1) = 4\pi|p_1(\omega, \epsilon_1)|/\hbar\epsilon_1|^2$ [cf. Eq. (22.39)], is plotted versus ω for different temperatures.[3] At high temperatures, $\eta_1(\omega)$ is peaked at $\omega = 0$, and there is only little structure in the frequency dependence. As the temperature is decreased, resonances are formed at fractional values of the static bias, $\omega = \epsilon_0/n$ $(n = 1, 2, \ldots)$. At these frequencies, the driving-induced coherent motion is amplified.

22.2   Markovian regime

When the system's characteristic motion is slow on the time scale τ on which the kernels $K_z^{(s/a)}(t, t-\tau)$ decay, the incoherent dynamics at long times is described by a time-local master equation. The GME (21.73) simplifies to the form

$$
\dot{P}(t) = \mathcal{M}_z^{(a)}(t) - \mathcal{M}_z^{(s)}(t)P(t),
\tag{22.21}
$$

with

$$
\mathcal{M}_z^{(s)}(t) = \int_0^\infty d\tau\,K_z^{(s)}(t, t-\tau) = \Delta^2\int_0^\infty d\tau\,e^{-Q'(\tau)}\cos[Q''(\tau)]\cos[\vartheta(t, t-\tau)],
$$

$$
\mathcal{M}_z^{(a)}(t) = \int_0^\infty d\tau\,K_z^{(a)}(t, t-\tau) = \Delta^2\int_0^\infty d\tau\,e^{-Q'(\tau)}\sin[Q''(\tau)]\sin[\vartheta(t, t-\tau)].
$$

The respective second forms are the NIBA expressions, in which $\vartheta(t_2, t_1)$ is given in Eq. (22.2). The master equation (22.21) is easily integrated and yields

---

[3]Figs. 22.1 – 22.4 are by courtesy of M. Grifoni, P. Hänggi and L. Hartmann.

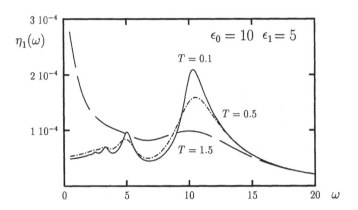

Figure 22.1: The spectral amplitude η_1 is plotted versus ω for different temperatures. See text for details. Frequencies and temperature are in units of $\gamma = \pi\Delta^2/2\omega_c$.

$$P(t) = \exp\left\{-\int_0^t dt'\, \mathcal{M}_z^{(s)}(t')\right\} + \int_0^t dt'\, \mathcal{M}_z^{(a)}(t') \exp\left\{-\int_{t'}^t dt''\, \mathcal{M}_z^{(s)}(t'')\right\} .$$

The adiabatic limit consists in the additional approximation $\vartheta(t, t-\tau) = \epsilon(t)\,\tau$. Substituting this form, the kernels are found as linear combinations of the forward/backward Golden Rule rates $k^\pm$ introduced in Subsection 20.2.1. We have

$$\mathcal{M}_z^{(s/a)}(t) = k^+[\epsilon(t)] \pm k^-[\epsilon(t)] , \tag{22.22}$$

in which the time-dependence enters solely through the bias $\epsilon(t)$. The adiabatic rates obey detailed balance with respect to the momentary bias $\epsilon(t)$.

22.3   High-frequency regime

When the driving frequency ω is very large compared to the characteristic frequencies of the damped TSS, the system is sluggish on the time scale $\mathcal{T}_\omega = 2\pi/\omega$. Then a coarse-grained description in which the dynamics is averaged over the period $\mathcal{T}_\omega = 2\pi/\omega$ is adequate. Hence we may substitute

$$\widehat{K}_\lambda^{(s/a)}(t) \;\Longrightarrow\; \langle \widehat{K}_\lambda^{(s/a)}\rangle \equiv \frac{1}{\mathcal{T}_\omega}\int_0^{\mathcal{T}_\omega} dt\, \widehat{K}_\lambda^{(s/a)}(t) = k_0^{(s/a)}(\lambda) . \tag{22.23}$$

Thus, the averaged dynamics is described by the $m = 0$ Fourier component of the kernels in the series (22.8). In this approximation, Eq. (22.6) can be solved easily for $\widehat{P}(\lambda)$. The resulting expression is analogous in form to Eq. (21.79),

$$\widehat{P}(\lambda) = \frac{1 + k_0^{(a)}(\lambda)/\lambda}{\lambda + k_0^{(s)}(\lambda)} . \tag{22.24}$$

We can deduce from this form the averaged transient and the long-time behaviour.

The characteristics of the transient dynamics is determined by the zeros of the equation $\lambda + k_0^{(s)}(\lambda) = 0$. Upon substituting into the form (22.14) for $k_0^{(s)}(\lambda)$ the decomposition

$$J_0(2z \sin \alpha) = J_0^2(z) + 2 \sum_{n=1}^{\infty} J_n^2(z) \cos(2n\alpha), \qquad (22.25)$$

the pole condition of Eq. (22.24) is found to read [$\widehat{K}_z^{(s)}(\lambda)$ is defined in Eq. (21.81)]

$$\lambda + \sum_{n=-\infty}^{\infty} J_n^2(\epsilon_1/\omega) \widehat{K}_z^{(s)}(\lambda + in\omega) = 0. \qquad (22.26)$$

Consider now first the incoherent regime, in which we disregard all poles except for the relaxation pole at $\lambda = -\gamma_r \equiv -k_0^{(s)}(\lambda = 0)$. In the relaxation limit, we have

$$P(t) = P_\infty + [1 - P_\infty] e^{-\gamma_r t}, \qquad P_\infty = k_0^{(a)}(0)/k_0^{(s)}(0). \qquad (22.27)$$

Here, P_∞ is the average equilibrium value, and the decay rate is given by the series[4]

$$\gamma_r(\epsilon_0; \omega, \epsilon_1) = \sum_{n=-\infty}^{\infty} J_n^2\left(\frac{\epsilon_1}{\omega}\right) \left[k^+(\epsilon_0 + n\omega) + k^-(\epsilon_0 + n\omega)\right]. \qquad (22.28)$$

The formula (22.28) describes the inclusive relaxation rate γ_r as a sum over all possible escape channels. The individual channels describe tunneling under emission and absorption of a fixed number of quanta of the fundamental frequency ω, and the weight factor for n quanta is $J_n^2(\epsilon_1/\omega)$. The exclusive channel rates are given in terms of the relaxation rates $k^\pm$ for a static bias $\epsilon_n = \epsilon_0 + n\omega$ for the dissipative mechanism under consideration. These rates have been discussed in Section 20.2. The inclusive tunneling rate γ_r sensitively depends on the parameters of the driving field. In Fig. 22.2, we see that the rate is enhanced as compared with the static case when the resonance condition $\epsilon_0 = \pm n\omega$ is met and when the weight does not fall on a zero of the respective Bessel function. Upon tuning ϵ_1/ω to a zero of one of the Bessel functions in Eq. (22.28), the corresponding decay channel is missing.

The striking dependence of γ_r on the external field parameters may be utilized, e.g., to control electron transfer rates in chemical reaction processes.

The fast ac field also leads to an asymptotic population $P_\infty = k_0^{(a)}(0)/k_0^{(s)}(0)$ which may differ drastically from the static case, $P_\infty = \tanh(\hbar\beta\epsilon_0/2)$. Even inverted population can be reached by means of a suitable choice of the parameters ϵ_1 and ω, as predicted in the works in Ref. [447].

Next, consider a symmetric Ohmic TSS. Using the NIBA expression (21.90) with (21.91), the pole condition (22.26) for $\omega \gg 2\pi K k_B T/\hbar$ reads [443]

$$\lambda + 2g(\lambda) \left\{ J_0^2(\epsilon_1/\omega) + \left(\frac{2\pi K + \hbar\beta\lambda}{\hbar\beta\omega}\right)^2 \sum_{n \neq 0} \frac{J_n^2(\epsilon_1/\omega)}{n^2} \right\} = 0. \qquad (22.29)$$

---

[4]Representations of the form (22.28) for transport quantities have a long tradition since the pioneering work by Tien and Gordon [446].

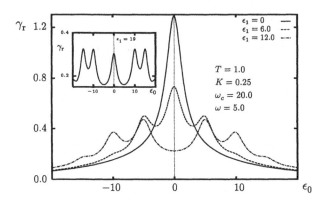

Figure 22.2: The inclusive rate γ_r is shown as a function of ϵ_0 for different amplitudes ϵ_1. The frequency parameters are given in units of $\Delta/2$. In the inset, the amplitude ϵ_1 is tuned on a zero of the $n = 1$ Bessel function, $J_1(\epsilon_1/\omega) = 0$ so that the first side band is absent.

The first term in the curly bracket dominates for large ω when the ratio ϵ_1/ω is away from a zero of the J_0-Bessel function. For $K \ll 1$, we then find weakly-damped coherent dynamics in the transient regime. Upon substituting a driving-renormalized tunneling matrix element for Δ,

$$\Delta \;\longrightarrow\; J_0(\epsilon_1/\omega)\,\Delta\,, \qquad (22.30)$$

all the results given in Subsection 21.3.2 directly apply to the present case.

The coherent dynamics is suppressed when ϵ_1/ω is tuned to $J_0(\epsilon_1/\omega) \approx 0$. Then the system relaxes incoherently with the rate

$$\gamma_r \;=\; 2g(0)\left(\frac{2\pi K}{\hbar\beta\omega}\right)^2 \sum_{n\neq 0} \frac{J_n^2(\epsilon_1/\omega)}{n^2}\,. \qquad (22.31)$$

The condition $J_0(\epsilon_1/\omega) = 0$ was found as a necessary criterion for the phenomenon of "coherent destruction of tunneling" for a driven undamped TSS [448, 449].

Consider now the coherent regime for $K \ll 1$ and $\epsilon_1/\omega \ll 1$. Then we may perform a systematic weak-damping study as explained in Section 21.4. For large ω, the leading effects of the driving force are taken into account by expanding the Bessel functions in Eq. (22.26) up to terms of order $(\epsilon_1/\omega)^2$. In this approximation, the *undamped* driven system is characterized by three bias frequencies $\mu_1 = \epsilon_0$, $\mu_2 = \epsilon_0 + \omega$, and $\mu_3 = \epsilon_0 - \omega$ and three tunneling frequencies ν_j. The squared ν_j are the solutions of a cubic equation in ν^2,

$$\nu^6 - a_4\nu^4 + a_2\nu^2 - a_0 = 0\,, \qquad (22.32)$$

$$a_0 = \mu_1^2\mu_2^2\mu_3^2 + [\,\Delta_1^2\mu_2^2\mu_3^2 + \text{cycl.}\,]\,, \qquad a_2 = \Delta_1^2(\mu_2^2 + \mu_3^2) + \mu_2^2\mu_3^2 + \text{cycl.}\,,$$

$$a_4 = \mu_1^2 + \Delta_1^2 + \text{cycl.}\,, \qquad \Delta_1^2 = (1 - \epsilon_1^2/2\omega^2)\Delta^2\,, \qquad \Delta_2^2 = \Delta_3^2 = (\epsilon_1^2/2\omega^2)\Delta^2\,.$$

The undamped system performs a superposition of coherent oscillations

$$P_{\text{undamped}}(t) \;=\; p_0 + \sum_{j=1}^{3} p_j \cos(\nu_j t)\,, \qquad (22.33)$$

with the amplitudes (p_2, p_3 cycl.)

$$p_1 \;=\; \frac{(\nu_1^2 - \mu_1^2)(\nu_1^2 - \mu_2^2)(\nu_1^2 - \mu_3^2)}{\nu_1^2(\nu_1^2 - \nu_2^2)(\nu_1^2 - \nu_3^2)}\,, \qquad p_0 \;=\; \prod_{j=1}^{3} \frac{\mu_j^2}{\nu_j^2}\,. \qquad (22.34)$$

For weak damping, the shift of the poles can be calculated as explained in Section 21.4. Disregarding irrelevant frequency shifts, one finds

$$P(t) \;=\; \sum_{j=1}^{3} p_j \cos(\nu_j t)\, e^{-\gamma_j t} + (p_0 - P_\infty)\, e^{-\gamma_0 t} + P_\infty\,. \qquad (22.35)$$

The rates are linear combinations of one-phonon emission and absorption processes [cf. Eq. (21.178)] for the possible transition frequencies.[5] One finds again by comparison of the exact weak-damping expressions with the NIBA results that the NIBA disregards the frequency shifts in the transition frequencies and amplitudes of the one-phonon processes resulting from the tunneling coupling.

One final remark is in order. On a closer look, the steady state exhibits oscillations about the mean value P_∞ given in Eq.(22.27). We find from Eq. (22.11) for large ω

$$P^{(\text{as})}(t) \;=\; P_\infty + \sum_{m \neq 0} \frac{1}{-im\omega} \left(k_m^{(\text{a})}(-im\omega) - k_m^{(\text{s})}(-im\omega) P_\infty \right) e^{-im\omega t}\,. \qquad (22.36)$$

With increasing driving-frequency, the oscillations of $P^{(\text{as})}(t)$ about P_∞ become suppressed since the $m \neq 0$ Fourier components get less important.

22.4 Quantum stochastic resonance

The stochastic resonance phenomenon refers to the amplification of the response to an applied periodic signal at a certain optimal value of the noise strength and is a cooperative effect of friction, noise, and periodic driving in a bistable system. This phenomenon has been dubbed *stochastic resonance* (SR) since classically the maximal output signal occurs when the thermal hopping rate is in resonance with the frequency of the driving force. It has been argued that this phenomenon is of fundamental importance in biological evolution. Several comprehensive reviews on classical stochastic resonance are available [450]. Qualitatively new signatures of stochastic resonance appear in the deep quantum regime [451, 452].

In the deep quantum regime, the double well can be truncated to the two-state system. The quantity of interest in SR is the power spectrum

---

[5]The various rate expressions are given in Ref. [413].

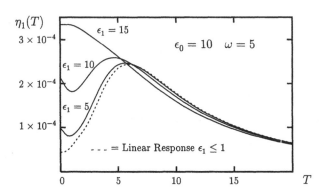

Figure 22.3: Depiction of QSR in the deep quantum regime for Ohmic damping $K = \frac{1}{2}$. The power amplitude η_1 of the fundamental frequency is plotted as a function of temperature for different driving amplitudes. See text for details. The units are the same as in Fig. 22.1.

$$S(\nu) = \int_{-\infty}^{\infty} d\tau\, e^{i\nu\tau}\, \overline{C}^{(as)}(\tau)\,, \qquad (22.37)$$

where $\overline{C}^{(as)}(\tau)$ is the time-averaged steady-state population correlation function,

$$\overline{C}^{(as)}(\tau) \equiv \lim_{\tau \to \infty} \frac{\omega}{2\pi}\, \mathrm{Re} \int_0^{2\pi/\omega} dt\, \langle\, \sigma_z(t+\tau)\sigma_z(t)\,\rangle = \sum_{m=-\infty}^{\infty} |p_m(\omega,\epsilon_1)|^2\, e^{-im\omega\tau}\,.$$

The amplitude $p_m(\omega,\epsilon_1)$ is given in Eq. (22.11). The power spectrum takes the form

$$S(\nu) = 2\pi \sum_{m=-\infty}^{\infty} |p_m(\omega,\epsilon_1)|^2\, \delta(\nu - m\omega)\,. \qquad (22.38)$$

A quantitative study of the power amplitude

$$\eta_m(\omega,\epsilon_1) = 4\pi\, |p_m(\omega,\epsilon_1)/\hbar\epsilon_1|^2 \qquad (22.39)$$

can be performed by numerically computing the kernels $k_m^{(s/a)}(\lambda)$ for the environment of interest and solving the recursive relations (22.11). In the sequel, we confine ourselves to an Ohmic coupling with a large cutoff.

In classical SR, the spectral amplification is maximal for a *symmetric* system [450], whereas in the deep quantum regime and $K < 1$, QSR is only effective in the presence of a static bias. One finds that QSR is most striking when the static bias ϵ_0 is larger than the driving amplitude ϵ_1. For $K = \frac{1}{2}$, the power amplitudes can be calculated from Eq. (22.20). In Fig. 22.3, the power amplitude $\eta_1(T)$ for $K = \frac{1}{2}$ is plotted as a function of temperature for different driving amplitudes. When $\epsilon_1 > \epsilon_0$, the power amplitude decreases monotonously with increasing T. As ϵ_1 is decreased, a minimum at low T followed by a QSR maximum at $T \approx \hbar\omega/k_B$ is formed. For $\epsilon_1 < 5\gamma$, the QSR can be studied within linear response theory.

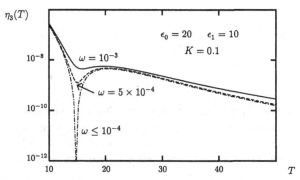

Figure 22.4: Noise-induced suppression of higher harmonics is illustrated for the third power amplitude η_3. Frequencies and temperature are given in units of Δ_{eff}.

In the linear response regime, we find from Eqs. (22.17) and (22.39) the relation $p_1 = -\hbar\epsilon_1\tilde{\chi}(\omega)/4$. In the regime $\hbar\omega < 2\pi\alpha k_B T$ and $k_B T > \hbar\Delta$ and/or $K > 1$, the NIBA form (22.18) is correct. From this we obtain in the Lorentzian approximation for the quasi-elastic peak centered at $\omega = 0$ the expression

$$\eta_1(\omega) = \frac{\pi}{(2k_B T)^2} \frac{1}{\cosh^4(\hbar\epsilon_0/2k_B T)} \frac{\gamma_r^2}{\omega^2 + \gamma_r^2} , \qquad (22.40)$$

where $\gamma_r = k^+ + k^- = [1 + \exp(-\hbar\beta\epsilon)]k^+$ is the width, and where $k^+(T, \epsilon_0)$ is the forward tunneling rate (20.78). Since the quasi-elastic peak reflects exponential relaxation, it is not surprising that the same form is also found in the classical case, e.g. for the bistable double well, in which γ_r is the thermal relaxation rate [450].

For large amplitude ϵ_1 of the driving force, the response of higher harmonics may become significant. One finds from the recursive relations (22.11) that quantum noise can enhance or suppress higher harmonics in $P^{(\text{as})}(t)$. In Fig. 22.4, the temperature dependence of the power amplitude η_3 is shown for different ω. As the driving frequency is decreased, noise-induced suppression of this amplitude occurs at a temperature where the fundamental power amplitude η_1 has a maximum.

22.5 Driving-induced symmetry breaking

Consider the cooperative effect of monochromatic fields modulating both the bias energy of a TSS with zero static bias,

$$\epsilon(t) = \epsilon_1 \sin(\omega t) , \qquad (22.41)$$

and the tunneling coupling (TC),

$$\Delta(t) = \Delta \exp[\mu \sin(\nu t)] . \qquad (22.42)$$

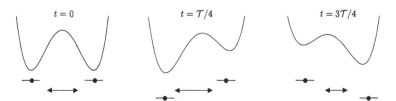

Figure 22.5: Bistable potential (top) and related TSS (bottom) for the case $\nu = \omega$. At time $t = 0$, the TSS is symmetric. At time $t = \mathcal{T}/4$, the left state is lower and the TC has the maximum value. At time $t = 3\mathcal{T}/4$, the right state is lower, but the TC has the minimum value. Thus, in the presence of damping, relaxation towards the left state is preferred.

If either the bias or the coupling energy is modulated, the left-right symmetry is dynamically broken. However, on average over a period, one finds asymptotically again equal occupation of both states.

When both parameters are modulated with commensurable frequencies (n and m integer),

$$\omega = m\Omega \qquad \text{and} \qquad \nu = n\Omega\,, \tag{22.43}$$

the TSS Hamiltonian has a discrete time translation symmetry. However, when n and m are odd, the left-right symmetry is broken even on average. The symmetry breaking is maximal when $n = m = 1$. This case is pictorially sketched and qualitatively explained in Fig. 22.5, and it is also chosen below in Eq. (22.45). Thus we expect that at long times the occupation of the left state is preferred on average.

The average dynamics is again described by the expression (22.24), in which the kernels $k_0^{(\text{s/a})}(\lambda)$ must be calculated for time-periodic bias and tunneling coupling. For simplicity, we now restrict the attention to the average equilibrium state

$$P_\infty = k_0^{(\text{a})}(0)/k_0^{(\text{s})}(0) = [\gamma^+ - \gamma^-]/[\gamma^+ + \gamma^-]\,. \tag{22.44}$$

Here, $\gamma^\pm$ are inclusive forward/backward tunneling rates. The exponentials in the bias phase term $e^{\pm i\vartheta(t,t-\tau)} = e^{\mp i(\epsilon_1/\omega)\cos(\omega t)}\,e^{\pm i(\epsilon_1/\omega)\cos[\omega(t-\tau)]}$ can be expanded in a series in J-Bessel functions, and the oscillatory factors $e^{\mu\sin(\omega t)}\,e^{\mu\sin[\omega(t-\tau)]}$ of the tunneling coupling in a series in I-Bessel functions. Then we find the inclusive rates $\gamma^\pm$ as

$$\gamma^\pm = \sum_{\ell,m,n=-\infty}^{\infty} I_\ell(\mu)I_m(\mu)J_n(\epsilon_1/\omega)J_{\ell+n-m}(\epsilon_1/\omega)k^\pm[\,(n-m)\omega\,]\,, \tag{22.45}$$

where $k^\pm(\epsilon)$ is the exclusive forward/backward rate for a static bias ϵ. The expression (22.45) is the generalization of Eq. (22.28) to the case of an additional periodic modulation of the tunneling coupling. It reveals that the forward/backward symmetry is dynamically broken, $\gamma^+ \neq \gamma^-$, when both the driving amplitude of the bias (ϵ_1) and the modulation amplitude of the tunneling coupling (μ) are nonzero.

PART V

THE DISSIPATIVE MULTI–STATE SYSTEM

A quantum Brownian particle in a multi-well potential coupled to a dissipative environment is archetypal for many problems in physics and chemistry. Examples include superionic conductors, atoms on surfaces, and interstitials in dielectrics and metals. Also the current-voltage characteristics of a Josephson junction, and charge transport in a quantum wire hindered by impurity scattering are described by this model. At high temperatures, the particle moves forward or backward from well to well by incoherent tunneling or thermally activated transitions. As the temperature is lowered, coherent tunneling across many wells may become significant, and the competing different tunneling paths may interfere constructively or destructively. The model has a far reaching duality symmetry between the weak-binding and strong-binding representation. In view of the broad area of applications, the understanding of quantum transport in multi-well systems is a central issue.

23. Quantum Brownian particle in a washboard potential

23.1 Introduction

In the preceding part, we have considered the dynamics of a damped quantum system which is effectively restricted to a two-dimensional Hilbert space. Many of the concepts and approximation schemes developed in Part IV can be generalized for a dissipative system with N tight-binding sites. A system with three sites, $N = 3$, for instance, is of particular interest for the study of the ultrafast primary electron transfer in bacterial photosynthesis [453]. In many systems in chemical and biological physics, the transfer of a particle or a charge from a donor to an acceptor state occurs via a bridge of few or many intermediate tight-binding states. The RDM for a discrete N-site system can be visualized as a square lattice with $N \times N$ lattice points. The path sum for the conditional propagating function (5.12) includes all paths on this lattice for given end points. Each path consists of a sequence of segments in which the path dwells for some time on a lattice site and then undertakes a sudden flip to a neighbouring site. A general path can be divided into a sequence of clusters: each cluster is a path section between two successive visits of a diagonal state. A natural generalization of the noninteracting-blip approximation to the case of N states is the

"noninteracting-cluster" approximation (NICA) developed in Ref. [454]. In the NICA, the intracluster correlations are taken into account while the intercluster correlations are disregarded. It is then easy to write down master equations for the populations. Each kernel is given by the sum of all clusters which have the same end points. For example, in the three-state system, there are two types of cluster functions:
(1) The clusters k_{12} and k_{23} describe "sequential" transport of the particle between the neighbouring sites 1 and 2, and between the sites 2 and 3, respectively.
(2) The "superexchange" cluster k_{13} describes direct transfer of the particle between site 1 and site 3 without any real population of the intermediate site 2.
In the sequel, we turn our attention to the case where the number of sites is infinity.

Quantum Brownian motion (QBM) in a periodic potential is a key model for many transport phenomena in condensed matter [455]. Early work was mainly concerned with the significance of polaronic effects [123] – [125]. Recent interest has been focussed on the influence of frequency-independent damping. For instance, the electron-hole drag of charged particles in metals, as well as quasiparticle tunneling in Josephson junctions [456, 143] give rise to Ohmic dissipation [cf. Subsections 4.2.8 and 4.2.10]. The duality symmetry between the weak- and strong-binding representations of the QBM model discovered by Schmid [136] corresponds to the charge-phase duality for the Josephson junction [143]. Transport of charge through impurities in quantum wires [457] and tunneling of edge currents through constrictions in fractional quantum Hall devices [458] are also described by the QBM model. The collective excitations of the correlated fermions away from the barrier are represented by a Luttinger harmonic liquid which corresponds to the thermal reservoir in the QBM model. Short-range electron interaction is equivalent to Ohmic damping in the related QBM model, whereas unscreened long-range Coulomb repulsion [459] corresponds to a sub-Ohmic reservoir coupling [460]. Many other physical and chemical systems involve transport of charge through barriers under Ohmic and super-Ohmic dissipation [89, 166].

23.2   Weak- and tight-binding representation

We now consider a quantum Brownian particle moving in a tilted corrugated potential. The dynamics is described by the translational-invariant global model (3.11) with a bilinear coordinate coupling of the particle to a bath of harmonic oscillators,

$$H_{\mathrm{WB}} = \frac{P^2}{2M} + V(X,t) + \sum_{\alpha} \left[\frac{p_{\alpha}^2}{2m_{\alpha}} + \frac{m_{\alpha}\omega_{\alpha}^2}{2} \left(x_{\alpha} - \frac{c_{\alpha}X}{m_{\alpha}\omega_{\alpha}^2} \right)^2 \right] , \qquad (23.1)$$

where the label WB indicates "weak binding" representation. We choose a trigonometric form for the corrugation with period X_0 and a global tilting force $F = 2\pi V_{\mathrm{tilt}}/X_0 = \hbar\epsilon_{\mathrm{WB}}/X_0$. To study the particle's response in thermal equilibrium, we assume that this force is acting only for $t \geq 0$. We then have [cf. Eq.(17.46)]

$$V(X,t) = -V_0 \cos\left(2\pi X/X_0\right) - \Theta(t)FX . \qquad (23.2)$$

There are many different physical situations in which an underlying discrete translational symmetry provides the periodic potential. In real systems, there are often complications due to perturbations of the periodic order by impurities or by local strain fields. Nevertheless, the model (23.1) with (23.2) provides an ideal description and does show already interesting nontrivial behaviour.

In the tight-binding regime $k_B T/\hbar\omega_0$, $FX_0/\hbar\omega_0 \ll 1 \ll V_0/\hbar\omega_0$, where $\hbar\omega_0$ is the frequency of small oscillations about the potential minima, only the lowest state in each well is occupied so that the potential system is effectively reduced to a single-band tight-binding lattice. The Hilbert space of the isolated system is spanned by the (localized) ground states of the wells. These states are coupled by a transfer matrix element Δ which represents quantum mechanical overlap of nearest-neighbour localized states. The tight-binding (TB) model with transfer matrix element Δ and translational-invariant coupling is

$$H_{\rm TB} = -\hbar\Delta \cos\left(\frac{q_0 p}{\hbar}\right) - \Theta(t)Fq + \sum_\alpha \left[\frac{\pi_\alpha^2}{2M_\alpha} + \frac{M_\alpha\Omega_\alpha^2}{2}\left(u_\alpha - \frac{d_\alpha q}{M_\alpha\Omega_\alpha^2}\right)^2\right], \quad (23.3)$$

$$q = q_0 \sum_n n\, a_n^\dagger a_n = q_0 \sum_n n\, |n\rangle\langle n|, \qquad e^{iq_0 p/\hbar} = \sum_n a_n^\dagger a_{n+1} = \sum_n |n\rangle\langle n+1|,$$

where q_0 is the TB lattice constant. Here we have given two equivalent forms for the discrete representation. The operators $a_n^\dagger$ and a_n create and annihilate a particle at the site n, and $|n\rangle$ denotes the localized state at this site. In the sequel, we express the tilting force F in terms of the potential drop $\hbar\epsilon$ between neighbouring sites,

$$F = \hbar\epsilon/q_0, \quad (23.4)$$

As far as we are interested in properties of the particle, the coupling constants $\{c_\alpha\}$ and the parameters of the bath are relevant only via the spectral density of the coupling. We define the spectral densities of the two models as in Eq. (3.30),

$$J_{\rm WB}(\omega) = \frac{\pi}{2}\sum_\alpha \frac{c_\alpha^2}{m_\alpha\omega_\alpha}\delta(\omega - \omega_\alpha), \qquad J_{\rm TB}(\omega) = \frac{\pi}{2}\sum_\alpha \frac{d_\alpha^2}{M_\alpha\Omega_\alpha}\delta(\omega - \Omega_\alpha). \quad (23.5)$$

The tight-binding and weak-binding models are related by a duality symmetry which becomes an exact self-duality in the Ohmic scaling limit (see Section 25.1). Explicit computations are conveniently performed in the discrete TB representation. Subsequently, upon using the duality transformation, the results can easily be transferred to the dual weak-binding model.

24. Multi-state dynamics

24.1 Quantum transport and quantum-statistical fluctuations

The reduced density matrix describes all properties pertaining to the Brownian particle. In the TB limit, the originally continuous coordinate q becomes discrete, $q = nq_0$ ($n = 0, \pm1, \pm2, \cdots$), and the density matrix is labelled by discrete indices numbering the wells. We now study the dynamics for two different kinds of initial states.

24.1.1 Product initial state

The first kind is a product initial state of the form discussed in Subsection 21.1.1. Suppose that the particle was prepared to start out at time zero from the site $n = 0$ of the discrete lattice. The dynamical quantity of interest is then the probability $P_n(t)$ for finding the particle at site n at a later time $t > 0$. To formulate the evolution of $P_n(t)$, we use again the real-time influence functional method for a product initial state of the system-plus-reservoir complex discussed in Section 5.1. The populations $P_n(t)$ are the diagonal elements of the reduced density matrix. They can be written as a double path integral for the propagating function,

$$P_n(t) \; = \; J(nn, t; 00, 0) \; = \; \int \mathcal{D}q \int \mathcal{D}q' \, \mathcal{A}[q] \mathcal{A}^*[q'] \mathcal{F}[q, q'] \,. \tag{24.1}$$

Here, $q(t')$, $q'(t')$ are discontinuous paths propagating vertically and horizontally in steps of length q_0 in the (q, q')-plane. The paths obey the end-point conditions

$$q(0) = q'(0) = 0 \,, \quad \text{and} \quad q(t) = q'(t) = n\, q_0 \,. \tag{24.2}$$

The functional $\mathcal{A}[q]$ is the probability amplitude to go from $q = 0$ to $q = nq_0$ along the path $q(t')$, and $\int \mathcal{D}q$ means summation over all possible paths on the TB-lattice with the end-points (24.2). Again, all the influences of the coupling to the reservoir are contained in the influence functional $\mathcal{F}[q, q']$ in the form of self-interactions of the paths $q(t')$ and $q'(t')$ and interactions between the paths $q(t')$ and $q'(t')$.

24.1.2 Characteristic function of moments and cumulants

The Fourier transform of the populations is the central dynamical quantity,

$$\mathcal{Z}(\rho, t) \; = \; \sum_{n=-\infty}^{\infty} e^{i\rho n q_0} P_n(t) \; = \; \langle e^{i\rho q(t)} \rangle \; = \; \sum_{m=1}^{\infty} \frac{(i\rho)^m}{m!} \langle q^m(t) \rangle \,. \tag{24.3}$$

The *characteristic function* $\mathcal{Z}(\rho, t)$ is the *moment generating function* (MGF), where ρ is the counting field. The Nth derivative of the MGF at $\rho=0$ gives the Nth moment

$$\langle q^N(t) \rangle \; \equiv \; q_0^N \sum_{n=-\infty}^{\infty} n^N P_n(t) \; = \; \left(-i \frac{\partial}{\partial \rho} \right)^N \mathcal{Z}(\rho, t) \bigg|_{\rho=0} \,. \tag{24.4}$$

The reducible moments of $P_n(t)$ can be written in terms of the irreducible moments or cumulants $\langle q^N(t) \rangle_c$ as

$$\begin{aligned}
\langle q(t) \rangle &= \langle q(t) \rangle_c \,, \\
\langle q^2(t) \rangle &= \langle q(t) \rangle_c^2 + \langle q^2(t) \rangle_c \,, \\
\langle q^3(t) \rangle &= \langle q(t) \rangle_c^3 + 3 \langle q^2(t) \rangle_c \langle q(t) \rangle_c + \langle q^3(t) \rangle_c \,.
\end{aligned} \tag{24.5}$$

The cumulant generating function (CGF) $\ln \mathcal{Z}(\rho, t)$ yields the cumulant expansion

$$\ln \mathcal{Z}(\rho, t) \;=\; \sum_{m=1}^{\infty} \frac{(i\rho)^m}{m!} \, \langle q^m(t) \rangle_{\rm c} \,. \tag{24.6}$$

Thus, the Nth cumulant can be found from the CGF as

$$\langle q^N(t) \rangle_{\rm c} \;=\; \left(-i \frac{\partial}{\partial \rho} \right)^N \ln \mathcal{Z}(\rho, t) \Big|_{\rho=0} \,. \tag{24.7}$$

Evidently, the cumulant generating function $\ln \mathcal{Z}(\rho, t)$ carries all statistical properties of the quantum transport process. The transport dynamics is referred to as diffusive when $\ln \mathcal{Z}(\rho, t)$, and hence all irreducible moments or cumulants, grow linearly with t at long time. This is the case when the dynamics takes place by incoherent tunneling events between the system's diagonal states (see Section 24.2).

The first cumulant gives the average current. From this one may deduce the mobility or conductance. The second cumulant yields the diffusion constant or, in connection with charge transport, the (nonequilibrium) dc current noise. All higher cumulants are zero when the statistics is Gaussian. The third cumulant, referred to as skewness, gives information about the leading asymmetric deviation from the Gaussian distribution. Experimentally, it can be discriminated from Gaussian noise by inverting the current. The fourth cumulant, called curtosis or sharpness, is a measure for the flatness of the distribution compared to the standard distribution. When the curtosis is positive (negative), the distribution is sharp (flat).

24.1.3 Thermal initial state and correlation functions

The second kind of initial state is a thermal initial state: the system is prepared by letting the particle equilibrate with the reservoir, and then at time $t = 0$ a measurement of the observable of interest of the particle is performed. This leads to a reduction of the canonical density operator according to $\hat{W}_0 = \hat{P}\hat{W}_\beta\hat{P}$, where $\hat{W}_\beta$ is the equilibrium density operator of the global system, and the operator $\hat{P}$ projects onto the measured interval of the observable in question. Upon measuring the same observable again at a later time $t > 0$, we obtain information about the equilibrium autocorrelations of this observable. Here we restrict our attention to position correlation functions that can be deduced from the generating function

$$\mathcal{Z}_{\rm th}(\rho, \kappa, \mu; t) \;\equiv\; \langle\, e^{i\rho q(0)} \, e^{i\kappa q(t)} \, e^{i\mu q(0)} \,\rangle_\beta \,, \tag{24.8}$$

where $\langle \cdots \rangle_\beta$ means thermal average of the states of the global system. We are mainly interested in the evolution of the average position $\langle q(t) \rangle_\beta$, in the mean square displacement

$$C(t) \;\equiv\; \langle\, [\,q(t) - q(0)\,]^2 \,\rangle_\beta \,, \tag{24.9}$$

and in the antisymmetrized correlation function

$$A(t) \;\equiv\; \frac{1}{2i} \, \langle\, [\,q(t)q(0) - q(0)q(t)\,]\,\rangle_\beta|_{F=0} \,. \tag{24.10}$$

These functions are directly found from $\mathcal{Z}_{\text{th}}$ upon differentiation,

$$
\langle q(t) \rangle_\beta \;=\; -i\, \frac{\partial}{\partial \kappa} \mathcal{Z}_{\text{th}}\Big|_{\rho=\kappa=\mu=0}\,,
$$

$$
C(t) \;=\; \left[-\frac{\partial^2}{\partial \kappa^2} - \frac{\partial}{\partial \rho}\frac{\partial}{\partial \mu} + \frac{\partial}{\partial \kappa}\frac{\partial}{\partial \mu} + \frac{\partial}{\partial \kappa}\frac{\partial}{\partial \rho} \right] \mathcal{Z}_{\text{th}}\Big|_{\rho=\kappa=\mu=0}\,, \qquad (24.11)
$$

$$
A(t) \;=\; \frac{i}{2}\left[\frac{\partial}{\partial \kappa}\frac{\partial}{\partial \mu} - \frac{\partial}{\partial \kappa}\frac{\partial}{\partial \rho} \right] \mathcal{Z}_{\text{th}}\Big|_{\rho=\kappa=\mu=0,\,F=0}\,.
$$

The generating function $\mathcal{Z}_{\text{th}}$ can be expressed in terms of three-time conditional propagating functions. For an ergodic system-plus-reservoir complex, these can be written as real-time path integrals evoluting from the infinite past (see Section 5.5 and Subsection 21.2.2). Suppose that the particle has been prepared at some negative time t_{p} in the diagonal state $(0,0)$. Then the particle will have relaxed at time zero to the thermal equilibrium state if we send t_p to the infinite past. Performing now at time $t = 0$ a measurement of the observable of interest, the desired thermal correlated initial state is prepared. We have

$$
\mathcal{Z}_{\text{th}}(\rho, \kappa, \mu; t) \;=\; \lim_{t_{\text{p}} \to -\infty} \sum_{n,m,r} J(n\,n, t;\, m\,r, 0;\, 0\,0, t_{\text{p}})\, e^{iq_0(\rho m + \kappa n + \mu r)}\,. \qquad (24.12)
$$

The propagating function may be expressed as a double path integral of the form (24.1) in which the paths on the (q, q')-lattice are constrained as

$$
\begin{aligned}
q(t_{\text{p}}) &= 0\,, & q(0) &= m\,q_0\,, & q(t) &= n\,q_0\,, \\
q'(t_{\text{p}}) &= 0\,, & q'(0) &= r\,q_0\,, & q'(t) &= n\,q_0\,.
\end{aligned} \qquad (24.13)
$$

The effects of different initial preparation are now easily visualized. For the product initial state discussed in Subsection 24.1.1, the dynamics of the particle at negative times is quenched. For the thermal initial state discussed here, the particle undergoes dynamics since the infinite past, and the effects of the correlations at time zero are represented by the interactions between the negative- and positive-time branch.

24.2 Poissonian quantum transport

24.2.1 Dynamics by incoherent nearest-neighbour tunneling moves

If the temperature is high and/or the bath coupling is strong, the particle tunnels incoherently between neighbouring TB states.[1] The system is then in a diagonal state of the RDM after every second transition it makes. Hence the paths are restricted to the tridiagonal for which the off-diagonal charges p_j are ± 1 or zero. In terms of our walker, every path consists of sequences of three elementary two-step processes:

---

[1]The precise conditions for this regime are discussed in connection with the dissipative two-state system in Chapters 20 and 21.

(1) The walker may step forward to the next diagonal state

(2) The walker may walk back to the preceding diagonal state

(3) The walker may return to the same diagonal state

The elementary processes (1) and (2) have each two variants, whereas the third has four variants, as counted from the number of possible intermediate off-diagonal states. According to the noninteracting-blip approximation, the elementary processes are assumed to be noninteracting and thus statistically independent. We denote the weights of the double jumps per unit time by w^+, w^-, and w, respectively. The weights are of order Δ^2. They are directly related to the nonadiabatic forward and backward tunneling rates in the two-state system discussed in Section 20.2. We have $w^+ = k^+$, $w^- = k^-$, and $w = -k$, where $k = k^+ + k^-$. As the walker finally reaches the diagonal state $(n\,n)$, there is an excess of n step-down processes over the step-up processes. Thus the grand-canonical sum of the two-step processes is given by

$$P_n(t) = \sum_{j,\ell,m=0}^{\infty} \delta_{j,\ell+n} \frac{1}{j!\,\ell!\,m!} (k^+)^j (k^-)^\ell (-k)^m \, t^{j+\ell+m} , \tag{24.14}$$

which can be summed upon using the detailed balance relation (20.32) to the form

$$P_n(t) = \exp(n\hbar\beta\epsilon/2 - kt) \, I_{|n|}[\, kt/\cosh(\hbar\beta\epsilon/2)\,] , \tag{24.15}$$

where $I_n(z)$ is a modified Bessel function. A simple interpretation of this result may be obtained by observing that the nearest-neighbour hopping dynamics is alternatively described in terms of the master equation [391, 455]

$$\dot{P}_n(t) = k^+ P_{n-1}(t) + k^- P_{n+1}(t) - kP_n(t) . \tag{24.16}$$

Upon using a recursion formula of the modified Bessel function, it is straightforward to see that the expression (24.15) solves the master equation. Next we calculate the Fourier transform of the expression (24.15) upon observing that the Fourier series can be summed up to the generating function of the modified Bessel functions. The resulting expression for $\mathcal{Z}(\rho, t)$ depends exponentially on time,

$$\mathcal{Z}(\rho, t) = \exp\left\{ [\cos(\rho q_0) - 1] \, kt + i\sin(\rho q_0)[\, k^+ - k^-] \, t \right\} , \tag{24.17}$$

which tells us that the dynamics is diffusive. Substituting this form into Eq. (24.7), the first and second cumulants are found to read

$$\langle q(t)\rangle_c = q_0 [\, k^+(\epsilon) - k^-(\epsilon)] \, t , \qquad \langle q^2(t)\rangle_c = q_0^2 \, k(\epsilon) \, t . \tag{24.18}$$

If the forward and backward rate are related by detailed balance [cf. Eq. (20.32)], we have $k^+(\epsilon) - k^-(\epsilon) = \tanh(\hbar\beta\epsilon/2) \, k(\epsilon)$.

Next, we employ the definining expressions for the nonlinear mobility $\mu(T, \epsilon)$ and the diffusion coefficient $D(T, \epsilon)$,

$$\mu = \frac{q_0}{\hbar\epsilon} \lim_{t\to\infty} \frac{\langle q(t)\rangle}{t} , \qquad D = \lim_{t\to\infty} \frac{\langle q^2(t)\rangle_c}{2t} . \tag{24.19}$$

With these definitions, we finally find from the forms (24.18) and the defining expressions (24.19) the nonlinear mobility and the diffusion coefficient as

$$
\begin{aligned}
\mu(T,\epsilon) &= \frac{\tanh(\hbar\beta\epsilon/2)}{\hbar\epsilon}\, q_0^2\, k(T,\epsilon)\,, \\[2mm]
D(T,\epsilon) &= \tfrac{1}{2}q_0^2\, k(T,\epsilon) = k_{\mathrm{B}}T\,\frac{\hbar\beta\epsilon/2}{\tanh(\hbar\beta\epsilon/2)}\,\mu(T,\epsilon)\,.
\end{aligned}
\tag{24.20}
$$

The nonadiabatic forward/backward tunneling rates $k^{\pm}(T,\epsilon)$ in a TB double-well system have been discussed in Section 20.2. With these results we then get explicit expressions for $\mu(T,\epsilon)$ and $D(T,\epsilon)$. The resulting expression for the nonlinear mobility in the Ohmic scaling limit is given below in Subsection 24.5.1.

24.2.2   The general case

We now generalize the discussion to a TB model in which the transport of mass or charge between the sites $n = 0, \pm1, \pm2, \cdots$ takes place via direct forward and backward transitions by ℓ sites, $\ell = 1, 2, \cdots$. We denote the respective transitions weights (transition "rates") by $k_\ell^{\pm}$. Assuming statistically independent transitions, the dynamics of the population probability $P_n(t)$ of site n is governed by the master equation

$$
\dot{P}_n(t) = \sum_{\ell=1}^{\infty} \left[\, k_\ell^+ P_{n-\ell}(t) + k_\ell^- P_{n+\ell}(t) - (k_\ell^+ + k_\ell^-)P_n(t) \,\right]\,.
\tag{24.21}
$$

The moment generating function defined in Eq. (24.3) is found from this equation as

$$
\mathcal{Z}(\rho,t) = \prod_{n=1}^{\infty} \exp\left\{\, t(\,\mathrm{e}^{i\rho q_0 n} - 1)k_n^+ + t(\,\mathrm{e}^{-i\rho q_0 n} - 1)k_n^- \,\right\}\,.
\tag{24.22}
$$

The cumulants are obtained from the cumulant generating function $\ln \mathcal{Z}(\rho,t)$ by differentiation. With use of Eq. (24.7) we get

$$
\langle q^N(t)\rangle_{\mathrm{c}} = q_0^N \sum_{n=1}^{\infty} n^N\,[\,k_n^+ + (-1)^N k_n^-\,]\,t\,.
\tag{24.23}
$$

The characteristic function is conveniently rewritten in terms of partial forward/backward currents $I_n^{\pm} = nk_n^{\pm}$ as

$$
\mathcal{Z}(\rho,t) = \prod_{n=1}^{\infty} Z_n^+(\rho,t)\, Z_n^-(\rho,t)\,,
\tag{24.24}
$$

where

$$
Z_n^{\pm}(\rho,t) = \sum_{\ell=0}^{\infty} \mathrm{e}^{\pm i\rho q_0 n\ell}\frac{(tI_n^{\pm}/n)^\ell}{\ell!}\,\mathrm{e}^{-tI_n^{\pm}/n} = \exp[\,t(\mathrm{e}^{\pm i\rho q_0 n} - 1)I_n^{\pm}/n\,]\,.
\tag{24.25}
$$

The physical meaning of this expression can be elucidated as follows. Suppose that the mass or charge transferred per unit time in forward direction were the results of a Poisson process for particles of unit mass propagating via uncorrelated nearest-neighbour forward steps, contributing a current $I_1^+ = \gamma_1^+$, plus a Poisson process of uncorrelated forward moves via next-to-nearest-neighbour transitions contributing a current $I_2^+ = 2\gamma_2^+$, etc. Suppose also that the total backward current were the result of independent Poisson processes with partial backward currents $I_n^- = n\gamma_n^-$, $n = 1, 2, \cdots$. The final form of the characteristic function would then be the expression (24.24) with (24.25).

The transport model (24.21) finds application, e.g., to charge transport through a weak link or through a quantum impurity in a 1D quantum wire, and to tunneling of edge currents in the fractional quantum Hall regime [cf. Chapter 26]. In these cases, the interpretation would be that the expression (24.24) with (24.25) describes independent Poisson processes for particles of charge one going across the impurity's barrier in forward/backward direction, contributing a current $I_1^\pm$, plus a Poisson process for particles of charge two (or for joint transport of a pair of particles of charge one) contributing a forward/backward current $I_2^\pm$, etc.

When the tunneling entities are coupled to a thermal reservoir, then the forward and backward transition weights are connected by detailed balance. Assuming that the potential drop per lattice period in forward direction is $\hbar\epsilon$, we have

$$k_n^-(\epsilon) = e^{-n\hbar\beta\epsilon} k_n^+(\epsilon) . \qquad (24.26)$$

We remark that the nonequilibrium relation between mobility and diffusion (24.20) does not hold when different pairs of transfer rates, say $k_1^\pm$, $k_2^\pm$, $\cdots$, contribute to the transport process, since they have different detailed balance factors, $e^{-\hbar\beta\epsilon}$, $e^{-2\hbar\beta\epsilon}$, $\cdots$.

In a classical process, the transition weights $k_n^\pm$, $n = 1, 2, \cdots$, have the usual meaning of rates, since they are all positive. When the Poisson processes come about quantum mechanically, the sign of transition weights may be negative, e.g., the sign of $k_n^\pm$ may alternate as a function of n. This does not spoil conservation of probability, since the master equation (24.21) provides $\sum_n \dot{P}_n(t) = 0$ regardles of the particular form chosen for the set $\{k_n^\pm\}$. In any case, each population $P_n(t)$, $N = 1, 2, \cdots$, must be non-negative at all times.

Below we shall see that the quantum mechanical transition weights $k_n^\pm(\epsilon)$ can be given in analytic form for particular models in ample regions of the parameter space.

24.3 Exact formal expressions for the system dynamics

In the tight-binding model (23.3), the system makes sudden transitions between neighbouring states of the density matrix with a probability amplitude $\pm i\Delta/2$ per unit time. The sequence of states of the density matrix visited in succession can be visualized in terms of a paths of a walker on an infinite square lattice spanned by the sites of the discrete (q, q') double path. The walker starts on some site on the principal diagonal, say at $(0,0)$, and then randomly makes vertical or horizontal steps along the q- or q'-axis. At each site of the square lattice there are four possible directions to continue. Return to a site on the principal diagonal is possible with an even number of steps.

A TB path on the square lattice is conveniently parametrized in terms of charges $u_j = \pm 1$ and $v_j = \pm 1$. A path with k steps in $q(\tau)$ and ℓ steps in $q'(\tau)$ is given by

$$q^{(k)}(\tau) \;=\; q_0 \sum_{j=1}^{k} u_j \Theta(\tau - t_j)\,, \qquad q'^{(\ell)}(\tau) \;=\; q_0 \sum_{i=1}^{\ell} v_i \Theta(\tau - t'_i)\,. \qquad (24.27)$$

A double path from the state $(0,0)$ to the state (n,n) is subject to the constraints

$$\sum_{j=1}^{k} u_j \;=\; n\,, \qquad \sum_{i=1}^{\ell} v_i \;=\; n\,. \qquad (24.28)$$

The environmental coupling manifests itself in complex-valued interactions between the charges. The influence action is conveniently split into the self-interactions of the paths $q(\tau)$ and $q'(\tau)$ and into the interaction between the paths $q(\tau)$ and $q'(\tau)$. Upon substituting the expression (24.27) into Eq. (5.22) we find

$$\mathcal{F}[q^{(k)}, q'^{(\ell)}] \;=\; \exp\left(\Phi_1[q^{(k)}] + \Phi_1^*[q'^{(\ell)}] + \Phi_2[q^{(k)}, q'^{(\ell)}] \right)\,, \qquad (24.29)$$

$$\Phi_1[q^{(k)}] \;=\; \sum_{j=2}^{k}\sum_{i=1}^{j-1} u_j u_i Q(t_j - t_i)\,, \qquad \Phi_1^*[q'^{(\ell)}] \;=\; \sum_{j=2}^{\ell}\sum_{i=1}^{j-1} v_j v_i Q^*(t'_j - t'_i)\,,$$

$$\Phi_2[q^{(k)}, q'^{(\ell)}] \;=\; -\sum_{i=1}^{\ell}\sum_{j=1}^{k} v_i u_j Q(t'_i - t_j)\,, \qquad (24.30)$$

where $Q(t)$ is the pair correlation function introduced in Section 18.2. We shall see that the (q, q')-representation (24.29) is advantageous if one aims at derivation of fluctuation-dissipation theorems, e.g., detailed balance and the Einstein relation (see Subsection 24.4.2).

For computational purposes it is again useful to introduce symmetric and antisymmetric spin paths,

$$\eta(\tau) \equiv [q(\tau) + q'(\tau)]/q_0\,, \qquad \xi(\tau) \;=\; [q(\tau) - q'(\tau)]/q_0\,. \qquad (24.31)$$

The path $\eta(\tau)$ describes propagation on the square lattice in direction parallel to the principal diagonal, whereas the path $\xi(\tau)$ measures moves in perpendicular direction. A general path with $2m$ steps on the square lattice, which is released at time t_p from the diagonal state $\xi = 0$, $\eta = 0$, is represented as

$$\eta^{(2m)}(\tau) \;=\; \sum_{k=1}^{2m} \eta_k\, \theta(\tau - t_k)\,, \qquad \xi^{(2m)}(\tau) = \sum_{k=1}^{2m} \xi_k\, \theta(\tau - t_k)\,, \qquad (24.32)$$

where $t_k > t_p$ for all k. The charges $\xi_k = \pm 1$ and $\eta_k = \pm 1$ label the four possibilities to move at time t_k instaneously from a site to a neighbouring site. Every individual u– or v–charge is associated with a $\{\eta, \xi\}$–pair. We have the correspondences

$$
\begin{aligned}
u_j = \pm 1 &\longleftrightarrow \{\eta_j, \xi_j\} = \{\pm 1, \pm 1\}\,, \\
v_k = \pm 1 &\longleftrightarrow \{\eta_k, \xi_k\} = \{\pm 1, \mp 1\}\,.
\end{aligned}
\tag{24.33}
$$

The influence function in (η, ξ)-representation takes the form

$$
\mathcal{F}^{(m)} = G_m H_m\,,
$$

$$
G_m = \exp\left[\sum_{j=2}^{2m}\sum_{k=1}^{j-1}\xi_j\xi_k Q'(t_j - t_k)\right],
\tag{24.34}
$$

$$
H_m = \exp\left[i\sum_{j=1}^{2m-1}\chi_{j,2m}\eta_j\right]\,, \quad \text{with} \quad \chi_{j,2m} = \sum_{k=j+1}^{2m}\xi_k Q''(t_k - t_j)\,.
$$

For nonzero tilting force in the Hamiltonian (23.3) at times $t > 0$, $F = \Theta(t)\hbar\epsilon/q_0$, the path (24.32) is weighted at positive times with a bias phase factor . For k transitions at negative times and $j = 2m - k$ transitions at positive times, the bias factor reads

$$
B_{j,k} = \exp(i\varphi_{j,k}) \qquad \varphi_{j,k} = \epsilon\int_0^t d\tau\,\xi^{(k+j)}(\tau) = \epsilon\sum_{i=k+1}^{k+j}\xi_i(t - t_i)\,.
\tag{24.35}
$$

With these preliminaries, it is straightforward to write down the path sum for all dynamical quantities of interest.

24.3.1 Product initial state

Consider now first the case of a product initial state of the global system. The path sum $\int \mathcal{D}q \int \mathcal{D}q'$ in Eq. (24.1) is expressed for the discrete system as follows. First, $P_n(t)$ is represented as an expansion in even numbers of steps or tunneling transitions. Secondly, in a given order of steps, say $2m$, we have to sum over all possible arrangements of labels $\{\xi_j = \pm 1\}$, $\{\eta_j = \pm 1\}$ which fulfill the boundary conditions (24.2). This gives the constraints

$$
\sum_j \eta_j = 2n\,, \qquad \sum_j \xi_j = 0\,.
\tag{24.36}
$$

Thirdly, we must take into account that the individual steps may occur at any time. The corresponding time-ordered integrations for ℓ steps in the negative time-branch and k steps in the positive-time branch are compactly expressed as

$$
\int_{t_p}^t \mathcal{D}_{k,\ell}\{t_j\} \times \cdots \equiv \int_0^t dt_{\ell+k}\cdots\int_0^{t_{\ell+2}} dt_{\ell+1}\int_{t_p}^0 dt_\ell\cdots\int_{t_p}^{t_2} dt_1\,\Delta^{k+\ell} \times \cdots\,.
\tag{24.37}
$$

We have included in the definition the product of the tunneling amplitudes. The various components are combined to yield for $P_n(t)$ the form

$$P_n(t) = \sum_{m=|n|}^{\infty} (-1)^{m-n} \left(\frac{1}{2}\right)^{2m} \int_0^t \mathcal{D}_{2m,0}\{t_j\} \sum_{\{\xi_j\}'} B_{2m,0} G_m \sum_{\{\eta_j\}'} H_m . \qquad (24.38)$$

The prime in $\{\xi_j\}'$ and $\{\chi_j\}'$ denotes summation under the constraints (24.36). With this form, the characteristic function (24.3) for a product initial state (pis) reads

$$\mathcal{Z}_{\text{pis}}(\rho, t) = \sum_{m=0}^{\infty} (-1)^m \int_0^t \mathcal{D}_{2m,0}\{t_j\} \sum_{\{\xi_j\}'} B_{2m,0} G_m F_{2m}(\rho) . \qquad (24.39)$$

The weighted average over the populations $P_n(t)$ is in the function

$$F_{2m}(\rho) = \frac{1}{2^{2m}} \sum_{n=-m}^{+m} (-1)^n e^{i\rho n q_0} \sum_{\{\eta_j\}'} H_m = \sin\left(\frac{\rho q_0}{2}\right) \prod_{j=1}^{2m-1} \sin\left(\frac{\rho q_0}{2} + \chi_{j,2m}\right) .$$

To obtain the second form, we have substituted the Fourier representation for the Kronecker function which brings in the constraint $\sum_j \eta_j = 2n$, and then performed the sum over n and $\{\eta_j = \pm 1\}$. Observe that every individual path together with its charge-conjugate counter part gives a real contribution to the generating function.

It is now easy to find the expressions for the moments of the probability distribution. For the most interesting cases $N = 1$ and $N = 2$ we obtain using Eq. (24.4)

$$\langle q(t) \rangle_{\text{pis}} = q_0 \sum_{m=1}^{\infty} \int_0^t \mathcal{D}_{2m,0}\{t_j\} \sum_{\{\xi_j\}'} b_{2m,0} B_{2m,0} G_m , \qquad (24.40)$$

$$\langle q^2(t) \rangle_{\text{pis}} = q_0^2 \sum_{m=1}^{\infty} \int_0^t \mathcal{D}_{2m,0}\{t_j\} \sum_{\{\xi_j\}'} c_{2m,0} B_{2m,0} G_m , \qquad (24.41)$$

where

$$b_{2m,0} = \frac{i}{2} (-1)^{m-1} \prod_{\ell=1}^{2m-1} \sin(\chi_{\ell,2m}) , \qquad (24.42)$$

$$c_{2m,0} = \frac{1}{2} (-1)^{m-1} \sum_{k=1}^{2m-1} \cos(\chi_{k,2m}) \prod_{\ell=1,\ell\neq k}^{2m-1} \sin(\chi_{\ell,2m}) . \qquad (24.43)$$

Eqs. (24.38) – (24.43) are exact series expressions for a product initial state. The phase correlations resulting from the friction action are in the functions $\chi_{k,2m}$. The correlations due to the noise action are captured by the functions G_m.

We also give the corresponding Laplace transforms. These are conveniently expressed in terms of the intervals the system spends in a particular state, $\tau_\ell = t_{\ell+1} - t_\ell$. We introduce for the corresponding integrations the compact symbol

$$\int_0^{\infty} \widehat{\mathcal{D}}_{j,0}(\lambda, \{\tau_\ell\}) \times \cdots \equiv \Delta^j \int_0^{\infty} \prod_{k=1}^{j-1} d\tau_k\, e^{-\lambda \tau_k} \times \cdots , \qquad \tau_k = t_{k+1} - t_k . \quad (24.44)$$

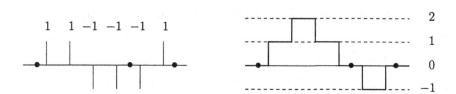

Figure 24.1: A sequence of six steps falling into two clusters separated by a sojourn ($\bullet$) is sketched. In the charge picture (left), the spikes or charges $\{\xi_j = \pm 1\}$ give the direction of the moves perpendicular to the diagonal of the RDM. The right diagram shows for the same walk the distance from the main diagonal in terms of the cumulative charges $\{p_j\}$.

The Laplace transforms of $\langle q(t) \rangle_{\text{pis}}$ and $\langle q^2(t) \rangle_{\text{pis}}$ are then given by

$$\langle \hat{q}(\lambda) \rangle_{\text{pis}} = i \frac{q_0}{\lambda^2} \sum_{m=1}^{\infty} \int_0^{\infty} \widehat{\mathcal{D}}_{2m,0}(\lambda, \{\tau_\ell\}) \sum_{\{\xi_j\}'} b_{2m,0} G_m \sin \varphi_{2m,0} , \qquad (24.45)$$

$$\langle \hat{q}^2(\lambda) \rangle_{\text{pis}} = \frac{q_0^2}{\lambda^2} \sum_{m=1}^{\infty} \int_0^{\infty} \widehat{\mathcal{D}}_{2m,0}(\lambda, \{\tau_\ell\}) \sum_{\{\xi_j\}'} c_{2m,0} G_m \cos \varphi_{2m,0} . \qquad (24.46)$$

The bias phase $\varphi_{2m,0}$ is conveniently expressed in terms of the cumulative charges

$$p_\ell = \sum_{k=1}^{\ell} \xi_k = - \sum_{k=\ell+1}^{2m} \xi_k . \qquad (24.47)$$

The second form is obtained with use of the constraint (24.36). The cumulative charge p_ℓ measures how far the system is off-diagonal after ℓ transitions. The meaning of the charges $\{\xi_j\}$ and the cumulative charges $\{p_j\}$ is illustrated in Fig. 24.1. With the cumulative charges $\{p_j\}$, the bias phase (24.35) can be written as

$$\varphi_{j,k} = \epsilon\, p_k(t_{k+1} - t) + \epsilon \sum_{\ell=k+1}^{k+j-1} p_\ell \tau_\ell , \qquad \varphi_{2m,0} = \epsilon \sum_{\ell=1}^{2m-1} p_\ell \tau_\ell . \qquad (24.48)$$

The dynamics at long times is found from the limit $\lambda \to 0$ in Eqs. (24.45) and (24.46).

24.3.2 Thermal initial state

To derive the dynamical expressions for a thermal initial state, we follow the discussion in Section 5.5 and Subsection 21.2.2. We then obtain the propagating function, which is a constituent of the expression (24.12), in the form

$$J(n\,n, t; m\,r, 0; 0\,0, t_{\text{p}}) = \sum_{\substack{k=|m|+|r| \\ j=|n-m|+|n-r|}}^{\infty} (-1)^n \left(\frac{i}{2} \right)^{j+k} \int_{t_{\text{p}}}^{t} \mathcal{D}_{j,k}\{t_\ell\} \sum_{\{\xi_\ell\}''} B_{j,k} G_{j+k} \sum_{\{\eta_\ell\}''} H_{j+k} .$$

$$(24.49)$$

The double prime in $\{\xi_\ell\}''$ and $\{\eta_\ell\}''$ is affixed to indicate the constraints

$$
\sum_{\ell=1}^{k}\eta_\ell = m+r\,, \qquad \sum_{\ell=k+1}^{k+j}\eta_\ell = 2n-(m+r)\,,
$$
$$
\sum_{\ell=1}^{k}\xi_\ell = m-r\,, \qquad \sum_{\ell=k+1}^{k+j}\xi_\ell = r-m\,. \tag{24.50}
$$

The summations $\{\xi_\ell=\pm1\}''$ and $\{\eta_\ell=\pm1\}''$ create all possible paths on the (q,q')-lattice to go in k steps in the negative time interval $t_{\rm p}<\tau<0$ from $(0,0)$ to (m,r) and in j steps at positive times $0<\tau<t$ from (m,r) to (n,n). Substituting the form (24.49) into Eq. (24.12), the generating function $\mathscr{Z}_{\rm th}$ takes the form

$$
\mathscr{Z}_{\rm th}(\rho,\kappa,\mu;t) = \sum_{j=0}^{\infty}\sum_{k=0}^{\infty}\int_{-\infty}^{t}\mathcal{D}_{j,k}\{t_\ell\}\sum_{\{\xi_\ell\}'}B_{j,k}G_{j+k}F_{j,k}(\rho,\kappa,\mu)\,, \tag{24.51}
$$

where the prime in $\{\xi_\ell\}'$ denotes again summation over all charge sequences obeying the neutrality condition (24.36). The function $F_{j,k}(\rho,\kappa,\mu)$ is given by

$$
F_{j,k}(\rho,\kappa,\mu) = \left(\frac{1}{2}\right)^{j+k}\sum_{n,m,r=-\infty}^{+\infty}\Theta(k-|m|-|r|)\,\theta(j-|n-m|-|n-r|)
$$
$$
\times (-1)^{(j+k-2n)/2}\,e^{iq_0(n\kappa+m\rho+r\mu)}\sum_{\{\eta_\ell\}''}H_{j+k}\,. \tag{24.52}
$$

Following the strategy employed for $F_{2m}(\rho)$, it is easy to perform the constrained sums. The result is

$$
F_{j,k}(\rho,\kappa,\mu) = i^{j+k}\exp\left(i\frac{q_0}{2}(\rho-\mu)p_k\right)\sin\left(\frac{\kappa q_0}{2}\right) \tag{24.53}
$$
$$
\times \prod_{\ell=1}^{k}\sin\left(\frac{q_0}{2}(\kappa+\rho+\mu)+\chi_{\ell,j+k}\right)\prod_{m=k+1}^{j+k-1}\sin\left(\frac{q_0}{2}\kappa+\chi_{m,j+k}\right)\,.
$$

Finally, performing the differentiations of $\mathscr{Z}_{\rm th}$ as given in Eq. (24.11) we find

$$
\langle q(t)\rangle_{\rm th} = \langle q(t)\rangle_{\rm pis}+R(t)\,, \tag{24.54}
$$
$$
C(t) = \langle q^2(t)\rangle_{\rm pis}+S(t)\,. \tag{24.55}
$$

The effects of the different initial conditions for $\langle\cdots\rangle_{\rm pis}$ and $\langle\cdots\rangle_{\rm th}$ are described by

$$
R(t) = q_0\sum_{j=1}^{\infty}\sum_{k=1}^{\infty}\int_{-\infty}^{t}\mathcal{D}_{j,k}\{t_\ell\}\sum_{\{\xi_\ell\}'}b_{j,k}B_{j,k}G_{j+k}\,, \tag{24.56}
$$
$$
S(t) = q_0^2\sum_{j=1}^{\infty}\sum_{k=1}^{\infty}\int_{-\infty}^{t}\mathcal{D}_{j,k}\{t_\ell\}\sum_{\{\xi_\ell\}'}c_{j,k}B_{j,k}G_{j+k}\,. \tag{24.57}
$$

The coefficients $b_{j,k}$ and $c_{j,k}$ are given by

$$b_{j,k} = i\frac{1}{2}(-1)^{(j+k-2)/2}\prod_{\ell=1}^{j+k-1}\sin(\chi_{\ell,j+k}) , \qquad (24.58)$$

$$c_{j,k} = \frac{1}{2}(-1)^{(j+k-2)/2}\prod_{\ell=1}^{k}\sin(\chi_{\ell,j+k})\left[\sum_{i=k+1}^{j+k-1}\cos(\chi_{i,j+k})\prod_{m=k+1,m\neq i}^{j+k-1}\sin(\chi_{m,j+k})\right] .$$

Further, the antisymmetrized correlation function is found as

$$A(t) = -\frac{i}{2}q_0^2\sum_{j=1}^{\infty}\sum_{k=1}^{\infty}\int_{-\infty}^{t}\mathcal{D}_{j,k}\{t_\ell\}\sum_{\{\xi_\ell\}'}p_kb_{j,k}G_{j+k} . \qquad (24.59)$$

Finally, consider the Laplace transforms of $R(t)$ and $S(t)$. In generalization of the expression (24.44), we define

$$\int_0^{\infty}\widehat{\mathcal{D}}_{j,k}(\lambda,\{\tau_\ell\})\times\cdots \equiv \Delta^{j+k}\prod_{i=k+1}^{k+j-1}\int_0^{\infty}d\tau_i\,e^{-\lambda\tau_i}\prod_{\ell=1}^{k-1}\int_0^{\infty}d\tau_\ell\,e^{-0^+\tau_\ell}\times\cdots . \qquad (24.60)$$

The additional two intervals not included in Eq. (24.60) are

$$r_k = -t_k ; \qquad s_k = t_{k+1} . \qquad (24.61)$$

The interval r_k is the time passed between the last step in the negativ-time branch and time zero, and s_k is the interval between time zero and the first step at positive time. The Laplace transforms of the functions $R(t)$ and $S(t)$ are

$$\hat{R}(\lambda) = i\frac{q_0}{\lambda}\sum_{j=1}^{\infty}\sum_{k=1}^{\infty}\int_0^{\infty}\widehat{\mathcal{D}}_{j,k}(\lambda,\{\tau_\ell\})\int_0^{\infty}dr_k\,ds_k\,e^{-(\lambda s_k+0^+r_k)}\sum_{\{\xi_\ell\}'}b_{j,k}G_{j+k}\sin\varphi_{j,k} ,$$

$$\hat{S}(\lambda) = \frac{q_0^2}{\lambda}\sum_{j=1}^{\infty}\sum_{k=1}^{\infty}\int_0^{\infty}\widehat{\mathcal{D}}_{j,k}(\lambda,\{\tau_\ell\})\int_0^{\infty}dr_k\,ds_k\,e^{-(\lambda s_k+0^+r_k)}\sum_{\{\xi_\ell\}'}c_{j,k}G_{j+k}\cos\varphi_{j,k} ,$$

and the Laplace transform of the antisymmetric correlation function for zero bias is

$$\hat{A}(\lambda) = \frac{-iq_0^2}{2\lambda}\sum_{j=1}^{\infty}\sum_{k=1}^{\infty}\int_0^{\infty}\widehat{\mathcal{D}}_{j,k}(\lambda,\{\tau_\ell\})\int_0^{\infty}dr_k\,ds_k\,e^{-(\lambda s_k+0^+r_k)}\sum_{\{\xi_\ell\}'}p_kb_{j,k}G_{j+k} .$$

The functions $R(t)$ and $S(t)$ describe the effects of the correlated initial state on the dynamics. Finally, $A(t)$ is related to the response function $\chi(t)$ as in Eq. (6.15).

24.4   Mobility and Diffusion

24.4.1   Exact formal series expressions for transport coefficients

The mobility and the diffusion coefficient are found from the limits $\lim_{\lambda \to 0} \lambda^2 \langle \hat{q}(\lambda) \rangle_{\text{pis}}$ and $\lim_{\lambda \to 0} \lambda^2 \langle \hat{q}^2(\lambda) \rangle_{\text{pis}}$ in the expressions (24.45) and (24.46).

In the charge picture, an offdiagonal (blip) interval with cumulative charge p_k separates a sequence of charges with total charge zero into two sections with charge $\pm p_k$. When the blip length τ_k is effectively large compared to the length of the sections, the blip interaction factor is $\exp\left[-p_k^2 Q'(\tau_k) \right]$. For the form (18.44) of the spectral density $G_{\text{lf}}(\omega)$, the function $Q(\tau)$ behaves asymptotically for nonzero temperature as

$$Q'(\tau \to \infty) \propto \tau^{2-s} ; \qquad Q''(\tau \to \infty) \propto \tau^{1-s} . \qquad (24.62)$$

Hence the integral over the length of a blip is convergent also in the limit $\lambda \to 0$ when $s < 2$. For $s \geq 2$ and $\lambda = 0$, the interaction between charged clusters can not suppress long blips anymore, and the integral is only convergent when the biasing force F is nonzero.

For a sojourn state, the respective charge label, say p_j, is zero. Then, convergence of the integral over the sojourn interval τ_j can not be provided by the charge interaction $Q'(\tau_j)$ so that, in the limit $\lambda \to 0$, convergence of the sojourn integral must be provided by the phase factors $b_{2m,0}$ or $c_{2m,0}$. Inspection of the phase factor $b_{2m,0}$ shows that it provides convergence of the integral over the sojourn time in a diagonal state when $s \geq 1$. Under convergence of the internal blip and sojourn integrals in a neutral charge cluster we then obtain at long times

$$\lim_{t \to \infty} \langle q(t) \rangle_{\text{pis}} = F\mu t + q_\infty , \qquad (24.63)$$

where we have added the next to leading order term. The quantity μ is the mobility. It is called nonlinear mobility when it depends on the bias, $\mu = \mu(T, \epsilon)$, and linear mobility in the linear response regime, $\mu_{\text{lin}}(T) = \mu(T, 0)$. We obtain the mobility from Eq. (24.45) in the form of a series in even numbers of transitions,

$$\mu(\epsilon, T) = \lim_{\lambda \to 0^+} i\frac{q_0^2}{\hbar \epsilon} \sum_{m=1}^{\infty} \int_0^\infty \widehat{\mathcal{D}}_{2m,0}(\lambda, \{\tau_\ell\}) \sum_{\{\xi_l\}'} b_{2m,0} G_m \sin\left(\epsilon \sum_{j=1}^{2m-1} p_j \tau_j \right) . \quad (24.64)$$

This series expression is formally exact. With the above arguments we see that the nonlinear mobility is well defined for $s \geq 1$ and the linear mobility for $T > 0$ in the range $2 > s \geq 1$. At zero temperature, $Q'(\tau)$ approaches a constant as $\tau \to \infty$ for $s > 1$. Thus the linear mobility becomes divergent as $T \to 0$ also in the super-Ohmic regime $1 < s < 2$. This indicates that the mean position moves faster than linear in t. The sub-Ohmic case is special and is discussed in Section 25.8.

Consider next the second moment $\langle \hat{q}^2(\lambda) \rangle_{\text{pis}}$ for the untilted washboard ($\epsilon = 0$). The phase factor (24.43) is nonzero for a sojourn k, $p_k = 0$, because of the term $\cos(\chi_{k,2m})$. In addition, the interactions in the factor G_m across the sojourn are not strong enough for $s \geq 1$ to regularize the integral over the sojourn length. Thus the sojourn's breathing integral diverges in the limit $\lambda \to 0$. However, the divergent contributions cancel out in the sum over all paths which visit the sojourn k. In the end one finds that the limiting expression $\lim_{\lambda \to 0} \lambda^2 \langle \hat{q}^2(\lambda) \rangle_{\mathrm{pis}}$ approaches a constant in the regime $1 \leq s < 2$ and $T > 0$. We obtain

$$\lim_{t \to \infty} \langle q^2(t) \rangle_{\mathrm{pis}} = 2Dt , \qquad (24.65)$$

with the diffusion coefficient

$$D(T) = \lim_{\lambda \to 0^+} \frac{q_0^2}{2} \sum_{m=1}^{\infty} \int_0^{\infty} \widehat{\mathcal{D}}_{2m,0}(\lambda, \{\tau_\ell\}) \sum_{\{\xi_j\}'} c_{2m,0} G_m . \qquad (24.66)$$

In conclusion, the TB model shows diffusive behaviour in the regime $1 < s < 2$. This is different from free Brownian motion which is superdiffusive in this regime, as we have discussed in Section 7.4.

24.4.2 Einstein relation

The Einstein relation links the diffusion coefficient to the linear mobility,

$$D(T) = k_{\mathrm{B}} T \mu_{\mathrm{lin}}(T) . \qquad (24.67)$$

As emphasized in the preceeding subsection, the static transport coefficients D and μ_{lin} in this relation are well-defined in the parameter regime $1 \leq s < 2$. We have seen already in Section 7.3 for the case of free Brownian motion that the relation (24.67) is a special version of the fluctuation-dissipation theorem. The relation was proven rigorously for Brownian motion in a corrugated potential in the classical regime [461]. There have been raised concerns about the general validity of the Einstein relation in the quantum regime [206]. The questions are related to possible problems with the definition of μ_{lin} since the stationary state in which the particle moves with constant velocity is not normalizable and therefore the standard Green-Kubo method might not be applicable. On the other hand, if one defines the mobility as in Eq. (24.63), it is not obvious a priori whether the Einstein relation is valid in all orders in Δ^2.

To prove the relation (24.67) for the TB model, we rewrite μ_{lin} and D as moments of the distribution function, and return to the parametrization (24.27). We have

$$\mu_{\mathrm{lin}} = q_0^2 \sum_n n X_n , \qquad D = \tfrac{1}{2} q_0^2 \sum_n n^2 Y_n , \qquad (24.68)$$

$$X_n = \sum_{k,\ell=|n|}^{\infty} i^{\ell-k} \left(\frac{\Delta}{2}\right)^{\ell+k} \int_0^{\infty} \prod_{j=1}^{k-1} d\rho_j \prod_{i=1}^{\ell-1} d\rho_i' \int_{-\infty}^{\infty} d\tau \sum_{\{u_k,v_\ell\}'} \frac{i\varphi_{k,\ell}}{\hbar\epsilon} \mathcal{F}[q^{(k)}, q'^{(\ell)}] ,$$

$$Y_n = \sum_{k,\ell=|n|}^{\infty} i^{\ell-k} \left(\frac{\Delta}{2}\right)^{\ell+k} \int_0^{\infty} \prod_{j=1}^{k-1} d\rho_j \prod_{i=1}^{\ell-1} d\rho_i' \int_{-\infty}^{\infty} d\tau \sum_{\{u_k,v_\ell\}'} \mathcal{F}[q^{(k)}, q'^{(\ell)}] ,$$

where $\rho_j = t_{j+1} - t_j$ and $\rho_i' = t_{i+1}' - t_i'$ are the intervals in the paths $q(\tau)$ and $q'(\tau)$, and $\tau = t_1' - t_k$. The bias phase is $\varphi_{k,\ell} = \epsilon \left(\sum_i v_i t_i' - \sum_j u_j t_j \right)$. The τ-integration introduces all possibilities of mixing up the steps in $q^{(k)}(\tau)$ with those in $q'^{(\ell)}(\tau)$.

To proceed, we assign to the double path $\{ q^{(k)}(\tau), q'^{(\ell)}(\tau) \}$ a conjugate double path $\{ q^{(\ell)}(\tau), q'^{(k)}(\tau) \}$ in which the charges are in reverse order. Since the function $Q(z)$ is analytic in the strip $0 \geq \operatorname{Im} z > -\hbar\beta$, the τ-contour can be shifted in this strip as explained below Eq. (20.29).[2] It is convenient to choose $\widetilde{\tau} = \tau - i\hbar\beta$. Using the reflection symmetry (18.43) of $Q(z)$, the influence function $\widetilde{\mathcal{F}}$ calculated for a conjugate pair of double paths on the shifted contour $\widetilde{\tau}$ has the property

$$\widetilde{\mathcal{F}}[q^{(k)}, q'^{(\ell)}] + \widetilde{\mathcal{F}}[q^{(\ell)}, q'^{(k)}] = \mathcal{F}^*[q^{(k)}, q'^{(\ell)}] + \mathcal{F}^*[q^{(\ell)}, q'^{(k)}], \qquad (24.69)$$

and the bias phase on the shifted contour is given by $\widetilde{\varphi}_{k,\ell} = \varphi_{k,\ell} - i\beta\hbar\epsilon n$. Upon writing X_n for the shifted contour, then employing Eqs. (24.69) and the shift of the bias phase on the shifted contour, and comparing with the expression for Y_n, we find for $1 \leq s < 2$ the relation [328]

$$X_n = -X_n + n\beta Y_n. \qquad (24.70)$$

This relation is valid in all orders in Δ^2. Substituting Eq. (24.70) into the expressions (24.68), we find the Einstein relation (24.67) (See also the discussion in Ref. [462]).

Since a conjugate pair of double paths visit the same sites, the relation (24.70) and hence the Einstein relation holds also in the presence of disorder.

24.5 The Ohmic case

We now consider the mobility for Ohmic spectral densities in various limits. We base the treatment on the scaling form (18.51) for the pair interaction with the usual dimensionless damping parameter [cf. Eq. (18.19)]

$$K = \frac{\eta q_0^2}{2\pi\hbar}. \qquad (24.71)$$

Substituting Eq. (18.51) into Eq. (24.64), the normalized TB mobility is found as

$$\frac{\mu^{(\mathrm{TB})}(T, \epsilon)}{\mu_0} = \frac{\pi K}{\epsilon} \sum_{m=1}^{\infty} (-1)^{m-1} \Delta^{2m} \int_0^\infty d\tau_1 \, d\tau_2 \cdots d\tau_{2m-1} \qquad (24.72)$$

$$\times \sum_{\{\xi_j\}'} G_m \left(\{\tau_j\}; K, \{\xi_j\} \right) \sin \left(\sum_{i=1}^{2m-1} \epsilon p_i \tau_i \right) \prod_{k=1}^{2m-1} \sin(\pi K p_k),$$

with the interaction factor (we put $\tau_{jk} \equiv t_j - t_k = \sum_{\ell=k}^{j-1} \tau_\ell$)

$$G_m \left(\{\tau_j\}; K, \{\xi_j\} \right) = \exp \left(2K \sum_{j>k=1}^{2m} \xi_j \xi_k \ln \left[(\hbar\beta\omega_c/\pi) \sinh(\pi\tau_{jk}/\hbar\beta) \right] \right). \qquad (24.73)$$

_____

[2]In the absence of a bias, $\exp[-Q(z)]$ must vanish for $|z| \to \infty$. This gives the bound $s < 2$.

Here, $\mu_0 = 1/\eta = q_0^2/2\pi\hbar K$ is the mobility of a free Brownian particle [cf. Eq. (7.25)]. Due to the product of sine factors, every double-path contributing to the mobility does not return to the diagonal at intermediate times, i.e., the affiliated cumulative charges p_k have all the same sign. In charge picture, the form (24.72) is the grand-canonical sum of all charge sequences which can not be separated into neutral subclusters.

24.5.1 Weak-tunneling regime

The $m = 1$ term of the series (24.72) is found by combining previous results. With the form (18.51) for $Q(t)$, the nonadiabatic forward tunneling rate k^+ takes the form (20.78). Substituting this result into the mobility expression (24.20), we readily get

$$\frac{\mu^{(\mathrm{TB})}(T,\epsilon)}{\mu_0} = \pi^2 K \frac{|\Gamma(K + i\hbar\epsilon/2\pi k_{\mathrm{B}}T)|^2}{\Gamma(2K)} \frac{\sinh(\hbar\epsilon/2k_{\mathrm{B}}T)}{\hbar\epsilon/2k_{\mathrm{B}}T} \left(\frac{\hbar\Delta_{\mathrm{r}}}{2\pi k_{\mathrm{B}}T}\right)^{2-2K} . \qquad (24.74)$$

This gives at zero temperature

$$\frac{\mu^{(\mathrm{TB})}(0,\epsilon)}{\mu_0} = \frac{\pi^2 K}{\Gamma(2K)} \left(\frac{\Delta_{\mathrm{r}}}{\epsilon}\right)^{2-2K} , \qquad (24.75)$$

and for the linear mobility at finite temperatures

$$\frac{\mu^{(\mathrm{TB})}(T,0)}{\mu_0} = \frac{\pi^2\sqrt{\pi}}{2} \frac{\Gamma(1+K)}{\Gamma(1/2+K)} \left(\frac{\hbar\Delta_{\mathrm{r}}}{\pi k_{\mathrm{B}}T}\right)^{2-2K} . \qquad (24.76)$$

These expressions describe the mobility for an Ohmic environment in the nearest-neighbour tunneling regime discussed in Subsection 24.2.1. The corresponding expression for the diffusion coefficient $D(T,\epsilon)$ follows with use of Eq. (24.20).

24.5.2 Weak-damping limit

When the Kondo parameter is very small, $K \ll 1$, the Gaussian filter G_m in the series expression (24.39) for $\mathcal{Z}_{\mathrm{pis}}(\rho,t)$ is wide so that paths with visits of far off-diagonal states of the RDM and long sojourn time in these states may give significant contributions to the long-time dynamics. Under this condition, we may use for the noise correlation function $Q'(\tau)$ the asymptotic long-time form [cf. Eq. (21.113)]

$$Q'(\tau) = 2K[\pi|\tau|/\hbar\beta + \ln(\hbar\beta\omega_c/2\pi)] . \qquad (24.77)$$

With this expression for $Q'(\tau)$, the filter function G_m, which is given in Eq. (24.73), separates into a product of exponential filters for the individual time interval. We get

$$G_m = \left(\frac{2\pi}{\hbar\beta\omega_c}\right)^{2Km} \prod_{k=1}^{2m-1} \exp\left(-2\pi K p_k^2 \frac{\tau_k}{\hbar\beta}\right) . \qquad (24.78)$$

With this form, the time integrals in the Laplace representation $\langle \dot{\hat{q}}^N(\lambda)\rangle_{\mathrm{fc}}$ separate into integrals of the exponential function. The mobility is found from Eq. (24.72) as

$$\frac{\mu^{(\text{TB})}(T,\epsilon)}{\mu_0} = -\frac{2\pi K}{\hbar\beta\epsilon}\sum_{m=1}^{\infty}(-u)^m\sum_{\{\xi_j=\pm 1\}'}\text{Im}\prod_{k=1}^{2m-1}\frac{\sin(\pi K p_k)}{\pi K p_k[p_k - i\,\hbar\beta\epsilon/2\pi K]}\,,\quad (24.79)$$

where

$$u \equiv (\hbar\beta\Delta_T/2)^2\,,\qquad \Delta_T = \Delta_{\text{r}}\,(2\pi k_{\text{B}}T/\hbar\Delta_{\text{r}})^K\,.\qquad (24.80)$$

Here, the adiabatic Franck-Condon factor $(2\pi/\hbar\beta\omega_{\text{c}})^K$ is included in the frequency scale Δ_T [cf. Eq. (21.114)]. The series (24.79) resembles the grand-canonical partition function of a one-dimensional Coulomb gas [463]. The summation over all arrangements of charges $\xi_j = \pm 1$ satisfying the neutrality condition (24.36) leads to the continued fraction expression [463, 456]

$$\frac{\mu^{(\text{TB})}(T,\epsilon)}{\mu_0} = \frac{4\pi K}{\hbar\beta\epsilon}\,\text{Im}\,\cfrac{a_1 u}{1+\cfrac{a_1 a_2 u}{1+\cfrac{a_2 a_3 u}{1+\cdots}}}\qquad (24.81)$$

with coefficients

$$a_n = \frac{\sin(\pi K n)}{\pi K n}\,\frac{1}{n - i\,\hbar\beta\epsilon/2\pi K}\,.\qquad (24.82)$$

For $K \ll 1$, the sine function in Eq. (24.82) may be linearized. This finally yields

$$a_n = 1/(n - i\nu)\qquad \text{with}\qquad \nu \equiv \hbar\beta\epsilon/2\pi K\,.\qquad (24.83)$$

With this form, the expression (24.81) matches the continued fraction (9.1.73) in Ref. [88] which can be written in terms of modified Bessel functions $I_\mu(z)$ with complex order μ [464]. Upon adjusting the parameters, we find the mobility in analytic form for all T and ϵ as

$$\frac{\mu^{(\text{TB})}(T,\epsilon)}{\mu_0} = \frac{\hbar\beta\Delta_T}{\nu}\,\text{Im}\left(\frac{I_{1-i\nu}(\hbar\beta\Delta_T)}{I_{-i\nu}(\hbar\beta\Delta_T)}\right)\,.\qquad (24.84)$$

Employing recursion relations of the modified Bessel function, this expression can be rewritten as

$$\frac{\mu^{(\text{TB})}(T,\epsilon)}{\mu_0} = 1 - \frac{\sinh(\pi\nu)}{\pi\nu}\,\frac{1}{|I_{i\nu}(\hbar\beta\Delta_T)|^2}\,.\qquad (24.85)$$

Upon performing the limit $\epsilon \to 0$ either in Eq. (24.84) or in Eq. (24.85), we get for the linear mobility the expression

$$\frac{\mu_{\text{lin}}^{(\text{TB})}(T)}{\mu_0} = 1 - \frac{1}{I_0^2(\hbar\beta\Delta_T)}\,.\qquad (24.86)$$

Consider next the case $T = 0$ and take the limit $K \to 0$ in the series expression (24.72). Now, thermal noise is absent and quantum noise is so weak that the only effect of the filter function G_m is to regularize the time integrals. Further, we may approximate each factor $\sin(\pi K p_k)$ by $\pi K p_k$. Putting $G_m = 1$, the time integrals are easily done, and we find that the cumulative charges p_j drop out. Thus, for a fixed number of charges in a neutral cluster, every individual sequence of the charges gives the same contribution. The series expression for the nonlinear mobility takes the form

$$\frac{\mu^{(\mathrm{TB})}(0,\epsilon)}{\mu_0} = \sum_{m=1}^{\infty} \left(\frac{\pi K\Delta}{\epsilon}\right)^{2m} \sum_{\{\xi_j\}'} 1 \,. \tag{24.87}$$

The number of possibilities for 2m charges with total charge zero to form a single irreducible cluster, i.e. all $p_j \neq 0$, is $2^{2m-1}\Gamma(m-1/2)/\sqrt{\pi}\,\Gamma(m+1)$. Thus we find

$$\frac{\mu^{(\mathrm{TB})}(0,\epsilon)}{\mu_0} = \frac{1}{2\sqrt{\pi}} \sum_{m=1}^{\infty} \frac{\Gamma(m-1/2)}{\Gamma(m+1)} \left(\frac{2\pi K\Delta}{\epsilon}\right)^{2m} \,. \tag{24.88}$$

This series is convergent in the regime

$$\epsilon > \epsilon_0 \equiv 2\pi K\Delta \,, \tag{24.89}$$

in which it is summed to the square root expression

$$\frac{\mu^{(\mathrm{TB})}(0,\epsilon)}{\mu_0} = 1 - \sqrt{1 - (\epsilon_0/\epsilon)^2} \,. \tag{24.90}$$

Two remarks seem to be in order. First, the expression (24.90) is found also from the continued fraction expression (24.81) if we substitute $a_n = i/\nu$ and put $u = (\hbar\beta\Delta/2)^2$. Secondly, in the zero temperature limit of the expression (24.84), or of the expression (24.85), the uniform asymptotic expansion of the modified Bessel function for large complex order [cf. Eq. (9.7.7) in Ref. [88]] is appropriate. Upon employing the leading term and disregarding the adiabatic Franck-Condon dressing factor, we attain again the expression (24.90).

24.6 Exact solution in the Ohmic scaling limit at $K = \frac{1}{2}$

The quantum transport for the particular case $K = \frac{1}{2}$ can be solved in analytic form in two different ways. In the first, one employs in the Coulomb gas representation of the TB model presented in Section 24.3 the concept of collapsed dipoles, which has been introduced already in Subsection 21.6.1. Alternatively, following the treatment given in Subsections 19.2.2 and 21.6.6, one transforms to a representation of noninteracting fermionic quasiparticles.

24.6.1 Current and mobility

In the Ohmic scaling limit, the TB series (24.72) for the mobility and the series (24.45) for $\langle \hat{q}(\lambda)\rangle_{\mathrm{pis}}$ can be calculated in analytic form at the coupling value $K = \frac{1}{2}$ using the concept of collapsed dipoles developed in Subsection 21.6.1 [465, 328]. The calculation is done by putting $K = \frac{1}{2} - \kappa$ and eventually taking the limit $\kappa \to 0$. In this limit, the phase factors $\sin(\pi K p_k)$ are reduced to factors $1, 2\pi\kappa, -1, -4\pi\kappa, \cdots$ for $p_k = 1, 2, 3, 4, \cdots$. On the other hand, the breathing mode integral of a dipole is $\propto 1/\kappa$, as $\kappa \to 0$. The singularity originates from the $\tau^{-1+2\kappa}$ short-distance behaviour of the intra-dipole interaction $\exp[-Q'(\tau)]$. Altogether, only those charge sequences, in which the occurring κ-factors are compensated by $1/\kappa$-singularities, contribute to the mobility. These are just the formations which have cumulative charge sequence $p_k = 1, 2, 1, 2, \cdots, 2, 1$ and $p_k = -1, -2, -1, -2, \cdots, -2, -1$. The respective charge sequences are

$$
\begin{aligned}
+ \ (+ \ -) \ (+ \ -) \ \cdots \ (+ \ -) \ (+ \ -) \ - \ , \\
- \ (- \ +) \ (- \ +) \ \cdots \ (- \ +) \ (- \ +) \ + \ .
\end{aligned}
\tag{24.91}
$$

Here, the round brackets indicate the charge pairs which form collapsed dipoles.

For later convenience (see Subsection 25.2.1), we now use the renormalized tunneling amplitude introduced below in Eq. (25.113). At $K = \frac{1}{2}$, we have

$$
\tilde{\epsilon}_0 \ = \ 2\gamma \ = \pi\Delta^2/\omega_c \, ,
\tag{24.92}
$$

where $\gamma = \Delta_{\text{eff}}(K = \frac{1}{2})$ is the frequency scale familiar from the spin-boson system at the Toulouse point $K = \frac{1}{2}$ (see Subsection 19.2.2 and Section 21.6).

The grand-canonical sum of collapsed noninteracting dipoles in the interval τ between the two peripheral charges yields the form factor (see Subsection 21.6.1)

$$
U_{\text{CD}}(\tau) \ = \ e^{-\tilde{\epsilon}_0\tau} \, .
\tag{24.93}
$$

The exponent in Eq. (24.93) differs from the exponent in the CS form factor (21.198) by a factor four. In this, a factor two stems from the cumulative charge $p = 2$ in the dipole phase compared with $p = 1$ in the sojourn phase, and another factor two comes from the different multiplicity of a collapsed dipole in the multi-well compared with a collapsed sojourn in the TSS. From these considerations we see that the two single dressed extended dipoles shown in (24.91) determine the mean particle current $\langle I \rangle = (\hbar\epsilon/q_0)\,\mu = q_0(\epsilon/\pi)\,\mu/\mu_0$, and hence also the mobility. The breathing mode integral yields

$$
\langle I \rangle \ = \ \lim_{t\to\infty}\langle q(t)\rangle_c/t \ = \ q_0\,\Delta^2\int_0^\infty d\tau\, e^{-Q'(\tau)-\tilde{\epsilon}_0\tau}\sin(\epsilon\tau) \, ,
\tag{24.94}
$$

which is a representation of the digamma function,

$$
\langle I \rangle \ = \ q_0\,\frac{\tilde{\epsilon}_0}{\pi}\,\text{Im}\,\psi\Big(\frac{1}{2} + \frac{\hbar\beta\tilde{\epsilon}_0}{2\pi} + i\,\frac{\hbar\beta\epsilon}{2\pi}\Big) \, .
\tag{24.95}
$$

This yields for the normalized nonlinear mobility the analytic expression

$$
\frac{\mu}{\mu_0} \ = \ \frac{\tilde{\epsilon}_0}{\epsilon}\,\text{Im}\,\psi\Big(\frac{1}{2} + \frac{\hbar\beta\tilde{\epsilon}_0}{2\pi} + i\,\frac{\hbar\beta\epsilon}{2\pi}\Big) \, .
\tag{24.96}
$$

At high temperatures, $\hbar\tilde{\epsilon}_0/2\pi k_{\mathrm B}T \ll 1$, we get in agreement with expression (24.74)

$$\mu/\mu_0 = \pi\left(\tilde{\epsilon}_0/2\epsilon\right)\tanh(\hbar\beta\epsilon/2) . \tag{24.97}$$

In the opposite limit $T \to 0$, we obtain the equivalent forms

$$\frac{\mu}{\mu_0} = \frac{\tilde{\epsilon}_0}{\epsilon}\arctan\left(\frac{\epsilon}{\tilde{\epsilon}_0}\right) = \frac{\pi\tilde{\epsilon}_0}{2\epsilon} - \frac{\tilde{\epsilon}_0}{\epsilon}\arctan\left(\frac{\tilde{\epsilon}_0}{\epsilon}\right) . \tag{24.98}$$

Thus, we find in agreement with the previous argumentation $\mu_{\mathrm{lin}}(T = 0) = \mu_0$.

It has been realized by Guinea that the TB model (23.3) in the Ohmic scaling limit at $K = \frac{1}{2}$ can be mapped on free fermions [466] (see also Ref. [457], Sec. VIII). This correspondence is analogous to that of the spin-boson model, which maps at $K = \frac{1}{2}$ on the Toulouse point of the anisotropic Kondo model (cf. Subsection 19.4.1) and on the resonance level model with vanishing repulsive contact interaction (cf. Subsection 19.4.2). The equivalence of the bosonic Coulomb gas representation (presented in Subsection 24.3.1) at $K = \frac{1}{2}$ with the respective fermionic representation directly follows from an integral representation of the charge interaction factor,

$$e^{-Q'(\tau,K=\frac{1}{2})} = \frac{i}{2\omega_{\mathrm c}}\int_{-\infty}^{\infty} d\omega\,\tanh(\hbar\beta\omega/2)\,e^{-i\omega\tau} . \tag{24.99}$$

With use of this representation the integral expression (24.94) for the current can be transformed into the frequency integral

$$\langle I\rangle = 2q_0\int_0^{\infty}\frac{d\omega}{2\pi}\,\mathcal{T}(\omega)[\mathcal{N}_+(\omega,\epsilon) - \mathcal{N}_-(\omega,\epsilon)] . \tag{24.100}$$

The interpretation of this expression is straightforward. The quantity

$$\mathcal{N}_\pm(\omega,\epsilon) = f(\omega \mp \epsilon)[1 - f(\omega \pm \epsilon)] , \tag{24.101}$$

where $f(\omega) = 1/(e^{\hbar\beta\omega} + 1)$, represents the spectral weight for right/left-moving fermions for an occupied state on the left/right side and an empty state on the right/left side of the barrier. It is convenient to introduce the linear combinations

$$\begin{aligned}
F_1(\omega,\epsilon) &\equiv \mathcal{N}_+(\omega,\epsilon) - \mathcal{N}_-(\omega,\epsilon) = f(\omega - \epsilon) - f(\omega + \epsilon) , \\
F_2(\omega,\epsilon) &\equiv \mathcal{N}_+(\omega,\epsilon) + \mathcal{N}_-(\omega,\epsilon) = \coth(\hbar\epsilon/k_{\mathrm B}T)\,F_1(\omega,\epsilon) .
\end{aligned} \tag{24.102}$$

In the second line, we have used the explicit form of the Fermi function $f(\omega)$.

The functions $\mathcal{T}(\omega)$ and $\mathcal{R}(\omega)$ represent the spectral transmission and reflection probability of the fermionic entity at frequency ω, $\mathcal{T}(\omega) + \mathcal{R}(\omega) = 1$,

$$\mathcal{T}(\omega) = \frac{\tilde{\epsilon}_0^2}{\tilde{\epsilon}_0^2 + \omega^2} , \qquad \mathcal{R}(\omega) = \frac{\omega^2}{\tilde{\epsilon}_0^2 + \omega^2} . \tag{24.103}$$

In terms of these functions, the first cumulant (24.100) may be written as

$$\langle I \rangle = 2q_0 \int_0^\infty \frac{d\omega}{2\pi} \, \mathcal{T}(\omega) F_1(\omega, \epsilon) = q_0 \frac{\epsilon}{\pi} - 2q_0 \int_0^\infty \frac{d\omega}{2\pi} \, \mathcal{R}(\omega) F_1(\omega, \epsilon) . \quad (24.104)$$

The first form expresses the current in terms of the transmission through the barrier. The first term in the second form is the maximum current, which appears in the absence of the barrier. The second term diminishes the maximum current because of the possibility that the particle is reflected at the barrier.

One remark about small deviations from $K = \frac{1}{2}$ is in order. For $K = \frac{1}{2} - \kappa$, $|\kappa| \ll 1$, a leading log summation of all diagrams contributing to the current can be performed [467]. The resulting expression for the mobility is again of the form (24.96), though with a bias- and temperature-dependent renormalization of the frequency scale,

$$\tilde{\epsilon}_0 \ \rightarrow \ \tilde{\epsilon}(\epsilon, T) = \bar{\epsilon} \left[(\epsilon/\bar{\epsilon})^2 + (2\pi k_{\mathrm{B}} T / \hbar \bar{\epsilon})^2 \right]^{-\kappa} \quad (24.105)$$

with $\bar{\epsilon} = \tilde{\epsilon}_0 (\omega_c / \tilde{\epsilon}_0)^{2\kappa}$. We remark that the expression (24.96) with (24.105) is consistent with the scaling forms at $K = \frac{1}{2} - \kappa$ of the weak- and strong-tunneling expansions of the mobility discussed below in Section 25.4.

24.6.2 Diffusion and skewness

The calculation of the series (24.46) for the second moment $\langle \hat{q}^2(\lambda) \rangle_{\mathrm{pis}}$ is more intricate because of the cosine factor in the phase term (24.43). This factor allows the particle to visit a diagonal state once at intermediate times. The charge sequences in the Coulomb gas representation which contribute to $\langle \hat{q}^2(\lambda) \rangle_{\mathrm{pis}}$ are those configurations in which all the cumulative charges p_j are confined to $|p_j| \leq 3$.

For zero bias, the resulting "bosonic" integral expression at long times reads [328]

$$\langle q^2(t) \rangle_c = t \, q_0^2 \frac{\tilde{\epsilon}_0}{2} \left\{ 1 - \frac{4\tilde{\epsilon}_0 \, \omega_c^2}{\pi^2} \int_0^\infty d\tau_1 \, d\tau_2 \, d\tau_3 \, e^{-\tilde{\epsilon}_0(\tau_1 + \tau_3)} \, e^{-Q'(\tau_1 + \tau_2) - Q'(\tau_2 + \tau_3)} \right\} .$$

Evaluation of the triple integral yields for the diffusion coefficient the analytic form

$$D(T) = (\hbar \tilde{\epsilon}_0 / 2\pi) \, \psi' \left(1/2 + \hbar \tilde{\epsilon}_0 / 2\pi k_{\mathrm{B}} T \right) . \quad (24.106)$$

It is evident that the linear mobility resulting from Eq. (24.96) and the diffusion coefficient (24.106) satisfy the Einstein relation (24.67), $D(T) = k_{\mathrm{B}} T \mu_{\mathrm{lin}}(T)$.

The diffusion coefficient for zero bias vanishes at $T = 0$, which indicates that the dispersion is subdiffusive. As we shall see in Subsection 24.7.2, for zero bias and zero temperature the second moment spreads out only logarithmically with at asymptotic times, $\langle q^2(t \rightarrow \infty) \rangle = (2/\pi^2) q_0^2 \ln(\tilde{\epsilon}_0 t)$ as $t \rightarrow \infty$. The logarithmic behaviour of the position dispersion at long times generally holds for all translational-invariant systems with Ohmic dissipation which have nonzero linear mobility at $T = 0$.

The discussion of the diffusion coefficient in the Coulomb gas representation for the biased case is reported in Ref. [468].

The bosonic forms of the cumulants, extracted from the formally exact series (24.39) [e.g. Eq. (24.41)], may be transformed with use of the integral representation (24.99) into fermionic representation. The resulting expressions for the second cumulant (diffusion) and third cumulant (skewness) in the biased case are

$$\langle q^2(t)\rangle_c = t\,(2q_0)^2 \int_0^\infty \frac{d\omega}{2\pi} \left[T(\omega)F_2(\omega,\epsilon) - T^2(\omega)F_1^2(\omega,\epsilon) \right], \qquad (24.107)$$

$$\langle q^3(t)\rangle_c = t\,(2q_0)^3 \int_0^\infty \frac{d\omega}{2\pi} T(\omega)F_1(\omega,\epsilon)\left[1 - 3T(\omega)F_2(\omega,\epsilon) + 2T^2(\omega)F_1^2(\omega,\epsilon) \right].$$

Upon employing the expressions (24.102) and (24.103), the cumulants (24.107) can be expressed in terms of cumulants of lower order. One finds

$$\langle q^2(t)\rangle_c = q_0 \left\{ \coth(\hbar\beta\epsilon)\,\tilde{\epsilon}_0\frac{d}{d\tilde{\epsilon}_0} + \left(2 - \tilde{\epsilon}_0\frac{d}{d\tilde{\epsilon}_0}\right)\frac{1}{\hbar\beta}\frac{d}{d\epsilon} \right\} \langle q(t)\rangle_c\,, \qquad (24.108)$$

$$\langle q^3(t)\rangle_c = q_0 \left\{ \coth(\hbar\beta\epsilon)\,\tilde{\epsilon}_0\frac{d}{d\tilde{\epsilon}_0} + \left(2 - \frac{1}{2}\tilde{\epsilon}_0\frac{d}{d\tilde{\epsilon}_0}\right)\frac{1}{\hbar\beta}\frac{d}{d\epsilon} \right\} \langle q^2(t)\rangle_c \qquad (24.109)$$

$$- q_0^2 \left\{ \coth(\hbar\beta\epsilon)\frac{1}{\hbar\beta}\frac{d}{d\epsilon} + \frac{1}{\sinh^2(\hbar\beta\epsilon)} \right\} \tilde{\epsilon}_0\frac{d}{d\tilde{\epsilon}_0} \langle q(t)\rangle_c\,.$$

With use of the analytic expression (24.95) for the current, it is now straightforward to derive analytic expressions for diffusion and skewness.

24.7 The effects of a thermal initial state

24.7.1 Mean position and variance

We have seen in Subsection 24.3.2 that the difference between $\langle q(t)\rangle_{\rm th}$ and $\langle q(t)\rangle_{\rm pis}$ is described by the function $R(t)$, which is given in Eq. (24.56). In the parameter regime, in which $\langle q(t\to\infty)\rangle_{\rm pis}$ grows linear with t at long times, Eq. (24.63), all the integrals over the blip and sojourn lengths in $\langle \hat{q}(\lambda)\rangle_{\rm pis}$, Eq. (24.45), and in $\hat{R}(\lambda)$, given below Eq. (24.61), are convergent in the limit $\lambda \to 0$. Hence the function $R(t)$ approaches a constant, $R(t \to \infty) = R_\infty$,

$$\lim_{t\to\infty}\left[\langle q(t)\rangle_{\rm th} - \langle q(t)\rangle_{\rm pis} \right] = R_\infty\,. \qquad (24.110)$$

As a result, the leading asymptotic behaviour of the mean values is exactly the same for both kinds of initial conditions, while subleading terms are different.

In the presence of a bias, the variances are defined by

$$\sigma_{\rm pis}^2(t) \equiv \langle q^2(t)\rangle_{\rm pis} - \langle q(t)\rangle_{\rm pis}^2\,, \quad \text{and} \quad \sigma_{\rm th}^2(t) \equiv C(t) - \langle q(t)\rangle_{\rm th}^2\,. \qquad (24.111)$$

Consider the term of order Δ^{2m} in Eq. (24.46) with (24.43). The leading contribution of $\langle \hat{q}^2(\lambda) \rangle_{\text{pis}}$ in the limit $\lambda \to 0$ comes from the terms in which those intervals τ_k which are decorated with a factor $\cos(\chi_{k,2m})$ [cf. Eq. (24.43)] are sojourns ($p_k = 0$). The τ_k-integral in question gives a factor $1/\lambda$. The two branches to the left and right of this interval are easily identified with terms of the series expansion of $\langle \hat{q}(\lambda) \rangle_{\text{pis}}$. In the end we find using Eq. (24.63)

$$\lim_{\lambda \to 0} \lambda^3 \langle \hat{q}^2(\lambda) \rangle_{\text{pis}} = 2(F\mu)^2 + 2\lambda F\mu \, q_\infty + 2\lambda D \,. \qquad (24.112)$$

With this form and with the definition of $\sigma_{\text{pis}}^2(t)$, we find in the diffusive regime

$$\lim_{t \to \infty} \sigma_{\text{pis}}^2(t) = 2Dt \,. \qquad (24.113)$$

The effects of the different initial conditions are in the function

$$\sigma_{\text{th}}^2(t) - \sigma_{\text{pis}}^2(t) = S(t) - R^2(t) - 2\langle q(t) \rangle_{\text{pis}} R(t) \,, \qquad (24.114)$$

as follows from Eqs. (24.111), (24.54), and (24.55). In the limit $\lambda \to 0$, the leading contribution in order Δ^{i+j} to $\hat{S}(\lambda)$ comes again from terms with the factor $\cos(\chi_{k,i+j})$ in which τ_k is a sojourn. The branch to the left of this interval can be identified with a contribution of $\hat{R}(\lambda)$, and the branch to the right is a term of the series expression for $\langle \hat{q}(\lambda) \rangle_{\text{pis}}$. Readily we find that the combined expression $S(t) - 2\langle q(t) \rangle_{\text{pis}} R(t)$ approaches asymptotically a constant,

$$\lim_{t \to \infty} \left[S(t) - 2 \langle q(t) \rangle_{\text{pis}} R(t) \right] = S_\infty \,. \qquad (24.115)$$

The leading behaviour at long times is the same for $\sigma_{\text{th}}^2(t)$ and $\sigma_{\text{pis}}^2(t)$, Eq. (24.113), while subleading terms are different,

$$\lim_{t \to \infty} \left[\sigma_{\text{th}}^2(t) - \sigma_{\text{pis}}^2(t) \right] = S_\infty - R_\infty^2 \,. \qquad (24.116)$$

Thus in the diffusive regime, the variances differ by a constant at asymptotic times.

24.7.2   Linear response

In this subsection, we focus the attention on the linear response to an external force $F = \hbar\epsilon/q_0$ switched on at time $t = 0$. One may think that the definition of the mobility within Kubo's linear-response theory is problematic [206] since the stationary state at asymptotic times is not normalizable. We now draw our attention to this question.

The Kubo formalism yields for the mean velocity at time t the expression

$$\bar{v}_{\text{lin}}(t) = -(2\epsilon/q_0) \, \Theta(t) \, A(t) = \frac{\hbar\epsilon}{q_0} \chi(t) \,. \qquad (24.117)$$

Here, $A(t)$ is the antisymmetric equilibrium correlation function (24.10) at zero bias. Alternatively, we may define the mean velocity $\bar{v}_{\mathrm{lin}}(t)$ as

$$\bar{v}_{\mathrm{lin}}(t) = \epsilon \lim_{\epsilon \to 0} \left[\langle \dot{q}(t) \rangle_{\mathrm{th}} / \epsilon \right] . \tag{24.118}$$

It is now interesting to see, using Eqs. (24.54) and (24.59), whether the two expressions coincide. Equating the expressions (24.117) and (24.118) and switching to the Laplace transforms, we obtain

$$\lim_{\epsilon \to 0} \left[\langle \hat{q}(\lambda) \rangle_{\mathrm{pis}} + \hat{R}(\lambda) \right] / \epsilon = -(2/q_0) \hat{A}(\lambda)/\lambda . \tag{24.119}$$

To prove the validity of this relation, we use for the time integrals occurring in the series expression for $\hat{R}(\lambda)$ and $\hat{A}(\lambda)$ the integral identities

$$\int_0^\infty dr_k \, ds_k \, e^{-\lambda s_k} f(\tau_k + s_k) = \int_0^\infty d\tau \, f(\tau) \frac{1}{\lambda} \left(1 - e^{-\lambda \tau} \right) ,$$

$$\int_0^\infty dr_k \, ds_k \, s_k \, e^{-\lambda s_k} f(\tau_k + s_k) = \int_0^\infty d\tau \, f(\tau) \left(\frac{1}{\lambda^2} \left(1 - e^{-\lambda \tau} \right) - \frac{\tau}{\lambda} e^{-\lambda \tau} \right) ,$$

where $\tau = r_k + s_k$, and where the factor $f(\tau)$ contains the interactions between the two sets of charges to the left and right of the interval τ. With these forms, it is straightforward to see that the relation (24.119) is indeed satisfied for all λ.

Using Eqs. (24.63) and (24.119), we find that the *linear* mobility is found from the absorptive part of the dynamical susceptibility as

$$\mu_{\mathrm{lin}} = \lim_{\omega \to 0} \omega \, \mathrm{Im} \, \tilde{\chi}(\omega) = - \lim_{\omega \to 0} \omega \frac{2}{\hbar} \mathrm{Im} \int_0^\infty dt \, A(t) \, e^{i\omega t} . \tag{24.120}$$

Consider next the mean square displacement. In the absence of the bias, $\epsilon = 0$, the variance $\sigma_{\mathrm{th}}^2(t)$ coincides with the equilibrium correlation function $C(t)$. Using the fluctuation-dissipation theorem (6.24), the Laplace transform can be written as

$$\hat{C}(\lambda) = \frac{\hbar}{\pi \lambda} \int_{-\infty}^{+\infty} d\omega \, \frac{\omega^2}{\lambda^2 + \omega^2} \coth \left(\frac{\hbar \beta \omega}{2} \right) \mathrm{Im} \, \tilde{\chi}(\omega) . \tag{24.121}$$

From this we find using Eq. (24.120)

$$\lim_{\lambda \to 0} \lambda^2 \, \hat{C}(\lambda) = 2 \, k_{\mathrm{B}} T \, \mu_{\mathrm{lin}} . \tag{24.122}$$

On the other hand, with the findings of the preceding subsection, we have in the diffusive regime

$$\lim_{\lambda \to 0} \lambda^2 \, \hat{C}(\lambda) = \lim_{\lambda \to 0} \lambda^2 \, \langle \hat{q}^2(\lambda) \rangle_{\mathrm{fc}} = 2D . \tag{24.123}$$

Upon equating the expression (24.122) with (24.123), we recover indeed the Einstein relation (24.67).

At zero temperature, we find from Eq. (24.121)

$$\lim_{\lambda \to 0} \hat{C}(\lambda) = -(2\hbar/\pi)\,\mu_{\text{lin},0}\,\ln(\lambda t_0)/\lambda\,, \qquad (24.124)$$

and hence

$$\lim_{t \to \infty} C(t) = (2\hbar/\pi)\mu_{\text{lin},0}\,\ln(t/t_0)\,, \qquad (24.125)$$

where $\mu_{\text{lin},0} = \mu_{\text{lin}}(T=0)$, and t_0 is a reference time. The derivation shows that the logarithmic dependence at zero temperature and long times holds for any system for which the linear mobility at $T=0$ is nonzero.

We can also study the behaviour at long times directly from the exact formal expression (24.46). The leading contribution in order Δ^{2m} for $T=0$ in the limit $\lambda \to 0$ comes again from the terms in which the cosine factor in the coefficient $c_{2m,0}$ falls on a sojourn ($p_k = 0$). This sojourn divides the charge sequence into two neutral clusters, and the average sojourn length $\bar{\tau}_k$ is very large compared to the lengths of the clusters. On this assumption, we have $\cos(\chi_{k,2m}) = 1$ and

$$c_{2m,0}G_m^{(0)} = \frac{q_0^2(-1)^{m-1}}{2}\prod_{l=1,l\neq k}^{2m-1}\sin\left(\chi_{\ell,2m}\right)G_{m-k}^{(0)}G_k^{(0)}\left(1 - \ddot{Q}'(\tau_k)\sum_{j=1}^{2k-1}p_j\tau_j\sum_{\ell=2k+1}^{2m-1}p_\ell\tau_\ell\right).$$

The terms $G_k^{(0)}$ and $G_{m-k}^{(0)}$ are the interaction factors (24.34) at $T=0$ for the two clusters without their mutual interaction across the interval τ_k. The second term in the square bracket represents the dipole-dipole interaction between the two neutral clusters. Higher-order multipole interactions are negligible for a long sojourn.

Substituing the first term into the expression (24.46), we obtain the first term in Eq. (24.112). This term vanishes for zero bias. The additional contribution to $\langle \hat{q}^2(\lambda)\rangle_{\text{fc}}$ stemming from the second (dipole-dipole interaction) term is found as well in product form. Performing the summation over all possible charge sequences on both sides of the interval τ_k, we find that the two distinguished branches for $\epsilon = 0$ can be identified with the series (24.64) for the zero temperature linear mobility $\mu_{\text{lin},0}$. Thus we find

$$\lim_{\lambda \to 0}\lambda^2\langle \hat{q}^2(\lambda)\rangle_{\text{fc}} = -2\left(\frac{\hbar}{q_0}\mu_{\text{lin},0}\right)^2\int_{\tau_{\text{min}}}^{\infty}d\tau\,\ddot{Q}'(\tau)\,e^{-\lambda\tau} = -\frac{2\hbar}{\pi}\frac{\mu_{\text{lin},0}^2}{\mu_0}\lambda\ln(\lambda t_0')\,.$$

Here we have put $\epsilon = 0$ and used $\ddot{Q}'(\tau) = -2K/\tau^2$, and we have extracted the λ-dependence of the integral from the region $\tau \gg \tau_{\text{min}}$. Hence we have at long times

$$\lim_{t \to \infty}\langle q^2(t)\rangle_{\text{fc}} = (2\hbar/\pi)\,(\mu_{\text{lin},0}^2/\mu_0)\,\ln(t/t_0')\,. \qquad (24.126)$$

For an ergodic system, as e.g. the Ohmic one, the prefactor of the logarithmic law in Eqs. (24.125) must coincide with that in Eq. (24.126). As a surprising consequence of this requirement, we find that the linear mobility at $T=0$ must be given by

$$\mu_{\text{lin},0} = \mu_0\,. \qquad (24.127)$$

This result is in correspondence with the duality relation (25.25) discussed below.

24.7.3   The exactly solvable case $K = \frac{1}{2}$

In the Ohmic scaling limit at the Toulouse point $K = \frac{1}{2}$, the function $\langle q(t) \rangle_{\mathrm{th}}$ and the correlation functions $C(t)$ and $A(t)$ can be calculated in analytic form [468] using again the concept of collapsed blips explained in Subsection 21.6.1. The Laplace transform of $A(t)$ is obtained in the form

$$\hat{A}(\lambda) = (q_0^2 \tilde{\epsilon}_0 / 4\pi) \hat{\Psi}(\lambda) , \tag{24.128}$$

$$\hat{\Psi}(\lambda) \equiv -\frac{2\omega_c}{\lambda} \int_0^\infty d\tau \, ds \, e^{-\lambda s} \, e^{-\tilde{\epsilon}_0(\tau+s)} \, e^{-Q'(\tau+s)} = \frac{2}{\lambda^2} \left[\psi(x) - \psi\left(x + \frac{\hbar\beta\lambda}{2\pi} \right) \right] ,$$

where $x = \frac{1}{2} + \hbar\beta\tilde{\epsilon}_0/2\pi$, $\tilde{\epsilon}_0 = \pi\Delta^2/\omega_c$, and $\psi(z)$ is the digamma function. We see from the integral form of $\hat{\Psi}(\lambda)$ that this function is represented by an extended dipole of length $\tau + s$ with τ being the length in the negative time branch, and the factors $e^{-\tilde{\epsilon}_0 \tau}$ and $e^{-\tilde{\epsilon}_0 s}$ describe grand-canonical sums of collapsed dipoles. The calculation of the function $\hat{C}(\lambda)$ is more difficult than of $\hat{A}(\lambda)$, since there are contributions from paths which cross the diagonal (cf. Ref. [468]). The final result for $\epsilon = 0$ is

$$\hat{C}(\lambda) = \frac{2q_0^2 \tilde{\epsilon}_0 \omega_c}{\pi^2 \lambda} \int_0^\infty d\omega \, \frac{\coth(\hbar\beta\omega/2)}{\omega^2 + \lambda^2} \int_0^\infty d\tau \, e^{-Q'(\tau) - \tilde{\epsilon}_0 \tau} \sin(\omega\tau) . \tag{24.129}$$

The dynamical susceptibility is found from Eq. (24.128) as

$$\chi(\omega) = -(q_0^2 \tilde{\epsilon}_0 / 2\pi\hbar) \, \hat{\Psi}(\lambda = -i\omega) . \tag{24.130}$$

Using Eqs. (24.128) and (24.130), we see that the τ-integral in Eq. (24.129) is just $(\hbar\omega^2/\Delta^2) \operatorname{Im} \chi(\omega)$. With this form, we find from Eq. (24.129) the relation

$$\operatorname{Re} \hat{C}(\lambda = -i\omega) = -\hbar \coth(\hbar\beta\omega/2) \operatorname{Im} \chi(\omega) , \tag{24.131}$$

which is the fluctuation-dissipation theorem (6.24) [note that $\widetilde{C}(\omega) = -\widetilde{S}(\omega)$]. Finally, it is recommended to compare these results with those of Subsection 21.6.4.

25. Duality symmetry

Duality is by now an old concept. In electromagnetism, the simplest form of duality is the invariance of the source free Maxwell equations under interchange of electric and magnetic field, $B \to E$, $E \to -B$. In the presence of sources, the product of electric and magnetic charges obeys the Dirac quantization condition. In general, duality maps a theory with strong coupling to one with weak coupling. Thus, if a duality symmetry exists, we can study the strong-coupling regime from the perturbative analysis of the weak-coupling regime. A theory is self-dual when there is an exact map from the strong-coupling sector to the weak-coupling sector of the same theory.

25.1 Duality for general spectral density

The model of the Brownian particle in a cosine potential exhibits a duality symmetry. Schmid [136] demonstrated that the equilibrium density matrix in the Ohmic case at $T = 0$ is self-dual in the scaling limit [cf. Eq. (18.54)]. The weak-binding representation can be mapped exactly on the tight-binding representation, and vice versa. The duality between a Brownian particle in a periodic potential and a discrete TB model was shown in an imaginary-time path integral approach by comparing the perturbation expansion in the corrugation strength V_0 with the multi-kink expansion. The analysis resulted in a transformation for the dc mobility in which diffusive and localized behaviour are interchanged. The duality was generalized in Ref. [455] to real time and finite temperature. It was shown that, when the scaling limit is dropped, there is still a duality symmetry, but the transformation involves a change of the spectral densities at high frequencies. Recently, it was shown that the Hamiltonians (23.1) and (23.3) are related by duality for general density $J(\omega)$ with power $s < 2$ at low frequencies [469]. In the duality, super-Ohmic and sub-Ohmic bath coupling are interchanged, while Ohmic low-frequency behaviour maps on Ohmic. The corresponding duality transformation is discussed in the following subsection. We shall expound in Section 25.3 that a corresponding duality symmetry holds between the charge and phase representation for a Josephson junction. An interesting correspondence between impurity scattering in a Tomonaga-Luttinger liquid and Brownian motion in a periodic potential is discussed in Chapter 26. In this case, there is an exact self-duality map between weak and strong backscattering.

25.1.1 The map between the TB and WB Hamiltonian

We begin with demonstrating that the Hamiltonians (23.1) and (23.3) can directly be mapped onto each other. For simplicity, we confine ourselves to the case $F = 0$.[1] In the first step, we perform the canonical transformation

$$
\begin{aligned}
p_\alpha &\rightarrow -m_\alpha \omega_\alpha x_\alpha\,, & x_\alpha &\rightarrow p_\alpha/m_\alpha\omega_\alpha + pc_\alpha/\kappa m_\alpha \omega_\alpha^2\,, \\
X &\rightarrow p/\kappa\,, & P &\rightarrow -\kappa q + \sum_\alpha c_\alpha x_\alpha/\omega_\alpha\,.
\end{aligned}
\tag{25.1}
$$

To map the cosine potential in Eq. (23.1) on the hopping term in Eq. (23.3), we choose

$$
V_0 = \hbar\Delta\,, \qquad\qquad \kappa = 2\pi\hbar/X_0 q_0\,.
\tag{25.2}
$$

---

[1]The case $F \neq 0$ is more complicated since the duality transformation changes a force coupled to a coordinate into a force coupled to a momentum. In the Ohmic scaling limit, the final result is as discussed in Subsection 25.1.3.

The transformed Hamiltonian reads

$$
\begin{aligned}
H &= -\hbar\Delta\cos\left(\frac{q_0 p}{\hbar}\right) + \frac{\kappa^2 q^2}{2M} - \frac{\kappa q}{M}\sum_\alpha \frac{c_\alpha x_\alpha}{\omega_\alpha} + H_{\mathrm R}\,, \\
H_{\mathrm R} &= \sum_\alpha\left(\frac{p_\alpha^2}{2m_\alpha} + \frac{m_\alpha\omega_\alpha^2 x_\alpha^2}{2}\right) + \frac{1}{2M}\left(\sum_\alpha \frac{c_\alpha x_\alpha}{\omega_\alpha}\right)^2 .
\end{aligned}
\tag{25.3}
$$

Now the modes of the environment are coupled with each other. Next, we diagonalize $H_{\mathrm R}$, thereby introducing new canonical variables π_α and u_α, and new parameters M_α and Ω_α. By the transformation, the interaction term $-(\kappa q/M)\sum_\alpha c_\alpha x_\alpha/\omega_\alpha$ is changed into $-q\sum_\alpha d_\alpha u_\alpha$. The mapping is completed by imposing spatial invariance of the system-bath coupling which requires that the q^2 term in Eq. (25.3) becomes the counter term of the model (23.3). This is achieved by imposing the condition

$$
\kappa^2/M = \sum_\alpha d_\alpha^2/M_\alpha\Omega_\alpha^2 .
\tag{25.4}
$$

A relation between $J_{\mathrm{WB}}(\omega)$ and $J_{\mathrm{TB}}(\omega)$ results from the dynamical matrix of $H_{\mathrm R}$,

$$
A_{\alpha\beta}(\omega) = B_{\alpha\beta}(\omega) + c_\alpha c_\beta/M\omega_\alpha\omega_\beta\,, \qquad B_{\alpha\beta}(\omega) = \delta_{\alpha\beta}m_\alpha(\omega_\alpha^2 - \omega^2)\,.
\tag{25.5}
$$

Since the coupling term of the bath modes is in the form of an exterior vector product, the ratio of the determinants of the matrices $\boldsymbol A(\omega)$ and $\boldsymbol B(\omega)$ gives

$$
\frac{\det \boldsymbol A(\omega)}{\det \boldsymbol B(\omega)} = 1 + \frac{1}{M}\sum_\alpha \frac{c_\alpha^2}{m_\alpha\omega_\alpha^2(\omega_\alpha^2 - \omega^2)} = 1 + i\,\frac{\gamma_{\mathrm{WB}}(\omega)}{\omega}\,.
\tag{25.6}
$$

In the last form, we have introduced the spectral damping function $\gamma(\omega)$ for the weak binding model. In the continuum limit, we have

$$
\gamma(\omega) = \lim_{\varepsilon\to 0}\frac{-i\omega}{M}\frac{2}{\pi}\int_0^\infty d\omega'\,\frac{J(\omega')}{\omega'(\omega'^2 - \omega^2 - i\varepsilon\,\mathrm{sgn}\,\omega)}\,.
\tag{25.7}
$$

On the other hand, we may resolve $A_{\alpha\beta}$ for $B_{\alpha\beta}$ and then switch to the unitarily equivalent form in which $\boldsymbol A$ is diagonal. We then get

$$
\tilde B_{\alpha\beta}(\omega) = \tilde A_{\alpha\beta}(\omega) - (M/\kappa^2)\,d_\alpha d_\beta\,, \qquad \tilde A_{\alpha\beta}(\omega) = \delta_{\alpha\beta}M_\alpha(\Omega_\alpha^2 - \omega^2)\,,
\tag{25.8}
$$

$$
\frac{\det \boldsymbol B(\omega)}{\det \boldsymbol A(\omega)} = 1 - \frac{M}{\kappa^2}\sum_\alpha \frac{d_\alpha^2}{M_\alpha(\Omega_\alpha^2 - \omega^2)} = -i\,\frac{M^2}{\kappa^2}\,\omega\gamma_{\mathrm{TB}}(\omega)\,.
\tag{25.9}
$$

To obtain the second form, we have used Eq. (25.4) and we have introduced the damping function $\gamma_{\mathrm{TB}}(\omega)$ of the TB model. Combining Eq. (25.6) with Eq. (25.9), we find an exact relation between the damping functions of the two models [469],

$$
\gamma_{\mathrm{TB}}(\omega)\,[\gamma_{\mathrm{WB}}(\omega) - i\omega] = \kappa^2/M^2 .
\tag{25.10}
$$

Using $J(\omega) = M\omega\,\mathrm{Re}\,\gamma(\omega)$, we then have

$$J_{\mathrm{WB}}(\omega) = (\kappa^2/M^2)J_{\mathrm{TB}}(\omega)/|\gamma_{\mathrm{TB}}(\omega)|^2 \,, \tag{25.11}$$

$$J_{\mathrm{TB}}(\omega) = (\kappa^2/M^2)J_{\mathrm{WB}}(\omega)/|\omega + i\gamma_{\mathrm{WB}}(\omega)|^2 \,. \tag{25.12}$$

Thus, the spectral density of the one model can be calculated for any form of the spectral density of the other model.

Consider now the WB model with the spectral power-law behaviour

$$J_{\mathrm{WB}}(\omega) = M\gamma_s\omega_{\mathrm{ph}}(\omega/\omega_{\mathrm{ph}})^s \,, \qquad 0 < s < 2 \,, \tag{25.13}$$

$$\gamma_{\mathrm{WB}}(\omega) = \lambda_s(-i\omega/\omega_{\mathrm{ph}})^{s-1} \,, \qquad \lambda_s \equiv \gamma_s/\sin(\pi s/2) \,. \tag{25.14}$$

Here we have introduced for $s \neq 1$ the usual phononic reference frequency ω_{ph}, as in Eq. (3.58). Let us now check the consistency of the transformation. We see from Eq. (25.9) that the constraint (25.4) is satisfied if $\omega\gamma_{\mathrm{TB}}(\omega)$ vanishes in the limit $\omega \to 0$. This in turn implies that $\gamma_{\mathrm{WB}}(\omega)/\omega$ must diverge in this limit which is the case in the parameter range $0 < s < 2$, as we see using Eq. (25.14). For $s \geq 2$, we have $\gamma_{\mathrm{WB}}(\omega \to 0) \propto \omega$ and hence the condition (25.4) cannot be fulfilled. For $s \geq 2$, damping becomes ineffective in the zero-frequency or long-time limit, and the only remaining effect is mass renormalization [86] (cf. Section 7.1).

Thus we have proven an exact duality between the weak corrugation model (23.1) and the tight-binding model (23.3) for the case $s < 2$. In the mapping, the continuous coordinate X in the WB model is identified, up to a scale factor $1/\kappa$, with the quasimomentum p in the dual TB model. Hence the duality is a sort of a Fourier transformation between real and momentum space [456]. A non-zero system-bath coupling is essential since otherwise the scale factor $1/\kappa$ is infinity. Strictly speaking, the mapping holds under the condition $\lim_{\omega\to 0} \omega^2/J(\omega) \to 0$. There are no restrictions on the form of $J(\omega)$ at finite frequencies except those enforced on physical grounds. There follows from Eq. (25.11) or Eq. (25.12) that the spectral density $J_{\mathrm{WB}}(\omega) \propto \omega^s$ of the weak corrugation model maps on the spectral density $J_{\mathrm{TB}}(\omega \to 0) \propto \omega^{2-s}$ of the dual TB model. Thus the power s in the spectral density is mapped on the power $2 - s$. This means that sub-Ohmic and super-Ohmic friction are interchanged in the transformation, while Ohmic friction is mapped on Ohmic friction.

Substituting Eq. (25.13) into Eq. (25.7) and using Eq. (25.12), the spectral density of the dual TB model takes the form with a soft cutoff,

$$J_{\mathrm{TB}}(\omega) = \frac{\kappa^2}{\eta_s} \frac{\omega_{\mathrm{ph}}(\omega/\omega_{\mathrm{ph}})^{2-s}}{1 + \left[\cot(\pi s/2) - (M\omega_{\mathrm{ph}}/\eta_s)(\omega/\omega_{\mathrm{ph}})^{2-s}\right]^2} \,. \tag{25.15}$$

Physically, the inertia of the particle provides a natural cutoff for the bath oscillators. In the Ohmic case, $s = 1$, the density $J_{\mathrm{TB}}(\omega)$ reduces to the Drude form [455]

$$J_{\mathrm{TB}}(\omega) = M\gamma\omega/[1 + (\omega/\omega_c)^2] \qquad \text{with} \qquad \omega_c = \gamma \,. \tag{25.16}$$

In the scaling limit (18.54), the particular form of the cutoff is disregarded. Let us finally briefly compare the path integral formulations of these models. In the WB model, the influence function (5.29) is calculated for a steplike path in which the steps are rounded because of the inertia term [455]. In the Ohmic case, the smearing scale is $1/\gamma$. In the dual TB representation, the paths are sharp tight-binding trajectories as in Eq. (24.32). Formally, the two influence functions are reconciled by the change of the spectral densities in the kernel $L(t)$, as given by the relations (25.11) and (25.12).

25.1.2 Frequency-dependent linear mobility

We now turn to a study of the respective linear ac mobilities. First, we observe that by the above unitary transformations a coordinate autocorrelation function of the WB model is transformed into a momentum autocorrelation function of the associated TB model. From this we find for the linear mobility of the WB model

$$\mu^{(WB)}(\omega) = -i\omega X_{\mathrm{TB}}(\omega)/\kappa^2 \,, \tag{25.17}$$

where $X_{\mathrm{TB}}(\omega)$ is the Fourier transform of the retarded momentum response function of the TB model, $X_{\mathrm{TB}}(t) = (i/\hbar)\,\Theta(t)\langle p(t)p(0) - p(0)p(t)\rangle_\beta$. On the other hand, the ac mobility of the TB model is related to the Fourier transform $Y_{\mathrm{TB}}(\omega)$ of the respective retarded coordinate response function by

$$\mu^{(TB)}(\omega) = -i\omega Y_{\mathrm{TB}}(\omega) \,. \tag{25.18}$$

To relate $X_{\mathrm{TB}}(\omega)$ to $Y_{\mathrm{TB}}(\omega)$, we use the equations of motion resulting from the Hamiltonian (23.3). We then obtain

$$\omega^2 X_{\mathrm{TB}}(\omega) = iM\omega\gamma_{\mathrm{TB}}(\omega) - M^2\omega^2\gamma_{\mathrm{TB}}^2(\omega)Y_{\mathrm{TB}}(\omega) \,. \tag{25.19}$$

Using this, and Eqs. (25.17), (25.18) and (25.10), we find the exact relations

$$\mu^{(WB)}(\omega, V_0/\hbar) = \frac{M}{\kappa^2}\,\gamma_{\mathrm{TB}}(\omega) - \frac{M^2}{\kappa^2}\,\gamma_{\mathrm{TB}}^2(\omega)\mu^{(TB)}(\omega, \Delta) \,, \tag{25.20}$$

$$\mu^{(TB)}(\omega, \Delta) = \frac{M}{\kappa^2}\,[\,\gamma_{\mathrm{WB}}(\omega) - i\omega\,] - \frac{M^2}{\kappa^2}\,[\,\gamma_{\mathrm{WB}}(\omega) - i\omega\,]^2\mu^{(WB)}(\omega, V_0/\hbar) \,. \tag{25.21}$$

The expressions (25.10) – (25.12),(25.20) and (25.21) are the central duality relations. Most remarkably, they hold for any spectral form of the damping, with the only restriction that damping remains effective at zero frequency, and no restrictions were placed on T. Here, the correspondence is shown for the frequency-dependent linear mobility. Clearly, the mapping can be extended to different kinds of dynamical variables. So far, we have considered the scales X_0 and q_0 as free parameters. In order to simplify the relation between the ac mobilities, we now make the specific choice

$$M\lambda_s X_0 q_0/2\pi\hbar \equiv M\lambda_s/\kappa = 1 \,. \tag{25.22}$$

Then the standard dimensionless coupling parameters $K_{\mathrm{WB}} \equiv M\lambda_s X_0^2/2\pi\hbar$ and $K_{\mathrm{TB}} \equiv M\lambda_s q_0^2/2\pi\hbar$ are related by

$$X_0/q_0 = K_{\mathrm{WB}} = 1/K_{\mathrm{TB}} \,. \tag{25.23}$$

In the low-frequency regime $\omega/\gamma_{\rm WB}(\omega) \ll 1$, the relations (25.20) and (25.21) become symmetric. Using Eqs. (25.10), (25.14) and (25.22), we then obtain the relation

$$\mu^{(\rm WB)}(\omega, V_0/\hbar; s) = 1/M\gamma_{\rm WB}(\omega) - (i\omega_{\rm ph}/\omega)^{2s-2} \mu^{(\rm TB)}(\omega, \Delta; 2-s) . \qquad (25.24)$$

For the Ohmic case, $s = 1$, the mobility of the free Brownian particle is $\mu_0 = 1/M\gamma$. Then the duality relation reads

$$\mu^{(\rm WB)}(\omega, V_0/\hbar; K) = \mu_0 - \mu^{(\rm TB)}(\omega, \Delta; 1/K) . \qquad (25.25)$$

In this form, the mobility still carries the label (WB) or (TB) because of the dependence on $V_0/\hbar$ or Δ. However, there is a universal frequency scale with which the mobility loses differentiation between WB- and TB-model. This frequency scale $\tilde{\epsilon}_0$ is introduced below in Subsection 25.5.1, and the relation with the TB-frequency Δ and WB-frequency $V_0/\hbar$ is given in Eq. (25.113). With the frequency scale $\tilde{\epsilon}_0$, the self-duality relation takes the concise form

$$\mu(\omega, \tilde{\epsilon}_0, K) = \mu_0 - \mu(\omega, \tilde{\epsilon}_0, 1/K) . \qquad (25.26)$$

This is equally a relation within the TB- or WB- model or between these models.

25.1.3   Nonlinear static mobility

In the presence of an external constant force F, the potential drop per lattice period X_0 in the continuous WB potential (23.2) is $\hbar\epsilon_{\rm WB} = FX_0$, whereas the potential drop in the dual TB model per length q_0 is $\hbar\epsilon_{\rm TB} = Fq_0$. Using the relation (25.23), the energy drop in the two models is related by $\epsilon_{\rm WB} = \epsilon_{\rm TB}/K_{\rm TB}$. Thus we find in generalization of Eq. (25.25) that the nonlinear static mobilities in the Ohmic case are related by

$$\mu^{(\rm WB)}(V_0/\hbar, T, \epsilon, K) = \mu_0 - \mu^{(\rm TB)}(\Delta, T, \epsilon/K, 1/K) , \qquad (25.27)$$

which holds for all T and ϵ. The mapping described by the self-duality relation (25.27) is pictorially sketched in Fig. 25.1 (left).

We may get again rid of the labels (TB) and (WB) upon introducing the universal frequency scale $\tilde{\epsilon}_0$ given below in Eq. (25.113),

$$\mu(\tilde{\epsilon}_0, T, \epsilon, K) = \mu_0 - \mu(\tilde{\epsilon}_0, T, \epsilon/K, 1/K) . \qquad (25.28)$$

At this point, a remark is expedient. The self-duality relation (25.28) is free of explicit dependence on the particular model. In the mapping, both the Ohmic viscosity η and the biasing force stay constant. Because the lattice period transforms as $X_0 \leftrightarrow q_0/K$, where $K = \eta q_0^2/(2\pi\hbar)$, we have in the duality $K \leftrightarrow 1/K$ and $\epsilon \leftrightarrow \epsilon/K$.

At $T = 0$, we may eliminate the change of the bias $\epsilon \leftrightarrow \epsilon/K$ by employing the modified frequency scale ϵ_0 given in Eqs. (25.111) and (25.112), as explained in Subsection 25.5.1. With the scaled bias $v = \epsilon/\epsilon_0$, we then have

$$\mu(v, K) = \mu_0 - \mu(v, 1/K) . \qquad (25.29)$$

As an example, we now translate the TB expression (24.85), which holds for $K \ll 1$, into a solution of the WB model. The duality relation $\omega_c = \gamma$ maps the scaling limit of the TB model on the high-friction limit of the WB model, $\hbar\gamma \gg V_0$, $\hbar\epsilon_{\mathrm{WB}}$. We see from the definition $K_{\mathrm{WB}} = M\gamma X_0^2/2\pi\hbar$ and the duality $K_{\mathrm{WB}} = 1/K_{\mathrm{TB}}$ that the condition $K_{\mathrm{TB}} \to 0$ corresponds to the classical limit in the dual WB model.

Using the self-duality relation (25.27), the mobility of the WB model in the classical limit $K_{\mathrm{WB}} \to \infty$ is found from the expression (24.85) to read

$$\frac{\mu^{(\mathrm{WB})}(T,\epsilon)}{\mu_0} = \frac{\sinh(\hbar\beta\epsilon/2)}{\hbar\beta\epsilon/2} \frac{1}{|I_{i\hbar\beta\epsilon/2\pi}(\beta V_0)|^2} . \tag{25.30}$$

This expression reduces in the special cases $\epsilon = 0$ and $T = 0$, respectively, to

$$\frac{\mu^{(\mathrm{WB})}(T.0)}{\mu_0} = \frac{1}{I_0^2(\beta V_0)} , \tag{25.31}$$

$$\frac{\mu^{(\mathrm{WB})}(0,\epsilon)}{\mu_0} = \Theta(\epsilon - \epsilon_0) \sqrt{1 - (\epsilon_0/\epsilon)^2} . \tag{25.32}$$

In the classical limit, the Kondo frequency scale ϵ_0 is related to the corrugation strength V_0 of the weak-binding model by

$$\epsilon_0 = 2\pi V_0/\hbar . \tag{25.33}$$

To our satisfaction, the results (25.30) - (25.32) coincide with solutions of the Smoluchowski diffusion equation (11.19) in corresponding limits (see Ref. [270]). For $\epsilon \le \epsilon_0$ or equivalently $V_{\mathrm{tilt}} \equiv FX_0/2\pi \le V_0$, the mobility is zero in the classical limit because the sliding motion of the overdamped particle comes to a stop at a point with zero slope and cannot move anymore since there are neither quantal nor thermal fluctuations. For $V_{\mathrm{tilt}} > V_0$, the potential (23.2) is sloping downwards everywhere, so that the particle is not trapped. This results in a nonzero mobility at zero temperature in the classical limit for $V_{\mathrm{tilt}} > V_0$.

As already emphasized, the mobility is most easily calculated in the TB representation. A major advantage of the duality now is that one can shift the numerical problem of calculating or simulating the dynamics of the WB model to the discrete TB model. For instance in quantum Monte Carlo simulations of the dynamics, the discrete variables of the latter model significantly reduce the relevant configuration space subject to Monte Carlo sampling compared to the continuous model, and meaningful real-time simulations in the interesting nonperturbative low-temperature regime are possible [471]. In many interesting cases, e. g. charge transfer in chemical reactions, the density $J_{\mathrm{WB}}(\omega)$ is not in the simple power-law form (25.13), and also the band width ω_c may be of the same order as the other frequencies. Nevertheless, the associated density $J_{\mathrm{TB}}(\omega)$ is given by (25.12), and one may now take advantage of performing numerical simulations of the equivalent TB model. Since the mapping is exact, the entire regime from quantum tunneling to thermal hopping across the barrier can be studied in the equivalent TB model.

In conclusion, we have discussed the exact duality symmetry between quantum Brownian motion in a continuous cosine potential and in a discrete tight-binding lattice. Because the tight-binding approximation ignores excited states at each lattice site, one might think that a general mapping between the two models is impossible. However, the TB model is fully sensitive to all aspects of both the quantum and the thermal hopping dynamics of the continuous model. In fact, the tight-binding model is reconciled with the continuous model by taking into account the change of the spectral density.

25.2 Self-duality in the exactly solvable cases $K = \frac{1}{2}$ and $K = 2$

In Section 24.6 we have discussed the three lowest cumulants of the tight-binding transport model (23.3) in the Ohmic scaling regime at the Toulouse point $K = \frac{1}{2}$. Now, we first extend the discussion to the full counting statistics. Employing self-duality, we then demonstrate that the resulting cumulant generating function can be transformed into the CGF of the TB model at $K = 2$. We show that the CGFs also describe the full-counting statistics of the WB model (23.1) in corresponding regimes.

25.2.1 Full counting statistics at $K = \frac{1}{2}$

In Subsection 24.6.1 we have seen that the current of the bosonic transport problem at $K = \frac{1}{2}$ can be mapped on one of free fermions with the transmission probability (24.103). Accordingly, it is quite natural to conjecture that the cumulant generating function is given by the Levitov-Lesovik formula [472]

$$
\begin{aligned}
\ln \mathcal{Z}(\rho, t) \;=\; t \int_0^\infty \frac{d\omega}{2\pi} \ln \Big\{ 1 + \mathcal{T}(\omega) \big[\mathcal{N}_+(\omega, \epsilon) \, (e^{i 2 q_0 \rho} - 1) \\
+ \; \mathcal{N}_-(\omega, \epsilon) \, (e^{-i 2 q_0 \rho} - 1) \big] \Big\}
\end{aligned}
\tag{25.34}
$$

with the spectral transmission probability at frequency ω [see Eq. (24.103)]

$$
\mathcal{T}(\omega) \;=\; \frac{\tilde{\epsilon}_0^2}{\tilde{\epsilon}_0^2 + \omega^2} \; .
\tag{25.35}
$$

Indeed, the CGF (25.34) reproduces the cumulants $\langle q(t) \rangle_c$, $\langle q^2(t) \rangle_c$ and $\langle q^3(t) \rangle_c$ as given in Eqs. (24.104) and (24.107).

An important limiting case is zero temperature at which up-hill particle flow is absent, $\mathcal{N}_+(\omega, \epsilon) = \Theta(\epsilon - |\omega|)$, and $\mathcal{N}_-(\omega, \epsilon) = 0$. In this limit, the CGF (25.34) yields for the cumulants the form

$$
\langle q^N(t) \rangle_c \;=\; t \, q_0^N \int_0^\epsilon \frac{d\omega}{2\pi} \, \mathcal{T}_N(\omega) \; .
\tag{25.36}
$$

Here

$$
\begin{aligned}
T_1(\omega) &= T(\omega)\,, \\
T_2(\omega) &= 2T(\omega)[1 - T(\omega)]\,, \\
T_3(\omega) &= 4T(\omega)[1 - T(\omega)][1 - 2T(\omega)]\,,
\end{aligned}
\tag{25.37}
$$

etc. The function $T_N(\omega)$ can be generated by differentiation,

$$
T_N(\omega) = [\tilde{\epsilon}_0(d/d\tilde{\epsilon}_0)]^{N-1} T(\omega)\,.
\tag{25.38}
$$

This helps us to express every single cumulant in terms of derivatives of the current,

$$
\langle q^N(t)\rangle_{\mathrm c} = t\left(q_0\,\tilde{\epsilon}_0\frac{d}{d\tilde{\epsilon}_0}\right)^{N-1}\langle I\rangle\,.
\tag{25.39}
$$

On the other hand, the CGF may be written in the tight-binding form (24.22), which at zero temperature is

$$
\ln \mathcal{Z}(\rho, t) = t\sum_{n=1}^{\infty}\left(e^{i\rho q_0 n} - 1\right)k_n^{+}\,.
\tag{25.40}
$$

This yields for the cumulants the form [cf. Eq. (24.23)]

$$
\langle q^N(t)\rangle_{\mathrm c} = t\,q_0^N\sum_{n=1}^{\infty}n^N k_n^{+}\,.
\tag{25.41}
$$

We now see upon comparing the form (25.39) with the expression (25.41) that $k_n^{+} \propto \tilde{\epsilon}_0^n$. This property allows us to extract all the transition weights k_n^{+} from the series expressions of one of the cumulants, e.g. from the series resulting from the second form of the mobility given in Eq. (24.98). This yields for the transition weight k_n^{+} at zero temperature the analytic expression

$$
k_n^{+} = \frac{\epsilon}{\pi}\frac{(-1)^{n-1}}{n}\frac{\sin[\frac{\pi}{2}(n-1)]}{n-1}\left(\frac{\tilde{\epsilon}_0}{\epsilon}\right)^n\,.
\tag{25.42}
$$

So we have $k_1^{+} = \tilde{\epsilon}_0/2$, while the transition weights for all n odd with $N > 1$ vanish. For n even, we have

$$
k_{2n}^{+} = \frac{\epsilon}{\pi}\frac{(-1)^n}{2n}\frac{1}{2n-1}\left(\frac{\tilde{\epsilon}_0}{\epsilon}\right)^{2n}\,.
\tag{25.43}
$$

Hence the sign of the transition weights turns out to be not strictly positive. This is a phenomenon of quantum interference (see also Section 25.6).

The expressions given here have lost explicit dependence on the particular representation. In application to the TB representation (23.3), the length q_0 is the lattice constant, ϵ is the bias frequency and the renormalized tunneling amplitude is [cf. Eq. (24.92)]

$$
\tilde{\epsilon}_0 = \pi\Delta^2/\omega_{\mathrm c}\,.
\tag{25.44}
$$

Passage to the WB representation (23.1) with lattice period X_0 and bias frequency ϵ_{WB} implies the substitutions $q_0 \to X_0/2$ and $\epsilon \to \epsilon_{\mathrm{WB}}/2$, as follows from the relation

(25.23). The dressed tunneling amplitude is given in terms of the WB parameters by

$$\tilde{\epsilon}_0 = \frac{1}{2\pi} \frac{\omega_c^2}{V_0/\hbar} , \tag{25.45}$$

as follows from the relation (25.113) at $K = \frac{1}{2}$.

With these assignments, the above expressions for the CGF and the cumulants are equally valid both for the WB and the TB representation at $K = \frac{1}{2}$. The former is appropriate in the weak-backscattering or strong-tunneling regime, and the latter is expedient at strong backscattering or weak tunneling.

25.2.2 Full counting statistics at $K = 2$

In the preceding section, we have seen that the full counting statistics at the Toulouse point $K = \frac{1}{2}$ is available in analytic form. Equipped with this solution, the case $K = 2$ can now be tackled without splendid calculation upon employing self-duality of the model under consideration. The dual mapping $K = \frac{1}{2} \leftrightarrow K = 2$ in the TB representation implies

$$\tilde{\epsilon}_0 \leftrightarrow \tilde{\epsilon}_0 , \qquad \epsilon \leftrightarrow \epsilon/2 , \qquad q_0 \leftrightarrow q_0/2 , \tag{25.46}$$

as follows from the relation (25.28). Here, the correspondence $q_0 \leftrightarrow q_0/2$ is because the force $F = \hbar\epsilon/q_0$ is constant in the mapping.

Current and mobility:

Employing the mapping (25.28) for $K = \frac{1}{2}$, we find from the expression (24.95) that the normalized mobility at $K = 2$ takes the form

$$\frac{\mu}{\mu_0} = 1 - \frac{2\tilde{\epsilon}_0}{\epsilon} \operatorname{Im} \psi\left(\frac{1}{2} + \frac{\hbar\beta\tilde{\epsilon}_0}{2\pi} + i\frac{\hbar\beta\epsilon}{4\pi}\right) , \tag{25.47}$$

and the mean current is

$$\langle I \rangle = q_0 \frac{\epsilon}{4\pi} - q_0 \frac{\tilde{\epsilon}_0}{2\pi} \operatorname{Im} \psi\left(\frac{1}{2} + \frac{\hbar\beta\tilde{\epsilon}_0}{2\pi} + i\frac{\hbar\beta\epsilon}{4\pi}\right) . \tag{25.48}$$

Here, q_0 and $\epsilon = Fq_0/\hbar$ are the lattice constant and bias frequency in the TB representation (23.3). The renormalized scattering amplitude $\tilde{\epsilon}_0$ is connected with the TB parameters Δ and ω_c by the relation

$$\tilde{\epsilon}_0 = \frac{1}{2\pi} \frac{\omega_c^2}{\Delta} , \tag{25.49}$$

as follows from Eq. (25.113) at $K = 2$. Because $\tilde{\epsilon}_0 \propto 1/\Delta$, the weak-tunneling series of current and mobility emerge from the asymptotic expansion of the digamma function in the expressions (25.48) and (25.47). The leading term of the weak tunneling series resulting from the expression (25.47) coincides indeed with the weak-tunneling expression (24.74) at $K = 2$,

$$\frac{\mu}{\mu_0} = \frac{\pi^2}{3} \frac{\Delta^2}{\omega_c^4} \left(\epsilon^2 + \left(\frac{2\pi}{\hbar\beta} \right)^2 \right) . \tag{25.50}$$

Passage to the WB representation implies the substitutions $q_0 \to 2X_0$ and $\epsilon \to 2\epsilon_{\text{WB}}$, and the relation of the dressed amplitude with the WB parameters is

$$\tilde{\epsilon}_0 = \pi (V_0/\hbar)^2 / \omega_c . \tag{25.51}$$

With these assignments, we may switch easily between the dual representations.

In the limit $T \to 0$, the expression (25.48) reduces to the equivalent forms

$$\langle I \rangle = q_0 \frac{\epsilon}{4\pi} - q_0 \frac{\tilde{\epsilon}_0}{2\pi} \arctan \left(\frac{\epsilon}{2\tilde{\epsilon}_0} \right) = q_0 \frac{\epsilon}{4\pi} - q_0 \frac{\tilde{\epsilon}_0}{4} + q_0 \frac{\tilde{\epsilon}_0}{2\pi} \arctan \left(\frac{2\tilde{\epsilon}_0}{\epsilon} \right) . \tag{25.52}$$

The Taylor expansion of the first form yields the weak-tunneling series of the current, and the Taylor expansion of the second form gives the strong-tunneling series.

By taking the line pursued in Subsection 24.6.1, we may transform the expression (25.48) into fermionic representation. The resulting expression is

$$\langle I \rangle = q_0 \frac{\epsilon}{4\pi} - q_0 \int_0^\infty \frac{d\omega}{2\pi} \mathcal{R}(\omega) F_1(\omega, \epsilon/2) = q_0 \int_0^\infty \frac{d\omega}{2\pi} \mathcal{T}(\omega) F_1(\omega, \epsilon/2) , \tag{25.53}$$

where $\mathcal{T}(\omega)$ and $\mathcal{R}$ are the spectral transition and reflexion probabilities at $K = 2$,

$$\mathcal{T}(\omega) = \frac{\omega^2}{\omega^2 + \tilde{\epsilon}_0^2} , \qquad \mathcal{R}(\omega) = \frac{\tilde{\epsilon}_0^2}{\omega^2 + \tilde{\epsilon}_0^2} . \tag{25.54}$$

Comparison of the expressions (25.53) with the forms (24.104) shows that in the mapping $K = \frac{1}{2} \leftrightarrow K = 2$ transmission and reflexion perform role reversal

$$\mathcal{T}(\omega) = \overline{\mathcal{R}}(\omega) , \qquad \text{and} \qquad \mathcal{R}(\omega) = \overline{\mathcal{T}}(\omega) . \tag{25.55}$$

The meaning of the expression (25.53) is as in the case $K = \frac{1}{2}$. The first term in the first form is the maximum current in the absence of the barrier. The second term diminishes the current because of the possibility that the particle is reflected at the barrier. The second form expresses the current in terms of the transmission through the barrier.

Full counting statistics:

With the relations (25.46) and the substitution $\mathcal{T}(\omega) \to \overline{\mathcal{T}}(\omega)$, we may also deduce the cumulant generating function at $K = 2$ from the CGF expression at $K = \frac{1}{2}$, Eq. (25.34) . The resulting expression is

$$\ln \mathcal{Z}(\rho, t) = t \int_0^\infty \frac{d\omega}{2\pi} \ln \Big\{ 1 + \overline{\mathcal{T}}(\omega) \big[\mathcal{N}_+(\omega, \epsilon/2) \, (e^{iq_0\rho} - 1)$$
$$+ \mathcal{N}_-(\omega, \epsilon/2) \, (e^{-iq_0\rho} - 1) \big] \Big\} . \tag{25.56}$$

The CGF yields for the second and third cumulant the fermionic frequency integrals analogous to the expressions (24.107),

$$\langle q^2(t)\rangle_c = t q_0^2 \int_0^\infty \frac{d\omega}{2\pi} \left[\overline{T}(\omega)F_2(\omega,\epsilon/2) - \overline{T}^2(\omega)F_1^2(\omega,\epsilon/2) \right], \qquad (25.57)$$

$$\langle q^3(t)\rangle_c = t q_0^3 \int_0^\infty \frac{d\omega}{2\pi} \overline{T}(\omega)F_1(\omega,\epsilon/2) \left[1 - 3\overline{T}(\omega)F_2(\omega,\epsilon/2) + 2\overline{T}^2(\omega)F_1^2(\omega,\epsilon/2) \right].$$

Again, these forms can be rewritten in terms of derivatives of lower-order cumulants,

$$\langle q^2(t)\rangle_c = \frac{q_0}{2} \left\{ -\coth\left(\tfrac{1}{2}\hbar\beta\epsilon\right) \tilde{\epsilon}_0 \frac{d}{d\tilde{\epsilon}_0} + \left(2 + \tilde{\epsilon}_0 \frac{d}{d\tilde{\epsilon}_0}\right) \frac{2}{\hbar\beta} \frac{d}{d\epsilon} \right\} \langle q(t)\rangle_c, \qquad (25.58)$$

$$\langle q^3(t)\rangle_c = \frac{q_0}{2} \left\{ -\coth\left(\tfrac{1}{2}\hbar\beta\epsilon\right) \tilde{\epsilon}_0 \frac{d}{d\tilde{\epsilon}_0} + \left(2 + \frac{\tilde{\epsilon}_0}{2} \frac{d}{d\tilde{\epsilon}_0}\right) \frac{2}{\hbar\beta} \frac{d}{d\epsilon} \right\} \langle q^2(t)\rangle_c \qquad (25.59)$$

$$+ \frac{q_0^2}{4} \left\{ \coth\left(\tfrac{1}{2}\hbar\beta\epsilon\right) \frac{2}{\hbar\beta} \frac{d}{d\epsilon} + \frac{1}{\sinh^2(\tfrac{1}{2}\hbar\beta\epsilon)} \right\} \tilde{\epsilon}_0 \frac{d}{d\tilde{\epsilon}_0} \langle q(t)\rangle_c.$$

At zero temperature, we have $\mathcal{N}_+(\omega) = \Theta(\epsilon - |\omega|)$ and $N_-(\omega) = 0$. We then obtain from Eq. (25.56) a relation which expresses cumulants of higher order again in terms of derivatives of the mean current,

$$\langle q^N(t)\rangle_c = t \left(-\frac{q_0}{2} \tilde{\epsilon}_0 \frac{d}{d\tilde{\epsilon}_0} \right)^{N-1} \langle I \rangle. \qquad (25.60)$$

Furthermore, the weak-tunneling representations of the CGF and the cumulants may be written again in the forms (25.40) and (25.41), respectively, but now the transition weight k_n^+ is

$$k_n^+ = \frac{\epsilon}{4\pi} \frac{(-1)^{n-1}}{n(2n+1)} \left(\frac{\epsilon}{2\tilde{\epsilon}_0} \right)^{2n}. \qquad (25.61)$$

Like in the case $K = \tfrac{1}{2}$, Eq. (25.42), the sign of the transition weight k_n^+ alternates.

We conclude the discussion of the special cases $K = \tfrac{1}{2}$ and $K = 2$ with a prospect. First we anticipate that the full counting statistics at $T = 0$ can be found in analytic form for general K (see Section 25.6). The functional relations (25.39) and (25.60) are consistent with the general functional relation (25.125). Similarly, the rate expressions (25.42) and (25.61) are special cases of the general rate expression (25.126). The indefinite sign of the transition weights is a signature of quantum interference.

25.3 Duality and supercurrent in Josephson junctions

25.3.1 Charge-phase duality

In Subsection 3.4.3 we have introduced the model for the Josephson junction in the charge representation which is appropriate for weak Josephson coupling, $E_J \ll E_c$.

In this limit, the charge on the junction is nearly sharp, and the phase fluctuations are described by the correlation function (3.179).

In the opposite limit of large Josephson coupling, $E_J \gg E_c$, the jump of the phase at the junction is nearly sharp. The relevant Hamiltonian is

$$H = -U_0 \cos\left(\frac{2\pi\bar{Q}}{2e}\right) + \sum_\alpha \left[\frac{\bar{q}_\alpha^2}{2\bar{C}_\alpha} + \left(\frac{\hbar}{2e}\right)^2 \frac{1}{2\bar{L}_\alpha}(\bar{\psi} - \bar{\psi}_\alpha)^2\right], \qquad (25.62)$$

$$\bar{\psi} = 2\pi \sum_n n a_n^\dagger a_n = 2\pi \sum_n n |n\rangle\langle n|, \qquad e^{i2\pi\bar{Q}/2e} = \sum_n a_n^\dagger a_{n+1} = |n\rangle\langle n+1|.$$

Here we have introduced a basis of discrete phase states. The Hamiltonians (3.178) and (25.62) correspond to the WB and TB Hamiltonians, Eq. (23.1) and Eq. (23.3), respectively. Following the discussion given in Subsection 25.1.1, we now show that the Hamiltonian (3.178) can be transformed into the form (25.62). First, we put $E_J \to U_0$, and we perform the unitary transformation (we put $\lambda_\alpha = (\hbar/2e)\sqrt{C_\alpha/L_\alpha}$),

$$q_\alpha \to -\lambda_\alpha \psi_\alpha, \qquad \psi_\alpha \to q_\alpha/\lambda_\alpha + (\pi/e)\bar{Q},$$

$$\psi \to \pi\bar{Q}/e, \qquad Q \to -(e/\pi)\bar{\psi} + \sum_\alpha \lambda_\alpha\psi_\alpha. \qquad (25.63)$$

We then obtain a representation in which the electromagnetic modes are coupled,

$$H = -U_0 \cos\left(\frac{2\pi\bar{Q}}{2e}\right) + \left(\frac{e}{\pi}\right)^2 \frac{\bar{\psi}^2}{2C} - \frac{e\bar{\psi}}{\pi C}\sum_\alpha \lambda_\alpha\psi_\alpha + H_R,$$

$$H_R = \sum_\alpha \left[\frac{q_\alpha^2}{2C_\alpha} + \left(\frac{\hbar}{2e}\right)^2 \frac{\psi_\alpha^2}{2L_\alpha}\right] + \frac{1}{2C}\left(\sum_\alpha \lambda_\alpha\psi_\alpha\right)^2. \qquad (25.64)$$

The mapping is completed by equating the $\bar{\psi}^2$ terms in Eqs. (25.64) and (25.62),

$$\left(\frac{e}{\pi}\right)^2 \frac{1}{C} = \left(\frac{\hbar}{2e}\right)^2 \sum_\alpha \frac{1}{\bar{L}_\alpha}, \qquad (25.65)$$

and by switching to the diagonal basis of the bath modes. The ratio of the determinants of the dynamical matrices $\mathbf{A}(\omega)$ and $\mathbf{B}(\omega)$,

$$B_{\alpha\beta}(\omega) = \delta_{\alpha\beta}C_\alpha (\hbar/2e)^2 (\omega_\alpha^2 - \omega^2), \qquad A_{\alpha\beta}(\omega) = B_{\alpha\beta}(\omega) + \lambda_\alpha\lambda_\beta/C, \qquad (25.66)$$

reads
$$\frac{\det \mathbf{A}(\omega)}{\det \mathbf{B}(\omega)} = 1 + \frac{1}{C}\sum_\alpha \frac{1}{L_\alpha}\frac{1}{\omega_\alpha^2 - \omega^2} = \frac{-i\omega C + Y^*(\omega)}{-i\omega C}. \qquad (25.67)$$

In the second form we have introduced the admittance $Y(\omega)$ of the weak-coupling model (3.156). On the other hand, in the eigen basis of the matrix $\mathbf{A}(\omega)$ we have

$$\bar{A}_{\alpha\beta} = \delta_{\alpha\beta}\bar{C}_\alpha \left(\frac{\hbar}{2e}\right)^2 (\bar{\omega}_\alpha^2 - \omega^2), \qquad \bar{B}_{\alpha\beta} = \bar{A}_{\alpha\beta} - \frac{\pi^2 C}{e^2}\left(\frac{\hbar}{2e}\right)^2 \lambda_\alpha\bar{\omega}_\alpha\lambda_\beta\bar{\omega}_\beta.$$

With this we find using Eq. (25.65)

$$\frac{\det \boldsymbol{B}(\omega)}{\det \boldsymbol{A}(\omega)} = -i\omega C \left(\frac{\pi\hbar}{2e^2}\right)^2 \sum_\alpha \frac{1}{\bar{L}_\alpha} \frac{-i\omega}{\bar{\omega}_\alpha^2 - \omega^2} = -i\omega C\, R_Q^2 \bar{Y}^*(\omega)\,, \qquad (25.68)$$

where $R_Q = h/4e^2$ is the resistance quantum for Cooper pairs. Solving Eq. (25.68) for $\det \boldsymbol{A}(\omega)/\det \boldsymbol{B}(\omega)$, and substituting this into Eq. (25.67), we find a simple relation between the admittance $Y(\omega)$ of the model for weak Josephson coupling and the admittance $\bar{Y}(\omega)$ of the strong-coupling model,

$$\bar{Y}(\omega)\,[\,i\omega C + Y(\omega)\,] = 1/R_Q^2\,. \qquad (25.69)$$

Using this correspondence, we can express quantitites of the one model in terms of those of the other. The phase-charge duality transformations are [143, 146]

$$\psi \;\longleftrightarrow\; \pi Q/e\,, \qquad \frac{1}{R_Q[\,i\omega C + Y(\omega)\,]} \;\longleftrightarrow\; R_Q Y(\omega)\,. \qquad (25.70)$$

The phase autocorrelation function $\langle[\,\psi(0) - \psi(t)\,]\psi(0)\rangle_\beta$ takes the form (3.179). The spectral density of the coupling reads

$$G_\psi(\omega) = 2\omega R_Q \,\mathrm{Re}\,\bar{Y}(\omega) = 2\omega\, \mathrm{Re}\frac{1}{R_Q[\,i\omega C + Y(\omega)\,]}\,. \qquad (25.71)$$

In the strong-coupling limit, the influence of the environment is described by the charge autocorrelation function $Q_Q(t) \equiv \langle[\,Q(0) - Q(t)\,]Q(0)\rangle_\beta$. Using the correspondence (25.70), we find the function $Q_Q(t)$ in the form (3.179) with the spectral density

$$G_Q(\omega) = (e/\pi)^2\, 2\omega R_Q \,\mathrm{Re}\, Y(\omega)\,. \qquad (25.72)$$

In the Ohmic scaling limit, we then obtain with

$$\rho = R/R_Q \qquad (25.73)$$

the forms
$$G_\psi(\omega) = 2\rho\omega\,, \qquad \text{and} \qquad G_Q(\omega) = (e/\pi)^2\,(2/\rho)\,\omega\,, \qquad (25.74)$$

which indicate exact self-duality between the charge and phase representation.

These results can directly be transferred to electron tunneling through a normal junction with the substitution $2e \to e$ and $R_Q \to R_K$. The respective spectral densities are in correspondence with the earlier results (3.157) and (3.160) with (3.161).

Nano-electronic devices based on low-capacitance Josephson junctions appear to be suitable for large-scale integration and physical realization of quantum bits. In the regime $E_c \gg E_J$, the Coulomb blockade effect allows the controll of individual charges [106, 473]. Single- and two-bit operations may be performed upon applying a sequences of gate voltages. When the gate voltage is chosen such that neighbouring charge states are degenerate, the system reduces to two states which are weakly coupled by E_J [cf. Subsection 3.2.1]. One may also think of quantum-logic Josephson elements with flux states instead of charge states. The effects of the electromagnetic environment are described by the phase correlation function (3.179) with (3.180).

In Subsection 20.3.4 we have studied the supercurrent through an ultrasmall Josephson junction with Josephson coupling energy E_J in the weak-tunneling limit, in which $I \propto E_J^2$. In the region $\rho < 1$, we found divergence of the expression (20.178) for $\rho < 1$ in the low-energy regime $k_B T$, $eV_x \ll E_J (E_J / E_c)^{\rho/(1-\rho)}$. This indicates that tunneling processes of higher order in E_J become relevant in this parameter regime. As we shall see, the higher-order terms smooth out the singularity in question.

Interestingly enough, in the particular cases $\rho \ll 1$, $\rho = \frac{1}{2}$ and $\rho = 2$, the tunneling dynamics of Cooper pairs can be solved in analytic form in all orders in E_J.

25.3.2 Supercurrent-voltage characteristics for $\rho \ll 1$

By virtue of the equivalence relations (20.173) between tunneling of bosons under Ohmic friction and tunneling of Cooper pairs in a resistive electromagnetic environment, we can translate the Ivanchenko-Zil'berman expression (24.84) into the supercurrent-voltage characteristics of a small capacitance Josephson junction [470],

$$I(V_x) = \frac{V_x}{R} \frac{\beta E_J^*}{\nu} \, \mathrm{Im} \left(\frac{I_{1-i\nu}(\beta E_J^*)}{I_{-i\nu}(\beta E_J^*)} \right) \tag{25.75}$$

where $\nu = \beta eV_x / \pi \rho$. Furthermore, E_J^* is the equivalent of $\hbar \Delta_T$ defined in Eq. (24.80),

$$E_J^* = E_J \, e^{-\rho \zeta} \left(\frac{2\pi^2 \rho}{\beta E_c} \right)^\rho , \tag{25.76}$$

where ζ is given in Eqs. (20.176) and (20.149). The formula (25.75) can be rewritten as [see Eqs. (24.84 and (24.85)]

$$I(V_x) = \frac{V_x}{R} \left(1 - \frac{\sinh(\pi\nu)}{\pi\nu} \frac{1}{|I_{i\nu}(\beta E_J^*)|^2} \right) . \tag{25.77}$$

The supercurrent-voltage characteristics (25.75) or (25.77) covers the entire regime ranging from weak to strong Cooper pair tunneling. It shows a peak at small voltage, as explained below Eq. (20.178).

Upon expanding the expression (25.75) to order E_J^2, we recover the weak-tunneling expression (20.181). Furthermore, we get from the expression (25.77) the linear superconductance at general T

$$G_{\mathrm{lin}}(T) = \frac{1}{R} \left[1 - \frac{1}{I_0^2(\beta E_J^*)} \right] , \tag{25.78}$$

and the supercurrent-voltage characteristics at zero temperature

$$I(V_x) = \Theta(V_x - RI_c) \frac{V_x}{R} \left[1 - \sqrt{1 - (RI_c/V_x)^2} \right] . \tag{25.79}$$

These results hold for small capacitance Josephson junctions in the domain $R \ll R_Q$.

Upon employing duality, it is straightforward to translate the expression (25.77) into the I-V characteristics of a large capacitance Josephson junction ($E_J \gg E_c$) coupled to a very strong resistive environment, $R/R_Q \gg 1$. The relevant Hamiltonian is given in Eq. (25.62), and the spectral density of the charge fluctuations is $G_Q(\omega) = (e/\pi)^2 2(R_Q/R)\,\omega$. The changeover to the latter case is analogous to the passage from the expression (24.85) to the form (25.30). In the classical limit $R/R_Q \to \infty$, the dc voltage drop V at the current bias junction with bias current $I_{\text{ext}} = I_b$ due to the backscattering current is (we put $\sigma = e\beta R I_b/\pi$) [474]

$$V(I_b) \;=\; R I_b \Big(1 \;-\; \frac{\sinh(\pi\sigma)}{\pi\sigma}\,\frac{1}{|I_{i\sigma}(\beta U_0)|^2}\Big)\,. \tag{25.80}$$

As $T \to 0$, this reduces to $V(I_b) = \Theta(I_b - \pi U_0/eR)\,R I_b[\,1 - \sqrt{1 - (\pi U_0/eRI_b)^2}\,]$.

25.3.3 Supercurrent-voltage characteristics at $\rho = \frac{1}{2}$

With the spadework set out in Section 24.6, it is only a small step to give the supercurrent-voltage characteristics for the case $\rho = \frac{1}{2}$, or equivalently $R = \pi\hbar/4e^2$. Employing the correspondence relations (20.173) - (20.175), the renormalized tunneling amplitude (25.44) is expressed in terms of the device parameters as

$$\tilde{\epsilon}_0 \;=\; \frac{\pi^2}{2\hbar}\frac{E_J^2\,e^{-\varsigma}}{E_c} \;=\; \frac{e\pi}{\hbar}\frac{E_J\,e^{-\varsigma}}{E_c}\,R I_c\,. \tag{25.81}$$

Putting $q_0 \,\hat{=}\, 2e$, the expression (24.95) transforms itself into the supercurrent

$$I(V_x) \;=\; I_c\,\frac{\hbar\tilde{\epsilon}_0}{\pi E_J}\,\operatorname{Im}\,\psi\Big(\frac{1}{2} + \frac{\hbar\beta\tilde{\epsilon}_0}{2\pi} + i\,\frac{\beta e V_x}{\pi}\Big)\,. \tag{25.82}$$

At $T = 0$, this expression reduces to [475]

$$I(V_x) \;=\; I_c\,\frac{\hbar\tilde{\epsilon}_0}{\pi E_J}\,\arctan\Big(\frac{2e V_x}{\hbar\tilde{\epsilon}_0}\Big) \;=\; I_c\,\frac{\hbar\tilde{\epsilon}_0}{2E_J} - I_c\,\frac{\hbar\tilde{\epsilon}_0}{\pi E_J}\,\arctan\Big(\frac{\hbar\tilde{\epsilon}_0}{2e V_x}\Big)\,. \tag{25.83}$$

In contrast to the regime $\rho < \frac{1}{2}$, this result does not show a peak structure anymore. The first term in the second form is the plateau value. It is equal to the Coulomb blockade expression (20.179) at $\rho = \frac{1}{2}$. This term is the leading one for weak Cooper pair tunneling, i.e., when the bias energy $2e V_x$ is large. The opposite regime of strong Cooper pair tunneling is captured, when the bias energy $2e V_x$ is small compared to $\hbar\tilde{\epsilon}_0$.

25.3.4 Supercurrent-voltage characteristics at $\rho = 2$

At $\rho = 2$, which corresponds to $R = \pi\hbar/e^2$, the renormalized frequency (25.49) with the correspondence relations (20.173) - (20.175) takes the form

$$\tilde{\epsilon}_0 \;=\; \frac{e}{\hbar}\Big(\frac{E_c\,e^\varsigma}{4\pi^2 E_J}\Big)^2 R I_c\,. \tag{25.84}$$

The supercurrent-voltage characteristics dual to the expression (25.82) is

$$I(V_x) = \frac{V_x}{R} - \left(\frac{E_c \, e^\varsigma}{4\pi^2 E_J}\right)^2 I_c \, \mathrm{Im}\,\psi\left(\frac{1}{2} + \frac{\hbar\beta\tilde\epsilon_0}{2\pi} + i\,\frac{\beta e V_x}{2\pi}\right). \qquad (25.85)$$

In the zero temperature limit, this expressions simplifies to [475]

$$I(V_x) = \frac{V_x}{R} - \left(\frac{E_c \, e^\varsigma}{4\pi^2 E_J}\right)^2 I_c \, \arctan\left[\left(\frac{4\pi^2 E_J}{E_c \, e^\varsigma}\right)^2 \frac{V_x}{RI_c}\right]. \qquad (25.86)$$

The perturbative expansion in V_x yields the weak-tunneling series. The leading term $\propto V_x^3$ coincides with the Coulomb blockade expression (20.179) at $\rho = 2$.

In the opposite limit of very large V_x, we have strong Cooper pair tunneling. The respective weak-backscattering series may be deduced from the complementary supercurrent-voltage relation

$$I(V_x) = \frac{V_x}{R} - \frac{\pi}{2}\left(\frac{E_c \, e^\varsigma}{4\pi^2 E_J}\right)^2 I_c + \left(\frac{E_c \, e^\varsigma}{4\pi^2 E_J}\right)^2 I_c \, \arctan\left(\left(\frac{E_c \, e^\varsigma}{4\pi^2 E_J}\right)^2 \frac{RI_c}{V_x}\right). \qquad (25.87)$$

The leading backscattering term turns out to be independent of the applied voltage.

With the correspondence relations (20.173) - (20.175), it is easy to translate the expressions for higher order cumulants and the CGF, which for $K = \frac{1}{2}$ are given in Eqs. (24.108), (24.109), and (25.34), and for $K = 2$ in Eqs. (25.58), (25.59) and (25.56), into the corresponding expressions for Cooper pair tunneling at $\rho = \frac{1}{2}$ and at $\rho = 2$.

25.4 Self-duality in the Ohmic scaling limit

In the WB model (23.1), the scaling limit (18.54) corresponds to the high-friction (Smoluchowski) limit $\gamma \gg V_0/\hbar$, ϵ_{WB}. Upon expanding the WB mobility in a power series in the corrugation strength V_0, one obtains the series [136]

$$\frac{\mu^{(\mathrm{WB})}(T,\epsilon)}{\mu_0} = 1 - \frac{\pi}{\epsilon}\,\mathrm{Im}\sum_{m=1}^{\infty}(-1)^{m-1}\left(\frac{V_0}{\hbar}\right)^{2m}\int_0^\infty d\tau_1\,d\tau_2\cdots d\tau_{2m-1} \qquad (25.88)$$

$$\times \sum_{\{\xi_j\}'} G_m\left(\{\tau_j\}; 1/K, \{\xi_j\}\right)\sin\left(\sum_{i=1}^{2m-1}\epsilon\,p_i\tau_i/K\right)\prod_{k=1}^{2m-1}\sin(\pi p_k/K).$$

The interaction factor G_m is given in Eq. (24.73) with K replaced by $1/K$. On the other hand, the TB series for the mobility in the Ohmic scaling limit is given by the series (24.72). With the expansions (24.72) and (25.88), the duality relation (25.27) is easily verified.

25.4.1   Linear mobility at finite T

We see from Eq. (24.72) by simply counting dimensions that, for zero bias, T is combined with the frequencies Δ and ω_c in the form $\Delta^2 T^{2K-2} \omega_c^{-2K}$. On the other hand, we find from the series expression (25.88) that, again for $\epsilon = 0$, V_0 is combined with T and ω_c as $V_0^2 T^{2/K-2} \omega_c^{-2/K}$. Fom this we see that there is a Kondo-type temperature which is related to the parameters of the TB and WB model by

$$k_{\mathrm{B}}T_0 = a_{\mathrm{TB}}(K)\,(\Delta/\omega_c)^{K/(1-K)}\,\hbar\Delta = a_{\mathrm{WB}}(K)(V_0/\hbar\omega_c)^{1/(K-1)}V_0 \,. \tag{25.89}$$

Upon introducing the Kondo-scaled temperature

$$\vartheta = T/T_0 \equiv k_{\mathrm{B}}T/\hbar\tilde{\epsilon}_0 \,, \tag{25.90}$$

the TB series (24.72) and the WB series (25.88) for the dimensionless linear mobility $\mathcal{M}_{\mathrm{lin}}(\vartheta, K) \equiv \mu_{\mathrm{lin}}/\mu_0$ can be rewritten in the self-dual scaling forms

$$\mathcal{M}_{\mathrm{lin}}(\vartheta, K) = \sum_{m=1}^{\infty} d_m(K)\,\vartheta^{2(K-1)m} \,, \tag{25.91}$$

$$\mathcal{M}_{\mathrm{lin}}(\vartheta, K) = 1 - \sum_{m=1}^{\infty} d_m(1/K)\,\vartheta^{2(1/K-1)m} \,. \tag{25.92}$$

In the Coulomb gas representation (24.72), the dimensionless coefficient $d_m(K)$ is given in terms of a $(2m - 1)$-fold integral.

Self-duality means that the expressions (25.91) and (25.92) are the Taylor and the asymptotic expansion of the same mobility function. By the self-duality transformation $K \to 1/K$, the two expansions are interchanged. Equating Eq. (25.91) with Eq. (25.92), we obtain the nonperturbative self-duality

$$\mathcal{M}_{\mathrm{lin}}(\vartheta, K) = 1 - \mathcal{M}_{\mathrm{lin}}(\vartheta, 1/K) \,, \tag{25.93}$$

which is valid for all values of K. By this relation, the Taylor expansion is mapped on the asymptotic expansion and vice versa. With the series expressions (25.91) and (25.92), the mapping holds term by term. Observe that the forms (25.92) and (25.93) have lost explicit dependence on the TB and on the WB model. Therefore, the two series (25.91) and (25.92) describe the same physical system in complementary regions of the parameter space.

Since T is invariant in the self-duality transformation, the coefficients in the definition (25.90) of the Kondo temperature scale are related by

$$a_{\mathrm{WB}}(K) = a_{\mathrm{TB}}(1/K) \,. \tag{25.94}$$

The radius of convergence of the power series (25.91) and (25.92) is conveniently expressed in terms of the temperature scale

$$T_{\mathrm{cr}}(K) \equiv \lim_{m\to\infty} |d_m(K)|^{1/[2(1-K)m]} T_0 = T_{\mathrm{cr}}(1/K) \,. \tag{25.95}$$

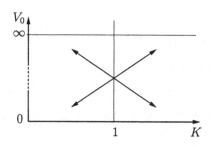
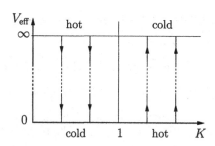

Figure 25.1: Regarding the TB lattice as the large barrier limit of the WB model, both models can be represented in a single diagram of V_0 versus K. The duality maps a WB model with small V_0 and Kondo parameter $K \lesssim 1$ to a TB model with small Δ (large V_0) and $K \gtrsim 1$ (left diagram). In the right diagram, the flow of the effective barrier height with decreasing temperature is sketched for the linear mobility. For $K < 1$ the system flows towards a vanishing barrier whereas it becomes localized for $K > 1$ at zero temperature. For the nonlinear mobility at $T = 0$, the flows as a function of the bias are similar.

The equality is a consequence of self-duality. The scale T_{cr} is a crossover temperature analogous to the Kondo temperature in Kondo models. For $K < 1$, the series (25.91) absolutely converges for $T > T_{cr}$, and the series (25.92) for $T < T_{cr}$. For $K > 1$, the regions of convergence of these two series are interchanged.

For $K < 1$, we have the following picture holding both for the TB and WB model. At $T \gg T_{cr}$, we are in the perturbative regime of the series (25.91), the $m = 1$ term being the leading one. As T is lowered, the barrier gets effectively weaker so that higher-order terms of the series become increasingly important. Physically, this means that coherent tunneling through two, three, $\cdots$, many barriers give relevant contributions to the mobility. At $T = T_{cr}$, the region of convergence of the series (25.91) is left. For $T < T_{cr}$, the asymptotic series (25.92) applies. As temperature is lowered further, convergence of the series is improved. At $T = 0$, all terms $m \geq 1$ have dropped to zero, and we are left with $\mu_{lin} = \mu_0$. Thus by cooling down the system from high to zero temperature, the originally strong barrier gradually fades away.

In the characteristic behaviour for $K > 1$, high and low temperatures are interchanged. Thus, as T approaches zero, the barrier becomes infinitely high and the particle is localized. The pertinent flows of the effective barrier height are depicted in Fig. 25.1 (right).

25.4.2 Nonlinear mobility at $T = 0$

Similar scaling behaviour is found for the nonlinear mobility at $T = 0$. In this case, the bias takes the role of temperature. By counting dimensions in the series expressions (24.72) and (25.88), we see that the Kondo energy scale is given in terms of the parameters of the TB and WB model as

$$\epsilon_0 \;=\; b_{\rm TB}(K)\, \Delta^{1/(1-K)} \omega_c^{-K/(1-K)} \;=\; b_{\rm WB}(K)(V_0/\hbar)^{K/(K-1)} \omega_c^{-1/(K-1)} \,. \qquad (25.96)$$

It is now convenient to introduce the Kondo-scaled dimensionless bias frequency

$$v \;=\; \epsilon/\epsilon_0 \,. \qquad (25.97)$$

Since v is chosen invariant, while ϵ maps on ϵ/K in the duality transfromation (25.27), we must have in difference to the relation (25.94)

$$b_{\rm WB}(K) \;=\; K\, b_{\rm TB}(1/K) \,. \qquad (25.98)$$

Explicit expressions for the prefactors $b_{\rm TB}(K)$ and $b_{\rm WB}(K)$, which satisfy the relation (25.98), can be extracted from the relations given below in Eqs. (25.111) and (25.112).

Then the TB series (24.72) for the dimensionless nonlinear mobility at zero temperature, $\mathcal{M}(v,K) \equiv \mu(T=0,\epsilon,K)/\mu_0$, takes the scaling form

$$\mathcal{M}(v,K) \;=\; \sum_{m=1}^{\infty} c_m(K)\, v^{2(K-1)m} \,. \qquad (25.99)$$

Likewise, the WB series (25.92) at $T=0$ can be written as

$$\mathcal{M}(v,K) \;=\; 1 - \sum_{m=1}^{\infty} c_m(1/K)\, v^{2(1/K-1)m} \,. \qquad (25.100)$$

Thus we find again nonperturbative self-duality, but now of the form

$$\mathcal{M}(v,K) \;=\; 1 - \mathcal{M}(v,1/K) \,. \qquad (25.101)$$

In the Coulomb gas representation, the coefficient $c_m(K)$ is again given in terms of a $(2m-1)$-fold integral. The convergence radius of the expansions (25.99) and (25.100) is determined by the critical frequency

$$\epsilon_{\rm cr}(K) \;\equiv\; \lim_{m\to\infty} |c_m(K)|^{1/[\,2(1-K)m\,]} \epsilon_0 \;=\; \epsilon_{\rm cr}(1/K) \,. \qquad (25.102)$$

For $K < 1$, the series (25.99) converges absolutely for $\epsilon > \epsilon_{\rm cr}$, wheras the series (25.100) converges for $\epsilon < \epsilon_{\rm cr}$. Again for $K > 1$, the regions of convergence are interchanged.

The dual series expressions (25.99) and (25.100) are again the Taylor and asymptotic expansion of the same physical quantity, and vice versa, depending on whether K is smaller or larger than one and whether the bias ϵ is larger or smaller than $\epsilon_{\rm cr}$. They hold both for the TB and WB model. Upon utilizing the self-duality property of the model, we determine in Subsection 25.5.1 the coefficients $c_m(K)$ for general K. There, we also relate the dressed frequencies ϵ_0 and $\epsilon_{\rm cr}$ to the parameters of the original WB and TB model. Finally, the relation (25.96) will make possible to express the parameters of the one model in terms of those of the other [see Eq. (25.114)].

25.5   Exact scaling function at $T = 0$ for arbitrary K

25.5.1   Construction of the self-dual scaling solution

The dual expansions (25.99) and (25.100) can be derived from the contour integral representation [476]

$$\mathcal{M}(v, K) \;=\; \frac{1}{2\pi i} \int_C dz \, P(z) F(z) \, v^{2z} \,. \tag{25.103}$$

Here we conceive the function $P(z)$ such that it provides the analytic structure of the power series (25.99) and (25.100),

$$P(z) \;=\; \frac{1}{z} \Gamma\!\left(1 + \frac{zK}{K-1}\right) \Gamma\!\left(1 + \frac{z}{1-K}\right). \tag{25.104}$$

The function $P(z)$ is analytic over the entire complex plane save for the points

$$
\begin{aligned}
z &= 0 \,, \\
z &= z_n^{(1)} \equiv (1/K - 1)n \,, & n &= 1, 2, \ldots \,, \\
z &= z_m^{(2)} \equiv (K - 1)m \,, & m &= 1, 2, \ldots \,,
\end{aligned}
\tag{25.105}
$$

where it possesses simple poles. The function $F(z)$ serves as an adaptive function. We require that $F(z)$ is an analytic function on the entire complex z-plane. The contour C starts at infinity, then circles the origin such that the pole at $z = 0$ and the set of poles $\{z_n^{(1)}\}$ lie to the left, and the set of poles $\{z_m^{(2)}\}$ lie to the right of the integration path, and finally returns to infinity (cf. Fig. 25.2). We close the contour, in which we require that the integrand in Eq. (25.103) tends to zero faster than $1/|z|$ on the chosen semicircle. In the regime $(\epsilon_{cr}/\epsilon)^{2(1-K)} < 1$, the contour is closed such that the set of poles $\{z_m^{(2)}\}$ is circled. The resulting series expression is in the form (25.99), and the coefficients are given by

$$c_m(K) \;=\; (-1)^{m-1} \Gamma(Km + 1) \, F[(K-1)m]/\Gamma(m+1) \,. \tag{25.106}$$

For $(\epsilon_{cr}/\epsilon)^{2(1-K)} > 1$, the contour is closed in the opposite direction of rotation, so that the pole at $z = 0$ and the set of poles $\{z_n^{(1)}\}$ are circled. We find upon comparing the corresponding residua with the terms of the series (25.100) that $F(z)$ is independent of K at the origin, $F(0) = 1$, and the coefficients with $m \geq 1$ are given by Eq. (25.106) with $1/K$ substituted for K.

    The entire analytic structure of the integral representation (25.103) required by the self-duality property is carried by the function $P(z)$. The only additional requirement imposed by self-duality is that the yet unknown function $F(z)$ is invariant under the substitution $K \to 1/K$. It is tempting to assume that self-dual models are "minimal theories" in the sense that the whole dependence on the coupling constant K is determined only by the analytic properties of the function $P(z)$. Thus, in a minimal theory, the function $F(z)$ does not depend on K at all. Under this condition it is

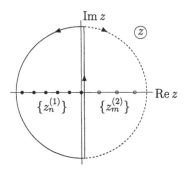

Figure 25.2: Sketch of the integration path $\mathcal{C}$ in the contour integral (25.103) for $K > 1$. Closing the contour in the counter-clockwise and clockwise sense gives the series (25.100) and (25.99), respectively. For $K < 1$, the contour and the poles are reflected at the origin.

possible to determine $F(z)$ provided that the series (25.99) or (25.100) is known in analytic form for a particular value of K different from unity. Fortunately, the TB mobility is known in analytic form for the case $K = \frac{1}{2}$ and in the limit $K \to 0$ as discussed in Subsections 24.5.2 and 24.6.1. The latter case is equivalent to the classical limit $K \to \infty$ in the WB model. Indeed, anyone of these special cases may be employed in order to determine the function $F(z)$.

To obtain $F(z)$, we equate the terms of the series (24.88) with those of the series (25.99) for $K = 0$ and use Eq. (25.106). Alternatively, we may match the power series of (25.32) with the series (25.100) for $K \to \infty$. In both ways, we unambiguously get

$$F(z) = \frac{\Gamma(3/2)}{\Gamma(z + 3/2)} \,. \tag{25.107}$$

The matching function (25.107) is analytic for all z, as postulated. Thus, we have established indeed an exact contour integral representation of the scaling function for general K at zero temperature [476],

$$\mathcal{M}(v, K) \;=\; \frac{\Gamma(3/2)}{2\pi i} \int_{\mathcal{C}} dz \, \frac{\Gamma[\,1 + zK/(K - 1)\,]\,\Gamma[\,1 + z/(1 - K)\,]\, v^{2z}}{z\,\Gamma(3/2 + z)} \,. \tag{25.108}$$

With the form (25.107) for $F(z)$, the series coefficient (25.106) is found to read

$$c_m(K) \;=\; \frac{(-1)^{m-1}}{m!} \, \frac{\sqrt{\pi}}{2} \, \frac{\Gamma(Km + 1)}{\Gamma[\,(K - 1)m + 3/2\,]} \,. \tag{25.109}$$

The series expressions (25.99) and (25.100) with (25.109) constitute the exact solution for the mobility for general values of the parameters K and ϵ at zero temperature.

Readily the crossover scale (25.102) delimiting the regions of convergence of the expansions (25.99) and (25.100) is found to be given by

$$\epsilon_{\mathrm{cr}}(K) \;=\; \epsilon_{\mathrm{cr}}(1/K) \;=\; \sqrt{|1 - K|}\, K^{K/[2(1-K)]}\, \epsilon_0 \,. \tag{25.110}$$

For instance, we find in the limit $K \rightarrow \infty$ the radius $\epsilon_{\mathrm{cr}} = 2\pi V_0/\hbar$. This is in agreement with the convergence radius of the power series of the expression (25.32).

To relate the frequency scale ϵ_0 of the scaling solution to the parameters of the WB model, we equate the $m = 1$ term of the series (25.100) with the $m = 1$ term of the series (25.88) at zero temperature. Thus we find

$$\epsilon_0^{2-2/K} = K^{2-2/K} 2^{2-2/K} [\pi/\Gamma(1/K)]^2 (V_0/\hbar)^2 \omega_c^{-2/K} . \tag{25.111}$$

Likewise, we can express ϵ_0 in terms of the parameters of the TB model. Equating the $m = 1$ term of the series (25.99) with the expression (24.75), we find

$$\epsilon_0^{2-2K} = 2^{2-2K} [\pi/\Gamma(K)]^2 \Delta^2 \omega_c^{-2K} . \tag{25.112}$$

The contour integral (25.108) with (25.111) and with (25.112) is an exact representation of the nonlinear dc-mobility of the WB-model and TB-model, respectively, holding at $T = 0$ in the scaling limit for general bias ϵ and general friction K.

The expressions (25.111) and (25.112) are symmetric under substitution $K \rightarrow 1/K$ except for the factor $K^{2-2/K}$ in (25.111). This factor appears because of the correspondence $\epsilon \leftrightarrow K\epsilon$ in the duality relation (25.27) [see relation (25.98)].

A convenient renormalized frequency $\tilde{\epsilon}_0$, which is fully symmetric with respect to the TB and WB model, is $\tilde{\epsilon}_0 = 2^{-3} K^{1/(K-1)} \epsilon_0$. We have

$$\tilde{\epsilon}_0^{2-2K} = 2^{4K-4} \frac{\pi^2}{\Gamma^2(1+K)} \frac{\Delta^2}{\omega_c^{2K}} , \qquad \tilde{\epsilon}_0^{2-\frac{2}{K}} = 2^{\frac{4}{K}-4} \frac{\pi^2}{\Gamma^2(1+\frac{1}{K})} \frac{(V_0/\hbar)^2}{\omega_c^{2/K}} . \tag{25.113}$$

At this point we wish to emphasize the following conclusions [476]:

(1) Using the relations (25.111) and (25.112) [or equivalently the relations (25.113)], we can express Δ in terms of V_0,

$$\Delta^2 = [\Gamma(1+K)]^2 [\Gamma(1+1/K)]^{2K} (\omega_c/\pi)^{2+2K} (V_0/\hbar)^{-2K} . \tag{25.114}$$

Resolution of this relation for V_0^2 gives the same functional form except that $1/K$ is substituted for K. It is interesting to compare the expression (25.114) with the corresponding relation (17.74) obtained by use of the bounce method for large K. Employing Stirling's asymptotic formula for $\Gamma(K)$ and identifying the cutoff ω_c with the damping frequency γ, the expression (25.114) indeed coincides with the result (17.74) of the bounce method. However the form (25.114) is more general since it holds for all K. Interestingly enough, here we have found the single-bounce (and all multi-bounce contributions) to the TB mobility solely by use of duality and analytic properties, without calculating bounce actions and fiddly fluctuation determinants.

(2) Besides duality and analytic properties we have used in the construction of the scaling solution only knowledge of the solution in the classical limit. Thus, the entire quantum regime of this transport problem is fully determined by self-duality.

(3) One should expect that also the scaling function $\mathcal{M}_{\mathrm{lin}}(\vartheta, K)$ for the linear mobility can be found from the above "minimal" assumptions. However, a formidable

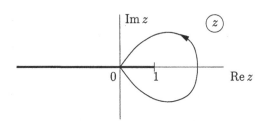

Figure 25.3: The path of integration $\mathcal{C}$ in Eq. (25.117) starts at the origin, circles around the branch point at $z = 1$ in the counter-clockwise sense, and returns to the origin.

practical complication arises: an ansatz equivalent to (25.103) does not work since the adaptive function $F(z)$ turns out to be nonanalytic in the complex z-plane.

(4) The existence of self-duality is a direct consequence of integrability in the equivalent boundary sine-Gordon model [477]. Integrability determines the low-energy properties of this model and shows itself in the above scaling behaviour.

25.5.2 Supercurrent-voltage characteristics at $T = 0$ for arbitrary ρ

We have seen in Section 25.3 that the charge-phase duality in the Josephson junction dynamics is an exact self-duality for strictly Ohmic damping. Thus, with the correspondence relations (20.173) - (20.175), the findings of the previous subsection can directly be applied to Cooper pair tunneling in a resistive environment. The supercurrent-voltage characteristics at $T = 0$ and arbitrary $\rho = R/R_{\rm Q}$ takes the scaling-invariant form [475]

$$I(V_{\rm x}) \;=\; \mathcal{M}(v, \rho)\, \frac{V_{\rm x}}{R}\,. \tag{25.115}$$

Here, $\mathcal{M}(v, \rho)$ is the scaling function just discussed, but now it represents the normalized conductance of the Josephson contact. The various energy scales of the system are aggregated in the scaling variable

$$v \;=\; \frac{eV_{\rm x}}{\pi E_{\rm J}}\left[\, \Gamma(\rho)\left(\frac{{\rm e}^{\varsigma}}{\pi^2\rho}\,\frac{E_{\rm c}}{E_{\rm J}}\right)^{\rho}\,\right]^{\frac{1}{1-\rho}}\,. \tag{25.116}$$

Taking the leading term of the weak Cooper-pair-tunneling series (25.99), we get the Coulomb blockade expression (20.179). On the other hand, the leading terms of the dual strong-tunneling series (25.100) for $\rho \ll 1$ agree with the expression (17.81).

25.5.3 Connection with Seiberg-Witten theory

It is possible to establish an interesting connection of the duality with supersymmetric Seiberg-Witten theory [478]. Consider the series (25.100) with (25.109) and substitute for the ratio of gamma functions in the coefficient $c_m(1/K)$ the contour integral representation [479] (cf. Fig. 25.3)

$$\frac{\Gamma(x)}{\Gamma(x+y)} = \Gamma(1-y)\frac{1}{2\pi i}\int_C dz \, z^{x-1}(z-1)^{y-1} . \tag{25.117}$$

Since the series is absolutely convergent within the circle of convergence, we can interchange the order of integration and summation. This yields

$$\mathcal{M}(v,K) = \frac{1}{4\sqrt{\pi}i}\int_C dz \frac{1}{\sqrt{z-1}}\Bigg\{\sum_{n=0}^{\infty}\frac{(-1)^n}{n!}\Gamma(n+\tfrac{1}{2})\left(\frac{z^{1/K}v^{2/K-2}}{z-1}\right)^n\Bigg\} . \tag{25.118}$$

The series in the curly bracket can be summed to a square root. With the substitution $z = t/v^2$, we finally obtain for the scaling function the integral form [477]

$$\mathcal{M}(v,K) = \frac{1}{4vi}\int_{C'} dt \frac{1}{\sqrt{t+t^{1/K}-v^2}} . \tag{25.119}$$

One can now prove the self-duality relation (25.101) without the series expansions since the integral form (25.119) is analytic in K and therefore holds for all K. With the substitution $t^K = x$ in Eq. (25.119), we obtain

$$M(v,1/K) = \frac{1}{4vi}\int_{C'} dx \frac{1}{\sqrt{x+x^{1/K}-v^2}}\left[\left(\frac{x^{1/K-1}}{K}+1\right)-1\right] , \tag{25.120}$$

where C' changes accordingly. The integrand with the round bracket term is a total differential yielding simply the term 1, whereas the residual integral is $-\mathcal{M}(v,K)$. Thus we have proven the self-duality (25.101) directly from the integral form (25.119). The representation (25.119) bears resemblance with representations for mass gaps in SU(2) supersymmetric Yang-Mills theory [478]. The parameter v corresponds to an order parameter related to the expectation value of the Higgs field. A geometrical picture in terms of tori for the form (25.119) is discussed in Ref. [480].

25.5.4 Special limits

At $K = \frac{1}{2}$, the renormalized frequency scale is $\epsilon_0 = 2\tilde{\epsilon}_0 = 2\pi\Delta^2/\omega_c$. We recover from the scaling function $\mathcal{M}(v, K = \frac{1}{2})$ previous results given in Subsection 24.6.1. Namely, upon putting $K = \frac{1}{2}$ in the series (25.100), it is summed to the first form given in Eq. (24.98), wheras the dual series (25.99) gives the second form in Eq. (24.98).

In the narrow regime $K = 1/2 - \kappa$, with $|\kappa| \ll 1$, the scaling function is consistent with the former result (24.105) at $T = 0$.

For $K = 1 + \kappa$ with $|\kappa| \ll 1$, the series (25.100) is summed to the expression

$$\mu_{\text{WB}}(\epsilon, K = 1+\kappa)/\mu_0 = 1/[1+(\pi V_0/\hbar\omega_c)^2(2\omega_c/\epsilon)^{2\kappa}] , \tag{25.121}$$

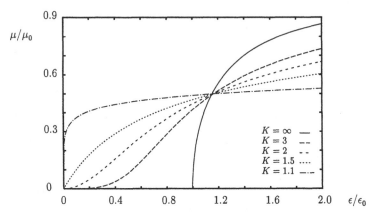

Figure 25.4: The normalized mobility μ/μ_0 at $T = 0$ is plotted as a function of ϵ/ϵ_0 for different K. In the regime $\epsilon < \epsilon_{cr}$ and $K > 1$, the weak-tunneling (TB) series (25.99) with coefficient (25.109) converges, whereas for $\epsilon > \epsilon_{cr}$ and $K > 1$ the strong-tunneling (WB) series (25.100) with coefficient (25.109) converges. The curve for $K = \infty$ shows the square root singularity of the classical case, Eq. (25.32). For finite K, the mobility is nonzero in the bias regime $\epsilon/\epsilon_0 < 1$ due to quantum mechanical tunneling.

where we have employed the relation (25.111). This form agrees with a result obtained by a leading-log summation in a related impurity scattering problem (discussed below in Section 26.1) with Luttinger parameter $g = 1 - \kappa$, where $g = 1/K$ [481].

The normalized mobility $\mu(\epsilon)/\mu_0$ is plotted in Fig. 25.4 as a function of ϵ/ϵ_0 for different values of the damping parameter K. The mobility shows a smooth transition from the square root singular behaviour (25.32) in the classical limit $K \to \infty$ via the kink-like shape $\mu/\mu_0 = 1 - (\epsilon_0/2\epsilon) \arctan(2\epsilon/\epsilon_0)$ at $K = 2$ to the constant behaviour (25.121) at $K = 1$. The curves for $K < 1$ are found from the self-duality relation (25.101). All curves cross the line $\mu/\mu_0 = 1/2$ in the interval $\epsilon/\epsilon_0 = 2/\sqrt{3} \pm 0.01$. As $K \to 1$, the frequency scale ϵ_0 drops to zero and μ/μ_0 becomes unity. Because of the scale employed in Fig. 25.4, this behaviour does not show up in the figure.

25.6 Full counting statistics at zero temperature

We have seen in Section 25.5 that the scaling function for the mobility at $T = 0$ can be given in analytic form for general damping strength K. Surprisingly, not only the mobility but also all statistical properties of this transport process at zero temperature can be found in analytic form for general K both in the weak- and in the strong-tunneling representation [482].

In the theory of full counting statistics (FCS), the key quantity of interest is the moment generating function $\mathcal{Z}(\rho, t)$ introduced in Eq. (24.3). For weak tunneling or strong backscattering, the TB model (23.3) applies. Observing that at $T = 0$ there are no transition from the energetically lower to the higher state, $k_n^- = 0$, the expression (24.22) reduces to

$$\ln \mathcal{Z}(\rho, t) \;=\; t \sum_{n=1}^{\infty} \left(e^{i\rho q_0 n} - 1 \right) k_n^+ \,. \tag{25.122}$$

This form can be discovered indeed from a systematic cluster decomposition of the series expression (24.39) in the limit of very large t. The clusters are defined in terms of the irreducible influence functions $\widetilde{F}_n^{\pm}$ discussed in Section 21.2.6. In Laplace representation, the clusters are the λ–independent (irreducible) path sections which start and end in diagonal states of the reduced density matrix. By definition, the clusters are noninteracting, and therefore each time interval which separates neighbouring clusters gives a factor $1/\lambda$. Path segments with intermediate visits of diagonal states are reducible, i.e., they factorize into clusters of lower order. After subtraction of the reducible components always an irreducible part is left. The transition rates $k_n^{\pm}$ can be identified as the sum of all clusters which interpolate between the (arbitrary) diagonal state m and the diagonal state $m \pm n$. The clusters are formally given in terms of a power series in the number of tunneling transitions. In the Ohmic scaling limit at $T = 0$, the tunneling matrix element Δ is united with the bias ϵ and the cut-off ω_c into a dimensionless parameter x. With the choice $x \propto \Delta$, the weak-tunneling expansion , which is a power series in x, takes the form

$$k_n^+ \;=\; \frac{\epsilon}{2\pi} \sum_{\ell=0}^{\infty} f_n^{(\ell)} \, x^{2n+2\ell} \,, \qquad x = \frac{\Delta}{\epsilon} \left(\frac{\epsilon}{\omega_c} \right)^K \propto v^{K-1} \,. \tag{25.123}$$

Here, v is determined by the relations (25.97) and (25.112). The coefficients $f_n^{(\ell)}$ are given in terms of a sum of $2(n + \ell) - 1$-fold integrals. Each of these integrals represents a particular sequence $2(n + \ell) - 1$ cumulative charges p_j. The cluster expressions at $T = 0$ have been analyzed in Ref. [433]. It was found that there are formidable cancellations between the various path contributions so that altogether only paths with the minimal number of transitions contribute. Consequently, the rate $k_n^{\pm}$ is exactly of order Δ^{2n}, and therefore in the series (25.123) only the $\ell = 0$ term remains,

$$k_n^+ \;=\; \frac{\epsilon}{2\pi} \, f_n^{(0)} \, x^{2n} \,. \tag{25.124}$$

Now, taking into consideration the definition (24.23) of the cumulants and the property $k_n^+ \propto x^{2n}$, Eq. (25.124), we immediately get the simple cumulant relation

$$\langle q^N(t) \rangle_c \;=\; \left(q_0 \, \frac{\Delta}{2} \, \frac{\partial}{\partial \Delta} \right)^{N-1} \langle q(t) \rangle \,, \tag{25.125}$$

which helps us to reduce every single cumulant to a calculation of derivatives of the current $\langle q(t) \rangle = t \langle I \rangle$ with respect to the tunneling coupling Δ. In addition, the central property (25.124) allows us to determine the transition rates k_n^+ unambiguously from one of the cumulants, e.g., from the current. Observing that the series for the current $\langle I \rangle = \langle q(t) \rangle / t = q_0 \sum_n n k_n^+$ corresponds to the backscattering series $\langle q(t) \rangle = t \, (\epsilon q_0 / 2\pi K) \mathcal{M}(v, K)$, where $\mathcal{M}(v, K)$ is given by the series (25.99), we get

$$k_n^+ = \frac{(-1)^{n-1}}{n!} \frac{\Gamma(\tfrac{3}{2})\Gamma(Kn)}{\Gamma[\tfrac{3}{2} + (K-1)n]} \frac{\epsilon}{2\pi} \left(\frac{\epsilon}{\epsilon_0}\right)^{(2K-2)n}. \qquad (25.126)$$

The expression (25.122) with (25.126) is the weak-tunneling series representation of the full counting statistics for general dissipation parameter K.

When the Poissonian transport process is classical, all the individual transition rates k_n^+ in Eq. (25.122) are positive. The subtle point now is that the rates (25.126) are not. Certainly, the first contribution k_1^+ represents indeed a classical Poisson process for all K (in reality, it is a quantum mechanical tunneling process); but the coherent tunneling through two barriers with transition weight k_2^+ comes with a negative sign when $K > \tfrac{1}{4}$, which is an effect of quantum coherence. Similar behaviour is found for tunneling transition with $n > 2$, depending on the particular value of K.

In the opposite limit of weak backscattering or strong tunneling, the series expression for $\ln \mathcal{Z}(\rho, t)$ reads

$$\ln \mathcal{Z}(\rho, t) = t \left(i\rho q_0 \frac{\epsilon}{2\pi K} + \sum_{n=1}^{\infty} \left(e^{-i\rho q_0 n / K} - 1 \right) \widetilde{k}_n^+ \right), \qquad (25.127)$$

where

$$\widetilde{k}_n^+ = \frac{(-1)^{n-1}}{n!} \frac{\Gamma(\tfrac{3}{2})\Gamma(n/K)}{\Gamma[\tfrac{3}{2} + (1/K - 1)n]} \frac{\epsilon}{2\pi K} \left(\frac{\epsilon}{\epsilon_0}\right)^{(2/K-2)n}. \qquad (25.128)$$

It is immediately obvious from these forms together with the relation (25.111) that there holds a cumulant relation similar to the form (25.125) in the strong-tunneling representation,

$$\langle q^N(t) \rangle_c = \left(-\frac{q_0}{K} \frac{V_0}{2} \frac{\partial}{\partial V_0} \right)^{N-1} \langle q(t) \rangle. \qquad (25.129)$$

The expression (25.127) is quite similar to the form (25.122), but there are subtle differences. The first term in the expression (25.127) represents the current in the absence of the barrier. The exponential factor $e^{-i\rho q_0 n / K}$ indicates that now we have tunneling over distance q_0 / K and multiples thereof, and the minus sign in the exponent means that the tunneling diminishes the current instead of building it up as in the weak-tunneling limit. Using the reflection formula for gamma functions, one finds that the sign of the tunneling rate $\widetilde{k}_n^+$ is the one of $\cos(n\pi / K)$ when $K > 2$. Therefore, the perception of a classical Poisson process for the tunneling dynamics is quite appropriate for transition weights with modest n, when K is large. As K goes to infinity, all the transition weights become amenable to classical interpretation but the quantum fluctuations $\langle q^n(t) \rangle_c$ with $n > 1$ disappear, as one reaches the classical limit [cf. Eq. (25.32)],

$$\ln \mathcal{Z}(\rho, t) = it\rho \frac{q_0 \epsilon}{2\pi K} \left[1 - \sum_{n=1}^{\infty} \frac{\Gamma(n - \tfrac{1}{2})}{2\sqrt{\pi} n!} \left(\frac{\epsilon_0}{\epsilon}\right)^2 \right] = it\rho \frac{q_0 \epsilon}{2\pi K} \sqrt{1 - \left(\frac{\epsilon_0}{\epsilon}\right)^2}. \qquad (25.130)$$

In the classical limit, the Kondo scale ϵ_0 is related to the corrugation strength V_0 of the weak-binding model (23.1) with (23.2) by the relation $\epsilon_0 = 2\pi V_0/\hbar$.

25.7 Low temperature behaviour of the characteristic function

For general damping strength K and general temperature, the evaluation of the Coulomb gas integral representation given in Subsection 24.3.1 is not possible. To make progress, one has to employ thermodynamic Bethe ansatz techniques to the related BSG model [483, 484]. In Ref. [485] a general procedure for the calculation of cumulants of any order is sketched and the third moment is explicitly calculated.

Interestingly, the leading low-temperature contribution to the zero temperature cumulant generating function can be calculated in analytic form. To this aim, we work in the Coulomb gas representation. In the first step, we perform the low temperature expansion of the charge interaction,

$$Q'(\tau) = Q_0'(\tau) + K\frac{\pi^2}{3}\left(\frac{\tau}{\hbar\beta}\right)^2 + \mathcal{O}\left[\left(\frac{\tau}{\hbar\beta}\right)^4\right], \tag{25.131}$$

where $Q_0'(\tau) = 2K\ln(\omega_c\tau)$ is the interaction at $T = 0$. Next, we determine the leading low-temperature dependence of the charge interaction factor G_m given in Eq. (24.34). Upon substituting the form (25.131), we obtain

$$G_m = G_m^{(0)}\left\{1 - K\frac{\pi^2}{3}\left(\frac{k_B T}{\hbar}\right)^2\left(\sum_{j=1}^{2m-1}p_j\tau_j\right)^2 + \mathcal{O}(T^4)\right\}, \tag{25.132}$$

where $G_m^{(0)}$ is the full interaction factor at $T = 0$. The crucial point now is that the squared sum in the curly bracket in Eq. (25.132) can be expressed in terms of the bias phase $\varphi_{2m,0}$ given in Eq. (24.48), $\sum_j p_j\tau_j = \varphi_{2m,0}/\epsilon$. Hence this annoying term can be generated by taking the second derivative of the bias factor $B_{2m,0} = e^{i\varphi_{2m,0}}$ with respect to the bias. The asymptotic low-temperature expansion of $G_m B_{2m,0}$ reads

$$G_m B_{2m,0} = G_m^{(0)}\left\{1 + K\frac{\pi^2}{3}\left(\frac{k_B T}{\hbar}\right)^2\frac{\partial^2}{\partial\epsilon^2} + \mathcal{O}(T^4)\right\}B_{2m,0} . \tag{25.133}$$

Upon employing the relation (25.133), we find that the T^2- contribution to the cumulant generating function can be universally written in terms of the second derivative of the zero temperature expression with respect to the bias. Examination of subleading contributions $\propto T^4$, T^6, $\cdots$ yields that they are nonuniversal, and therefore they can not be expressed in simple terms.

Substituting the form (25.133) into the series expression (24.39), we readily obtain the cumulant generating function in the TB representation in the form (25.122), in which the rate k_n^+ is given by

$$k_n^+(\epsilon, T) = \left\{1 + K\frac{\pi^2}{3}\left(\frac{k_B T}{\hbar}\right)^2\frac{\partial^2}{\partial\epsilon^2} + \mathcal{O}(T^4)\right\}k_n^+(\epsilon, 0) . \tag{25.134}$$

Observing from the form (25.126) that $k_n^+(\epsilon, 0) \propto \epsilon^{(2K-2)n+1}$, we readily get

$$k_n^+(\epsilon, T) = \left\{ 1 + \frac{\pi^2}{3} K\, (2K-2) n [\, 1 + (2K-2) n\,] \left(\frac{k_{\mathrm{B}} T}{\hbar \epsilon} \right)^2 \right\} k_n^+(\epsilon, 0)\,, \quad (25.135)$$

where terms of order T^4 are disregarded.

The dual weak-backscattering or strong-tunneling representation of the CGF is found in generalization of the expression (25.127) as

$$\ln \mathcal{Z}(\rho, t) = \tau \left[i\rho \frac{q_0 \epsilon}{2\pi K} - \rho^2 \frac{q_0^2 k_{\mathrm{B}} T}{2\pi K \hbar} + \sum_{n=1}^{\infty} \left(\mathrm{e}^{-i\rho n q_0 / K} - 1 \right) \widetilde{k}_n^+(\epsilon, T) \right]\,, \quad (25.136)$$

where

$$\widetilde{k}_n^+(\epsilon, T) = \left\{ 1 + \frac{\pi^2}{3} \frac{(2-2K)\, n [\, K + (2-2K)\, n\,]}{K} \left(\frac{k_{\mathrm{B}} T}{\hbar \epsilon} \right)^2 \right\} \widetilde{k}_n^+(\epsilon, 0)\,. \quad (25.137)$$

The expressions (25.122) with (25.135) and (25.136) with (25.137) are the dual series representations of the cumulant generating function $\mathcal{Z}(\rho, t)$. It is straightforward to derive from these forms the weak- and strong-tunneling expansions of all cumulants. Observe that the second term in Eq. (25.136) gives the remaining contribution to the second cumulant in the asymptotic limit in which the corrugation is fully irrelevant.

The T^2-contribution to the full counting statistics is a distinctive signature of Ohmic dissipation inherent in the quantum transport process. A related phenomenon is the universal T^2-behaviour observed at low temperatures in other open quantum systems subject to Ohmic dissipation, e.g. the T^2-enhancement in macroscopic quantum tunneling (cf. Section 17.2). It is also the origin of the universal Wilson ratio occurring e.g. in Kondo systems and in the related Ohmic two-state system (cf. Subsection 19.2.1 and Section 21.9). Other examples are the T^2 enhancement of the free energy and static susceptibility of the two-state system [cf. Eqs. (21.281) and (21.282) for $s = 1$]. In these cases, the prefactor is determined by the susceptibility at zero temperature and its second derivative, respectively. The physical origin of the T^2 corrections is the low-frequency thermal noise of Ohmic dissipation [327].

We finally remark that with the same strategy one may calculate also finite temperature corrections in the non-Ohmic case $s \neq 1$, e.g. for the nonlinear mobility. It is straightforward to see that in this case the power-law of the leading thermal enhancement is T^{1+s} [208].

25.8 The sub- and super-Ohmic case

Consider first the TB model in the sub-Ohmic case. For $s < 1$, a long blip interval τ_k with cumulative charge label p_k is exponentially suppressed by the interaction factor $\exp[-p_k^2 Q'(\tau_k)]$, as discussed before Eq. (24.62). In contrast, every integral over a sojourn length ρ is regularized only by the Laplace factor $\mathrm{e}^{-\lambda \rho}$. Thus, in the limit

$\lambda \to 0$, the charges in the series (24.45) arrange themselves in sequences of dense neutral clusters which are separated by long sojourn intervals. The sojourn factor $f(\lambda)$ can be determined upon expanding the influence phase term $b_{2m,0}$ for short cluster and long sojourn lengths. One then finds that the first moment to linear order in the bias $\langle \hat{q}(\lambda) \rangle_{\mathrm{lin}}$ takes the form of a geometrical series of sojourn-cluster pairs $f(\lambda)k(\lambda)$ with an additional cluster of exactly the same form [328],

$$\lim_{\lambda \to 0} \langle \hat{q}(\lambda) \rangle_{\mathrm{lin}} = \lim_{\lambda \to 0} \frac{q_0 \epsilon}{2} \frac{k(\lambda)}{\lambda^2 [1 + f(\lambda)k(\lambda)]} = \lim_{\lambda \to 0} \frac{q_0 \epsilon}{2} \frac{1}{\lambda^2 f(\lambda)} . \tag{25.138}$$

In obtaining the second expression we have used that $k(\lambda)$ is regular and the sojourn factor function diverges in the limit $\lambda \to 0$. We obtain using Eq. (18.45)

$$f(\lambda) = \int_0^\infty d\rho \, \dot{Q}''(\rho) \, e^{-\lambda \rho} = [\pi \delta_s / \sin(\pi s/2)] \omega_{\mathrm{ph}}^{1-s} \lambda^{s-1} , \tag{25.139}$$

We thus find in the asymptotic time regime the sluggish subdiffusive behaviour

$$\langle q(t) \rangle_{\mathrm{lin}} = \frac{q_0 \sin(\frac{1}{2}\pi s)}{2\pi \delta_s \Gamma(1+s)} \frac{\epsilon}{\omega_{\mathrm{ph}}} (\omega_{\mathrm{ph}} t)^s . \tag{25.140}$$

Thus, the dependence on T and on Δ fades away at long times. Using for large t the relation $\chi(t) = \langle \dot{q}(t) \rangle_{\mathrm{lin}} q_0 / \hbar \epsilon$, we find that the asymptotic motion in the TB lattice is fully equivalent, including the prefactor, to free Brownian motion, Eq. (7.51). The second moment is more difficult to evaluate. In the end one finds the simple relation

$$\langle q^2(t) \rangle_{\epsilon=0} = (2q_0 / \hbar \beta \epsilon) \langle q(t) \rangle_{\mathrm{lin}} \qquad \text{for} \qquad \omega_{\mathrm{ph}} t \gg 1 . \tag{25.141}$$

Similarly, the linear ac mobility coincides in the limit $\omega \to 0$ with free Brownian motion, Eq. (7.55). We find $\tilde{\mu}^{(\mathrm{TB})}(\omega \to 0) = 1/M\gamma_{\mathrm{TB}}(\omega) \propto (-i\omega/\omega_{\mathrm{ph}})^{1-s}$ and the subleading correction is $\propto (1/\Delta^2)(-i\omega/\omega_{\mathrm{ph}})^{2-2s}$. As the temperature is decreased, the effective transfer matrix element of the sub-Ohmic TB model scales to lower values for fixed ω, and at $T = 0$ it vanishes with an essential singularity $\propto \exp(-\mathrm{const}/\omega^{s/(1-s)})$ as $\omega \to 0$ [cf. Eq. (20.50)]. Hence the mobility vanishes faster than any power of ω in the zero-frequency limit, and the particle becomes strictly localized.

Utilizing the duality relation (25.24), we can now deduce results for the mobility of the WB model in the super-Ohmic range $s' = 2 - s$. At finite T, the superdiffusive term $\propto \omega^{1-s'}$ of free Brownian motion in Eq. (25.24) is cancelled and the next to leading order term describes slower *diffusive* behaviour $\mu^{(\mathrm{WB})}(\omega \to 0) = \mathrm{const}$. This is a nonperturbative result. At $T = 0$, we find, using duality, to leading order the superdiffusive behaviour of the free Brownian particle, $\mu^{(\mathrm{WB})}(\omega \to 0) = 1/M\gamma_{\mathrm{WB}}(\omega) \propto (-i\omega/\omega_{\mathrm{ph}})^{1-s'}$, and the potential is an irrelevant perturbation.

Consider next the TB model for super-Ohmic damping, $1 < s < 2$. The TB model is perturbative in Δ^2 at finite T and it has the diffusive limit $\mu_{\mathrm{TB}}(\omega \to 0) = \bar{\mu}$, where $\bar{\mu}$ depends on temperature and the leading term is of order Δ^2 [cf. Eq. (24.20)]. With decreasing temperature, the effective transfer matrix element of the super-Ohmic TB model scales to higher values for fixed ω, and for $T = 0$ and $\omega \to 0$ it even scales to infinity. Hence the TB model becomes nonperturbative in Δ^2 in this limit. Using a cluster expansion analogous to Eq. (25.138), we find that the tight-binding lattice becomes completely dissolved in this limit, and the mobility is that of a free super-Ohmic Brownian particle, $\mu^{(\mathrm{TB})}(\omega \to 0) = 1/M\gamma_{\mathrm{TB}}(\omega) \propto (-i\omega/\omega_{\mathrm{ph}})^{1-s}$. The related mean position grows superdiffusively [86], $\langle q(t) \rangle \propto t^s$.

Using again the duality relation (25.24), we can infer the implications for the dual sub-Ohmic WB model with $s' = 2 - s < 1$. The linear mobility of this model at finite T is $\mu^{(\mathrm{WB})}(\omega \to 0) = 1/M\gamma_{\mathrm{WB}}(\omega) \propto (-i\omega/\omega_{\mathrm{ph}})^{1-s'}$, and the subleading correction is $-(-i\omega/\omega_{\mathrm{ph}})^{2-2s'}\bar{\mu}$. The potential V_0 represents an irrelevant perturbation in this regime. The mean position $\langle X(t) \rangle$ grows subdiffusively $\propto t^{s'}$, i.e., just as in the absence of the potential, and the leading potential corrections grow as $t^{2s'-1}$. This is in agreement with the findings in Ref. [486]. At $T = 0$, the term $\propto \omega^{1-s}$ in Eq. (25.24) is again cancelled by a corresponding term from the TB mobility. As a result, the mean position of the WB particle grows even slower than in the free sub-Ohmic case.

Considering the fact that the system becomes increasingly sensitive to its low-frequency properties as T is decreased, the above characteristics are fully consistent with physical intuition. For sub-Ohmic damping and decreasing temperature, the mobility is progressively suppressed at constant low frequency in both models. This is due to the diverging spectral damping function $\gamma(\omega)$ for $\omega \to 0$. On the contrary, for super-Ohmic damping the mobility is enhanced with decreasing temperature in both models since the respective spectral damping function vanishes at zero frequency.

26. Charge transport in quantum impurity systems

The physics of interacting particles in one dimension is drastically different from the usual physics of interacting particles in two or three dimensions. The theoretical methods and techniques relevant to quantum physics in one dimension have been carefully reviewed by Gogolin, Nersesyan, and Tsvelik [487], and by Giamarchi [488]. Signatures of many-body correlations have attracted a great deal of interest in recent years. Many investigations have been concentrated on one-dimensional (1D) electron systems, in which the usual Fermi liquid behaviour is destroyed by the interaction. The generic features of many 1D interacting fermion systems are well described in terms of the Tomanaga-Luttinger liquid (TLL) model [489, 366]. In the TLL model, all effects of the electron-electron interaction are captured by a dimensionless parameter g. A sensitive experimental probe of a Luttinger liquid state is the tunneling conductance through a point contact in a 1D quantum wire, as studied by Kane and Fisher (KF) [457]. Of interest are also the dc nonequilibrium current noise [490] and higher cumulants. The generic model is a quantum impurity embedded in a Luttinger liquid environment (QI-TLL model). Tunneling of edge currents in the fractional quantum Hall (FQH) regime provides another realization of a Luttinger phase. As shown by

Wen [458], the edge state excitations are described by a (chiral) Luttinger liquid with Luttinger parameter $g = \nu$, where ν is the fractional filling factor.

26.1   Generic models for transmission of charge through barriers

In this section, we discuss the conductance of 1D interacting spinless electrons in the presence of a barrier. First, we consider a pure interacting electron gas and give the two-terminal conductance. Then we discuss transport through a single barrier. We approach this problem perturbatively in two limits: a very weak barrier (strong tunneling) and a very large barrier (weak tunneling). We find that the QI-TLL model is closely related to the model of an Ohmic Brownian particle in a tilted washboard potential, discussed in the preceding sections. As we shall see in Subsection 26.1.4, the model is also equivalent to a one-channel coherent conductor in a resistive electromagnetic environment.

26.1.1   The Tomonaga-Luttinger liquid

The characteristic feature of a Fermi liquid is the discontinuity of the momentum density of states at the Fermi surface. In one dimension, electron-electron interaction is so strong that this discontinuity is dissolved and replaced by a power law essential singularity at the Fermi points. The ground-state of the interacting 1D electron gas is a Tomonaga-Luttinger liquid which is distinguished by a gapless collective sound mode [366]. At low energies and long wave lengths, the strength of the interaction of spinless electrons is characterized by a single dimensionless parameter g.

The low-energy modes of the 1D interacting electron liquid are conveniently treated in the framework of bosonization [489, 366, 487]. This approach is appropriate for low temperatures where only excitation near the Fermi points are relevant. The creation operator for spinless fermions can equivalently be expressed in terms of boson phase fields $\theta(x)$ and $\phi(x)$, which obey the equal-time commutation relation

$$[\phi(x,t), \theta(x',t)] = -(i/2)\,\text{sgn}(x-x') . \tag{26.1}$$

Thus $\partial_x \phi(x)$ is the canonically conjugate momentum density to $\theta(x)$, and $\partial_x \theta(x)$ is conjugate to $\phi(x)$. In terms of these bosonic fields, the fermion field operator may be written as

$$\psi^\dagger(x) \propto \sum_{n\,\text{odd}} \exp\{in[k_F x + \sqrt{\pi}\theta(x)]\} \exp[i\sqrt{\pi}\phi(x)] . \tag{26.2}$$

At long wavelengths, only the terms $n = \pm 1$ are important, corresponding to the right- and left-moving parts of the electron field. The boson representation of the electron density operator is then given by

$$\rho(x) = \frac{k_F}{\pi} + \frac{1}{\sqrt{\pi}}\partial_x \theta(x) + \frac{k_F}{\pi}\cos[2k_F x + 2\sqrt{\pi}\theta(x)] , \tag{26.3}$$

where $k_{\rm F}$ is the Fermi momentum. The first term is the background charge, the second term represents the density fluctuations of the right- and left-movers, and the last term describes interference between right- and left-movers.

The interaction between these bosons emerges from a combination of the hard-core condition of the bosons and the electron-electron interaction of the original fermions. Taking short-range electron-electron interactions and disregarding backscattering, the TLL liquid is described by the generic harmonic Hamiltonian

$$H_{\rm L} = \frac{\hbar v}{2} \int dx \left[(\partial_x \theta)^2/g + g(\partial_x \phi)^2 \right] , \qquad (26.4)$$

where g is the interaction parameter, and $v = v_{\rm F}/g$ is the sound velocity. For the noninteracting Fermi gas, we have $g = 1$, and the case $g < 1$ corresponds to repulsive interaction. We assume a sharp cutoff ω_c in the band width for the linear dispersion relation implicit in $H_{\rm L}$ and take ω_c as the largest frequency of the problem.

By switching from the Hamiltonian to the Lagrangian in either θ or ϕ, we get two equivalent representations of the Luttinger liquid. The respective Euclidean actions read

$$S_{\rm L}^{(E)} = \begin{cases} \dfrac{\hbar v}{2g} \displaystyle\int dx\, d\tau \left((\partial_x \theta)^2 + \dfrac{1}{v^2}(\partial_\tau \theta)^2 \right) , \\[2ex] \dfrac{\hbar v g}{2} \displaystyle\int dx\, d\tau \left((\partial_x \phi)^2 + \dfrac{1}{v^2}(\partial_\tau \phi)^2 \right) . \end{cases} \qquad (26.5)$$

It is well-known that the "two-terminal" conductance for a single-channel Fermi liquid is $G_0(g = 1) = e^2/2\pi\hbar$. For interacting electrons, the conductance may be calculated from the current-current correlation function in the zero frequency limit, where the current is $J = ie\dot\theta/\sqrt{\pi}$. One then finds that the conductance is renormalized by the electron-electron interaction [457],[1]

$$G_0(g) = g\frac{e^2}{2\pi\hbar} . \qquad (26.6)$$

Thus we may regard the Luttinger parameter g as a dimensionless measure of the conductance of the pure Luttinger liquid.

26.1.2 Transport through a single weak barrier

The weak impurity is modelled by a barrier Hamiltonian $H_{\rm sc} = \int dx\, V(x)\psi^\dagger(x)\psi(x)$, where $V(x)$ is a scattering potential [457]. Omitting multiple-electron backscattering processes, one finds for a short-ranged impurity potential, which is suppposed to be centered at $x = 0$, the form

$$H_{\rm sc} = -V_0 \cos[\, 2\sqrt{\pi}\,\bar\theta\,] , \qquad (26.7)$$

---

[1]The conductance considered here is a low-frequency microwave conductance. A two-terminal setup with reservoirs held at fixed chemical potentials would lead to modifications [491].

where $\bar{\theta}(t) = \theta(x = 0, t)$, and where V_0 is the Fourier transform of $V(x)$ at momentum $2k_{\mathrm{F}}$. The scattering Hamiltonian originates from the interference between left- and right-movers and represents $2k_{\mathrm{F}}$-backscattering. An applied voltage drop V_{a} at the impurity gives rise to the contribution

$$H_{\mathrm{V}} = eV_{\mathrm{a}}\bar{\theta}/\sqrt{\pi} . \tag{26.8}$$

The weak barrier or θ model is then given by

$$H_{\theta} = H_{\mathrm{L}} + H_{\mathrm{sc}} + H_{\mathrm{V}} . \tag{26.9}$$

In the model (26.9), the nonlinear tunneling degree of freedom $\bar{\theta}(t)$ is coupled to a harmonic field which represents the modes in the leads away from $x = 0$. This is the convenient starting point if one wishes to study backscattering off a weak impurity, in particular with regard to computations of conductance and statistical fluctuations.

Since the backscattering term H_{sc} only acts at $x = 0$, we may integrate out fluctuations of $\theta(x)$ for all x away from the origin. This can be done without approximation because of the harmonic nature of the pure Luttinger liquid action (26.5). With the Fourier ansatz $\theta(x, \tau) = (1/\hbar\beta) \sum_n \theta(x, \nu_n) e^{-i\nu_n t}$, where $\nu_n = (2\pi/\hbar\beta)n$, the Euclidean action is minimized when $\theta(x, \nu_n) = \bar{\theta}(\nu_n) \exp(-|\nu_n x|/v)$. Here, $\bar{\theta}(\nu_n)$ is the Fourier coefficient of $\bar{\theta}(\tau)$. The resulting influence action [457] includes all effects of the modes in the leads on the tunneling degree of freedom $\bar{\theta}(t)$,

$$S_{\mathrm{infl}}^{(\mathrm{E})}[\bar{\theta}]/\hbar = \frac{1}{g} \frac{1}{\hbar\beta} \sum_n |\nu_n| \, |\bar{\theta}(\nu_n)|^2 . \tag{26.10}$$

It is convenient to consider the nonlinear conductance given in units of the conductance $G_0(g)$ of a pure 1D quantum wire,

$$\mathcal{G}(T, V_{\mathrm{a}}, g) = I(T, V_{\mathrm{a}}, g)/[V_{\mathrm{a}}G_0(g)] . \tag{26.11}$$

For a weak barrier, the θ-representation (26.7) is appropriate. Then the conductance is determined by

$$\mathcal{G}_{\theta}(T, V_{\mathrm{a}}, g) = e\langle \dot{\bar{\theta}}(t \to \infty) \rangle_{\beta} /[\sqrt{\pi}V_{\mathrm{a}}G_0(g)] , \tag{26.12}$$

where $\langle \cdots \rangle_{\beta}$ denotes the thermal average over all modes of H_{L} away from the impurity. As we have just seen, this means average with the weight function $\exp\{-S_{\mathrm{infl}}[\bar{\theta}]/\hbar\}$.

An analytical expression for $\mathcal{G}_{\theta}$ can be derived as follows. First, we expand the formal path integral expression into a power series in V_0^2. Then, in each term of the series, we integrate out the Gaussian field $\theta(x, \tau)$ away from $x = 0$. The resulting expression has the form of a statistical, grand-canonical ensemble of interacting discrete charges, analogous to the corresponding expressions given in Section 24.3. Because of the analogy of the interaction term (26.7) with the interaction (23.2), the charge conditions are as specified in Subsection 24.3.1. As a result of the elimination of the harmonic modes in the leads, the charges are interacting with each other. The pair interaction in imaginary time $W_\theta(\tau)$ is related to the correlator of $\bar{\theta}(\tau)$ by

$$\left\langle \mathcal{T} e^{i2\sqrt{\pi}[\bar{\theta}(\tau)-\bar{\theta}(0)]} \right\rangle_\beta = e^{-W_\theta(\tau)}. \tag{26.13}$$

Here, the thermal average $\langle \cdots \rangle$ denotes again average with the weight function $\exp[-S_{\text{infl}}^{(\text{E})}[\bar{\theta}]/\hbar]$. Upon completing the square, we get

$$W_\theta(\tau) = 2g\pi\frac{1}{\hbar\beta}\sum_n \frac{1}{|\nu_n|}\left[1 - e^{i\nu_n\tau}\right]. \tag{26.14}$$

We see from the expression (4.70) with (4.57) and (4.56) that $W_\theta(\tau)$ is the charge interaction for an Ohmic spectral density $G_\theta(\omega) = 2g\,\omega$. Thus, if we perform analytic continuation to real time $\tau = it$, the charge interaction $Q_\theta(t) = W_\theta(\tau = it)$ is given by the integral representation (18.42) and takes the Ohmic scaling form [cf. Eq. (18.51)]

$$Q_\theta(t) = 2g\ln\left[\frac{\hbar\beta\omega_c}{\pi}\sinh\left(\frac{\pi t}{\hbar\beta}\right)\right] + i\pi g\,\text{sgn}(t). \tag{26.15}$$

The strong-tunneling series for the normalized conductance is found to read

$$\mathcal{G}_\theta(T,V_a,g) = 1 - \frac{\pi\hbar}{eV_a}\sum_{m=1}^\infty (-1)^{m-1}\left(\frac{V_0}{\hbar}\right)^{2m}\int_0^\infty d\tau_1\,d\tau_2\cdots d\tau_{2m-1} \tag{26.16}$$

$$\times \sum_{\{\xi_j\}'} G_m(\{\tau_j\}; g, \{\xi_j\})\sin\left(\sum_{i=1}^{2m-1} geV_a p_i\tau_i/\hbar\right)\prod_{k=1}^{2m-1}\sin(\pi g p_k),$$

where p_j is the cumulative charge defined in Eq. (24.47). The interaction factor $G_n(\{\tau_j\}; g, \{\xi_j\})$ is defined in Eq. (24.73).

26.1.3 Transport through a single strong barrier

Next, we discuss the opposite limit of a large barrier or equivalently a weak link. For an infinitely high barrier, we have two disconnected semi-infinite Luttinger leads described in terms of independent bosonic fields. To describe electron tunneling across a weak link, a hopping term of the form

$$H_I' \propto -\hbar\Delta[\psi^\dagger(x=0^+)\psi(x=0^-) + \text{h.c.}] \tag{26.17}$$

is added. This term depends on the phase jump $\bar{\phi}$ of the ϕ-field at the point-like barrier $\bar{\phi} \equiv [\phi(x=0^+) - \phi(x=0^-)]/2$. Hence the ϕ-representation of the action (26.5) is appropriate. In terms of $\bar{\phi}$, the tunneling term takes the form

$$H_I' = -\hbar\Delta\cos[2\sqrt{\pi}\bar{\phi} + eV_a t/\hbar]. \tag{26.18}$$

A voltage drop at the barrier induces a jump of the chemical potential. The jump induces a phase shift, which we have included in the expression (26.18). Upon adding the terms describing the harmonic liquid in the left $(-)$ and right $(+)$ lead, we then arrive at the Hamiltonian of the weak link problem (ϕ–model)

$$H_\phi = H_{\mathrm{L},+} + H_{\mathrm{L},-} + H_{\mathrm{I}}' . \tag{26.19}$$

This model describes again a nonlinear tunneling degree of freedom coupled to a harmonic field. We may again integrate out the harmonic modes away from the barrier, this time in the ϕ-representation. Proceeding as in the previous subsection, the Euclidean influence action of the tunneling mode $\bar{\phi}$ resulting from the Luttinger liquid environment takes the form

$$S_{\mathrm{infl}}^{(\mathrm{E})}[\bar\phi]/\hbar = g\,\frac{1}{\hbar\beta}\sum_n |\nu_n|\,|\bar\phi(\nu_n)|^2 . \tag{26.20}$$

Because of the substitution $g \to 1/g$, when switching from the θ- to the ϕ-representation, the action (26.20) is the dual of the action (26.10).

With employment of the Heisenberg equation of motion for $\bar\theta(t)$, we may express the normalized conductance (26.12) in terms of the ϕ-field. This yields the expression

$$\mathcal{G}_\phi(T,V_{\mathrm{a}},g) = e\Delta\langle\, \sin[\,2\sqrt{\pi}\bar\phi(t\to\infty)\,]\,\rangle_\beta /V_{\mathrm{a}} , \tag{26.21}$$

Thermal average $\langle\cdots\rangle_\beta$ over all Luttinger liquid modes away from the barrier now means average with the weight function $\exp[-S_{\mathrm{infl}}^{(\mathrm{E})}[\bar\phi]/\hbar]$.

The proceeding is now as in the preceeding subsection. We formally expand the path integral in powers of Δ^2 and, in view of the formal similarity of the expression (26.18) with (26.7), we introduce again a charge representation. The effects of the Luttinger liquid modes are included in the charge interaction, which is again a pair correlation function, but now for $\bar\phi(t)$. Following the lines (26.13) – (26.15), we get for the real-time correlator

$$\Big\langle\, \mathcal{T}\, e^{i2\sqrt{\pi}[\bar\phi(t)-\bar\phi(0)]}\,\Big\rangle_\beta = \exp\Big\{ -\frac{2}{g}\ln\Big[\frac{\hbar\beta\omega_c}{\pi}\sinh\Big(\frac{\pi t}{\hbar\beta}\Big)\Big] - i\frac{\pi}{g}\,\mathrm{sgn}(t)\Big\} . \tag{26.22}$$

The weak-tunneling series expansion for the normalized conductance is readily found as the dual of the series (26.16)

$$\mathcal{G}_\phi(T,V_{\mathrm{a}},g) = \frac{\pi\hbar}{geV_{\mathrm{a}}}\sum_{m=1}^{\infty}(-1)^{m-1}\Delta^{2m}\int_0^\infty d\tau_1\,d\tau_2\cdots d\tau_{2m-1} \tag{26.23}$$

$$\times \sum_{\{\xi_j\}'} G_m(\{\tau_j\};1/g,\{\xi_j\})\,\sin\Big(\sum_i^{2m-1} eV_{\mathrm{a}}p_i\tau_i/\hbar\Big)\prod_{k=1}^{2m-1}\sin(\pi p_k/g) .$$

26.1.4 Coherent conductor in an Ohmic environment

A mesoscopic conductor embedded in an electromagnetic environment forms a quantum system violating Ohm's law. The electrons in the conductor induce electromagnetic modes in the electrical circuit and therefore undergo inelastic scattering processes. By this, the current at low voltage is reduced, as we have discussed already for weak tunneling in Subsection 20.3.1. This picture of dynamical Coulomb blockade changes in the opposite limit of a good conductor. The description of tunneling of discrete charges is then no longer valid. The Luttinger parameter of the coherent conductor in the absence of the electromagnetic environment is $g = 1$. According to convenience, we may use either the θ- or the ϕ-representation of the action (26.5).

As we have seen already in Section 20.3 in the weak-tunneling limit, an Ohmic environment can well simulate the electronic interactions in a coherent conductor. We now fully extend the analogy of a one-channel conductor in an Ohmic environment to impurity scattering in a Luttinger liquid (see Ref. [492]).

Let us take the Hamiltonian (3.167) as a starting-point. For spinless electrons and a constant real tunneling amplitude T_{T}, it can be written as

$$H_{\mathrm{wt}} = H_1 + H_2 + T_{\mathrm{T}}[\psi_2^\dagger(0^+)\psi_1(0^-)\,e^{-i\varphi(t)} + \text{h.c.}] + H_{\mathrm{env}}(\mathcal{Q},\varphi)\,. \qquad (26.24)$$

The label "wt" stands for *weak tunneling*, and $H_{1,2}$ is the electronic part for the left/right electrode with the voltage drop V_{a} across the barrier included in H_1, as specified in Eq. (3.166). The term H_{env} is supposed to describe a resistive environment with impedance R, and V_a is the applied voltage drop at the barrier. The last term couples the phase fluctuations $\varphi(t)$ induced by the impedance to the local electronic fields $\psi_{1,2}(0)$ at the edges of the barrier. Since the barrier is point-like, electron tunneling across the barrier is effectively one-dimensional. Therefore, the electronic part can be bosonized, as discussed in Subsection 26.1.1.

The tunneling term in Eq. (26.24) can thus be written as [cf. Eq. (26.18]

$$H_I = -\tfrac{1}{2}\hbar\Delta'\left[e^{i(2\sqrt{\pi}\bar\phi - \varphi)} + \text{h.c.}\right]\,. \qquad (26.25)$$

where $\Delta' \propto T_{\mathrm{T}}$. The phase correlation functions for the electromagnetic phase $\varphi(t)$ and the electronic phase jump at the barrier $2\sqrt{\pi}\bar\phi(t)$, respectively, are [cf. Eq. (26.22) and Eq. (20.132) with Eq. (20.147)]

$$\langle\,[\varphi(0) - \varphi(t)]\,\varphi(0)\,\rangle_\beta \;=\; Q_{\mathrm{ohm}}(t;\alpha) + \Delta Q_{\mathrm{em}}(\alpha)\,, \qquad (26.26)$$

$$4\pi\langle\,[\bar\phi(0) - \bar\phi(t)]\,\bar\phi(0)\,\rangle_\beta \;=\; Q_{\mathrm{ohm}}(t;1)\,, \qquad (26.27)$$

where $Q_{\mathrm{ohm}}(t;g)$ is given in (20.142) and $\Delta Q_{\mathrm{em}}(\alpha)$ in (20.147). Since $\bar\phi(t)$ and $\varphi(t)$ commute and the Hamiltonian $H_1 + H_2 + H_{\mathrm{env}}$ is quadratic in the combined phase $\chi = 2\sqrt{\pi}\bar\phi + \varphi$, the equilibrium correlation function of the auxiliary phase $\chi(t)$ is

$$\langle\,[\chi(0) - \chi(t)]\,\chi(0)\,\rangle_\beta \;=\; Q_{\mathrm{ohm}}(t;1+\alpha) + \Delta Q_{\mathrm{em}}(\alpha)\,. \qquad (26.28)$$

The constant term $\Delta Q_{\rm em}(\alpha)$ is conveniently absorbed into a dressed tunneling amplitude, $\Delta = \Delta' e^{-\Delta Q_{\rm em}(\alpha)/2}$. The relation between Δ' and the original tunneling amplitude $T_{\rm T}$ follows from the expressions (20.161) and (20.128).

Thus, the Hamiltonian $H_{\rm wt}$ is equivalent to the impurity Hamiltonian H_ϕ, Eq. (26.19), if we identify

$$1 + \alpha = 1/g \,. \tag{26.29}$$

The correspondence holds in all orders of Δ^2 in the series expression (26.23). The standard first order for the current has been dicussed already in Subsection 20.3.2.

We may also think of the "dual limit" of weak backscattering by the scattering potential $H_{\rm sc}$ given in bosonized form in Eq. (26.7). The electromagnetic environment causes a fluctuating potential drop $\hbar\dot\varphi(t)$ at the barrier which is coupled to the scattering mode $\bar\theta$ as in Eq. (26.8). Thus, the Hamiltonian of the coherent conductor plus electromagnetic environment takes the form

$$H_{\rm st} = H_{{\rm L},g=1} + H_{\rm sc} + H_{\rm V} + \dot\varphi\,\bar\theta/\sqrt{\pi} + H_{\rm env}(\mathcal{Q},\varphi) \,. \tag{26.30}$$

where the last term represents the electromagnetic environment as given in Eq. (3.154). It is straightforward to integrate out the modes $\{\varphi_\alpha\}$ of the electromagnetic environment. The resulting Euclidean action in Fourier representation for an Ohmic environment, $Y(\omega) = 1/R$, reads [cf. Eq. (4.204)]

$$S_{\rm em}^{\rm (E)}[\bar\theta,\varphi]/\hbar = \frac{1}{\hbar\beta} \sum_n \left\{ \frac{1}{4\pi} \frac{R_{\rm K}}{R} |\nu_n||\varphi_n|^2 - i\nu_n\varphi_n \frac{\bar\theta_n}{\sqrt{\pi}} \right\} \,. \tag{26.31}$$

In the next step, we also integrate out the phase $\varphi(t)$. Upon completing the square, we find that the effects of the electromagnetic environment on the tunneling degree of freedom $\bar\theta$ are described by the Euclidean influence action $(\alpha = R/R_{\rm K})$

$$S_{\rm infl,\,em}^{\rm (E)}[\bar\theta]/\hbar = \alpha \frac{1}{\hbar\beta} \sum_n |\nu_n|\,|\bar\theta_n|^2 \,. \tag{26.32}$$

On the other hand, the electronic fluctuations in the leads give rise to the action

$$S_{\rm infl,\,el}^{\rm (E)}[\bar\theta/]/\hbar = \frac{1}{\hbar\beta} \sum_n |\nu_n|\,|\bar\theta_n|^2 \,, \tag{26.33}$$

as we can read off from the expression (26.10). In the sum of both terms, the resistive environment leads to a renormalization of the electronic modes, $|\nu_n| \to (1+\alpha)|\nu_n|$.

Thus, there is again a formal equivalence to backscattering off an impurity in TLL, this time in the weak-backscattering limit. The relation of the effective coupling parameter $1 + \alpha$ to the fictitious TLL parameter g is the same as that found in the tunneling regime, Eq. (26.29), $1+\alpha = 1/g$. The correspondence of the model (26.30) with the QI-TLL model (26.9) is again on the level of the effective action.

26.1.5   Equivalence with quantum transport in a washboard potential

After having established a formal equivalence between the models of a quantum impurity in a TLL and a coherent one-channel conductor in a resistive environment, we now discuss the correspondence of the QI-TLL model with that of a Brownian particle in a tilted cosine potential.

The model described by H_θ, Eq. (26.9), directly corresponds to the weak-binding Brownian particle model (23.1). The correspondence can be shown by employing canonical transformations and by a study of the equations of motion of the coordinate and momentum autocorrelation functions (cf. the discussion in Subsections 25.1.1 and 25.1.2). The tunneling degree of freedom $\bar{\theta}$ corresponds to $\sqrt{\pi} X/X_0$. Ohmic damping of the mode $\bar{\theta}$ is provided by excitation of the TLL liquid away from the barrier, and the parameter g is related to the viscosity η by $g = \eta X_0^2/2\pi\hbar$. In the correspondence, the cutoff frequency ω_c of the liquid modes is identified with η/M. The equivalence becomes exact when the force of inertia is negligibly small compared with the friction force, i.e., when ω_c is the largest frequency of the problem. For $k_{\rm B}T \ll V_0$ and $eV_{\rm a} \ll V_0$, this means $\eta^2 \gg MV_{WB}''(0)$, or equivalently $\hbar\omega_c \gg 2\pi g V_0$. As a result of the mapping, the nonlinear conductance in the θ–model is directly related to the nonlinear mobility in the WB–model.

Similarly, the high-barrier ϕ-model (26.19) is equivalent to the TB-model (23.3), as follows again by unitary transformations.

Alternatively, the correspondences of the θ– and ϕ–model with the dissipative WB– and TB–model can also be seen directly from the exact formal series expressions for the conductance and mobility, respectively. The series (26.16) agrees with the WB mobility (25.88) if we employ the correspondences

$$g = 1/K , \quad \text{and} \quad eV_{\rm a}/\hbar = \epsilon . \tag{26.34}$$

Likewise, the series expression (26.23) for the weak-link conductance concurs with the series (24.72) for the mobility of the TB model if we identify again $1/g$ with the Ohmic damping parameter K.

The relations (26.34) are valid on the level of the effective actions. This means that they hold not only for the conductance but also for all statistical fluctuations.

26.2   Self-duality between weak and strong tunneling

The results of the preceding subsection can be condensed into an important correspondence between the normalized mobility of the Brownian particle and the normalized conductance for the QI-TLL model,

$$\mathcal{G}(T, V_{\rm a}, g) = \mu(T, \epsilon = eV_{\rm a}/\hbar, K = 1/g)/\mu_0 , \tag{26.35}$$

which relates the θ-model to the WB model, and the ϕ-model to the TB model. Upon employing the self-duality relation (25.27) between the TB and WB representation we immediately infer from the correspondence (26.35) that the weak-link conductance $\mathcal{G}_\phi$ is related to the weak-barrier conductance $\mathcal{G}_\theta$ by

$$\mathcal{G}_\theta(T, V_a, g) = 1 - \mathcal{G}_\phi(T, gV_a, 1/g) . \qquad (26.36)$$

For repulsive electron interaction, the Luttinger parameter g is restricted to the domain $g < 1$, whereas in the related Brownian particle model the parameter regime is $0 < K < \infty$. The entire expansions around weak and strong backscattering are related to each other term by term by the correspondence (26.36). Upon introducing the scaling variables ϑ and v, the self-duality relations for $\mathcal{G} = \mathcal{M}$ take the form (25.93) and (25.101), respectively, where $g = 1/K$.

With use of the correspondence (26.35), the analytical results obtained in Chapter 25.4 for the Brownian particle can now immediately be transferred to charge transport across a quantum impurity in a Luttinger liquid. The exact scaling solutions at $T = 0$, Eqs. (25.99) and (25.100) with Eq. (25.109), agree with expressions derived by Fendley et al. [483]. They utilized a suitable basis of interacting quasiparticles in which the model is integrable, and they employed sophisticated thermodynamic Bethe-ansatz (TBA) technology to calculate the non-Fermi distribution function and the density of states of the quasiparticles, which then determine the conductance by a Boltzmann-type rate expression. The Kondo frequency ϵ_0 directly corresponds to the temperature scale T'_B used in Ref. [483], $T'_B = \hbar\epsilon_0/k_B$. The simple derivation given here in Section 25.5 sheds additional light on the underlying symmetries of these models.

26.3 Full counting statistics of charge transfer through impurities

The full counting statistics of charge transport through an impurity is encapsulated in the moment generating function $\chi(\rho, t)$, which is the Fourier transform of the probability distribution $P(Q, t)$ of the charge Q crossing the impurity during time t, $\chi(\rho, t) = \sum_Q e^{i\rho Q} P(Q, t)$, where ρ is the counting field. The function $\chi(\rho, t)$ generates moments of the charge $Q_t = \int_0^t dt'\, I(t')$ transferred during time t,

$$\chi(\rho, t) = \sum_k \frac{(i\rho)^k}{k!} \langle Q_t^k \rangle = \exp\left[t \sum_k \frac{(i\rho)^k}{k!} \langle \delta^k Q \rangle \right] . \qquad (26.37)$$

Here, $t \langle \delta^k Q \rangle$ is the kth cumulant of the distribution, and $I(t)$ is the time-dependent current through the scattering region.

26.3.1 Charge transport at low temperature for arbitrary Luttinger parameter g

An important limiting case is the zero temperature regime in which the strong- and weak-backscattering expansions of all cumulants can be found in analytic form [482]. There are clear physical pictures in these different limits.

For strong backscattering or weak tunneling, the ϕ–model (26.19) applies, which is the equivalent of the TB model (23.3) in the Ohmic scaling limit. Here, the true ground state is that of two completely disconnected leads. Then, evidently, only quasiparticles with integer unit charge, i.e. electrons, may tunnel through the barrier between the leads. Since at $T = 0$ there are no transition from the energetically lower to the higher lead, we have in correspondence with the expression (25.122)

$$\ln \chi(\rho, t) = t \sum_{n=1}^{\infty} \left(e^{i\rho n e} - 1 \right) \frac{I_n^+}{n} , \qquad (26.38)$$

where $I_n^+ = n k_n^+$. The physical meaning of this expression is quite illuminating. Suppose that k_n^+ is the probability per unit time to transfer a particle of charge ne through the impurity barrier. Then the charge transferred in the time interval t is the result of a Poisson process for particles of charge e crossing the barrier, contributing a current I_1^+, plus a Poisson process for particles of charge $2e$ contributing a current I_2^+, etc. All these Poisson processes are represented by $\ln \chi(\rho, t)$.

Observing that the correspondence of the TB-model (23.3) with the ϕ–model (26.19) does hold not only for the conductance but for all cumulants, we immediately get from the expression (25.126) for the partial current I_n^+ at $T = 0$

$$I_n^+(\epsilon, 0) = \frac{(-1)^{n-1}}{\Gamma(n)} \frac{\Gamma(\frac{3}{2})\Gamma(n/g)}{\Gamma[\frac{3}{2} + (1/g - 1)n]} \frac{\epsilon}{2\pi} \left(\frac{\epsilon}{\epsilon_0} \right)^{(2/g-2)n} . \qquad (26.39)$$

This yields for the first cumulant or mean current the expression

$$\langle \delta Q \rangle = \langle I(V_a) \rangle = e \sum_{n=1}^{\infty} I_n(eV_a/\hbar, 0) . \qquad (26.40)$$

While in a classical Poisson process all the partial currents I_n^+ in Eq. (26.38) are positive, the subtle point now is that the partial currents (26.39) are not. Certainly, the first contribution I_1^+ is indeed a Poisson process for the tunneling of electrons; but the joint tunneling of pairs of electrons (and of multiples thereof), $m = 2, 4, \cdots$, comes with a negative sign, which is an effect of quantum interference.

Following the arguments given in Section 25.7, we find that the expression (26.38) also holds at low temperature, in which the universal leading low-temperature contribution is included in the partial current I_n^+,

$$I_n^+(\epsilon, T) = \left\{ 1 + \frac{\pi^2}{3g} \frac{2 - 2g}{g} n \left[1 + \frac{2 - 2g}{g} n \right] \left(\frac{k_B T}{\hbar \epsilon} \right)^2 \right\} I_n^+(\epsilon, 0) . \qquad (26.41)$$

On the other hand, in the opposite limit of strong-tunneling, a collective state between the edges is formed, which has charge ge for elementary excitations. The resulting expression for $\ln \chi(\rho, t)$ at low temperatures reads

$$\ln \chi(\rho, t) = t \left(i\rho g e \frac{\epsilon}{2\pi} - \frac{\rho^2}{2} \frac{g e^2 k_B T}{\pi \hbar} + \sum_{n=1}^{\infty} \left(e^{-i\rho n g e} - 1 \right) \frac{\tilde{I}_n^+}{n} \right) , \qquad (26.42)$$

where

$$\tilde{I}_n^+(\epsilon, T) = \left\{ 1 + \frac{\pi^2}{3g}(2g - 2)\, n[\, 1 + (2g - 2)\, n\,]\left(\frac{k_{\mathrm{B}}T}{\hbar\epsilon}\right)^2 \right\} \tilde{I}_n^+(\epsilon, 0) \qquad (26.43)$$

with

$$\tilde{I}_n^+(\epsilon, 0) = \frac{(-1)^{n-1}}{\Gamma(n)} \frac{\Gamma(\frac{3}{2})\Gamma(ng)}{\Gamma[\frac{3}{2} + (g-1)n]} \frac{ge}{2\pi} \left(\frac{\epsilon}{\epsilon_0}\right)^{(2g-2)n}. \qquad (26.44)$$

The form (26.42) is quite similar to Eq. (26.38), but there are subtle differences. The first two terms in the expression (26.42) represent the current and noise of fractionally charged quasiparticles in the absence of the barrier. The exponential factor $e^{-i\rho nge}$ indicates that now we have tunneling of quasiparticles of charge ge and of multiples thereof, and the minus sign in the exponent means that the tunneling diminishes the current instead of building it up as in the strong-backscattering limit. Since the sign of the partial current $\tilde{I}_n^+$ is the one of $\cos(ng\pi)$ for $g < \frac{1}{2}$, the perception of clusters of quasiparticles with fractional charge ge tunneling independently with a classical Poisson process is quite appropriate for bundles with modest n, when g is small. The mean current at zero temperature is

$$\langle \delta Q \rangle = \langle I \rangle = g\frac{V_{\mathrm{a}}}{R_{\mathrm{K}}} - ge \sum_{n=1}^{\infty} \tilde{I}_n^+(eV_{\mathrm{a}}/\hbar, 0). \qquad (26.45)$$

As g goes to zero, all the partial currents become positive, but the quantum fluctuations $\langle \delta^n Q \rangle$ with $n > 1$ disappear at $T = 0$, as one reaches the classical limit,

$$\ln \chi(\rho) = i\rho\tau\epsilon\frac{ge}{2\pi}\left[1 - \sum_{n=1}^{\infty} \frac{\Gamma(n - \frac{1}{2})}{2\sqrt{\pi}n!}\left(\frac{\epsilon_0}{\epsilon}\right)^2 \right] = i\rho\tau\epsilon\frac{ge}{2\pi}\sqrt{1 - \left(\frac{\epsilon_0}{\epsilon}\right)^2}, \qquad (26.46)$$

where $\epsilon_0 = 2\pi V_0/\hbar$. The first term is the current in the absence of the barrier, and the second is the backscattering contribution made up of partial backscattering currents.

The T^2-contribution to the cumulants vanishes at $g = 1$. Hence T^2-variation of the current and of its fluctuations is a distinctive signature of 1D interacting electrons.

Finally, we remark that cumulant relations analogous to the expressions (25.125) and (25.129) hold. The corresponding relation in the weak-tunneling representation is

$$\langle \delta^N Q \rangle = \left(e\frac{\Delta}{2}\frac{\partial}{\partial\Delta} \right)^{N-1} \langle \delta Q \rangle, \qquad (26.47)$$

whereas in the strong-tunneling representation

$$\langle \delta^N Q \rangle = \left(-ge\frac{V_0}{2}\frac{\partial}{\partial V_0} \right)^{N-1} \langle \delta Q \rangle. \qquad (26.48)$$

These relations help us to reduce every single cumulant to a calculation of derivatives of the current $\langle \delta Q \rangle$ with respect to the tunneling coupling Δ or to the corrugation strength V_0, respectively. They are in agreement with the findings from the integrable approach to the BSG model [482].

In the case of a fractional quantum Hall bar with $g = \nu = \frac{1}{3}, \frac{1}{5}, \cdots$, the collective excitations are Laughlin quasiparticles and the impurity corresponds to a point contact. In the weak-backscattering limit, Laughlin quasiparticles with fractional charge

νe are tunneling. In the strong-backscattering limit, the system consists of two different Hall devices which are weakly coupled by the interaction (26.18), and only electrons can tunnel. Strict duality means that the entire expansions around weak and strong backscattering are related. In the FQHE system, the crossover from weak to strong backscattering comes along with a crossover from Laughlin quasiparticle tunneling to electron tunneling.

For general g and general temperature, the analytical calculation of higher-order Coulomb integrals in the bosonic representation is not possible. Then one may resort to thermodynamic Bethe ansatz techniques in the related BSG model [483, 484]. In Ref. [485] a general scheme for the calculation of cumulants of any order is proposed.

26.3.2 *Full counting statistics at $g = \frac{1}{2}$ and general temperature*

In the weak-tunneling limit (strong impurity), the mean current at $T = 0$ is $\propto V_a^{2/g-1}$, as we see from the leading term of the series (26.40). On the other hand, the current I_B that is backscattered from a weak barrier (weak impurity) varies as V_a^{2g-1}. Hence the backscattering current in leading order becomes independent of the applied voltage at $g = \frac{1}{2}$. This is a sign that $g = \frac{1}{2}$ is a special point. By virtue of the mapping (26.29), the case $g = \frac{1}{2}$ corresponds to a coherent conductor (e.g. a quantum dot) with series resistance $R = R_K$ or $\alpha = 1$. An open point contact is a realization of such series resistance, as discussed in Ref. [493].

Employing the nonequilibrium Keldysh formalism and refermionization techniques, which map a Luttinger liquid at $g = \frac{1}{2}$ on noninteracting fermions, the cumulant generating function has been calculated in Ref. [493]. With a transformation set out in Appendix C of Ref. [494], the expression of the CGF given in Ref. [493] can be brought into the form

$$\ln \chi(\rho, t) = t \int_0^\infty \frac{d\omega}{2\pi} \ln \left\{ 1 + \overline{\mathcal{T}}(\omega) \left[\mathcal{N}_+(\omega, eV_a/2\hbar) \left(e^{ie\rho} - 1 \right) \right. \right.$$
$$\left. \left. + \mathcal{N}_-(\omega, eV_a/2\hbar) \left(e^{-ie\rho} - 1 \right) \right] \right\}, \tag{26.49}$$

with the spectral transition probability $\overline{\mathcal{T}}(\omega) = \omega^2/(\omega^2 + \tilde{\epsilon}_0^2)$ [cf. Eq.(25.54)]. The renormalized frequency $\tilde{\epsilon}_0$ is related to the bare impurity strength V_0 by [see the relation (25.113) at $g = \frac{1}{K} = \frac{1}{2}$ or the relation (25.51)]

$$\tilde{\epsilon}_0 = \pi \frac{(V_0/\hbar)^2}{\omega_c}. \tag{26.50}$$

The functions $\mathcal{N}_\pm(\omega, \epsilon)$ and relations of them are given in Eqs. (24.101) and (24.102).

The CGF (26.49) of the QI-TLL model (26.9) at $g = \frac{1}{2}$ also describes the FCS of a coherent conductor in series with an Ohmic resistor of resistance $R = R_K$. The expression (26.49) is in agreement with the CGF (25.56) of the quantum Brownian particle model (23.1) at $K = 2$, and with the CGF of Cooper pair tunneling in a resistive environment with resistance $R = 2R_Q$. Finally, the leading cumulants are easily found from the expressions (25.57) - (25.59).

Bibliography

[1] G. Baym, *Lectures on Quantum Mechanics* (Benjamin, Reading, 1969).

[2] D. Chandler, *Introduction to Modern Statistical Mechanics* (Oxford University Press, New York, 1987).

[3] R. P. Feynman and A. R. Hibbs, *Quantum Mechanics and Path Integrals* (Mc Graw-Hill, New York, 1965).

[4] R. P. Feynman, *Statistical Mechanics* (Benjamin, Reading, Mass., 1972).

[5] L. S. Schulman, *Techniques and Applications of Path Integration* (Wiley, 1981).

[6] H. Kleinert, *Path Integrals in Quantum Mechanics, Statistics, Polymer Physics, and Financial Markets*, (World Scientific, Singapore, 3rd edition, 2004).

[7] S. Chandrasekhar, Rev. Mod. Phys. **15**, 1 (1943).

[8] M. S. Green, J. Chem. Phys. **20**, 1281 (1952);
R. Kubo, Rep. Progr. Phys. (London) **29**, 255 (1966).

[9] P. Caldirola, Il Nuovo Cim. **18**, 393 (1941).

[10] E. Kanai, Progr. Theor. Phys. **3**, 440 (1948).

[11] W. H. Louisell, *Quantum Statistical Properties of Radiation* (Wiley, N.Y., 1973).

[12] H. Dekker, Phys. Rev. A **16**, 2116 (1977).

[13] M. D. Kostin, J. Chem. Phys. **57**, 3589 (1972).

[14] K. Yasue, Ann. Phys. (N.Y.) **114**, 479 (1978).

[15] E. Nelson, Phys. Rev. **150**, 1079 (1966).

[16] S. Nakajima, Progr. Theor. Phys. **20**, 948 (1958).

[17] R. Zwanzig, J. Chem. Phys. **33**, 1338 (1960);
R. Zwanzig, in: *Lectures in Theoretical Physics* (Boulder), Vol. 3, ed. by W. E. Brittin, B. W. Downs, and J. Down (Interscience, New York, 1961).

[18] J. Prigogine and P. Resibois, Physica **27**, 629 (1961).

[19] J. R. Senitzky, Phys. Rev. **119**, 670 (1960).

[20] G. W. Ford, M. Kac, and P. Mazur, J. Math. Phys. **6**, 504 (1965).

[21] H. Mori, Progr. Theor. Phys. **33**, 423 (1965).

[22] F. Haake, in *Quantum Statistics in Optics and Solid State Physics*, Springer Tracts in Modern Physics, Vol. 66, ed. by G. Höhler (Springer, Berlin, 1973).

[23] H. Haken, Rev. Mod. Phys. **47**, 67 (1975).

[24] H. Spohn, Rev. Mod. Phys. **52**, 569 (1980).

[25] H. Dekker, Phys. Rep. **80**, 1 (1981).

[26] K. Blum, *Density Matrix Theory and Applications* (Plenum Press, 1981).

[27] H. Grabert, *Projection Operator Techniques in Nonequilibrium Statistical Mechanics*, Springer Tracts in Modern Physics, Vol. 95, (Springer, Berlin, 1982).

[28] P. Talkner, Ann. Phys. (N.Y.) **167**, 390 (1986).

[29] R. Alicki and K. Lendi, in *Quantum Dynamical Semigroups and Applications*, Lecture Notes in Physics Vol. 286, ed. H. Araki *et al.*, (Springer, Berlin, 1987).

[30] C. W. Gardiner, *Quantum Noise* (Springer, Berlin, 1991).

[31] A.G. Redfield, IBM J. Res. Dev. **1**, 19 (1957); Adv. Magn. Reson. **1**, 1 (1965).

[32] G. C. Schatz and M. A. Ratner, *Quantum Mechanics in Chemistry* (Prentice Hall, Englewood Ciffs, New Jersey, 1993).

[33] M. Mehring, *Principles of High-Resolution NMR in Solids* (Springer, 1983).

[34] R. R. Ernst, G. Bodenhausen, and A. Wokaun, *Principles of Nuclear Magnetic Resonance in One and Two Dimensions* (Clarendon Press, Oxford, 1990).

[35] C. P. Slichter, *Principles of Magnetic Resonance* (Springer, Berlin, 1990).

[36] L. Allen and Z. H. Eberly, *Optical Resonance and Two-Level Atoms* (Wiley, New York, 1975).

[37] Y. R. Shen, *The Principles of Nonlinear Optics* (Wiley, New York, 1984).

[38] J. M. Jean, J. Chem. Phys **101**, 10464 (1994); A. K. Felts, W. T. Pollard, and R. A. Friesner, J. Phys. Chem. **99**, 2929 (1995); J. M. Jean and G. R. Fleming, J. Chem. Phys. **103**, 2092 (1995).

[39] O. Kühn, V. May, and M. Schreiber, J. Chem. Phys.**101**, 10404 (1994).

[40] V. Sidis, Adv. Chem. Phys. **82**, 73 (1992).

[41] T. Pacher, L. S. Cederbaum, and H. Köppel, Adv. Chem. Phys. **84**, 293 (1993).

[42] W. Domcke and G. Stock, Adv. Chem. Phys. **100**, 1 (1997).

[43] G. Lindblad, Commun. Math. Phys. **48**, 119 (1976).

[44] A. Isar *et al.*, Int. J. Mod. Phys. E **3**, 635 (1994).

[45] Sh. Gao, Phys. Rev. Lett. **79**, 3101 (1997).

[46] a.: Ph. Pechukas, Phys. Rev. Lett. **73**, 1060 (1994); b.: A. Suárez, R. Silbey, and I. Oppenheim, J. Chem. Phys. **97**, 5101 (1992); c.: W. J. Munro and C. W. Gardiner, Phys. Rev. A **53**, 2633 (1996).

[47] R. Karrlein and H. Grabert, Phys. Rev. E **55**, 153 (1997).

[48] E. Fick and G. Sauermann, *The Quantum Statistics of Dynamic Processes*, Springer Series in Solid-State Sciences, Vol. 88, (Springer, Berlin, 1990).

[49] C. W. Gardiner, IBM J. Res. Dev. **32**, 127 (1988).

[50] G. W. Ford and M. Kac, J. Stat. Phys. **46**, 803 (1987).

[51] G. W. Ford. J. T. Lewis, and R. F. O'Connell, Phys. Rev. A **37**, 4419 (1988).

[52] R. Benguria and M. Kac, Phys. Rev. Lett. **46**, 1 (1981).

[53] H. Kleinert and S. V. Shabanov, Phys. Lett. A **200**, 224 (1995).

[54] A. Schmid, J. Low Temp. Phys. **49**, 609 (1982).

[55] U. Eckern, W. Lehr, A. Menzel-Dorwarth, F. Pelzer, and A. Schmid, J. Stat. Phys. **59**, 885 (1990).

[56] R. H. Koch, D. J. van Harlingen, and J. Clarke, Phys. Rev. Lett. **45**, 2132 (1980); *ibid.* **47**, 1216 (1981).

[57] D. Giulini, E. Joos, C. Kiefer, J. Kupsch, I.-O. Stamatescu, and H. D. Zeh, *Decoherence and the Appearance of a Classical World in Quantum Theory* (Springer, Berlin, 1996).

[58] M. H. A. Davis, *Markov Models and Optimization* (Chapman, London, 1993).

[59] N. Gisin, Phys. Rev. Lett. **52**, 1657 (1984).

[60] N. Gisin and I. C. Percival, J. Phys. A **25**, 5677 (1992); I. C. Percival, Proc. R. Soc. London A **447**, 189 (1994).

[61] I. Percival, *Quantum State Diffusion*, (Cambridge University Press, Cambridge, U.K., 1998).

[62] H. P. Breuer and F. Petruccione, J. Phys. A: Math. Gen. **31**, 33 (1998).

[63] N. G. van Kampen, *Stochastic Processes in Physics and Chemistry* (North-Holland, Amsterdam, 1992).

[64] L. Diósi, J. Phys. Math. Gen. **21**, 2885 (1988).

[65] J. Dalibard, Y. Castin, and K. Mølmer, Phys Rev. Lett. **68**, 580 (1992).

[66] R. Dum, P. Zoller, and H. Ritsch, Phys. Rev. A **45**, 4879 (1992); C. W. Gardiner, A. S. Parkins, and P. Zoller, *ibid.* A **46**, 4363 (1992); R. Dum, A. S. Parkins, P. Zoller, and C. W. Gardiner, *ibid.* A **46**, 4382 (1992).

[67] H. J. Carmichael, S. Singh, R. Vyas, and P. R. Rice, Phys. Rev. A **39**, 1200 (1989); H. J. Carmichael, *An Open System Approach to Quantum Optics* (Springer, Berlin, 1993).

[68] M. Naraschewski and A. Schenzle, Z. Physik A **25**, 5677 (1992).

[69] K. Mølmer, Y. Castin, and J. Dalibard, J. Opt. Soc. Am. B **10**, 524 (1993).

[70] H. M. Wiseman and G. J. Milburn, Phys. Rev A **47**, 1652 (1992).

[71] H. P. Breuer and F. Petruccione, Phys. Rev. Lett. **74**, 3788 (1995); Phys. Rev. E **51**, 4041 (1995); *ibid.* E **52**, 428 (1995).

[72] H.-P. Breuer and F. Petruccione, *The Theory of Open Quantum Systems*, (Oxford University Press, Oxford, U.K., 2002).

[73] R. J. Cook, *Quantum Jumps* in Progress in Optics, Vol. XXVIII, ed. by E. Wolf (Elsevier, Amsterdam, 1990).

[74] W. Nagourney, J. Sandberg, and H. Dehmelt, Phys. Rev. Lett.**56**, 2797 (1986); Th. Sauter, W. Neuhauser, R. Blatt, and P. E. Toschek, Phys. Rev. Lett. **57**, 1699 (1986); Th. Basché, S. Kummer, and C. Bräuchle, Nature **373**, 132 (1995).

[75] P. Ullersma, Physica (Utrecht) **32**, 27, 56, 74, 90 (1966).

[76] R. Zwanzig, J. Stat. Phys. **9**, 215 (1973).

[77] A. O. Caldeira and A. J. Leggett, Phys. Rev. Lett. **46**, 211 (1981); Ann. Phys. (N.Y.) **149**, 374 (1983); *ibid.* **153**, 445(E) (1983).

[78] K. H. Stevens, J. Phys. C **16**, 5765 (1983).

[79] V. Ambegaokar and U. Eckern, Z. Physik B **69**, 399 (1987).

[80] V. B. Magalinskiĭ, Sov. Phys.–JETP **9**, 1381 (1959).

[81] R. J. Rubin, J. Math. Phys. **1**, 309 (1960); *ibid.* **2**, 373 (1961).

[82] W. Bez, Z. Physik B **39**, 319 (1980).

[83] P. Hänggi, in *Stochastic Dynamics*, Lecture Notes in Physics, Vol.484, ed. by L. Schimansky-Geier and Th. Pöschel (Springer, Berlin, 1997), p. 15.

[84] A. J. Leggett, Phys. Rev. B **30**, 1208 (1984).

[85] H. Grabert and U. Weiss, Z. Physik B **56**, 171 (1984).

[86] H. Grabert, P. Schramm, and G.-L. Ingold, Phys. Rep. **168**, 115 (1988); P. Schramm and H. Grabert, J. Stat. Phys. **49**, 767 (1987).

[87] I. S. Gradshteyn and I. M. Ryzhik, *Tables of Integrals, Series and Products* (Academic Press, London, 1965).

[88] M. Abramowitz and I. Stegun, *Handbook of Mathematical Functions* (Dover, New York, 1971).

[89] A. J. Leggett, S. Chakravarty, A. T. Dorsey, M. P. A. Fisher, A. Garg, and W. Zwerger, Rev. Mod. Phys. **59**, 1 (1987); *ibid.* **67**, 725 (1995) [erratum].

[90] R. J. Rubin, Phys. Rev. **131**, 964 (1963).

[91] H. Wipf, D. Steinbinder, K. Neumaier, P. Gutsmiedl, A. Magerl, and A. J. Dianoux, Europhys. Lett. **4**, 1379 (1989); D. Steinbinder, H. Wipf, A. Magerl, A. D. Dianoux, and K.Neumaier, *ibid.* **6**, 535 (1988); *ibid.* **16**, 211 (1991). For a review see H. Grabert and H. Wipf, in: Festkörperprobleme/Advances in Solid State Physics, Vol. 30, p. 1, ed. by U. Rössler (Vieweg, Braunschweig, 1990).

[92] G. M. Luke *et al.*, Phys. Rev. B **43**, 3284 (1991); O. Hartmann *et al.*, Hyperfine Interactions **64**, 641 (1990), and references therein; I. S. Anderson, Phys. Rev. Lett. **65**, 1439 (1990).

[93] *Hydrogen in Metals III*, Topics in Applied Physics, Vol. 73, ed. by H. Wipf (Springer, Berlin, 1997).

[94] A. Würger, *From Coherent Tunneling to Relaxation*, Springer Tracts in Modern Physics, Vol. 135 (Springer, Berlin, 1997).

[95] K. Chun and N. O. Birge, Phys. Rev. B **48**, 11 500 (1993); B **54**, 4629 (1996).

[96] B. Golding, N. M. Zimmermann, and S. N. Coppersmith, Phys. Rev. Lett. **68**, 998 (1992).

[97] R. Marcus, J. Chem. Phys. **24**, 966 (1956).

[98] R. A. Marcus and N. Sutin, Biochim. Biophys. Acta **811**, 265 (1985).

[99] S. Coleman, Phys. Rev. D **15**, 2929 (1977); S. Coleman, in *The Whys of Subnuclear Physics*, ed. by A. Zichichi (Plenum, New York, 1979) p. 805.

[100] U. Weiss and W. Häffner, Phys. Rev. D **27**, 2916 (1983).

[101] A. Barone and G. Paterno, *Physics and Application of the Josephson Effect* (Wiley, New York, 1982).

[102] J. R. Friedman *et al.*, Nature (London) **406**, 43 (2000).

[103] C. H. van der Wal *et. al.*, Science **290**, 773 (2000).

[104] J. E. Mooij *et al.*, Science **285**, 1036 (1999).

[105] M. V. Feigelman *et al.*, J. Low Temp. Phys. **118**, 805 (2000).

[106] Yu. Makhlin, G. Schön, and A. Shnirman, Rev. Mod. Phys. **73**, 357 (2001).

[107] E. Paladino, L. Faoro, G. Falci, and R. Fazio, Phys. Rev. Lett. **88**, 228304 (2002).

[108] A. Shnirman, G. Schön, I. Martin, and Yu. Makhlin, Phys. Rev. Lett. **94**, 127002 (2005).

[109] Yu. Makhlin, and A. Shnirman, Phys. Rev. Lett. **92**, 178301 (2004).

[110] P. W. Anderson, B. I. Halperin, and C. M. Varma, Phil. Mag. **25**, 1 (1972).

[111] W. A. Phillips, J. Low Temp. Phys. **7**, 351 (1972).

[112] J. L. Black, in *Glassy Metals I*, Topics in Applied Physics, Vol. 46, ed. by H.-J. Güntherodt and H. Beck (Springer, Berlin, 1981).

[113] Y. Nakamura, Yu. A. Pashkin, and J. S. Tsai, Nature **398**, 786 (1999).

[114] D. Vion *et al.*, Science **296**, 886 (2002).

[115] J. M. Martinis *et al.*, Phys. Rev. Lett. **89**, 117901 (2002).

[116] I. Chiorescu, Y. Nakamura, C. Harmans, and J. E. Mooij, Science **299**, 1869 (2003).

[117] A. Wallraff *et al.*, Nature **431**, 162 (2004).

[118] A. O. Niskanen, K. Harrabi, F. Yoshihara, Y. Nakamura, S. Lloyd, and J. S. Tsai, Science **316**, 723 (2007).

[119] J. Yamashita and T. Kurosawa, J. Chem. Solids **5**, 34 (1958).

[120] a.: T. Holstein, Ann. Phys. (N.Y.) **8**, 325 (1959); b.: *ibid.* **8**, 343 (1959).

[121] H. B. Shore and L. M. Sander, Phys. Rev. B **12**, 1546 (1975).

[122] M. Wagner, J. Phys. A **18**, 1915 (1986).

[123] C. P. Flynn and A. M. Stoneham, Phys. Rev. B **1**, 3966 (1970).

[124] Y. Kagan and M. I. Klinger, J. Phys. C **7**, 2791 (1974).

[125] H. Teichler and A. Seeger, Phys. Lett. **82** A, 91 (1981).

[126] H. Fröhlich, Adv. Phys. **3**, 325 (1954).

[127] J. Kondo, in *Fermi Surface Effects*, Vol. 77 of Springer Series in Solid State Sciences, eds. J. Kondo and A. Yoshimori (Springer, Berlin, 1988).

[128] T. Regelmann, L. Schimmele, and A. Seeger, Z. Physik B **95**, 441 (1994).

[129] G. D. Mahan, Many-Particle Physics (Plenum Press, New York, 1981).

[130] G. D. Mahan, in *Fermi Surface Effects*, Vol. 77 of Springer Series in Solid State Sciences, eds. J. Kondo and A. Yoshimori (Springer, Berlin, 1988).

[131] K. Ohtaka and Y. Tanabe, Rev. Mod. Phys. **62**, 929 (1990).

[132] A. M. Tsvelik and P. B. Wiegmann, Adv. Phys. **32**, 453 (1983).

[133] S. Chakravarty, Phys. Rev. Lett. **49**, 681 (1982).

[134] A. J. Bray and M. A. Moore, Phys. Rev. Lett. **49**, 1546 (1982).

[135] V. Hakim, A. Muramatsu, and F. Guinea, Phys. Rev. B **30**, 464 (1984).

[136] A. Schmid, Phys. Rev. Lett. **51**, 1506 (1983).

[137] S. Bulgadaev, Sov. Phys.–JETP Lett. **39**, 314 (1985).

[138] F. Guinea, V. Hakim, and A. Muramatsu, Phys. Rev. Lett. **54**, 263 (1985).

[139] P. G. de Gennes, *Superconductivity of Metals and Alloys* (Wesley, N.Y, 1989).

[140] D. J. Scalapino, in *Superconductivity*, Vol. 1, ed. by R. D. Parks (Marcel Dekker, New York, 1969).

[141] a.: V. Ambegaokar, U. Eckern, and G. Schön, Phys. Rev. Lett. **48**, 1745 (1982); b.: U. Eckern, G. Schön, and V. Ambegaokar, Phys. Rev. B **30**, 6419 (1984).

[142] A. I. Larkin and Yu. N. Ovchinnikov, Phys. Rev. B **28**, 6281 (1983).

[143] G. Schön and A. D. Zaikin, Phys. Rep. **198**, 237 (1990).

[144] Special Issue on *Single Charge Tunneling*, Z. Physik B **85** (3), 317-468 (1991).

[145] *Single Charge Tunneling*, ed. by H. Grabert and M. H. Devoret, NATO ASI Series B: Physics Vol. 294 (Plenum Press, New York, 1992).

[146] G.-L. Ingold and Yu. V. Nazarov, in Ref. [145], p. 21-107.

[147] D. V. Averin and K. K. Likharev, J. Low Temp. Phys. **62**, 345 (1986).

[148] *Quantum Tunneling of Magnetization - QTM 94*, ed. by L. Gunther and B. Barbara (Kluwer, Dordrecht, 1995).

[149] P. C. E. Stamp in Ref. [150].

[150] *Tunneling in Complex Systems*, Proceedings from the Institute for Nuclear Theory - Vol. 5, ed. by S. Tomsovic (World Scientific, Singapore,1998).

[151] J.L. van Hemmen and A. Suto, Europhys. Lett. **1**, 481 (1986); and in Ref. [148].

[152] M. Enz and R. Schilling, J. Phys. C **19**, L711 and 1765 (1986); and in Ref. [148].

[153] D. Gatteschi, R. Sessoli, and J. Villain, *Molecular Nanomagnets* (Oxford University Press, Oxford, 2006).

[154] I. S. Tupitsyn, N. V. Prokof'ev and P. C. E. Stamp, Int. J. Mod. Phys. B **11**, 2901 (1997).

[155] A. O. Caldeira, A. H. Castro Neto, and T. O. de Carvalho, Phys. Rev. B **48**, 13 974 (1993).

[156] G. Mahler and V. A. Weberruß, *Quantum Networks* (Springer, Berlin, 1998).

[157] J. Allinger and U. Weiss, Z. Physik B **98**, 289 (1995).

[158] D. Cohen, Phys. Rev. E **55**, 1422 (1997); Phys. Rev. Lett. **78**, 2878 (1997).

[159] S. Doniach and E. H. Sondheimer, *Green's Functions for Solid State Physicists* (Benjamin, Reading, Ma., 1974).

[160] J. P. Sethna, Phys. Rev. B **24**, 698 (1981); *ibid.* B **25**, 5050 (1982).

[161] A. A. Louis and J. P. Sethna, Phys. Rev. Lett. **74**, 1363 (1995).

[162] H. Sugimoto and Y. Fukai, Phys. Rev. B **22**, 670 (1980); A. Klamt and H. Teichler, Phys. Stat. Sol. (B) **134**, 103 (1986).

[163] A. Sumi and Y. Toyozawa, J. Phys. Soc. Jpn. **35**, 137 (1973).

[164] F. M. Peeters and J. T. Devreese, Phys. Rev. B **32**, 3515 (1985).

[165] B. Gerlach and H. Löwen, Rev. Mod. Phys. **63**, 63 (1991).

[166] Yu. Kagan, J. Low Temp. Phys. **87**, 525 (1992).

[167] *Quantum Tunnelling in Condensed Media*, ed. by Yu. Kagan and A. J. Leggett, (Elsevier Publishers, Amsterdam, 1992).

[168] Yu. Kagan and N. V. Prokov'ev, in Ref. [167].

[169] F. Napoli, M. Sassetti, and U. Weiss, Physica B **202**, 80 (1994).

[170] J. A. Stroscio and D. M. Eigler, Science **254**, 1319 (1991).

[171] D. M. Eigler, C. P. Lutz, and W. E. Rudge, Nature **352**, 600 (1991).

[172] M.F. Crommie *et al.*, Nature (London) **363** 524 (1993); Science **262**, 218 (1993).

[173] J. E. Artacho and L. M. Falicov, Phys. Rev. B **47**, 1190 (1993).

[174] M. Sassetti, E. Galleani d'Agliano, and F. Napoli, Physica B **154**, 359 (1989).

[175] E. G. d'Agliano, P. Kumar, W. Schaich, H. Suhl, Phys. Rev. B **11**, 2122 (1975).

[176] L.-D. Chang and S. Chakravarty, Phys. Rev. B **31**, 154 (1985).

[177] P. Nozières and C. De Dominicis, Phys. Rev. **178**, 1097 (1969).

[178] F. Guinea, V. Hakim, and A. Muramatsu, Phys. Rev. B **32**, 4410 (1985).

[179] F. Sols and F. Guinea, Phys. Rev. B **36**, 7775 (1987).

[180] K. Schönhammer, Phys. Rev. B **43**, 11323 (1991).

[181] P. W. Anderson, Phys. Rev. Lett. **18**, 1049 (1967).

[182] K. Yamada and K. Yosida, Progr. Theor. Phys. **68**, 1504 (1982); K. Yamada, A. Sakurai, S. Miyazima, and H. S. Wang, Progr. Theor. Phys. **75**, 1030 (1986).

[183] A. Oguchi and K. Yosida, Progr. Theor. Phys. **75**, 1048 (1986); T. Kitamura, A. Oguchi and K. Yosida, Progr. Theor. Phys. **78**, 583 (1987).

[184] N. R. Wertheimer, Phys. Rev. **147**, 255 (1966).

[185] G. Schön and A. D. Zaikin, Phys. Rev. B **40**, 5231 (1989).

[186] F. Guinea and G. Schön, Physica B **152**, 165 (1988).

[187] F. W. J. Hekking, L. I. Glazman, K. A. Matveev, and R. I. Shekhter, Phys. Rev. Lett. **70**, 4138 (1993).

[188] F. W. J. Hekking and Yu V. Nazarov, Phys. Rev. Lett. **71**, 1625 (1993).

[189] E. Pollak, Chem. Phys. Lett. **127**, 178 (1986).

[190] W. P. Schleich, *Quantum Optics in Phase Space* (Wiley-VCH, 2001).

[191] H. Weyl, Z. Physik **46**, 1 (1927).

[192] M. Hillery, R. F. O'Connell, M. O. Scully, and E. P. Wigner, Phys. Rep. **106**, 122 (1984).

[193] E. Wigner, Phys. Rev. **40**, 749 (1932).

[194] U. Weiss, Z. Physik B **30**, 429 (1978).

[195] R. P. Feynman and F. L. Vernon, Ann. Phys. (N.Y.) **24**, 118 (1963).

[196] A. O. Caldeira and A. J. Leggett, Physica **121 A**, 587 (1983).

[197] A. Stern, Y. Aharonov, and Y. Imry, Phys. Rev. A **41**, 3436 (1990).

[198] D. Loss and K. Mullen, Phys. Rev. B **43**, 13252 (1991).

[199] B. d'Espagnat, *Conceptual Foundations of Quantum Mechanics* (Benjamin, Reading, Ma., 1976).

[200] P. Grigolini, *Quantum Mechanical Irreversibility and Measurement* (World Scientific, 1993).

[201] S. Dattagupta, *Relaxation Phenomena in Condensed Matter Physics* (Academic Press, New York, 1987).

[202] L. P. Kadanoff and G. Baym, *Quantum Statistical Mechanics* (Benjamin, 1962).

[203] L. V. Keldysh, Sov. Phys.–JETP **20**, 1018 (1965).

[204] K. C. Chou, Z. B. Su, B. L. Hao, and L. Yu, Phys. Rep. **118**, 1 (1985).

[205] J. Rammer and H. Smith, Rev. Mod. Phys. **58**, 323 (1986).

[206] Y.-C. Chen, J. L. Lebowitz, and C. Liverani, Phys. Rev. B **40**, 4664 (1989).

[207] M. Sassetti and U. Weiss, Phys. Rev. A **41**, 5383 (1990).

[208] M. Sassetti and U. Weiss, Phys. Rev. Lett. **65**, 2262 (1990).

[209] H. Svensmark and K. Flensberg, Phys. Rev. A **47**, R23 (1993).

[210] K. S. Chow, D. A. Browne, and V. Ambegaokar, Phys. Rev. B **37**, 1624 (1988); K. S. Chow and V. Ambegaokar, Phys. Rev. B **38**, 11 168 (1988).

[211] D. S. Golubev, J. König, H. Schoeller, G. Schön, and A. D. Zaikin, Phys. Rev. B **56**, 15 782 (1997).

[212] W. T. Strunz, L. Diósi, and N. Gisin, Phys. Rev. Lett. **82**, 1801 (1999).

[213] J. T. Stockburger and C. H. Mak, Phys. Rev. Lett. **80**, 2657 (1998); J. Chem. Phys. **110**, 4983 (1999).

[214] J.T. Stockburger and H. Grabert, Phys. Rev. Lett. **88**, 170407 (2002).

[215] H. Grabert, U. Weiss, and P. Talkner, Z. Physik B **55**, 87 (1984).

[216] H. Metiu and G. Schön, Phys. Rev. Lett. **53**, 13 (1984).

[217] C. Aslangul, N. Pottier, and D. Saint-James, J. Stat. Phys. **40**, 167 (1985).

[218] P. S. Riseborough, P. Hänggi, and U. Weiss, Phys. Rev. A **31**, 471 (1985).

[219] R. Jung, G.-L. Ingold, and H. Grabert, Phys. Rev. A **32**, 2510 (1985).

[220] K. Lindenberg and B. J. West, Phys. Rev. A **30**, 568 (1984).

[221] F. Haake and R. Reibold, Phys. Rev. A **32**, 2462 (1985).

[222] A. Einstein, Ann. Phys. (Leipzig) **17**, 549 (1905).

[223] J. B. Johnson, Phys. Rev. **32**, 97 (1928).

[224] H. Nyquist, Phys. Rev. **32**, 110 (1928).

[225] H. B. Callen and T. A. Welton, Phys. Rev. **83**, 34 (1951).

[226] P. Talkner, Z. Physik B **41**, 365 (1981).

[227] A. Hanke and W. Zwerger, Phys. Rev. E **52**, 6875 (1995).

[228] V. Hakim and V. Ambegaokar, Phys. Rev. A **32**, 423 (1985).

[229] A. Erdélyi, *Higher Transcendental Functions*, Vol. 3 (McGraw-Hill, N.Y., 1955).

[230] R. P. Feynman, Phys. Rev. **97**, 660 (1955).

[231] R. Giachetti and V. Tognetti, Phys. Rev. Lett. **55**, 912 (1985); Phys. Rev. B **33**, 7647 (1986).

[232] R. P. Feynman and H. Kleinert, Phys. Rev. A **34**, 5080 (1986).

[233] H. Leschke, in *Path Summations: Achievements and Goals*, ed. by S. Lundquist *et al.* (World Scientific, Singapore, 1987).

[234] W. Janke, in *Path Integrals from meV to MeV*, ed. by V. Sa-yakanit *et al.* (World Scientific, Singapore, 1989).

[235] R. Giachetti, V. Tognetti, and R. Vaia, in *Path Summations: Achievements and Goals*, ed. by S. Lundquist *et al.* (World Scientific, Singapore, 1987).

[236] A. Cuccoli, V. Tognetti, and R. Vaia, in *Quantum Fluctuations in Mesoscopic and Macroscopic Systems*, ed. by H. A. Cerdeira, F. Guinea Lopez, and U. Weiss, (World Scientific, Singapore, 1991).

[237] R. Giachetti and V. Tognetti, Phys. Rev. A **36**, 5512 (1987); R. Giachetti, V. Tognetti, R. Vaia, Phys. Rev. A **37**, 2165 (1988); A **38**, 1521, 1638 (1988).

[238] G. Falci, R. Fazio, and G. Giaquinta, Europhys. Lett. **14**, 145 (1991); S. Kim and M. Y. Choi, Phys. Rev. B **42**, 80 (1990).

[239] A. Cuccoli *et al.*, Phys. Rev. A **45**, 8418 (1992); A. Cuccoli, R. Giachetti, V. Tognetti, R. Vaia, and P. Verrucchi, J. Phys.: Condens. Matter **7**, 7891 (1995).

[240] H. Kleinert, Phys. Lett. A **174**, 332 (1992).

[241] H. Kleinert, W. Kürzinger, and A. Pelster, J. Phys. A **31**, 8307 (1998).

[242] A. Cuccoli, A. Rossi, V. Tognetti, and R. Vaia, Phys. Rev. E **55**, 4849 (1997).

[243] D. M. Larsen, Phys. Rev. B **32**, 2657 (1985), *ibid.* B **33**, 799 (1986); S. N. Gorshkov, A. V. Zabrodin, C. Rodriguez, V. K. Fedyanin, Theor. Math. Phys. **62**, 205 (1985); K. M. Broderix, N. Heldt, H. Leschke, Z. Physik B **66**, 507 (1987).

[244] J. T. Devreese and F. Brosens, Phys. Rev. B **45**, 6459 (1992).

[245] W. H. Zurek, Phys. Today **44** (10), 36 (1991); B. L. Hu, J. P. Paz, and Y. Zhang, Phys. Rev. D **45**, 2843 (1992).

[246] S. Chakravarty and A. Schmid, Phys. Rep. **140**, 193 (1986).

[247] S. Washburn and R. A. Webb, Adv. Phys. **35**, 375 (1986).

[248] P. Mohanty, E. M. Q. Jariwala, R. A. Webb, Phys. Rev. Lett. **78**, 3366 (1997).

[249] P. Mohanty and R. A. Webb, Phys. Rev. B **55**, R13 452 (1997).

[250] I. L. Aleiner, B. L. Altshuler, and M. E. Gershenson, Waves in Random Media **9**, 201 (1999); I. L. Aleiner, B. L. Altshuler, and M. G. Vavilov, J. Low Temp. Phys. **126**, 1377 (2002).

[251] D.S. Golubev and A. D. Zaikin, Phys. Rev. B **59**, 9195 (1999); Phys. Rev. B **62**, 14061 (2000).

[252] F. Guinea, Phys. Rev. B **65**, 205317 (2002).[-7mm]

[253] D. S. Golubev, A. D. Zaikin, and G. Schön, J. Low Temp. Phys. **126**, 1355 (2002).

[254] B. Altshuler, A. G. Aronov, and D. E. Khmelnitskii, J. Phys. C **15**, 7367 (1982).

[255] D. Cohen, J. Phys. A **31**, 8199 (1998).

[256] D. Golubev and A. D. Zaikin, Phys. Rev. Lett. **81**, 1074 (1998).

[257] R. Schuster *et al.*, Nature (London) **385**, 417 (1997).

[258] I. L. Aleiner, N. S. Wingreen, and Y. Meir, Phys. Rev. Lett. **79**, 3740 (1997).

[259] S. Arrhenius, Z. Phys. Chem. (Leipzig) **4**, 226 (1889).

[260] H. A. Kramers, Physica (Utrecht) **7**, 284 (1940).

[261] A. J. Leggett, Contemp. Phys. **25**, 583 (1984).

[262] A. J. Leggett, in *Directions in Condensed Matter Physics*, Vol. 1, ed. by G. Grinstein and G. Mazenko (World Scientific, Singapore, 1986), p. 187.

[263] H. Grabert, P. Olschowski, and U. Weiss, Phys. Rev. B **36**, 1931 (1987).

[264] P. Hänggi, P. Talkner, and M. Borkovec, Rev. Mod. Phys. **62**, 251 (1990).

[265] J. Ankerhold, *Quantum Tunneling in Complex Systems*, Springer Tracts in Modern Physics, Vol. 224 (Springer Verlag, Berlin, 2007).

[266] A. N. Cleland, J. M. Martinis, and J. Clarke, Phys. Rev. B **37**, 5950 (1988).

[267] D. M. Brink, J. M. Neto, and H. A. Weidenmüller, Phys. Lett. B **80**, 170 (1979).

[268] K. Möhring and U. Smilansky, Nucl. Phys. A **338**, 227 (1980).

[269] D. Emin and T. Holstein, Ann. Phys. (N.Y.) **53**, 439 (1969).

[270] H. Risken, *The Fokker-Planck Equation* (Springer Verlag, Berlin, 1984).

[271] F. Hund, Z. Physik **43**, 805 (1927).

[272] J. R. Oppenheimer, Phys. Rev. **31**, 80 (1928).

[273] G. Gamow, Z. Physik **51**, 204 (1928).

[274] R. W. Gurney and E. U. Condon, Nature (London) **122**, 439 (1928).

[275] E. P. Wigner, Z. Phys. Chem. B **19**, 203 (1932).

[276] W. H. Miller, J. Chem. Phys. **62**, 1899 (1975).

[277] W. H. Miller, S. D. Schwartz, J. W. Tromp, J. Chem. Phys. **79**, 4889 (1983).

[278] W. H. Miller, J. Chem. Phys. **61**, 1823 (1974).

[279] T. Yamamoto, J. Chem. Phys. **33**, 281 (1960).

[280] E. Pollak and J.-L. Liao, J. Chem. Phys. **108**, 2733 (1998); G. Gershinsky and E. Pollak, J. Chem. Phys. **108**, 2756 (1998).

[281] F. Matzkies and U. Manthe, J. Chem. Phys. **106**, 2646 (1997).

[282] W. H. Thompson and W. H. Miller, J. Chem. Phys. **102**, 7409 (1995); *ibid.* **106**, 142 (1997).

[283] F. J. Mc Lafferty and Ph. Pechukas, Chem. Phys. Lett. **27**, 511 (1974).

[284] M. C. Gutzwiller, J. Math. Phys. **8**, 1979 (1967).

[285] M. C. Gutzwiller, *Chaos in Classical and Quantum Mechanics*, Interdisciplinary Applied Mathematics, Vol. 1 (Springer, Berlin, 1990).

[286] F. Haake, *Quantum Signatures of Chaos* (Springer, Berlin, 2nd edition, 2000).

[287] I. Affleck, Phys. Rev. Lett. **46**, 388 (1981).

[288] J. S. Langer, Ann. Phys. (N.Y.) **41**, 108 (1967); *ibid.* **54**, 258 (1969).

[289] J. S. Langer, in *Systems far from Equilibrium*, Lecture Notes in Physics, Vol. 132, ed. by L. Garrido (Springer, Berlin, 1980), p. 12.

[290] T. Nakamura, A. Ottewill, and S. Takagi, Ann. Phys. (N.Y.) **260**, 9 (1997).

[291] R. F. Dashen, B. Hasslacher, and A. Neveu, Phys. Rev. D **10**, 4114 (1974).

[292] R. P. Bell, *The Tunnel Effect in Chemistry* (Chapman and Hall, London, 1980).

[293] V. I. Goldanskii, Dokl. Acad. Nauk SSSR **124**, 1261 (1959); **127**, 1037 (1959).

[294] P. Reimann, M. Grifoni, and P. Hänggi, Phys. Rev. Lett. **79**, 10 (1997).

[295] S. Keshavamurthy and W. H. Miller, Chem. Phys. Lett. **218**, 189 (1994).

[296] N. T. Maitra and E. J. Heller, Phys. Rev. Lett. **78**, 3035 (1997).

[297] M. J. Gillan, J. Phys. C **20**, 3621 (1987); see also P. G. Wolynes, J. Chem. Phys. **87**, 6559 (1987).

[298] G. A. Voth, D. Chandler, and W. H. Miller, J. Chem. Phys. **91**, 7749 (1989).

[299] J. Cao and G. A. Voth, J. Chem. Phys. **105**, 6856 (1996).

[300] D. Makarov and M. Topaler, Phys. Rev. E **52**, 178 (1995).

[301] M. C. Gutzwiller, J. Math. Phys. **12**, 343 (1971); M. C. Gutzwiller, Physica D (Utrecht) **5**, 183 (1982).

[302] T. Banks, C. M. Bender, and T. T. Wu, Phys. Rev. D **8**, 3346 (1973).

[303] R. F. Grote and J. T. Hynes, J. Chem. Phys. **73**, 2715 (1980); P. Hänggi and F. Mojtabai, Phys. Rev. A **29**, 1168 (1982).

[304] P. G. Wolynes, Phys. Rev. Lett. **47**, 968 (1981); see also: V. I. Mel'nikov and S. V. Meshkov, Sov. Phys.–JETP Lett. **60**, 38 (1983).

[305] H. Grabert, U. Weiss, and P. Hänggi, Phys. Rev. Lett. **52**, 2193 (1984).

[306] H. Grabert and U. Weiss, Phys. Rev. Lett. **53**, 1787 (1984).

[307] A. I. Larkin and Yu. N. Ovchinnikov, Sov. Phys.–JETP Lett. **37**, 382 (1983).

[308] A. I. Larkin and Yu. N. Ovchinnikov, Sov. Phys.–JETP **59**, 420 (1984).

[309] A. M. Levine, M. Shapiro, and E. Pollak, J. Chem. Phys. **88**, 1959 (1988).

[310] H. Grabert, Phys. Rev. Lett. **61**, 1683 (1988).

[311] E. Pollak, H. Grabert, and P. Hänggi, J. Chem. Phys. **91**, 4073 (1989).

[312] V. I. Mel'nikov and S. V. Meshkov, J. Chem. Phys. **85**, 1018 (1986).

[313] E. Hershkovitz and E. Pollak, J. Chem. Phys. **106**, 7678 (1997).

[314] P. Hänggi, H. Grabert, G.-L. Ingold, and U. Weiss, Phys. Rev. Lett. **55**, 761 (1985). For an experimental confirmation of Eq. (15.16) see J. B. Bouchaud, E. Cohen de Lara, and R. Kahn, Europhys. Lett. **17**, 583 (1992).

[315] J. Ankerhold, Ph. Pechukas, and H. Grabert, Phys. Rev. Lett. **87**, 086802 (2001).

[316] W. T. Coffey, Yu. P. Kalmykov, S. V. Titov, and B. P. Mulligan, J. Phys. A: Math. Theor. **40**, F91 (2007); *ibid.* **40**, 12505 (2007).

[317] L. Machura *et al.*, Phys. Rev. E **70**, 031107 (2004); J. Luczka, R. Rudnicki, and P. Hänggi, Physica A 351, 60 (2005).

[318] D. Waxman and A. J. Leggett, Phys. Rev. B **32**, 4450 (1985).

[319] P. Hänggi and W. Hontscha, J. Chem. Phys. **88**, 4094 (1988); Ber. Bunsenges. Phys. Chem. **95**, 379 (1991).

[320] C. G. Callan and S. Coleman, Phys. Rev. D **16**, 1762 (1977).

[321] M. Stone, Phys. Lett. **67B**, 186 (1977).

[322] A. Schmid, Ann. Phys. (N.Y.) **170**, 333 (1986).

[323] U. Eckern and A. Schmid, in Ref. [167].

[324] H. Grabert, P. Olschowski, and U. Weiss, Z. Physik B **68**, 193 (1987).

[325] E. Freidkin, P. S. Riseborough, and P. Hänggi, Z. Physik B **64**, 237 (1986).

[326] H. Grabert and U. Weiss, Z. Physik B **56**, 171 (1984).

[327] J. M. Martinis and H. Grabert, Phys. Rev. B **38**, 2371 (1988).

[328] U. Weiss, M. Sassetti, Th. Negele, M. Wollensak, Z. Physik B **84**, 471 (1991).

[329] S. Washburn, R. A. Webb, R. F. Voss, and S. M. Faris, Phys. Rev. Lett. **54**, 2712 (1985); D. B. Schwartz, B. Sen, C. N. Archie, and J. E. Lukens, **55**, 1547 (1985); A. N. Cleland, J. M. Martinis, J. Clarke, Phys. Rev. B **36**, 58 (1987).

[330] S. E. Korshunov, Sov. Phys.–JETP **65**, 1025 (1987).

[331] A. Erdélyi, *Higher Transcendental Functions*, Vol. 1 (McGraw-Hill, N.Y., 1955).

[332] L.-D. Chang and S. Chakravarty, Phys. Rev. B **29**, 130 (1984); *ibid.* B **30**, 1566(E) (1984).

[333] M. H. Devoret, D. Esteve, C. Urbina, J. Martinis, A. N. Cleland, and J. Clarke, in Ref. [167].

[334] S. Tagaki, *Macroscopic Quantum Tunneling*, (Cambridge, 2002).

[335] G. Careri and G. Consolini, Ber. Bunsenges. Phys. Chem. **95**, 376 (1991).

[336] W. Kleemann, V. Schönknecht, D. Sommer, Phys. Rev. Lett. **66**, 762 (1991).

[337] F. Bruni, G. Consolini, and G. Careri, J. Chem. Phys. **99**, 538 (1993).

[338] B. Golding, J. E. Graebner, A. B. Kane, and J. L. Black, Phys. Rev. Lett. **41**, 1487 (1978).

[339] J. L. Black and P. Fulde, Phys. Rev. Lett. **43**, 453 (1979).

[340] G. Weiss, W. Arnold, K. Dransfeld, and H.-J. Güntherodt, Solid State Comm. **33**, 111 (1980).

[341] J. Stockburger, U. Weiss, and R. Görlich, Z. Physik B **84**, 457 (1991).

[342] J. Stockburger, M. Grifoni, M. Sassetti, U. Weiss, Z. Physik B **94**, 447 (1994).

[343] P. Esquinazi, R. König, and F. Pobell, Z. Physik B **87**, 305 (1992).

[344] *Tunneling Systems in Amorphous and Crystalline Solids*, ed. by P. Esquinazi (Springer Verlag, Berlin, 1998).

[345] G. Cannelli, R. Cantelli, F. Cordero, and F. Trequattrini, in Ref. [344].

[346] H. Grabert and H. R. Schober, in Ref. [93].

[347] J. Kondo, Physica **125** B, 279 (1984); *ibid.* **126** B, 377 (1984).

[348] C. D. Tesche, Ann. N. Y. Acad. Sci. **480**, 36 (1986); S. Chakravarty, *ibid.* **480**, 25 (1986).

[349] H. Spohn and R. Dümcke, J. Stat. Phys. **41**, 389 (1985).

[350] U. Weiss, H. Grabert, P. Hänggi, P. Riseborough, Phys. Rev. B **35**, 9535 (1987).

[351] S. Chakravarty and S. Kivelson, Phys. Rev. B **32**, 76 (1985).

[352] A. T. Dorsey, M. P. A. Fisher, and M. Wartak, Phys. Rev. A **33**, 1117 (1986).

[353] R. Silbey and R. A. Harris, J. Chem. Phys. **80**, 2615 (1983); J. Phys. Chem. **93**, 7062 (1989).

[354] F. Wegner, Ann. Physik (Leipzig) **3**, 77 (1994).

[355] S. Kehrein, *The Flow Equation Approach to Many-Particle Systems*, Springer Tracts in Modern Physics, Vol. 217 (Springer Verlag, Berlin, 2006).

[356] S. K. Kehrein, A. Mielke, and P. Neu, Z. Physik B **99**, 269 (1996).

[357] S. K. Kehrein and A. Mielke, Ann. Physik (Leipzig) **6**, 90 (1997); J. Stat. Phys. **90**, 889 (1997).

[358] S. K. Kehrein and A. Mielke, Phys. Lett. A **219**, 313 (1996).

[359] R. Görlich and U. Weiss, Phys. Rev. B **38**, 5254 (1988).

[360] R. Görlich and U. Weiss, Il Nuovo Cim. **11** D, 123 (1989).

[361] P. B. Vigmann and A. M. Finkel'steĭn, Sov. Phys.–JETP **48**, 102 (1978).

[362] P. Nozières, J. Low Temp. Phys. **17**, 31 (1974).

[363] P. Schlottmann, J. Magn. Mater. **7**, 72 (1978); Phys. Rev. B **25**, 4805 (1982).

[364] P. W. Anderson, Phys. Rev. Lett. **18**, 1049 (1967).

[365] K. D. Schotte and U. Schotte, Phys. Rev. **182**, (1969); K. Schönhammer, Z. Phys. B **45**, 23 (1981).

[366] V. J. Emery, in *Highly Conducting One-Dimensional Solids*, ed. by J. T. Devreese *et al.* (Plenum, New York, 1979); F. D. M. Haldane, Phys. Rev. Lett. **47**, 1840 (1981).

[367] D. L. Cox and A. Zawadowski, Adv. Phys. **47**, 599 (1998).

[368] G. Yuval and P. W. Anderson, Phys. Rev. B **1**, 1522 (1970).

[369] P. W. Anderson and G. Yuval, J. Phys. C **4**, 607 (1971).

[370] J. Cardy, Journ. of Physics A **14**, 1407 (1981).

[371] S. Chakravarty and J. Rudnick, Phys. Rev. Lett. **75**, 501 (1995).

[372] K. Völker, Phys. Rev. B **58**, 1862 (1998).

[373] H. Spohn, Comm. Math. Phys. **123**, 277 (1989).

[374] B. Carmeli and D. Chandler, J. Chem. Phys. **82**, 3400 (1985); D. Chandler in *Liquids, Freezing and Glass Transition*, ed. by D. Levesque, J. P. Hansen, and J. Zinn-Justin (Elsevier Science Publishers, 1990).

[375] a.: J. Ulstrup, *Charge Transfer in Condensed Media* (Springer, 1979); b.: B. Fain, *Theory of Rate Processes in Condensed Media* (Springer, 1980).

[376] P. Ao and J. Rammer, Phys. Rev. Lett. **62**, 3004 (1989).

[377] A. Garg, J. N. Onuchic, and V. Ambegaokar, J. Chem. Phys. **83**, 4491 (1985).

[378] I. Rips and J. Jortner, J. Chem. Phys. **87**, 2090 (1987); M. Sparpaglione and S. Mukamel, J. Chem. Phys. **88**, 3263 (1987).

[379] J. N. Gehlen and D. Chandler, J. Chem. Phys. **97**, 4958 (1992); J. N. Gehlen, D. Chandler, H. J. Kim, and J. T. Hynes, J. Phys. Chem. **96**, 1748 (1992).

[380] X. Song and A. A. Stuchebrukhov, J. Chem. Phys. **99**, 969 (1993); A. A. Stuchebrukhov and X. Song, J. Chem. Phys. **101**, 9354 (1994).

[381] J. Cao, C. Minichino, and G. A. Voth, J. Chem. Phys. **103**, 1391 (1995).

[382] K. Huang and A. Rhys, Proc. Roy. Soc. A **204**, 406 (1950); M. Lax, J. Chem. Phys. **20**, 1752 (1952); R. Kubo and Y. Toyozawa, Progr. Theor. Phys. **13**, 160 (1955); V. G. Levich and R. R. Dogonadze, Coll. Czech. Chem. Comm. **26**, 193 (1961). See also V. G. Levich in *Advances in Electrochemistry and Electrochemical Engineering*, ed. by P. Delahay and C. W. Tobias, Vol. 4, 249 (Interscience, 1965).

[383] M. H. Devoret, D. Esteve, H. Grabert, G.-L. Ingold, H. Pothier, and C. Urbina, Phys. Rev. Lett. **64**, 1824 (1990); Physica B **165 & 166**, 977 (1990); S. M. Girvin, L. I. Glazman, M. Jonson, D. R. Penn, and M. D. Stiles, Phys. Rev. Lett. **64**, 3183 (1990).

[384] R. Bruinsma and P. Bak, Phys. Rev. Lett. **56**, 420 (1986).

[385] J. Jortner, J. Chem. Phys. **64**, 4860 (1975).

[386] P. Minnhagen, Phys. Lett. A **56**, 327 (1976).

[387] G. Falci, V. Bubanja, and G. Schön, Z. Physik B **85**, 451 (1991).

[388] K. Ando, J. Chem. Phys. **106**, 116 (1997).

[389] J. S. Bader, R. A. Kuharski, and D. Chandler, J. Chem. Phys. **93**, 230 (1990).

[390] P. Siders and R. A. Marcus, J. Am. Chem. Soc. **103**, 741 (1981).

[391] U. Weiss and H. Grabert, Phys. Lett. **108A**, 63 (1985).

[392] a.: H. Grabert and U. Weiss, Phys. Rev. Lett. **54**, 1605 (1985).
       b.: M. P. A. Fisher and A. T. Dorsey, Phys. Rev. Lett. **54**, 1609 (1985).

[393] C. Aslangul, N. Poitier, and D. Saint-James, J. Phys. (Paris) **47**, 1671 (1986).

[394] R. Egger, C. H. Mak, and U. Weiss, J. Chem. Phys. **100**, 2651 (1994).

[395] H. Grabert, Phys. Rev. B **46**, 12 753 (1992).

[396] A. Würger, in Ref. [344]; Phys. Rev. Lett. **78**, 1759 (1997).

[397] Q. Niu, J. Stat. Phys. **65**, 317 (1991).

[398] H. Grabert, U. Weiss, and H. R. Schober, Hyperfine Interactions **31**, 147 (1986).

[399] R. Pirc and P. Gosar, Phys. Kondens. Mater. **9**, 377 (1969).

[400] A. Würger, Physics Letters A **236**, 571 (1998).

[401] A. Würger, Solid State Comm. **106**, 63 (1998).

[402] M. Sassetti and U. Weiss, Europhys. Lett. **27**, 311 (1994).

[403] M. Sassetti, F. Napoli, and U. Weiss, Phys. Rev. B **52**, 11 213 (1995).

[404] *Mesoscopic Electron Transport*, ed. by L. L. Sohn, L. P. Kouwenhoven, and G. Schön, NATO ASI Series E, Vol. 345 (Kluwer, Dordrecht, 1997).

[405] H. Schoeller and G. Schön, Phys. Rev. B **50**, 18 436 (1994); J. König, J. Schmid, H. Schoeller, and G. Schön, Phys. Rev. B **54**, 16 820 (1996); H. Schoeller, in Ref. [404].

[406] G.-L. Ingold, H. Grabert, and U. Eberhardt, Phys. Rev. B **50**, 395 (1994).

[407] R. D. Coalson, D. G. Evans, and A. Nitzan, J. Chem. Phys. **101**, 436 (1994).

[408] A. Lucke, C. H. Mak, R. Egger, J. Ankerhold, J. Stockburger, and H. Grabert, J. Chem. Phys. **107**, 8397 (1997).

[409] C. Cohen-Tannoudji, B. Diu, and F. Laloë, *Quantum Mechanics*, Vol. 1 (Wiley, New York).

[410] M. Grifoni, M. Winterstetter, and U. Weiss, Phys. Rev. E **56**, 334 (1997).

[411] F. Guinea, Phys. Rev. B **32**, 4486 (1985).

[412] G. Lang, E. Paladino, and U. Weiss, Europhys. Lett. **43**, 117 (1998); Phys. Rev. E **58**, 4288 (1998).

[413] M. Grifoni, M. Sassetti, and U. Weiss, Phys. Rev. E **53**, R2033 (1996).

[414] H. Dekker, Phys. Rev. A **35**, 1436 (1987).

[415] F. Lesage and H. Saleur, Phys. Rev. Lett. **80** 4370 (1998).

[416] J. Cardy, Nucl. Phys. B **324**, 581 (1989); I. Affleck and A. W. W. Ludwig, Nucl. Phys. B **360**, 641 (1991); *ibid.* B **428**, 545 (1994).

[417] P. Fulde and I. Peschel, Adv. Phys. **21**, 1 (1972).

[418] U. Weiss and H. Grabert, Europhys. Lett. **2**, 667 (1986); U. Weiss, H. Grabert, and S. Linkwitz, J. Low Temp. Phys. **68**, 213 (1987).

[419] A. Garg, Phys. Rev. B **32**, 4746 (1985).

[420] S. Dattagupta, H. Grabert, and R. Jung, J. Phys.: Cond. Mat. **1**, 1405 (1989).

[421] U. Weiss and M. Wollensak, Phys. Rev. Letters **62**, 1663 (1989); R. Görlich, M. Sassetti, and U. Weiss, Europhys. Lett. **10**, 507 (1989).

[422] A. Würger, Phys. Rev. B **57**, 347 (1998).

[423] D. A. Parshin, Z. Physik B **91**, 367 (1993).

[424] J. Stockburger, M. Grifoni, and M. Sassetti, Phys. Rev. B **51**, 2835 (1995).

[425] M. Winterstetter and U. Weiss, Chem. Phys. **217**, 155 (1997).

[426] R. Egger, H. Grabert, and U. Weiss, Phys. Rev. E **55**, R3809 (1997).

[427] H. Shiba, Progr. Theor. Phys. **54**, 967 (1975).

[428] T. A. Costi and C. Kieffer, Phys. Rev. Lett. **76**, 1683 (1996).

[429] T. A. Costi, Phys. Rev. B **55**, 3003 (1997); Phys. Rev. Lett.**80**, 1038 (1998).

[430] F. Lesage and H. Saleur, Nucl. Phys. B **490**, 543 (1997).

[431] S. P. Strong, Phys. Rev. E **55**, 6636 (1997).

[432] J. T. Stockburger and C. H. Mak, J. Chem. Phys. **105**, 8126 (1996).

[433] H. Baur, A. Fubini, and U. Weiss, Phys. Rev. B **70**, 024302 (2004).

[434] R. N. Silver, D. S. Sivia, and J. E. Gubernatis in: *Quantum Simulations of Condensed Matter Phenomena*, ed. by J. D. Doll and J. E. Gubernatis (World Scientific, Singapore, 1990).

[435] P. Fendley, F. Lesage, and H. Saleur, J. Stat. Phys. **79**, 799 (1995).

[436] F. Grossmann, T. Dittrich, and P. Hänggi, Phys. Rev. Lett. **67**, 516 (1991).

[437] N. Makri, J. Chem. Phys. **106**, 2286 (1997); N. Makri and L. Wei, Phys. Rev. E **55**, 2475 (1997).

[438] a.: D. G. Evans, R. D. Coalson, H. J. Kim, and Yu. Dakhnovskii, Phys. Rev. Lett. **75**, 3649 (1995).
b.: M. Morillo and R. I. Cukier, Phys. Rev. B **54**, 13 962 (1996).

[439] M. Grifoni, L. Hartmann, and P. Hänggi, Chem. Phys. **217**, 167 (1997).

[440] Special issue on *Dynamics of Driven Quantum Systems*, ed. by W. Domcke, P. Hänggi, and D. Tannor, Chem. Phys. **217** (2,3), 117-416 (1997).

[441] M. Grifoni and P. Hänggi, *Driven Quantum Tunneling*, Phys. Rep. **304**, 229 (1998).

[442] S. Han, J. Lapointe, and J. E. Lukens, Phys. Rev. Lett. **66**, 810 (1991); Phys. Rev. B **46**, 6338 (1992).

[443] M. Grifoni, Phys. Rev. E **54**, R3086 (1996).

[444] M. Grifoni, M. Sassetti, P. Hänggi, and U. Weiss, Phys. Rev. E **52**, 3596 (1995).

[445] M. Grifoni, M. Sassetti, J. Stockburger, U. Weiss, Phys. Rev. E **48**, 3497 (1993).

[446] P. K. Tien and J. P. Gordon, Phys. Rev. **129**, 647 (1963).

[447] Yu. Dakhnovskii, Phys. Rev. B **49**, 4649 (1994); Yu. Dakhnovskii and R. D. Coalson, J. Chem. Phys. **103**, 2908 (1995); I. A. Goychuk, E. G. Petrov, and V. May, Chem. Phys. Lett. **253**, 428 (1996). See also A. Lück, M. Winterstetter, U. Weiss, and C.H. Mak, Phys. Rev. E **58**, 5565 (1998).

[448] F. Grossmann and P. Hänggi, Europhys. Lett. **18**, 571 (1992).

[449] J. M. Gomez-Llorente, Phys. Rev. A **45**, R6958 (1992); erratum Phys. Rev. E **49**, 3547 (1994).

[450] P. Jung, Phys. Rep. **234**, 175 (1993); Proc. NATO Workshop on SR in Physics and Biology, F. Moss *et al.* (Eds.), J. Stat. Phys. **70**, 1 (1993); L. Gammaitoni, P. Hänggi, P. Jung, and F. Marchesoni, Rev. Mod. Phys. **70**, 223 (1998).

[451] R. Löfstedt and S. N. Coppersmith, Phys. Rev. Lett. **72**, 1947 (1994); Phys. Rev. E **49**, 4821 (1994).

[452] M. Grifoni and P. Hänggi, Phys. Rev. Lett. **76**, 1611 (1996); Phys. Rev. E **54**, 1390 (1996).

[453] *The Photosynthetic Reaction Center*, Vol. 1 and 2, ed. by J. Deisenhofer and J. R. Norris (Academic Press, New York, 1993).

[454] R. Egger, C. H. Mak, and U. Weiss, Phys. Rev. E **50**, R655 (1994).

[455] M. P. A. Fisher and W. Zwerger, Phys. Rev. B **32**, 6190 (1985).

[456] W. Zwerger, Phys. Rev. B **35**, 4737 (1987).

[457] C. L. Kane and M. P. A. Fisher, Phys. Rev. Lett. **68**, 1220 (1992); Phys. Rev. B **46** 15 233 (1992).

[458] X. G. Wen, Phys. Rev. B **41**, 12 838 (1990); **43** 11 025 (1991); **44**, 5708 (1991).

[459] H. J. Schulz, Phys. Rev. Lett. **71**, 1864 (1993).

[460] M. Fabrizio, A. O. Gogolin, and S. Scheidl, Phys. Rev. Lett. **72**, 2235 (1994).

[461] H. Rodenhausen, J. Stat. Phys. **55**, 1065 (1989).

[462] Y.-C. Chen, J. Stat. Phys. **65**, 761 (1991).

[463] A. Lenard, J. Math. Phys. **2**, 682 (1961).

[464] Yu. M. Ivanchenko and L. A. Zil'berman, Sov. Phys.–JETP **28**, 1272 (1969).

[465] U. Weiss and M. Wollensak, Phys. Rev. B **37**, 2729 (1988).

[466] F. Guinea, Phys. Rev. B **32**, 7518 (1985).

[467] U. Weiss, R. Egger, and M. Sassetti, Phys. Rev. B **52**, 16 707 (1995).

[468] M. Sassetti, M. Milch, and U. Weiss, Phys. Rev. A **46**, 4615 (1992).

[469] M. Sassetti, H. Schomerus, and U. Weiss, Phys. Rev. B **53**, R2914 (1996).

[470] H. Grabert, G.-L. Ingold, and B. Paul, Europhys. Lett. **44**, 360 (1998).

[471] K. Leung, R. Egger, and C. H. Mak, Phys. Rev. Lett. **75**, 3344 (1995).

[472] L. S. Levitov and G. B. Lesovik, JETP Lett. **58**, 230 (1993).

[473] D. V. Averin, Solid State Comm. **105**, 659 (1998).

[474] I. S. Beloborodov, F. W. J. Hekking and F. Pistolesi, in *New directions in Mesoscopic Physics*, ed. by R. Fazio et al. (Kluwer Academic Publisher, 2002).

[475] G.-L. Ingold and H. Grabert, Phys. Rev. Lett. **83**, 3721 (1999).

[476] U. Weiss, Solid State Comm. **100**, 281 (1996);
U. Weiss, in *Tunneling and its Implications*, ed. by D. Mugnai, A. Ranfagni, and L. S. Schulman, p. 134 (World Scientific, Singapore, 1997).

[477] P. Fendley and H. Saleur, Phys. Rev. Lett. **81**, 2518 (1998).

[478] N. Seiberg and E. Witten, Nucl. Phys. B **426**, 19 (1994); B **431**, 484 (1994).

[479] N. M. Temme, *Special Functions* (Wiley, New York, 1996), p. 49.

[480] L. Álvarez-Gaumé and F. Zamora, hep-th/9709180 (1997).

[481] K. A. Matveev, D. Yue, and L. I. Glazman, Phys. Rev. Lett. **71**, 3351 (1993).

[482] H. Saleur and U. Weiss, Phys. Rev. B **63**, 201302(R) (2001).

[483] P. Fendley, A. W. W. Ludwig, and H. Saleur, Phys. Rev. B **52**, 8934 (1995).

[484] P. Fendley and H. Saleur, Phys. Rev. B **54**, 10 845 (1996).

[485] A. Komnik and H. Saleur, Phys. Rev. Lett. **96**, 216406 (2006).

[486] Y.-C. Chen and J. L. Lebowitz, Phys. Rev. B **46**, 10 743 (1992).

[487] A. O. Gogolin, A. A. Nersesyan, and A. M. Tsvelik, *Bosonization and Strongly Correlated Systems* (Cambridge University Press, Cambridge, 1998).

[488] T. Giamarchi, *Quantum Physics in One Dimension* (Clarendon, Oxford, 2004).

[489] A. Luther and I. Peschel, Phys. Rev. B **9**, 2911 (1974).

[490] P. Fendley, A. W. W. Ludwig, and H. Saleur, Phys. Rev. Lett. **75**, 2196 (1995).

[491] R. Egger and H. Grabert, Phys. Rev. Lett. **77**, 538 (1996), and Ref. 10 therein.

[492] I. Safi and H. Saleur, Phys. Rev. Lett. **93**, 126602 (2004).

[493] M. Kindermann and B. Trauzettel, Phys. Rev. Lett. **94**, 166803 (2005).

[494] A. O. Gogolin and A. Komnik, Phys. Rev. B **73**, 195301 (2006).

Index